Frederick Manson Bailey

The Queensland flora

Part 1

Frederick Manson Bailey

The Queensland flora
Part 1

ISBN/EAN: 9783337271589

Printed in Europe, USA, Canada, Australia, Japan

Cover: Foto ©berggeist007 / pixelio.de

More available books at **www.hansebooks.com**

THE
QUEENSLAND FLORA:

BY

F. MANSON BAILEY, F.L.S.,
COLONIAL BOTANIST OF QUEENSLAND.

WITH PLATES ILLUSTRATING SOME RARE SPECIES.

PART I.

RANUNCULACEÆ TO ANACARDIACEÆ.

PUBLISHED UNDER THE AUTHORITY OF THE QUEENSLAND
GOVERNMENT.

QUEENSLAND:
PRINTED BY H. J. DIDDAMS & CO., ELIZABETH STREET, BRISBANE.

1899.

H. J. DIDDAMS AND CO., PRINTERS,
HARPER'S BUILDINGS, ELIZABETH STREET, BRISBANE.

PREFACE.

The issue of the "Synopsis of the Queensland Flora" having for some time been exhausted, it has been deemed advisable to publish another and more complete work concerning the plants of the Colony. This will be issued in six parts, and each part will be accompanied by a few lithographic plates. Part I., now issued, contains the Orders Ranunculaceæ to Anacardiaceæ.

The arrangement is that of Bentham and Hooker's "Genera Plantarum." Where plants are described in the "Flora Australiensis" such descriptions are reproduced in full, with any needful further descriptive notes which may have come to our knowledge since. In all cases where Mr. Bentham gave notes of the genera or species such notes I have considered far too important to leave out, so they will be found generally intact.

The explanation of systematic names, and the inclusion of aboriginal and local names will doubtless prove acceptable; and it is hoped that the notes upon the economic and other properties will be found of service. The authorities and references have been curtailed to little more than works dealing with Australian plants. In a large number of cases the names of works in which illustrations of certain plants may be found, have been quoted; and frequent references will be found to fungus blights which have been observed upon the plants. With regard to the localities where each plant has been found, only a few are recorded for plants having a wide range, and seldom other than the names of the earlier collectors are given.

As in most works of this kind the right to republish Mr. Bentham's excellent "Outlines of Botany" has been acquired from Messrs. L. Reeve & Co., of London.

The naturalised plants will be found numbered in with the indigenous species, but will be marked with an asterisk by way of distinction. Several strays from cultivation, which at present can scarcely be considered as naturalised, are also included.

F.M.B.

December, 1899.

LIST OF PLATES.

To face page

PLATE I.—Legnephora Moorii 29
(A portion of a branch of the male plant in flower)
Fig. 1, a pedicellated male flower; Fig. 2, the same, expanded, *both natural size*; Fig. 3, the same magnified; Fig. 4, the bract of the pedicele; Fig. 5, the three outer sepals; Fig. 6, the three inner sepals; Fig. 7, the six petals; Fig. 8, the six stamens, *all magnified to the same scale*; Fig. 9, a petal in different positions, showing the gland on its margins; Fig. 10, a stamen in different positions, before and after dehiscence, *all more magnified*.

PLATE II.—Legnephora Moorii 29
(A portion of a branch of the female plant in flower and in fruit)
Fig. 11, a female flower on its bracteolated pedicel, *natural size*; Fig. 12, the bracteole, *magnified*; Fig. 13, the three outer sepals; Fig. 14, the three inner sepals; Fig. 15, the six sterile stamens; Fig. 16, the three ovaries, *all magnified to the same scale*; Fig. 17, a sterile stamen; Fig. 18, an ovary, showing the form of the stigma, *both more magnified*; Fig. 19, a drupe, seated on the receptacle, from which the two others have fallen; Fig. 20, its putamen; Fig. 21, the same, seen edgeways, *all natural size*; Fig. 22, the putamen, seen on its face, showing the three series of imbricating, laciniated, flat, pergamaneous scales which surround the flat, concave, scutiform condyle; Fig. 23, the same, shown endways, *both magnified*; Fig. 24, one of the scales, seen in front and edgeways, *more magnified*; Fig 25, a cross section of the putamen, showing the seed and the hollow between the two plates of the condyla along the sutural line of division; Fig. 26, inner view of half the putamen, showing the hippocrepiform cell of the seed and the groove from the basal hilum to the point of the attachment of the seed, at its sinus between the plates of the condyle; Fig. 27, the seed extracted; Fig. 28, a cross section of the same, showing the embryo imbedded in albumen, *all equally magnified*. *All from Miers' Contributions*, iii pl. 129.

PLATE III.—Pachygone longifolia 34
Fig. 1, leaf; Fig. 2. whole drupe; Fig. 3, inside view of same; Fig. 4, endocarp, outside view.

PLATE IV.—Pittosporum setigerum 69
(Flowering branch, expanded flower and bud, stamen, and calyx and ovary)

PLATE V.—Saurauja Andreana 106
(Flowering branch and expanded flower)

PLATE VI.—Sterculia Garrawayæ.. 136
(Leaf, flower and fruit)

PLATE VII.—Sterculia vitifolia 137
(Leaf, flower and fruit)

PLATE VIII.—Asterolasia Woombye 195
(Flowering branch and expanded flower)

PLATE IX.—Geijera Helmsiæ 206
Fig. 1, flowering branch; Fig. 2, flower, enlarged; Fig. 3, longitudinal section of flower, enlarged; Fig. 4, anther, with portion of filament, enlarged; Fig. 5, stigma with portion of style, enlarged; Fig. 6, transverse section of ovary, enlarged; Fig. 7, open cocci, natural size; Fig. 8, endocarp, natural size—*From drawing by Mrs. R. Helms*.

PLATE X.—Citrus inodora 215
(Branchlet and fruit)

PLATE XI.—Rhodosphæra rhodanthema 321
Fig. 1, petal; Fig. 2, gynæcium and stamens; Fig. 3, longitudinal section of gynæcium; Fig. 4, drupe; Fig. 5, longitudinal section of drupe; Fig. 6, transverse section of drupe; Fig. 7, embryo—*From Engl. in A. and C., Dl. Mon. Phane.*, iv. pl. xii. Fig. 8, inflorescence.

PLATE XII.—Pleiogynium Solandri 324
Fig. 1, flower-bud, with bracts; Fig. 2, expanded flower seen from below; Fig. 3, andrœcium, with disk; Fig. 4, anthers, seen from back and front; Fig. 5, gynæcium; Fig. 6, longitudinal section of same; Fig. 7, immature fruit; Fig. 8, mature fruit; Fig. 9, longitudinal section; Fig. 10, transverse section—*From Endl. in A. and C. Dl. Mon. Phane.*, iv. pl. vii. Fig. 11, male inflorescence; Fig. 12, female inflorescence.

CONTENTS.

	Page
OUTLINES OF BOTANY, WITH SPECIAL REFERENCE TO LOCAL FLORAS	i
CHAP. I. DEFINITIONS AND DESCRIPTIVE BOTANY	i
§ 1. The Plant in General	i
§ 2. The Root	ii
§ 3. The Stock	iii
§ 4. The Stem	iv
§ 5. The Leaves	v
§ 6. Scales, Bracts, and Stipules	vii
§ 7. Inflorescence and its Bracts	viii
§ 8. The Flower in General	ix
§ 9. The Calyx and Corolla or Perianth	xi
§ 10. The Stamens	xii
§ 11. The Pistil	xiii
§ 12. The Receptacle and Relative Attachment of the Floral Whorls	xv
§ 13. The Fruit	xvi
§ 14. The Seed	xvii
§ 15. Accessory Organs	xvii
CHAP. II. CLASSIFICATION, OR SYSTEMATIC BOTANY	xix
CHAP. III. VEGETABLE ANATOMY AND PHYSIOLOGY	xx
§ 1. Structure of the Elementary Tissues	xx
§ 2. Arrangement of the Elementary Tissues, or Structure of the Organs of Plants	xxi
§ 3. Growth of the Organs	xxiii
§ 4. Functions of the Organs	xxiv
CHAP. IV. COLLECTION, PRESERVATION, AND DETERMINATION OF PLANTS	xxv
INDEX OF TERMS, OR GLOSSARY	xxix
THE QUEENSLAND FLORA.	
CLASS I. DICOTYLEDONS	1
Subclass I. Polypetalæ	1
Order I. Ranunculaceæ	4
II. Dilleniaceæ	8
III. Magnoliaceæ	18
IV. Anonaceæ	20
V. Menispermaceæ	27
VI. Nymphæaceæ	36
VII. Papaveraceæ	40
VIII. Cruciferæ	42
IX. Capparideæ	53
X. Violarieæ	61
XI. Bixineæ	65

CONTENTS.

		Page
Order XII.	Pittosporeæ	67
XIII.	Tremandreæ	75
XIV.	Polygaleæ	76
XV.	Frankeniaceæ	83
XVI.	Caryophylleæ	84
XVII.	Portulaceæ	92
XVIII.	Elatineæ	98
XIX.	Hypericineæ	100
XX.	Guttiferæ	102
XXI.	Ternstrœmiaceæ	105
XXII.	Malvaceæ	106
XXIII.	Sterculiaceæ	134
XXIV.	Tiliaceæ	151
XXV.	Lineæ	165
XXVI.	Malpighiaceæ	168
XXVII.	Zygophylleæ	170
XXVIII.	Geraniaceæ	176
XXIX.	Rutaceæ	181
XXX.	Simarubeæ	216
XXXI.	Ochnaceæ	221
XXXII.	Burseraceæ	222
XXXIII.	Meliaceæ	225
XXXIV.	Olacineæ	243
XXXV.	Ilicineæ	251
XXXVI.	Celastrineæ	252
XXXVII.	Stackhousieæ	262
XXXVIII.	Rhamneæ	265
XXXIX.	Ampelideæ	278
XL.	Sapindaceæ	285
XLI.	Anacardiaceæ	319

OUTLINES OF BOTANY

WITH SPECIAL REFERENCE TO LOCAL FLORAS.

(From Bentham's *Flora Australiensis.*)

Chap. I. Definitions and Descriptive Botany.

1. The principal object of a **Flora** of a country, is to afford the means of *determining* (i.e. ascertaining the name of) any plant growing in it, whether for the purpose of ulterior study or of intellectual exercise.

2. With this view, a Flora consists of descriptions of all the wild or native plants contained in the country in question, so drawn up and arranged that the student may identify with the corresponding description any individual specimen which he may gather.

3. These descriptions should be *clear, concise, accurate* and *characteristic,* so as that each one should be readily adapted to the plant it relates to, and to no other one; they should be as nearly as possible arranged under *natural* (184) divisions, so as to facilitate the comparison of each plant with those nearest allied to it; and they should be accompanied by an *artificial key* or index, by means of which the student may be guided step by step in the observation of such peculiarities or *characters* in his plant, as may lead him, with the least delay, to the individual description belonging to it.

4. For descriptions to be clear and readily intelligible, they should be expressed as much as possible in ordinary well-established language. But, for the purpose of accuracy, it is necessary not only to give a more precise technical meaning to many terms used more or less vaguely in common conversation, but also to introduce purely technical names for such parts of plants or forms as are of little importance except to the botanist. In the present chapter it is proposed to define such *technical* or *technically limited* terms as are made use of in these Floras.

5. At the same time mathematical accuracy must not be expected. The forms and appearances assumed by plants and their parts are infinite. Names cannot be invented for all; those even that have been proposed are too numerous for ordinary memories. Many are derived from supposed resemblances to well-known forms or objects. These resemblances are differently appreciated by different persons, and the same term is not only differently applied by two different botanists, but it frequently happens that the same writer is led on different occasions to give somewhat different meanings to the same word. The botanist's endeavours should always be, on the one hand, to make as near an approach to precision as circumstances will allow, and on the other hand to avoid that prolixity of detail and overloading with technical terms which tends rather to confusion than clearness. In this he will be more or less successful. The aptness of a botanical description, like the beauty of a work of imagination, will always vary with the style and genius of the author.

§ 1. *The Plant in General.*

6. The **Plant,** in its botanical sense, includes every being which has *vegetable life,* from the loftiest tree which adorns our landscapes, to the humblest moss which grows on its stem, to the mould or fungus which attacks our provisions, or the green scum that floats on our ponds.

7. Every portion of a plant which has a distinct part or *function* to perform in the operations or phenomena of vegetable life is called an **Organ.**

8. What constitutes *vegetable life,* and what are the functions of each organ, belong to *Vegetable Physiology*; the microscopical structure of the tissues composing the organs, to *Vegetable Anatomy*; the composition of the substances of which they are formed, to *Vegetable Chemistry*; under *Descriptive and Systematic Botany* we have chiefly to consider the forms of organs, that is, their *Morphology,* in the proper sense of the term, and their general structure so far as it affects classification and specific resemblances and differences. The terms we shall now define

belong chiefly to the latter branch of Botany, as being that which is essential for the investigation of the Flora of a country. We shall add, however, a short chapter on Vegetable Anatomy and Physiology, as a general knowledge of both imparts an additional interest to and facilitates the comparison of the characters and affinities of the plants examined.

9. In the more perfect plants, their organs are comprised in the general terms **Root, Stem, Leaves, Flowers,** and **Fruit.** Of these the three first, whose function is to assist in the growth of the plant, are *Organs of Vegetation*; the flower and fruit, whose office is the formation of the seed, are the *Organs of Reproduction.*

10. All these organs exist, in one shape or another, at some period of the life of most, if not all, *flowering plants*, technically called *phænogamous* or *phanerogamous plants*; which all bear some kind of flower and fruit in the botanical sense of the term. In the lower classes, the ferns, mosses, fungi, moulds, or mildews, seaweeds, etc., called by botanists *cryptogamous plants*, the flowers, the fruit, and not unfrequently one or more of the organs of vegetation, are either wanting, or replaced by organs so different as to be hardly capable of bearing the same name.

11. The observations comprised in the following pages refer exclusively to the flowering or phænogamous plants. The study of the cryptogamous classes has now become so complicated as to form almost a separate science. They are therefore not included in these introductory observations, nor, with the exception of ferns, in the present Flora.

12. **Plants** are
Monocarpic, if they die after one flowering-season. These include *Annuals*, which flower in the same year in which they are raised from seed; and *Biennials*, which only flower in the year following that in which they are sown.

Caulocarpic, if, after flowering, the whole or part of the plant lives through the winter and produces fresh flowers another season. These include *Herbaceous perennials*, in which the greater part of the plant dies after flowering, leaving only a small perennial portion called the Stock or Caudex, close to or within the earth; *Undershrubs, suffruticose* or *suffrutescent* plants, in which the flowering branches, forming a considerable portion of the plant, die down after flowering, but leave a more or less prominent perennial and woody base; *shrubs (frutescent* or *fruticose plants)*, in which the perennial woody part forms the greater part of the plant, but branches near the base, and does not much exceed a man's height; and *Trees (Arboreous* or *arborescent plants)* when the height is greater and forms a woody *trunk*, scarcely branching from the base. *Bushes* are low, much branched shrubs.

13. The terms *Monocarpic* and *Caulocarpic* are but little used, but the other distinctions enumerated above are universally attended to, although more useful to the gardener than to the botanist, who cannot always assign to them any precise character. Monocarpic plants, which require more than two or three years to produce their flowers, will often, under certain circumstances, become herbaceous perennials, and are generally confounded with them. Truly perennial herbs will often commence flowering the first year, and have then all the appearance of annuals. Many tall shrubs and trees lose annually their flowering branches like undershrubs. And the same botanical species may be an annual or perennial, a herbaceous perennial or an undershrub, an undershrub or a shrub, a shrub or a tree, according to climate, treatment, or variety.

14. Plants are usually *terrestrial*, that is, growing on earth, or *aquatic, i.e.* growing in water; but sometimes they may be found attached by their roots to other plants, in which case they are *epiphytes* when simply growing upon other plants without penetrating into their tissue, *parasites* when their roots penetrate into and derive more or less nutriment from the plant to which they are attached.

15. The simplest form of the perfect plant, the annual, consists of—

(1) The **Root,** or descending axis, which grows downward from the stem, divides and spreads in the earth or water, and absorbs food for the plant through the extremities of its branches.

(2) The **Stem,** or ascending axis, which grows upwards from the root, branches and bears first one or more leaves in succession, then one or more flowers, and finally one or more fruits. It contains the tissues or other channels (217) by which the nutriment absorbed by the roots is conveyed in the form of *sap* (192) to the leaves or other points of the surface of the plant, to be *elaborated* or *digested* (218), and afterwards redistributed over different parts of the plant for its support and growth.

(3) The **Leaves,** usually flat, green, and horizontal, are variously arranged on the stem and its branches. They *elaborate* or *digest* (218) the nutriment brought to them through the stem, absorb carbonic acid gas from the air, exhaling the superfluous oxygen, and returning the assimilated sap to the stem.

(4) The **Flowers,** usually placed at or towards the extremities of the branches. They are destined to form the future seed. When perfect and complete they consist: 1st, of a *pistil* in the centre, consisting of one or more *carpels*, each containing the germ of one or more seeds; 2nd, of one or more *stamens* outside the pistil, whose action is necessary to *fertilize* the pistil or enable it to ripen its seed; 3rd, of a *perianth* or *floral envelope*, which usually encloses the stamens and pistil when young, and expands and exposes them to view when fully formed. This complete

perianth is double; the outer one called *Calyx*, is usually more green and leaf-like; the inner one, called the *Corolla*, more conspicuous and variously coloured. It is the perianth, and especially the Corolla, as the most showy part, that is generally called the flower in popular language.

(5) The **Fruit**, consisting of the pistil or its lower portion, which persists or remains attached to the plant after the remainder of the flower has withered and fallen off. It enlarges and alters more or less in shape or consistence, becomes a *seed-vessel*, enclosing the seed until it is ripe, when it either opens to discharge the seed or falls to the ground with the seed. In popular language the term *fruit* is often limited to such seed-vessels as are or look juicy and eatable. Botanists give that name to all seed-vessels.

16. The herbaceous perennial resembles the annual during the first year of its growth; but it also forms (usually towards the close of the season), on its *stock* (the portion of the stem and root which does not die), one or more *buds*, either exposed, and then popularly called *eyes*, or concealed among leaves. These buds, called *leaf-buds*, to distinguish them from *flower-buds* or unopened flowers, are future branches as yet undeveloped; they remain dormant through the winter, and the following spring grow out into new stems bearing leaves and flowers like those of the preceding year, whilst the lower part of the stock emits fresh roots to replace those which had perished at the same time as the stems.

17. Shrubs and trees form similar leaf-buds either at the extremity of their branches, or along the branches of the year. In the latter case these buds are usually *axillary*, that is, they appear in the *axil* of each leaf, *i.e.* in the angle formed by the leaf and the branch. When they appear at any other part of the plant they are called *adventitious*. If these buds by producing roots (19) become distinct plants before separating from the parent, or if adventitious leaf-buds are produced in the place of flowers or seeds, the plant is said to be *viviparous* or *proliferous*.

§ 2. *The Root.*

18. **Roots** ordinarily produce neither buds, leaves, nor flowers. Their branches, called *fibres* when slender and long, proceed irregularly from any part of their surface.

19. Although roots proceed usually from the base of the stem or stock, they may also be produced from the base of any bud, especially if the bud lie along the ground, or is otherwise placed by nature or art in circumstances favourable for their development, or indeed occasionally from almost any part of the plant. They are then often distinguished as *adventitious*, but this term is by some applied to all roots which are not in prolongation of the original radicle.

20. **Roots** are

fibrous, when they consist chiefly of slender fibres.

tuberous, when either the main root or its branches are thickened into one or more short fleshy or woody masses called *tubers* (25).

taproots, when the main root descends perpendicularly into the earth, emitting only very small fibrous branches.

21. The stock of a herbaceous perennial, or the lower part of the stem of an annual or perennial, or the lowest branches of a plant, are sometimes underground and assume the appearance of a root. They then take the name of *rhizome*. The rhizome may always be distinguished from the true root by the presence or production of one or more buds, or leaves, or scales.

§ 3. *The Stock.*

22. The **Stock** of a herbaceous perennial, in its most complete state, includes a small portion of the summits of the previous year's roots, as well as of the base of the previous year's stems. Such stocks will increase yearly, so as at length to form dense tufts. They will often preserve through the winter a few leaves, amongst which are placed the buds which grow out into stems the following year, whilst the under side of the stock emits new roots from or amongst the remains of the old ones. These perennial stocks only differ from the permanent base of an undershrub in the shortness of the perennial part of the stems and in their texture usually less woody.

23. In some perennials, however, the stock consists merely of a branch, which proceeds in autumn from the base of the stem either aboveground or underground, and produces one or more buds. This branch, or a portion of it, alone survives the winter. In the following year its buds produce the new stem and roots, whilst the rest of the plant, even the branch on which these buds were formed, has died away. These *annual stocks*, called sometimes *hybernacula*, *offsets*, or *stolons*, keep up the communication between the annual stem and root of one year and those of the following year, thus forming altogether a perennial plant.

24. The stock, whether annual or perennial, is often entirely underground or root-like. This is the *rootstock*, to which some botanists limit the meaning of the term *rhizome*. When the stock is entirely root-like, it is popularly called the *crown* of the root.

25. The term *tuber* is applied to a short, thick, more or less decumbent rootstock or rhizome, as well as to a root of that shape (20), although some botanists propose to restrict its meaning to the one or to the other. An Orchis tuber, called by some a *knob*, is an annual tuberous rootstock with one bud at the top. A potato is an annual tuberous rootstock with several buds.

26. A *bulb* is a stock of a shape approaching to globular, usually rather conical above and flattened underneath, in which the bud or buds are concealed, or nearly so, under *scales*. These scales are the more or less thickened bases of the decayed leaves of the preceding year, or of the undeveloped leaves of the future year, or of both. Bulbs are annual or perennial, usually underground or close to the ground, but occasionally buds in the axils of the upper leaves become transformed into bulbs. Bulbs are said to be *scaly* when their scales are thick and loosely imbricated, *tunicated* when the scales are thinner, broader, and closely rolled round each other in concentric layers.

27. A *corm* is a tuberous rootstock, usually annual, shaped like a bulb, but in which the bud or buds are not covered by scales, or of which the scales are very thin and membranous.

§ 4. *The Stem.*

28. **Stems** are

erect, when they ascend perpendicularly from the root or stock ; *twiggy* or *virgate*, when at the same time they are slender, stiff, and scarcely branched.

sarmentose, when the branches of a woody stem are long and weak, although scarcely climbing.

decumbent or *ascending*, when they spread horizontally, or nearly so, at the base, and then turn upwards and become erect.

procumbent, when they spread along the ground the whole or the greater portion of their length ; *diffuse*, when at the same time very much and rather loosely branched.

prostrate, when they lie still closer to the ground.

creeping, when they emit roots at their nodes. This term is also frequently applied to any rhizomes or roots which spread horizontally.

tufted or *cæspitose*, when very short, close, and many together from the same stock.

29. Weak climbing stems are said to *twine*, when they support themselves by winding spirally round any object ; such stems are also called *voluble*. When they simply climb without twining, they support themselves by their leaves, or by special clasping organs called *tendrils* (169), or sometimes, like the Ivy, by small root-like excrescences.

30. *Suckers*, are young plants formed at the end of creeping, underground rootstocks. *Scions, runners*, and *stolons*, or *stoles*, are names given to young plants formed at the end or at the nodes (31) of branches or stocks creeping wholly or partially aboveground, or sometimes to the creeping stocks themselves.

31. A *node* is a point of the stem or its branches at which one or more leaves, branches, or leaf-buds (16) are given off. An *internode* is the portion of the stem comprised between two nodes.

32. **Branches** or **leaves** are

opposite, when two proceed from the same node on opposite sides of the stem.

whorled or *verticillate* (in a *whorl* or *verticil*), when several proceed from the same node, arranged regularly round the stem ; *geminate, ternate, fascicled*, or *fasciculate*, when two, three, or more proceed from the same node on the same side of the stem. A tuft of fasciculate leaves is usually in fact an axillary leafy branch, so short that the leaves appear to proceed all from the same point.

alternate, when one only proceeds from each node, one on one side and the next above or below on the opposite side of the stem.

decussate, when opposite, but each pair placed at right-angles to the next pair above or below it ; *distichous*, when regularly arranged one above another in two opposite rows, one on each side of the stem ; *tristichous*, when in three rows, etc. (92).

scattered, when irregularly arranged round the stem ; frequently, however, botanists apply the term *alternate* to all branches or leaves that are neither opposite nor whorled.

secund, when all start from or are turned to one side of the stem.

33. **Branches** are *dichotomous*, when several times forked, the two branches of each fork being nearly equal ; *trichotomous*, when there are three nearly equal branches at each division instead of two ; but when the middle branch is evidently the principal one, the stem is usually said to have two opposite branches ; *umbellate*, when divided in the same manner into several nearly equal branches proceeding from the same point. If however the central branch is larger than the two or more lateral ones, the stem is said to have opposite or whorled branches, as the case may be.

34. A *culm* is a name sometimes given to the stem of Grasses, Sedges, and some other Monocotyledonous plants.

§ 5. *The Leaves.*

35. The ordinary or perfect **Leaf** consists of a flat *blade* or *lamina*, usually green, and more or less horizontal, attached to the stem by a stalk called a *footstalk* or *petiole*. When the form or dimensions of a leaf are spoken of, it is generally the blade that is meant, without the petiole or stalk.

36. The end by which a leaf, part of the flower, a seed, or any other organ, is attached to the stem or other organ, is called its *base*, the opposite end is its *apex* or summit, excepting sometimes in the case of anther-cells (115).

37. **Leaves** are
sessile, when the blade rests on the stem without the intervention of a petiole.
amplexicaul or *stem-clasping*, when the sessile base of the blade clasps the stem horizontally.
perfoliate, when the base of the blade not only clasps the stem, but closes round it on the opposite side, so that the stem appears to pierce through the blade.
decurrent, when the edges of the leaf are continued down the stem so as to form raised lines or narrow appendages, called *wings*.
sheathing, when the base of the blade, or of the more or less expanded petiole, forms a vertical sheath round the stem for some distance above the node.

38. Leaves and flowers are called *radical*, when inserted on a rhizome or stock, or so close to the base of the stem as to appear to proceed from the root, rhizome, or stock; *cauline*, when inserted on a distinct stem. Radical leaves are *rosulate* when they spread in a circle on the ground.

39. **Leaves** are
simple and *entire*, when the blade consists of a single piece, with the margin nowhere indented, *simple* being used in opposition to *compound*, *entire* in opposition to *dentate*, *lobed*, or *divided*.
ciliate, when bordered with thick hairs or fine hair-like teeth.
dentate or *toothed*, when the margin is only cut a little way in, into what have been compared to teeth. Such leaves are *serrate*, when the teeth are regular and pointed like the teeth of a saw; *crenate*, when regular and blunt or rounded (compared to the battlements of a tower); *serrulate* and *crenulate*, when the serratures or crenatures are small; *sinuate*, when the teeth are broad, not deep, and irregular (compared to bays of the coast); *wavy* or *undulate*, when the edges are not flat, but bent up and down (compared to the waves of the sea).
lobed or *cleft*, when more deeply indented or divided, but so that the incisions do not reach the midrib or petiole. The portions thus divided take the name of *lobes*. When the lobes are narrow and very irregular, the leaves are said to be *laciniate*. The spaces between the teeth or lobes are called *sinuses*.
divided or *dissected*, when the incisions reach the midrib or petiole, but the parts so divided off, called *segments*, do not separate from the petiole, even when the leaf falls, without tearing.
compound, when divided to the midrib or petiole, and the parts so divided off, called *leaflets*, separate, at least at the fall of the leaf, from the petiole, as the whole leaf does from the stem, without tearing. The common stalk upon which the leaflets are inserted is called the *common petiole* or the *rhachis*; the separate stalk of each leaflet is a *petiolule*.

40. Leaves are more or less marked by *veins*, which, starting from the stalk, diverge or branch as the blade widens, and spread all over it more or less visibly. The principal ones, when prominent, are often called *ribs* or *nerves*, the smaller branches only then retaining the name of *veins*, or the latter are termed *veinlets*. The smaller veins are often connected together like the meshes of a net, they are then said to *anastomose*, and the leaf is said to be *reticulate* or *net-veined*. When one principal vein runs direct from the stalk towards the summit of the leaf, it is called the *midrib*. When several start from the stalk, diverge slightly without branching, and converge again towards the summit, they are said to be *parallel*, although not mathematically so. When 3 or 5 or more ribs or nerves diverge *from the base*, the leaf is said to be *3-nerved*, *5-nerved*, etc., but if the lateral ones diverge from the midrib a little above the base, the leaf is *triplinerved*, *quintuplinerved*, etc. The arrangement of the veins of a leaf is called their *venation*.

41. The **Leaflets, Segments, Lobes,** or **Veins** of leaves are
pinnate, (feathered), when there are several succeeding each other on each side of the midrib or petiole, compared to the branches of a feather. A pinnately lobed or divided leaf is called *lyrate* when the the terminal lobe or segment is much larger and broader than the lateral ones, compared, by a stretch of imagination, to a lyre; *runcinate*, when the lateral lobes are curved backwards towards the base of the leaf; *pectinate*, when the lateral lobes are numerous, narrow, and regular, like the teeth of a comb.
palmate or *digitate*, when several diverge from the same point, compared to the fingers of the hand.
ternate, when three only start from the same point, in which case the distinction between the palmate and pinnate arrangement often ceases, or can only be determined by analogy with allied plants. A leaf with ternate lobes is called *trifid*. A leaf with three leaflets is sometimes improperly called a ternate leaf; it is the leaflets that are ternate; the whole leaf is *trifoliolate*. Ternate leaves are leaves growing three together.
pedate, when the division is at first ternate, but the two outer branches are forked, the outer ones of each fork again forked, and so on, and all the branches are near together at the base, compared vaguely to the foot of a bird.

42. Leaves with pinnate, palmate, pedate, etc., leaflets, are usually for shortness called *pinnate*, *palmate*, *pedate*, etc., *leaves*. If they are so cut into segments only, they are usually said to be *pinnatisect*, *palmatisect*, *pedatisect*, etc., although the distinction between segments and leaflets, is often unheeded in descriptions, and cannot indeed always be ascertained. If the leaves are so cut only into lobes, they are said to be *pinnatifid*, *palmatifid*, *pedatifid*, etc.

43. The teeth, lobes, segments, or leaflets, may be again toothed, lobed, divided, or compounded. Some leaves are even three or more times divided or compounded. In the latter case they are termed *decompound*. When twice or thrice pinnate (*bipinnate* or *tripinnate*), each primary or secondary division, with the leaflets it comprises, is called a *pinna*. When the pinna of a leaf or the leaflets of a pinna are in pairs, without an odd terminal pinna or leaflet, the leaf or pinna so divided is said to be *abruptly pinnate*; if there is an odd terminal pinna or leaflet, the leaf or pinna is *unequally pinnate* (*imparipinnatum*).

44. The number of leaves or their parts is expressed adjectively by the following numerals, derived from the Latin:—

uni-,	bi-,	tri-,	quadri-,	quinque-,	sex-,	septem-,	octo-,	novem-,	decem-,	multi-
1.,	2.,	3.,	4.,	5.,	6.,	7.,	8.,	9.,	10.,	many.

prefixed to a termination indicating the particular kind of part referred to. Thus—

unidentate, bidentate, multidentate, mean one-toothed, two-toothed, many-toothed, etc.

bifid, trifid, multifid, mean two-lobed, three-lobed, many-lobed, etc.

unifoliolate, bifoliolate, multifoliolate, mean having one leaflet, two leaflets, many leaflets, etc.

unifoliate, bifoliate, multifoliate, mean having one leaf, two leaves, many leaves, etc.

biternate and *triternate,* mean twice or thrice ternately divided.

unijugate, bijugate, multijugate, etc., pinnæ or leaflets, mean that there are in one, two, many, etc., pairs (*juga*).

45. **Leaves** or their parts, when **flat,** or any other flat organs in plants, are

linear, when long and narrow, at least four or five times as long as broad, falsely compared to a mathematical line, for a linear leaf has always a perceptible breadth.

lanceolate, when about three or more times as long as broad, broadest below the middle, and tapering towards the summit, compared to the head of a lance.

cuneate, when broadest above the middle, and tapering towards the base, compared to a wedge with the point downwards; when very broadly cuneate and rounded at the top, it is often called *flabelliform* or *fan-shaped.*

spathulate, when the broad part near the top is short, and the narrow tapering part long, compared to a spatula or flat ladle.

ovate, when scarcely twice as long as broad, and rather broader below the middle, compared to the longitudinal section of an egg; *obovate* is the same form, with the broadest part above the middle.

orbicular, oval, oblong, elliptical, rhomboidal, etc., when compared to the corresponding mathematical figures.

transversely oblong, or *oblate,* when conspicuously broader than long.

falcate, when curved like the blade of a scythe.

46. Intermediate forms between any two of the above are expressed by combining two terms. Thus, a *linear-lanceolate* leaf is long and narrow, yet broader below the middle, and tapering to a point; a *linear-oblong* one is scarcely narrow enough to be called linear, yet too narrow to be strictly oblong, and does not conspicuously taper either towards the summit or towards the base.

47. The *apex* or *summit* of a leaf is

acute or *pointed,* when it forms an acute angle or tapers to a point.

obtuse or *blunt,* when it forms a very obtuse angle, or more generally when it is more or less rounded at the top.

acuminate or *cuspidate,* when suddenly narrowed at the top, and then more or less prolonged into an *acumen* or *point,* which may be acute or obtuse, linear or tapering. Some botanists make a slight difference between the *acuminate* and *cuspidate* apex, the acumen being more distinct from the rest of the leaf in the latter case than in the former; but in general the two terms are used in the same sense, some preferring the one and some the other.

truncate, when the end is cut off square.

retuse, when very obtuse or truncate, and slightly indented.

emarginate or *notched,* when more decidedly indented at the end of the midrib; *obcordate,* if at the same time approaching the shape of a heart with its point downwards.

mucronate, when the midrib is produced beyond the apex in the form of a small point.

aristate, when the point is fine like a hair.

48. The base of the leaf is liable to the same variations of form as the apex, but the terms more commonly used are *tapering* or *narrowed* for acute and acuminate, *rounded* for obtuse, and *cordate* for emarginate. In all cases the petiole or point of attachment prevent any such absolute termination at the base as at the apex.

49. A leaf may be *cordate* at the base whatever be its length or breadth, or whatever the shape of the two lateral lobes, called *auricles* (or *little ears*), formed by the indenture or notch, but the term *cordiform* or *heart-shaped* leaf is restricted to an ovate and acute leaf, cordate at the base, with rounded auricles. The word auricles is more particularly used as applied to sessile and stem-clasping leaves.

50. If the auricles are pointed, the leaf is more particularly called *auriculate;* it is moreover said to be *sagittate,* when the points are directed downwards, compared to an arrow-head; *hastate,* when the points diverge horizontally, compared to a halbert.

51. A *reniform* leaf is broader than long, slightly but broadly cordate at the base, with rounded auricles, compared to a kidney.

52. In a *peltate* leaf, the stalk, instead of proceeding from the lower edge of the blade, is attached to the under surface, usually near the lower edge, but sometimes in the very centre of the blade. The peltate leaf has usually several principal nerves radiating from the point of attachment, being, in fact, a cordate leaf, with the auricles united.

53. All these modifications of division and form in the leaf pass so gradually one into the other that it is often difficult to say which term is the most applicable—whether the leaf be toothed or lobed, divided or compound, oblong or lanceolate, obtuse or acute, etc. The choice of the most apt expression will depend on the skill of the describer.

54. **Leaves**, when **solid, Stems, Fruits, Tubers,** and other parts of plants, when not flattened like ordinary leaves, are

setaceous or *capillary*, when very slender like bristles or hairs.
acicular, when very slender, but stiff and pointed like needles.
subulate, when rather thicker and firmer like awls.
linear, when at least four times as long as thick; *oblong*, when from about two to about four times as long as thick, the terms having the same sense as when applied to flat surfaces.
ovoid, when egg-shaped, with the broad end downwards, *obovoid* if the broad end is upwards; these terms corresponding to *ovate* and *obovate* shapes in flat surfaces.
globular or *spherical* when corresponding to *orbicular* in a flat surface. *Round* applies to both.
turbinate, when shaped like a top.
conical, when tapering upwards; *obconical* when tapering downwards, if in both cases a transverse section shows a circle.
pyramidal, when tapering upwards; *obpyramidal*, when tapering downwards, if in both cases a transverse section shows a triangle or polygon.
fusiform, or spindle-shaped, when tapering at both ends; *cylindrical*, when not tapering at either end, if in both cases the transverse section shows a circle, or sometimes irrespective of the transverse shape.
terete, when the transverse section is not angular; *trigonous, triquetrous*, if the transverse section shows a triangle, irrespective in both cases of longitudinal form.
compressed, when more or less flattened laterally; *depressed*, when more or less flattened vertically, or at any rate at the top; *obcompressed* (in the achenes of *Compositæ*), when flattened from front to back.
articulate or *jointed*, if at any period of their growth (usually when fully formed and approaching their decay, or in the case of fruits when quite ripe) they separate, without tearing, into two or more pieces placed end to end. The joints where they separate are called *articulations*, each separate piece an *article*. The name of *joint* is, in common language, given both to the articulation and the article, but more especially to the former. Some modern botanists, however, propose to restrict it to the article, giving the name of *joining* to the articulation.
didymous, when slightly two-lobed, with rounded obtuse lobes.
moniliform, or beaded, when much contracted at regular intervals, but not separating spontaneously into articles.

55. In their consistence **Leaves** or other organs are

fleshy, when thick and soft; *succulent* is generally used in the same sense, but implies the presence of more juice.
coriaceous, when firm and stiff, or very tough, of the consistence of leather.
crustaceous, when firm and brittle.
membranous, when thin and not stiff.
scarious or *scariose*, when very thin, more or less transparent and not green, yet rather stiff.

56. The terms applied botanically to the consistence of solids are those in general use in common language.

57. The mode in which unexpanded leaves are disposed in the leaf-bud is called their *vernation præfoliation;* it varies considerably, and technical terms have been proposed to express some of its varieties, but it has been hitherto rarely noticed in descriptive botany.

§ 6. *Scales, Bracts, and Stipules.*

58. **Scales** (*Squamæ*) are leaves very much reduced in size, usually sessile, seldom green or capable of performing the respiratory functions of leaves. In other words, they are organs resembling leaves in their position on the plant, but differing in size, colour, texture, and functions. They are most frequent on the stock of perennial plants, or at the base of annual branches, especially on the buds of future shoots, when they serve apparently to protect the dormant living germ from the rigour of winter. In the latter case they are usually short, broad, close together, and more or less *imbricated*, that is, overlapping each other like the tiles of a roof. It is this arrangement, as well as their usual shape that has suggested the name of *scales*, borrowed from the scales of a fish. Imbricated scales, bracts, or leaves, are said to be *squarrose*, when their tips are pointed and very spreading or recurved.

59. Sometimes, however, most or all the leaves of the plant are reduced to small scales, in which case they do not appear to perform any particular function. The name of *scales* is also given to any small broad scale-like appendages or reduced organs, whether in the flower or any other part of the plant.

60. **Bracts** (*Bracteæ*) are the upper leaves of a plant in flower (either all those of the flowering branches, or only one or two immediately under the flower), when different from the stem-leaves in size, shape, colour, or arrangement. They are generally much smaller and more sessile. They often partake of the colour of the flower, although they very frequently also retain the green colour of the leaves. When small they are often called *scales*.

61. *Floral leaves* or *leafy bracts* are generally the lower bracts on the upper leaves at the base of the flowering branches, intermediate in size, shape, or arrangement, between the stem-leaves and the upper bracts.

62. *Bracteoles* are the one or two last bracts under each flower, when they differ materially in size, shape, or arrangement from the other bracts.

63. **Stipules** are leaf-like or scale-like appendages at the base of the leaf-stalk, or on the node of the stem. When present there are generally two, one on each side of the leaf, and they sometimes appear to protect the young leaf before it is developed. They are, however, exceedingly variable in size and appearance, sometimes exactly like the true leaves except that they have no buds in their axils, or looking like the leaflets of a compound leaf, sometimes apparently the only leaves of the plant; generally small and narrow, sometimes reduced to minute scales, spots or scars, sometimes united into one opposite the leaf, or more or less united with, or *adnate* to the petiole, or quite detached from the leaf, and forming a ring or sheath round the stem in the axil of the leaf. In a great number of plants they are entirely wanting.

64. *Stipellæ*, or secondary stipules, are similar organs, sometimes found on compound leaves at the points where the leaflets are inserted.

65. When scales, bracts, or stipules, or almost any part of the plant besides leaves and flowers are stalked, they are said to be *stipitate*, from *stipes*, a *stalk*.

§ 7. *Inflorescence and its Bracts.*

66. The **Inflorescence** of a plant is the arrangement of the flowering branches, and of the flowers upon them. An *Inflorescence* is a flowering branch, or the flowering summit of a plant above the last stem-leaves, with its branches, bracts, and flowers.

67. A single flower, or an inflorescence, is *terminal* when at the summit of a stem or leafy branch, *axillary* when in the axil of a stem-leaf, *leaf-opposed* when opposite to a stem-leaf. The inflorescence of a plant is said to be *terminal* or *determinate* when the main stem and principal branches end in a flower or inflorescence (not in a leaf-bud), *axillary* or *indeterminate* when all the flowers or inflorescence are axillary, the stem or branches ending in leaf-buds.

68. A *Peduncle* is the stalk of a solitary flower, or of an inflorescence; that is to say, the portion of the flowering branch from the last stem-leaf to the flower, or to the first ramification of the inflorescence, or even up to its last ramifications; but the portion extending from the first to the last ramifications or the axis of inflorescence is often distinguished under the name of *rhachis*.

69. A *Scape* or *radical Peduncle* is a leafless peduncle proceeding from the stock, or from near the base of the stem, or apparently from the root itself.

70. A *Pedicel* is the last branch of an inflorescence, supporting a single flower.

71. The branches of inflorescences may be, like those of stems, opposite, alternate, etc. (32, 33), but very often their arrangement is different from that of the leafy branches of the same plant.

72. **Inflorescence** is

centrifugal, when the terminal flower opens first, and those on the lateral branches are successively developed.

centripetal, when the lowest flowers open first, and the main stem continues to elongate, developing fresh flowers.

73. Determinate inflorescence is usually centrifugal. Indeterminate inflorescence is always centripetal. Both inflorescences may be combined on one plant, for it often happens that the main branches of an inflorescence are centripetal, whilst the flowers on the lateral branches are centrifugal; or *vice versâ*.

74. An **Inflorescence** is

a *Spike*, or *spicate*, when the flowers are sessile along a simple undivided axis or rhachis.

a *Raceme*, or *racemose*, when the flowers are borne on pedicels along a single undivided axis or rhachis.

a *Panicle*, or *paniculate*, when the axis is divided into branches bearing two or more flowers.

a *Head*, or *capitate*, when several sessile or nearly sessile flowers are collected into a compact head-like cluster. The short, flat, convex or conical axis on which the flowers are seated is called the *receptacle*, a term also used for the torus of a single flower (135). The very compact flower-heads of *Compositæ* are often termed *compound flowers*.

an *Umbel*, or *umbellate*, when several branches or pedicels appear to start from the same point and are nearly of the same length. It differs from the head, like the raceme from the spike, in that the flowers are not sessile. An umbel is said to be *simple* when each of its branches or *rays* bears a single flower; *compound*, when each ray bears a *partial umbel* or *umbellule*.

a *Corymb*, or *corymbose*, when the branches and pedicels, although starting from different points, all attain the same level, the lower ones being much longer than the upper. It is a flat-topped or *fastigiate* panicle.

a *Cyme* or *cymose* when branched and centrifugal. It is a centrifugal panicle, and is often corymbose. The central flower opens first. The lateral branches successively developed are usually forked or opposite (dichotomous or trichotomous), but sometimes after the first forking the branches are no longer divided, but produce a succession of pedicels on their upper side, forming apparently unilateral centripetal racemes; whereas if attentively examined it will be found that each pedicel is at first terminal, but becomes lateral by the development of one outer branch only, immediately under the pedicel. Such branches, when in bud, are generally rolled back at the top, like the tail of a scorpion; and are thence called *scorpioid*.

a *Thyrsus*, or *thyrsoid*, when cymes, usually opposite, are arranged in a narrow pyramidal panicle.

75. There are numerous cases where inflorescences are intermediate between some two of the above, and are called by different botanists by one or the other name, according as they are guided by apparent or by theoretical similarity. A spike-like panicle, where the axis is divided into very short branches forming a cylindrical compact inflorescence, is called sometimes a spike, sometimes a panicle. If the flowers are in distinct clusters along a simple axis, the inflorescence is described as an *interrupted* spike or raceme, according as the flowers are nearly sessile or distinctly pedicellate; although when closely examined the flowers will be found to be inserted not on the main axis, but on a very short branch, thus, strictly speaking, constituting a panicle.

76. The *catkins (amenta)* of *Amentaceæ*, the *spadices* of several Monocotyledons, the *ears* and *spikelets* of Grasses are forms of the spike.

77. **Bracts** are generally placed singly under each branch of the inflorescence, and under each pedicel; bracteoles are usually two, one on each side, on the pedicel or close under the flower, or even upon the calyx itself; but bracts are also frequently scattered along the branches without axillary pedicels; and when the differences between the bracts and bracteoles are trifling or immaterial, they are usually all called bracts.

78. When three bracts appear to proceed from the same point, they will, on examination, be found to be really either one bract and two stipules, or one bract with two bracteoles in its axils. When two bracts appear to proceed from the same point, they will usually be found to be the stipules of an undeveloped bract, unless the branches of the inflorescence are opposite, when the bracts will of course be opposite also.

79. When several bracts are collected in a whorl, or are so close together as to appear whorled, or are closely imbricated round the base of a head or umbel, they are collectively called an *Involucre*. The bracts composing an involucre are described under the names of *leaves*, *leaflets*, *bracts*, or *scales*, according to their appearance. *Phyllaries* is a useless term, lately introduced for the bracts or scales of the involucre of *Compositæ*. An *Involucel* is the involucre of a partial umbel.

80. When several very small bracts are placed round the base of a calyx or of an involucre, they have been termed a *calycule*, and the calyx or involucre said to be *calyculate*, but these terms are now falling into disuse, as conveying a false impression.

81. A *Spatha* is a bract or floral leaf enclosing the inflorescence of some Monocotyledons.

82. *Paleæ*, *Pales*, or *Chaff* are the inner bracts or scales in *Compositæ*, *Gramineæ*, and some other plants, when of a thin yet stiff consistence, usually narrow and of a pale colour.

83. *Glumes* are the bracts enclosing the flowers of *Cyperaceæ* and *Gramineæ*.

§ 8. *The Flower in General.*

84. A *complete* **Flower** (15) is one in which the calyx, corolla, stamens, and pistils are all present; a *perfect* flower, one in which all these organs, or such of them as are present, are capable of performing their several functions. Therefore, properly speaking, an *incomplete* flower is one in which any one or more of these organs is wanting; and an *imperfect* flower, one in which any one or more of these organs is so altered as to be incapable of properly performing its functions. These imperfect organs are said to be *abortive* if much reduced in size or efficiency, *rudimentary* if so much so as to be scarcely perceptible. But, in many works, the term *incomplete* is specially applied to those flowers in which the perianth is simple or wanting, and *imperfect* to those in which either the stamens or pistils are imperfect or wanting.

85. A **Flower** is
dichlamydeous, when the perianth is double, both calyx and corolla being present and distinct.
monochlamydeous, when the perianth is single, whether by the union of the calyx and corolla, or the deficiency of either.

asepalous, when there is no calyx.
apetalous, when there is no corolla.
naked, when there is no perianth at all.
hermaphrodite or *bisexual*, when both stamens and pistil are present and perfect.
male or *staminate*, when there are one or more stamens, but either no pistil at all or an imperfect one.
female or *pistillate*, when there is a pistil, but either no stamens at all, or only imperfect ones.
neuter, when both stamens and pistil are imperfect or wanting.
barren or *sterile*, when from any cause it produces no seed.
fertile, when it does produce seed. In some works the terms *barren*, *fertile*, and *perfect* are also used respectively as synonyms of *male*, *female*, and *hermaphrodite*.

86. The flowers of a plant or species are said collectively to be *unisexual* or *declinous* when the flowers are all either male or female.
monœcious, when the male and female flowers are distinct, but on the same plant.
diœcious, when the male and female flowers are on distinct plants.
polygamous, when there are male, female, and hermaphrodite flowers on the same or on distinct plants.

87. A head of flowers is *heterogamous* when male, female, hermaphrodite, and neuter flowers, or any two or three of them, are included in one head; *homogamous*, when all the flowers included in one head are alike in this respect. A spike or head of flowers is *androgynous* when male and female flowers are mixed in it. These terms are only used in the case of very few Natural Orders.

88. As the scales of buds are leaves undeveloped or reduced in size and altered in shape and consistence, and bracts are leaves likewise reduced in size, and occasionally altered in colour; so the parts of the flower are considered as leaves still further altered in shape, colour, and arrangement round the axis, and often more or less combined with each other. The details of this theory constitute the comparatively modern branch of botany called *Vegetable Metamorphosis*, or *Homology*, sometimes improperly termed *Morphology* (8).

89. To understand the arrangement of the floral parts, let us take a *complete* flower, in which moreover all the parts are *free* from each other, *definite* in number, *i.e.* always the same in the same species, and *symmetrical* or *isomerous*, *i.e.* when each whorl consists of the same number of parts.

90. Such a complete symmetrical flower consists usually of either four or five whorls of altered leaves (88), placed immediately one within the other.

The **Calyx** forms the outer whorl. Its parts are called *sepals*.

The **Corolla** forms the next whorl. Its parts, called *petals*, usually *alternate* with the sepals; that is to say, the centre of each petal is immediately over or within the interval between two sepals.

The **Stamens** form one or two whorls within the petals. If two, those of the outer whorl (the *outer stamens*) alternate with the petals, and are consequently opposite to, or over the centre of the sepals; those of the inner whorl (the *inner stamens*) alternate with the outer ones, and are therefore opposite to the petals. If there is only one whorl of stamens, they most frequently alternate with the petals; but sometimes they are opposite the petals and alternate with the sepals.

The **Pistil** forms the inner whorl; its carpels usually alternate with the inner row of stamens.

91. In an axillary or lateral flower the *upper* parts of each whorl (sepals, petals, stamens, or carpels) are those which are next to the main axis of the stems or branch, the *lower* parts those which are furthest from it; the intermediate ones are said to be *lateral*. The words *anterior* (front) and *posterior* (back) are often used for lower and upper respectively, but their meaning is sometimes reversed if the writer supposes himself in the centre of the flower instead of outside of it.

92. The number of parts in each whorl of a flower is expressed adjectively by the following numerals derived from the Greek:—

mono-, di-, tri-, tetra-, penta-, hexa-, hepta-, octo-, ennea-, deca-, etc., poly-
1., 2., 3., 4., 5., 6., 7., 8., 9., 10., many.

prefixed to a termination indicating the whorl referred to.

93. Thus, a **Flower** is
disepalous, *trisepalous*, *tetrasepalous*, *polysepalous*, etc., according as there are 2, 3, 4, or many (or an indefinite number of) sepals.
dipetalous, *tripetalous*, *polypetalous*, etc., according as there are 2, 3, or many petals.
diandrous, *triandrous*, *polyandrous*, etc., according as there are 2, 3, or many stamens.
digynous, *trigynous*, *polygynous*, etc., according as there are 2, 3, or many carpels.

And generally (if symmetrical), *dimerous*, *trimerous*, *polymerous*, etc., according as there are 2, 3, or many (or an indefinite number of) parts to each whorl.

94. Flowers are *unsymmetrical* or *anisomerous*, strictly speaking, when any one of the whorls has a different number of parts from any other; but when the pistils alone are reduced in number, the flower is still frequently called symmetrical or isomerous, if the calyx, corolla, and staminal whorls have all the same number of parts.

95. Flowers are *irregular* when the parts of any one of the whorls are unequal in size, dissimilar in shape, or do not spread regularly round the axis at equal distances. It is however more especially irregularity of the corolla that is referred to in descriptions. A slight inequality in size or direction in the other whorls does not prevent the flower being classed as *regular*, if the corolla or perianth is conspicuous and regular.

§ 9. *The Calyx and Corolla, or Perianth.*

96. The **Calyx** (90) is usually green, and smaller than the corolla; sometimes very minute, rudimentary, or wanting, sometimes very indistinctly whorled, or not whorled at all, or in two whorls, or composed of a large number of sepals, of which the outer ones pass gradually into bracts, and the inner ones into petals.

97. The **Corolla** (90) is usually coloured, and of a more delicate texture than the calyx, and, in popular language, is often more specially meant by the *flower*. Its petals are more rarely in two whorls, or indefinite in number, and the whorl more rarely broken than in the case of the calyx, at least when the plant is in a natural state. *Double flowers* are in most cases an accidental deformity or monster in which the ordinary number of petals is multiplied by the conversion of stamens, sepals, or even carpels into petals, by the division of ordinary petals, or simply by the addition of supernumerary ones. Petals are also sometimes very small, rudimentary, or entirely deficient.

98. In very many cases, a so-called *simple perianth* (15) (of which the parts are usually called *leaves* or *segments*) is one in which the sepals and petals are similar in form and texture, and present apparently a single whorl. But if examined in the young bud, one half of the parts will generally be found to be placed outside the other half, and there will frequently be some slight difference in texture, size, and colour, indicating to the close observer the presence of both calyx and corolla. Hence much discrepancy in descriptive works. Where one botanist describes a simple perianth of six segments, another will speak of a double perianth of three sepals and three petals.

99. The following terms and prefixes, expressive of the modifications of form and arrangement of the corolla and its petals, are equally applicable to the calyx and its sepals, and to the simple perianth and its segments.

100. The Corolla is said to be *monopetalous* when the petals are united, either entirely or at the base only, into a cup, tube, or ring; *polypetalous* when they are all free from the base. These expressions, established by a long usage, are not strictly correct, for *monopetalous* (consisting of a single petal) should apply rather to a corolla really reduced to a single petal, which would then be on one side of the axis; and *polypetalous* is sometimes used more appropriately for a corolla with an indefinite number of petals. Some modern botanists have therefore proposed the term *gamopetalous* for the corolla with united petals, and *dialypetalous* for that with free petals; but the old-established expressions are still the most generally used.

101. When the petals are partially united, the lower entire portion of the corolla is called the *tube*, whatever be its shape, and the free portions of the petals are called the *teeth*, *lobes*, or *segments* (39), according as they are short or long in proportion to the whole length of the corolla. When the tube is excessively short, the petals appear at first sight free, but their slight union at the base must be carefully attended to, being of importance in classification.

102. The **Æstivation** of a corolla, is the arrangement of the petals, or of such portion of them as is free, in the unexpanded bud. It is

valvate, when they are strictly whorled in their whole length, their edges being placed against each other without overlapping. If the edges are much inflexed, the æstivation is at the same time induplicate; *involute*, if the margins are rolled inward; *reduplicate*, if the margins project outwards into salient angles; *revolute*, if the margins are rolled outwards; *plicate*, if the petals are folded in longitudinal plaits.

imbricate, when the whorl is more or less broken by some of the petals being outside the others, or by their overlapping each other at least at the top. Five-petalled imbricate corollas are *quincuncially* imbricate when one petal is outside, and an adjoining one wholly inside, the three others intermediate and overlapping on one side; *bilabiate*, when two adjoining ones are inside or outside the three others. Imbricate petals are described as *crumpled* (*corrugate*) when puckered irregularly in the bud.

twisted, *contorted*, or *convolute* when each petal overlaps an adjoining one on one side, and is overlapped by the other adjoining one on the other side. Some botanists include the twisted æstivation in the general term *imbricate*; others carefully distinguish the one from the other.

103. In a few cases the overlapping is so slight that the three æstivations cannot easily be distinguished one from the other; in a few others the æstivation is variable, even in the same species, but, in general, it supplies a constant character in species, in genera, or even in Natural Orders.

104. In general shape the **Corolla** is

tubular, when the whole or the greater part of it is in the form of a tube or cylinder.

campanulate, when approaching in some measure the shape of a cup or bell.

urceolate, when the tube is swollen or nearly globular, contracted at the top, and slightly expanded again in a narrow rim.

rotate or *stellate*, when the petals or lobes are spread out horizontally from the base, or nearly so, like a wheel or star.

hypocrateriform or *salver-shaped*, when the lower part is cylindrical and the upper portion expanded horizontally. In this case the name of *tube* is restricted to the cylindrical part, and the horizontal portion is called the *limb*, whether it be divided to the base or not. The orifice of the tube is called its *mouth* or *throat*.

infundibuliform or *funnel-shaped*, when the tube is cylindrical at the base, but enlarged at the top into a more or less campanulate limb, of which the lobes often spread horizontally. In this case the campanulate part, up to the commencement of the lobes, is sometimes considered as a portion of the tube, sometimes as a portion of the limb, and by some botanists again described as independent of either, under the name of *throat (fauces)*. Generally speaking, however, in campanulate, infundibuliform, or other corollas, where the lower entire part passes gradually into the upper divided and more spreading part, the distinction between the *tube* and the *limb* is drawn either at the point where the lobes separate, or at the part where the corolla first expands, according to which is the most marked.

105. Irregular corollas have received various names according to the more familiar forms they have been compared to. Some of the most important are the

bilabiate or *two-lipped* corolla, when, in a four or five-lobed corolla, the two or three upper lobes stand obviously apart, like an *upper lip*, from the two or three lower ones, or *under lip*. In Orchideæ and some other families the name of lip, or *labellum*, is given to one of the divisions or lobes of the perianth.

personate, when two-lipped, and the orifice of the tube closed by a projection from the base of the upper or lower lip, called a *palate*.

ringent, when very strongly two-lipped, and the orifice of the tube very open.

spurred, when the tube or the lower part of the petal has a conical hollow projection, compared to the spur of a cock ; *saccate*, when the spur is short and round like a little bag ; *gibbous*, when projecting at any part into a slight swelling ; *foveolate*, when marked in any part with a slight glandular or thickened cavity.

resupinate or *reversed*, when a lip, spur, etc., which in allied species is usually lowest, lies uppermost, and *vice versâ*.

106. The above terms are mostly applied to the forms of monopetalous corollas, but several are also applicable to those of polypetalous ones. Terms descriptive of the special forms of corolla in certain Natural Orders, will be explained under those Orders respectively.

107. Most of the terms used for describing the forms of leaves (39, 45) are also applicable to those of individual petals ; but the flat expanded portion of a petal, corresponding to the blade of the leaf, is called its *lamina*, and the stalk, corresponding to the petiole, its *claw (unguis)*. The stalked petal is said to be *unguiculate*.

§ 10. *The Stamens.*

108. Although in a few cases the outer stamens may gradually pass into petals, yet, in general, **Stamens** are very different in shape and aspect from leaves, sepals, or petals. It is only in a theoretical point of view (not the less important in the study of the physiological economy of the plant) that they can be called altered leaves.

109. This usual form is a stalk, called the *filament*, bearing at the top an *anther* divided into two pouches or *cells*. These anther-cells are filled with *pollen*, consisting of minute grains, usually forming a yellow dust, which, when the flower expands, is scattered from an opening in each cell. When the two cells are not closely contiguous, the portion of the anther that unites them is called the *connectivum*.

110. The filament is often wanting, and the anther sessile, yet still the stamen is perfect ; but if the anther, which is the essential part of the stamen, is wanting, or does not contain pollen, the stamen is imperfect, and is then said to be *barren* or *sterile* (without pollen), *abortive*, or *rudimentary* (84), according to the degree to which the imperfection is carried. Imperfect stamens are often called *staminodia*.

111. In unsymmetrical flowers, the stamens of each whorl are sometimes reduced in number below that of the petals, even to a single one, and in several Natural Orders they are multiplied indefinitely.

112. The terms *monandrous* and *polyandrous* are restricted to flowers which have really but one stamen, or an indefinite number respectively. Where several stamens are united into one, the flower is said to be *synandrous*.

113. **Stamens** are

monadelphous, when united by their filaments into one cluster. This cluster either forms a tube round the pistil, or, if the pistil is wanting, occupies the centre of the flower.

diadelphous, when so united into two clusters. The term is more especially applied to certain *Leguminosæ,* in which nine stamens are united in a tube slit open on the upper side, and a tenth, placed in the slit, is free. In some other plants the stamens are equally distributed in the two clusters.

triadelphous, pentadelphous, polyadelphous, when so united into three, five, or many clusters.

syngenesious, when united by their anthers in a ring round the pistil, the filaments usually remaining free.

didynamous, when (usually in a bilabiate flower) there are four stamens in two pairs, those of one pair longer than those of the other.

tetradynamous, when (in *Cruciferæ*) there are six, four of them longer than the two others.

exserted, when longer than the corolla, or even when longer than its tube, if the limb be very spreading.

114. An **Anther** (109) is

adnate, when continuous with the filament, the anther-cells appearing to lie their whole length along the upper part of the filament.

innate, when firmly attached by their base to the filament. This is like an adnate anther, but rather more distinct from the filament.

versatile, when attached by their back to the very point of the filament, so as to swing loosely.

115. Anther-cells may be *parallel* or *diverging* at a less or greater angle, or *divaricate,* when placed end to end so as to form one straight line. The end of each anther-cell placed nearest to the other cell is generally called its *apex* or *summit,* and the other end its *base* (36); but some botanists reverse the sense of these terms.

116. Anthers have often, on their connectivum or cells, appendages termed *bristles* (setæ), *spurs, crests, points, glands,* etc., according to their appearance.

117. Anthers have occasionally only one cell; this may take place either by the disappearance of the partition between two closely contiguous cells, when these cells are said to be *confluent;* or by the abortion or total deficiency of one of the cells, when the anther is said to be *dimidiate.*

118. Anthers will open, or *dehisce,* to let out the pollen, like capsules, in *valves, pores,* or *slits.* Their dehiscence is *introrse,* when the opening faces the pistil; *extrorse,* when towards the circumference of the flower.

119. Pollen (109) is not always in the form of dust. It is sometimes collected in each cell into one or two little wax-like masses. Special terms used in describing these masses or other modifications of the pollen will be explained under the Orders where they occur.

§ 11. *The Pistil.*

120. The carpels (91) of the **Pistil,** although they may occasionally assume, rather more than stamens, the appearance and colour of leaves, are still more different in shape and structure. They are usually sessile; if stalked, their stalk is called a *podocarp.* This stalk, upon which each separate carpel is supported above the receptacle, must not be confounded with the *gynobasis* (143), upon which the whole pistil is sometimes raised.

121. Each carpel consists of three parts:

1. The **Ovary,** or enlarged base, which includes one or more cavities or *cells,* containing one or more small bodies called *ovules.* These are the earliest condition of the future seeds.

2. the **Style,** proceeding from the summit of the ovary, and supporting—

3. the **Stigma,** which is sometimes a point (or *punctiform* stigma) or small head (a *capitate* stigma) at the top of the style or ovary, sometimes a portion of its surface more or less lateral and variously shaped, distinguished by a looser texture, and covered with minute protuberances called *papillæ.*

122. The style is often wanting, and the stigma is then sessile on the ovary, but in the perfect pistil there is always at least one ovule in the ovary, and some portion of stigmatic surface. Without these the pistil is imperfect, and said to be *barren* (not setting seed), *abortive,* or *rudimentary* (84), according to the degree of imperfection.

123. The ovary being the essential part of the pistil, most of the terms relating to the number, arrangement, etc., of the carpels, apply specially to their ovaries. In some works each separate carpel is called a pistil, all those of a flower constituting together the *gynæcium;* but this term is in little use, and the word *pistil* is more generally applied in a collective sense. When the ovaries are at all united, they are commonly termed collectively a compound ovary.

124. The number of carpels or ovaries in a flower is frequently reduced below that of the parts of the other floral whorls, even in flowers otherwise symmetrical. In a very few genera, however, the ovaries are more numerous than the petals, or indefinite. They are in that case either arranged in a single whorl, or form a head or spike in the centre of the flower.

125. The terms *monogynous, digynous, polygynous,* etc. (with a pistil of one, two or more parts), are vaguely used, applying sometimes to the whole pistil, sometimes to the ovaries alone, or to the styles or stigmas only. Where a more precise nomenclature is adopted, the flower is

monocarpellary, when the pistil consists of a single simple carpel.

bi-, tri-, etc., to *poly-carpellary,* when the pistil consists of two, three, or an indefinite number of carpels, whether separate or united.

syncarpous, when the carpels or their ovaries are more or less united into one compound ovary.

apocarpous, when the carpels or ovaries are all free and distinct.

126. A *compound* ovary is

unilocular or *one-celled*, when there are no partitions between the ovules, or when these partitions do not meet in the centre so as to divide the cavity into several cells.

plurilocular or *several-celled*, when completely divided into two or more cells by partitions called *dissepiments* (*septa*), usually vertical and radiating from the centre or axis of the ovary to its circumference.

bi-, tri-, etc., to *multi-locular*, according to the number of these cells, two, three, etc., or many.

127. In general the number of cells or of dissepiments, complete or partial, or of rows of ovules, corresponds with that of the carpels, of which the pistil is composed. But sometimes each carpel is divided completely or partially into two cells, or has two rows of ovules, so that the number of carpels appears double what it really is. Sometimes again the carpels are so completely combined and reduced as to form a single cell, with a single ovule, although it really consist of several carpels. But in these cases the ovary is usually described as it appears, as well as such as it is theoretically supposed to be.

128. In apocarpous pistils the styles are usually free, each bearing its own stigma. Very rarely the greater part of the styles or the stigmas alone, are united, whilst the ovaries remain distinct.

129. Syncarpous flowers are said to have

several styles, when the styles are free from the base.

one style, with several branches, when the styles are connected at the base, but separate below the point where the stigmas or stigmatic surfaces commence.

one simple style, with several stigmas, when united up to the point where the stigmas or stigmatic surfaces commence, and then separating.

one simple style, with a branched, lobed, toothed, notched, or entire stigma (as the case may be), when the stigmas also are more or less united. In many works, however, this precise nomenclature is not strictly adhered to, and considerable confusion is often the result.

130. In general the number of styles, or branches of the style or stigma, is the same as that of the carpels, but sometimes that number is doubled, especially in the stigmas, and sometimes the stigmas are dichotomously or pinnately branched, or *penicillate*, that is, divided into a tuft of hair-like branches. All these variations sometimes make it a difficult task to determine the number of carpels forming a compound ovary, but the point is of considerable importance in fixing the affinities of plants, and by careful consideration, the real as well as the apparent number has now in most cases been agreed upon.

131. The *Placenta* is the part of the inside of the ovary to which the ovules are attached, sometimes a mere point or a line on the inner surface, often more or less thickened or raised. *Placentation* is therefore the indication of the part of the ovary to which the ovules are attached.

132. Placentas are

axile, when the ovules are attached to the axis or centre, that is, in plurilocular ovaries, when they are attached to the inner angle of each cell; in unilocular simple ovaries, which have almost always an excentrical style or stigma, when the ovules are attached to the side of the ovary nearest to the style; in unilocular compound ovaries, when the ovaries are attached to a central protuberance, column, or axis rising up from the base of the cavity. If this column does not reach the top of the cavity, the placenta is said to be *free* and *central*.

parietal, when the ovules are attached to the inner surface of the cavity of a one-celled compound ovary. Parietal placentas are usually slightly thickened or raised lines, sometimes broad surfaces nearly covering the inner surface of the cavity, sometimes projecting far into the cavity, and constituting partial dissepiments, or even meeting in the centre, but without cohering there. In the latter case the distinction between the one-celled and the several-celled ovary sometimes almost disappears.

133. Each **Ovule** (121), when fully formed, usually consists of a central mass or *nucleus* enclosed in two bag-like *coats*, the outer one called *primine*, the inner one *secundine*. The *chalaza* is the point of the ovule at which the base of the nucleus is confluent with the coats. The *foramen* is a minute aperture in the coats over the *apex* of the nucleus.

134. **Ovules** are

orthotropous or *straight*, when the chalaza coincides with the base (86) of the ovule, and the foramen is at the opposite extremity, the axis of the ovule being straight.

campylotropous or *incurved*, when the chalaza still coinciding with the base of the ovule, the axis of the ovule is curved, bringing the foramen more or less towards that base.

anatropous or *inverted*, when the chalaza is at the apex of the ovule, and the foramen next to its base, the axis remaining straight. In this, one of the most frequent forms of the ovule, the chalaza is connected with the base by a cord, called the *raphe*, adhering to one side of the ovule, and becoming more or less incorporated with its coats, as the ovule enlarges into a seed.

amphitropous, or *half-inverted*, when the ovule being as it were attached laterally, the chalaza and foramen at opposite ends of its straight or curved axis are about equally distant from the base or point of attachment.

§ 12. *The Receptacle and Relative Attachment of the Floral Whorls.*

135. The **Receptacle** or *torus* is the extremity of the peduncle (above the calyx), upon which the corolla, stamens, and ovary are inserted. It is sometimes little more than a mere point or minute hemisphere, but it is often also more or less elongated, thickened, or otherwise enlarged. It must not be confounded with the receptacle of inflorescence (74).

136. A *Disk*, or *disc*, is a circular enlargement of the receptacle, usually in the form of a cup (*cupular*), of a flat disk or quoit, or of a cushion (*pulvinate*). It is either immediately at the base of the ovary within the stamens, or between the petals and stamens, or bears the petals or stamens or both on its margin, or is quite at the extremity of the receptacle, with the ovaries arranged in a ring round it or under it.

137. The disk may be *entire*, or *toothed*, or *lobed*, or *divided* into a number of parts, usually equal to or twice that of the stamens or carpels. When the parts of the disk are quite separate and short, they are often called glands.

138. *Nectaries* are either the disk, or small deformed petals, or abortive stamens, or appendages at the base of petals or stamens, or any small bodies within the flower which do not look like petals, stamens, or ovaries. They were formerly supposed to supply bees with their honey, and the term is frequently to be met with in the older Floras, but is now deservedly going out of use.

139. When the disk bears the petals and stamens, it is frequently adherent to, and apparently forms part of, the tube of the calyx, or it is adherent to, and apparently forms part of, the ovary, or of both calyx-tube and ovary. Hence the three following important distinctions in the relative insertion of the floral whorls.

140. Petals, or as it is frequently expressed, flowers, are

hypogynous (*i.e.* under the ovary), when they or the disk that bears them are entirely free both from the calyx and ovary. The ovary is then described as *free* or *superior*; the calyx as *free* or *inferior*, the petals as being *inserted on the receptacle*.

perigynous (*i.e.* round the ovary), when the disk bearing the petals is quite free from the ovary, but is more or less combined with the base of the calyx-tube. The ovary is then still described as *free* or *superior*, even though the combined disk and calyx-tube may form a deep cup with the ovary lying in the bottom; the calyx is said to be *free* or *inferior*, and the petals are described as *inserted on the calyx*.

epigynous (*i.e.* upon the ovary), when the disk bearing the petals is combined both with the base of the calyx-tube and the base outside of the ovary; either closing over the ovary so as only to leave a passage for the style, or leaving more or less of the top of the ovary free, but always adhering to it above the level of the insertion of the lowest ovule (except in a very few cases where the ovules are absolutely suspended from the top of the cell). In epigynous flowers the ovary is described as *adherent* or *inferior*, the calyx as *adherent* or *superior*, the petals as *inserted on* or *above the ovary*. In some works, however, most epigynous flowers are included in the perigynous ones, and a very different meaning is given to the term *epigynous* (144), and there are a few cases where no positive distinction can be drawn between the epigynous and perigynous flowers, or again between the perigynous and hypogynous flowers.

141. When there are no petals, it is the insertion of the stamens that determines the difference between the hypogynous, perigynous, and epigynous flowers.

142. When there are both petals and stamens,

in hypogynous flowers, the petals and stamens are usually free from each other, but sometimes they are combined at the base. In that case, if the petals are distinct from each other, and the stamens are monadelphous, the petals are often said to be *inserted on* or *combined with the staminal tube*; if the corolla is gamopetalous and the stamens distinct from each other, the latter are said to be *inserted in the tube of the corolla*.

In perigynous flowers, the stamens are usually inserted immediately within the petals, or alternating with them on the edge of the disk, but occasionally much lower down within the disk, or even on the unenlarged part of the receptacle.

In epigynous flowers, when the petals are distinct, the stamens are usually inserted as in perigynous flowers; when the corolla is gamopetalous, the stamens are either free and hypogynous, or combined at the base with (inserted in) the tube of the corolla.

143. When the receptacle is distinctly elongated below the ovary, it is often called a *gynobasis*, *gynophore*, or *stalk of the ovary*. If the elongation takes place below the stamens or below the petals, these stamens or petals are then said to be *inserted on the stalk of the ovary*, and are occasionally, but falsely, described as *epigynous*. Really epigynous stamens (*i.e.* when the filaments are combined with the ovary) are very rare, unless the rest of the flower is epigynous.

144. An *epigynous disk* is a name given either to the thickened summit of the ovary in epigynous flowers, or very rarely to a real disk or enlargement of the receptacle closing over the ovary.

145. In the relative position of any two or more parts of the flower, whether in the same or in different whorls, they are

connivent, when nearer together at the summit than at the base.

divergent, when further apart at the summit than at the base.

coherent, when united together, but so slightly that they can be separated with little or no laceration; and one of the two cohering parts (usually the smallest or least important) is said to be *adherent* to the other. Grammatically speaking, these two terms convey nearly the same meaning, but require a different form of phrase; practically however it has been found more convenient to restrict *cohesion* to the union of parts of the same whorl, and *adhesion* to the union of parts of different whorls.

connate, when so closely united that they cannot be separated without laceration. Each of the two connate parts, and especially that one which is considered the smaller or of the least importance, is said to be *adnate* to the other.

free, when neither coherent nor connate.

distinct is also used in the same sense, but is also applied to parts distinctly visible or distinctly limited.

§ 13. *The Fruit.*

146. The **Fruit** (15) consists of the ovary and whatever other parts of the flower are *persistent* (i.e. persist at the time the seed is ripe), usually enlarged, and more or less altered in shape and consistence. It encloses or covers the seed or seeds till the period of maturity, when it either opens for the seed to escape, or falls to the ground with the seed. When stalked, its stalk has been termed a *carpophore*.

147. Fruits are, in elementary works, said to be *simple* when the result of a single flower, *compound* when they proceed from several flowers closely packed or combined in a head. But as a fruit resulting from a single flower, with several distinct carpels, is compound in the sense in which that term is applied to the ovary, the terms *single* and *aggregate*, proposed for the fruit resulting from one or several flowers, may be more appropriately adopted. In descriptive botany a fruit is always supposed to result from a single flower unless the contrary be stated. It may, like the pistil, be syncarpous or apocarpous (125); and as in many cases carpels united in the flower may become separate as they ripen, an apocarpous fruit may result from a syncarpous pistil.

148. The involucre or bracts often persist and form part of aggregate fruits, but very seldom so in single ones.

149. The receptacle becomes occasionally enlarged and succulent; if when ripe it falls off with the fruit, it is considered as forming part of it.

150. The adherent part of the calyx of epigynous flowers always persists and forms part of the fruit; the free part of the calyx of epigynous flowers or the calyx of perigynous flowers, either persists entirely at the top of or round the fruit, or the lobes alone fall off, or the lobes fall off with whatever part of the calyx is above the insertion of the petals, or the whole of what is free from the ovary falls off, including the disk bearing the petals. The calyx of hypogynous flowers usually falls off entirely or persists entirely. In general a calyx is called *deciduous* if any part falls off. When it persists it is either enlarged round or under the fruit, or it withers and dries up.

151. The corolla usually falls off entirely; when it persists it is usually withered and dry (*marcescent*), or very seldom enlarges round the fruit.

152. The stamens either fall off, or more or less of their filaments persists, usually withered and dry.

153. The style sometimes falls off, or dries up and disappears; sometimes persists, forming a point to the fruit, or becomes enlarged into a wing or other appendage to the fruit.

154. The *Pericarp* is the portion of the fruit formed of the ovary, and whatever adheres to it exclusive of and outside of the seed or seeds, exclusive also of the persistent receptacle, or of whatever portion of the calyx persists round the ovary without adhering to it.

155. Fruits have often external appendages, called *wings* (alæ), *beaks*, *crests*, *awns*, etc., according to their appearance. They are either formed by persistent parts of the flower more or less altered, or grow out of the ovary or the persistent part of the calyx. If the appendage be a ring of hairs or scales round the top of the fruit, it is called a *pappus*.

156. Fruits are generally divided into *succulent* (including *fleshy*, *pulpy*, and *juicy* fruits) and *dry*. They are *dehiscent* when they open at maturity to let out the seeds, *indehiscent* when they do not open spontaneously but fall off with the seeds. Succulent fruits are usually indehiscent.

157. The principal kinds of succulent fruits are

the *Berry*, in which the whole substance of the pericarp is fleshy or pulpy, with the exception of the outer skin or rind, called the *Epicarp*. The seeds themselves are usually immersed in the pulp; but in some berries, the seeds are separated from the pulp by the walls of the cavity or cells of the ovary, which forms as it were a thin inner skin or rind, called the *Endocarp*.

the *Drupe*, in which the pericarp, when ripe, consists of two distinct portions, an outer succulent one called the *Sarcocarp* (covered like the berry by a skin or epicarp), and an inner dry endocarp called the *Putamen*, which is either *cartilaginous* (of the consistence of parchment) or hard and woody. In the latter case it is commonly called a *stone*, and the drupe a *stone-fruit*. When the putamen consists of several distinct stones or nuts, each enclosing a seed, they are called *pyrenes*, or sometimes *kernels*.

158. The principal kinds of dry fruits are
the *Capsule* or *Pod*,* which is dehiscent. When ripe the pericarp usually slits longitudinally into as many or twice as many pieces, called *valves*, as it contains cells or placentas. If these valves separate at the line of junction of the carpels, that is, along the line of the placentas or dissepiments, either splitting them or leaving them attached to the axis, the dehiscence is termed *septicidal*; if the valves separate between the placentas or dissepiment, the dehiscence is *loculicidal*, and the values either bear the placentas or dissepiments along their middle line, or leave them attached to the axis. Sometimes also the capsule discharges its seeds by *slits*, *chinks*, or *pores*, more or less regularly arranged, or bursts irregularly, or separates into two parts by a horizontal line; in the latter case it is said to be *circumsciss*.

The *Nut* or *Achene*, which is indehiscent and contains but a single seed. When the pericarp is thin in proportion to the seed it encloses, the whole fruit (or each of its lobes) has the appearance of a single seed, and is so called in popular language. If the pericarp is thin and rather loose, it is often called an *Utricle*. A *Samara* is a nut with a wing at its upper end.

159. Where the carpels of the pistil are distinct (125) they may severally become as many distinct berries, drupes, capsules, or achenes. Separate carpels are usually more or less compressed laterally, with more or less prominent inner and outer edges, called *sutures*, and, if dehiscent, the carpel usually opens at these sutures. A *Follicle* is a carpel opening at the inner suture only. In some cases where the carpels are united in the pistil they will separate when ripe; they are then called *Cocci* if one-seeded.

160. The peculiar fruits of some of the large Orders have received special names, which will be explained under each Order. Such are the *siliqua* and *silicule* of Cruciferæ, the *legume* of Leguminosæ, the *pome* of *Pyrus* and its allies, the *pepo* of Cucurbitaceæ, the *cone* of Coniferæ, the *grain* or *caryopsis* of Gramineæ, etc.

§ 14. *The Seed.*

161. The **Seed** is enclosed in the pericarp in the great majority of flowering plants, called therefore *Angiosperms* or *angiospermous plants*. In *Coniferæ* and a very few allied genera, called *Gymnosperms* or *gymnospermous plants*, the seed is naked, without any real pericarp. These truly gymnospermous plants must not be confounded with *Labiatæ*, *Boragineæ*, etc., which have also been falsely called gynospermous, their small nuts having the appearance of seeds (158).

162. The seed when ripe contains an *embryo* or young plant, either filling or nearly filling the cavity, but not attached to the outer skin or the seed, or more or less immersed in a mealy, oily, fleshy, or horn-like substance, called the *albumen* or *perisperm*. The presence or absence of this albumen, that is, the distinction between *albuminous* and *exalbuminous* seeds, is one of great importance. The embryo or albumen can often only be found or distinguished when the seed is quite ripe, or sometimes only when it begins to germinate.

163. The shell of the seed consists usually of two separable *coats*. The outer coat, called the *testa*, is usually the principal one, and in most cases the only one attended to in descriptions. It may be hard and *crustaceous*, woody or bony, or thin and *membranous* (skin-like), dry, or rarely succulent. It is sometimes expanded into *wings*, or bears a tuft of hair, cotton or wool, called a *coma*. The inner coat is called the *tegmen*.

164. The *funicle* is the stalk by which the seed is attached to the placenta. It is occasionally enlarged into a membranous, pulpy, or fleshy appendage, sometimes spreading over a considerable part of the seed, or nearly enclosing it, called an *aril*. A *strophiole* or *caruncle* is a similar appendage proceeding from the testa by the side of or near the funicle.

165. The *hilum* is the scar left on the seed where it separates from the funicle. The *micropyle* is a mark indicating the position of the foramen of the ovule (133).

166. The **Embryo** (162) consists of the *Radicle* or base of the future root, one or two *Cotyledons* or future seed-leaves, and the *Plumule* or future bud within the base of the cotyledons. In some seeds, especially where there is no albumen, these several parts are very conspicuous, in others they are very difficult to distinguish until the seed begins to germinate. Their observation, however, is of the greatest importance, for it is chiefly upon the distinction between the embryo with one or with two cotyledons that are founded the two great classes of phænogamous plants, *Monocotyledons* and *Dicotyledons*.

167. Although the embryo lies loose (unattached) within the seed, it is generally in some determinate position with respect to the seed or to the whole fruit. This position is described by stating the direction of the radicle next to or more or less remote from the *hilum*, or it is said to be *superior* if pointing towards the summit of the *fruit*, *inferior* if pointing towards the base of the *fruit*.

§ 15. *Accessory Organs.*

168. Under this name are included, in many elementary works, various external parts of plants which do not appear to act any essential part either in the vegetation or reproduction of the plant. They may be classed under four heads: *Tendrils* and *Hooks*, *Thorns* and *Prickles*, *Hairs* and *Glands*.

---

* In English descriptions, *pod* is more frequently used when it is long and narrow; *capsule*, or sometimes *pouch*, when it is short and thick or broad.

A

169. **Tendrils** (*cirrhi*) are usually abortive petioles, or abortive peduncles, or sometimes abortive ends of branches. They are simple or more or less branched, flexible, and coil more or less firmly round any objects within their reach, in order to support the plant to which they belong. *Hooks* are similar holdfasts, but of a firmer consistence, not branched, and less coiled.

170. **Thorns** and **Prickles** have been fancifully called the weapons of plants. A *Thorn* or *Spine* is the strongly pointed extremity of a branch, or abortive petiole, or abortive peduncle. A *Prickle* is a sharply pointed excrescence from the epidermis, and is usually produced on a branch, on the petiole or veins of a leaf, or on a peduncle, or even on the calyx or corolla. When the teeth of a leaf or the stipules are pungent, they are also called *prickles* not *thorns*. A plant is *spinous* if it has thorns, *aculeate* if has prickles.

171. **Hairs**, in the general sense, or the *indumentum* (or clothing) of a plant, include all those productions of the epidermis which have, by a more or less appropriate comparison, been termed *bristles, hairs, down, cotton* or *wool*.

172. Hairs are often branched. They are said to be *attached by the centre*, if parted from the base, and the forks spread along the surface in opposite directions ; *plumose*, if the branches are arranged along a common axis, as in a feather ; *stellate*, if several branches radiate horizontally. These stellate hairs have sometimes their rays connected together at the base, forming little flat circular disks attached by the centre, and are then called *scales*, and the surface is said to be *scaly* or *lepidote*.

173. The *Epidermis*, or outer skin, of an organ, as to its surface and indumentum, is
smooth, when without any protuberance whatever.
glabrous, when without hairs of any kind.
striate, when marked with parallel longitudinal lines, either slightly raised or merely discoloured.
furrowed (sulcate) or *ribbed (costate)* when the parallel lines are more distinctly raised.
rugose, when wrinkled or marked with irregular raised or depressed lines.
umbilicate, when marked with a small round depression.
umbonate, when bearing a small boss like that of a shield.
viscous, viscid, or *glutinous*, when covered with a sticky or clammy exudation.
scabrous, when rough to the touch.
tuberculate or *warted*, when covered with small, obtuse, wart-like protuberances.
muricate, when the protuberances are more raised and pointed but yet short and hard.
echinate, when the protuberances are longer and sharper, almost prickly.
setose or *bristly*, when bearing very stiff erect straight hairs.
glandular-setose, when the setæ or bristles terminate in a minute resinous head or drop. In some works, especially in the case of *roses* and *robus*, the meaning of *setæ* has been restricted to such as are glandular.
glochidiate, when the setæ are hooked at the top.
pilose, when the surface is thinly sprinkled with rather long simple hairs.
hispid, when more thickly covered with rather stiff hairs.
hirsute, when the hairs are dense and not so stiff.
downy or *pubescent*, when the hairs are short and soft ; *perbulent*, when slightly pubescent.
strigose, when the hairs are rather short and stiff, and lie close along the surface all in the same direction ; *strigillose*, when slightly strigose.
tomentose or *cottony*, when the hairs are very short and soft, rather dense and more or less intricate, and usually white or whitish.
woolly (lanate), when the hairs are long and loosely intricate, like wool. The wool or tomentum is said to be *floccose* when closely intricate and readily detached, like fleece.
mealy (farinose), when the hairs are excessively short, intricate and white, and come off readily, having the appearance of meal or dust.
canescent or *hoary*, when the hairs are so short as not readily to be distinguished by the naked eye, and yet give a general whitish hue to the epidermis.
glaucous, when of a pale bluish-green, often covered with a fine bloom.

174. The meanings here attached to the above terms are such as appear to have been most generally adopted, but there is much vagueness in the use practically made of many of them by different botanists. This is especially the case with the terms *pilose, hispid, hirsute, pubescent*, and *tomentose*.

175. The name of **Glands** is given to several different productions, and principally to the four following :—

1. Small wart-like or shield-like bodies, either sessile or sometimes stalked, of a fungous or somewhat fleshy consistence, occasionally secreting a small quantity of oily or resinous matter, but more frequently dry. They are generally few in number, often definite in their position and form, and occur chiefly on the petiole or principal veins of leaves, on the branches of inflorescences, or on the stalks or principal veins of bracts, sepals, or petals.

2. Minute raised dots, usually black, red, or dark-coloured, of a resinous or oily nature, always superficial, and apparently exudations from the epidermis. They are often numerous on leaves, bracts, sepals, and green branches, and occur even on petals and stamens, more rarely on pistils. When raised upon slender stalks they are called *pedicellate* (or *stipitate*) *glands*, or *glandular hairs*, according to the thickness of the stalk,

3. Small, globular, oblong or even linear vesicles, filled with oil imbedded in the substance itself of leaves, bracts, floral organs, or fruits. They are often very numerous, like transparent dots, sometimes few and determinate in form and position. In the pericarp of *Umbelliferæ* they are remarkably regular and conspicuous, and take the name of *vittæ*.

4. Lobes of the disk (137), or other small fleshy excrescences within the flower, whether from the receptacle, calyx, corolla, stamens, or pistil.

CHAP. II. CLASSIFICATION, OR SYSTEMATIC BOTANY.

176. It has already been observed (3) that descriptions of plants should, as nearly as possible, be arranged under natural divisions, so as to facilitate the comparison of each plant with those most nearly allied to it. The descriptions of plants here alluded to are descriptions of *species*; the natural divisions of the Flora refer to natural *groups of species*.

177. A **Species** comprises all the individual plants which resemble each other sufficiently to make us conclude that they are all, or *may have been* all, descended from a common parent. These individuals may often differ from each other in many striking particulars, such as the colour of the flower, size of the leaf, etc., but these particulars are such as experience teaches us are liable to vary in the seedlings raised from one individual.

178. When a large number of the individuals of a species differ from the others in any striking particular they constitute a **Variety**. If the variety generally comes true from seed, it is often called a *Race*.

179. A *Variety* can only be propagated with certainty by grafts, cuttings, bulbs, tubers, or any other method which produces a new plant by the development of one or more buds taken from the old one. A *Race* may with care be propagated by seed, although seedlings will always be liable, under certain circumstances, to lose those particulars which distinguish it from the rest of the species. A real *Species* will always come true from seed.

180. The known species of plants (now near 100,000) are far too numerous for the human mind to study without classification, or even to give distinct single names to. To facilitate these objects, an admirable system, invented by Linnæus, has been universally adopted, viz. one common substantive name is given to a number of species which resemble each other more than they do any other species; the species so collected under one name are collectively called a **Genus**, the common name being the *generic* name. Each species is then distinguished from the others of the same genus by the addition of an adjective epithet or *specific name*. Every species has thus a botanical name of two words. In Latin, the language usually used for the purpose, the first word is a substantive and designates the genus; the second, an adjective, indicates the species.

181. The genera thus formed being still too numerous (above 6,000) for study without further arrangement, they have been classed upon the same principles; viz. genera which resemble each other more than they do any other genera, have been collected together into groups of a higher degree called **Families** and **Natural Orders,** to each of which a common name has been given. This name is in Latin an adjective plural, usually taken from the name of some one *typical* genus, generally the best known, the first discovered, or the most marked (e.g. *Ranunculaceæ* from *Ranunculus*). This is however for the purpose of study and comparison. To speak of a species, to refer to it and identify it, all that is necessary is to give the generic and specific names.

182. Natural Orders themselves (of which we reckon near 200) are often in the same manner collected into **Classes**; and where orders contain a large number of genera, or genera a large number of species, they require further classification. The genera of an Order are then collected into minor groups called *Tribes*, the species of a genus into *Sections*, and in a few cases this intermediate classification is carried still further. The names of these several groups the most generally adopted are as follows, beginning with the most comprehensive or highest:—

Classes.	Genera.
Subclasses or *Alliances*	*Subgenera.*
Natural Orders or Families.	Sections.
Suborders.	*Subsections.*
Tribes.	Species.
Subtribes.	Varieties.
Divisions.	
Subdivisions.	

183. The characters (3) by which a species is distinguished from all other species of the same genus are collectively called the *specific character* of the plant; those by which its genus is distinguished from other genera of the Order, or its order from other Orders, are respectively called the *generic* or *ordinal* character, as the case may be. The *habit* of a plant, of a species, a genus, etc., consists of such general characters as strike the eye at first sight, such as size, colour, ramification, arrangement of the leaves, inflorescence, etc., and are chiefly derived from the organs of vegetation.

184. Classes, Orders, Genera, and their several subdivisions, are called *natural* when, in forming them, all resemblances and differences are taken into account, valuing them according to their evident or presumed importance; *artificial*, when resemblances and differences in some one or very few particulars only are taken into account independently of all others.

185. The number of species included in a genus, or the number of genera in an Order, is very variable. Sometimes two or three or even a single species may be so different from all others as to constitute the entire genus; in others, several hundred species may resemble each other so much as to be all included in one genus; and there is the same discrepancy in the number of genera to a Family. There is, moreover, unfortunately, in a number of instances, great difference of opinion as to whether certain plants differing from each other in certain particulars are varieties of one species or belong to distinct species; and again, whether two or more groups of species should constitute as many sections of one genus, or distinct genera, or tribes of one Order, or even distinct Natural Orders. In the former case, as a species is supposed to have a real existence in nature, the question is susceptible of argument, and sometimes of absolute proof. But the place a group should occupy in the scale of degree is very arbitrary, being often a mere question of convenience. The more subdivisions upon correct principles are multiplied, the more they facilitate the study of plants, provided always the main resting-points for constant use, the Order and the Genus, are comprehensive and distinct. But if every group into which a genus can be divided be erected into a distinct genus, with a substantive name to be remembered whenever a species is spoken of, all the advantages derived from the beautiful simplicity of the Linnæan nomenclature are gone.

CHAP. III. VEGETABLE ANATOMY AND PHYSIOLOGY.

§ 1. *Structure and Growth of the Elementary Tissues.*

186. If a very thin slice of any part of a plant be placed under a microscope of high magnifying power, it will be found to be made up of variously shaped and arranged ultimate parts, forming a sort of honeycombed structure. These ultimate parts are called *cells*, and form by their combination the *elementary tissues* of which the entire plant is composed.

187. A cell in its simplest state is a closed membranous sac, formed of a substance permeable by fluids, though usually destitute of visible pores. Each cell is a distinct individual, separately formed and separately acting, though cohering with the cells with which it is in contact, and partaking of the common life and action of the tissue of which it forms a part. The membranes separating or enclosing the cells are also called their *walls*.

188. Botanists usually distinguish the following tissues:—

(1) *Cellular tissue* or *parenchyma*, consists usually of thin-walled cells, more or less round in form, or with their length not much exceeding their breadth, and not tapering at the ends. All the soft parts of the leaves, the pith of stems, the pulp of fruits, and all young growing parts, are formed of it. It is the first tissue produced, and continues to be formed while growth ntinues, and when it ceases to be active the plant dies.

(2) *Woody tissue* or *prosenchyma*, differs in having its cells considerably longer than broad, usually tapering at each end into points and overlapping each other. The cells are commonly thick walled; the tissue is firm, tenacious, and elastic, and constitutes the principal part of wood, of the inner bark, and of the nerves and veins of leaves, forming, in short, the framework of the plant.

(3) *Vascular tissue*, or the *vessels* or *ducts* of plants, so called from the mistaken notion that their functions are analogous to those of the vessels (veins and arteries) of animals. A *vessel* in plants consists of a vertical row of cells, which have their transverse partition-walls obliterated so as to form a continuous tube. All phænogamous plants, as well as ferns and a few other cryptogamous plants, have vessels, and are therefore called *vascular plants*; so the majority of cryptogams having only cellular tissue are termed *cellular plants*. Vessels have their sides very variously marked; some called *spiral vessels*, have a spiral fibre coiled up their inside, which unrolls when the vessel is broken; others are marked with longitudinal slits, cross bars, minute dots or pits, or with traverse rings. The size of vessels is also very variable in different plants; in some they are of considerable size and visible to the naked eye in cross sections of the stem, in others they are almost absent or can only be traced under a strong magnifier.

189. Various modifications of the above tissues are distinguished by vegetable anatomists under names which need not be enumerated here as not being in general practical use. *Air-vessels, cysts, turpentine-vessels, oil-reservoirs*, etc., are either cavities left between the cells, or large cells filled with peculiar secretions.

190. When tissues are once formed, they increase, not by the general enlargement of the whole of the cells already formed, but by *cell-division*, that is, by the division of young and vitally active cells, and the enlargement of their portions. In the formation of the embryo, the first cell of the new plant is formed, not by division, but around a segregate portion of the contents of a previously existing cell, the embryo-sac. This is termed *free cell-formation*, in contradistinction to cell-division.

191. A young and vitally active cell consists of the *outer wall*, formed of a more or less transparent substance called *cellulose*, permeable by fluids, and of ternary chemical composition (carbon, hydrogen, and oxygen); and of the *cell-contents* usually viscid or mucilaginous, consisting of *protoplasm*, a substance of quaternary chemical composition (carbon, hydrogen, oxygen,

and nitrogen), which fills an important part in cell-division and growth. Within the cell (either in the centre or excentrical) is usually a minute, soft, subgelatinous body called the *nucleus*, whose functions appear to be intimately connected with the first formation of the new cell. As this cell increases in size, and its walls in thickness, the protoplasm and watery cell-sap become absorbed or dried up, the firm cellulose wall alone remaining as a permanent fabric, either empty or filled with various organised substances produced or secreted within it.

192. The principal organised contents of cells are
sap, the first product of the digestion of the food of plants; it contains the elements of vegetable growth in a dissolved condition.

sugar, of which there are two kinds, called *cane-sugar* and *grape-sugar*. It usually exists dissolved in the sap. It is found abundantly in growing parts, in fruits, and in germinating seeds.

dextrine, or vegetable mucilage, a gummy substance, between mucilage and starch.

starch or *fecula*, one of the most universal and conspicuous of cell-contents, and often so abundant in farinaceous roots and seeds as to fill the cell-cavity. It consists of minute grains called *starch-granules*, which vary in size and are marked with more or less conspicuous concentric lines of growth. The chemical constitution of starch is the same as that of cellulose; it is unaffected by cold water, but forms a jelly with boiling water, and turns blue when tested by iodine. When fully dissolved it is no longer starch, but dextrine.

chlorophyll, very minute granules, containing nitrogen, and coloured green under the action of sunlight. These granules are most abundant in the layers of cells immediately below the surface or epidermis of leaves and young bark. The green colouring matter is soluble in alcohol, and may thus be removed from the granules.

chromule, a name given to a similar colouring matter when not green.

wax, *oils*, *camphor*, and *resinous* matter, are common in cells or in cavities in the tissues between the cells, also various mineral substances, either in an amorphous state or as microscopic crystals, when they are called *Raphides*.

§ 2. *Arrangement of the Elementary Tissues, or Structure of the Organs of Plants.*

193. Leaves, young stems, and branches, and most parts of phænogamous plants, during the first year of their existence consist anatomically of

1, a *cellular system*, or continuous mass of cellular tissue, which is developed both vertically as the stem or other parts increase in length, and horizontally or laterally as they increase in thickness or breadth. It surrounds or is intermixed with the fibrovascular system or it may exist alone in some parts of phænogamous plants, as well as in cryptogamous ones.

2, a *fibro-vascular system*, or continuous mass of woody and vascular tissue, which is gradually introduced vertically into, and serves to bind together the cellular system. It is continued from the stem into the petioles and veins of the leaves, and into the pedicels and parts of the flowers, and is never wholly wanting in any phænogamous plant.

3, An *epidermis*, or outer skin, formed of one or more layers of flattened (horizontal), firmly coherent, and usually empty cells, with either thin and transparent or thick and opaque walls. It covers almost all parts of plants exposed to the outward air, protecting their tissues from its immediate action, but is wanting in those parts of aquatic plants which are constantly submerged.

194. The epidermis is frequently pierced by minute spaces between the cells, called *Stomates*. They are oval or mouth shaped, bordered by *lips*, formed of two or more elastic cells so disposed as to cause the stomate to open in a moist and to close up in a dry state of the atmosphere. They communicate with intercellular cavities, and are obviously designed to regulate evaporation and respiration. They are chiefly found upon leaves, especially on the under surface.

195. When a phænogamous plant has outlived the first season of its growth, the anatomical structure of its stem or other perennial parts becomes more complicated and very different in the two great classes of phænogamous plants called *Exogens* and *Endogens*, which correspond with very few exceptions to the two classes Dicotyledons and Monocotyledons (167), founded on the structure of the embryo. In Exogens (Dicotyledons) the woody system is placed in concentric layers between a central *pith* (198, 1) and an external separable *bark* (198, 5). In Endogens (Monocotyledons) the woody system is in separate small bundles or fibres running through the cellular system without apparent order, and there is usually no distinct central pith, nor outer separable bark.

196. The anatomical structure is also somewhat different in the different organs of plants. In the **Root**, although it is constructed generally on the same plan as the stem, yet the regular organisation, and the difference between Exogens and Endogens is often disguised or obliterated by irregularities of growth, or by the production of large quantities of cellular tissue filled with starch or other substances (192). There is seldom, if ever, any distinct pith, the concentric circles of fibro-vascular tissue in Exogens are often very indistinct or have no relation to seasons of growth, and the epidermis has no stomates.

197. In the **Stem** or branches, during the first year or season of their growth, the difference between Exogens and Endogens is not always very conspicuous. In both there is a tendency to a circular arrangement of the fibro-vascular system, leaving the centre either vacant or filled with cellular tissue (pith) only, and a more or less distinct outer rind is observable even in

several Endogens. More frequently, however, the distinction is already very apparent the first season, especially towards its close. The fibro-vascular bundles in Endogens usually anastomose but little, passing continuously into the branches and leaves. In Exogens the circle of fibro-vascular bundles forms a more continuous cylinder of network emitting lateral offsets into the branches and leaves.

198. The Exogenous stem, after the first year of its growth, consists of

1, the *pith*, a cylinder of cellular tissue, occupying the centre or longitudinal axis of the stem. It is active only in young stems or branches, becomes dried up and compressed as the wood hardens, and often finally disappears, or is scarcely distinguishable in old trees.

2, the *medullary sheath*, which surrounds and encases the pith. It abounds in spiral vessels (188, 3), and is in direct connection, when young, with the leaf-buds and branches, with the petioles and veins of leaves, and other ramifications of the system. Like the pith, it gradually disappears in old wood.

3, the *wood*, which lies immediately outside the medullary sheath. It is formed of woody tissue (188, 2), through which in most cases, vessels (188, 3) variously disposed are interspersed. It is arranged in annual concentric circles (211), which usually remain active during several years, but in older stems the central and older layers become hard, dense, comparatively inactive and usually deeper colored, forming what is called *heart-wood* or *duramen*, the outer, younger, and usually paler colored living layers constituting the *sap-wood* or *alburnum*.

4, the *medullary rays*, which form vertical plates, originating in the pith, and radiating from thence, traverse the wood and terminate in the bark. They are formed of cellular tissue, keeping up a communication between the living portion of the centre of the stem and its outer surface. As the heart-wood is formed, the inner portion of the medullary rays ceases to be active, but they usually may still be seen in old wood, forming what carpenters call the *silver grain*.

5, the *bark*, which lies outside the wood, within the epidermis. It is, like the wood, arranged in annual concentric circles (211), of which the outer older ones become dry and hard, forming the *corky layer* or *outer bark*, which, as it is distended by the thickening of the stem, either cracks or is cast off with the epidermis, which is no longer distinguishable. Within the corky layer is the *cellular*, or *green*, or *middle bark*, formed of loose thin-walled pulpy cells containing chlorophyll (192); and which is usually the layer of the preceding season. The innermost and youngest circle, next the young wood, is the *liber* or *inner bark*, formed of long tough woody tissue called *bast-cells*.

199. The Endogenous stem, as it grows old, is not marked by the concentric circles of Exogens. The wood consists of a *matrix* of cellular tissue irregularly traversed by vertical cords, or bundles of woody and vascular tissue, which are in connection with the leaves. These vascular bundles change in structure and direction as they pass down the stem, losing their vessels, they retain only their bast or long wood-cells, usually curving outwards towards the rind. The old wood becomes more compact and harder towards the circumference than in the centre. The epidermis or rind either hardens so as to prevent any increase of diameter in the stem, or it distends, without increasing in thickness or splitting or casting off any outer layers.

200. In the **Leaf,** the structure of the petioles and principal ribs or veins is the same as that of the young branches of which they are ramifications. In the expanded portion of the leaf the fibro-vascular system becomes usually very much ramified, forming the smaller veins. These are surrounded and the interstices filled up by a copious and very active cellular tissue. The majority of leaves are horizontal, having a differently constructed upper and under surface. The cellular stratum forming the upper surface consists of closely set cells, placed vertically, with their smallest ends next the surface, and with few or no stomates in the epidermis. In the stratum forming the under surface, the cells are more or less horizontal, more loosely placed, and have generally empty spaces between them, with stomates in the epidermis communicating with these intercellular spaces. In vertical leaves (as in a large number of Australian plants) the two surfaces are nearly similar in structure.

201. When leaves are reduced to scales, acting only as protectors of young buds, or without taking any apparent part in the economy of vegetable life, their structure, though still on the same plan, is more simple; their fibro-vascular system is less ramified, their cellular system more uniform, and there are few or no stomates.

202. Bracts and floral envelopes, when green and much developed, resemble leaves in their anatomical structure, but in proportion as they are reduced to scales or transformed into petals, they lose their stomates, and their systems, both fibro-vascular and cellular, become more simple and uniform, or more slender and delicate.

203. In the stamens and pistils the structure is still nearly the same. The fibro-vascular system, surrounded by and intermixed with the cellular tissue, is usually simple in the filaments and style, more or less ramified in the flattened or expanded parts, such as the anther cases, the walls of the ovary, or carpellary leaves, etc. The pollen consists of granular cells variously shaped, marked, or combined, peculiar forms being constant in the same species, or often in large genera, or even Orders. The stigmatic portion of the pistil is a mass of loosely cellular substance, destitute of epidermis, and usually is in communication with the ovary by a channel running down the centre of the style.

204. Tubers, fleshy thickenings of the stem or other parts of the plant, succulent leaves or branches, the fleshy, woody, or bony parts of fruits, the albumen, and the thick fleshy parts of embryos, consist chiefly of largely developed cellular tissue, replete with starch or other substances (192), deposited apparently in most cases for the eventual future use of the plant or its parts when recalled into activity at the approach of a new season.

205. Hairs (171) are usually expansions or processes of the epidermis, and consist of one or more cells placed end to end. When thick or hardened into prickles, they still consist usually of cellular tissue only. Thorns (170) contain more or less of a fibro-vascular system, according to their degree of development.

206. Glands, in the primary sense of the word (175, 1), consist usually of a rather loose cellular tissue without epidermis, and often replete with resinous or other substances.

§ 3. *Growth of the Organs.*

207. Roots grow in length constantly and regularly at the extremities only of their fibres, in proportion as they find the requisite nutriment. They form no buds containing the germ of future branches, but their fibres proceed irregularly from any part of their surface without previous indication, and when their growth has been stopped for a time, either wholly by the close of the season, or partially by a deficiency of nutriment at any particular spot, it will, on the return of favourable circumstances, be resumed at the same point, if the growing extremities be uninjured. If during the dead season, or at any other time, the growing extremity is cut off, dried up, or otherwise injured, or stopped by a rock or other obstacle opposing its progress, lateral fibres will be formed on the still living portion; thus enabling the root as a whole to diverge in any direction, and travel far and wide when lured on by appropriate nutriment.

208. This growth is not however by the successive formation of terminal cells attaining at once their full size. The cells first formed on a fibre commencing or renewing its growth, will often dry up and form a kind of terminal cap, which is pushed on as cells are formed immediately under it; and the new cells, constituting a greater or less portion of the ends of the fibres, remain some time in a growing state before they have attained their full size.

209. The roots of Exogens, when perennial, increase in thickness like stems by the addition of concentric layers, but these are usually much less distinctly marked; and in a large number of perennial Exogens and most Endogens the roots are annual, perishing at the close of the season, fresh adventitious roots springing from the stock when vegetation commences the following season.

210. The Stem, including its branches and appendages (leaves, floral organs, etc.), grows in length by additions to its extremity, but a much greater proportion of the extremity and branches remains in a growing and expanding state for a much longer time than in the case of the root. At the close of one season, leaf-buds or seeds are formed, each containing the germ of a branch or young plant to be produced the following season. At a very early stage of the development of these buds or seeds, a commencement may be found of many of the leaves it is to bear; and before a leaf unfolds, every leaflet of which it is to consist, every lobe or tooth which is to mark its margin, may often be traced in miniature, and thenceforth till it attains its full size, the branch grows and expands in every part. In some cases however the lower part of a branch and more rarely (*e.g.* in some *Meliaceæ*) the lower part of a compound leaf attains its full size before the young leaves or leaflets of the extremity are yet formed.

211. The perennial stem, if exogenous (198), grows in thickness by the addition every season of a new layer or ring of wood between the outermost preceding layer and the inner surface of the bark, and by the formation of a new layer or ring of bark within the innermost preceding layer and outside the new ring of of wood, thus forming a succession of concentric circles. The sap elaborated by the leaves finds its way, in a manner not as yet absolutely ascertained, into the *cambium-region*; a zone of tender thin-walled cells connecting the wood with the bark, by the division and enlargement of which new cells (190) are formed. These cells separate in layers, the inner ones constituting the new ring of wood, and the outer ones the new bark or liber. In most exogenous trees, in temperate climates, the seasons of growth correspond with the years, and the rings of wood remain sufficiently distinct to indicate the age of the tree; but in many tropical and some evergreen trees, two or more rings of wood are formed in one year.

212. In endogenous perennial stems (199), the new wood or woody fibre is formed towards the centre of the stem, or irregularly mingled with the old. The stem consequently either only becomes more dense without increasing in thickness, or only increases by gradual distension, which is never very considerable. It affords therefore no certain criterion for judging of the age of the the tree.

213. Flowers have generally all their parts formed, or indicated by protuberances or growing cells at a very early stage of the bud. These parts are then usually more regularly placed than in the fully developed flower. Parts which afterwards unite are then distinct, many are present in this rudimentary state which are never further developed, and parts which are afterwards very unequal or dissimilar are perfectly alike at this early period. On this account flowers in this very early stage are supposed by some modern botanists to be more *normal*, that is, more in

conformity to a supposed type; and the study of the early formation and growth of the floral organs, called *Organogenesis*, has been considered essential for the correct appreciation of the affinities of plants. In some cases, however, it would appear that modifications of development, not to be detected in the very young bud, are yet of great importance in the distinction of large groups of plants, and that Organogenesis, although it may often assist in clearing up a doubtful point of affinity, cannot nevertheless be exclusively relied on in estimating the real value of peculiarities of structure.

214. The flower is considered as a *bud (flower-bud, alabastrum)* until the perianth expands. The *period of flowering (anthesis)* is that which elapses from the first expanding of the perianth, till the pistil is set or begins to enlarge, or, when it does not set, until the stamens and pistil wither or fall. After that, the enlarged ovary takes the name of *young fruit*.

215. At the close of the season of growth, at the same time as the leaf-buds or seeds are formed containing the germ of future branches or plants, many plants form also, at or near the bud or seed, large deposits, chiefly of starch. In many cases—such as the tubers of a potato or other root-stock, the scales or thickened base of a bulb, the albumen or the thick cotyledons of a seed—this deposit appears to be a store of nutriment, which is partially absorbed by the young branch or plant during its first stage of growth, before the roots are sufficiently developed to supply it from without. In some cases, however, such as the fleshy thickening of some stems or peduncles, the pericarps of fruits which perish long before *germination* (the first growth of the eed), neither the use nor the cause of these deposits has as yet been clearly explained.

§ 4. *Functions of the Organs.*

216. The functions of the Root are:—1. To fix the plant in or to the soil or other substance on which it grows. 2. To absorb nourishment from the soil, water, or air, into which the fibres have penetrated (or from other plants in the case of parasites), and to transmit it rapidly to the stem. The absorption takes place through the young growing extremities of the fibres, and through a peculiar kind of hairs or absorbing organs which are formed at or near those growing extremities. The transmission to the stem is through the tissues of the root itself. The nutriment absorbed consists chiefly of carbonic acid and nitrogen or nitrogenous compounds dissolved in water. 3. In some cases roots secrete or exude small quantities of matter in a manner and with a purpose not satisfactorily ascertained.

217. The stem and its branches support the leaves, flowers, and fruit, transmit the crude sap, or nutriment absorbed by the roots and mixed with previously organised matter, to the leaves, and re-transmit the assimilated or elaborated sap from the leaves to the growing parts of the plant, to be there used up, or to form deposits for future use (204). The transmission of the ascending crude sap appears to take place chiefly through the elongated cells associated with the vascular tissues, passing from one cell to another by a process but little understood, but known by the name of *endosmose*.

218. Leaves are functionally the most active of the organs of vegetation. In them is chiefly conducted digestion or *Assimilation*, a name given to the process which accomplishes the following results:—1. The chemical decomposition of the oxygenated matter of the sap, the absorption of carbonic acid, and the liberation of pure oxygen at the ordinary temperature of the air. 2. A counter operation by which oxygen is absorbed from the atmosphere and carbonic acid is exhaled. 3. The transformation of the residue of the crude sap into the organized substances which enter into the composition of the plant. The exhalation of oxygen appears to take place under the influence of solar heat and light, chiefly from the under surface of the leaf, and to be in some measure regulated by the stomates; the absorption of oxygen goes on always in the dark, and in the daytime also in certain cases. The transformation of the sap is effected within the tissues of the leaf, and continues probably more or less throughout the active parts of the whole plant.

219. The Floral Organs seldom contribute to the growth of the plant on which they are produced; their functions are wholly concentrated on the formation of the seed with the germ of a future plant.

220. The Perianth (calyx and corolla) acts in the first instance in protecting the stamens and pistils during the early stages of their development. When expanded, the use of the brilliant colours which they often display, of the sweet or strong odours they emit, has not been adequately explained. Perhaps they may have great influence in attracting those insects whose concurrence has been shown in many cases to be necessary for the due transmission of the pollen from the anther to the stigma.

221. The pistil, when stimulated by the action of the pollen, forms and nourishes the young seed. The varied and complicated contrivances by which the pollen is conveyed to the stigma, whether by elastic action of the organs themselves, or with the assistance of wind, of insects, or other extraneous agents, have been the subject of numerous observations and experiments of the most distinguished naturalists, and are yet far from being fully investigated. Their details, however, as far as known, would be far too long for the present outline.

222. The fruit nourishes and protects the seed until its maturity, and then often promotes its dispersion by a great variety of contrivances or apparently collateral circumstances, *e.g.* by an elastic dehiscence which casts the seed off to a distance; by the development of a pappus, wings, hooked or other appendages, which allows them to be carried off by winds, or by animals, etc., to which they may adhere; by their small specific gravity, which enables them to float down streams; by their attractions to birds, etc., who, taking them for food, drop them often at great distances, etc. Appendages to the seeds themselves also often promote dispersion.

223. Hairs have various functions. The ordinary indumentum (171) of stems and leaves indeed seems to take little part in the economy of the plant besides perhaps some occasional protection against injurious atmospheric influences, but the root-hairs (216) are active absorbents, the hairs on styles and other parts of flowers appear often materially to assist the transmission of pollen, and the exudations of glandular hairs (175, 2) are often too copious not to exercise some influence on the phenomena of vegetation. The whole question however of vegetable exudations and their influence on the economy of vegetable life, is as yet but imperfectly understood.

CHAP. IV. COLLECTION, PRESERVATION, AND DETERMINATION OF PLANTS.

224. Plants can undoubtedly be most easily and satisfactorily examined when freshly gathered. But time will rarely admit of this being done, and it is moreover desirable to compare them with other plants previously observed or collected. *Specimens* must, therefore, be selected for leisurely observation at home, and preserved for future reference. A collection of such specimens constitutes a *Herbarium*.

225. A botanical **Specimen,** to be perfect, should have *root, stem, leaves, flowers* (both open and in the bud), and *fruit* (both young and mature). It is not, however, always possible to gather such complete specimens, but the collector should aim at completeness. Fragments, such as leaves without flowers, or flowers without leaves, are of little or no use.

226. If the plant is small (not exceeding 15in.) or can be reduced to that length by folding, the specimen should consist of the whole plant, including the principal part of the root. If it be too large to preserve the whole, a good flowering-branch should be selected, with the foliage as low down as can be gathered with it; and one or two of the lower stem-leaves or radical leaves, if any, should be added, so as to preserve as much as possible of the peculiar aspect of the plant.

227. The specimens should be taken from healthy uninjured plants of a medium size. Or if a specimen be gathered because it looks a little different from the majority of those around it, apparently belonging to the same species, a specimen of the more prevalent form should be taken from the same locality for comparison.

228. For bringing the specimens home, a light portfolio of pasteboard, covered with calico or leather, furnished with straps and buckles for closing, and another for slinging on the shoulder, and containing a few sheets of stout coarse paper, is better than the old-fashioned tin box (except, perhaps, for stiff prickly plants and a few others). The specimens as gathered are placed between the leaves of paper, and may be crowded together if not left long without sorting.

229. If the specimen brought home be not immediately determined when fresh, but dried for future examination, a note should be taken of the time, place, and situation in which it was gathered; of the stature, habit, and other particulars relating to any tree, shrub, or herb of which the specimen is only a portion; of the kind of root it has; of the colour of the flower; or of any other particulars which the specimen itself cannot supply, or which may be lost in the process of drying. These memoranda, whether taken down in the field, or from the living specimen when brought home, should be written on a label attached to the specimen or preserved with it.

230. To dry specimens, they are laid flat between several sheets of bibulous paper, and subjected to pressure. The paper is subsequently changed at intervals, until they are dry.

231. In laying out the specimen, care should be taken to preserve the natural position of the parts as far as consistent with the laying flat. In general, if the specimen is fresh and not very slender, it may be simply laid on the lower sheet, holding it by the stalk and drawing it slightly downwards; then, as the upper sheet is laid over, if it be slightly drawn downwards as it is pressed down, it will be found, after a few trials, that the specimen will have retained a natural form with very little trouble. If the specimen has been gathered long enough to have become flaccid, it will require more care in laying the leaves flat and giving the parts their proper direction. Specimens kept in tin boxes, will also often have taken unnatural bends which will require to be corrected.

232. If the specimen is very bushy, some branches must be thinned out, but always so as to show where they have been. If any part, such as the head of a thistle, the stem of an *Orobanche*, or the bulb of a Lily, be very thick, a portion of what is to be the under side of the specimen may be sliced off. Some thick specimens may be split from top to bottom before drying.

233. If the specimen be succulent or tenacious of life, such as a *Sedum* or an *Orchis*, it may be dipped in boiling water *all but the flowers*. This will kill the plant at once, and enable it to be dried rapidly, losing less of its colour or foliage than would otherwise be the case. Dipping in boiling water is also useful in the case of Heaths and other plants which are apt to shed their leaves during the process of drying.

234. Plants with very delicate corollas may be placed between single leaves of very thin unglazed tissue-paper. In shifting these plants into dry paper the tissue-paper is not to be removed, but lifted with its contents on to the dry paper.

235. The number of sheets of paper to be placed between each specimen or sheet of specimens, will depend, on the one hand, on the thickness and humidity of the specimens; on the other hand, on the quantity and quality of the paper one has at command. The more and the better the paper, the less frequently will it be necessary to change it, and the sooner the plants will dry. The paper ought to be coarse, stout, and unsized. Common blotting-paper is much too tender.

236. Care must be taken that the paper used is well dried. If it be likewise hot, all the better: but it must then be very dry; and wet plants put into hot paper will require changing very soon, to prevent their turning black, for hot damp without ventilation produces fermentation and spoils the specimens.

237. For pressing plants, various more or less complicated and costly presses are made. None is better than a pair of boards the size of the paper, and a stone or other heavy weight upon them if at home, or a pair of strong leather straps round them if travelling. Each of these boards should be double, that is made of two layers of thin boards, the opposite way of the grain, and joined together by a row of clenched brads round the edge, without glue. Such boards, in deal, rather less than half an inch thick (each layer about $2\frac{1}{4}$ lines) will be found light and durable.

238. It is useful also to have extra boards or pasteboards the size of the paper, to separate thick plants from thin ones, wet ones from those nearly dry, etc. Open wooden frames with cross-bars, or frames of strong wirework lattice, are still better than boards for this purpose, as accelerating the drying by promoting ventilation.

239. The more frequently the plants are shifted into dry paper the better. Excepting for very stiff or woody plants, the first pressure should be light, and the first shifting, if possible, after a few hours. Then, or at the second shifting, when the specimens will have lost their elasticity, will be the time for putting right any part of a specimen which may have taken a wrong fold or a bad direction. After this the pressure may be gradually increased, and the plants left from one to several days without shifting. The exact amount of pressure to be given will depend on the consistence of the specimens and the amount of paper. It must only be borne in mind that too much pressure crushes the delicate parts, too little allows them to shrivel, in both cases interfering with their future examination.

240. The most convenient specimens will be made if the drying-paper is the same size as that of the herbarium in which they are to be kept. That of writing-demy, rather more than 16in. by $10\frac{1}{2}$in., is a common and very convenient size. A small size reduces the specimens too much, a large size is both costly and inconvenient for use.

241. When the specimens are quite dry and stiff, they may be packed up in bundles with a single sheet of paper between each layer, and this paper need not be bibulous. The specimens may be placed very closely on the sheets, but not in more than one layer on each sheet, and care must be taken to protect the bundles by sufficient covering from the effects of external moisture or the attacks of insects.

242. In laying the specimens into the herbarium, no more than one species should ever be fastened on one sheet of paper, although several specimens of the same species may be laid side by side. And throughout the process of drying, packing, and laying in, great care must be taken that the labels be not separated from the specimens they belong to.

243. To examine or dissect flowers or fruits in dried specimens it is necessary to soften them. If the parts are very delicate, this is best done by gradually moistening them in cold water; in most cases, steeping them in boiling water or in steam is much quicker. Very hard fruits and seeds will require boiling to be able to dissect them easily.

244. For dissecting and examining flowers in the field, all that is necessary is a penknife and a pocket lens of two or three glasses from 1 to 2in. focus. At home it is more convenient to have a mounted lens or simple microscope, with a stage holding a glass plate, upon which the flowers may be laid; and a pair of dissectors, one of which should be narrow and pointed, or a mere point, like a thick needle, in a handle; the other should have a pointed blade, with a sharp edge, to make clean sections across the ovary. A compound microscope is rarely necessary, except in cryptogamic botany and vegetable anatomy. For the simple microscope, lenses of $\frac{1}{4}$, $\frac{1}{2}$, 1, and $1\frac{1}{2}$in. focus are sufficient.

245. To assist the student in *determining* or ascertaining the name of a plant belonging to a Flora, analytical tables should be prefixed to the Orders, Genera, and Species. These tables should be so constructed as to contain, under each bracket, or equally indented, two (rarely three or more) alternatives as nearly as possible contradictory or incompatible with each other, each alternative referring to another bracket, or having under it another pair of alternatives further indented. The student having a plant to determine, will first take the general table of Natural Orders, and examining his plant at each step to see which alternative agrees with it, will be led

on to the Order to which it belongs; he will then compare it with the detailed character of the Order given in the text. If it agrees, he will follow the same course with the table of the genera of that Order, and again with the table of species of the genus. But in each case, if he finds that his plant does not agree with the detailed description of the genus or species to which he has thus been referred, he must revert to the beginning and carefully go through every step of the investigation before he can be satisfied. A fresh examination of his specimen, or of others of the same plant, a critical consideration of the meaning of every expression in the characters given, may lead him to detect some minute point overlooked or mistaken, and put him into the right way. Species vary within limits which it is often very difficult to express in words, and it proves often impossible, in framing these analytical tables, so to divide the genera and species that those which come under one alternative should absolutely exclude the others. In such doubtful cases both alternatives must be tried before the student can come to the conclusion that his plant is not contained in the Flora or that it is erroneously described.

246. In those Floras where analytical tables are not given, the student is usually guided to the most important or prominent characters of each genus or species, either by a general summary prefixed to the genera of an Order or to the species of the genus, for all such genera or species; or by a a special summary immediately preceding the detailed description of each genus or species. In the latter case this summary is called a *diagnosis*. Or sometimes the important characters are only indicated by italicizing them in the detailed description.

247. It may also happen that the specimen gathered may present some occasional or accidental anomalies peculiar to that single one, or to a very few individuals, which may prevent the species from being at once recognized by its technical characters. It may be useful here to point out a few of these anomalies which the botanist may be most likely to meet with. For this purpose we may divide them into two classes, viz.:

1. *Aberrations from the ordinary type or appearance of a species for which some general cause may be assigned.*

A bright, light, and open situation, particularly at considerable elevations above the sea, or at high latitudes, without too much wet or drought, tends to increase the size and heighten the colour of flowers, in proportion to the stature and foliage of the plant.

Shade, on the contrary, especially if accompanied by richness of soil and sufficient moisture, tends to increase the foliage and draw up the stem, but to diminish the number, size, and colour of the flowers.

A hot climate and dry situation tends to increase the hairs, prickles, and other productions of the epidermis, to shorten and stiffen the branches, rendering thorny plants yet more spinous. Moisture in a rich soil has a contrary effect.

The neighbourhood of the sea, or a saline soil or atmosphere, imparts a thicker and more succulent consistence to the foliage and almost every part of the plant, and appears not unfrequently to enable plants usually annual to live through the winter. Flowers in a maritime variety are often much fewer, but not smaller.

The luxuriance of plants growing in a rich soil, and the dwarf stunted character of those crowded in poor soils, are too well known to need particularizing. It is also an everyday observation how gradually the specimens of a species become dwarf and stunted as we advance into the cold damp regions of the summits of high mountain-ranges, or into high northern latitudes; and yet it is frequently from the want of attention to these circumstances that numbers of false species have been added to our Enumerations and Floras. Luxuriance entails not only increase of size to the whole plant, or of particular parts, but increase of number in branches, in leaves, or leaflets of a compound leaf; or it may diminish the hairiness of the plant, induce thorns to grow out into branches, etc.

Capsules which, while growing, lie close upon the ground, will often become larger, more succulent, and less readily dehiscent, than those which are not so exposed to the moisture of the soil.

Herbs eaten down by sheep or cattle, or crushed underfoot, or otherwise checked in their growth, or trees or shrubs cut down to the ground, if then exposed to favourable circumstances of soil and climate, will send up luxuriant side-shoots, often so different in the form of their leaves, in their ramification and inflorescence, as to be scarcely recognizable for the same species.

Annuals which have germinated in spring, and flowered without check, will often be very different in aspect from individuals of the same species, which, having germinated later, are stopped by summer droughts or the approach of winter, and only flower the following season upon a second growth. The latter have often been mistaken for perennials.

Hybrids, or crosses between two distinct species, come under the same category of anomalous specimens from a known cause. Frequent as they are in gardens, where they are artificially produced, they are probably rare in nature, although on this subject there is much diversity of opinion, some believing them to be very frequent, others almost denying their existence. Absolute proof of the origin of a plant found wild, is of course impossible; but it is pretty generally agreed that the following particulars must always co-exist in a *wild hybrid*. It partakes of the characters of its two parents; it is to be found isolated, or almost isolated, in places where the two parents are abundant; if there are two or three, they will generally be dissimilar from each other, one partaking more of one parent, another of the other; it seldom ripens good seed; it will never be found where one of the parents grows alone.

Where two supposed species grow together, intermixed with numerous intermediates bearing good seed, and passing more or less gradually from the one to the other, it may generally be concluded that the whole are mere varieties of one species. The beginner, however, must be very cautious not to set down a specimen as intermediate between two species, because it appears to be so in some, even the most striking characters, such as stature and foliage. Extreme varieties of one species are connected together by transitions in all their characters, but these transitions are not all observable in the same specimens. The observation of a single intermediate is therefore of little value, unless it be one link in a long series of intermediate forms, and, when met with, should lead to the search for the other connecting links.

2. *Accidental aberrations from the ordinary type, that is, those of which the cause is unknown.*

These require the more attention, as they may sometimes lead the beginner far astray in his search for the genus, whilst the aberrations above-mentioned as reducible more or less to general laws, affect chiefly the distinction of species.

Almost all species with coloured flowers are liable to occur occasionally with them all white.

Many may be found even in a wild state with double flowers, that is, with a multiplication of petals.

Plants which have usually conspicuous petals will occasionally appear without any at all, either to the flowers produced at particular seasons, or to all the flowers of individual plants, or the petals may be reduced to narrow slips.

Flowers usually very irregular, may, on certain individuals, lose more or less of their irregularity, or appear in some very different shape. Spurs, for instance, may disappear, or be produced on all instead of one only of the petals.

One part may be occasionally added to, or subtracted from, the usual number of parts in each floral whorl, more especially in regular polypetalous flowers.

Plants usually monœcious or diœcious may become occasionally hermaphrodite, or hermaphrodite plants may produce occasionally unisexual flowers by the abortion of the stamens or of the pistils.

Leaves cut or divided where they are usually entire, variegated or spotted where they are usually of one colour, or the reverse, must also be classed amongst those accidental aberrations which the botanist must always be on his guard against mistaking for specific distinctions.

INDEX OF TERMS, OR GLOSSARY.

The Figures refer to the Paragraphs of the Outlines.

Term	Par.
Aberrations	247
Abortive	84
Abruptly pinnate	43
Accessory organs	168
Acicular	54
Achene	158
Aculeate	170
Acuminate, acumen	47
Acute	47
Adherent	140, 145
Adnate	63, 145
Adnate anther	114
Adventitious	17, 19
Aerial=growing in the air.	
Æstivation	102
Aggregate fruit	147
Alabastrum (bud)	214
Alæ (wings)	37, 155
Alate=having wings.	
Albumen, albuminous	162
Alburnum	198
Alliances	182
Alternate	32, 90
Amentum=catkin	76
Amphitropous	134
Amplexicaul	37
Amygdaloid=almond-like.	
Amyloid	192
Anastomose	40
Anatropous	134
Androgynous	87
Angiospermous	161
Anisomerous	94
Annuals	12
Anterior	91
Anther	109, 114
Anthesis (flowering period)	214
Apetalous	85
Apex	36, 47, 115
Apiculate=with a little point.	
Apocarpous	125
Aquatic=growing in water	14
Aboreous or arborescent plants	12
Aril, arillus	164
Arillate (having an aril)	164
Aristate	47
Article, articulate, articulation	54
Artificial divisions and characters	184
Ascending	28
Asepalous	85
Assimilation	218

Term	Par.
Auricle	49
Auriculate=having auricles	50
Axil, axillary	17
Axile (in the axis)	132
Bark	198
Barren	85, 110
Base	36, 48, 115
Bast-cells	198
Berry	157
Bi- (2 in composition)	44
Bicarpellary	125
Bidentate	44
Biennials	12
Bifid	44
Bifoliolate	44
Bijugate	44
Bilabiate (two-lipped)	102, 105
Bilocular	126
Bipinnate	43
Bisexual	85
Biternate	44
Blade	35
Bracts, bracteæ	60, 77, 202
Bracteate=having bracts.	
Bracteoles	62
Bristles, bristly	173
Bud	16
Bulb	26
Bush	12
Cæspitose=tufted	28
Callous=hardened and usually thickened.	
Calycule, calyculate	80
Calyx	15, 90, 96
Cambium-region	211
Campanulate	104
Campylotropous	134
Canescent	173
Capillary=hair-like	54
Capitate	74
Capsule	158
Carpel	15, 123
Carpophore	146
Cartilaginous=of the consistence of cartilage or parchment.	
Caruncule, carunculate	164
Caryopsis	160
Catkins	76
Cauline (on the stem)	38
Caulocarpic	12
Cells (elementary)	186

Term	Par.
Cells (of anthers)	109
Cells (of the ovary)	121
Cellular system	198
Cellular tissue	188
Cellulose	191
Centrifugal	72
Centripetal	72
Chaff	82
Chalaza	133
Character	183
Chlorophyll	192
Chromule	192
Ciliate	39
Circumsciss	158
Cirrhus=tendril	169
Class	182
Claw (of a petal)	107
Climbing stem	29
Coats of the ovule	133
Coats of the seed	163
Coccus	159
Coherent	145
Collateral=inserted one by the side of the other.	
Collection of specimens	224
Coma	163
Common petiole	39
Complete flower	89
Compound fruit	147
Compound leaf	39
Compound flower	74
Compound ovary	126
Compound umbel	74
Compressed	54
Cone	160
Confluent	117
Conical	54
Connate	145
Connective, connectivum	109
Connivent	145
Contorted, convolute	102
Cordate	49
Cordiform	49
Coriaceous	55
Corky layer	198
Corm	27
Corolla	15, 90, 97
Corrugate (crumpled)	102
Corymb, corymbose	74
Costate	173
Cotton, cottony	173
Cotyledons	166
Creeping	28
Crenate, crenulate	39

GLOSSARY OF TERMS.

Term	Par.
Cristate=having a crest-like appendage.	
Crown of the root	24
Crumpled	102
Crustaceous	55
Cryptogamous plants	10
Culm	34
Cuneate	45
Cupular (cup-shaped)	136
Cuspidate	47
Cylindrical	54
Cyme, cymose	74
Deca- or decem- (10 in composition)	44, 92
Deciduous calyx	152
Decompound	43
Decumbent	28
Decurrent	37
Decussate	32
Definite	89
Definitions	(p. i.)
Dehiscence, dehiscent	118, 156
Dentate	39
Depressed	54
Descriptive Botany	(p. i.)
Determinate	67
Determination of plants	245
Dextrine	192
Di- (2 in composition)	92
Diadelphous	113
Diagnosis	246
Dialypetalous	100
Diandrous	93
Dichlamydeous	85
Dichotomous	33
Diclinous	86
Dicotyledonous plants	167
Didymous	54
Didynamous	113
Diffuse	28
Digitate	41
Dignous	93, 125
Dimerous	93
Dimidiate	117
Diœcious	86
Dipetalous	93
Disepalous	93
Disk	136
Dissepiment	126
Dissected	39
Distichous	32
Distinct	145
Divaricate	115
Diverging, divergent	115, 145
Divided	39
Dorsal=on the back.	
Double flowers	97
Down, downy	173
Drupe	157
Dry fruits	158
Ducts	188
Duramen	198
Ear	76
Echinate	173
Elaborated sap	217
Elementary cells and tissues	186

Term	Par.
Elliptical	45
Emarginate	47
Embryo	162, 166
Endocarp	157
Endogens, endogenous plants	195
Endogenous stem	199
Endosmose	217
Ennea- (9 in composition)	92
Entire	39
Epicarp	157
Epidermis	173, 193
Epigynous	140
Epigynous disk	144
Epiphyte	14
Erect	28
Exalbuminous (without albumen)	162
Examination of plants	234
Exogens, exogenous plants	195
Exogenous stem	198
Exserted	113
Extrorse	118
Falcate	45
Families	181
Farinose	173
Fascicled, fasciculate	32
Fastigiate	74
Fecula	192
Female	85
Fertile	85
Fibre	18
Fibrous root	20
Fibro-vascular system	193
Filament	109
Filiform=thread-like.	
Fimbriate=fringed.	
Flabelliform=fan-shaped	45
Fleshy	55
Floccose	173
Floral Envelope	15
Floral leaves	61
Flowers	15, 84, 213, 219
Flowering plants	10
Foliaceous=leaf-like.	
Follicle	159
Foramen	133
Forked	33
Foveolate	105
Free	89, 132, 140, 145
Fruit	15, 146, 222
Frutescent, fruticose	12
Function	7
Funicle (funiculus)	164
Funnel-shaped	104
Furrowed	173
Fusiform=spindle-shaped	54
Gamopetalous	100
Geminate	32
Genus, genera	180
Germ, germination	215
Gibbous	105
Glabrous	173
Glands	175, 206
Glandular-setose	173
Glaucous	173

Term	Par.
Globose, globular	54
Glochidiate	173
Glume	83
Glutinous	173
Grain	160
Gymnospermous	161
Gynobasis, gynophore	143
Habit	183
Hairs	171, 205, 223
Hastate	50
Head	74
Heart-wood	198
Hepta- (7 in composition)	92
Herbaceous perennials	12
Herbarium	224
Hermaphrodite	85
Heterogamous	87
Hexa- (6 in composition)	92
Hilum	162
Hirsute	173
Hispid	173
Hoary	173
Homogamous	87
Hooks	169
Hybernaculum	23
Hybrids	247
Hypocrateriform (salver-shaped)	104
Hypogynous	140
Imbricate, imbricated	58, 102
Imparipinnate	43
Imperfect	84
Incomplete	84
Indefinite	92
Indehiscent	156
Indeterminate	67
Indumentum	171
Induplicate	102
Inferior	140
Inferior radicle	167
Inflorescence	66
Infundibuliform (funnel-shaped)	104
Innate anther	114
Insertion	140
Internode	31
Interrupted spike or raceme	75
Introrse	118
Involucre, involucel	79
Involute	102
Irregular	95
Isomerous	89
Joint, joining	54
Jugum, juga=pairs	44
Kernel	157
Knob	25
Labellum	105
Laciniate	39
Lamina	35, 107
Lanate=woolly	173
Lanceolate	45
Lateral	91

GLOSSARY OF TERMS.

Term	Par.
Leaf, leaves	15, 35, 200, 218
Leaf-bud	16
Leaflet	39
Leaf-opposed	67
Legume	160
Lepidote	172
Liber	198, 211
Ligulate=strap-shaped.	
Limb	104
Linear	45, 54
Lip, lipped	105
Lobe, lobed	39
Loculicidal	158
Lower	91
Lunate=crescent-shaped.	
Lyrate	41
Male	85
Marescent	151
Mealy	173
Medullary rays and sheath	198
Membranous	55
Micropyle	165
Midrib	40
Monadelphous	113
Monandrous	112
Moniliform	54
Mono- (1 in composition)	92
Monocarpellary	125
Monocarpic	12
Monochlamydeous	85
Monocotyledonous plants	167
Monœcious	86
Monogynous	125
Monopetalous	100
Morphology	8, 88
Mucronate	47
Multi- (many, or an indefinite number, in composition)	44
Muricate	173
Naked	85, 161
Natural divisions and characters	184
Natural Order	181
Navicular=boat-shaped.	
Nectary	138
Nerve	40
Net-veined	40
Neuter	85
Node	31
Novem- (9 in composition)	44
Nucleus of a cell	191
Nucleus of the ovule	133
Nut	158
Obcompressed	54
Obconical	54
Obcordate	47
Oblate	45
Oblong	45, 54
Obovate	45
Obovoid	54
Obpyramidal	54
Obtuse	47
Oct- or octo- (8 in composition)	44, 92
Offset	23
Opposite	32
Orbicular	45
Order	181
Organ	7
Organogenesis	213
Organs of vegetation and reproduction	9
Orthotropous	134
Oval	45
Ovary	121
Ovate	45
Ovoid	54
Ovule	121, 133
Palate	105
Palea, paleæ	82
Paleaceous=of a chaffy consistence.	
Palmate	41, 42
Palmatifid, palmatisect	42
Panicle, paniculate	74
Papillæ	122
Pappus	155
Parallel veins	40
Parasite	14
Parenchyma	188
Parietal	132
Pectinate	41
Pedate	41, 42
Pedatifid, pedatisect	42
Pedicel	70
Pedicellate=on a pedicel.	
Peduncle	68
Pedunculate=on a peduncle.	
Peltate	52
Penicillate	130
Penta- (5 in composition)	92
Pepo	160
Perennials	12
Perfect flower	84
Perfoliate	37
Perianth	15, 98, 202, 230
Pericarp	154
Perigynous	140
Perisperm	162
Persistent	146
Personate	105
Petal	90
Petiole	35
Petiolule	39
Phænogamous, phanerogamous	10
Phyllaries	79
Phyllodium=a flat petiole with no blade.	
Pilose	173
Pinna	43
Pinnate	41, 42
Pinnatifid, pinnatisect	42
Pistil	15, 90, 120, 203, 221
Pistillate	85
Pith	198
Placenta, placentation	131
Plant	6
Plicate	102
Plumose	172
Plumule	166
Pluri=several, in composition.	
Plurilocular	126
Pod	158
Podocarp	120
Pollen	109, 119
Poly- (many, or an indefinite number, in composition)	92
Polyadelphous	113
Polyandrous	92, 112
Polygamous	86
Polygynous	92, 125
Polypetalous	100
Pome	160
Posterior	91
Præfoliation	57
Preservation of specimens	224
Prickles	170
Primine	133
Procumbent	28
Proliferous	17
Prosenchyma	188
Prostrate	28
Protoplasm	191
Pubescent, puberulent	173
Pulvinate (cushion-shaped)	136
Punctiform=like a point or dot.	
Putamen	157
Pyramidal	54
Pyrenes	157
Quadri- (4 in composition)	44
Quincuncial	102
Quinque- (5 in composition)	44
Quintuplinerved	40
Race	178
Raceme, racemose	74
Rachis	39, 68
Radical	38
Radicle	166
Raphe	134
Raphides	192
Receptacle	74, 185
Reduplicate	102
Regular	95
Reniform	51
Resupinate	105
Reticulate	40
Retuse	47
Revolute	102
Rhachis	39, 68
Rhaphe	134
Rhizome	21, 24
Rhomboidal	45
Ribs	40
Ribbed	173
Ringent	105
Root	15, 18, 196, 207, 216
Rootstock	24
Rostrate=beaked.	
Rosulate	38
Rotate	104
Rudimentary	84

GLOSSARY OF TERMS.

Term	Par.	Term	Par.	Term	Par.
Rugose	173	Stem	15, 28, 197, 210, 217	Triplinerved	40
Runcinate	41	Stem-clasping	37	Triquetrous	54
Runner	30	Sterile	85	Tristichous	32
		Stigma	121	Truncate	47
Saccate	105	Stipella	64	Trunk	12
Sagittate	50	Stipes, stipitate	65	Tube	101, 104
Salver-shaped	104	Stipules	63	Tuber, tuberous	20, 25, 204
Samara	158	Stock	16, 22	Tuberculate	173
Sap	192	Stole, stolon	23, 30	Tubular	104
Sapwood	198	Stomates	194	Tufted	28
Sarcocarp	157	Stone, stone-fruit	157	Tunicated bulb	27
Sarmentose	28	Striate	173	Turbinate=top-shaped	54
Scabrous	173	Strigose, strigillose	173	Twiner	29
Scales	58, 59, 172, 201	Strophiole, strophiolate	164	Twisted	102
Scaly bulb	26	Style	121	Type, typical	181
Scaly surface	172	Sub=*almost*, or *under*, in composition.			
Scape	69			Umbel, umbellate, umbellule	33, 74
Scariose, scarious	55	Subclass, suborder	182		
Scattered	32	Submerged=under water.		Umbilicate	173
Scion	30	Subulate	54	Umbonate	173
Scorpioid cyme	74	Succulent	55	Uncinate=hooked.	
Section	182	Succulent fruits	157	Undershrubs	12
Secund	32	Sucker	30	Undulate	39
Secundine	133	Suffrutescent, suffruticose	12	Unequally pinnate	43
Seed	161	Sugar	192	Unguiculate	107
Segment	39	Sulcate	173	Unguis (claw)	107
Sepals	90	Superior	140	Uni- (1 in composition)	44
Septem- (7 in composition)	44	Superior radicle	167	Unilateral (one-sided) racemes	74
Septicical	158	Superposed=inserted one above the other.		Unilocular	126
Septum=partition	126			Unisexual	86
Serrate, serrulate	39	Suture	159	Unsymmetrical	94
Sessile	37	Symmetrical	89	Upper	91
Seta, setæ (bristles)	173	Synandrous	112	Urceolate	104
Setaceous (bristle-like)	54	Syncarpous	125	Utricle	158
Setose (bearing bristles)	173	Syngenesious	113		
Sex- (6 in composition)	44	Systematic Botany	(p. xix)	Valvate	102
Sheathing	37			Valves	158
Shrubs	12	Taproot	20	Variety	178
Silicule, siliqua	196	Teeth	39, 101	Vascular tissue	188
Silver grain	198	Tegmen	163	Vegetable Anatomy	8, 186
Simple	39	Tendril	29, 169	Vegetable Chemistry	8
Sinuate	39	Terete	54	Vegetable Homology or Metamorphosis	88
Sinus	39	Ternate	32, 41		
Smooth	173	Terrestrial=growing on the earth	14	Vegetable Physiology	8, 207
Spadix	76			Veins, veinlets, venation	40
Spatha	81	Testa	163	Vernation	57
Spathulate	45	Tetra (4 in composition)	92	Versatile anther	114
Species	177	Tetradynamous	113	Verticil, verticillate	32
Specimen	225	Thorns	170	Vessels	188
Spherical	54	Throat	104	Virgate=twiggy	28
Spike, spicate	74	Thyrsus, thyrsoid	74	Viscid, viscous	173
Spikelet	76	Tissues (elementary)	186	Vitta, vittæ	175
Spinous	170	Tomentose	173	Viviparous	17
Spiral vessels	188	Toothed	39	Voluble	29
Spur, spurred	105	Torus	135		
Squamæ=scales	58	Trees	12		
Squarrose	58	Tri- (3 in composition)	44, 92	Wart, warted	173
Stamens	15, 90, 108, 203	Tribe	182	Wavy	39
Staminate	85	Trichotomous	33	Whorl, whorled	32
Staminodia	110	Trifid	41	Wing, winged	37, 155
Starch	192	Trifoliolate	41	Wood	198
Stellate	104	Trigonous	54	Woody tissue	188
Stellate hairs	172	Tripinnate	43	Wool, woolly	173

THE QUEENSLAND FLORA.

Class I. **DICOTYLEDONS.**

Stem, when perennial, consisting of a pith in the centre, of one or more concentric circles of woody tissue, and of the bark on the outside. Embryo with two cotyledons, the young stem in germination proceeding from between the two lobes of the embryo or from a notch at its summit.

The above characters are the most constant to separate *Dicotyledons* from *Monocotyledons*; these two great classes have, however, each a peculiar habit, which in most cases is easily recognised. All Queensland trees and shrubs, except *Palms*, a few *Ferns*, and *Bamboos*, and a few others with linear grass-like leaves, are *Dicotyledons*; so also are almost all plants with opposite, or whorled, or netted-veined leaves, or with the parts of the flower in fours, fives, or eights, or with indefinite stamens, all these characters being very rare in *Monocotyledons*. Benth. Fl. Austr.

(The following short ordinal characters given are not absolute, nor without exception, and are inserted for the purpose of calling attention to one or two of the most striking or most important features of each Order.) *Benth. l. c.*

SUBCLASS I. POLYPETALÆ.

Petals several, distinct (wanting in a few genera, very rarely united).

Series I. Thalamifloræ.—Torus small or elongated, rarely expanded into a disk. Ovary superior. Stamens definite or more frequently indefinite.

Alliance (Cohors) **I. Ranales.**—*Stamens indefinite, or if definite, opposite the petals. Carpels distinct or united at the base only, superior, or rarely enclosed in a fleshy torus. Embryo small, in fleshy albumen.*
(Carpels united in *Eupomatia* and *Nymphæa*. Embryo large, without albumen in some *Menispermaceæ* and in *Nelumbium*.)

I. Ranunculaceæ. Herbs with radical or alternate leaves, or climbers with opposite leaves. No stipules. Sepals usually coloured and deciduous. Petals in a single series or none. Stamens indefinite. No arillus.

II. Dilleniaceæ. Trees, shrubs, or undershrubs with alternate leaves. No stipules. Sepals usually herbaceous and persistent. Petals in a single series. Stamens usually indefinite. Seeds with an arillus or strophiola.

III. Magnoliaceæ. Shrubs or trees, with alternate leaves. Petals indefinite. Stamens indefinite. No arillus. (Calyx entire in the bud, irregularly split.)

IV. Anonaceæ. Shrubs, trees, or woody climbers, with alternate leaves. No stipules. Sepals 3. Petals in two series of 3 each (excepting *Eupomatia*, where sepals and petals are combined in a mass). Stamens indefinite. Carpels indefinite. Albumen ruminate.

V. Menispermaceæ. Twiners, with alternate leaves. No stipules. Flowers small, diœcious. Sepals in 2 or more series of 3 or 2 each. Petals smaller than the inner sepals, or none. Stamens definite opposite the petals. Carpels 6 or fewer.

VI. Nymphæaceæ. Aquatic herbs. Leaves usually peltate. Sepals or petals indefinite, or rarely in threes. Stamens indefinite. Carpels free or united, the ovules not in the inner angle,

DICOTYLEDONS.

Alliance II. Parietales.—*Stamens definite or indefinite. Ovary syncarpous, with 2 or more parietal placentas, either 1-celled, or incompletely divided by the placentas protruding in the cavity, or divided by false dissepiments connecting the placentas. Ovules usually several to each placenta, rarely solitary.*

VII. PAPAVERACEÆ. Herbs with alternate leaves. No stipules. Sepals 2. Petals 4. Flowers regular, with indefinite stamens, or irregular, with diadelphous definite stamens. Albumen copious. Embryo small.

VIII. CRUCIFERÆ. Herbs with alternate leaves. No stipules. Sepals 4. Petals 4. Stamens 6, tetradynamous or rarely 4. Placentas 2, connected by a false dissepiment. No albumen. Embryo curved.

IX. CAPPARIDEÆ. Herbs, shrubs, or trees. Stipules often prickly. Sepals 4 (2 outer ones sometimes united). Petals 4 (rarely more, or none, or united). Stamens indefinite, or if few, not tetradynamous. Placentas 2 or more. No albumen. Embryo curved.

X. VIOLARIEÆ. Herbs or shrubs. Stipules herbaceous or small. Sepals 5. Petals 5 (often irregular). Anthers 5, on short filaments, connivent or connected in a ring round the pistil. Placentas usually 3. Albumen fleshy. Embryo rather large.

XI. BIXINEÆ. Trees or shrubs. Stipules none. Sepals 5 or fewer. Petals various, often none. Stamens indefinite. Placentas 2, 3, or more (meeting in the axis in *Cochlospermum*). Albumen fleshy. Embryo rather large.

Alliance III. Polygalineæ.—*Sepals and petals 5 each, rarely fewer. Stamens the same number or twice as many, or fewer when the flowers are irregular. Ovary usually 2-merous (although in most genera occasionally 3–5-merous), partially or completely divided into as many cells. Ovules indefinite, or solitary with a superior micropyle. Albumen fleshy.*

XII. PITTOSPOREÆ. Trees, shrubs, undershrubs, or twiners, with alternate leaves. No stipules. Flowers regular or oblique. Stamens as many as petals. Embryo minute.

XIII. TREMANDREÆ. Shrubs often heath-like, with alternate or whorled or opposite leaves. No stipules. Flowers regular. Stamens twice as many as petals. Embryo small or minute.

XIV. POLYGALEÆ. Herbs, undershrubs, or shrubs, with alternate leaves. No stipules. Flowers irregular. Stamens monadelphous. Embryo rather large, sometimes almost or quite without albumen.

Alliance IV. Caryophyllineæ.—*Sepals or calyx-lobes 5 or fewer. Petals 5 or fewer. Stamens as many or twice as many, or indefinite. Ovary 1-celled, with central placentas (except Frankenia). Albumen mealy. Embryo curved, or rarely straight when the albumen is scanty.*

(Ovary half-inferior in *Portulaca*.)

XV. FRANKENIACEÆ. Small or prostrate undershrubs, or herbs, with small opposite leaves. No stipules. Calyx angular, toothed. Petals isomerous with the calyx. Stamens definite. Placentas parietal.

XVI. CARYOPHYLLEÆ. Herbs rarely undershrubs, with opposite entire leaves. Stipules none or scarious. Calyx toothed or sepals free. Petals isomerous with the calyx. Stamens definite. Placentas central.

XVII. PORTULACEÆ. Herbs, often succulent, with alternate or opposite leaves. Stipules scarious or changed into hairs. Sepals 2. Petals more numerous than the sepals. Stamens indefinite or rarely definite. Placentas central.

Alliance V. Guttiferales.—*Sepals imbricate. Petals as many as sepals, or rarely more. Stamens indefinite (except Elatineæ). Ovary divided into cells, with axile placentas.*

XVIII. ÆLATINEÆ. Herbs or undershrubs, with small opposite leaves. Stipules small. Flowers hermaphrodite. Stamens definite.

XIX. HYPERICINEÆ. Herbs or shrubs, with opposite leaves. No stipules. Flowers hermaphrodite. Stamens indefinite.

XX. GUTTIFERÆ. Trees or shrubs with opposite leaves. No stipules. Flowers polygamous or unisexual. Stamens indefinite.

XXI. TERNSTRŒMIACEÆ. Trees or shrubs, with often alternate, coriaceous, undivided leaves, very rarely opposite or digitate. Stipules usually absent. Flowers hermaphrodite or rarely unisexual. Stamens often indefinite.

Alliance VI. Malvales.—*Sepals valvate (except Echinocarpus). Petals as many as sepals, or none. Stamens indefinite or monadelphous (except Lasiopetaleæ). Ovary divided into cells with axile placentas.*

XXII. MALVACEÆ. Herbs, shrubs, or trees, with alternate leaves. Stipules usually present. Stamens monadelphous. Anthers 1-celled.

XXIII. STERCULIACEÆ. Herbs, shrubs, or trees, with alternate leaves. Stipules usually present. Stamens monadelphous, or, if free, definite and alternating with the petals. Anthers 2-celled.

XXIV. TILIACEÆ. Trees or shrubs. rarely herbs, with alternate leaves. Stipules usually present. Stamens indefinite, free, or scarcely united at the base. Anthers 2-celled.

SERIES II. DISCIFLORÆ.—Torus usually thickened or expanded into a disk, either free or adnate to the ovary, or to the calyx, or to both, rarely reduced to glands, or wanting. Stamens as many or twice as many as petals, or fewer. Ovary superior, or partially immersed in the disk, divided into cells with axile placentas, or the carpels distinct.

(Stamens indefinite in a very few exceptional species. Ovary inferior or enclosed in the calyx-tube in most *Rhamneæ*; 1-celled in some *Olacineæ*.)

Alliance VII. Geraniales.—*Disk within the stamens, or confluent with the staminal tube, or reduced to glands, or obsolete. Gynœcium lobed or apocarpous, or sometimes entire. Ovules usually 1 or 2 in each cell, 1 or both pendulous with a ventral raphe.*

XXV. LINEÆ. Herbs or shrubs, with undivided alternate leaves. Stipules often present Disk small, glandular, or none. Ovary entire. Ovules usually 2 in each cell. Albumen fleshy rarely wanting.

XXVI. MALPIGHIACEÆ. Woody climbers (rarely trees or shrubs), with opposite (rarely alternate) leaves. Stipules present. Two glands on the outside of some or all the calyx-lobes (wanting in the Australian genera.) Disk large. Gynœcium lobed or apocarpous. Ovules solitary in each cell. No albumen.

XXVII. ZYGOPHYLLEÆ. Herbs or shrubs, usually articulate or succulent, without glandular dots. Leaves 2-foliolate or pinnate, rarely simple. Stipules present. Disk fleshy. Ovary angular or lobed. Ovules 2 or rarely more in each cell. Albumen fleshy or none.

XXVIII. GERANIACEÆ. Herbs or shrubs, articulate or not, with toothed, divided, or compound leaves without glandular dots. Stipules usually present. Disk reduced to 5 glands or obsolete. Ovary angular or lobed. Ovules 1, 2 or rarely more in each cell. Albumen none or rarely fleshy.

XXIX. RUTACEÆ. Trees or shrubs, very rarely herbs, with compound or rarely simple leaves, always marked with pellucid glandular dots. No stipules. Disk within the stamens. Ovary rarely entire, usually lobed or the carpels distinct, with the styles connate or gynœcium entirely apocarpous. Ovules 2 in each cell. Albumen fleshy or none.

XXX. SIMARUBEÆ. Characters of *Rutaceæ*, except that the leaves are not dotted and the ovules are usually solitary in each cell. Taste generally bitter.

XXXI. OCHNACEÆ. Shrubs or trees, with alternate, simple, glabrous, penninerved leaves. Stipules various. Sepals 4—5, free, often scarious or rigid. Torus enlarging after flowering. Stamens definite or indefinite. Anthers linear, often elongate. Ovary often lobed. Ovules 1, 2, or more in each cell.

XXXII. BURSERACEÆ. Trees or shrubs, not dotted, but with a balsamic juice. Leaves pinnately or ternately compound. No stipules. Disk free or adnate to the calyx-tube. Ovary entire. Ovules usually 2 in each cell. Albumen none. Cotyledons much folded or rarely thick and fleshy.

XXXIII. MELIACEÆ. Trees or shrubs, with compound or rarely simple leaves. No stipules. Stamens monadelphous. Anthers sessile or rarely stipitate within or on the top of the staminal tube. Ovary entire. Ovules 2 in each cell. Albumen none or fleshy.

Alliance VIII. Olacales.—*Disk various or none. Ovary entire. Ovules 1 to 3 in a solitary cell or 1 in each cell, pendulous with a dorsal raphe, the integuments not distinct from the nucleus. Seeds solitary in the fruit or in the cells. Albumen copious.*

XXXIV. OLACINEÆ. Trees or shrubs, rarely undershrubs or climbers. No stipules. Petals or corolla-lobes valvate (except *Villaresia*). Ovary 1-celled or incompletely 3- to 5-celled. Fruit 1-seeded.

XXXV. ILICINEÆ. Trees or shrubs. No stipules. Petals or corolla-lobes imbricate. Ovary 3- or more celled.

Alliance IX. Celastrales.—*Disk thick and fleshy or adnate to the calyx, the stamens outside or upon it. Ovary entire (except* Stackhousia). *Ovules 1 or 2 in each cell, erect with a ventral raphe.*

XXXVI. CELASTRINEÆ. Trees or shrubs, with simple leaves. Stipules none, or minute and deciduous. Calyx-lobes imbricate. Petals spreading. Stamens alternating with the petals or fewer. Ovary entire.

XXXVII. STACKHOUSIEÆ. Herbs or undershrubs, with simple leaves. Calyx-lobes imbricate. Petals erect, usually connate. Stamens alternating with the petals. Ovary lobed.

XXXVIII. RHAMNEÆ. Trees or shrubs, with simple leaves. Stipules usually present. Calyx-lobes valvate. Petals small, concave (or none). Stamens opposite the petals. Ovary entire, often inferior.

XXXIX. AMPELIDEÆ. Climbers, with simple or compound leaves, the petiole usually expanded into a stipule. Calyx-lobes imbricate. Petals valvate. Stamens opposite the petals. Ovary entire. Albumen cartilaginous. Embryo small.

Alliance X. Sapindales.—*Disk fleshy or adnate to the calyx, within or under or outside the stamens. Gynæcium entire, lobed or apocarpous. Ovules 1 or 2 in each cell, ascending with a ventral raphe, or reversed, or suspended from an erect funiculus, or pendulous with an inferior micropyle.*

XL. SAPINDACEÆ. Trees, shrubs, or climbers, with compound or simple leaves. Stamens anisomerous with the petals, or twice as many as petals, or of the same number. Often (but not always) within the disk. Style 1. Ovules ascending.

XLI. ANACARDIACEÆ. Trees or shrubs, with compound or simple leaves. Stamens as many or twice as many as petals, never within the disk. Ovules suspended from an erect funicle or from the top or side of the cell with an inferior micropyle.

ORDER I. **RANUNCULACEÆ.**

Sepals 3 or more, most frequently 5, usually petal-like and deciduous. Petals of the same number or more, or sometimes none, or very small and deformed. Stamens indefinite, hypogynous, free. Anthers innate. Gynœcium of several carpels, usually free; ovules anatropous, either solitary and ascending, with a ventral raphe, or pendulous with a dorsal raphe, or several. Fruit of one or more indehiscent achenes or berries, or follicular capsules, the distinct styles usually persistent as short points, or lengthened into long, often bearded tails. Seeds without any arillus. Embryo very small, near the base of a copious albumen.—Herbs either annual, or with a perennial rootstock, or creeping stolons, with radical or alternate leaves, or climbers with opposite leaves. Leaves entire, or palmately or pinnately lobed or divided, the petiole often dilated and sheathing at the base, or rarely accompanied by stipular appendages. Hairs, when present, simple. Flowers regular (or in a few genera, not Australian, irregular), terminal or leaf-opposed, rarely axillary, solitary paniculate or racemose.

The Order is chiefly numerous in the temperate regions of the northern hemisphere, rare within the tropics, and not represented by many species in the southern hemisphere. The Australian ones are nearly all extra tropical, and belong to genera more numerously represented in the north.—*Benth.*

TRIBE I. **Clematideæ.**—*Sepals valvate. Carpels indehiscent, with 1 pendulous ovule or seed in each. Stems often climbing. Leaves opposite.*
Petals none . . . . 1. CLEMATIS.

TRIBE II. **Anemoneæ.**—*Sepals imbricate. Carpels indehiscent, with 1 pendulous ovule or seed in each. Herbs. Leaves radical or alternate or forming an involucre below the flower.*
Petals minute, narrow. No involucre. Achenes very numerous, in a long, close, slender spike . . . . 2. MYOSURUS.

TRIBE III. **Ranunculeæ.**—*Sepals imbricate. Carpels indehiscent, with 1 ascending ovule or seed in each. Herbs. Leaves radical or alternate.*
Sepals deciduous. Petals 3, 5, or more . . . . 3. RANUNCULUS.

1. CLEMATIS, Linn.
(From the Greek, alluding to the twisting branches.)

Sepals 4, or rarely 5 to 8, petal-like, valvate in the bud. Petals none, or smaller than the sepals, and passing gradually into the stamens. Carpels many, with one pendulous ovule in each. Achenes capitate, sessile, or scarcely stipitate, terminating in a plumose or simple tail, formed by the persistent and enlarged style.—Stem woody and climbing, or rarely dwarf or prostrate. Leaves opposite, pinnately or ternately divided into three or more petiolulate segments, or rarely simple, the petiole often twisted or twining. Flowers axillary or terminal solitary, or in panicles, which are shortened branches with the leaves reduced to small bracts, and often polygamous or diœcious.

A large genus, dispersed over the temperate regions both of the New and the Old World, rare within the tropics. The Australian species are all endemic, although one is closely connected with a South Pacific one. They have all simple or once- or twice-ternately divided leaves, diœcious, apetalous, white or cream-coloured flowers, the males usually without any ovaries, the females with a few imperfect stamens, and the carpels of all have plumose tails.—*Benth.*

Anthers linear or oblong, tipped by a subulate or oblong appendage.
  Woody climbers. Leaflets mostly once or twice ternate.
    Anther-points slender. Leaflets almost coriaceous, when
     large usually toothed, when small twice ternate . . . 1. *C. aristata.*
    Anther-points very short. Leaflets usually 3, rather large,
     thin, and entire . . . . 2. *C. glycinoides.*
Anthers short, without any appendage.
  Leaflets ternate, rather large, loosely pubescent underneath . . . . var. *submutica.*
  Leaflets mostly twice ternate, small or narrow, glabrous
   or closely pubescent . . . . 3. *C. microphylla*
                                              var. *Fawcettii.*

1. **C. aristata** (awned), *R. Br.*, *Benth. Fl. Austr.* i. 6. A woody climber, trailing over rocks and bushes, or ascending into tall trees, glabrous, or softly pubescent, especially on the inflorescence. Leaves mostly on long petioles, and divided into 3-petiolulate segments or leaflets, varying from ovate-cordate to narrow-lanceolate, obtuse or acute, 1 to 2 or even 3in. long, usually irregularly toothed when large, entire when small, and of a firm consistence when full grown, but some of the leaves near the base of the flowering branches are occasionally simple, and others have often twice-ternate leaflets. Flowers white or yellowish, usually in short panicles or clusters in the upper axils. Sepals 4, or very rarely 5, oblong or linear-lanceolate, usually ¾ to 1in. long when fully out, glabrous or pubescent. Anthers oblong-linear, tipped by a subulate appendage, often as long as the cells, usually rather shorter, the outer anthers on long filaments, the inner ones almost sessile. Achenes numerous, ovate or lanceolate, pubescent or glabrous, with a plumose tail often attaining 1½in.—*F. v. M. Pl. Vict.*, i. 3; *Bot. Reg.* t. 238.

Hab.: Killarney. Flowering in Oct.
Var. *longiseta*, *Bail. Bot. Bull.* vii. A climber, glabrous except the young shoots and inflorescence. Leaves on slender petioles, leaflets 3, ovate-lanceolate, attaining the length of about 2in. and mostly under ½in. broad at the base, where they slightly taper to the rather long petiolules, margins bordered by distant setaceous teeth. Flowers yellowish, tomentose, in short racemes in the axils of the leaves. Pedicels rather long and slender. Sepals 4, about 5 lines long, linear-lanceolate. Anthers often more ovate than oblong, and usually upon short filaments, the terminal awn frequently exceeding in length that of both anther and filament and often three times the length of the anther, and tapering to a hair-like point. Female flowers and achenes not to hand.
Hab.: Upper Nerang, *H. Schneider.* Flowering in Nov.

2. **C. glycinoides** (resembling a Glycine), *DC. Benth. in Fl. Austr.* i. 7. A woody climber, very near to those forms of *C. aristata* which have simply ternate rather large ovate-lanceolate or cordate leaflets, but these leaflets are usually of a thinner consistence, often broader, and quite entire or rarely with a single tooth near the base. Flowers usually smaller, the sepals narrow, from ½ to ¾in., pubescent or rarely glabrous. Anthers rather shorter, with a very short obtuse and almost gland-like appendage. Achenes glabrous or pubescent, usually narrower than in *C. aristata*, with tails of about 2in.—*C. stenosepala*, DC.

Var. *submutica.* Leaf-segments loosely pubescent underneath, sepals shorter, broader, and more villous than in the other forms, anthers short, tipped by a minute gland or entirely without appendage, as in *C. microphylla.*
Hab.: Brisbane River to Rockingham Bay and beyond. Flowering in Aug.

3. **C. microphylla** (small-leaved), *DC. Benth. in Fl. Austr.* i. 7. A tall woody climber, with the habit of the smaller-leaved varieties of *C. aristata*. Leaflets mostly twice ternate, narrow, from ovate-lanceolate or oblong to nearly linear, ½ to 1in. long, but sometimes simply ternate and larger and broader, or three times ternate and much smaller. Flowers rather smaller than in *C. aristata*, usually numerous in short panicles. Sepals cream-coloured, from oblong-lanceolate to narrow-linear, mostly about ½in. rarely near 1in. long, glabrous or pubescent. Stamens with unequal filaments as in *C. aristata*, but the anthers are always very shortly oblong or ovate and very obtuse, without any terminal appendage. Achenes of *C. aristata*, but usually with thicker, often wrinkled or warted margins and longer tails.—F. Muell. Pl. Vict. i. 4; *C. linearifolia*, Steud.; Hook. f. Fl. Tasm. i. 4, t. 1; *C. stenophylla*, Fras.; Hook. in Mitch. Trop. Aust. 363.

Hab.: Darling Downs, Springsure.

Var. *Fawcettii*. Leaflets twice ternate, membranous, ½ to 1¼in. long, broad or rhomboid-lanceolate, incised and acutely toothed, the petioles and petiolules rather long. Peduncles bearing 3 to 5 flowers on somewhat long pedicels. Sepals about 1in. long, and 1 line broad, acute, margins somewhat tomentose. Anthers oblong, about 1 line long. Styles plumose at the ends.—*C. Fawcetti*, F. v. M. Fragm. x. 1.

Var. *colorata* (coloured). This variety differs from the last mentioned in its dull-purple flowers, and in the segments of the leaf being usually narrow-linear.
Hab.: Killarney.

2. MYOSURUS, Linn.
(Inflorescence resembling tail of mouse).

Sepals usually 5, produced below their insertion into a small spur. Petals 5, small and very narrow, almost tubular at the top, often wanting. Carpels numerous, with one pendulous ovule in each. Achenes closely packed in a long slender spike, flat on the back, or with a raised nerve ending in the short persistent style.—Small annuals, with linear radical entire leaves. Flowers very small, on leafless scapes.

A genus comprising, besides the following, only one other species, *M. aristatus*, Geyer, distinguished by the more prominent and spreading points of the achenes, which, although originally described from North America and from Chili, has also been found in New Zealand, and may not improbably appear in Australia.—*Benth.*

1. **M. minimus** (very small), *Linn., Benth. Fl. Austr.* i. 8. Mouse tail. Leaves sometimes not an inch long, sometimes attaining 2 or even 3in., including their long petiole. Scapes shorter or longer than the leaves. Sepals yellowish or pale green, very small; petals rarely longer than the calyx, and in the Australian specimens often deficient. Stamens usually 4 or 5, and seldom above 10. Achenes sometimes near 300, the head lengthening into a spike of 1 to 2in.—*M. australis*, F. v. M. in Trans. Phil. Soc. Vict. i. 6.

Hab.: Southern Queensland, near the N.S.W. border.—*Rev. Dr. Wm. Woolls.*

3. RANUNCULUS, Linn.
(From *rana*, a frog, many species being found in boggy places.)

Sepals usually 5, deciduous. Petals as many or more, usually marked with a small nectariferous pit, or a minute scale near the base. Carpels several, with a single ascending ovule in each. Achenes in a globular or ovoid head or oblong spike, tipped or beaked by the persistent hooked or straight style.—Herbs either annual or with a perennial rootstock, and tufted entire or variously cut radical leaves. Flowering stems either a leafless scape, or several-flowered, bearing few leaves and chiefly at the base of the peduncles. Flowers yellow, white, or red.

A large genus abounding in the temperate and colder regions of both the northern and southern hemispheres, but more especially in the former, and almost confined in the tropics to the higher mountain ranges. *Benth.*

I. RANUNCULACEÆ.

SECT. 1. **Hecatonia.**—*Carpels smooth. Perennials (in Australia) with a tufted rootstock, or creeping or floating stolons. Flowers white or yellow.*

Radical leaves pinnate, with flat segments or digitate. Flowers yellow.
Stems tufted or erect or decumbent, without stolons. Petals usually 5.
 Calyx appressed or spreading, not reflexed.
  Carpels with a much recurved point. Plant hispid, or silky hairy, or nearly glabrous. Leaves pinnatisect, or 3- to 5-lobed, or entire . . 1. *R. lappaceus.*
 Calyx reflexed. Stem weak, hirsute. Leaves not pinnate. Flowers small . . . . 2. *R. plebeius.*
Stems creeping, floating, or stoloniferous. Plant glabrous or nearly so.
Leaves digitate. Petals usually 6 to 10 . . . . 3. *R. rivularis.*

SECT. 2. **Echinella.**—*Carpels tuberculate or muricate or hispid on the sides. Annuals. Flowers yellow.*

Flowers lateral, sessile, or on peduncles shorter than the leaves.
Hairy plant, with very small flowers, often sessile. Carpels usually about 1 line long, with a small recurved point . . . . 4. *R. parviflorus.*

1. **R. lappaceus** (burdock-like), *Sm.; DC. Prod.* i. 39; *Benth. Fl. Austr.* i. 12. A perennial, more or less clothed with soft spreading or rarely silky and appressed hairs. Rootstock short, with long fibres and no stolons. Leaves chiefly radical, on long petioles, usually divided into 3 or 5 deep lobes or segments, ovate or rhomboid-cuneate, either pinnately distinct or, if confluent, almost palmate, although the middle lobe is generally longer than the lateral ones, each lobe or segment is often again lobed or toothed and sometimes much cut into narrow lobes, more rarely the leaves are all entire or shortly 3-lobed. Flowering stems either a leafless 1-flowered scape or branching and erect or decumbent, bearing several flowers and a few leaves, smaller and less divided than the radical ones. Flowers of a rich yellow. Sepals hairy or rarely glabrous, usually much shorter than the petals, appressed or open, but not closely reflexed. Petals usually 5, broadly obovate and rather large, with a small glandular pit near the base. Carpels in a globular head, compressed or rarely turgid, glabrous and smooth, with a recurved style, usually short, but longer and slender in some western specimens.—Hook. f. Fl. Tasm. i. 6; F. Muell. Pl. Vict. i. 7; *R. colonorum*, Endl. in Hueg. Enum. 1; *R. discolor*, Steud. in Pl. Preiss. i. 263 (calyx certainly not reflexed).

Hab.: On the ranges about the Brisbane River. Flowering during the winter and spring months.

Var. *pimpinellifolius, forma multiplex.* Double buttercup.
This form of our common buttercup has been found by Miss Cameron near Ormiston, Cleveland railway line, and by Miss Schneider, Nerang. The meeting with so-called double flowers amongst wild plants is by no means common, and when of compact habit and well formed flowers such as the one now under notice are a real boon to the horticulturist. The flowers of the present plant closely resemble those of the Bachelor's Button, so common in the gardens around London, which is a form of *R. bulbosus*.

2. **R. plebeius** (common), *R. Br. in DC. Syst. Veg.* i. 288; *Benth. Fl. Austr.* i. 13. Hirsute with spreading or rarely nearly appressed hairs. Radical leaves on long petioles, digitately divided into 3 deeply lobed and toothed cuneate or rhomboid segments. Stems weak, decumbent or erect, often above a foot long and branched, with a few leaves, the lower ones more divided than the radical ones, with the primary segments petiolate, the others smaller, more sessile, and less cut. Flowers several, small, on long peduncles. Calyx reflexed, shorter than the petals, very deciduous. Petals obovate or oblong, seldom above 2 lines long. Achenes few or numerous, more or less compressed, rather small, with a hooked or recurved slender style.—Steud. in Pl. Preiss. i. 263: *R. hirtus*, Banks and Sol. in DC. Syst. Veg. i. 289; F. Muell. Pl. Vic. i. 8.

Hab.: Southern Queensland.

3. **R. rivularis** (river kind), *Banks and Sol. in DC. Syst. Veg.* i. 270; *Benth. Fl. Austr.* i. 13. Stems creeping or stoloniferous, producing at every node tufts of radical leaves and erect scapes, or weak slightly branched flowering stems, rarely forming short thick rhizomes. Leaves on long petioles, digitately divided

into 3, 5, or 7 segments, varying from cuneate to narrow-linear, rarely entire, usually 3-lobed, and sometimes much cut, but never pinnate, either quite glabrous, as well as the whole plant, or rarely with a very few appressed hairs. Flowers yellow, usually small, the sepals not reflexed. Petals 6 to 10, about twice as long as the sepals, or 5 only in small-flowered varieties, narrow-oblong. Achenes rather small and broad, with a firm or slender recurved or rarely nearly straight point, not tubercled or muricate.—*F. Muell. Pl. Vict.* i. 8.

Hab.: Brisbane River.

Var. *major*. Tufts erect. Leaf-segments ½ to 1in. long or more, often very narrow and much cut, on petioles of 2 to 6 inches. Flowers rather large.—*R. inundatus*, R. Br. in DC. Syst. Veg. i, 269. *R. glabrifolius*, Hook. Journ. Bot. i. 243; Hook. f. Fl. Tasm. i. 9. *R. incisus*, Hook. f. Fl. Nov. Zeal. 1, 10 t. 4.

Hab.: Watery places, Main Range.

4. **R. parviflorus** (small-flowered), *Linn.; DC. Prod.* i. 42 : var. *australis; Benth. Fl. Austr.* i. 14. A slender hairy annual, either with tufted, erect stems of a few inches, or weak, procumbent, and lengthening to a foot or even more. Leaves small, orbicular, the lower ones often only 3- or 5-lobed, but mostly divided into three segments, either entire or 3-lobed, or again cut into narrow segments. Flowers small, leaf-opposed, sessile, or on short, slender peduncles. Sepals rarely above 1 line long and very deciduous. Petals 5 or fewer, seldom much longer than the calyx. Achenes in a small globular head, much compressed, with a smooth margin, seldom much exceeding a line in breadth in Australian specimens, the sides covered with short hairs, or tubercles, or short hooked bristles, the style forming usually a very short recurved point, more rarely rigid and dilated at the base.—*F. v. M. Pl. Vict.* i. 9; *R. sessiliflorus*, R. Br.; *R. collinus*, R. Br.; *R. pumilio*, R. Br.; *R. leptocaulis*, Hook.; *R. pilulifer*, Hook. Ic. Pl. t. 600.

Hab.: A common weed of moist land in Southern Queensland.

Order II. **DILLENIACEÆ**.

Sepals usually 5, persistent, imbricate in the bud. Petals 5 or rarely fewer, deciduous, imbricate in the bud. Stamens hypogynous, indefinite, few or numerous, or rarely definitely 10, free or rarely united in clusters. Anthers innate or adnate. Gynœcium of carpels several, free and distinct or cohering at the base, or rarely single and excentrical, 1-celled, with 1 or more ovules in each. Styles quite distinct and diverging. Fruit-carpels either indehiscent and succulent, or opening along the inner edge, or in two valves. Seeds furnished with an arillus; testa crustaceous. Embryo very small, at the base of a fleshy albumen.—Trees, shrubs, climbers, or herbs. Leaves alternate or very rarely opposite. Stipules minute or none. Flowers usually yellow or white.

A considerable Order, of which rather the larger portion, with regularly pinnate veins prominent on the under side of the leaves, is entirely tropical.

Tribe I. **Delimeæ**.—*Stamens with the filaments more or less dilated upwards. Anthers short, cells divergent or rarely parallel, leaves with parallel lateral nerves, often scabrous.*

Sepals 5. Spreading. Carpels 3—5 (rarely 1—2?), acuminate, ovules many.
2-seriate. Panicles terminal . . . . 1. Tetracera.

Tribe II. **Dillenieæ**.—*Stamens with the filaments not dilated upwards. Anthers linear or rarely oblong, cells parallel contiguous. Leaves large, with parallel lateral nerves.*

Sepals 5. Anthers biporous. Carpels 5—10, ovules many, scarcely cohering, dehiscent at maturity. Trees . . . . 2. Wormia.

II. DILLENIACEÆ.   9

TRIBE III.—**Hibbertieæ**.—*Stamens with the filaments not at all or slightly dilated upwards. Anthers often oblong, cells parallel, contiguous rarely a little divergent. Leaves small uninerved or reticulately penniveined, sometimes absent.*

Perfect stamens free or nearly so, more than 10, or, if fewer, on one side of
the pistil . . . . 3. HIBBERTIA.
Perfect stamens 10 or fewer, in a complete ring round the pistil. No
staminodia within the perfect stamens . . . . 4. ADRASTÆA.

1. TETRACERA, Linn.
(Supposed resemblance of carpels to 4 horns.)
(Euryandra, *Forst.*)

Sepals 4 to 6, spreading. Petals just as many as sepals or rarely fewer. Stamens with the filaments dilated upwards. Anthers small, cells distant more or less divergent. Carpels 3 to 5, acuminate, ovules many, biseriate, at maturity coriaceous, shining, folliculate or dehiscing in two valves. Seeds 1 to 5, with a fimbriated or toothed aril. Trees or climbing shrubs, smooth scabrous, or pubescent. Leaves with parallel lateral veins. Flowers in terminal or in the upper axils, in loose panicles, hermaphrodite or partially unisexual.—B. & H. Gen. Pl. i. 12.

Leaves pilose on the under, scabrous on the upper side, primary veins close
 margins dentate. Sepals 4 . . . . 1. *T. Nordtiana.*
Leaves glossy, scabrous on both sides, margins sharply dentate. Sepals 4 . 2. *T. Cowleyana.*
Leaves glabrous except the midrib and primary veins, margins usually entire.
 Sepals 4 . . . . 3. *T. Wuthiana.*
Leaves glabrous, primary veins distant, margins entire. Sepals 5 . . . 4. *T. Dæmeliana.*

1. **T. Nordtiana** (after a lady horticulturist), *F. v. M. Fragm.* v. 1. A tall climbing evergreen shrub with a smooth bark and hard wood. Branchlets densely clothed with stellate and scattered longer simple hairs. Leaves ovate, 3 to 5in. long, 1¼ to about 3in. broad, the upper surface scabrous, the under clothed with short stellate hairs giving a hoary appearance, the lateral nerves parallel and rather close, often projecting beyond the margin and forming glandular teeth, more or less decurrent upon the petiole, which latter is from ½in. to 1in. long. Panicles loose and straggling in the upper axils and terminal, flowers fragrant. Bracts and bracteoles small, silky. Sepals 4, scabrous-pilose on the outside, nerveles are of unequal size, the 2 outer ones subrotund, 1 to 1½ lines long, the 2 inner ones rotund-ovate, 2 or 3 lines long. Petals 3, white, not much exceeding the sepals, cuneate-obovate, emarginate, ciliolate, and soon deciduous. Stamens numerous, glabrous, capillary, with a cuneate expanded apex, thus separating the anther-cells, but less so than in other Australian species. Carpels 3, densely hairy, styles very short, glabrous. Ripe carpels, obliquely ovate, about 3 lines long. Seeds sub-globose, of a dark or chestnut brown. Arillus 1½ to 2 lines long, fringed.

Hab.: Rockingham Bay, *J. Dallachy.*

2. **T. Cowleyana** (after E. Cowley), *Bail. Bot. Bull.* v. *Teeweeree*, Barron River, *Cowley.* A coarse climber, the branches appearing angular from the bark peeling and rolling back from longitudinal fissures, chestnut brown and scabrous. leaves scabrous, ovate-lanceolate, often 6in. long and 3in. broad in the centre, the apex sometimes sharply acuminate; petiole 1in. or more long, and often slender, hispid with appressed hairs, with which the costa and primary nerves on the under side are usually clothed; the primary parallel nerves numerous, regular, extending beyond the margin in the form of mucronate teeth. Panicle scabrous, from 6 to 9in. long, bracts narrow linear-lanceolate, silky. Pedicles slender. Sepals obtuse, velvety, with ciliate edges, the inner ones twice the size of the

outer. Petals veined, 3 lines long, obovate, velvety, with the margin ciliate like the sepals. Filaments much dilated, and more or less bifid at the apex. The anther-cells thus being widely separated. Carpels usually 3, hirsute, 3½ lines long. Seeds black, glossy, enveloped in a fringed crimson arillus, which when expanded has a diameter of 4 or 5 lines.

Hab.: Herbert River, *H. G. Eaton;* Cairns, *E. Cowley.*

3. **T. Wuthiana** (after D. E. Wuth), *F. v. M. Fragm.* x. 49. A tall climber with a smooth bark, and hard wood. Leaves on rather long petioles, ovate entire, somewhat acute, 3 to 5in. long 1½ to 2in. broad, texture somewhat thick-chartaceous, smooth, shining on both sides, the primary veins somewhat distant and prominent, rarely exserted beyond the margin in minute teeth. Panicle pilose with appressed hairs, from a few inches to a foot long, pedicels 2 to 10 lines long. Bracts lanceolate to subulate-linear 1½ line or less long. Flowers for the most part bisexual. Bractioles minute silky. Sepals 4, glabrous inside, unequal in length and nerveless, 2 or 3 lines long. Petals 3, scarcely equalling the sepals, slightly ciliate, white. Filaments suddenly much dilated at the upper end; anther-cells thus widely separated. Ovary sericeus, carpels 3. Style glabrous 1 line long. Stigma dilated. Ripe fruit not yet obtained.

Hab.: Daintree River, *E. Fitzalan;* Rockingham Bay, *J. Dallachy.*

The fungus *Dimerosporium Tetracerae,* Cke., sometimes infests the leaves.

4. **T. Dæmeliana** (after E. Dæmel), *F. v. M. Fragm.* v. 191. A tall glabrous climber. Leaves ovate-lanceolate, 4 to 7in. long and about 1½ to 2¼in. broad, decurrent upon short petioles, smooth, shining, and remotely-reticulate between distant primary veins. Panicle about 7in. long, upon peduncles of moderate length. Flowers bisexual. Bracteoles ciliolate. Sepals 5, glabrous, 2 or 3 lines long, almost ovate, obtuse, nerveless. Petals 4 or 5, glabrous, fugaceous, scarcely exceeding the calyx. Stamens numerous. Carpels 3, glabrous, tapering into short styles. Ripe fruit as yet not collected.

Hab.: Cape York.—*E. Dæmel.*

2. WORMIA, Rottb.
(After O. Wormius, a Dane.)

Sepals 5, spreading. Petals 5. Stamens numerous, with erect linear anthers opening at the summit in two pores, the inner ones often longer and recurved. Carpels 5 to 10, scarcely cohering, with several ovules in each, dehiscent when ripe. Seeds with an arillus.—Trees often very lofty. Leaves large, with raised parallel veins diverging from the midrib, the petioles often bordered with narrow deciduous wings. Flowers large, in loose terminal panicles.

A tropical genus, extending over tropical Asia and the Indian Archipelago, with one Madagascar species. The only Australian one is endemic. *Benth.*

1. **W. alata** (winged), *R. Br. in DC. Syst. Veg.* i. 484; *Benth. Fl. Austr.* i. 16; *F. v. M. Fragm.* vii. 124. Attaining the height of 60ft., with a stem diameter of 2ft. Glabrous, or the young parts very slightly hoary. Bark loose, papery, of a reddish colour. Leaves oval or nearly orbicular, rounded at both ends, 4 to 8in. long, entire or slightly sinuate, rather rough to the touch, with about 9 prominent veins on each side of the midrib and transversely reticulate veinlets, the petiole 1in. long or more, with longitudinal wings about 2 or 3 lines broad, which fall off in the greater part of their length. Peduncles terminal, not usually exceeding the leaves, bearing 2 or 3 large flowers on pedicels of nearly 1in. Sepals 6 to 8 lines long, ovate, concave, ciliate. Petals obovate, 1¼in. long, narrowed at the base.

Stamens very numerous, the inner ones long and recurved, the others shorter, and the outermost sometimes small and barren. Gynœcium of 5 to 8 glabrous, deep-crimson carpels, tapering into long recurved styles. Ovules 6 to 8 in each carpel. Seeds enclosed in a waxy-white arillus.

Hab.: Tropical coast.
Wood of a dark colour. Cut one way it shows a pretty red "clash," differing in colour but somewhat resembling that of English oak. It is close in grain and easy to work—a good cabinetmaker's wood Bailey's Cat. Ql. Woods, No. 1.

2. HIBBERTIA, Andr.
(After Dr. Hibbert).

(Hemmistemma, Pleurandra, *and* Hibbertia, *DC.* ; Ochrolasia, *Turcz.* ; Hemistephus, *Drummond*).

Sepals 5, spreading, sometimes shortly united at the base. Petals 5. Stamens indefinite, rarely fewer than 12, and then usually all on one side of the carpels, either all perfect or some of them reduced to staminodia, all free or the filaments shortly and irregularly united at the base. Anthers erect, oblong, or rarely ovate or orbicular, opening in longitudinal slits. Carpels usually 2 to 5, rarely solitary or more than 5, free or shortly cohering on their inner edge, with 2 to 6 or rarely only 1 or more than 6 ovules in each. Styles filiform, diverging, terminal, or almost dorsal. Fruit-carpels usually dehiscent at the top. Seeds reniform or nearly globular, with an entire or divided arillus.—Shrubs or undershrubs, usually much branched and low, erect or procumbent, sometimes almost herbaceous or climbing, rarely 5 or 6 feet high. Leaves usually small, alternate in all the Australian species, with a midrib prominent underneath, the lateral veins reticulate and rarely prominent. Flowers yellow or white, solitary and terminal, or (owing to the shortness or abortion of the flowering shoot) apparently axillary sessile in a tuft of floral leaves or pedunculate.

SECT. I. **Hemmistemma.**—*Perfect stamens and staminodia all on one side of the carpels, the staminodia outside. Peduncles mostly 2- or more-flowered.*

Leaves oblong or lanceolate, flat or the margins slightly recurved.
  Leaves obtuse.
    Leaves with recurved margins, narrowed into a petiole, rusty-brown
      underneath. Sepals obtuse . . . . 1. *H. Banksii.*
    Leaves flat, closely sessile with a rounded base, white underneath.
      Sepals acute . . . . 2. *H. Brownei.*
  Leaves acute or mucronate, white underneath.
    Peduncles lateral, 2- or 3-, rarely 1-flowered . . . . 3. *H. candicans.*
    Flowers rather large, midrib of leaf not red . . . . 4. *H. Millari.*

SECT. III. **Pleurandra.**—*Stamens all on one side of the carpels without any staminodia. Peduncle 1-flowered or none.*

Flowers subsessile. Stamens cohering at the base . . . . 5. *H. synandra.*
Flowers pedunculate. Leaves narrow-linear, rigid, glabrous or scabrous.
  Calyx glabrous, stellate-tomentose, or, if hirsute, pedicels very short . 6. *H. stricta.*
Flowers pedunculate. Leaves obovate, oblong or shortly-linear.
  Peduncles usually short. Ovules 2—4 . . . . 7. *H. Billardieri.*
Leaves nearly flat, rigidly pungent . . . . 8. *H. acicularis.*

SECT. IV. **Euhibbertia.**—*Stamens placed all round the carpels, with occasionally small staminodia outside.*

§ 1. *Tomentosæ.*—Carpels usually tomentose or scaly and 2-ovulate. Stamens numerous, without any or rarely with small staminodia outside. Leaves flat or the margins slightly revolute, usually stellately tomentose or scaly. Flowers pedunculate, axillary.

Tomentum soft and velvety. Leaves oblong, 1—2in. long . . . . . . 9. *H. velutina.*
                                                                          10. *H. melhanoides.*
Leaves narrow-linear. Tomentum of peltate scales. Peduncles 1 to 3
  lines . . . . 11. *H. lepidota.*

§ 2. *Vestitæ.*—Carpels (usually 3) villous, 4—6-ovulate. Stamens with or without staminodia outside. Leaves small, narrow, with revolute margins.

Flowers sessile, or pedunculate not exceeding the leaves . . . . . . 12. *H. vestita.*

12   II. DILLENIACEÆ.   [*Hibbertia*.

§ 3. *Fasciculatæ.*—Carpels glabrous. 2—6-ovulate. No staminodia. Leaves very narrow, convex underneath, the margins not revolute. Bracts small. Flowers sessile.
   Ovules 2, or rarely 3 or 4 in each carpel. Leaves usually fine, much
      clustered, often hirsute or pubescent . . . . 13. *H. fasciculata.*
§ 4. *Bracteatæ.*—Carpels glabrous, 1—2-ovulate. No staminodia. Leaves flat or convex underneath. Flowers closely sessile within broad brown shining bracts, like those of some of the *Hemihibbertiæ.*
Leaves very narrow, convex underneath.
   Leaves obtuse.
      Glabrous and green. Leaves not dilated at the top . . . . 14. *H. virgata.*
§ 5. *Subsessiles.*—Carpels glabrous. Stamens usually numerous, without staminodia. Leaves flat or the margins slightly re-curved. Bracts small or passing into the sepals. Flowers sessile or nearly so.
   Carpels 1—2-ovulate. Stems erect or diffuse.
      Leaves mostly under 1in. long.
         Leaves linear-oblong or scarcely enlarged above the middle. Stems
           usually erect or ascending . . . . 15. *H. linearis.*
         Leaves obovate or cuneate. Stems usually diffuse or prostrate . . 16. *H. diffusa.*
      Leaves 1 to 3in. long. Plant softly hairy.
         Leaves obovate-oblong, obtuse . . . . 15. *H. linearis,* var.
   Carpels 6—8-ovulate. Stems twining or trailing. Leaves large . . . 17. *H. volubilis.*
§ 6. *Hemihibbertiæ.*—Carpels glabrous or rarely villous. Stamens very numerous, with several small, subulate or clavate staminodia outside. Leaves flat. Flowers pedunculate.
Leaves distinctly petiolate, ovate, or oblong, mostly toothed.
   Carpels 3, glabrous, 6- to 8-ovulate . . . . 18. *H. dentata.*
   Leaves oblong-lanceolate, tapering at the base, and half stem-clasping . 19. *H. glaberrima.*
   Leaves narrow-elongate to linear-lanceolate, 6in. long, 2 to 6 lines
      broad . . . . 20. *H. longifolia.*
   Leaves narrow linear-lanceolate, 2 to 3in. long, 2 to 4 lines broad,
      staminodia numerous, carpels 4 . . . . 21. *H. œnotheroides*
   Leaves linear-lanceolate 1½ to 3½in. long, 2 to 4 lines broad, staminodia
      few, carpels 3 . . . . 22. *H. Bennettii.*

    1. **H. Banksii** (after Sir J. Banks), *Benth. Fl. Austr.* i. 20. Young branches and under side of the leaves densely clothed with a short, soft, rusty tomentum. Leaves oblong, obtuse, 2 to 3in. long, ½ to near 1in. broad, the margins more or less recurved, narrowed into a short petiole, glabrous above and somewhat shining when old, the pinnate and anastomosing veins prominent underneath. Spikes terminal, 1-sided, rusty-villous, about 1in. long, the flowers closely sessile. Sepals about 4 lines long. Petals longer. Stamens about 20, obtuse, with half as many staminodia outside, about one-third shorter.—*Hemistemma Banksii,* R. Br. in DC. Syst. Veg. i. 414.

    Hab.: Endeavour River, *Banks.*

    2. **H. Brownei** (after Robert Brown), *Benth. Fl. Austr.* i. 21. Young branches clothed with a short rusty down. Leaves oblong-lanceolate, obtuse or scarcely pointed, 2 to 3in. long, closely sessile and very obtuse or rounded at the base, the margins flat, glabrous, and at length almost shining above, white underneath, with the midrib alone prominent and rust-coloured. Spikes terminal, 1-sided, silky-villous. Sepals scarcely 4 lines long, acute. Stamens nearly as in *H. Banksii.*

    Hab.: Recorded for Queensland by *F. v. M.*

    3. **H. candicans** (whitish), *Benth. Fl. Austr.* i. 21. Like *H. dealbata* in the white tomentum that covers the under side of the leaves, but it is rather more silky or rusty on the peduncles and calyx, the leaves are rather narrower, and the inflorescence is very different; peduncles all axillary, ½ to 1in. long, bearing at their extremity 1 to 3 sessile flowers, and bracts and sepals usually broader. Stamens and carpels the same as in *H. Banksii.*—*Hemistemma candicans,* Hook. f. in Kew Journ. Bot. ix. 48, t. 2.

    Hab.: Cape York, *M'Gillivray*; Albany Island, *F. Mueller.*

II. DILLENIACEÆ.

4. H. Millari (after T. Barclay Millar), *Bail. 2nd Suppl. Syn. Ql. Fl.* 5. Branches slender, reddish-brown, silky-hoary or more or less clothed with white silky hairs. Leaves linear, acute, and apiculate, 2 to 3in. long, 1 to 1½ line broad, margins revolute, the upper surface glabrous, hoary-white on the under side, the midrib prominent but not rusty as in *H. angustifolia*. Spikes ¾in. long, terminal or in the upper axils bearing 1 or 2 flowers. Sepals ovate, about 4 to 5 lines long, silky-hairy outside. Petals cuneate, about 6 or 7 lines long, the end deeply emarginate. Stamens about 20, with a few filiform staminodia outside. Anthers oblong, longer than the filaments. Carpels 2, villous.

This species is very closely allied to *Hemistemma angustifolia*, R. Br., but of a more robust habit, with fewer and larger flowers in the spike, and wanting the prominent rusty-red midrib of the leaf of that species.—*Bail.*

Hab.: Musgrave, *T. Barclay Millar.*

5. H. synandra (anthers close together), *F. v. M. Fragm.* iv. 151. An erect branching shrub. Leaves subcoriaceous, lanceolate or broad-linear, 4 to 6 lines long, 1 to 1½ line broad, margins revolute, deep glossy-green above, slightly canescent underneath. Flowers solitary, subsessile. Sepals 2 to 4 lines long, sparsely puberulous, outer ones oblong-lanceolate, inner ones broad or orbicular. Petals obcordate-ovate, about 5 lines long. Stamens about 20, unilaterals, filaments about 1 line long, cohering at the base, anthers linear, no staminodia. Carpels slightly silky, oblique ovate, ovules often 3. Seeds glossy-brown. Arillus white, thin.

Hab.: Rockingham Bay. *J. Dallachy.*

6. H. stricta (erect), *R. Br. Herb.*; *F. Muell. Pl. Vict.* i. 15; *Benth. Fl. Austr.* i. 2. Erect, spreading, or diffuse, but scarcely prostrate, sometimes throwing up almost simple stems of 6in. from a thick rhizome, sometimes attaining several feet in height, more or less hoary or scabrous, with a minute stellate tomentum, although sometimes appearing glabrous at first sight. Leaves narrow-linear, erect or spreading, rather obtuse, mostly ¼ to ½in. long, the closely revolute margins disclosing little more than the midrib underneath. Flowers nearly sessile, or on pedicels of 2 or 3 lines in length. Sepals usually about 3 lines long, oblong, lanceolate, or the inner ones ovate. Stamens usually 8 to 12. Carpels tomentose, or very rarely glabrous, with 4 to 6, or very rarely more ovules in each. Arillus usually very small.—*Pleurandra stricta*, R. Br. in DC. Syst. Veg. i. 422; *P. riparia*, R. Br. in DC. l. c. i. 419; *P. ericifolia*, DC. l. c. i. 420; Hook. f. Fl. Tasm. i. 17; *P. cistiflora*, Sieb. in Spreng. Syst. Cur. Post. 191; Reichb. Icon. Exot. t. 79.

Hab.: Various parts of Southern Queensland.

Var. *canescens.* Leaves and calyx more or less hoary with stellate hairs. Flowers pedunculate or more rarely nearly sessile. Ovules usually 4.

Var. *hirtiflora.* Leaves nearly as in the var. *canescens.* Calyx usually large, more sessile, and hirsute with spreading hairs. Ovules usually 6 to 8 or more.

7. H. Billardieri (after Dr. J. J. Labillardière), *F. v. M. Pl. Vict.* i. 14; *Benth. Fl. Austr.* i. 28. Stems weak, sometimes short and erect, but more frequently trailing to the length of 2 or 3 feet or more over other shrubs, the branches clothed with stellate hairs, often mixed with long spreading ones. Leaves from obovate, ovate or oval-oblong to oblong-cuneate or narrow-oblong, the larger ones ½ to 1in. long, but in the commoner slender varieties not half that size, the margins recurved, more or less stellately pubescent, especially underneath, and scabrous above, but becoming glabrous with age. Pedicels terminating short, leafy shoots, or apparently axillary, slender, and recurved, ¼ to ½in. long. Sepals 2 to 3 lines long, or in some varieties rather shorter or longer, the outer ones usually pointed, the inner broader and more obtuse, glabrous, or nearly so. Petals broad. Stamens usually 10 to 12. Carpels downy or villous, with 2 to 4 ovules. Arillus sometimes almost enveloping the seed, sometimes very short.—*Pleurandra ovata*, Labill. Pl. Nov. Holl. ii. 5, t. 143; Hook. f. Fl. Tasm. i. 16.

Hab.: Southern Queensland, common.

8. **H. acicularis** (needle-like), *F. Muell. Pl. Vict.* i. 17; *Benth. Fl. Austr.* i. 29. Nearly or quite glabrous, procumbent or diffuse, with a thick woody stock, and numerous branches, short and intricate, or lengthened to a foot. Leaves narrow-linear, rigid, with a stiff, often pungent point, about 3 to 6 lines long, the margins recurved. Pedicels terminal or axillary, often on very short shoots, with a few leaves at the base sometimes reduced to minute bracts, recurved, ¼ to ½in. long. Sepals glabrous, or very slightly downy, about 2 lines long. Stamens usually 8, or fewer. Carpels downy, or rarely glabrous, with 2, or very rarely 4 ovules.—*Pleurandra acicularis*, Labill. Pl. Nov. Holl. ii. 6, t. 144; Hook. f. *Fl. Tasm.* i. 15.

Hab.: Common on ironbark forest land in southern Queensland.

9. **H. velutina** (velvety) *R. Br.*; *Benth. Fl. Austr.* i. 30. Shrub, all parts clothed with a whitish velvety tomentum; height 4 to 5ft., and spreading; branches prominently angled. Leaves oblong-ovate, 1 to nearly 2in. long, 3 to 7 lines broad, much tapering towards a petiole of about a line long; midrib very prominent, lateral veins patent and distinct, margins recurved. Peduncles axillary near the ends of the branches, flattened, ½ to ¾in. long, with a narrow bract close under the calyx. Sepals 3 or 4 lines long, outer ones acute, inner ones obtuse, and wider. Petals broadly obovate, about ½in. long. Stamens numerous. Carpels 3, at first silky then rather echinate, 2-ovulate.

Hab.: On summit of Mount Harold, off Tringilburra Creek, and Walsh's Pyramid, Mulgrave River.

10. **H. melhanoides** (Melhania-like), *F. v. M. Fragm.* iv. 116. An erect shrub of 3 to 4ft.; thinly clothed with a stellate pubescence, branches angular. Leaves 1½ to 3in. long, 5 to 9 lines broad, oblong-lanceolate, quite entire, smooth, glaucous, and somewhat scabrous on the under surface.—Petioles very short. Peduncles solitary, very short, slender, angular. Bracts 1½ to 2½ lines long. Outer sepals 4 to 6 lines, 1-nerved, almost lanceolate, inner one ovate or roundish. Petals obcordate, stamens 40 to 60, filaments 1 to 1¼ line long. Staminodia few. Anthers ¾ to 1 line. Style scaly below the middle.

Hab.: Rockingham Bay.

11. **H. lepidota** (scaly), *R. Br. in DC. Syst. Veg.* i. 432; *Benth. Fl. Austr.* i. 31. Branches stiff but slender, covered as well as the leaves and sepals with a close silvery or slightly rusty tomentum, consisting of minute peltate scales with scarious edges. Leaves linear, rather acute, mostly ¼ to ⅜in. long, concave, the margins not revolute. Flowers rather small, on pedicels of 1 to 3 lines, solitary or 2 or 3 together in the axils. Sepals broad, very obtuse, about 2 lines long, or 3 when in fruit, the 2 outer rather shorter. Stamens about 12, mostly, but not all, on one side of the carpels, with several small staminodia outside. Carpels 2, scaly-tomentose, 2-ovulate.

Hab.: Northcote, *R. C. Burton.*

12. **H. vestita** (clothed), *A. Cunn. Herb.*; *Benth. Fl. Austr.* i. 31. Branches elongated, decumbent or erect, clothed as well as the young leaves with short spreading hairs. Leaves narrow-linear, obtuse, 3 to 4 lines long, rigid with recurved margins, often glabrous when full grown. Flowers nearly sessile, in clusters of floral leaves shorter than them, the inner ones passing into small linear

bracts. Sepals ovate-lanceolate, obtuse, or the outer ones scarcely acute, 3 or even 4 lines long, with rather silky hairs outside. Petals obovate, deeply emarginate. Stamens above 30, with several short filiform or clavate staminodia outside. Carpels 3, villous, 6-ovulate. The general aspect is sometimes that of *H. serpyllifolia*, but it is readily known by the stamens.

Hab.: Open forest land near Moreton Bay, *A. Cunningham*; Stradbroke Island, *Fraser*; Glasshouse mountains, *F. Mueller*; swamps towards Durval, *Leichhardt*.
Var. *thymifolia*. Leaves shorter, often recurved at the end.—Near Moreton Bay, *A. Cunningham*.

13. **H. fasciculata** (fascicled), *R. Br. in DC. Syst. Veg.* i. 428; *Benth. Fl. Austr.* i. 33. Stems erect, procumbent or prostrate. Leaves very narrow-linear, clustered and crowded, 2 to 3 lines or rarely ½in. long, hirsute with soft rather spreading hairs, or at length glabrous, obtuse, or scarcely pointed, the margins never revolute or recurved, but rather turned upwards so as to leave the under surface convex with the prominent midrib. Flowers sessile, on very short leafy shoots along the branches, with 2 or 3 small sepal-like bracts at their base. Sepals 2 to 3 lines long, broadly ovate, membranous at the edge, the outer ones narrower and less obtuse. Petals obcordate. Stamens usually 8 to 12, without staminodia. Carpels usually 3, glabrous, with two erect ovules in each.—Hook. f. Fl. Tasm. i. 13; *H. angustifolia* (partly), F. Muell. Pl. Vict. i. 18; *H. virgata*, Hook. Ic. Pl. t. 267, not R. Br.; *H. prostrata*, Hook. Journ. Bot. i. 246; *Pleurandra camforosma*, Sieb. in Spreng. Syst. Cur. Post. 191; *H. camphorosma*, A. Gray Bot. Amer. Expl. Exped. i. 21.

Hab.: Southern parts of the colony.

14. **H. virgata** (twiggy), *R. Br. in DC. Syst. Veg.* i. 428; *Benth. Fl. Austr.* i. 34. Diffuse or erect, glabrous, with numerous thin but stiff and often wiry branches. Leaves narrow-linear, obtuse or scarcely acute, mostly about ½in. long, but sometimes much longer, stiff and rather thick, the margins not revolute, and sometimes almost terete. Flowers sessile, surrounded by 2 or 3 very broad scarious pale brown bracts fully half as long as the calyx. Sepals about 4 lines long, obtuse or more frequently acute, or with a short sharp point, glabrous and more scarious than in any other species. Petals broadly obovate, scarcely emarginate. Stamens 10 to 15, without staminodia. Carpels 3, glabrous, 2-ovulate. —Hook. f. Fl. Tasm. i. 14; *H. angustifolia*, var., F. Muell. Pl. Vict. i. 19.

Hab.: Southern parts of the colony.

15. **H. linearis** (leaves linear), *R. Br. in DC. Syst. Veg.* i. 428; *Benth. Fl. Austr.* i. 36. Much branched, erect or divaricate, or rarely decumbent, glabrous in all its parts, or with a very minute pubescence on the young shoots. Leaves in the normal forms linear, rather acute or obtuse, with a short recurved point, 4 to 8 lines long, or nearly 1in. when luxuriant, the margins flat or slightly recurved, and not convex underneath. Flowers on very short peduncles, and usually surrounded by rather longer floral leaves, with small acuminate brown bracts at the base of the peduncle, and one or two at the summit passing into the sepals. Sepals all or the inner ones only obtuse, glabrous with thin margins, 2½ to 3 lines long. Petals obovate, scarcely notched. Stamens 15 to 20, without staminodia. Carpels usually 3, rarely 2 or 1, glabrous, 2-ovulate.

Hab.: Moreton Island, *M'Gillivray*, F. Mueller.
Var. *floribunda*. Sepals more acute and rather hairy. Stamens more numerous.—Peel's Island, *A. Cunningham*.
Var. ? *obtusifolia*. More rigid than the normal form, more frequently erect, and more or less hairy, with a minute crisped or shortly stellate tomentum, sometimes densely and softly pubescent, and very rarely glabrous. Leaves from linear to broadly oblong spathulate, very obtuse or truncate, in some southern specimens above 1½in. long, and mostly narrowed into a short petiole. Flowers rather larger than in the normal variety, with numerous stamens.—*H. obtusifolia*, DC. Syst. Veg. i. 429; *H. canescens*, Sieb. in Spreng. Syst. Cur. Post. 211.

16. **H. diffusa** (wide-spreading). *R. Br. in DC. Syst. Veg.* i. 429; *Benth. Fl. Austr.* i. 36. Stems low, usually diffuse or prostrate, with numerous short ascending branches, pubescent or at length glabrous. Leaves from obovate to linear-cuneate, very obtuse or truncate, seldom above ½in. long, and then often 2- or 3-toothed. Peduncles very short. Sepals broadly oblong, obtuse, about 4 lines long, the outer ones rather shorter and narrower. Petals obovate, entire. Stamens about 20 to 25, without staminodia. Carpels usually 3, or rarely 2 or 4, glabrous, 2-ovulate.

Hab.: Southern parts of the colony.

17. **H. volubilis** (twining), *Andr. Bot. Rep.* t. 126; *Benth. Fl. Austr.* i. 87. Stems woody, trailing, or twining and climbing to the height of 20 to 30ft. the young parts more or less clothed with silky hairs. Leaves from obovate to lanceolate, obtuse or acute, 1½ to 3in. long, narrowed below, but slightly enlarged and stem-clasping at the base, leaving a raised ring on the stem, as in most *Candolleas*, glabrous above, silky-hairy underneath. Flowers the largest of the genus, nearly sessile, the upper leaves passing into sepal-like bracts. Sepals 8 lines to 1in. long, ovate-acuminate, very silky-hairy outside. Petals obovate, entire. Stamens very numerous, without staminodia. Carpels usually 5, but sometimes up to 8, glabrous, 6- to 8-ovulate.—*Dillenia scandens*, Willd. Spec. ii. 1251; *Dillenia volubilis*, Vent. Choix. t. 11; *D. speciosa*, Bot. Mag. t. 449, not of Thumb.

Hab.: Both southern and northern parts of the colony.

18. **H. dentata** (leaves toothed), *R. Br. in DC. Syst. Veg.* i. 426; *Benth. Fl. Austr.* i. 38. Stems woody at the base only, trailing or twining, glabrous or the young branches pubescent. Leaves distinctly petiolate, oblong, obtuse or acute, 1¼ to 2½in. long, flat, marked with a few distant callous teeth, or slightly sinuate, rounded at the base, glabrous or pubescent when young. Flowers rather large, on short peduncles, with 1 or 2 small bracts at their base. Sepals ovate, ½in. long, the inner ones obtuse, the outer rather shorter and more acute, rarely all acuminate, pubescent or silky-hairy. Petals obovate, entire or scarcely notched. Stamens very numerous with slender filaments, the anthers short, although not so broad as in the *Brachyanthera*, and a considerable number of filiform or clavate staminodia outside. Carpels 3, glabrous, 6- to 8-ovulate.—F. Muell. Pl. Vict. i. 217; Bot. Reg. t. 282; Bot. Mag. t. 2338; Lodd. Bot. Cab. t. 317.

Hab.: Towards the N.S.W. border.

19. **H. glaberrima** (smooth without indumentum), *F. Muell. Fragm.* iii. 1; *Benth. Fl. Austr.* i. 39. Perfectly glabrous. Leaves (the upper ones only known) oblong-lanceolate, obtuse with a short glandular point, 1 to 1¼in. long, quite entire, tapering below the middle almost into a petiole, and slightly expanded so as to half-clasp the branch. Peduncles axillary or terminal, about 1¼in. long. Innermost sepals fully 6 to 7 lines long, and very broad, the others gradually diminishing to the outermost, which is lanceolate and about 3 lines. Petals not much longer than the calyx. Stamens very numerous (200 to 30 ¹), with numerous (2 or 3 dozen) short clavate staminodia outside. Carpels 3, glabrous, with about 8 ovules in each.

Hab.: Inland northern parts of the colony.

20. **H. longifolia** (long-leaved), *F. v. M. Fragm.* iv. 115. A perfectly glabrous shrub about 2ft. high. Leaves narrow-elongate to linear-lanceolate, quite entire, about 6in. long, and 2 to 6 lines broad, narrowly stem-clasping, the apex attenuated. Peduncles 1 to 1¾in. long. Bracts about ½in. long, linear-

subulate. Sepals 6 to 10 lines long, outer ones tenui-acuminate, inner ones subovate or lanceolate-ovate. Petals not incised at the top. Stamens 200 to 300, filaments about 2 lines long, slender, a few reduced to staminodia. Styles about 2 lines. Carpels acute.

Hab.: Rockingham Bay.
Specimens of this species have been several times forwarded to me as a suspected poisonous plant.

21. **H. œnotheroides** (like evening primrose), *F. v. M. Fragm.* vii. 87. Plant perfectly glabrous, branches angular. Leaves narrow or linear-lanceolate, 2 to 3in. long, 2 to 4 lines broad, rarely broader, pale underneath the base, not broadly stem-clasping, margins entire recurved, apex acute. Peduncles about 4 lines long. Bracts subulate-linear, about 5 lines long. Sepals about 8 lines long, the outer ones subulate-acuminate. Petals about 10 lines, obcordate, bilobed. Stamens about 140, filaments 2 lines long. Anthers scarcely 1 line. Staminodia capillare about 60. Carpels 4.

Hab.: Gilbert River, *R. Daintree.*

22. **H. Bennettii** (after F. Bennett), *Bail. Ql. Agri. Journ.* June 1899. A spreading shrub from 6 to 18in. high, quite glabrous, branches angular. Leaves linear-lanceolate, $1\frac{1}{2}$ to $3\frac{1}{2}$in. long, 2 to 4 lines broad, slightly expanding and shortly clasping the stem at the base, veins obscure, pale on the under side, margins revolute, apex subulate. Peduncles terminal on the branchlets, somewhat flattened, 6 to 12 lines longs, bracts at the base short, clasping the peduncle, the one close under the flower narrow lanceolate, 3 to 4 lines long, often patent. Sepals ovate-lanceolate, 6 to 8 lines long, 3 to 4 lines broad, the inner ones the broadest, with more scarious margins. Expanded flower nearly 2in. diameter. Petals obovate or broad-cuneate, with minute mucro in the centre but scarcely showing lobes, about 8 or 9 lines long and nearly as broad. Stamens numerous, filaments about 2 lines long. Anthers oblong, obtuse opening laterally, about $\frac{1}{2}$ line long; only a very few of the outer filaments wanting anthers. Carpels 3, glabrous, 3-ovulate. Seeds globose, brown, $1\frac{1}{2}$ line diameter.

Hab.: Irvinebank, *F. Bennett*, who says that the plant is known locally as the "Arsenic Plant" and that it is considered exceptionally poisonous to stock.

4. ADRASTÆA, DC.
(After the Goddess).

Sepals 5. Petals 5. Stamens 10, or occasionally fewer, in a single series, filaments dilated and regularly cohering in a short tube round the pistil. Carpels and fruit of *Hibbertia.*

The genus consists of only one species, with the habit of a *Hibbertia* or *Candollea.—Benth.*

1. **A. salicifolia** (willow-leaved), *DC. Syst. Veg.* i. 424; *Benth. Fl. Austr.* i. 46. Branches rather slender, apparently erect, the young ones silky-hairy. Leaves linear or linear-oblong, mostly with a minute fine point, $\frac{3}{4}$ to $1\frac{1}{2}$in. long, often bordered by a few remote and minute callous teeth, glabrous above when old, more or less silky underneath. Flowers small, sessile in clusters of small leaves in the older axils. Sepals lanceolate, very acute, nearly 3 lines long. Petals scarcely longer, obovate-oblong, obtuse. Anthers oblong, longer than the filaments. Carpels 2, glabrous, 1-ovulate.—*Hibbertia salicifolia,* F. Muell. Fragm. i. 161.

Hab.: In the southern swamps of the colony.

C

Order III. **MAGNOLIACEÆ**.

Sepals and petals several, imbricate, and often passing gradually from the one to the other, deciduous; or in the Australian genera the calyx exceptionally 2 or 3-cleft. Stamens indefinite, hypogynous; filaments often thickened or dilated, anthers adnate. Carpels indefinite, rarely solitary, free or partially cohering. Ovules 2 or more, attached to the inner angle of the cavity, or rarely ascending from the base. Stigma sessile. Ripe carpels opening in 2 valves or indehiscent. Seeds with a crustaceous testa, often succulent externally; albumen copious, oily. Embryo minute, near the hilum, with divaricate cotyledons.—Trees or shrubs, often aromatic. Leaves alternate, undivided, reticulately penninerved, entire or toothed, with or without stipules. Flowers axillary or terminal, solitary or fasciculate, often large.

An Order chiefly distributed over tropical and eastern temperate Asia and North America, and only represented by two somewhat anomalous genera in the southern hemisphere. *Benth. in part.*

Tribe I. **Wintereæ**.—*Fowers hermaphrodite or rarely polygamous-diœcious. Carpels-verticillate or solitary. Stipules none.*

Sepals 2 or 3, united in the bud in a globular calyx, irregularly split or separating when open. Carpels baccate . . . . 1. Drimys.
Sepals 2, at first entire, at length opening on one side to the base. Fruit globose, 8 or more celled . . . . 2. Galbulimima.

1. DRIMYS, Forst.
(Alluding to the acridity of the plants.)
(Tasmannia, R. Br.)

Sepals 2 or 3, membranous, united in the bud in a globular calyx, irregularly split or separating when open. Petals usually few. Filaments thick, the anthercells parallel or divergent. Carpels various in number, mostly solitary in the Australian species, containing several ovules. Berries indehiscent.—Glabrous and aromatic trees or shrubs. Leaves marked with pellucid dots. Peduncles (in the Australian species 1-flowered) arising from the axils of deciduous scales at the base of the new shoots, but as these shoots are rarely developed till the fruit has ripened, the flowers appear to be in terminal umbels with a central bud. Flowers of a greenish-yellow, white, or coloured.

Leaves on very short petioles, the lamina ending at the base in two minute auricles . . . . 1. *D. dipetala.*
Leaves on longer petioles, the lamina tapering much towards the base, without auricles . . . . 2. *D. membranea.*
Leaves large, subcoriaceous, very obtuse and tapering towards the rather long petiole, under side grey . . . . 3. *D. semecarpoides.*

1. **D. dipetala** (two petals), *F. v. M. Pl. Vict.* i. 21; *Benth. Fl. Austr.* i. 49. A tall shrub. Leaves oblong-lanceolate or rarely oval-oblong, acute or acuminate, usually 3 to 5in. long, narrowed towards the base, but all (except sometimes a few of the smaller leaves of lateral shoots) abruptly obtuse or minutely biauriculate at the very base, on an exceedingly short broad petiole, or almost sessile. Peduncles exceeding 1in. in length. Sepals and petals 2 each. Carpels often 2 or 3, but 1 only usually enlarges. Stigma short or linear, more or less unilateral. Berry ovoid, fully ½in. long, purple or white and succulent. Seed reniform black.—*Tasmannia insipida,* R. Br., and *T. dipetala,* R. Br.; also *T. monticola.* A. Rich. Sert. Astrolab. 50 t. 19.

Hab.: Mount Lindsay, Mount Mistake, and frequently met with on the low land along the North Coast railway line.

2 **D. membranea** (referring to the thin leaves), *F. v. M. Fragm.* v. 175. A small glabrous tree. Leaves lanceolate tapering much towards each end, 3 to 5in. long, ¼ to 1in. broad, without auricles at the base of lamina, somewhat

glaucous on the under side, the reticulate veins not so prominent as in *D. dipetala*, and the petioles longer than in that species. Peduncles very slender, about 1in. long. Petals very few, often 2 or 3. Anthers ovate-cordate ¼ line long. Carpels 2 or 3. Stigma decurrent.

Hab.: Hills about the Mulgrave River.

3. **D. semecarpoides** (like a Semecarpus), *F. v. M. Vict. Nat.*, March, 1891. A tree said to attain the height of about 25ft. Leaves on petioles often 1in. long, chartaceous, glabrous, from ovate to elongate-elliptic, but gradually narrowed into a cuneate base, rounded-blunt at the summit, attaining 8in. in length and 2½in. in breadth, very grey on the under side, punctular-rough, the costular veins very thin, venules much concealed. Peduncles about 2 or 3in. long, glabrous. Flowers unknown. Pedicels few or two, or even solitary. Sepals two, very small, roundish. Ripe carpels solitary, almost globular, ¼ to ½in. in diameter.

Hab.: Rockingham Bay, *Dallachy*, *F. v. M.*, Fragm. vii. 18; Russell Creek, *W. Sayer*, *F. v. M.*, *l.c.*

This species differs from *D. Howeana* in almost entire absence of aroma, in leaves of larger size, of thinner texture, of far less prominent venulation, and with the dots not transparent, in the perfect separation of the sepals, and probably also in characteristics of the flowers. It comes very near to *Drimys rivularis* Vieillard, of New Caledonia, but the petioles are much longer, the venules of the leaves more occult, the inflorescence is less ramified, the ovularies are fewer, and also in this case the flowers, which in an only specimen available here for comparison are not developed, may be different.—*F. v. M.*

2. GALBULIMIMA, Bail. Bot. Bull. ix. 5.

(Named from the resemblance of the fruit to a galbulus).

Sepals 2, deciduous, at first entire but at length opening on one side down to the base, 2-seriate. Petals none, except the single outer series of staminodia be regarded as such. Stamens numerous in many series, on a raised torus; filaments much flattened, linear, bearing on the back, nearer the base than the apex, 2 adnate oblong anthers. Ovary glandular hirsute with about 7 or 8 prominent angles; stigmas purplish, more or less recurved and papillose. Berry globose, 8 or more celled, 5 usually with matured seed. Seeds with a loose outer ragged coat; testa smooth, cartilaginous; albumen copious, oily. Embryo not particularly small near the hilum, apical with reference to the position of the seed in the berry. An evergreen tree of about 50ft., foliage and fruit possessing a strong resinous odour. The nearest ally of this new genus seems to be *Illicium*.

1. **G. baccata** (berry-like fruit), *Bail. Bot. Bull.* ix. 5. An evergreen tree of about 50ft. in height, having a stem diameter of about 1¼ft; the young branchlets with a bronzed appearance from numerous bright ferruginous scales. Leaves alternate, margins entire, oblong-lanceolate, attaining the length of 4½in. on petioles of about ¾in.; the upper face dark green, glossy; under side covered with minute scurfy glands or scales, pellucidly-dotted. Flowers axillary, solitary, on peduncles of about ¼in., bearing near the top 2 or 3 thick angular bracts; pedicel short; bud ovoid; sepals 2, one entirely overcovering the other; petals wanting; stamens numerous, the outer series without anthers; filaments much flattened, linear, bearing in the lower half; 2 parallel, oblong, sessile anthers. Ovary angular, sessile, clothed with ferruginous bright hairs. Fruit globose, crimson, resembling a fleshy *Callitris* fruit in its form and markings. Seeds compressed, embedded in the substance of the fruit.

Hab.: Eumundi, *E. H. Arundell*; Boar Pocket and Evelyn, Herberton district, *J. F. Bailey*. Wood of a light colour, centre brown, soft and light.

Order IV. **ANONACEÆ.**

Sepals usually 3, distinct, or more or less united in a 3-lobed or 3-toothed calyx (in *Eupomatia* united in one mass with the petals). Petals usually 6, hypogynous, in two rows, 3 outer ones alternating with the sepals, 3 inner ones alternating with the outer, sometimes all united in a ring at the base, those of each row valvate or imbricate in the bud. Stamens indefinite, usually very numerous, closely packed on the thickened torus, round or under the carpels, linear or wedge-shaped, with 2 adnate anther-cells on the back or edges, often concealed by the more or less dilated summit of the connectivum. Gynœcium of several, often very many carpels, distinct (except in *Eupomatia*), closely packed on the centre of the torus, terminating each in a capitate stigma, or in a thick oblong or rarely more slender style, stigmatic on the top or inner side. Ovules in each carpel either 1 or 2, ascending from the base, or two or more attached to the inner angle of the cavity, anatropous. Fruit either of several distinct carpels, sessile or stalked, indehiscent and fleshy or pulpy, sometimes opening along the inner edge, or the carpels more or less united in a single mass. Seeds with or without an arillus. Albumen copious, always ruminate. Embryo very small, near the hilum.—Trees, shrubs or woody climbers. Leaves alternate, simple, and quite entire, without stipules. Flowers sessile, or on 1-flowered pedicels, solitary, or few together, terminal, lateral, or axillary, usually of a greenish-yellow or purple colour.

A large Order, widely distributed over the New World as well as the Old, but chiefly confined to the tropics.—*Benth.*

Tribe I. **Uvarieæ.**—*Petals 2-seriate, one or both series imbricate in bud. Stamens many, closely packed; their anther-cells usually concealed by the overlapping connectives. Ovaries indefinite.*

Flowers 2-sexual; ovules many, rarely few; torus almost flat . . . . 1. Uvaria.
Sepals valvate in the bud; connective not concealing the anther-cells . 2. Fitzalania.

Tribe II. **Unoneæ.**—*Petals valvate or open in bud, spreading in flower, flat or concave at the base only, inner subsimilar or none. Stamens many, close-packed, their anther-cells concealed by the overlapping connectives. Ovaries usually indefinite.*

Petals lanceolate, flat; spreading from the base. Ovaries many, 2-seriate. Ripe carpels indehiscent . . . . 3. Cananga.
Petals subulate-linear. Ovaries 3. Ripe carpels or berries few-seeded 4. Ancana.
Petals flat, spreading from the base. Ripe carpels indehiscent. Ovules 1—2, basal or sub-basal . . . . 5. Polyalthia.

Tribe III. **Mitrephoreæ.**—*Petals valvate in the bud, outer spreading; inner dissimilar, concave, connivent, arching over the stamens and pistils. Stamens many, close-packed, anther-cells concealed by the overlapping connectives.*

Inner petals clawed, usually smaller than the outer .
Ovaries indefinite, ovules many . . . . 6. Mitrephora.
Petals connate towards the base Ovaries 6 . . . . 7. Haplostichanthus.

Tribe IV. **Xylopieæ.**—*Petals valvate in the bud, thick and rigid, connivent, inner similar but smaller, rarely none. Stamens many, close-packed, anther-cells concealed by the produced connectives. Ovaries indefinite.*

Outer petals broad; torus convex. Ovules 2—many . . . . . . 8. Melodorum.

Tribe V. **Miliuseæ.**—*Petals imbricate or valvate in bud. Stamens often definite, loosely imbricate, anther-cells not concealed by the overlapping connectives. Ovaries solitary or indefinite.*

Petals valvate, inner largest, ovules indefinite . . . . . . 9. Saccopetalum.
Petals and sepals united in a conical mass, which falls off entire . 10. Eupomatia.

1. UVARIA, Linn.
(Fruit resembling grapes.)

Sepals broad. Petals 6, imbricate in the bud in each row, spreading. Stamens numerous and closely packed, rather flat, the connective produced into a shortly ovoid or truncate appendage, concealing the cells in the normal species. Receptacle slightly raised. Carpels numerous, with a short truncate style, and several ovules in two rows along the inner angle. Berries distinct, sessile, or stalked, usually with several seeds. Stems climbing or trailing. Flowers usually rather large, leaf-opposed or axillary.

A considerable genus, chiefly Asiatic, with a few African species. The following Australian ones are both endemic.

Petals all broad. Anthers dilated at the top, concealing the lateral cells . 1. *U. membranacea.*
Outer petals ovate, contracted upwards, inner ones ovate-lanceolate . . . 2. *U. Goezeana.*

1. **U. membranacea** (membranous), *Benth. Fl. Austr.* i. 51. A tall woody climber, quite glabrous, except a slight tomentum on the petioles and buds. Leaves on short stalks, oval-oblong, obtuse, or with a very short, broad point, 6 to 10in. long, 3 to 3½in. broad, oblique, and somewhat cordate at the base, thin and membranous, with distant primary veins branching into the reticulate smaller venation. Flowers large, solitary, on peduncles of about ½in. Petals obovate, very obtuse, fully 1in. long, narrowed, and slightly united at the base. Connective truncate and dilated above the anther-cell. Ripe carpels numerous, upon a globose receptacle, oblong, 1½in. long, with a diameter of ⅜in., deep-scarlet on stipites of about 1½in. Seeds lenticular, 5 lines diameter, with a more or less prominent border.

Hab.: Somerset, *McGillivray,* flowering specimens. *Bailey,* fruit June, 1897.

At Somerset the leaves are attacked by the blight fungus *Phyllosticta Uvariæ. Berk.*

2. **U. Goezeana** (after E. Goeze), *F. v. M. Fragm.* vii. 125. A glabrescent climber, attaining the height of 80ft. Leaves on short petioles, elliptical or lanceolate-ovate to elongate-lanceolate, 3 to 6in. long, 1 to 2in. broad, upper side shining, under side glaucous. Pedicels leaf-opposed, slender, 1 to 1¼in. long, with a solitary bract below the middle. Flowers fragrant, sepals 3 to 4 lines long, rhomboid-ovate. Petals yellowish, thinly pilose, outer ones ovate, nearly 1in. long, imbricate at the base, inner ones ovate-lanceolate, not imbricate at the base. Stamens 70 to 80, scarcely ⅔ line long, anther connective truncate and dilated, concealing the cells. Stigma depressed, black. Carpels about 20, moniliform, of from 5 to 9 articles, each 4 to 5 lines long.

Hab.: Mountains around Rockingham Bay, J. *Dallachy* (F. v. M. l.c.

2. FITZALANIA, F. v. M.
(After E. Fitzalan.)

(Uvaria heteropetala, *F. v. M. Fragm.* iii. 1.)

Sepals small, distinct, deciduous, lanceolate-ovate. Petals 6, hypogynous, sessile, 2-seriate, the inner longer than the outer, which in the bud are sub-valvate at the base, and quite valvate at the upper part, inner ones imbricate in the bud. Stamens many-seriate, compressed, cuneate, indefinite. Anthers subsessile, connective not dilated, anther-cells dorsal. Torus depressed, tomentose. Carpels numerous, free, 2-seriate, with 6 to 8 ovules in each. Stigmas subsessile, depressed. Berries cylindric-globose, few-seeded and not constricted between the seeds. Seeds in 1 series.—*F. v. M. Fragm.* iv. 88.

1. **F. heteropetala** (petals not all alike), *F. v. M. Fragm.* iv. 88. A scrubby shrub of 8 to 10ft., the young branches densely pubescent. Leaves on very short petioles, broadly ovate, obtuse, or shortly acuminate, 2 to 4in. long, not coriaceous, glabrous above, loosely pubescent underneath. Flowers dark purple, solitary, on very short recurved terminal or lateral pedicels. Sepals ovate-lanceolate, villous, 3 to 4 lines long. Petals imbricate in each series, the outer ones broadly ovate, attaining at least 7 lines, and probably longer when full grown, silky-villous outside, glabrous inside, the inner ones narrower and

perhaps longer. Stamens numerous, the short triangular terminal appendage not dilated, showing the rather large dorsal parallel cells. Carpels numerous, densely hirsute; stigma small. Ovules 6 to 8 in each carpel, in 2 series. Fruit ½ to 1in. long, turged, obtuse, ¼in. thick. Seeds large, roundish.—*Uvaria heteropetala*, F. v. M. in Fl. Austr. i. 51.

Hab.: Port Denison and Burnett River. This plant differs from *Uvaria* in the stamens, which are those of *Saccopetalum*. The habit and foliage are also more those of the latter genus than of *Uvaria*, but the petals certainly appear to be imbricate in each row, and the outer ones are much more developed than is usual in *Saccopetalum*. The flowers in the specimens seen are, however, still young and insufficient for fixing the precise affinities of the species.—*Benth*.

8. CANANGA, Rumph.
(Malay name of one species.)

Sepals 3, ovate or triangular, valvate. Petals 6, 2-seriate, subequal or inner smaller, long, flat, valvate. Stamens linear, anther-cells approximate extrorse; connective produced into a lanceolate acute process. Ovaries many, style oblong, stigmas subcapitate; ovules numerous, 2-seriate. Ripe carpels many, berried, stalked or sessile. Seeds many, testa crustaceous, pitted, sending spinous processes into the albumen.—Tall trees. Leaves large. Flowers large, yellow, solitary or fascicled on short axillary peduncles.

1. **C. odorata** (fragrant), *H. F. and T. Fl. Ind.* A tall tree with straight trunk; bark smooth; shoots glabrous. Leaves ovate-oblong, 5 to 8in. long, 2 to 3in. broad, finely acuminate, puberulous on the under side, rounded at the base, margins wavy. Petiole about ½in. long. Peduncles solitary or several from old scars; pedicels about 1in., recurved horny, with a few basal and a median scaly bract. Flowers usually 3-nate, drooping, yellow, odorous. Petals about 3in. long, subequal, narrow-linear, silky when young. Carpels about 12, glabrous long-stalked, ovoid or obovoid, black, 6 to 12-seeded. *Uvaria odorata* Lamb. Ill. t. 495, f. 1; Blume Fl. Jav. Anon. t. 9.

Hab.: Scrubs of tropical parts of the colony.
Wood of a grey colour, close-grained and hard.

4. ANCANA, F. v. M.
(After Baron Anca).

Sepals 3, free, persistent, much shorter than the petals. Petals 6, equal subulate-linear, biseriate-valvate in the bud. Stamens numerous, 4-angular-cuneate. Torus hemispherical. Carpels 3, with capitate sessile stigmas, ovules several. Berries containing few seeds. F. v. M. Fragm. v. 27.

1. **A. stenopetala** (slender petals), *F. v. M. Fragm.* v. 27 and vii. 126. A scandent shrub, the branchlets more or less hairy. Leaves chartaceous, 3 to 4½ in. long, ¾ to 1¼in. broad, shining on both sides, copiously netted-veined, base obtuse, apex acuminate. The sparsely scattered hairs rigid, at length glabrate. Pedicels thick, 1½ to 2½ lines long. Bracts and bracteoles like the sepals but smaller, the tawny appressed hairs with which they are covered almost silky. Flowers fragrant. Sepals 2 lines long, acute, opposite to the interior petals. Petals 1in. long, narrow, and tapering from the base, somewhat thick, bearing more or less of a tawny, hoary covering. Anthers nearly sessile. Fruit yellow.

Hab.: Scrubs of the southern parts of the colony.

5. POLYALTHIA, Blume.

Sepals broad. Petals 6, valvate in the very young bud, in two rows, but spreading or open long before they have attained their full size, nearly equal and flat, usually narrow. Stamens numerous, narrow-wedge-shaped, the connective flattened at the top, concealing the cells. Torus slightly raised. Carpels several, with a short, oblong, or capitate style, and 1 or 2 erect ovules. Berries stalked, globular or ovoid.—Trees or shrubs. Flowers solitary or clustered, axillary or leaf-opposed.

A considerable genus, chiefly Asiatic, with one African species, one of the Queensland species extending to New Caledonia. *Benth.*

Carpel shortly stalked . . . . 1. *P. nitidissima.*
Carpels sessile or nearly so . . . . 2. *P. Armitiana.*

1. **P. nitidissima** (very bright), *Benth. Fl. Austr.* i. 51. A tree of 15 to 50 or 60ft., glabrous in all its parts. Leaves elliptical, or the upper ones almost lanceolate, obtuse or obtusely acuminate, 2 to 3in. long, narrowed into a petiole varying from 2 to 5 lines, smooth and shining, the veins fine and reticulate, but not numerous. Peduncles solitary, axillary, 3 to 6 lines long, or more when in fruit, with 2 or 3 small bracts near the base. Sepals short and broad. Petals linear, rather thick, 5 or 6 lines long when fully out, but spreading very early. Stamens very short, and closely packed. Carpels 10 to 20 in the flower, much fewer in the fruit, and then globular or shortly ovoid, 1-seeded, shortly stalked. —*Unona nitidissima*, Dun. Anon. 109, t. 28 ; *Unona fulgens*, Labill. Sert. Austr. Caled. 57, t. 56 ; *Unona nitens*, F. v. M. Fragm. iii. 2.

Hab.: Scrubs on islands in Moreton Bay, and northward. Also found in New Caledonia.

In some specimens the torus, after flowering, becomes thick and woody, enclosing several cavities, probably a deformity occasioned by the puncture of some insect. Labillardière describes and figures the carpels as having several ovules, but this is a mistake. His own specimens, quite similar to the Australian ones, have but one erect ovule in each.—*Benth.*

Wood of a dark grey colour, close-grained, nicely marked, and with a strong spice-like fragrance when fresh cut. *Bail. Cat. Ql. Woods*, No. 3.

2. **P. Armitiana** (after W. E. Armit), *F. v. M. in Austr. Journ. of Pharm. Jan.* 1887, 2nd *Syst Cens. Austr. Pl.* 5.

I can find no description of this plant in the Journal of Pharmacy above quoted. After the description of *Mitrephora Froggattii* Baron Mueller says:—" Mr. Armit's collection from tributaries of the Gilbert River contains an Anonaceous plant possibly conspecific with the one just described ; but the lateral nerves of the leaves are thinner, more approximated, and less curved, while the fruits are not unlike those of *Polyalthia Holtzeana*, but seem not provided with conspicuous stipes ; the flowers of Armit's plant are unknown. It has meanwhile been specifically designated under the discoverer's name."

6. MITREPHORA, Blume.
(Mitre-bearing.)

Sepals 3, orbicular or ovate. Petals 6, 2-seriate, valvate ; outer ovate, thin, veined ; inner clawed, vaulted and cohering. Stamens oblong-cuneate ; above anther-cells dorsal, remote. Carpels oblong ; style oblong or clavate, ventrally furrowed ; ovules 4 or more, 2-seriate. Ripe carpels globose or ovoid, stalked or subsessile.—Trees. Leaves coriaceous, strongly ribbed, plaited in vernation. Flowers usually terminal or leaf-opposed, sometimes 1-sexual.—Hook. Flora British India i. 76.

1. **M. Froggattii** (after Froggatt), *F. v. M. Austr. Journ. of Pharm.* ii. 3. So far as known, a tree of about 20ft. ; the branchlets soon glabrous. Leaves on very short petioles, as much as 8in. long and 3in. broad, chartaceous, elliptic-lanceolate, acuminate, nearly or quite blunt at the base, glabrous on both sides, slightly dotted, of a rather dark green and shining on both sides, particularly beneath ; distantly costate-nerved, prominent on the under side of leaf ; veins thin, reticulate. Peduncles obliterated ; pedicels axillary or lateral, solitary or 2 together, about twice as long as the petals, thin-downy, minutely scaly-bracteate at the base. All flowers seen

only staminate. Sepals minute, about 1 line long, nearly deltoid, membranous. Petals black-purplish, somewhat curved inwards, 2 or 3 lines long; outer ones almost orbicular and nearly sessile, the inner ones considerably longer than the outer, roundish or obcordate-rhomboid, attenuated into a stalk-like base, bicallous above the middle inside, all as well as the sepals downy on the outside, and valvate in the upper part before expansion. Head of stamens rather depressed. Anthers numerous, broadly cuneate, truncate, almost sessile; connectives dark, their flat summits forming an even surface for the head of stamens; cells pale. Torus very convex, velvety-hairy. Fruit unknown.

Hab.: Mossman River.—Collected by *Sayer* and *Froggatt*.

In the absence of pistils it remains uncertain whether this interesting plant should be placed in the genus *Mitrephora* or *Goniothalamus*; but it is not dissimilar to *M. reticulata*, differing in glabrous more distinctly petiolated leaves, in fewer and larger flowers, and in dark-coloured petals, the inner quite blunt. In some respects it reminds of the *Orophea*, particularly *O. zeylanica*, though the stamens are so different.—*F. v. M.* in Austr. Journ. of Pharmacy, Jan. 1887.

7. HAPLOSTICHANTHUS, F. v. M. in Vict. Nat. 1891.

Sepals 3, deltoid, early valvate; petals 6, uniseriate-valvate in bud, completely connate towards the base, thus forming a 6-lobed corolla, 3 of the lobes deltoid, 3 doubly as long and almost semi-elliptic, all remaining much connivent; torus depressed; stamens about 30, of pyramidate-cuneate, their connectives at the summit slightly convex, or almost truncate and somewhat peltate, concealing the cells; ovularies 6, with sessile depressed stigmas; fruit unknown. Shrub with comparatively small chartaceous leaves, and with short-stalked, solitary, dark-coloured flowers of remarkable smallness.

This new Anonaceous genus seems best placed in the tribe of *Mitrephoreæ*, but it agrees with the otherwise very different *Hexalobus* in the downward conspicuously-connate petals. As regards the 6 petioline parts, placed in a single row, this plant seems to stand alone in the whole order, large as it is. The circumscription, however, of many of the genera needs revision also in this order, as much new material has been obtained during later years, affecting the generic limits as drawn formerly. The style and stigma offer good notes for primary distinctions also.—*F.v.M.l.c.*

1. H. Johnsoni (after S. Johnson), *F. v. M.*, *Vict. Nat.* 1891. Young branchlets thinly pubescent. Leaves almost sessile, rather narrowly lanceolate, acuminate, but at the base obliquely rounded, when young scantily beset with appressed hairlets, subsequently glabrescent, paler green beneath, from 1 to 3in. long, ⅓ to nearly 1in. broad, thinly venulated. Peduncles recurved, measuring at flowering time ½in. or less, occasionally supported at the base by a spinescently indurated bud. Sepals about ¹⁄₁₀in. long, pale-brown. Flowers as small as those of *Bocagea pisocarpa*, *Polyalthia moonii*, and *Popowia australis*, measuring, even when flattened out, only ⅓in. in diameter. Corolla outside beset with minute appressed hairlets, the connate portion quite as long as the 3 deltoid lobes, without any sutural indications; the 3 longer lobes somewhat triangular at the summit. Stamens only about ¹⁄₂₄in. long. Ovularies silky, during anthesis not emerging beyond the stamens.—*F. v. M.* l.c.

Hab.: Mount Bartle-Frere, *Stephen Johnson*.

8. MELODORUM, Dun.
(Leaves of one species honey-scented.)

Sepals small, united at the base. Petals 6, valvate in the bud in 2 rows, the outer ones broad, thick, concave, connivent or scarcely open, the inner ones smaller. Stamens numerous, the connective ovate or truncate, concealing the cells. Torus convex or conical. Carpels several, with an oblong thick style and 2 or more ovules in each, attached to the inner angle. Berries distinct, sessile or stalked. Stems woody, usually climbing. Primary veins of the leaves prominent underneath. Flowers terminal or leaf-opposed.

IV. ANONACEÆ.

The genus comprises several species, dispersed over tropical Asia and the Indian Archipelago. The Australian species all endemic.

Woody climber. Outer petals about 6 lines. . . . . 1. *M. Leichhardtii*
Woody climber. Outer petals about 4 lines . . . . 2. *M. Uhrii.*
Small tree. Outer petals narrow-lanceolate, inner ones 4-angled subulate . 3. *M. Maccreai.*

1. **M. Leichhardtii** (after L. Leichhardt), *Benth. Fl. Austr.* i. 52; *Merangara*, N. Q'land, *Thozet*. A strong woody climber, the young growth more or less rusty-tomentose. Leaves on short petioles, oblong, obtuse or obtusely-acuminate, 3 to 6in. long, coriaceous, glabrous somewhat shining, sprinkled with a few almost microscopic fringed scales or stellate hairs on the under side, the veins not very prominent. Peduncles $\frac{1}{2}$ to 1in. long, rusty-tomentose. Flowers brown, fragrant, about 1in. in diameter. Sepals 3 lines long, spreading. Outer petals exceeding 6 lines, slightly tomentose, very obtuse, concave and connivent, the inner ones thicker and rather shorter. Stamens very numerous. Berries stipitate, either depressed-globose 4 or 5 lines diameter and 1-seeded, or oblong 2-seeded with a slight transverse furrow between the seeds, or moniliform, consisting of 2 depressed-globose 1-seeded or oblong 2-seeded portions. Ripe about January, edible.—*Unona Leichhardtii*, F. v. M. Fragm. iii. 41.

Hab.: Scrubs of the south and north.
Fruit eaten by natives, *Thozet*.

2. **M. Uhrii** (after — Uhr), *F. v. M. Fragm.* vi. 2. A climbing shrub, with ferruginous-tomentose branchlets. Leaves on very short petioles, 3 to 5in. long, $1\frac{1}{2}$ to 2in. broad, nerves of the underside prominent, glabrescent on the upper side, ovate, obtusely acuminate. Peduncles few-flowered, pedicels short. Sepals cordate-deltoid, 1 or 2 lines long, velvety. Petals coriaceous, outer ones deltoid-cordate, 4 lines long, interior ones rhomboid, ferruginous-velvety. Stamens very numerous, almost a line long, glabrous, clavate, connective of anther peltate. Styles compressed-cylindrical. Carpels ferruginous-silky.

Hab.: Scrubs of Rockingham Bay, *J. Dallachy* (F. v. M. l.c.)

3. **M. Maccreai** (after Dr. W. Maccre), *F. v. M. Fragm.* vi. 176. A tree of 40 to 50ft., branches clothed with brownish or hoary hairs. Leaves $1\frac{1}{2}$ to $4\frac{1}{2}$in. long, $\frac{3}{4}$ to $1\frac{1}{2}$in. broad, chartaceous, ovate to ovate-lanceolate, the upper side glabrous, under side slightly hairy, petioles very short. Peduncles short, 1-flowered. Calyx-lobes deltoid, about $1\frac{1}{4}$ line broad. Petals, outer ones narrow, lanceolate, thick, $\frac{1}{2}$in. long, inner ones tetragonus-subulate. Anthers numerous, $\frac{1}{2}$ line long, connective, lanceolate-acuminate, cell very slender. Styles very short and bearded at the apex. Berries about 10, stipitate, $\frac{3}{4}$ to $1\frac{1}{2}$in. long, pyriform, red inside. Seeds angular, ovate, 4 or 5 lines long, smooth and of a shiny brown outside.

Hab.: Rockingham Bay, *J. Dallachy* (F. v. M. l.c.)

9. SACCOPETALUM, Bennett.
(From bag-like form of petal.)

Sepals small. Petals 6, valvate in 2 rows, the outer one small and resembling the sepals, the inner large, erect, and very concave. Stamens numerous, but loosely imbricate, showing the anther-cells on their backs, just below the short tips. Torus nearly globular. Carpels several, with an ovoid or oblong thick style, and 6 or more ovules in each attached to the inner angle. Berries globular. Trees or shrubs, with deciduous leaves. Flowers usually appearing on the young shoots before or with the young leaves.

A small genus, dispersed over India and the Archipelago; the Australian species endemic.

Branchlets densely hairy, leaves 3 or 4in. long, carpels very hairy. . . . . . 1. *S. Bidwilli.*
Branchlets never densely hairy, leaves about $1\frac{1}{2}$in. or more long, carpels very slightly hairy . . . . 2. *S. Brahei.*

IV. ANONACEÆ. [*Saccopetalum.*

1. **S. Bidwilli** (after J. C. Bidwill), *Benth. Fl. Austr.* i 53. A shrub, the branchlets densely hirsute with short rusty hairs. Leaves very shortly stalked, oblong or obovate oblong, obtuse or very shortly acuminate, 3 to 4in. long, rounded at the base, glabrous above, hairy underneath. Flowers lateral, solitary or 2 together, on very short pedicels. Sepals thin, lanceolate, hairy, about 2 lines long. Outer petals similar, but twice as long. Inner petals when fully developed 1½in. long, not saccate at the base only, as in most other species of the genus, but hollowed into a broad boat-shape all the way up, with the upper end turned inward, thin, and very hairy both inside and out. Stamens numerous, the anther-cells contiguous and conspicuous, terminated by the small flat tip of the connectivum. Carpels very hairy in the flower, when ripe nearly sessile, oblong, 6 to 8 lines long, thick and hard, covered with rusty hairs, containing 3 to 6 flattened seeds.

Hab.: Wide Bay.

2. **S. Brahei** (after W. Brahe), *F. v. M. Fragm.* viii. 159. Branchlets only slightly, never densely, hirsute. Leaves 1½in. long, broad-lanceolate, acute or acuminate, narrowed to a very short petiole. Pedicels lateral or terminal, solitary or in pairs, slender, and 3 to 5 times as long as the small flowers. Sepals and exterior petals about 1 line long, interior petals about ⅓in. long, saccate at the base, the margins velvety. Anthers imbricate, cordate, about ⅙ of a line long. Carpels about 20, very slightly pilose.

Hab.: Port Denison.

10. EUPOMATIA, R. Br.

(*Eu*, well ; *poma*, a lid ; the calyptra consolidation.)

Sepals and petals completely consolidated into one mass, the upper part falling off in a conical lid, leaving the lower companulate tube (or enlarged peduncle) filled with the thick flat-topped torus. Stamens inserted on the margin of the torus, the inner one in many rows, converted into petal-like obovate staminodia, the outer ones in fewer rows, perfect, linear-lanceolate, curved, with acuminate tips and longitudinal dorsal anther-cells. Carpels many, immersed in the torus, appearing like the cells of a single inferior ovary, the stigmas adnate on the flat areolate surface; ovules several in each carpel or cell. Fruit several-celled, formed of the enlarged perianth-tube more or less enclosing the carpels, becoming turbinate or urceolate and succulent. Seeds 1 or 2 in each cell, irregularly angular; albumen ruminate, and embryo precisely as in the more normal *Anonaceæ*. Small trees, shrubs, or undershrubs, quite glabrous. Leaves alternate, entire, shortly petiolate. Peduncles short, 1-flowered, terminal or lateral.

The genus is confined to Australia.

Petioles shortly decurrent. Flowers terminal. Outer staminodia spreading
  and longer than the stamens. Fruit turbinate . . . . 1. *E. Bennettii.*
Petioles not decurrent. Flowers lateral. Staminodia all connivent, shorter
  than the stamens. Fruit urceolate . . . . 2. *E. laurina.*

1. **E. Bennettii** (after G. Bennett), *F. v. M. Fragm.* i 45, *Benth. Fl. Austr.* i 54. A shrub or undershrub. Roots fleshy, almost tuberous. 1 to 2ft. high and quite glabrous. Leaves oblong-lanceolate, acuminate or acute, 3 to 5in. long, narrowed at the base into a short petiole, which is again enlarged at the base and shortly decurrent on the stem, leaving oblique raised lines when they fall off. Flowers solitary, terminal, on a short peduncle above the last leaf, when fully expanded rather more than 1in. diameter. Petal-like staminodia very numerous, yellow, the outer ones stained with orange or blood-red, beset with stipitate glands and bordered with stellate hairs spreading and completely concealing the perfect stamens, which are reflexed on the peduncle, the inner staminodia shorter

and connivent. Fruit turbinate, about ¾in. diameter, the pericarp thin, the top convex, with the tips of the carpels distinctly prominent, the base of the perianth scarcely projecting as a slight ring round the edge.—*E. laurina*, Hook. in Bot. Mag. t. 4848.

Hab.: Scrub land north and south. Flowering from September to March.

2. **E. laurina** (laurel-like), *R. Br. in Flind. Voy.* ii. 597, *t.* 2; *Benth. Fl. Austr.* i. 54. An erect glabrous tall shrub or small tree with weak branches. Leaves evergreen, oblong or almost elliptical, shortly acuminate, 3, 4, or sometimes 5in. long, narrowed into a short petiole which is not decurrent on the branch. Flowers solitary, on short lateral or nearly axillary peduncles, the buds at first oblong, becoming nearly globular and about ½in. diameter before opening; when the bud has fallen the stamens expand to about 1in. diameter. Petal-like staminodia connivent or the outer ones scarcely open, glabrous or with a very few stipitate glands; perfect stamens longer, erect or spreading, the linear anthers tipped by a short fine point, the filaments dilated. Fruit urceolate-globular, nearly ½in. diameter, the persistent base of the perianth forming a narrow rim projecting above the nearly flat top.—F. v. M. Fragm. i. 45.

Hab.: In most southern and northern coast scrubs. Flowering about November.

Wood close-grained, of a light colour, and prettily marked.—*Bailey's Cat. Ql. Woods No.* 3a.

Order V. MENISPERMACEÆ.

Flowers dioecious. Sepals usually 6 in 2 series, rarely 9 or 12 in 3 or 4 series, or very rarely 5 or fewer, imbricate or very rarely valvate in each series, the inner ones the largest. Petals usually 6, smaller than the sepals (except in *Sarcopetalum*), nearly equal but imbricate in 2 series in the bud, rarely fewer or none. Male fl.: Stamens usually 6, free and opposite the petals, or united in a central column, rarely 9 or more or only 3. Female fl.: Staminodia usually 6, free. Carpels distinct, usually 3, sometimes 6 or more or only 1, containing 1 or very rarely 2 amphitropous ovules peltately attached to the inner angle. Style terminal, usually recurved, and often expanding into a short sessile stigma. Fruit-carpels drupaceous, nearly straight, or more frequently curved, so that the remains of the style are near the base, the putamen then becoming more or less horseshoe-shaped, with an inner projection of the endocarp bearing the placentæ. Seed taking the shape of the cavity, with a thin membranous testa. Albumen sometimes fleshy, entire or ruminate, sometimes thin or none. Embryo nearly as long as the albumen or occupying the whole seed, the radicle pointing to the remains of the style.—Climbers usually woody, or in a very few non-Australian species erect herbs or shrubs. Leaves alternate, without stipules, entire or rarely palmately lobed, usually with 3 or more palmate ribs at the base. Flowers small, in axillary panicles, racemes, or cymes.

A considerable tropical Order, both in the New and the Old World, a very few species extending into more temperate regions in North America, eastern Asia, and southern Africa.—*Benth.*

Tribe I. **Tinosporeæ.**—*Flowers 3-merous. Ovaries usually 3. Drupes with a subterminal rarely ventral or subbasal style-scar. Seed oblong or subglobose; albumen copious or scanty; cotyledons foliaceous, usually spreading laterally.*

Flowers in simple racemes. Inner sepals broad and thin.
Carpels of the fruit ovoid, the style at the top. Seed albuminous, nearly straight . . . . 1. Tinospora.
Male flowers paniculate, female spicate. Carpels of fruit oblique, ovate turgid, echinulate-scabrous . . . . 2. Fawcettia.

V. MENISPERMACEÆ.

TRIBE II. **Cocculeæ.**—*Flowers 3-merous. Ovaries usually 3. Drupes with a subbasal rarely subterminal style-scar. Seed horseshoe-shaped, albumen copious; embryo slender; cotyledons linear or slightly dilated.*

Flowers in much-branched cymes. Carpels of the fruit broad, the style near the base. Seeds albuminous . . . . 3. PERICAMPYLUS.
Petals 5 to 8. Ovaries 3. Styles compressed. Sepals 8 to 12, inner imbricate in subgenus *Hypserpa* . . . . 4. LIMACIA (HYPSERPA).
Sepals imbricate. Petals 3. Stamens 9 to 12. Carpels 3, 2-ovulate . 5. ADELIOPSIS.
Sepals 9, 3-seriate. Petals 6. Stamens 6, free. Anthers didymous-globose, almost 4-lobed. Drupe renate-ovate, turgid. Seed reniform 6. TRISTICOCALYX.

TRIBE III. **Cissampelideæ.**—*Flowers 3—5-merous. Ovaries usually solitary. Drupes with a subbasal style-scar; endocarp dorsally muricate or echinate. Seed horseshoe-shaped; albumen scanty; embryo linear; cotyledons oppressed.*

Sepals 2 to 5, very small. Petals 3 to 6, thick and fleshy, almost globular. Anthers 2 or 3. Carpels 3 to 6. Flowers racemose . . 7. SARCOPETALUM.
Sepals 6, membranous. Petals 3, somewhat fleshy. Stamens 3, connate in a very short column. Flowers in racemose-panicles . . 8. LEICHHARDTIA.
Sepals 6 to 10, free.
Petals free, smaller than the sepals, concave, of both male and female 3 to 5. Anthers 4 or 5. Carpels solitary. Flowers umbellate . . . . 9. STEPHANIA.
Sepals 4. free. Petals of male 4-connate, of female 1. Male flowers cymose; female racemose . . . . 10. CISSAMPELOS.

TRIBE IV. **Pachygoneæ.**—*Flowers usually 3-merous. Ovaries usually 3. Drupes with subbasal or ventral style-scar. Seed curved-hooked or inflexed; albumen none; cotyledons thick, fleshy.*

Sepals, petals, and stamens 6 each . . . . 11. PACHYGONE.
Sepals and petals 6 each. Stamens 9 . . . . 12. PYCNARRHENA.
Sepals 9. Petals 6. Stamens 3 . . . . 13. PLEOGYNE.
Sepals 9. Petals very minute, bilobed. Stamens 6 . . . . . . 14. HUSEMANNIA.

1. TINOSPORA, Miers.

(Small seeds.)

Sepals 6, in 2 series, the inner ones large. Petals 6, smaller than the sepals, nearly flat. Male flowers: Stamens 6, free, thickened towards the top, the anther-cells lateral. Female flowers: Staminodia 6. Carpels 3, stigma jagged. Drupes ovoid, the remains of the style nearly terminal. Putamen slightly concave on the inner face, the internal projection hemispherical and hollow, forming an empty cell. Seeds disk-shaped, albuminous. Cotyledons ovate, spreading laterally. Leaves cordate or truncate at the base. Flowers usually clustered in long, simple racemes.

A small genus, chiefly Asiatic, but extending also to tropical Africa. The Australian species endemic.

1. **T. smilacina** (Smilax-like), *Benth. Fl. Austr.* i. 55. A glabrous twiner, the branches somewhat succulent. Leaves ovate, deeply and broadly cordate at the base, or almost hastate with rounded auricles, obtuse or scarcely acuminate, 8 to 4in. long, 5-nerved, the smaller pinnate veins scarcely prominent, on petioles of about 1in. Flowers green, the male racemes 2 or 3in., the female about 1in. long; pedicels about 1 line. Sepals: 3 outer ones very small and triangular, 3 inner ones about 1 line long, ovate, thin, spreading. Petals about half as long as the inner sepals, obovate. Anthers terminal, ovoid, almost globular, the cells almost parallel. Drupe red, oblong, about 3 lines long. Ripe in June.

Hab.: Cape York and Thursday Island.

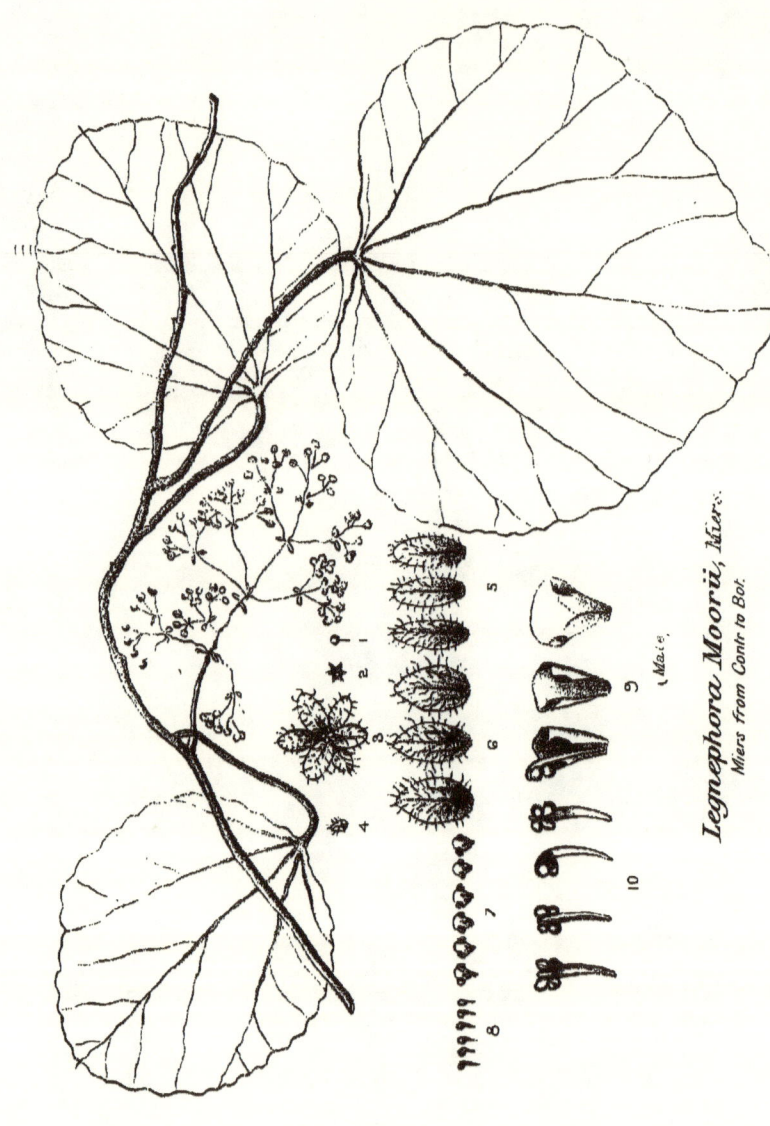

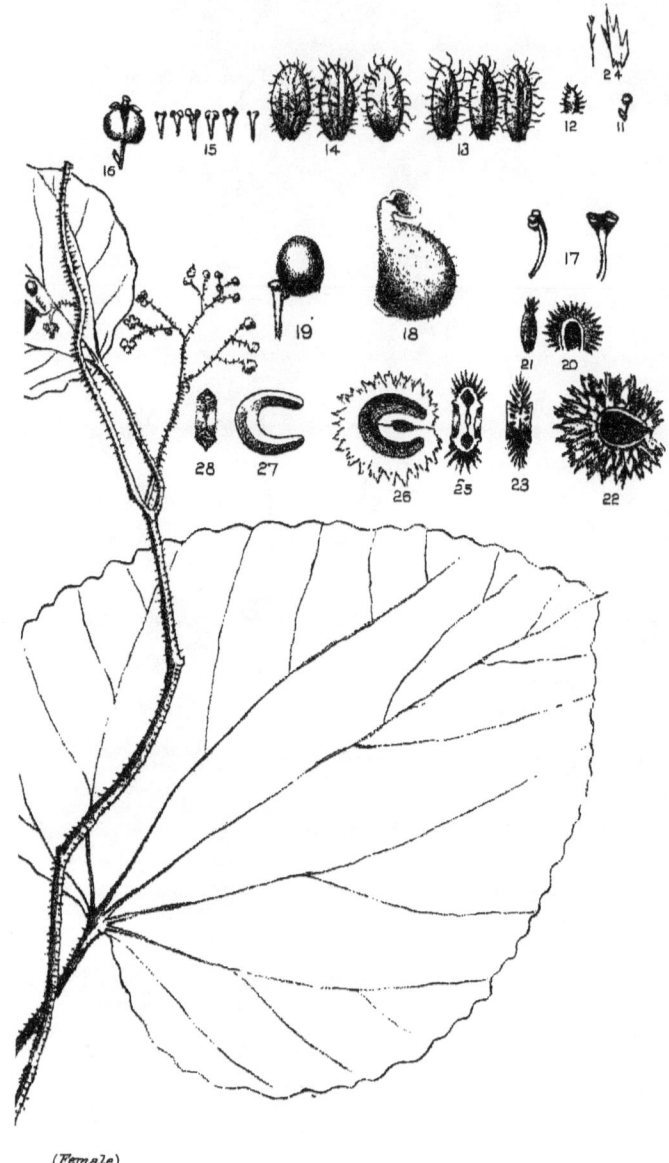

(Female)

Legnephora Moorii, Miers.
Miers from Contr. to Bot.

V. MENISPERMACEÆ.

2. FAWCETTIA, F. v. M.

(After C. Fawcett).

Sepals 6, membranous, 3 outer ones ovate-lanceolate, 3 inner ones longer, broadly or orbicularly-ovate, imbricate in the bud. Petals 6, a little longer than the outer sepals, obcordate or rhomboid-orbicular, membranous. Stamens 6, free, filaments planiuscula. Anthers cordate, bursting longitudinally. Stigma terminal. Endocarp echinulate-scabrous outside. Condyle (a name of Miers for the intrusion of the endocarp) ample, 1-celled. Seed longitudinally horseshoe-shaped. Albumen even. Cotyledons very broad, plain, divergent, quite entire; radicle slender, very short.—A climbing shrub. Leaves cordate or ovate-lanceolate. Fruit red.

1. **F. tinosporoides** (Tinospora-like), *F. v. M. Fragm.* x. 93. Leaves 2 to 3in. long, 1 to 2in. broad, glossy, 3—5-nerved, and reticulate-veined, on petioles of about 1in. Panicles from the old leafless branches; male about 6 or 7in. long, the branchlets about 1in. or less. Interior sepals entire, about $\frac{2}{3}$ of a line long. Petals green. Anthers yellow. Female flowers not seen. Carpels $\frac{2}{3}$ to 1in. long. Seeds very bitter. Endocarp dark, parchment-like.

Hab.: Southern scrubs.

LEGNEPHORA, Miers.

(Fringe or border bearing.)

Flowers diœcious. Male: Sepals 6, 2-seriate, subequal, elliptical scarcely acute, pilose outside, the outer ones a little narrower, imbricate in the bud. Petals 6, squamiform, opposite the sepals and 6 times shorter, cuneate-rotundate, the sides glandulously fleshy, glabrous. Stamens 6 of equal length and opposite to the petals; filaments terete, thickening upwards. Anthers subglobose, twice as broad at the filament, dorsifixed, introrse, 2-celled, cell-connective narrow, separation a trifle excurrent, fissures transverse on both sides, dehiscence 2-valved. Female: Sepals as in the male. Petals none. Stamens 6, sterile, opposite and half the length of the sepals, cuneate-linear, apex dilated and truncate, subcanaliculate, with semi-immersed concave glands. Ovaries 3, gibbose-globose, pilose, 1-celled, 1-ovulate. Style very short or obsolete. Stigma cordate-orbiculate, concave, entire, horizontally reflexed. Drupes 3 or less by abortion, compressed-globose, fleshy, stigma near the base; putamen cuneate-orbicular much compressed, thin bony, both faces with a prominent horseshoe-shaped scar, with 8 series of imbricating laciniated, flat, pergamenous scales, surrounding the flat, concave, scutiform condyle.—From Miers' Contri. to Bot. iii. 288.

1. **L. Moorei** (after C. Moore), *Miers Contri. to Bot.* iii. 289. A tall puberulent scrub-climber. Branches terete, striate, compressed at the nodes. Leaves 2¼ to 5½in. long, 2 to 5½in. broad, ovate, the lower ones very obtuse or truncate, narrowing above the centre, to an acute point, margins crispate-undulate, 5 to 7 nerves from the base, rigid and prominently reticulate, shining on the upper with a yellowish or whitish thin down on the underside; petiole 2 to 3in. long, thin puberulous. Panicles (male) axillary, tomentose, longer than the petioles, alternately branched; branches often 2 to 4 in a verticil, trichotomous at the apex, branches few-flowered. Sepals elliptical, both sides puberulent. Racemes (female) axillary or terminal of few flowers. Drupes subglobose, glabrous.—Miers l.c. *Cocculus Moorei*, F. v. M. Fragm. i. 162. *Pericampylus incanus*, Benth. Fl. Austr. i. 58.

Hab.. Scrubs of southern arts of Queensland.

V. MENISPERMACEÆ.

3. PERICAMPYLUS, Miers.

Sepals 6 in 2 series, the inner ones larger. Petals 6, smaller than the sepals, the edges embracing the stamens. Male flower: Stamens 6, free, the anther-cells lateral. Female flower: Staminodia 6. Carpels 3, the styles 2-cleft. Drupes globular, somewhat flattened, the remains of the style near the base. Putamen horseshoe-shaped, crested on the back, the sides concave. Seed horseshoe-shaped. Embryo in the axis of the albumen, with narrow cotyledons closed against each other. Leaves broad. Cymes dichotomously branched.

The genus is limited to the following species.

1. **P. incanus** (hoary), *Miers; Hook. and Thoms. Fl. Ind.* i. 194; *Benth. Fl. Austr.* i. 56. Achenium with the younger branches shortly tomentose or at length glabrous. Leaves nearly orbicular, sometimes slightly peltate, 2 to 4in. or sometimes above 5in. diameter, glabrous above, usually hoary underneath, on petioles of 1 to 2in. Flowers very small, in axillary dichotomous cymes, shorter than the leaves. Sepals hairy on the back. Drupes red.—*Cocculus Moorei*, F. v. M. Fragm. i. 162.

Hab.: Woody valleys, Moreton Bay and Wide Bay, *C. Moore. W. Hill, F. Mueller*.

The root-bark contains an active poisonous principle, *T. L. Bancroft*.

Miers in *Contri. Bot.* iii. p. 118 has the following note after his description of *Pericampylus incana*:—"The authors of the *Flora Indica* have absorbed in this the only species they acknowledge in the genus; and Mr. Bentham has done the same in his *Flora Australiensis* i. 58: a plant from Australia which is extremely different, not only in a specific, but in a generic point of view; it is the *Cocculus Moorei* of Dr. Mueller, which I have elsewhere described as the type of a distinct genus, under the name of *Legnephora*." As Mr. Miers disowns the Australian plant, being that named by him *P. incanus*, I have considered it advisable in the present work to insert both the notice given in the *Flora Australiensis* and almost all the generic and specific descriptions of the plant from *Miers Contri.* iii. 118 of the Australian plant under the name *Legnephora*; and also a copy of his excellent plate of the same.

4. LIMACIA, Lour.

(Fruit resembling shell of a snail.)

Sepals 6, 2-seriate, outer smaller. Petals 6, much smaller, auricled, embracing the stamens. Male flowers: Stamens 3, 6, or 9, free; anthers adnate, bursting vertically. Female flowers: Staminodia 6, clavate. Ovaries 3; style short, compressed. Drupes obovoid or reniform, style-scar subbasal; endocarp 3-celled, 2 lateral cells empty. Seeds elongate, embracing the intruded endocarp. Embryo slender, cotyledons elongate, $\frac{1}{2}$-terete, appressed.

Selwynia, *F. v. M.*—Sepals about 9, *outer* 2 *to* 4 *smaller, unequal, roundish-oval, inner* 5 *to* 6 *suborbicular, imbricate in the bud. Petals* 7 *to* 9, *shorter than the inner sepals, free, spathulate-cuneate, membranous. Stamens* 9 *to* 10, *clavate-cuneate, free. Anthers terminal, almost 4-lobed-globose, 2-celled, dehiscing longitudinally. Carpels drupaceous, pyriform-globose.*

1. **H. Selwynii** (after Bishop Selwyn), *F. v. M.* A tall climber, glabrous except for the white mealy substance which frequently covers the branches. Branchlets striate, often elongated and slender when bearing inflorescence. Leaves alternate, thin-chartaceous, from ovate-lanceolate to oblong or linear-lanceolate, 3 to 8in. long. 1 to 2½in. broad, the apex often elongated, rounded at the base, primary nerves distant, very oblique and looping distant from the margin, 3-5-nerved at the base, the transverse-reticulation fine, margins entire; petioles slender but thickening towards the top, $\frac{1}{2}$ to 1¼in. long. Inflorescence supra-axillary, very slender panicles. Male flowers pedicellate about 2 lines diameter; bracts minute orbicular, ciliate. Sepals orbicular, more than twice the size of the bracts, marked by 3 to 5 veins all starting from the base but not

reaching the apex, margins entire. Petals very minute, thick and somewhat triangular, shorter than the stamens and hidden by these organs in the expanded flower. Stamens 12, filaments cuneate. Anthers almost white. Drupe red when fresh, oval, compressed, about 5 lines long, drying a dark colour.

The above is written from the examination of specimens received from E. Cowley, Kamerunga, and these in general appearance seem identical with a small specimen since received by me from the late Baron Mueller; his description and mine in some respects differ, but not, in my opinion, sufficient to found distinct species upon. The Baron first mentioned his plant as a *Cocculus*, then gave it generic rank as *Selwynia*, and lastly as *Hypserpa* (a genus of Loureiro included by present botanists in *Limacia*). It will be observed from the generic characters, copied from *Hook. Fl. of Brit. Ind.*, of *Limacia*, and the brief generic characters of *Selwynia*, from *F. v. M. Fragm.* iv., that some alterations will have to be made in the generic characters of *Limacia* to admit the Queensland plant.

5. ADELIOPSIS, Benth.

(Named so on account of some doubt regarding the plant.)

Sepals 6 in 2 rows, the inner ones considerably larger, and 2 or 3 outer smaller bracts, all much imbricate in each row. Petals 3, smaller than the inner sepals, broad and slightly concave. Male flowers: Stamens 9 to 12; filaments linear-terete; anthers small, globose-didymous. Female flowers: Staminodia wanting. Carpels 3, with a large, incurved, broad and thick stigma, and 2 ovules in each carpel, inserted one above the other on the inner angles. Fruit unknown. Flowers clustered in short axillary spikes, or racemose panicles.

1. **A. decumbens** (spreading over the ground), *Benth. Fl. Austr.* i. 59. Branches rather thick, densely clothed with a soft velvety tomentum or almost hirsute, and, from the name given, probably decumbent and not climbing. Leaves ovate or oval-oblong, $1\frac{1}{4}$ to 2in. long, very obtuse, rounded at the base, thickly coriaceous, softly tomentose or velvety on both sides when young, becoming nearly glabrous above when old, the thickened revolute nerve-like margin terminating at the top of the midrib on the under side in a prominent hirsute gland or tuft of hairs. Flowers small, in little clusters along the rhachis of short axillary spikes or branches of the panicle, $\frac{1}{2}$ to 2in. broad, the outer bracts very small, acute, and hairy, the outer sepals also hairy, but rather larger and more obtuse, the inner sepals much larger, orbicular, and glabrous, except the ciliate edge, the petals about two-thirds as large as the inner sepals and quite glabrous.

Hab.: Cape York Peninsula.

6. TRISTICHOCALYX, F. v. M.

(Three rows of sepals).

Male flowers: Sepals 9, 3-seriate, outer ones small lanceolate, intermediate ones longer, inner ones larger, narrow-ovate. Petals 6, smaller than the inner sepals. Stamens 6, free, filaments thickened upwards; anthers subglobose-didymous, 4-celled. Female flowers: Drupes "renate-ovate turgid." Style-scar a little from the base; putamen reniform-globose, somewhat compressed, ventrically concave, with an intruding process. Seed reniform, albumen fleshy, uniform, embryo radicle short, cotyledons oblong.—Scandent pubescent shrubs. Leaves coriaceous, broadly ovate, not peltate. Male flowers in axillary racemes, usually in many-flowered clusters.

Leaves 3 to 4in. long. Flowers glabrous . . . . 1. *T. pubescens*.
Leaves 2in. long. Flowers pubescent . . . . 2. *T. diffusus*.

1. **T. pubescens** (downy), *F. v. M.* A woody climber, the young branches pubescent. Leaves petiolate, broadly ovate, shortly acuminate or rarely obtuse, 3 to 4 in. long, 5-nerved at the base, coriaceous, glabrous and shining or slightly scabrous above, pubescent underneath. Male racemes axillary, often 2 or 3

together, many-flowered but much shorter than the leaves, pubescent. Pedicels clustered, about 1 line long. Flowers glabrous, scarcely more than 1 line diameter when open. Sepals 9, in 3 series, the outer ones small and lanceolate, the next longer, the innermost still larger, narrow-ovate. Petals about half as long as the inner sepals. Stamens 6; anthers globose-didymous, almost 4-lobed. *Pachygone? pubescens.*—Benth. Fl. Austr. i. 58.

Hab.: Quail Island, — *Flood* (F. v. M.)

2. **T. diffusus** (diffuse), *Miers Contrib. to Bot.* iii. 286. Branches scandent, flexuose, slender, terete, ferruginous, tomentose. Leaves 2in. long, about 1in. broad, elliptical, acutely cuneate, apex shortly acuminate, sharply or obtusely attenuated, 5-nerved from the base, nerves branching from the outside, the branches looping within the margin, puberulent, upper side shining, under side hoary glaucous, texture somewhat thick, margins revolute and tomentose; on slender pubescent petioles about 8 lines long. Panicles (male) axillary, solitary or in twos, loosely branched, corymbose, trichotomously divided, tomentose, 1 to 1¼in. long and about 1in. broad; the short secondary branchlets each bearing about three alternate pedicellate flowers; pedicels about 1½ line long, the expanded flowers about 2 lines diameter. Sepals 9, ciliate, pubescent, rotate, 3 exterior ones lanceolate, 3 intermediate ones lanceolate-oblong, acute, 3 interior ones elliptical, of equal length but twice as broad as the intermediate ones. Petals 6, interior ones one-third the length of the interior sepals, glabrous, cuneate-subtrilobed, lobes rotund, lateral ones involute. Stamens 6, a little exceeding the petals.

Hab.: In the interior, *Sir Thos. Mitchell.*

7. SARCOPETALUM, F. v. M.
(Petals fleshy.)

Sepals 2 to 5, small. Petals 3 to 6, thickly fleshy, nearly globular. Male flowers: Stamens united in a column, divided at the top into two or three short horizontal lobes, each bearing a 2-celled anther. Female flowers: Carpels 3 to 6, with recurved lobed stigmas. Drupes flattened, the remains of the style near the base. Putamen horseshoe-shaped, the sides concave. Seed horseshoe-shaped. Embryo curved, linear, in rather copious albumen; cotyledons closed; racemes simple.

The genus is limited to the following species.

1. **S. Harveyanum** (after Dr. W. H. Harvey), *F. v. M. Pl. Vict.* i. 27 *and* 221, *t. suppl.* 3; *Benth. Fl. Austr.* i. 56. A tall woody climber, with thick terete stems. Leaves broadly ovate or orbicular, acuminate or rarely obtuse, and sometimes angular or lobed, attaining 4 to 6in. in breadth, deeply cordate at the base or sometimes slightly peltate, 7 to 9-nerved, quite glabrous, on a petiole of 1 to 3in. Racemes simple, axillary or mostly lateral below the leaves, solitary or clustered, 1 to 3in. long. Bracts small. Pedicels about 1 line long. Flowers reddish-yellow, scarcely 2 lines diameter, the sepals usually shorter than the thick almost gland-like petals. Drupes 3 or 4 lines diameter, almost pear-shaped.

Hab.: Scrubs of the south.
The root-bark contains an active poisonous principle.—*Dr. T. L. Bancroft.*

8. LEICHHARDTIA, F. v. M.
(After L. Leichhardt, the explorer.)

Sepals 6, membranous, of equal length, all free, imbricate in the bud, obovate. Petals 3, somewhat fleshy, obcordate-reniform, sometimes shorter than the calyx, opposite the outer sepals. Stamens 3, all connate in a very short column,

Anthers opposite the petals, close together forming a head. Female flowers and fruit unknown.—An evergreen glabrous climber. Leaves chartaceous, ovate-lanceolate. Inflorescence in racemose, axillary and terminal panicles.

1. **L. clamboides** (Clambus-like), *F. v. M. Fragm.* x. 68. Leaves 3 to 7in. long, deep-green and very glossy on the upper surface, the apex shortly acuminate, prominently reticulate. Panicles a few or several inches long; pedicels capillare 1½ to 3 lines long, scattered or fasciculate; bracts very minute. Sepals about 1 line long, staminal column and anthers of about an equal length.

Hab.: Tropical scrubs.

9. STEPHANIA, Lour.

(Alluding to the anthers being united and forming a crown.)

(Clypea, *Blume.*)

Male flowers: Sepals 6, 8, or 10, in 2 series. Petals 3, 4, or 5, shorter than the sepals, obovate. Stamens united in a column bearing a flat disk, with the sessile anthers confluent into a single ring round the margin. Female flowers: Sepals 3, 4, or 5. Petals as many. Carpel 1, with a divided stigma. Drupe compressed, the scar of the style not far from the base. Putamen horseshoe-shaped, with an open concavity on each side. Seed curved, with little albumen. Embryo linear, with closed cotyledons. Leaves mostly peltate. Flowers in simple or compound umbels.

A small genus, extending over tropical or subtropical Africa and Asia. One of the Queensland species common over the whole range, the other endemic.

Stems glabrous or pubescent . . . . 1. *S. hernandiæfolia.*
Stems prickly . . . . 2. *S. aculeata.*

1. **S. hernandiæfolia** (Hernandia-like), *Walp.; Hook. and Thoms. Fl. Ind.* i. 196; *Benth. Fl. Austr.* i. 56. A glabrous or more or less pubescent climber. Leaves broadly ovate, orbicular, or nearly triangular, usually more or less peltate at the base, the larger ones 3 or 4in. long, on a petiole of 2 or 3in., but often much smaller, glabrous or pubescent underneath. Peduncles axillary, shorter than or rather longer than the petioles, bearing an umbel of about 5 rays, each ray terminated by a head or partial umbel of 8 to 12 small sessile or shortly pedicellate flowers, or the partial umbel again compound.—*F. v. M. Pl. Vict.* i. 220; *Clypea hernandifolia,* W. and Arn. Prod. i. 14; Wight, Ic. t. 939.

Hab.: Coast lands, south and north.

The root of this plant is bitter, and an extract of it is extremely poisonous to frogs.—*Dr. T. L. Bancroft.*

2. **S. aculeata** (prickly), *Bail.* A prickly climber, the stems ribbed, prickles reflexed, of irregular length. Leaves broadly triangular, and more or less peltate at the base, 2 to 2½in. long, and the same broad at the base, apex glandular apiculate, lower angles rounded, 5 to 7-nerved, margins entire, glabrous, pale or glaucescent on the under side. Petioles slender, armed with reflexed prickles, 1 to 1½in. long. Panicles of male flowers axillary, very slender, 2 to 4in. long, the branches almost capillary, with few small lanceolate petiolate leaves or bracts, on lateral shoots; these bracts are often larger and of a similar shape to the stem leaves. Flowers minute, mostly under 1 line in diameter when expanded. Sepals 6, imbricate, obovate, prettily veined. Petals 6, scarcely half the length of the sepals, rotundate, imbricate. Stamens united in a very short column, bearing at the summit 3 rather large didymous anthers. Female flowers and fruit unknown.

Hab.: Mount Gravatt and Taylor's Range.

This hitherto undescribed plant is probably closely allied to the tropical African species, of which also the fruit is unknown—*S. lætificata,* Miers. The flowers of that plant, however, are said to have but 3 petals.

An extract of the root extremely poisonous.—*Dr. T. L. Bancroft.*

V. MENISPERMACEÆ.

10. CISSAMPELOS, Linn.
(Fanciful resemblance to the ivy and the vine.)

Sepals 4 (5 to 6), erose. Petals 4, connate, forming a 4-lobed cup. Anthers 4, connate, encircling the top of the staminal column, bursting transversely. Female flowers: Racemose, crowded in the axils of leafy bracts. Sepals 2 (or sepal and petal 1 each), 2-nerved, adnate to the bracts. No staminodes. Ovary 1; style short, 3-fid or 8-toothed. Drupe ovoid, style-scar subbasal; endocarp horseshoe-shaped, compressed, dorsally tubercled, sides excavated. Seed curved; embro slender; cotyledons narrow, half-terete, appressed. Suberect or climbing shrub. Leaves often peltate. Male flowers cymose.

The species of this genus are met with in all hot climates.

1. C. pareira (so named under the idea that it yielded the *pareira brava* of commerce), *Linn.* A lofty climber, the branchlets rarely glabrous. Leaves 1 to 4in. diameter, orbicular-reniform or cordate, usually peltate, obtuse and mucronate, petiole long as the leaf or longer. Male cymes $\frac{1}{2}$ to $1\frac{1}{2}$in. (sometimes replaced by a shoot with small leaves and small axillary cymes), axillary or nearly so, usually 2 to 3 superposed, decompound; bracts minute, rarely foliaceous; peduncles long, slender, pubescent, tomentose or hirsute. Female racemes with large reniform bracts, 1 or 2 axillary, the bracts lax or densely imbricate, usually hoary, sometimes petiolated; pedicels very short. Ovaries rarely glabrate. Drupes 2 lines diameter, subglobose, hairy, scarlet.

Hab.: Tropical scrubs. The root of this plant is employed in India as a mild tonic and diuretic.

11. PACHYGONE, Miers.

Sepals 6 or 9, in 2 or 3 series, the inner ones larger, imbricate. Petals 6, shorter than the sepals, embracing the stamens at the base. Male flowers: Stamens 6, free, incurved at the top, anthers small, globose-didymous. Female flowers: Staminodia 6; carpels 3, with thick horizontal stigmas. Drupes reniform, the scar of the style near the base; putamen slightly excavated, with an internal process. Seed horseshoe-shaped, without albumen, cotyledons semi-terete, almost horny, the radicle very short. Leaves ovate. Flowers in racemes, the males clustered along the rhachis, the females solitary.

Leaves broad ovate or ovate-cordate, pilose on under side, 2 to 5in. long, 3in. broad . . . . 1. *P. Hullsii.*
Leaves oblong-lanceolate, 6 to 9in. long, 3in. broad, rounded and slightly peltate at the base . . . . 2. *P. longifolia.*

1. P. Hullsii (after C. Hulls), *F. r. M. Fragm.* ix. 81. A tall climber clothed with a yellowish or brownish tomentum. Leaves chartaceous, broadly ovate or ovate-cordate, about 5in. long and 3in. broad, quintuplinerved or here and there triplinerved at the base, the apex acute, sparsely or the under side densely pilose or when old nearly glabrous and shining on both sides, and the reticulation prominent, petioles about $1\frac{1}{2}$in. long. Racemes solitary or in pairs, about 5in. long; pedicels sparsely tomentose, 1 to 2 lines long, bracts subtending the flowers narrow, silky tomentose. Sepals 6, glabrous. Petals 6, inflexed, stamens 6, free. Anthers cubical-rotund, slightly didymous, bursting longitudinally. Stigma somewhat broad. Carpels of female flowers 3, sessile. Staminodes 6, slender.

Hab.: Rockhampton and in the tropical scrubs northward.

2. P. longifolia (long-leaved), *Bail. n. sp.* A strong climber, almost or quite glabrous, the branches deeply striate. Leaves oblong-lanceolate, 6 to 9in. long, and about 3in. broad at the rounded slightly peltate base; thinly coriaceous, the lateral nerves distant, except those near the base, which are somewhat crowded,

Pl. III.

F.Elliott. Lith.   F.Wills.

Pachygone longifolia, Bail.

prominent as well as the reticulation; petioles from 1¼ to 2in. long, articulated near the top, and prominently swelled at each attachment. With the leaf-bearing shoot were received 2 panicles about 6in. long, bearing rigid branches of about 1½in. long, with few pedicellate flowers (those termed panicles were old and dry and possibly may be terminal leafless shoots and the supposed branches really short racemose panicles). Drupes red, broad pyriform, the style-scar very near the base, compressed, about 1¼in. long and nearly as broad, the transverse section 8 or 9 lines. Pericarp somewhat fleshy; endocarp roughly tuberculose.

Hab.: Mourilyan Harbour, *W. Mugford*.

I should not have published the above imperfect description had not Mr. Mugford left the district, and I have no one in the locality now to collect and forward me additional specimens of the species. The leaves of the present plant closely resemble those figured by Miers in *Cont. to Bot.* iii. pl. 144 of *Spirospermum penduliflorum* (Thouars).

12. PYCNARRHENA, Miers.

(Referring to the dense fascicles of male flowers).

Sub-erect or climbing shrubs. Flowers axillary, fascicled or shortly panicled, diœcious. Male flowers: Sepals 6, with 3 bracts, inner larger, orbicular; petals 6, small-lobed; stamens 9, filaments very short; anthers sub-didymous, bursting transversely. Female flowers unknown. Drupe broadly oblong, sub-gibbous, style-scar lateral; endocarp sub-reniform. Seed slightly concave ventrally, albumen none; cotyledons oblong, half-terete, very thick, slightly incurved; radicle minute, ventral.—Hook. Fl. of Brit. Ind. i. 105.

1. **P. australiana** (Australian), *F. v. M. Vict. Nat.*, Sept., 1886. A tall climber. Leaves glabrous (as far as seen), attaining 11in. in length with a breadth of 5in., thick-chartaceous, shining on both sides, hardly paler on the back, ovate or elliptical, protracted into a short and blunt apex, distinctly penninerved and finely net-veined, on short petioles. Inflorescence axillary or lateral, with minute appressed hairs; the peduncles few or many together, rather elongated, 1 to 1¼in. long, branching cymosely about 1in. wide. Pedicels 2 lines long, or scarcely any. Sepals almost orbicular, in 3 rows, the outer 3 considerably shorter, the other 6 nearly equal, about 1 line long; the petals much smaller and almost rhomboidal. Stamens very short, the filaments nearly cuneate, united at the base. Anthers of the genus. Female flowers unknown. Carpels about ½in. long, glabrous, obliquely ovate-globular, on often very short stipes; exocarp somewhat fleshy, endocarp thinly cartilaginous. Seeds obliquely ovate, about 4 lines long. Cotyledons very convex outward.—F. v. M., l.c.

Hab.: Endeavour River and near Trinity Bay.

The above species, Baron Mueller says, differs from *P. pleniflora* in shorter petioles, larger leaves, long peduncles, more distinctly pedicellate flowers, different proportionate size of sepals, and rather larger carpels; from *P. tumefacta* in leaves also dark-green underneath, not distinctly dilated petioles, 6 inner sepals, and perhaps also in fruit, but the disposition of the flowers is similar; from *P. lucida* and *P. manillensis*, the congener is far more removed. *P. novaguinensis*, as yet imperfectly known, is in some respects allied to the Queensland species.—F. v. M., l.c.

13. PLEOGYNE, Miers.

(Stigmas numerous.)

Outer sepals about 6, very small, 3 inner ones much larger, valvate in the bud, connivent at the base and recurved at the top when open. Petals 6, much shorter, the margins dilated and involute. Male flowers: Stamens 3; filaments linear-terete; anthers small, globose-didymous. Female flowers with 6 carpels (*Miers*). Drupes 3 to 6, reniform, with the scar of the style lateral, the putamen not excavated on the sides, nor with any internal process: Seed reniform, without albumen; cotyledons thick and fleshy, scarcely separable; radicle scarcely distinct.—Flowers in short axillary branching panicles.

Bentham remarks that this genus is distinguished from all. except the African *Triclisia*, by the remarkably valvate inner sepals.

1. **P. australis** (Australian), *Benth. Fl. Austr.* i. 59. A climber, with a soft pubescence like that of *Pericampylus*, sometimes very copious, sometimes quite disappearing from the upper surface of the leaves. Leaves from ovate to oblong, obtuse or scarcely acute, the large ones 8 to 4in. long, rounded but not cordate at the base, at length rather coriaceous and shining above, reticulate penninerved. Males cymes or single flowers in little axillary solitary or clustered panicles, seldom above 1in. long and softly pubescent; inner sepals about 1 line long, the outer ones very minute. Female inflorescence probably more simple. Drupes about 5 lines broad, glabrous, with a very thin endocarp.— *Microclisia*, Benth. in Benth. and Hook. Gen. Pl. i. addend.

Hab.: Coastal scrubs.

14. **HUSEMANNIA**, F. v. M. in Wing's So. Sc. Record, iii. 187.

(After Dr. Theodor Husemann, of Goettingen.)

Sepals 9, in 3 series; the 2 outer ones very minute, the 8 inner ones much longer, ovate-roundish, valvate and slightly induplicate in the bud. Petals very minute, flat, bilobed at the summit, much contracted at the base. Stamens of male flowers 6, free except at the base; filaments thickened upwards; anthers nearly globular but somewhat didymous, their cells opening by anterior almost semicircular slits; the connective narrow and not produced beyond the cells. Ovaries of female flowers 6; stigma of each awl-shaped, recurved, undivided, finally becoming nearly basal. Carpels on a conspicuous stipe, oblique-ovate, somewhat impressed on both sides, rather acutely margined; pericarp almost coriaceous; internal process erect, thin, flat, extending to somewhat beyond the middle of the cavity. Seeds nearly cylindrical, conduplicate by horseshoe-like curvature, no albumen, integument smooth; cotyledons for the greater part of their length turned dorsally towards the pericarp; radicle extremely short. A tall climber, with large, almost ovate, somewhat pointed and stiff leaves. Flowers in spicate-paniculate clusters of very small, dark, silky-hairy flowers, with short stamens and stipitate carpels.

1. **H. protensa** (referring to its extending habit), *F. v. M.* A tall climber. Leaves sometimes 15in. long and over 6in. in width, but often much smaller, glabrous and shining, distantly penninerved, the closely reticulate veins prominent on the under side. Petioles $\frac{1}{2}$ to 3in. long, thickened and velvety towards the top. Panicles about 15in. long, and thinly velvety. Flowers scarcely $\frac{1}{4}$in. long, the inside of the sepals and both sides of the petals glabrous. Carpels about $\frac{3}{8}$in. long, thinly brown-velvety, stipe about $\frac{1}{4}$in.—F. v. M. in Wing's So. Sc. Rec.

Hab.: The tropical scrubs.

Order VI. **NYMPHÆACEÆ**.

Sepals 3 to 5, petals 3 or more and stamens 6 or more, either all free and hypogynous, or the inner ones or all adnate at the base to the torus or ovary, or inserted on its summit. Anthers innate or adnate, the cells opening in longitudinal slits. Gynœcium of 3 or more carpels, either free and distinct, or immersed in the torus so as to form a several-celled ovary. Styles or stigmas

VI. NYMPHÆACEÆ.

free or adnate on an epigynous disk. Ovules solitary, and suspended from the apex of the cavity, or indefinite and attached to the sides of the cavity, not to its inner angle. Ripe carpels indehiscent, free or united in a fleshy or spongy fruit. Seeds immersed in a fleshy or pulpous arillus, or naked, the embryo either small, enclosed in the embryo-sac and half immersed in a cavity of a farinaceous albumen near the hilum, or without albumen, large, with thick fleshy cotyledons, and a remarkably developed plumule.—Aquatic herbs, with a submerged root or rhizome. Leaves carried by their long petioles to the surface of the water or raised above it, usually peltate or deeply cordate, or a few remaining under water and deeply cut. Flowers growing singly on long radical scapes or axillary peduncles, either on the surface of the water or raised above it.

The Order, although not numerous in species, is found in pure, quiet, or slowly-flowing waters nearly all over the globe. The three Australian species belong to the three genera considered as typical of as many tribes or sub-orders, raised by some botanists to the rank of distinct Orders. All three genera are common to the New and the Old World. They are absent, however, from the southern Australian colonies as well as from New Zealand.—*Benth.*

SUBORDER I. **Cabombeæ.**—*Sepals and petals 3 each, free. Carpels free. Ovules few. Seeds albuminous.*

Sepals and petals 3 each. Carpels 6 or more, free, on a small torus. Ovules few. Flowers small . . . . 1. BRASENIA.

SUBORDER II. **Nymphæa.**—*Sepals 4 to 6. Petals and stamens indefinite. Carpels confluent with one another or with the disk into one ovary. Ovules many. Seeds albuminous.*

Sepals 4 to 6. Petals and stamens numerous, the outer ones free, the inner more and more adnate to the torus. Carpels immersed in the torus in a ring round a central conical projection . . . . 2. NYMPHÆA.

SUBORDER III. **Nelumbieæ.**—*Sepals 4 to 5. Petals and stamens indefinite. Carpels irregular, scattered, sunk in pits of the turbinate disk. Ovules 1 to 2. Seeds albuminous.*

Sepals 4 or 5. Petals and stamens numerous, hypogynous. Carpels half immersed without order in the flat top of the torus. No albumen . . . . 3. NELUMBIUM.

1. BRASENIA, Schreb.

(Its name in Guiana.)

(Hydropeltis, *Mich.*)

Sepals 3, petal-like, and petals 3, hypogynous. Stamens 12 to 18, hypogynous; filaments subulate, anther-cells lateral. Carpels 6 to 18, free, on a small torus, attenuate at the top into short styles, stigmatic along the inner edge. Ovules 2 or 3, pendulous from the dorsal side of the cavity. Ripe carpels coriaceous, indehiscent. Seeds albuminous.

The genus is limited to the following species.

1. **B. peltata** (peltate), *Pursh. Fl. N. Amer.* 889; *Benth. Fl. Austr.* i. 60. Water shield. Rhizome prostrate at the bottom of the water. Stems forked, leafy, covered as well as other submerged parts, especially when young, with a thick coating of transparent jelly. Leaves floating on the surface of the water, peltately attached by their centre to long petioles, oval, entire, 3 to 4in. long and about half as broad. Peduncles axillary, bearing solitary flowers of a dull purple on the surface of the water. Sepals and petals very much alike, about 4 or 5 lines long when they first open, but lengthening to 7 or 8 lines. Carpels shorter. —A. Gray, Gen. Ill. t. 39.; *Hydropeltis purpurea*, Mich. Bot. Mag. t. 1147.

Hab.: Queensland lagoons. This species is also found in North America and the East Indies.

VI. NYMPHÆACEÆ.

2. NYMPHÆA, Linn.

(From *Nymphe*, a water nymph.)

Sepals 4, inserted near the base of the torus. Petals numerous, passing gradually from the sepals to the stamens, inserted on the torus or ovary, the outer petals near the base, the inner stamens almost at the top. Filaments of the outer stamens dilated and petal-like, with small lateral anther-cells, of the inner ones narrow or filiform, with longer anthers opening inwards. Carpels several, immersed in a ring in the fleshy torus, having the appearance of a several-celled ovary, with a conical or globular process in the centre. Styles thick, radiating, free or united at the base, often with an incurved appendage beyond the stigmatic portion. Ovules numerous, pendulous from the sides of the cavity. Fruit a spongy berry, breaking up irregularly when ripe. Seeds embedded in pulp, arillate, albuminous. Rhizome perennial. Leaves floating, peltate or very deeply cordate. Flowers large, solitary, floating on the surface of the water or slightly raised above it, on long radical peduncles.

The species of this genus are met with in most temperate and tropical regions.

Rhizome globose, rather large. Leaves sharply toothed, flowers white, petals obtuse, stamens without appendages beyond the anther . . . . 1. *N. lotus.*

Rhizome globose, rather large. Leaves more or less toothed, often 18in. in diameter, with much raised reticulations underneath. Flowers often 10 or 12in. across, anther appendage very short or none . . . . 2. *N. gigantea.*

Rhizome globose, small. Leaves usually quite entire, the reticulation on the underside not raised. Anther appendage very short or none, petals more acute than in *N. gigantea* . . . . 3. *N. Brownii.*

Rhizome not preserved. Leaves quite entire, cordate, 1 to 3in. long; flowers white, often stained with purple, sepals and petals acute. Expanded flowers about 1½ to 2in. diameter . . . . 4. *N. tetragona.*

Rhizome erect, prominently tuberculose. Leaves broadly oblong, often purplish on the underside. Flowers yellow . . . . 5. *N. flava.*

1. **N. lotus,** var. *australis* (Australian), *Syn. Ql. Flora* 10. This plant is of more compact growth and taller than other Australian species. Leaves in shape and size much like those of *N. gigantea*, thick, somewhat spongy on the under surface, where the reticulation is prominent, but the veins are not so much raised as in *N. gigantea*, margins bearing distant bristle-like teeth. Flowers 4 or 5in. diameter, very fragrant, often raised some distance above the water, white, but sometimes the sepals and outer petals tinged with pink or blue. Petals usually more or less obtuse. Anthers without appendages, except the somewhat subulate points be considered as such.

Hab.: Still waters off the Barron River.

2. **N. gigantea** (large), *Hook. Bot. Mag.* t. 4647; *Benth. Fl. Austr.* i. 61. Blue water lily ("Arnurna," Mitchell, *Palmer*; "Yako-kalo," Rockhampton, *Thozet*; "Kaooroo," Cleveland Bay, *Thozet*; "Moi-u," Brisbane; "Thindah" (root), "Thoolambool" (stalk), "Mille" (seedhead), Cloncurry (Mycoolon tribe), *Palmer*; "Thoongoon" (root), "Urgullathy" (seedstalk), "Irrpo" (seedhead), Mitchell, *Palmer*). Leaves orbicular or very broadly ovate, very deeply cordate, the early leaves as well as the flowers much smaller than the ultimate ones, but always with prominently raised reticulation on the under side, margins entire or with distant short teeth. Flowers blue, 5 to 12in. across; petals and stamens numerous, filaments filiform or the outer ones flattened, but narrowed under the anther; connective scarcely projecting beyond the cells. Stigmas thick, radiating, united at the base, either without any or with only a very short terminal appendage.

Hab.: Waters of the southern and northern parts of the colony. The tubers after preparation, the seedheads, and also the flower-stalks of the unexpanded flowers after being broken and deprived of their fibrous part, are eaten by natives.—*Thozet, Palmer.*

8. **N. Brownii** (after Dr. R. Brown), *Bail.* The small blue water lily of our tropics. Leaves smaller but much the shape of those of *N. gigantea*, margins slightly repand, entire, texture thin and the reticulation on the underside never raised. Flowers blue, 4 or 5in. diameter. Sepals spotted and as well as the petals often somewhat acute. Stamens numerous, filaments flattened. Anthers with very short or no terminal appendage. From note in Flora Australiensis i. 61, it seems that Mr. Bentham concurred with some other botanists in considering this northern plant only a form of *N. gigantea*. Baron Mueller gave it in his publications as *N. stellata*, and the author of the Flora Australiensis says that Robert Brown was in favour of placing it under this name. In the synopsis of the Queensland Flora I recorded it as *N. cærulea*, considering it agreeing best with the species so named, which at the present time is merged by botanists into *N. stellata*. It will be seen to differ from *N. stellata* in the want of prominent anther-appendages and from *N. gigantea* in the absence of the prominent raised reticulation on the underside of the leaves. Therefore it is here given as a species bearing the name of the botanist who probably first recorded it.

Hab.: Waters of the northern parts of the colony.

4. **N. tetragona** (four-angled), *Georgi*. A small plant with oblong-orbicular leaves 1 to 3in. long, the basal-lobes acute, sometimes spreading; petioles smooth, slender. Flowers white or more or less stained with purple. Sepals lanceolate, about 1in. long, the petals somewhat longer. Anther without appendages, but the cells not extending to the end of the obtuse apex. Seeds somewhat flask-shaped, bearing many longitudinal ciliate ribs.—*N. pygmæa*, Ait. Hook. Fl. Brit. Ind. i. 115; *N. minima*, Bail. Syn. Ql. Flora 10.

Hab.: Still shallow waters off the Barron River. The rootstock not preserved for examination; in the Indian specimens said to be woolly, with soft black hairs.

5. **N. flava** (yellow), *Leit. Hook. in Bot. Mag.* t. 6917. Rhizome erect, 7 or more inches long and often more than 1¼in. thick, having the appearance of being prominently tuberculous from the swelled persistent bases of fallen leaves; from near the crown are emitted long white stout fleshy running stems, which form fresh plants at distant intervals. Leaves orbicular, elliptic or broadly oblong, 6in. or more in diameter, on pale-green petioles; upper side deep-green, the under side more or less of a purplish colour, margins entire, slightly undulate, sinus narrow, basal lobes subacute, the nerves and veins not prominent on either side. Flowers pale yellow, about 4in. in diameter, each bloom remaining perfect for several days. Sepals lanceolate, more or less tinged on the back with purplish-red. Petals similar in shape but smaller than the sepals, the inner ones smaller than the outer, pale-yellow on both sides. Stamens numerous, sub-erect, outer filaments much the longest and dilated at the centre, inner filaments linear. Anthers linear, connective hardly produced, tips rounded, cells parallel. Stigmatic-rays 8, short, broad, obtuse incurved, inappendiculate. Berry globose, about 1in. in diameter, almost white, marked with short transverse scars. Seeds globose-oval, 3 lines long, silky-villous.

Besides Sir J. D. Hooker's account in Bot. Mag. l.c., there are two accounts with figures in The Garden xxiii. and xxvii.

Hab.: This beautiful Florida water-lily was introduced some 14 or 15 years ago (without name) by the Queensland Acclimatisation Society; soon after which it was planted out in one of the ponds at Bowen Park. It soon took possession of the pond, and is now said to have become quite naturalised in the still waters about Wellington Point.—*Colin Kifford*. According to Hook. l.c., this species was first figured in an American work on Ornithology about the year 1832. The present description refers to the plants growing in Queensland waters.

VI. NYMPHÆCEÆ.

8. NELUMBIUM, Juss.

(From Indian name *Nelumbo*.)

Sepals 4 or 5, free. Petals and stamens numerous, hypogynous. Anthers opening inwards, the connective produced in a club-shaped appendage. Carpels several, half-immersed in the flat top of an obconical torus, the styles shortly projecting, with somewhat dilated terminal stigmas. Ovules 1 or 2 in each carpel, suspended from the top of the cavity with a dorsal raphe. Nuts globose-oval, shortly protruding from the cells of the large flat-topped torus. Seeds with a spongy testa, without albumen; cotyledons thick and fleshy, enclosing a much-developed plumula; radicle very short. Leaves peltate, supported above the water on erect petioles. Flowers solitary, on erect scapes above the water.

Besides the following Asiatic and Australian species there is a second one from the West Indies.

1. **N. speciosum** (showy), *Willd.*: *Wight Ill. l. t. 9; Benth. Fl. Austr.* i. 62. Pink water lily, sacred lotus. *Aquaie*, N. Queensland, *Thozet*. An erect large water herb with a milky juice; rootstock stout, creeping. Leaves raised high above the water, orbicular, peltate, somewhat concave, 1 to 2ft. diameter, quite entire or slightly sinuate, glabrous and often somewhat glaucous. Peduncles and petioles 3 to 6ft. high, full of spiral vessels, smooth or more often bearing scattered prickles. Flowers pink, 4 to 8in. or more across, appendages to the anthers linear-clubshaped. Fruiting torus resembling a wasp's nest, 2 to 4in. diameter; the nuts oblong or roundish, about ½in. long, nearly black.

Hab.: Waters of northern Queensland, where the roots and seeds are eaten by the aborigines, as they were by the Egyptians and are by the native population of India at the present time.

Order VII. PAPAVERACEÆ.

Flowers hermaphrodite, regular, or in *Fumarieæ* irregular. Sepals 2 or 3, rarely 4, free, imbricate, very caducous. Petals 4, 6, or rarely 8 or 12, hypogynous, free, imbricate, and often crumpled in the bud, in 2 rarely 3 series, deciduous. Stamens hypogynous, indefinite, and free, or in *Fumarieæ* definite, with the filaments usually united. Anthers erect, the cells opening longitudinally. Ovary free, either 1-celled with parietal placentas often protruding into the cavity, or rarely completely several-celled by the placentas meeting in the axis, or 2-celled by a false dissepiment connecting 2 parietal placentas. Style short or none; stigmas as many as placentas, usually confluent and radiating on the disk-like or dilated top of the ovary or style. Ovules indefinite, anatropous, ascending with an inferior micropyle or horizontal. Fruit capsular, usually opening in pores or valves. Seeds globular or subreniform. Embryo minute, at the base of a fleshy albumen.—Herbs or rarely small shrubs, glabrous and often glaucous or hispid, the juice usually coloured. Leaves alternate or the floral ones almost opposite, entire, lobed or dissected without stipules. Flowers usually solitary on long peduncles, either terminal or in the upper axils.

The Order belongs almost entirely to the temperate or subtropical regions of the northern hemisphere.

Suborder I. **Papavereæ.**—*Petals alike. Stamens numerous. Capsule usually short, opening by short valves or pores.*

Stigmas 4 or more, radiating on a sessile disk . . . . 1. Papaver.
Stigmas 4—6, radiating from the top of a depressed style . . . . 2. Argemone.

Suborder II. **Fumarieæ.**—*Inner and outer petals dissimilar. Stamens definite.*

Stamens 6, diadelphous, outer petals spurred. Fruit indehiscent, 1-seeded . . 3. Fumaria.

VII. PAPAVERACEÆ.

1. PAPAVER, Linn.
(The old Latin name.)

Sepals 2, rarely 3. Petals 4, rarely 6. Stamens indefinite. Placentas of the ovary 4 or more, covered with ovules and projecting more or less into the cavity, rarely meeting in the centre; stigmas radiating on the convex or almost conical disk-like summit of the ovary. Capsule opening in transverse pores between the placentas under the disk, with very short opercular valves. Seeds furrowed.— Herbs with a milky juice. Leaves usually lobed or cut. Peduncles long, the buds nodding.

Capsule ovoid-oblong, smooth; stigmatic rays 6 to 8 . . . . 1. *P. horridum*.
Capsule subglobose, glabrous; stigmatic rays 8 to 12 . . . . 2. *P. rhæas*.

1. **P. horridum** (horrid), *DC. Benth. Flora Austr.* i. 68. An erect annual, beset with subulate prickles or stiff bristles, but otherwise glabrous and usually glaucous. Leaves narrow-oblong or lanceolate, irregularly pinnatifid and coarsely toothed, the radical ones contracted into a petiole, the stem ones sessile or partially stem-clasping. Flowers small for the genus, of a pale brick or red colour. Sepals hispid. Petals nearly ovate, about ½in. long. Capsule ovoid-oblong, perfectly smooth and glabrous, the terminal disk at first pyramidal, at length nearly flat, usually with 6, 7, or 8 stigmatic rays. Placentas as many, projecting in the cavity but not meeting in the centre.—*P. gariepinum*, DC. Bot. Mag. t. 8623.

Hab.: Southern parts of the colony, also in extratropical South Africa.

Dr. T. L. Bancroft states that this plant does not contain morphine, but an active principle quite as poisonous as morphine.

2. **P. *rhæas** (Greek name for wild poppy), Linn. The common corn poppy. A branching hispid annual. Leaves 1—2-pinnatifid, with the lobes more or less cut, ascending, awned. Scapes with spreading or appressed hairs. Flowers 3—4in. diameter, scarlet, the pairs of petals unequal; filaments filiform. Stigma convex, rays overlapping, 8 to 12. Capsule subglobose, glabrous.

Hab.: Only met with as a stray from garden culture. A plant of Europe met with also in W. Asia, N. Africa.

The petals readily impart their red color to water. The milky juice possesses sedative action.

2. *ARGEMONE, Linn.
(Named from supposed medicinal properties.)

Sepals 2 or 3. Petals 4—6. Stamens indefinite. Ovary 1-celled; style very short, stigma 4—7 lobed; ovules numerous, on 4—7 parietal placentas. Capsule short, dehiscing at the top by short valves that alternate with the stigmas and placentas. Seeds many. An erect prickly herb; juice yellow. Flowers yellow or white, showy.

1. **A. mexicana** (Mexican), Linn. *Figo del Inferno* of the Spaniards; prickly poppy. A robust plant almost woody at the base, 2 to 4ft. high, with spreading branches and sessile half-amplexicant sinuate-pinnatifid variegated green and white prickly leaves. Flowers 2 or 3in. diameter. Sepals horned at the top. Capsule about 1in. long, terete, usually bristly, elliptic or oblong.

Hab.: An American weed naturalised in many parts of the colony. Some years ago this was the one met with about Brisbane, but at present the one more generally met with is the variety *ochroleuca*.

VII. PAPAVERACEÆ.

8. *FUMARIA, Linn.

(From the smoky odour of the plants).

Sepals 2, small. Petals 4, erect or conniving; 2 outer ones dissimilar, anterior flat or concave, posterior gibbous or spurred at the base; 2 inner clawed, tips free or cohering, keeled. Stamens 6, diadelphous; posterior bundle with a basal spur enclosed in the petal-spur; mid-anther of each bundle 2-celled, lateral 6-celled. Ovary 1-celled; style filiform, stigma entire or shortly lobed; ovules 2, on 2 placentas. Fruit indehiscent, globose, 1-seeded.—Annual, rarely perennial herbs, usually branched, often scandent. Leaves much divided, segments very narrow. Flowers small, white, rose-coloured, or purplish, in terminal or leaf-opposed racemes. Generally to be met with as weeds in cultivation plots of the temperate regions of the old world.

1. **F. parviflora** (small-flowered), *Lam*. The common small-flowered fumitory. Plant diffused, pale-green, much branched. Leaves with flat segments. Racemes 1 to 2in. long, flowers white or rose-coloured, with purple tips, $\frac{1}{4}$ to $\frac{1}{3}$in.; sepals lanceolate, much smaller than the corolla-tube, pedicels exceeding the bracts. Fruit globose, rugose, dry, rounded at the top, with 2 pits.

This and some other species are met with as weeds of cultivation.

Order VIII. CRUCIFERÆ.

Flowers hermaphrodite, regular, or with the outer petals larger. Sepals 4, free, imbricate in 2 series, the outer ones often saccate at the base. Petals 4, rarely wanting, the laminæ spreading in the form of a cross; torus usually bearing 4 glands opposite the sepals. Stamens usually 6, of which 2 outer ones shorter or rarely wanting, 4 inner ones longer, in pairs alternating with the outer ones. Anthers 2-celled, attached by the base. Ovary 1-celled, with two parietal placentas or rarely a single one, or more frequently divided into two cells by a thin membranous septum connecting the two parietal placentas. Style simple, often very short or none; stigmas 2, erect, or divaricate, or united into a single capitate or minute stigma. Ovules 1, 2, or more in each cell, horizontal or pendulous from the parietal placenta. Fruit a pod, either long and narrow, and then called a *siliqua*, or short and broad, called a *silicule*, usually 2-celled, each cell opening by a deciduous valve, leaving persistent the thin septum surrounded by the nerve-like placentas, which form a rim called the *replum*; exceptionally the pod is 1-seeded and indehiscent, or separating into 2 indehiscent cocci or into 2 or more bead-like articles. Seeds attached in each cell in 2 rows, one proceeding from each edge of the septum, but when each seed is as broad as the cell they overlap each other, so as to appear to be and to be described as in a single row; testa cellular, sometimes winged, often exuding when soaked a thick coat of mucilage. Albumen usually none. Embryo usually curved, the cotyledons plano-convex with the radicle curved against their edge, when they are said to be *accumbent*, or over the back of one of them, when they are *incumbent*: in the latter case they are either flat or more or less folded over the radicle, or *conduplicate*.—Herbs or rarely undershrubs, without milky juice. Hairs simple, stellate or attached by the centre. Leaves simple, usually alternate, entire, lobed or pinnately divided, the radical ones often lyrate and the stem ones auricled. Stipules none. Flowers usually in terminal racemes, which are at first corymbose but lengthen out as the fruiting advances, and usually without bracts.

Cruciferæ form a very large Order, dispersed over nearly the whole globe, but most abundant in the temperate and cold regions of the northern hemisphere. They are rare within the tropics,

VIII. CRUCIFERÆ.

especially in districts where there are no high mountain ranges. The Order is one of the most easily recognised by the flowers or fruits, but to determine the genera and species it is absolutely necessary to have the pod and the seed in a good state.—*Benth.*

Series A.
Pods long or short, dehiscing throughout their length, terete, 4-angled or compressed dorsally (parallel to the septum).

Tribe I. **Arabideæ.**—*Pods narrow, long. Seeds usually 1-seriate. Cotyledons accumbent.*
Sepals spreading, not saccate; pods tumid. Seeds minute, 2-seriate.
Flowers usually yellow . . . . 1. Nasturtium.
Sepals not saccate; pods flat, usually acute. Stamens simple. Flowers white or purple . . . . 2. Cardamine.

Tribe II. **Abyssineæ.**—*Pods short, broad. Seeds usually 2-seriate. Cotyledons accumbent.*
Stamens often appendaged. Pods usually orbicular and 4-seeded. Hoary herbs . . . . 3. Alyssum.

Tribe III. **Sisymbrieæ.**—*Pods usually sessile, long, narrow. Seeds usually 1-seriate. Cotyledons straight, flat, incumbent.*
Sepals erect or spreading. Pods many-seeded, valves 1—3-nerved. Seeds usually 1-seriate. Hairs simple or none . . . . Sisymbrium.

Tribe IV. **Camelineæ.**—*Pods short or long. Seeds usually 2-seriate. Cotyledons flat, incumbent.*
Fruiting racemes erect. Petals obovate, or if narrow erect and short.
Septum broader than the transverse diameter of the pod . . . . 5. Blennodia.
Fruiting racemes erect. Petals tapering into a long subulate, often twisted, point . . . . 6. Stenopetalum.
Fruiting peduncle recurved, pod ripening underground . . . . . . 7. Geococcus.

Tribe V. **Brassiceæ.**—*Pods short or long. Cotyledons longitudinally folded or deeply grooved.*
Pods long. Seeds 1-seriate . . . . 8. Brassica.

Series B.
Pods short, dehiscing throughout their length, compressed laterally (at right angles to the septum).

Tribe VI. **Lepidineæ.**—*Cotyledons incumbent, straight, curved, or longitudinally folded.*
Pods many-seeded; valves not winged . . . . 9. Capsella.
Pods either indehiscent or separating into 2 indehiscent cocci . . . 10. Senebiera.
Pods few-seeded; valves winged or not, dehiscent . . . . . . 11. Lepidium.

Tribe VII. **Thlaspideæ.**—*Cotyledons accumbent, straight.*
Pods compressed, notched; valves winged or keeled . . . . . . 12. Thlaspi.

1. NASTURTIUM, R. Br.
(Very old name for cress plant.)

Sepals short, equal, spreading. Petals scarcely clawed. Pods nearly cylindrical, short or elongated, the valves convex, slightly 1-nerved, the septum transparent; style short or long, with an entire or 2-lobed stigma. Seeds usually distinctly ranged in 2 rows, small, turgid, with short free funicles. Cotyledons accumbent.—Herbs, either glabrous or pubescent, with simple hairs. Leaves entire, lobed, or pinnately divided. Flowers small, generally yellow.

A considerable genus, dispersed over the greater part of the globe, and very difficult both as to the discrimination of its species and as to its distinction from other genera. The Australian species is one of the most widely diffused.—*Benth.*

Flowers yellow . . . . 1. *N. palustre.*
Flowers white. Half-aquatic perennial. Petals obovate . . . . 2. *N. officinale.*

1. **N. palustre** (found in marshy places), *DC. Syst. Veg.* ii. 191; *Benth. Fl. Austr.* i. 65. An erect or decumbent or almost trailing annual or biennial, from a few inches to 2ft. or more in length, quite glabrous or very rarely pubescent.

44 VIII. CRUCIFERÆ. [*Nasturtium*.

Leaves toothed or pinnately lobed, or the lower ones sometimes lyrate, auriculate at the base, the lobes ovate, oblong, or rarely lanceolate, always irregular, confluent and usually sinuate or toothed. Racemes short, loose, without bracts. Flowers small, yellow, the petals scarcely exceeding the calyx. Style short. Pod sessile, turgid, oblong, obtuse, straight or slightly curved, generally 2 to 4 lines long and about 1½ line broad, but occasionally rather longer and narrower.—Reichb. Ic. Fl. Germ. ii. 53; *N. terrestre*, R. Br. in Ait. Hort. Kew. ed. 2, iv. 110; Hook. f. Fl. Tasm. i. 21; F. v. M. Fl. Vict. i. 31; *N. semipinnatifidum*, Hook. Journ. Bot. i. 246.

Hab.. Many parts of the colony, north and south.

2. **N. *officinale** (officinal), *R. Br.; Benth. Fl. Austr.* i. 65. European watercress. Stems creeping and floating, much branched. Leaves pinnate, the upper with 3 to 7 pairs of leaflets and a terminal one which is usually larger, varying from roundish to ovate or lanceolate, obtuse, sinuate, or dentate. Flowers white, small, in short racemes. Petals longer than the sepals. Pods 4 to 12 lines long, stalked, spreading or curved upwards. Seeds small, 2-seriate.

Hab.: Naturalised in many mountain streams.

2. CARDAMINE, Linn.

(Named from supposed medicinal qualities.)

Sepals equal at the base. Petals clawed. Pod elongated, linear, compressed; valves usually flat, without conspicuous nerves, opening elastically; septum transparent; style short or long; stigma entire or 2-lobed. Seeds flattened, not bordered, in a single row (except in *C. eustylis*).—Herbs, usually flaccid and glabrous. Leaves entire or more frequently pinnately divided, in a few species not Australian opposite or whorled. Flowers erect or nodding, white, purple, or lilac, not yellow. Pods usually slender.

A large genus, widely spread over the temperate and colder regions both of the northern and southern hemisphere.—*Benth.*

Seeds reticulate and pitted, rather large.
    Leaves entire or sinuate-toothed, the stem ones sagittate. Plant of 2 to 5ft. 1. *C. stylosa*.
    Petals very narrow, small, nearly erect.
        Seeds nearly the breadth of the septum, in a single row . . . . 2. *C. hirsuta*.
        Seeds numerous, small, almost biseriate. Valves of the pod convex . . 3. *C. eustylis*.

1. **C. stylosa** (style prominent), *DC. Syst. Veg.* ii. 248; *Benth. Fl. Austr.* i. 68. A rather coarse glabrous herb, branching and decumbent, or nearly erect, usually 2 to 3ft. high and sometimes attaining 5ft. Leaves oblong-lanceolate, entire or sinuate, and minutely but remotely toothed, the lower ones narrowed into a long petiole, the upper ones sessile but narrow below the middle and clasping the stem by their sagittate base, the longest 3 to 5in. long. Flowers small, white, with obovate spreading petals. Fruiting racemes long and rather rigid, the pedicels very spreading, 3 to 4 lines long. Pods 1 to 1½in. long and ¾ to 1 line broad, with a very faint nerve on the valves. Seeds oval, dark-coloured, reticulated with raised longitudinal nerves and transverse pits between them.—Hook. f. Fl. Tasm. i. 18; F. v. M. Pl. Vict. i. 34; *Arabis gigantea*, Hook. Ic. t. 259; *C. divaricata*, Hook. f. Fl. Nov. Z. i. 18.

Hab.: On ranges southern parts of the colony.

2. **C. hirsuta** (Hairy), *Linn.; DC. Prod.* i. 152; *Benth. Fl. Austr.* i. 70. A much-branched decumbent or tufted annual, seldom above 6in. high, either quite glabrous or slightly hirsute with short spreading hairs. Leaves pinnately divided, the lower ones with 1 ovate or rounded terminal segment and a few smaller petiolulate lateral ones, or sometimes reduced to the terminal one, the upper leaves few with narrow lobes. Flowers very small, the petals narrow and erect or scarcely spreading. Stamens often reduced to 4 (especially in European

specimens). Fruiting racemes usually short and rather dense, the pedicels not very spreading. Pods erect, slender, usually 7 to 9 lines long, and scarcely more than ½ line broad, the stigma sessile or on a style not longer than the breadth of the pod. Seeds smooth, as broad as the septum, and in a single row as in all the preceding species.—Reichb. Ic. Fl. Germ. ii. t. 26; Hook. f. Fl. Tasm. i. 20; *C. parviflora*, Linn.; DC. Prod. i. 152; also F. v. M. Pl. Vict. i. 36, partly; *C. debilis*, Banks in DC. Syst. Veg. ii. 265; *C. paucijuga*, Turcz. in Bull. Mosc. 1854 ii. 295.

Hab.: Not uncommon damp land in the southern parts of the colony.

3. **C. eustylis** (style prominent on fruit), *F. v. M. in Trans. Vict. Inst.* i. 114; *Pl. Vict.* i. 37; *Benth. Fl. Austr.* i. 70. An erect annual, much branched from the base, scarcely exceeding 6 to 8in. in height and quite glabrous. Leaves pinnately divided, the lower ones with ovate segments, the others with narrower ones, all usually with a few teeth or lobes. Flowers smaller than in *C. hirsuta*, the petals narrow, erect, and scarcely exceeding the calyx. Fruiting racemes short, leafless. Pods rather spreading, slender, 6 to 9 lines long, tipped by a style of ½ to near 1 line, the valves convex, smooth, without nerves. Seeds very numerous and small, much narrower than the septum, and showing 2 distinct rows.

Hab.: River banks in tropical parts of the colony.

3. ALYSSUM, Linn.

(Supposed by the ancients to have the power of allaying anger.)

(Meniocus, *Desv.*)

Sepals rather short, equal at the base. Petals rather short, entire or bifid. Stamens often bearing a tooth or small appendage on the filaments of some or all of them. Pod short, from nearly orbicular to oblong, very flat or turgid; the valves flat, concave, or turgid in the centre and flat on the margins, the septum membranous; style short or long, with an entire stigma. Seeds 2 to 10 in each cell. Cotyledons accumbent.—Branching herbs or small shrubs, usually hoary with stellate tomentum. Leaves undivided, usually linear. Racemes without bracts, with white or yellow flowers.

A large genus, dispersed over the temperate regions of the Old World, but chiefly in the Mediterranean region and western Asia. None are found in America, eastern Asia, or in the Pacific Islands. The only Australian species is identical with one common in the eastern Mediterranean region.—*Benth.*

1. **A. linifolium** (flax-leaved), *Steph. in Willd. Spec. Pl.* iii. 467; *Benth. Fl. Austr.* i. 71. A small, but hard, wiry, and much-branched erect annual, hoary, with a minute, close, stellate tomentum. Leaves linear, oblong, oblong-spathulate or almost obovate, mostly under ½in., but the longest sometimes nearly 1in. long, quite entire. Flowers white, very small. Pods orbicular or broadly ovate, 2 to 3 lines long, minutely hoary; the valves flat and without nerves; style small, subulate. Seeds 4 to 6 in each cell.—*Meniocus linifolius*, DC. Syst. Veg. ii. 325; Deless.' Ic. Sel. ii. t. 42; *M. Serpyllifolius*, Desv.; DC. l.c.; *M. australasicus*, Turcz. in Bull. Mosc. 1854, ii. 297.

Hab.: Inland downs country.

4. SISYMBRIUM, Linn.

(Name given by Greeks to some fragrant plant; not at present day recognised).

Sepals equal or the lateral ones slightly saccate. Petals usually elongated, with long claws. Pod linear-elongated, cylindrical or flattened, several-seeded, the valves usually convex and 3-nerved; septum membranous; style usually short, with an entire or slightly 2-lobed stigma. Seeds in a single row, not bordered, oblong, with filiform funicles. Cotyledons incumbent.—Herbs, usually annual or biennial, glabrous, hirsute or tomentose. Leaves entire or pinnately lobed or divided. Flowers yellow, or rarely white or pink.

A large genus, chiefly European and Asiatic, with a few North American and a very few Antarctic species.—*Benth.*

*1. **S. officinale** (officinal), *Scop.; DC. Prod.* i. 191; *Benth. Fl. Austr.* i. 71. An erect annual, more or less pubescent, a foot high or rather more, with very rigid spreading branches. Leaves deeply pinnatifid, with few lanceolate slightly toothed lobes, the terminal one 1 to 1¼in. long, the others smaller, often curved backwards towards the stem, the upper leaves sometimes undivided and hastate. Flowers very small, yellow. Pods about ½in. long, thick at the base, tapering to the point, more or less hairy, almost sessile, and closely pressed against the axis in long, slender, stiff racemes.—Reichb. Ic. Fl. Germ ii, t. 72.

Hab.: Southern parts of the colony, on roadsides and waste places about townships.

5. BLENNODIA, R. Br.

(Mucus emitted by seeds when soaked.)

Sepals short, open, equal at the base or slightly saccate. Petals obovate or short and narrow. Pod linear or linear-oblong (short in a variety of *B. trisecta*), terete or 4-angled, the valves very convex, without nerves or with a prominent midrib; septum membranous or almost spongy; stigma capitate, sessile or on a very short style. Seeds oblong or ovoid, more or less distinctly 2-rowed, not bordered, when soaked usually emitting a copious fibrous mucus; funicles free, filiform. Cotyledons incumbent.—Herbs or low undershrubs, glabrous or hoary-tomentose with simple or stellate hairs. Leaves entire or pinnatifid. Flowers white, yellow, or pink, the racemes without bracts.

A genus limited to extratropical or subtropical Australia, differing from *Sisymbrium*, to which some species have been referred, in the seeds never so completely overlapping each other as to form a single row, and generally in the copious mucus of the seeds, which is, however, not constant in all the species. From *Capsella* it differs in the longer pod and in the dissepiment broader in proportion to the transverse diameter of the pod.—*Benth.*

Glabrous undershrubs. Leaves or their lobes linear-filiform. Pods slender.
Leaves mostly 3-cleft . . . . 1. *B. trisecta.*
Annuals, glabrous or with simple hairs. Leaf-lobes narrow. Pods slender, scarcely contracted at the base.
Glabrous . . . . 2. *B. nasturtioides.*
Hoary, with simple hairs . . . . 3. *B. eremigera.*
Annuals, with stellate pubescence. Leaves pinnatifid or toothed. Pods acute at the top and at the base; valves very convex.
Pod rather slender, glabrous . . . . 4. *B. cardaminoides.*
Pod thicker in the middle, hirsute or stellately tomentose. Petals twice as long as the calyx, white or pink.
Calyx about 1 line long . . . . 5. *B. lasiocarpa.*
Calyx 2¼ lines long . . . . 6. *B. canescens.*
Perennials, with stellate pubescence. Leaves toothed or pinnatifid. Pods acute at the top and at the base; valves very convex.
Hoary. Pod at least 5 times as long as broad . . . . . . 7. *B. Cunninghamii.*

1. **B. trisecta** (referring to leaves), *Benth. Flora Austr.* i. 74. A perfectly glabrous often glaucous undershrub or almost a shrub, 1 to several feet high. Leaves numerous, often clustered, linear-filiform, sometimes rather thick, divided into 3 (rarely 2 or 5) unequal linear-filiform segments, the whole leaf seldom above 1in. long, except in very luxuriant specimens. Flowers white, scented. Sepals 1 to 1¼ line long. Petals obovate, spreading. Fruiting raceme 4 to 6in. long or rarely more, with slightly spreading pedicels of ¼ to ½in. Pod sessile on the pedicel, usually narrow-linear, 4 to 6 lines long, but sometimes very short,

straight or curved, the stigma sessile or nearly so; valves convex, with a slender longitudinal nerve. Seeds numerous, small, oblong-ovoid, those which I have soaked scarcely emitting any mucus.—*Sisymbrium trisectum*, F. v. M. in Trans. Vict. Inst. i. 114; Pl. Vict. i. 39.

Hab.: Downs of the southern parts of the colony.

2. **B. nasturtioides** (Nasturtium-like), *Benth. Fl. Austr.* i. 74. A glabrous annual, the central scape erect and leafless, the lateral branches decumbent at the base and leafy, from 2 or 3in. to nearly 1ft. long. Leaves usually pinnately divided into a few linear rather thick segments, the radical ones often 2in. long, the others much smaller. Flowers yellow, rather small. Fruiting racemes loose, 3 to 6in. long, with slender pedicels. Pod narrow, 4 to 7 lines long, nearly straight and scarcely contracted at the base; stigma sessile or nearly so; valves slightly convex, the longitudinal nerve very slender and sometimes quite inconspicuous. Seeds small, ovate, emitting a considerable mucus when soaked. —*Erysimum nasturtium*, F. v. M. in Linnæa xxv. 368; *Sisymbrium nasturtioides*, F. v. M. in Trans. Vict. Inst. i. 115; Pl. Vict. i. 39.

Hab.: Southern localities.

3. **B. eremigera** (desert, place of growth), *Benth. Fl. Austr.* i. 74. Annual and erect or branching and decumbent at the base, more or less hairy with short simple hairs, from a few inches to 1½ft. high. Leaves deeply and irregularly pinnatifid, with few oblong-linear or linear, sometimes falcate lobes. Flowers small, yellow. Fruiting racemes loose, 2 to 4in. long, with slender spreading pedicels. Pods like those of *B. nasturtioides*, mostly about ½in. long, slender, straight or curved, not contracted at the base; stigma sessile or nearly so; valves with a slender nerve. Seeds small, oblong-ovate, emitting mucus when soaked.—*Sisymbrium eremigerum*, F. v. M. Fragm. ii. 143.

Hab.: Maranoa.

4. **B. cardaminoides** (Cardamine-like), *F. v. M. Herb.* (as a *Sisymbrium*); *Benth. Fl. Austr.* i. 75. A slender or small annual like *B. nasturtioides*, but more or less clothed with a minute stellate pubescence, sometimes scarcely visible without a lens. Leaves pinnatifid, the radical ones with rather numerous small ovate triangular or lanceolate lobes, the terminal ones confluent, the lower ones becoming distinct segments along the petiole; stem-leaves few and small, with few short lobes. Flowers white (or pink?), the sepals barely 1 line long. Petals obovate, twice as long. Fruiting raceme loose and slender, 2 to 4in. long, with slender spreading pedicels. Pod 4 to 6 lines long, scarcely 1 line broad, usually curved, narrowed towards the base, glabrous or with a very minute stellate tomentum; valves very convex and keeled. Seeds small, ovate, emitting mucus when soaked.

Hab.: Southern localities, *F. v. M.*

5. **B. lasiocarpa** (hairy pods), *F. v. M. in Trans. Phil. Soc. Vict.* i. 100 and Pl. Vict. i. 40, t. 2; *Benth. Fl. Austr.* i. 76. ("Woombun Woombun," St. George, *Wedd.*) An annual, hoary with stellate pubescence, the central scape short and erect, the lateral stems decumbent and leafy at the base, branching and attaining 1ft. or more. Radical leaves petiolate, lyrate-pinnatifid, 1, 2, or even 3in. long; stem-leaves smaller, pinnatifid, or the upper ones toothed only. Flowers pink or white. Calyx about 1 line, petals obovate, fully twice as long. Fruiting racemes loose, 2 to 4in. long, with divaricate pedicels of 4 to 6 lines. Pods not above ½in. long, turgid, curved, tapering at the top with a short slender style, contracted at the base, hispid with simple or stellate hairs; valves very convex, with the midrib scarcely conspicuous. Seeds ovate, the mucus copious,— *Erysimum blennodioides*, F. v. M. in Linnæa, xxv. 367.

Hab.: St. George, *J. Wedd.*

6. **B. canescens** (hoary), *R. Br. in App. Sturt Exped.* 4; *Benth. Fl. Austr.* i. 76. Annual, but the lateral branching stems apparently harder at the base at the close of the season, so as to be almost woody; the whole plant hoary with a short, soft, stellate pubescence. Leaves lanceolate or oblong-linear, the radical ones about 2in. long, pinnatifid and narrowed into a petiole, the upper ones linear, toothed, or entire. Flowers large, pink, resembling those of a *Matthiola.* Calyx 2¼ lines long, hoary. Petals fully twice as long, with long claws. Fruiting racemes rather loose, 2 to 6in. long, the pedicels short, slightly spreading. Pod linear, 1 to 1½in. long, slightly pubescent, with convex valves, crowned by the large persistent stigma. Seeds oval-oblong, smooth.

Hab.: Inland localities, *F. v. M.*

7. **B. Cunninghamii** (after A. Cunningham), *Benth. Fl. Austr.* i. 76. A tufted herbaceous perennial, more or less hoary, with soft stellate hairs, occasionally mixed with simple ones; annual stems erect or decumbent at the base, from a few inches to 1ft. high, slightly branched. Radical leaves petiolate, 1 to 2in. long, oblong or lanceolate, coarsely toothed or shortly pinnatifid; stem-leaves few and small, from lanceolate to nearly obovate. Flowers small, apparently white. Fruiting racemes loose, 2 to 4in. long, with spreading pedicels. Pod 4 to 5 lines long, acute at the top and at the base, tipped by a very short subulate style, pubescent with simple or stellate hairs, or nearly glabrous; valves very convex, with a prominent midrib. Seeds oval-oblong, smooth, the mucus rather copious.

Hab.: Maranoa.

6. STENOPETALUM, R. Br.
(Narrow petals.)

Sepals narrow, erect, equal at the base. Petals shortly lanceolate above the claw, tapering to a point, often long and twisted. Pod globular, ovoid, or shortly linear, the valves very convex, usually without any conspicuous nerve; septum membranous; stigma globular, sessile or rarely on a very short style. Seeds several, small, in 2 rows, not bordered, with free filiform funicles; cotyledons incumbent.—Annuals, usually slender and glabrous, rarely tomentose and more rigid. Leaves linear. Flowers orange-yellow or white.

The genus is limited to Australia.—*Benth.*

Pods erect, 2 to 4 times as long as broad.
   Hoary tomentose. Pedicels as long as the pod. Petals 3 times as long as
      the calyx . . . . 1. *S. velutinum.*
   Glabrous or slightly tomentose. Pedicels shorter than the pod. Petals
      about twice as long as the calyx . . . . 2. *S. lineare.*
Pods spreading or pendulous.
   Sepals 1¼ line or more, petals more than twice as long. Pedicels slender,
      two or three times as long as the sepals, slightly hoary with appressed
      hairs. Leaves entire or remotely toothed . . . . 3. *S. nutans.*

1. **S. velutinum** (plant velvety), *F. v. M. Pl. Vict.* i. 49; *Benth. Fl. Austr.* i. 78. Erect and rather rigid, 1 to 1½ft. high, white or hoary, with a very short stellate tomentum, which disappears from the older leaves and the base of the stem. Leaves narrow-linear, rather thick, entire or with a few minute distant teeth, the lower ones 1¼ to 2in. long, the upper ones much shorter. Flowers erect, on pedicels about as long as the calyx. Sepals about 2 lines long, tomentose. Petals yellowish, the long slender point fully three times as long as the calyx. Fruiting pedicels erect, 3 to 5 lines long. Pod elliptical-oblong or almost ovoid, about 3 lines long, very turgid, glabrous; valves nerveless; ovules 8 to 12 in each cell.

Hab.: Amby Downs.

VIII. CRUCIFERÆ.

2. S. lineare (leaves linear), *R. Br. in DC. Syst. Veg.* ii. 513; *Benth. Fl. Austr.* i. 78. Usually erect, slender, little branched and quite glabrous, ¾ to 1½ft. high. Leaves few, narrow-linear, 1 to 1½in. long, entire or occasionally pinnatifid, with 1 or 2 short linear lobes on each side. Flowers small. Sepals not 1½ line long. Petals of a brownish-yellow, the narrow-linear exserted portion not longer than the sepals. Fruiting racemes slender but rigid, with erect pedicels not half so long as the pod. Pods erect, oblong, 2 to 3 lines long and scarcely 1 line broad, glabrous, the valves usually showing the midrib. Seeds 8 to 12 in each cell, small, ovate, smooth.—Hook. Ic. Pl. t. 618; Hook. f. Fl. Tasm. i. 22; F. v. M. Pl. Vict. i. 49.

Hab.: Interior of the colony, Darling Downs.

3. S. nutans (nodding), *F. v. M. Fragm.* iii. 27; *Benth. Fl. Austr.* i. 79. An erect annual, about 5in. high in the single specimen seen, slightly hoary with appressed hairs. Leaves linear, entire or remotely toothed, about 1in. long, narrowed at each end. Racemes loose. Pedicels much longer than the calyx, slender, erect when in flower, reflexed when in fruit. Sepals about 1½ line long. Petals with a filiform point of 4 or 5 lines. Pod broadly oval-oblong, about 4 lines long, very turgid, glabrous, ripening 3 or 4 seeds in each cell.

Hab.: Inland localities, Cooper's Creek, &c.

7. GEOCOCCUS, J. Drumm.

(From its burying its seed-vessels in the earth.)

Sepals short, spreading, equal at the base. Petals small. Pod oblong, slightly compressed, obtuse, the valves convex, with a prominent midrib; stigma sessile, entire. Seeds few, the two series rather distinct, oblong, not bordered, with long funicles; cotyledons incumbent.—A stemless herb, with radical pinnately-divided leaves, ripening its pods underground.

The genus is limited to the following species.

1. G. pusillus (small), *J. Drumm. in Hook. Kew Journ.* vii. 52; *Benth. Fl. Austr.* i. 80. A stemless, tufted annual. Leaves all radical, spreading, 1½ to 3 in. long, pinnately divided, with triangular or shortly lanceolate lobes, the lower ones distinct, the ultimate ones confluent. Flowers in our specimens imperfect, on short, erect, radical peduncles. Petals, according to Drummond, oblong, not clawed, shorter than the calyx. Fruiting peduncles lengthening to from ¼ to 1in., recurved so as to bury the pod in the ground. Our pods are irregularly ripened.

Hab.: Inland localities.

This curious little plant may possibly prove to be a condition of some species having usually dimorphous flowers, in which the more perfect ones are not developed. If so, it may very likely be a *Blennodia*, of some species of which it has the radical leaves.—*Benth*.

8. *BRASSICA, Linn.

(From the Celtic word for cabbage.)

Sepals erect or spreading, lateral, usually saccate at the base. Pods elongated, terete or angular, often with an indehiscent 1-seed beak; valves convex, 1—3-nerved, lateral nerves flexuose; style beaked or ensiform: stigma truncate or 2-lobed. Seeds 1-seriate, globose or sub-compressed; cotyledons incumbent, concave or conduplicate, the radicle within the longitudinal fold. Glabrous or hispid herbs; rootstock often woody. Leaves large, pinnatifid or lyrate, rarely entire. Flowers yellow, in long racemes.

Distributed over the temperate regions of the Old World,

1. **B. (Sinapis) nigra** (black), *Boiss.* Black mustard. Annual, hairy or quite glabrous, especially in the upper parts, the lower leaves and stems generally slightly hispid. Stems 2ft. or more high. Leaves mostly deeply divided, with one large terminal ovate or oblong lobe and a few small lateral ones, the upper leaves often small and entire. Flowers rather small. Pods on short pedicels, closely pressed against the axis of the long slender racemes, glabrous, seldom more than ½in. long, with a slender style, slightly conical at the base, the valves marked with a strong midrib.

Hab.: Europe. A stray weed from cultivation plots in southern Queensland.

9. CAPSELLA, Mœnch.
(A diminutive of *capsula*, a capsule.)
(Microlepidium, F. v. M.)

Sepals spreading, equal at the base. Petals short. Pod ovoid or oblong, laterally compressed or nearly terete, the valves very turgid or boat-shaped, keeled, the septum thin; style short or stigma sessile. Seeds several, in 2 rows, not bordered, on free funicles; cotyledons incumbent or rarely accumbent.— Small or weak annuals. Radical leaves rosulate, entire or lobed. Racemes slender, with small white flowers.

A small genus dispersed over the temperate regions of both the northern and southern hemispheres. Two of the following species are exclusively Australian. The genus is nearly allied to *Blennodia*, but the pod is shorter and more compressed laterally, the septum being usually narrower than the transverse diameter.—*Benth.*

Plant little branched ½ to 1½ft. Leaves pinnatifid. Pods cuneate-triangular, with numerous seeds . . . . 1. *C. bursa-pastoris.*
Plant dwarf, beset with short hairs. Leaves linear, blunt, entire. Pod ellipsoid turgid. Seeds generally 4 in each cell . . . . 2. *C. Andræana.*
Plant only a few inches high. Leaves divided into linear-lanceolate lobes. Pod rhomboid-rotund, about 4-seeded . . . . 3. *C. humistrata.*

1. **C. *bursa-pastoris** (shepherd's purse), *Mœnch.* Stems erect, from a few inches to 1ft. or more high, rather rough and often hairy with a few oblong or lanceolate entire or toothed leaves clasping the stem with projecting auricles. Radical leaves spreading on the ground, pinnatifid, with a large ovate or triangular lobe or sometimes entire. Root tapering often to a great depth. Flowers scarcely 1 line in diameter. Pods in a long loose raceme, usually triangular-truncate at the top, with the angles slightly rounded and narrowed at the base, sometimes notched at the top and almost obcordate; pedicels slender, style short, valves smooth. Seeds many, 10 to 12 in each cell, oblong punctate.—*Thlaspi bursa-pastoris*, Linn.

Hab.: Become naturalised near townships.

2. **C. Andræana** (after H. Andræ), *F. v. M.; Wing's Sou. Sc. Rec., Mar.* 1885. Annual, dwarf, erect. Stem as well as branches, flower-stalks, and stalklets beset with short papillular hair. Leaves short, linear, blunt, entire, glabrous; racemes short. Flowers minute, sepals soon spreading, petals white or yellowish, not or little longer than the sepals; filaments partly dilated at the base; anthers yellowish, cordate-roundish; stigma sessile. Pod small, ellipsoid, or globular-ovate, turgid, glabrous, not divided nor dilated at the summit, on a stalklet of usually the same length; valves subtilely 1-nerved, not keeled nor much compressed; septum lanceolar. Seeds generally 4 in each cell, ovate-roundish, compressed, brown-yellowish, margined by indurated through moisture much-expanding mucus.

Hab.: Southern localities.

In some respects allied to *C. pilulosa*, in others to *C. humistrata.*—*F. v. M., l.c.*

8. **C. humistrata** (found on damp spots), *F. v. M. Fragm.* xi. 25. An annual glabrous plant of a few inches high. Leaves divided into linear-lanceolate lobes. Racemes 1 to 4in. long, flowers numerous, pedicels spreading, very slender, from 1½ line under the flower to 4 lines under the fruit. Sepals ovate or oblong, ½ to ¾ line long. Petals oblong-ovate, attenuated at the base, yellow, about 1 line long, entire. Stamens 6, filaments free, subulate-linear; anthers yellow, almost round, introrse. Pods 1½ to 2 lines long, rhomboid-rotund, often 4-seeded, much compressed, with the base somewhat acute and the apex very shortly acuminate, entire. Seeds roundish-oval, ½ to ¾ line long, with pellucid margin.

Hab.: South Queensland.

10. SENEBIERA, Poir.

(After J. de Senebier).

Sepals short, spreading, equal at the base. Petals short. Pod laterally compressed, orbicular or broader than long, either indehiscent or separating into two nuts, each with a single seed. Embryo bent in a circle, or the radicle incumbent on the back of the cotyledons, but with the bend above the attenuated base of the cotyledons, not at their junction with the radicle.—Annuals or biennials, much branched and usually prostrate. Leaves entire or pinnately divided. Flowers very small, in short leaf-opposed racemes.

There are several species dispersed over the warm as well as the temperate regions both of the New and the Old World, and more especially near the sea, the following ones extending to Australia.—*Benth.*

Pods 1 line broad, slightly wrinkled, on slender pedicels.
Leaves linear, entire . . . . 1. *S. integrifolia.*
Leaves pinnate . . . . 2. *S. didyma.*

1. **S. integrifolia** (entire-leaved), *DC. in Mem. Soc. Hist. Nat. Par. ann.* 7, 144, *t.* 8, *and Syst. Veg.* ii. 522; *Benth. Fl. Austr.* i. 82. A rigid, glabrous, somewhat glaucous annual (or biennial?), usually decumbent, and very much branched. Leaves linear, usually acute, ¼ to 1in. long or rather more, narrowed into a petiole, quite entire or very rarely with 1 or 2 small teeth. Flowers very small and numerous, in terminal or leaf-opposed racemes usually much longer than the leaves; pedicels slender, rarely exceeding 1 line. Pods like those of *S. didyma*, of the same size, and reticulate when young, becoming often warted or even corky when old.—*S. linoides*, DC.; Harv. and Sond. Fl. Cap. i. 27.

Hab.: Bird Island, Wreck Reef, *Denham.*
The species has a wide range on the seacoasts of S. Africa and Madagascar, and we have it also from Pratas and other islands of the Chinese seas. *S mexicana*, Hook. and Arn. Bot. Beech. 276, is the same plant, but was probably gathered in the islands of Loo Choo and Bonin, and not in Mexico.—*Benth.*

2. **S. didyma** (double-podded), *Pers. Syn.* ii. 185; *Benth. Fl. Austr.* i. 83. Wart cress. A much-branched, prostrate annual, spreading on the ground from 6in. to 1ft. or more, glabrous, or with a few long loose hairs. Leaves pinnately divided into 7 to 11 narrow segments, which are usually again cut into 2 to 4 unequal linear or lanceolate lobes, the lower leaves often once pinnate, with oblong or obovate, entire or shortly lobed segments. Flowers very small and numerous, in leaf-opposed racemes, which seldom, even in fruit, exceed the leaves, the pedicels slender, 1 to 2 lines long. Pods about ¾ line long and 1 line broad, wrinkled, formed of 2 ovoid distinct lobes, which separate into 1-seeded nuts when ripe.—Reichb. Ic. Fl. Germ. ii. t. 9; *S. pinnatifida*, DC. Syst. Veg. ii. 523; Prod. i. 203.

A common weed in sandy soil, especially near the sea, in all warm countries, perhaps indigenous to N. Australia, and now established in the neighbourhood of towns in almost all the colonies.—*Benth.*

VIII. CRUCIFERÆ.

11. LEPIDIUM, Linn.
(Pod scale-like.)
(Monoploca, *Bunge.*)

Sepals short, equal at the base. Petals short, equal, sometimes wanting. Pod ovate or shortly oblong, rarely orbicular, usually much compressed laterally and notched at the top, the valves boat-shaped, keeled or winged, the septum narrow; style filiform or stigma sessile. Seeds solitary in each cell, suspended from the top of the septum with a free funicle; cotyledons incumbent in all except one species not Australian.—Herbs, undershrubs, or even small shrubs, very variable in habit. Leaves in the Australian species narrow or entire. Flowers small, white, the racemes without bracts.

A large genus, spread over the temperate and warmer regions of the globe, but not alpine and scarcely Arctic.

Leaves all quite entire. Pod usually conspicuously winged.
 Leaves broadly ovate or orbicular . . . . 1. *L. strongylophyllum.*
  Petals none. Stamens 4. Pod-wings almost united with the style   2. *L. monoplocoides.*
Leaves mostly toothed or lobed. Flowers very small. Pod-wings small or none, except in *L. papillosum.*
 Petals none. Leaves narrow-linear, the upper ones auricled.
  Stems papillose. Stamens 4. Pod about 2 lines long, with 2 short lobes or wings . . . . 3. *L. papillosum.*
  Stems glabrous. Leaves linear or cuneate, not auricled, the radical ones pinnatifid. Stamens 2. Pod about 1½ line, scarcely lobed . . . . 4. *L. ruderale.*

1. **L. strongylophyllum** (upper leaves round), *F. v. M. Herb.; Benth. Fl. Austr.* i. 84. Apparently shrubby, quite glabrous, with the branches denuded at the base. Leaves in the upper part of the branches, broadly ovate or nearly orbicular, or the upper ones elliptical-oblong, ½ to ¾in. long, entire, rather thick, narrowed into a short petiole. Flowers unknown. Fruiting raceme evidently dense, with spreading pedicels of about 2 lines, the thick rachis 1 to near 2in. long. Pods only known by the persistent replum, which is oblong-lanceolate, nearly 3 lines long, ¼ line broad in the centre, terminating in a subulate style of about 1 line, and the scars of a funicle on each side at the upper angle of the replum show that there had been a single pendulous seed in each cell as in other *Lepidia*.

Hab.: Inland localities.

2. **L. monoplocoides** (like a Monoploca), *F. v. M. in Trans. Phil. Soc. Vict.* i. 85 *and Pl. Vict.* i. 47; *Benth. Fl. Austr.* i. 85. An erect branching annual of about 6in., glabrous or slightly rough with minute papillæ. Leaves narrow-linear, entire and not auricled, the lower ones sometimes 2in. long, but mostly ½ to 1in. Flowers very minute, without petals and with only 4 stamens. Fruiting racemes 2 to 3in. long, with rigid, rather spreading, flattened pedicels, of 1½ to 2 lines. Pod orbicular, scarcely 2 lines long, flat, winged all round, the wings united with the style at the top, and projecting beyond it in 2 minute, connivent, acute lobes, forming a short point to the pod. Seeds with a viscid, clear mucus, as in several of the preceding species.

Hab.: Southern Queensland.

3. **L. papillosum** (Papillose), *F. v. M. in Linnæa* xxv. 870 *and Pl. Vict.* i. 46; *Benth. Fl. Austr.* i. 86. An erect branching annual, usually under 6in., but according to F. v. Mueller sometimes 1ft. high or more, the stems covered with little transparent papillæ, and exhaling an unpleasant scent. Radical leaves petiolate, often 2in. long or more, linear-oblong, coarsely toothed or irregularly pinnatifid, the upper ones lanceolate or linear-cuneate, with a few remote teeth, and clasping the stem by their auricled base, ½ to 1in. long and all glabrous. Flowers very small, without petals and with only 4 stamens. Fruiting racemes mostly 2 to 4in. long, with rigid, flattened, rather spreading pedicels, of about 2

lines. Pod obovate, about 2 lines long, the valves winged only above the middle, forming 2 rounded terminal lobes, a little more than ¼ line long, with the stigma sessile in the rather narrow sinus. Seeds exuding a viscid, clear mucilage in great abundance.

Hab.: Southern Queensland.

4. **L. ruderale** (found in waste places), *Linn.; DC. Prod.* i. 205; *Benth. Fl. Austr.* i. 86. An annual, biennial, or sometimes perennial, glabrous or with a few minute scattered hairs, commencing to flower when very small, but growing out to 1 or even 2ft., with hard stems and numerous divaricate, thin, wiry branches. Radical leaves once or twice pinnatifid, with narrow-linear lobes, but soon decaying; stem-leaves linear or rarely almost oblong-cuneate, usually with a few irregular teeth, especially towards the top, sometimes almost pinnatifid, the uppermost often linear and entire. Flowers minute, without petals and with only 2 stamens. Fruiting racemes usually rather loose but rigid, 2 to 3in. long, with slender stiff spreading pedicels of 2 or 8 lines, but sometimes the racemes remain short and dense as when in flower. Pods ovate, 1 to near 1½ line long, minutely 2-lobed at the top, with a short style between the lobes. Seeds ovate, usually exuding no mucus.—Reichb. Ic. Fl. Germ. ii. t. 10; Hook. f. Fl. Tasm. i. 25; F. v. M. Pl. Vict. i. 45; *L. puberulum*, Bunge, Pl. Preiss. i. 261; *L. hyssopifolium*, Desv. Journ. Bot. iii. 164 and 179; *L. fruticulosum*, Desv. l.c. 165 and 180 (a tall luxuriant form).

Hab.: Common in Southern Queensland along the fences around cultivation paddocks.

12. THLASPI, Linn.
(Pods compressed.)

Sepals erect, equal at the base. Petals obovate, equal. Pod short, ovate, obovate, obcuneate or oblong, much compressed laterally, notched or rarely acute at the top, the valves boat-shaped, keeled or winged, the septum narrow; style filiform or stigma sessile. Seeds 2 or rarely 3 or 4 in each cell, not winged; cotyledons accumbent.—Annual or perennial herbs, the radical leaves usually spreading, entire or toothed, those of the stem often auricled at the base. Flowers white, pink, or pale purple, rarely yellow.

A considerable genus spread over the temperate and colder regions of the northern hemisphere, with a very few S. American species, and none from S. Africa.

1. **T. cochlearinum** (like a cochlearia), *F. v. M. Pl. Vict.* i. 51; *Benth. Fl. Austr.* i. 88. An erect, rigid, branching annual, 6in. to 1ft. high, slightly pubescent, with a few short, mostly simple and reflexed hairs. Leaves lanceolate or linear-oblong, entire or with 1 or 2 coarse teeth or lobes on each side, narrowed into a petiole, the lower leaves about 2in. long, the upper ones few and smaller. Flowers white, rather large. Sepals open, 1¼in. long. Petals much larger. Fruiting racemes loose, about 2in. long, with half-spreading pedicels of 6 to 8 lines. Pod broadly oval, 4 to 5 lines long, obtuse at the top but not notched, pubescent with short, rigid, reflexed hairs; styles subulate, nearly 1 line long. Valves keeled, but not distinctly winged. Seeds 2 to 4 in each cell, flat, orbicular, emitting a clear, viscid mucus when soaked; cotyledons accumbent.—*Eunomia cochlearina*, F. v. M. in Linnæa, xxv. 869.

Hab.: Southern Queensland.

Order IX. CAPPARIDEÆ.

Flowers usually hermaphrodite. Sepals 4 to 8, either in a single series, free or united in a campanulate calyx, or 2 outer and 2 inner ones. Petals usually 4, imbricate, rarely 2 or none. Torus either small or expanded into a disk or lengthened into a straight or curved stalk to the ovary. Stamens inserted at the base or the summit of the torus or stalk of the ovary, definite or indefinite, all perfect or some reduced to staminodia. Ovary 1-celled, with 1 or usually several parietal placentas, which sometimes protrude so as to divide the ovary into imperfect cells. Stigma sessile or borne on a distinct style. Ovules usually numerous, rarely solitary, anatropous. Fruit either a capsule, with the valves separating from the persistent septum or placentas as in *Cruciferæ*, or indehiscent and succulent, or rarely dry. Seeds reniform or angular, without or with only a very thin albumen. Embryo curved, the cotyledons incumbent, folded, or convolute, very rarely flat.—Herbs or shrubs, rarely trees. Leaves alternate, or very rarely opposite, simple, or consisting of 1 to 5 digitate leaflets, with or without stipules, which when present are occasionally prickly. Flowers either solitary or clustered in the axis of the leaves, or more frequently in terminal racemes.

The Order is pretty generally distributed over the warmer and tropical regions of both the New and the Old World. Of the following genera two only, of one species each and both anomalous in the Order, are peculiar to Australia (one met with in Queensland, the other in West Australia). The other three are widely-spread tropical genera.—*Benth.*

TRIBE I. **Cleomeæ.**—*Herbs with a capsular fruit.*
Torus short, the stamens inserted immediately within the sepals and petals.
  Seeds several.
    Stamens 4 to 6, or rarely 8 . . . . 1. CLEOME.
    Stamens 8 to 16 . . . . 2. POLANISIA.
Torus elongated, bearing the stamens at the top under the ovary.
  Stamens all perfect, with long filaments. Leaves alternate, with digitate
    leaflets. Sepals 4. Seeds several . . . . 3. GYNANDROPSIS.

TRIBE II. **Cappareæ.**—*Shrubs or trees, with an indehiscent succulent fruit.*
Ovules and seeds many. Torus short without any basal appendage . . . . 4. CAPPARIS.
Ovules and seeds usually solitary. Leaves minute or none. Flowers diœcious.
  Sepals imbricate. Torus small. Filaments long . . . . 5. APOPHYLLUM.

1. CLEOME, Linn.

(Name used by a Latin physician to designate a plant unknown to modern botanists.)

Sepals 4, sometimes united in a 4-toothed calyx. Petals 4, nearly equal. Stamens 6, rarely 4 or 8, all or some only perfect, inserted on the short torus immediately within the petals. Ovary sessile or stalked, with many ovules, the stigma sessile or on a short subulate style. Capsule usually elongated, sessile or stipitate. Seeds many, reniform, usually rough or woolly.—Herbs, either glabrous or glandular-pubescent. Leaves with 3 to 7 digitate leaflets, or in some species not Australian simple. Flowers solitary or in terminal racemes.

A large genus, chiefly abundant in the warm parts of America, and in the hot sandy districts of N.E. Africa and S.W. Asia.

Stemless, with radical leaves and 1-flowered scapes . . . . 1. *C. oxalidea.*
Erect and leafy, with racemose flowers . . . . 2. *C. tetrandra.*
Plant, toothed prickly . . . . 3. *C. pungens.*

1. **C. oxalidea** (Oxalis-like), *F. v. M. Fragm.* i. 69; *Benth. Fl. Austr.* i. 90. A little glabrous, glaucous, almost stemless annual. Leaves radical, consisting of 3 obovate or orbicular leaflets, 2 to 4 lines long, on a slender petiole longer than themselves. Scapes filiform, 1-flowered, 1½ to 2in. long. Sepals about 1 line long. Petals of a pale pink, ovate, about 2 lines long. Stamens 6 to 8, with linear-oblong anthers attached near the base. Capsule sessile, linear-oblong or narrow-linear, ½ to 1in. long.

Hab.: Northern inland localities.

Cleome.] IX. CAPPARIDEÆ. 55

2. **C. tetrandra** (stamens often four), *Banks in DC. Prod.* i. 240; *Benth. Fl. Austr.* i. 90. An annual, either glabrous or sprinkled with a few short glandular hairs, the stems often several together, slender, ascending from a few inches to 1½ft. Leaves chiefly at the base of the stems on long petioles, with 3 or 5 linear-lanceolate or narrow-oblong leaflets sometimes above an inch long, the upper leaves few, small, with only 3 leaflets or simple. Raceme loose and slender, with filiform pedicels. Sepals ½ to 1 line long. Petals narrow, 3 to 6 lines long, nearly equal. Stamens 4 to 6. Capsule sessile, slender, 1 to 1½in. long, with a short subulate style, the valves thin and minutely striate. Seeds transversely wrinkled.

Hab.: Gulf of Carpentaria, *R. Brown.*

3. **C. *pungens** (pungent), *Willd.* Spider-flower. A robust annual clothed with a glandular pubescence, having a heavy scent, 2 to 5ft. high. Stipules spiny. Leaves petiolate, of from 5 to 7 lanceolate leaflets; petioles and midribs prickly. Stamens 6, long exserted from the corolla. Ovary much shorter than the gynophore. Ripe capsule about 4in. long.

Hab.: A South American plant now naturalised. Near towns, a stray from garden culture: considered a good bee plant

2. POLANISIA, Rafin.

(Stamens unequal.)

Sepals and petals 4 each, as in *Cleome.* Stamens usually 8 or more, inserted on the short torus. Ovary and capsule sessile or stalked, with many ovules and seeds, as in *Cleome.*—Herbs, with the habit of *Cleome,* from which the genus only differs in the increased number of stamens. Flowers in terminal racemes.

The genus is distributed over the warmer and tropical regions of both the New and the Old World. The only Australian species is a common tropical weed.

1. **P. viscosa** (viscid), *DC. Prod.* i. 242; *Benth. Fl. Austr.* i. 90. An erect branching annual or biennial, usually about 1ft. high, more or less covered with short, glandular, viscid hairs. Leaflets 3 or 5, very rarely 7, from obovate or oblong-cuneate to linear-lanceolate, the largest usually 1 to 1½in. long, but mostly much smaller. Flowers yellow, in terminal racemes. Sepals about 2 lines, petals twice or thrice as long, from narrow-oblong to almost ovate. Stamens from 8 to 16. Capsule from oblong-linear about 1in. long to narrow-linear and 3in. long strongly striate, the nerves very oblique and anastomosing in the short pods, nearly parallel in the long ones, and always glandular-pubescent. Seeds wrinkled.—*Cleome flava,* Banks, in DC. Prod. i. 241.

Hab.: Common in most parts of Queensland.

Var. *grandiflora.* Slightly pubescent. Leaflets narrow. Sepals about 4 lines, petals nearly 1in. long. Capsule above 4 in. long, N.W. coast, *Bynoe;* Sweers Island, *Henne.*

Some specimens from the gravelly bed of the Victoria River, *F. v. Mueller,* have shot out from the flowering racemes numerous branches crowded with small leaves, and very small axillary flowers almost without stamens, but producing small slender capsules, the whole plant assuming the appearance of the *P. micrantha,* Boj., from Madagascar. Other specimens from the same locality have all the leaves entire or 3-lobed, but these have no flowers to determine the species with certainty.

The species is a common weed throughout India, extending into tropical Africa.—*Benth.*

3. GYNANDROPSIS, DC.

(Stamens appearing to be on the style.)

(Rœperia, *F. v. M.*)

Sepals and petals 4 each, as in *Cleome.* Torus produced into a long slender gynophore, bearing at its summit about 6 stamens with filiform filaments. Ovary sessile or stalked within the stamens, with many ovules, the stigma sessile or on a subulate style, and the capsule sessile or stalked and many-seeded, as in *Cleome*.—Herbs, with the habit of *Cleome*, from which the genus only differs in the long stalk-like torus bearing the stamens. Flowers in terminal racemes.

Gynandropsis, like the last two genera, is dispersed over the tropical regions both of the New and the Old World.—*Benth.*

Flowers yellow. Capsule not striate . . . . 1. *G. Muelleri.*
Flowers white or purplish. Capsule striate . . . . 2. *G. pentaphylla.*

1. G. Muelleri (after Baron v. Mueller), *Benth. Fl. Austr.* i. 91. An erect annual, covered with a glandular viscid pubescence. Leaflets 3 or 5, lanceolate or oblong-linear, those of the upper leaves ¼ to 1in. long on a long petiole. Flowers yellow, on short pedicels in the upper axils, forming a terminal leafy raceme. Sepals ½ to near 1in. long, narrow, acuminate, unequal. Petals fully 3in. long, oblong, narrowed into a long claw. Stamens 5 to 7, the stipes or elongated torus often 1½in. long. Capsule linear, 2 to 2½in. long, not striate, but rough with short, glandular hairs, terminated by a slender style of nearly 1in.—*Rœperia cleomoides*, F. v. M. in Hook. Kew Journ. ix. 15.

Hab.: Gulf country.

2. G. pentaphylla (leaf of five parts), *DC.* An erect herb of 1 to 3ft., or sometimes shrubby below and taller, or reduced to 3 or 4in.; the extremities and young leaves usually thinly pilose or pubescent. Leaves 5-foliolate, the upper 3-foliolate; leaflets obovate or oblanceolate, acute-acuminate or obtuse, denticulate-serrulate or entire. Racemes glutinous, with simple or 3-foliolate bracts. Flowers white or purplish. Capsule narrow-linear, tapering into the style, usually puberulous or minutely setulose, 3 to 4in. long, gynophore ¾ to 2in., with the scar of the stamens near the middle style, variable in length or stigma sub-sessile.

Hab.: This Indian plant is now met with in most warm countries.

4. CAPPARIS, Linn.

(Name used by ancients for common caper plant, one of the genus.)

(Busbeckia, *Endl.*)

Sepals usually 4, rarely 5, free or the outer ones united in the bud into an entire calyx, which splits irregularly as the flower expands. Petals usually 4, imbricate. Stamens indefinite, inserted on the short torus, the filaments free, filiform. Ovary borne on a long stalk, 1 to 4-celled, with 2 to 6 placentas and several or many ovules; stigma sessile. Berry stalked, globose or elongated, very rarely dehiscent. Seeds several, immersed in pulp, with a hard or coriaceous testa and convolute embryo.—Trees or shrubs, sometimes climbing, unarmed or prickly. Leaves simple, membranous or coriaceous; stipules prickly or setaceous, often only on the young or barren shoots.

A large genus, distributed over the tropical and warm regions both of the New and the Old World; and divisible, chiefly from remarkable differences in the calyx, into several sections, of which two only are Australian—one (*Eucapparis*) comprises the greater number of the Asiatic and African species, but is not American; the other (*Busbeckia*) is confined to Australia and Norfolk Island. The Australian species of both sections are all endemic, and many of them are remarkable for producing slender barren shoots, with very prickly stipules, and small leaves so very differently shaped from those of the flowering-branches that where we have specimens of these barren branches only it is impossible to identify them.—*Benth.*

SECT. I. **Eucapparis.**—*Sepals 4, rather large, imbricate in 2 series. Berry globular or ovoid.*

Flowers on slender pedicels in terminal umbels. Outer sepals equal . . . 1. *C. umbellata.*
Flowers lateral or axillary, pedicels solitary or one above the other. One of
   the outer sepals larger and saccate or concave at the base.
  Stamens 12 or under. Flowers small.
    Pedicels usually 2, one over the other. Flowers very tomentose . . . 2. *C. lasiantha.*
    Pedicels 4 or 5, one above the other. Flowers slightly pubescent . . . 3. *C. quiniflora.*

Stamens numerous, or more than 15.
  Sepals very unequal, the largest ¾in. . . . . . . 4. *C. nummularia.*
  Sepals slightly unequal, about 3 lines . . . . . 5. *C. sarmentosa.*
Flowers small Sepals 2 lines. Petals 4 lines. . . . 6. *C. uberiflora.*

SECT. II. **Busbeckia.**—*Two outer sepals broad, very concave, completely united in the bud, and separating irregularly as the flower expands.*

Leaves mostly ovate or oblong.
  Leaves mostly 2 to 4in. long. Ovary glabrous. Fruit from ½ to a little
    more than 1in. diameter.
  Flowers mostly axillary, distant.
    Leaves ovate. Buds ovoid, acuminate, 1in. long, almost woody . . 7. *C. ornans.*
    Leaves ovate or oblong. Buds globular, ⅜in. long, coriaceous . . 8. *C. nobilis.*
    Leaves ovate. Buds 4-angled . . . . 9. *C. canescens.*
    Leaves oblong. Buds globose-pyramidal . . . . 10. *C. Shanesiana.*
  Flowers in a terminal corymb or short raceme. Buds globular . . 11. *C. lucida.*
  Leaves mostly 1 to 1½in. long. Ovary tomentose. Fruit 2in. diameter . 12. *C. Mitchelli.*
Leaves lanceolate or long and narrow.
  Leaves obtuse at the base. Petiole very short . . . . 13. *C. loranthifolia*
  Leaves coriaceous, narrowed into a rather long petiole . . . . 14. *C. umbonata.*
  Leaves ovate-lanceolate, membranous. Petioles short . . . . 15. *C. humistrata.*
  Leaves very narrow, on very short petioles . . . . 16. *C. Thozetiana.*

1. **C. umbellata** (form of inflorescence), *R. Br. in DC. Prod.* i. 247; *Benth. Fl. Austr.* i. 93. Shrubby, with the young branches tomentose. Stipulary spines small, nearly straight or recurved. Leaves from ovate to narrow-oblong, mostly 1½ to 2in., or when full grown 3in. long, at first membranous, softly pubescent or tomentose, at length stiff and usually glabrous, on petioles of about 2 lines. Pedicels slender, 6 to 9 lines long, usually 6 to 8 together in terminal umbels, sessile above the last leaves, or sometimes on short, lateral, leafless branches. Buds small, globular. Outer sepals thin, but stiff, equal, 2 to 2½ lines long, orbicular, concave, slightly imbricate, glabrous, inner ones scarcely longer, much imbricate. Petals about 3 lines long, pubescent. Stamens numerous. Ovary glabrous, with 8 to 10 ovules to each placenta. Berry globular, smooth, in our specimens not 1in. diameter, on a stipes of 1in. Seeds separated by spurious partitions.

Hab.: Coast scrubs from Port Denison to Cape York.
The species is mostly nearly allied to the common Indian *C. sepiaria*, differing chiefly in its sessile umbels and less numerous flowers.

2. **C. lasiantha** (alluding to clothing of flowers) *R. Br. in DC. Prod.* i. 247; *Benth. Fl. Austr.* i. 94. "Wyjeelah" or "Thulla-Kurbin," Cloncurry, *Palmer*. A much-branched shrub, clothed with a soft tomentum, usually rust-coloured on the young branches and inflorescence, afterwards paler, and sometimes disappearing on the old leaves. Leaves from ovate to narrow-oblong or almost lanceolate, obtuse, 1 to 2in. long, rounded at the base, with a very short petiole, thickly coriaceous when full grown, with very oblique primary nerves. Pedicels axillary, solitary or 2 together, one above the other, much shorter than the leaves. Outer sepals very concave and unequal, slightly imbricate, softly tomentose, the larger one about 3 lines long and almost saccate at the base ; inner sepals and petals ovate, 4 to 5 lines long, very tomentose outside. Stamens about 12. Ovary glabrous, with 10 to 12 ovules to each placenta. Young fruit ovoid, on a slender stipes of 1¼in.

Hab.: Many parts of the colony.
Pulp of fruit eaten by natives.—*Palmer*.

3. **C. quiniflora** (alluding to number of flowers together), *DC. Prod.* i. 247; *Benth. Fl. Austr.* i. 94. Branches weak and flexuose, the young ones and very young leaves rusty-tomentose, but soon becoming glabrous. Leaves ovate, obtuse or acuminate, 3 to 4in. long, rounded or almost cordate at the base, on petioles of 3 to 4 lines, rather coriaceous. Pedicels usually under ½in. long, 3 to

5 together, one above the other, in lateral clusters along the leafless tops of the side-branches, or above the upper axils. Outer sepals thin, slightly pubescent, unequal, the larger one saccate at the base and about 3 lines long; inner sepals and petals longer, oval-oblong, pubescent. Stamens few. Fruit glabrous, globular, ½ to 1in. diameter, on a stipes of about 1in. Some barren shoots, with very small ovate, rhomboid, or oblong leaves, assume a totally different aspect from the rest of the plant.

Hab.: Ranges about Cairns to Cape York, the Hammond Island, Torres Straits. Also in New Caledonia.

4. **C. nummularia** (leaves roundish, like a piece of coin), *DC.; Benth. Fl. Austr.* i. 94; "Longullah" and "Mijah," Cloncurry, *Palmer*. A dense or rambling shrub 5 or 6ft. high, or in some situations nearly prostrate or reclining on rocks, with hard tortuous branches. Stipular spines short, straight or recurved. Leaves broadly ovate or orbicular, very obtuse or sometimes emarginate, with a minute point in the notch, ½ to ¾in. long, rather thick, on petioles of 3 to 4 lines. Peduncles axillary, solitary, 1in. long or more. Outer sepals glabrous, very unequal, imbricate, the large one broadly hood-shaped, acuminate, ¾in. long, the other much narrower and concave. Inner sepals and petals apparently longer and glabrous, but very imperfect in our specimens. Stamens very numerous. Berry ovoid, succulent, fully 1½in. long, marked with longitudinal ribs, bursting when ripe like the fruit of a *Momordica*, on a stipes of at least 1½in.—F. v. M. Fragm. i. 143 and 244.

Hab.: Many parts of the colony. About Boulia, Burke River, it forms handsome large, dense bushes.
Fruit eaten by natives.

5. **C. sarmentosa** (branches straggling), *A. Cunn. Herb.*; *Benth. Fl. Austr.* i. 95. A slender tree, supporting itself on the branches of others, the younger branches slightly rusty-tomentose. Stipulary spines very short and hooked. Leaves almost sessile, broadly ovate, obovate or orbicular, obtuse, ½ to ¾in. long or sometimes much smaller, thin and glabrous when full grown. Flowers 1 or 2 together in the upper axils, on pedicels of 4 to 6 lines. Outer sepals glabrous, slightly unequal, about 3 lines long; inner sepals and petals rather longer, slightly tomentose or pubescent. Stamens 15 or more. Berry ovoid, not large, on a slender stipes of about an inch.

Hab.: Brisbane river, and many other southern localities.
The twigs sometimes infested with the fungus *Didymosphæria conoidella*. Sacc. and Berk.

6. **C. uberiflora** (flowers numerous), *F. v. M. Fragm.* ix. 172. A glabrous (except the petals) climbing shrub. Leaves deep-green and membranous, oblong-lanceolate, 2½ to 3in. long, 1¼ to 3in. broad, the apex usually with a minute sharp point, tapering to a slender petiole of about ½in. Stipular spines on the older branches, none on the flowering branchlets, in pairs, small, recurved. Peduncles axillary near the ends of the branchlets, often attaining 1½in. in length, thin and compressed, bearing at the end a dense or more or less elongated raceme of rather small white flowers. Pedicels filiform, about ½in. long. Sepals boat-like, about 2 lines long. Petals linear, 4 lines long and 1 broad, tomentose, densely so near the base. Stamens rather numerous, filaments flexuose, very slender, ¾in. long. Berry oval, on a slender stipes of ⅜in.

Hab.: Brook Island, *Dallachy;* Cairns, *L. J. Nugent*—the above from these specimens.

7. **C. ornans** (alluding to beauty of flowers), *F. v. M. Herb.; Benth. Fl. Austr.* i. 95. A woody climber, the branches hoary with a minute pubescence. Leaves ovate, obtuse, 2 to 3in. long, narrowed at the base, on petioles of ½ to 1in., glabrous on both sides. Stipulary spines conical, reflexed, often wanting on the

flowering branches. Pedicels solitary in the upper axils, 1½ to 2in. long. Flowers large and showy. Outer sepals united into an ovoid acuminate bud of above 1in. long, of a woody texture, and bursting irregularly; inner sepals orbicular, woolly inside, thick but petal-like. Petals (4 ?) obovate, more than 2in. long. Stamens numerous, about 3in. long. Ovary glabrous. Fruit not seen.

Hab.: Port Denison.

8. **C. nobilis** (referring to size of plant), *F. v. M. Herb.; Benth. Fl. Austr.* i. 95. "Rarum," N. Queensland, *Thozet.* A small tree, either perfectly glabrous or the young shoots and the under side of the leaves slightly covered with a close minute pubescence. Stipulary prickles short and conical, seldom seen on the flowering-branches. Leaves oval-oblong or oblong, acute, shortly acuminate or obtuse, 2 to 4in. long, coriaceous and often shining above, on petioles of 3 to 6 lines. Pedicels solitary in the upper axils or very rarely 2 together, about 1in. long. Buds globular, about ½in. diameter, often slightly emarginate at the top, showing the tips of the 2 outer sepals, which are perfectly united into a coriaceous calyx bursting or splitting irregularly; inner sepals broadly ovate, ½in. long, firm in the centre, thin on the edges. Petals 4, white, larger and thinner than the sepals, pubescent inside. Stamens very numerous. Fruit globular, about 1in. diameter, with a small protuberance at the top, the stipes ½in. to nearly 2in. long. Seeds numerous, embedded in a hard almost woody pulp.—*Busbeckia nobilis*, Endl. Prod. Fl. Norf. 64; *Busbeckia arborea*, F. v. M. Fragm. i. 163.

Hab.: From the Brisbane river scrubs to Rockhampton.
Wood of a light or whitish colour, close-grained, firm, should be useful for engraving.—*Bailey's Cat. Ql. Woods No. 4.*
Fruit eaten by natives, *Thozet.*
Var. *pubescens*. Petioles shorter, leaves more pubescent underneath, fruit scarcely umbonate. Brisbane river, *A. Cunningham.*
The same species is also found in Norfolk Island.

9. **C. canescens** (alluding to colour of foliage), *Banks in DC. Prod.* i. 246; Benth. Fl. Austr. i. 96. "Mondoleu," N. Queensland, *Thozet.* Habit and foliage so nearly that of *C. nobilis* that some specimens without the buds are difficult to distinguish from it, but in general they are of a paler more glaucous green, either minutely pubescent or glabrous. Stipulary prickles subulate, wanting on the flowering branches. Leaves as in *C. nobilis*, or more frequently broader and more obtuse, mostly 1½ to 2in. long, those of the barren shoots sometimes broadly ovate-cordate with a prickly point. Pedicels solitary or 2 together in the upper axils or terminal, 1 to 2in. long. Buds tomentose, larger than in *C. nobilis*, and prominently 4-angled. Flowers, of which I have only seen fragments, apparently like those of *C. nobilis*. Fruit (not yet ripe) as in *C. nobilis*, but on a longer stipes.

Hab.: Bay of Inlets, Northumberland islands and Keppel Bay, Burdekin and Lynd rivers.
Fruit eaten by natives, *Thozet.*
Var. *glauca*. Leaves 3 to 4in. long, very thick and glaucous. Between the Flinders and Lynd rivers.—*F. Mueller.*

10. **C. Shanesiana** (after P. A. O'Shanesy), *F. v. M. Fragm.* x. 94. A small tree with a rough bark. Branches spreading. Stipulary spines often absent, short, thick, and slightly curved. Leaves 3 or 4in. long, 1 to 1½in. broad, oblong to lanceolate-oval, the under side velvety pubescent; petioles 3 to 9 lines. Flowers large, mostly forming a terminal corymb, pedicels of flowers 1½ to 2in. long, lengthening under the fruit to 3in., and with the branchlets and calyx velvety. Buds about 1in. long, the lower portion globose, the upper pyramidal, longitudinally sulcate-angular. Petals 1½in.; style 2½ to 3in. long, slightly woolly, ovary glabrous. Fruit globose rugose, muricate-tuberculous umbonate, 2in. diameter.

Hab.: Brigalow scrubs, Rockhampton and Herbert's Creek.

11. **C. lucida** (leaves shiny), *R. Br. Herb.; Benth. Fl. Austr.* i. 96. Aboriginal name at Cloncurry "Thoogeer," *Palmer*. A shrub, very nearly allied to *C. nobilis*, but more often pubescent. Leaves ovate or oblong, obtuse, 2 to 3 or rarely 4in. long, coriaceous and shining when old, but often thinner than in *C. nobilis* and more reticulate. Flowers white, rather smaller than in *C. nobilis*, and usually several together in a terminal cluster or short raceme, the outer ones in the axils of the uppermost leaves. Buds globular, on pedicels of about 1in. Fruit globular, like that of *C. nobilis*.—*Thylacium lucidum*, DC. Prod. i. 254; *Busbeckia corymbiflora*, F. v. M. Fragm. i. 163.

Hab.: Burdekin river, Howitt's Isles, Hope islets, Port Molle and Port Denison.
Ripe fruit eaten by natives, *Palmer*.

12. **C. Mitchelli** (after Sir T. Mitchell), *Lindl. in Mitch. Three Exped.* i. 315; *Benth. Fl. Austr.* i. 96. Native pomegranate; Bumble; "Kam-doo-thal," Cloncurry, *Palmer*; "Mondo," N. Queensland, *Thozet*; "Eeger," St. George, *Wedd*. A much-branched shrub, more or less clothed with a minute yellowish or whitish tomentum, sometimes soft and dense, sometimes disappearing on the older leaves. Stipular prickles short, somewhat hooked, often wanting on the flowering branches. Leaves ovate or oblong, obtuse, 1 to 1¼in. long, narrowed into a petiole of 2 to 3 lines, coriaceous and rather thick, obscurely veined. Pedicels few, axillary, 1 to 1½in. long, thickened upwards. Buds ovoid-globular, usually acuminate, nearly ½in. long. Outer calyx thick, opening irregularly or sometimes into 2 valvate concave sepals. Inner sepals 4 to 8 lines long, more or less pubescent, especially at the base, thin and glabrous on the edges. Petals similar but larger. Ovary tomentose, on a long neary glabrous stipes. Berry globular, 2in. diameter when ripe. Seeds 4 to 5 lines long, embedded in a hard dry pulp.—*Busbeckia Mitchelli*, F. v. M. Pl. Vict. i. 53, t. suppl. 4.

Hab.: A common tree inland.
The aborigines and bushmen consider the bark to possess healing properties, and use it in cases of sores and piles.
Wood whitish, close-grained, hard; suitable for engraving or carving.—*Bailey's Cat. Ql. Woods No. 6.*
Fruit eaten by natives, *Thozet, Palmer*.
Twigs sometimes infested with the fungus *Calonectria Otagensis*.

13. **C. loranthifolia** (Loranthus-leaved), *Lindl. in Mitch. Trop. Austr.* 220; *Benth. Fl. Austr.* i. 97. A scrubby bush with more or less tomentose branches. Leaves from oblong-linear to broadly lanceolate, obtuse or acute, 1½ to 2½in. long, obtuse at the base, on a petiole of 1 or rarely 2 lines, coriaceous and at length glabrous. Pedicels in the upper axils about 1in. long, thickened upwards. Buds ovoid, scarcely acuminate, the outer calyx not so thick as in the other species of the section *Busbeckia*. Inner sepals larger, thickened in the centre. Petals longer, thinner, villous inside. Stamens numerous. Ovary glabrous.

Hab.: Not uncommon inland.

14. **C. umbonata** (form of fruit), *Lindl. in Mitch. Trop. Austr.* 257; *Benth. Fl. Austr.* i. 97. A shrub with tomentose branches like the last, but the leaves usually much longer, often 7 to 8in. long, and rarely under 3in., always lanceolate, and narrowed into a rather long petiole. Pedicels axillary, thickened upwards, 1 to 1½in. long. Buds ovoid, the outer calyx very thick and coriaceous. Petals as in *C. Mitchelli*. Fruit apparently small, glabrous, not always marked with the terminal protuberance which suggested the specific name; the stipes very long.

Hab.: Brigalow scrub on the Belyando, Dawson River, and other localities.

15. **C. humistrata** (spreading on the ground), *F. v. M. Fragm.* v. 156. A procumbent pubescent shrub, with terete, spreading, softly pubescent branches. Stipulary spines setaceous-subulate, scarcely curved. Leaves ¾ to 1¼in. long, ovate-lanceolate, mucronulate, chartaceous, margins slightly recurved, pale green; petioles short. Flower pedicel about ½in. long, axillary, solitary, buds about 4 lines long. Sepals roundish, contracted at the apex. Petals scarcely exceeding ½in., slightly pubescent near the base. Stamens about 30, ovary glabrous apiculate. Fruit?

Hab.: Near the town of Stanwell.

16. **C. Thozetiana** (after M. A. Thozet), *F. v. M. Fragm.* v. 104. An erect shrub resembling a Bossiæa, glabrous or nearly so; branchlets terete, slightly flexuose. Stipulary spines 1 to 1½ lines long, subulate, slightly curved. Leaves 1 to 2in. long, 1½ to 2 lines broad, linear, margins recurved, apex mucronulate, base obtuse, on very short petioles. Flowers axillary, solitary, on pedicels of about 1in.; bud before expanding about 3 lines diameter, globose, inner sepals obovate-cuneate; much shorter than the petals. Petals 5—6 lines long, velvety-pubescent beneath. Stamens 16 to 20. Ovary glabrous, very shortly obtuse-apiculate.

Hab.: Near Rockhampton.

5. APOPHYLLUM, F. v. M.
(Plant leafless.)

Flowers diœcious. Sepals 3 or 4, imbricate, 2 outside the others. Petals 2 or 4, sessile, imbricate. Maleflower: Stamens 8 to 16, inserted on the short torus with filiform filaments. Ovary none. Female flower: Stamens none, or rarely 1 to 3. Ovary stipitate with a sessile stigma; ovules 1 or 2, attached to the sides of the cavity above the middle. Berry shortly stipitate. Seeds 1 or 2, with a smooth testa and involute cotyledons. Leaves very few, small, alternate.

The genus is limited to the following species, and differs from *Capparis* only in its diœcious flowers and the usually solitary ovule.—*Benth.*

1. **A. anomalum** (strange appearance), *F. v. M. in Hook. Kew Journ.* ix. 307; *Benth. Fl. Austr.* i. 97. A shrub or tree, almost leafless, with cylindrical, often pendulous branches, silky-white when young but soon becoming glabrous. Leaves on the young shoots few, linear or linear-acute, 2 to 3 lines long and very deciduous, or rarely above ½in. long and more persistent. Flowers small, fragrant, either growing singly along the young shoots or in short lateral racemes or clusters. Petals 1 to 1¼ lines long. Sepals rather more than 1 line long, pubescent. Petals unequal, as long as or longer than the sepals, pubescent inside at the base. Fruit nearly globular, the size of a small pea.

Hab.: In the interior, *Mitchell.*

Order X. VIOLARIEÆ.

Flowers usually hermaphrodite. Sepals 5, imbricate. Petals 5, imbricate, equal or unequal, with the lower one larger, or spurred or otherwise dissimilar. Stamens 5, hypogynous or nearly so, the anthers erect and connivent or connate round the pistil, sessile or on short filaments, the connective often very broad, with the anther-cells opening inwards. Ovary free, sessile, 1-celled, with usually 3 parietal placentas, and several or rarely only 1 or 2 anatropous ovules to each placenta. Style usually simple, often thickened or curved at the top. Fruit a capsule, opening in as many valves as placentas, or rarely an indehiscent berry.

X. VIOLARIEÆ.

Seeds with a fleshy albumen; embryo axile, usually straight, the cotyledons usually broad and flat, the radicle next the hilum.—Herbs or shrubs. Leaves usually alternate, simple, and rarely lobed or cut, with lateral stipules. Flowers axillary, solitary, or in cymes or panicles, very rarely in racemes. Pedicels usually with 2 bracteoles. Capsules often opening elastically.

An Order generally dispersed over the globe. The two Queensland genera have a very wide geographical range.

Herbs or undershrubs, with very irregular flowers. Fruit capsular.
Sepals produced into a small appendage, or at least a protuberance below their
  insertion. Lower petal spurred or saccate . . . . 1. VIOLA.
Sepals not produced at the base. Lower petal saccate or gibbous at the base . 2. IONIDIUM.

1. VIOLA, Linn.
(Derived from its Greek name.)

Sepals produced into a small appendage or protuberance below the insertion. Petals spreading, the lowest usually larger, spurred or saccate at the base. Anthers nearly sessile, the connectives flat, produced into a membranous appendage beyond the cells, those of the 2 lower anthers usually bearing a small dorsal reflexed protuberance or spur. Style variously thickened or dilated at the top, straight with a terminal stigma, or incurved with the stigma in front. Capsule opening elastically in 3 valves. Seeds ovoid-globular with a crustaceous testa.—Herbs, with the stipules usually foliaceous and persistent. Peduncles axillary, 1-flowered. Most species, besides the perfect flowers, produce later in the season small apetalous but very prolific flowers.

A very large genus, most of the species natives of the temperate regions of the northern hemisphere, or of the high mountains of South America, with a very few dispersed over Africa, Australia, and New Zealand. The Australian species are either quite endemic or extend only to Norfolk Island and New Zealand. They are all perennials.—*Benth.*

Stemless, with a tufted or creeping rhizome.
  Leaves lanceolate, oblong, or scarcely ovate. No stolons. Stipules
    adnate . . . . 1. *V. betonicæfolia.*
  Leaves nearly orbicular. Stolons creeping. Spur reduced to a slight
    protuberance. Stipules free . . . . 2. *V. hederacea.*

1. **V. betonicæfolia** (Betony-leaved), *Sm.; DC. Prod.* i. 294; *Benth. Fl. Austr.* i. 99. Glabrous or pubescent, stemless, and without stolons, and often tufted, the stock either ending underneath abruptly, with thick spreading fibres, or tapering into a horizontal or descending root. Leaves radical, from lanceolate to oblong or nearly ovate, mostly obtuse, and 1 to 1½in. long, entire or slightly crenate, truncate or slightly cordate, rarely narrowed at the base, with the long petiole usually dilated at the top. Stipules linear, adnate to the petiole. Scapes of the perfect flowers usually considerably longer than the leaves, with the subulate bracts below the middle. Flowers violet, rather large. Sepals lanceolate, acute, 2¼ to nearly 3 lines long, with short blunt basal appendages. Lateral petals usually copiously bearded inside, the upper ones less so, the lowest not at all; spur broad and obtuse, much shorter than the sepals. Style thickened upwards, concave at the top, not winged. Apetalous flowers on very short scapes.—Hook. f. Fl. Tasm. i. 27; F. v. M. Pl. Vict. i. 64; *V. phyteumæfolia* and *V. longiscapa*, DC. in Herb. Lamb., from the char. in G. Don, Gen. Syst. i. 322.

Hab.: Near Brisbane.

Received also from Norfolk Island, *Backhouse*, and the species is nearly allied to *V. Patrinii,* DC., which is common in India, eastern Siberia, and China, and only appears to differ from *V. betonicæfolia* in the rather longer spur and the style usually broadly winged.

2. **V. hederacea** (like the English ground-ivy), *Labill. Pl. Nov. Holl.* i. 66, t. 91; *Benth. Fl. Austr.* i. 99. Glabrous or pubescent, densely tufted or widely creeping by its numerous stolons, very rarely emitting weak leafy stems. Leaves reniform, orbicular, or spathulate, usually under ½in. diameter, but when very luxuriant, 1 to 1½in., entire or irregularly and sometimes coarsely toothed. Stipules free, brown, lanceolate-subulate. Scapes usually longer than the leaves, the bracts about the middle. Flowers usually small, blue, rarely white, but sometimes fully ⅜in. broad. Sepals lanceolate, with only a slight protuberance below their insertion. Petals glabrous, or the lateral ones slightly pubescent inside, the spur of the lower one reduced to a slight concavity. Lower anthers with a very slight dorsal protuberance. Style bent at the base, the upper part cylindrical, truncate at the top, but not thickened. Seeds usually dark-coloured, but sometimes white.—*DC. Prod.* i. 305; *Hook. Exot. Fl.* iii. t. 225; *Reichb. Icon. Exot.* t. 110; *Hook. f. Fl. Tasm.* i. 26; *F. v. M. Pl. Vict.* i. 65; *V. Sieberiana*, Spreng. Syst. Cur. Post. 96; *Erpetion reniforme*, Sweet, *Brit. Fl. Gard.* ii. t. 170; *E. hederaceum, E. petiolare*, and *E. spathulatum*, G. Don, *Gen. Syst.* i. 335.

Hab.: Moreton Bay, *Fitzalan.*

2. IONIDIUM, Vent.

(Name from resemblance to violet.)

(Pigea, *DC.*)

Sepals not produced at the base. Petals spreading, the lowest sometimes slightly larger than the others, more frequently very much larger with a broad claw, gibbous or saccate at the base. Anthers nearly sessile, or on distinct filaments, the connectives flat, produced into a membranous appendage beyond the cells, those of the 2 lower ones bearing a dorsal reflexed protuberance, spur, or gland, the two rarely united into one. Style thickened and incurved at the top, with the stigma in front. Capsule opening elastically in 3 valves. Seeds ovoid-globular, with a crustaceous testa.—Herbs or small shrubs. Leaves alternate or rarely opposite, usually narrow. Stipules small and narrow. Peduncles axillary or in a terminal raceme, 1 or several-flowered.

A considerable genus, chiefly tropical, and the greater number of species American; four or five are found in tropical Asia and Africa, and one of these occurs in Australia, the others here enumerated are all endemic.—*Benth.*

Peduncles axillary, 1-flowered, or very rarely here and there 2-flowered.
  Lower petal more than twice as long as the calyx.
    Leaves entire, or rarely toothed. Appendages of the lower filaments
      nearly glabrous. Seeds striate . . . . 1. *I. suffruticosum.*
    Leaves toothed. Appendages of the lower filaments woolly-hairy.
      Seeds smooth . . . . 2. *I. aurantiacum.*
Peduncles 1-flowered in the upper axils, the upper ones longer than the
  leaves, and forming a terminal leafy raceme . . . . 3. *I. Vernonii.*
Peduncles slender, much longer than the leaves, with a leafless raceme of
  2 or more flowers.
    Upper leaves often opposite. Sepals lanceolate, shorter than the lateral
      petals . . . . 4. *I. filiforme.*

1. **I. suffruticosum** (shrubby), *Ging. in DC. Prod.* i. 311; *Benth. Fl. Austr.* i. 101. Much-branched, glabrous, or very slightly pubescent, and usually from 1 to 1½ft. high, and more or less woody at the base. Leaves alternate, narrow-linear, or rarely linear-oblong or lanceolate, entire or rarely toothed, mostly 1 to

2in. long. Peduncles axillary, filiform, 1-flowered, 2 to 4 lines long, with a pair of minute bracts under the pedicel. Sepals lanceolate, very acute, with a very prominent green midrib, 1½ to 2 lines long. Lateral petals rather longer than the calyx, with a broad-ovate falcate base, and a small, ciliate, obtuse extremity, sometimes expanded into a small lamina; upper petals smaller; lowest petal purple or rarely yellow, about ½in. long, the claw longer than the other petals, saccate at the base, the lamina broadly ovate and longer than the claw. Filaments at least half as long as the anthers, the 2 lower ones with a thick spur, either quite glabrous or with a minute tuft of hair. Seeds elegantly marked with longitudinal striæ.—Wight, Ic. t. 308; *Pigea Banksiana*, DC. Prod. i. 307.

Hab: Most parts of the colony.

The species is widely spread over tropical Asia and Africa.

2. **I. aurantiacum** (orange-coloured), *F. v. M. Herb.*; *Benth. Fl. Austr.* i. 102. Pubescent with short spreading hairs or rarely glabrous, often woody at the base, branched, 6in. to 1ft. high or rather more. Leaves linear or oblong-lanceolate, 1 to 1½in. long, bordered with small, distant, acute teeth. Flowers axillary, on peduncles of 3 to 4 lines, as in *I. suffruticosum*, and nearly similar in structure, but the lower petal is smaller and always yellow, the broad lamina usually shorter than the long narrow claw, which is scarcely saccate at the base, and the appendages of the filaments of the lower stamens are covered with long woolly hairs. Seeds, in the few capsules I have seen, smooth and not striate.

Hab.: Georgina River, *J. Coghlan*.

3. **I. Vernonii** (after W. Vernon), *F. v. M. Pl. Vict.* i. 228; *Benth. Fl. Austr.* i. 103. Glabrous, with erect, slender, but stiff stems, little branched, except at the base, and usually about 1ft. high, as in *I. filiforme*, but the branches more angular. Leaves all alternate, linear or narrow-lanceolate, rarely above 1in. long, and the upper ones much smaller and very narrow. Peduncles 1-flowered, as in *I. suffruticosum*, but only in the upper axils, and the upper ones longer than the small floral leaves, so as to form a terminal leafy raceme. Flowers blue, very much like those of *I. filiforme*, the lower petal of the same shape and size, except that the claw is distinctly spurred at the base, and the lateral petals are more obtuse than in that species; stamens the same, except that the subulate appendages at the top of the anther-cells are still more minute.

Hab.: What may be a form of this species has been gathered near the Pine River.

4. **I. filiforme** (thread-like), *F. v. M. Pl. Vict.* i. 66; *Benth. Fl. Austr.* i. 103. A perfectly glabrous herb, said by some collectors to be annual, but certainly in many instances forming a perennial rootstock. Stems slender, but stiff and wiry, simple or branched, usually 1 to 2ft. high, but when eaten down sending up numerous short erect branches. Leaves alternate or the upper ones opposite, narrow-linear, mostly 1 to 2in. long, entire, the lowest ones shorter, broader, and petiolate. Flowers blue, in slender leafless racemes, on terminal or axillary peduncles, always much longer than the leaves, the pedicels under a line long. Sepals shorter than the lateral petals, lanceolate, acute. Lower petal usually fully ½in. long, ovate, narrowed into a concave claw, saccate at the base, but varying considerably in size and breadth; lateral petals broadly falcate, acute, about 2 lines long; upper ones smaller. Anthers with an orange ovate appendage at the top of the connective, and two minute subulate appendages on the cells themselves; the two lowest have also a small glandular protuberance on the back at their base.—*Pigea filiformis*, DC. Prod. i. 307; *I. linarioides*, Presl. Bot. Bm. 12.

Hab.: Common in many localities.

Order XI. BIXINEÆ.

Flowers regular. Sepals 2 to 6, usually 4 or 5 and imbricate. Petals either none, or as many as the sepals, or indefinite, imbricate or contorted in the bud, deciduous. Stamens hypogynous or slightly perigynous, indefinite or very rarely definite. Anthers 2-celled, opening by longitudinal slits or rarely by terminal pores. Torus often bearing glands or a glandular disk. Ovary free, usually 1-celled, with 3 or more, rarely 2 or 1, parietal placentas. Styles or stigmas as many as placentas, free or united. Ovules 2 or more to each placenta, amphitropous or anatropous. Fruit succulent or dry, opening in valves, bearing the placentas in the middle, or indehiscent. Seeds usually few, with a copious and fleshy or rarely thin albumen. Embryo in the axis, straight or curved, the radicle next the hilum, the cotyledons usually broad.—Trees or shrubs, in one genus twiners. Leaves alternate, simple, and often toothed, or rarely palmately lobed or divided. Flowers axillary or terminal, solitary or in clusters, corymbs, racemes, or panicles.

A considerable Order, dispersed over the tropical or warm regions both of the Old and the New World.

Tribe I. **Bixeæ.**—*Petals broad, contorted, without a scale or basal appendage. Anthers bursting by pores or short slits.*
Capsule almost 3—5-celled. Seeds curved. Trees or shrubs. Leaves
digitate. Flowers large . . . . 1. Cochlospermum.

Tribe II. **Flacourtieæ.**—*Petals small, imbricate or none. Anthers short, bursting by slits.* Seeds straight. Trees or shrubs. Leaves simple. Flowers small.
Sepals 4 to 6. Petals as many. Anthers with an appendage . . . . 2. Scolopia.
Sepals 4 to 6. Petals none. Anthers without any appendage . . . . 3. Xylosma.

1. COCHLOSPERMUM, Kunth.
(Seeds twisted.)

Flowers hermaphrodite. Sepals 5, imbricate, deciduous. Petals 5, large. Stamens numerous. Anthers oblong or linear, opening in terminal pores or very short fissures. Placentas 3 to 5, projecting more or less into the cavity of the ovary, with numerous ovules. Style simple. Capsule 3 to 5-valved, the membranous endocarp separating from the pericarp. Seeds kidney-shaped or spirally curved, covered with wool or bordered by long hairs.—Trees, shrubs, or rarely undershrubs, usually yielding a yellow juice. Leaves palmately lobed or divided. Racemes loose, few-flowered, in the upper axils or in terminal panicles. Flowers large, yellow.

Besides the 2 following species of the 4 peculiar to Australia, there is 1 known from southern India, 2 from Africa, and about 5 from South America.
Calyx and inflorescence glabrous or slightly glandular pubescent.
Leaves glabrous, with deep ovate-lanceolate or oblong lobes . . . . 1. *C. Gillivræi.*
Leaves glabrous, divided to the base into narrow-oblong, pedate segments . 2. *C. Gregorii.*

1. **C. Gillivræi** (after John M'Gillivray), *Benth. Fl. Austr.* i. 106. The specimens are perfectly glabrous, except a very slight pubescence on the branches of the panicle and pedicels. Leaves palmately divided to within $\frac{1}{4}$ or $\frac{1}{2}$in. of the base, into 5 or 7 ovate-lanceolate or oblong-acuminate slightly toothed lobes, of which the central largest ones are usually 2 to 3in. long, the 2 outermost short and very acuminate. Panicles short and loose. Pedicels $\frac{1}{2}$ to 1in. long. Sepals very unequal, glabrous except at the base, the edges very thin. Anthers about $1\frac{1}{4}$ line long. Capsule obovoid-oblong, rarely 3in. long, truncate at the top, and very much depressed in the centre. Seeds enveloped in a very deciduous wood.

Hab.: Thursday and other islands off the N.E. coast, Burdekin River, Port Denison. Used by the natives for fibre.—*Roth,*

2. **C. Gregorii** (after Hon. A. C. Gregory, the explorer), *F. v. M. Fragm.* i. 71; *Benth. Fl. Austr.* i. 106. A small tree, quite glabrous except a very slight glandular pubescence on the branches of the inflorescence and pedicels. Leaves pedately divided to the base into about 7 narrow-lanceolate entire segments, the central ones 2 to 3in. long, the common petiole 3 to 6in. Panicles apparently short and not much divided, or reduced to a single raceme. Pedicels about ½in. long. Sepals and petals as in the last species. Style filiform, slightly thickened towards the top. Outer stamens, as in all the other species, on longer filaments than the inner ones, but the difference is rather more decided in this species. Placentas 5. Fruit not seen.

Hab.: Gilbert River.

2. SCOLOPIA, Schreb.
(Some species being very thorny.)
(Phoberos, *Lour.*)

Flowers hermaphrodite. Sepals 4 to 6, slightly imbricate when very young, but open long before flowering. Petals as many and nearly similar. Stamens indefinite, inserted on the thickened torus, with or without glands. Anthers short, the connective terminating in a thick process. Ovary with 3 or 4 placentas and few ovules. Style filiform, with an entire or lobed stigma. Fruit a berry. Seeds 2 to 4, with a hard testa. Cotyledons leafy.—Trees, often armed with axillary spines. Leaves simple, with pinnate veins, entire or toothed. Flowers small, in axillary racemes.

The genus is dispersed over southern and eastern Africa and tropical Asia. The Australian species is endemic.—*Benth.*

1. **S. Brownii** (after R. Brown), *F. v. M. Fragm.* iii. 11; *Benth. Fl. Austr.* i. 107. Perfectly glabrous in all its parts. Leaves from ovate to oblong-lanceolate, mostly acuminate, obtuse or almost acute, rarely rounded at the top, 1½ to 3in. long, always narrowed into a petiole of 3 to 4 lines, entire or slightly undulate-toothed, rather thick and smooth, obscurely triplinerved, but all the veins less conspicuous than in most species, either without glands or with 2 or 3 marginal glands underneath. Racemes short and axillary or forming a terminal panicle of 1 to 2in. Pedicels 2 to 3 lines. Calyx 4-cleft, smaller than in *S. crenata*, apparently persistent. Petals 4, rather longer than the calyx, deciduous. Stamens numerous, with slender filaments, surrounded by a ring of glands, either distinct and shortly club-shaped or irregularly connate. Anthers small, the process of the connective glabrous and usually as long as the cells. Placentas 3, with about 4 ovules to each. Stigma slightly 3-lobed.

Hab.: Coast scrubs south and north.

Wood pinkish, darkening towards the centres; close-grained, tough.—*Bailey's Cat. Ql. Woods No. 7A.*

This species has much the foliage of some forms of the Indian *C. crenata*, but is readily known by the glands of the disk.—*Benth.*

3. XYLOSMA, Forst.
(From Greek, on account of wood of one species being bitter.)

Flowers diœcious. Sepals 4 or 5, small, imbricate. Petals none. Male flowers: Stamens indefinite, often surrounded by a glandular disk; anthers short, without appendage. Female flowers: Ovary inserted on an annular disk, with 2 or rarely more placentas, and 2 or few ovules to each; style entire or divided, with dilated stigmas, or rarely stigma sessile. Berry small, indehiscent,

Seeds 2 to 8, with a smooth crustaceous testa. Cotyledons broad.—Trees, often thorny. Leaves toothed or rarely quite entire. Flowers small, axillary, clustered, or shortly racemose.

A genus widely dispersed over the tropical and subtropical regions of the New and the Old World. The only Australian species is endemic.

1. **X. ovatum** (from form of leaf), *Benth. Fl. Austr.* i. 108. Glabrous in all its parts, the branches short and slender, rough with lenticels, and in our specimens without thorns. Leaves mostly ovate, obtuse, about 1½in. long, quite entire, narrowed into a very short petiole, thinly coriaceous, with numerous fine reticulate veins; a few lower leaves short and almost orbicular, and the upper ones narrow. Male flowers not seen. Female flowers very small, 5 or 6 together in very short axillary racemes. Pedicels about 1 line long, in the axils of small, ovate, ciliate bracts. Sepals 4, orbicular, ciliate, about ½ line long. Disk deeply lobed or divided. Ovary ovoid, conical, but scarcely tapering into a distinct style, with a broad, thick, slightly 2-lobed stigma. Placentas 2, very prominent, forming a complete dissepiment above the insertion of the ovules, but far from meeting below. Ovules 2 to each placenta. Fruit a berry about 4 lines diameter, black, maturing 1 to 4 seeds. Testa brown outside, smooth.

Hab.: N.E. coast.

This appears to come nearest to *X. orbiculatum*, Forst., which, judging from Fiji Island specimens, has a similar almost sessile stigma, but its leaves are much larger and broader, and the ovary has 3 placentas, a 3-lobed stigma, and more than 2 ovules to each placenta.—*Benth.*

Order XII. PITTOSPOREÆ.

Flowers hermaphrodite, regular or oblique. Sepals 5, distinct and imbricate, or rarely connate at the base. Petals 5, imbricate, the claws or narrowed base usually erect and connivent or cohering in a tube, rarely spreading from the base. Stamens 5, hypogynous, free, alternating with the petals. Torus small, rarely produced into a short gynophore, sometimes bearing 5 glands. Ovary 1-celled, with 2 or rarely 3 to 5 parietal placentas, or divided into cells by the protrusion of the placentas, which often unite in the axis, at least after flowering. Style simple, with an entire, small, capitate, or dilated stigma. Ovules several, superposed in 2 rows on each placenta, horizontal. Fruit either a capsule opening loculicidally, the valves sometimes splitting also septicidally, or succulent and indehiscent. Seeds several or rarely solitary in each cell, dry or enveloped in pulp, with a thin testa, smooth or rarely muricate, and a hard albumen. Embryo very small, in a cavity of the albumen next the hilum.—Trees, erect shrubs, or undershrubs, with flexuose, decumbent or twining branches. Leaves alternate, entire, toothed, or rarely lobed, without stipules. Flowers white, blue, yellow, or rarely reddish, terminal or axillary, solitary and nodding, or in short racemes or corymbose panicles.

With the exception of *Pittosporum* itself, the genera are all limited to Australia.

* *Anthers ovate or oblong. Capsule dehiscent. Petals (except in Bursaria) erect at the base.*

Trees or erect shrubs. Petals erect at the base. Capsule thick or coriaceous. Seeds several.
    Seeds thick, not winged. Flowers usually small . . . . 1. PITTOSPORUM.
    Seeds flat, horizontal, winged. Flowers large, yellow . . . . . . 2. HYMENOSPORUM.
Erect shrubs, often prickly. Petals small, spreading from the base. Capsule thin, small, and flat. Seeds 1 or 2 in each cell, vertical, flat . 3. BURSARIA.
Undershrubs or twiners. Petals erect at the base. Capsule membranous or thinly coriaceous, Seeds thick or horizontal . . . . 4. MARIANTHUS.

68   XII. PITTOSPOREÆ.

** *Anthers ovate or oblong. Berry indehiscent. Petals erect at the base.*
Prickly shrub, with small leaves and small sessile solitary flowers. Berry
  globular . . . . 5. CITRIOBATUS.
Undershrubs or twiners. Flowers pedunculate. Berry ovoid or oblong . . 6. BILLARDIERA.

*** *Anthers linear, or longer than the filaments. Petals spreading from the base, or nearly so.*
Undershrubs or twiners.
Fruit dehiscent. Anthers turned to one side, opening in terminal pores . . 7. CHEIRANTHERA.

1. PITTOSPORUM, Banks.

(From the gummy matter surrounding the seeds.)

Petals usually connivent or cohering in a tube at their base or above the middle. Anthers ovate-oblong. Ovary sessile or shortly stipitate, incompletely or almost completely 2-celled, or rarely 3 to 5-celled; style short. Capsule globose, ovate or obovate, often laterally compressed; the valves coriaceous or thick and hard, bearing the placentas along their centre. Seeds thick or globular, not winged, often enveloped in a viscous liquor.—Shrubs or trees, glabrous or rarely tomentose. Leaves usually evergreen, entire or minutely toothed, the upper ones frequently collected into a false whorl. Flowers not large, axillary or terminal, solitary or in close corymbose panicles.

A large genus, dispersed over the warmer regions of Africa, Asia, the Pacific islands, and New Zealand. The Australian species are all endemic excepting one which is common to eastern tropical Asia and the eastern Archipelago.

Flowers numerous, small, in compound terminal corymbs, with the lower
  branches axillary.
  Leaves ovate-rhomboid, toothed. Sepals obtuse . . . . 1. *P. rhombifolium*.
  Leaves oblanceolate, aristate. Sepals broadly ovate . . . . 2. *P. setigerum*.
  Leaves from obovate to oblong or lanceolate, quite entire. Sepals
    subulate or subulate-pointed. Plant glabrous . . . . 3. *P. melanospermum*.
Peduncles all terminal, clustered, short, each bearing a short simple cyme
  or umbel.
  Glabrous, or the young shoots and inflorescence very slightly pubescent.
    Flowers about ½in. long . . . . 4. *P. undulatum*.
  Young shoots and inflorescence rusty-tomentose or hirsute
    Flowers about ½in. Capsule ¾in., very rough . . . . 5. *P. revolutum*.
    Veins of the leaf dark-coloured, 8 to 10 lines . . . . 6. *P. venulosum*.
    Flowers 3 to 4 lines. Capsule under ½in.
      Leaves on long petioles, ovate to oblong-lanceolate. Tomentum
        short and crisp, ferruginous . . . . 7. *P. ferrugineum*.
      Leaves of thin texture, petioles of medium length. Capsule
        velvety . . . . 8. *P. Wingii*.
      Leaves nearly sessile, oblong-lanceolate. Tomentum almost
        hirsute, very dense . . . . 9. *P. rubiginosum*.
Pedicels axillary, solitary or clustered, 1-flowered; the uppermost sometimes in a terminal cluster. Leaves glabrous, flat. Flowers yellow 10. *P. phillyræoides*.

1. **P. rhombifolium** (form of leaf), *A. Cunn. in Hook. Ic. Pl. t. 621; Benth. Fl. Austr.* i. 110. A tree attaining, according to A. Cunningham, 60 to 80ft., glabrous in all its parts. Leaves rhomboid-oval or rarely broadly oblong-lanceolate, mostly 3 to 4in. long, coarsely and irregularly toothed from the middle upwards, narrowed into a petiole of ½ to 1in., coriaceous and shining, but with the pinnate and netted veins prominent on both sides. Flowers white, numerous, and rather small, in a dense terminal compound corymb, the branches sometimes minutely glandular. Sepals obtuse, rather more than 1 line. Petals oblong, about 3 lines long, spreading from below the middle. Ovary shortly stipitate, the thick placentas nearly meeting, each bearing about 12 to 14 ovules. Capsule more or less obliquely pear-shaped or almost globular, usually about 3 lines long, and ripening 2 or 3 black seeds.

Hab.: Wide Bay, forests on the Brisbane; Araucaria range, between Brisbane and Dawson rivers and edge of the Killarney scrub, near Warwick.

Wood whitish, close-grained, tough, rather hard, considered suitable for carving and engraving.—*Bailey's Cat. Ql. Woods No. 8.*

Pittosporum setigerum, Bail.

2. **P. setigerum** (bristle-bearing), *Bail.* A small glabrous tree. Branchlets furrowed, with the bark often reddish. Leaves coriaceous, the reticulate veins close and raised, 2½ to 4in. long, ¾ to 1in. wide, tapering from above the middle to a rather long slender petiole, the apex terminating in a prominent bristle. Flowers, judging from the dried ones, seemingly light-yellow in a broad-spreading terminal panicle longer than the leaves. Pedicels slender. Sepals broadly ovate, minute. Petals free, patent, about 2½ lines long, obtuse, veins obscure. Stamens shortly exceeding the petals. Ovary on a short glabrous stipes, densely covered with a white tomentum. Capsule globose, 4 lines diameter, exuding an amber-coloured resin or gum. Seeds black, angular, from 2 to 6 in each capsule.

Hab.: Walsh River, *T. Barclay-Millar.*

3. **P. melanospermum** (black seed), *F. v. M. Fragm.* i. 70; *Benth. Fl. Austr.* i. 111. A small tree, quite glabrous, or with a scanty minute glandular pubescence on the inflorescence. Leaves from obovate to oblong or even lanceolate, shortly acuminate, mucronate or obtuse, 2 to 4in. long, entire and flat or slightly undulate on the margin, narrowed into a petiole of 4 to 5 lines, coriaceous, but not shining, of a pale hue and prominently veined. Corymbs compound, terminal, many-flowered, but shorter than the last leaves. Flowers small, the sepals subulate or lanceolate-subulate, the petals 3 or scarcely 4 lines long, spreading from about the middle. Ovary shortly stipitate, with 10 to 12 ovules to each placenta. Capsule obliquely globular or pear-shaped, somewhat compressed, with few or sometimes a single black seed.

Hab.: Tropical parts of the colony.
Var.(?) *lateralis.* Corymbs usually lateral. York Sound. *A. Cunningham;* Whitsunday Island, *Henne.*

4. **P. undulatum** (leaves wavy), *Vent. Hort. Cels. t.* 76; *Xenth. Fl. Austr.* i. 111. Mock orange. A tree, attaining in favourable situations 40ft., or according to M'Arthur 60 to 90ft., although in barren exposed localities it remains a shrub, quite glabrous, except a slight appressed pubescence on the young shoots and inflorescence. Leaves from oval-oblong to lanceolate, mostly 3 to 6in. long and acuminate, flat or undulate on the margin, narrowed into a petiole of ½ to ¾in., coriaceous and shining, with the veins little conspicuous, the upper ones often almost whorled. Peduncles several, in terminal clusters, much shorter than the leaves, mostly bearing a simple cyme or umbel of 3 or 4 rather large white flowers, and one or two often 1-flowered. Sepals lanceolate, acuminate, often connate at the base. Petals 5 to 6 lines long, spreading from the middle. Ovary almost sessile, hairy, the 2 placentas united at the base, each bearing numerous ovules. Capsule nearly globular, rarely attaining ½in., smooth, with thick coriaceous valves and numerous seeds.—*DC. Prod.* i. 846; *Andr. Bot. Rep.* t. 383; *Bot. Reg. t.* 16; *F. v. M. Pl. Vict.* i. 71 and 224.

Hab.: A common tree upon the ranges of southern Queensland.
Wood of a light colour, close in grain, and tough.—*Bailey's Cat. Ql. Woods No.* 9.
The bark contains a dye.

5. **P. revolutum** (petals rolled back), *Ait. Hort. Kew. ed.* 2 ii. 27; *Benth. Fl. Austr.* i. 111. A tall shrub, the young shoots tomentose. Leaves ovate-elliptical or elliptical-oblong, shortly acuminate, 2 to 4in. long, scarcely undulate, narrowed into a petiole, usually very short, but sometimes near ½in., coriaceous, glabrous above when full grown, clothed underneath with a loose rusty tomentum easily rubbed off, the upper ones often almost whorled. Peduncles terminal, few or solitary, usually decurved, bearing sometimes a single rather large flower, but more frequently a short dense ovate or corymbose raceme. Sepals lanceolate-subulate. Petals nearly ½in. long, often united to above the middle, shortly spreading or recurved at the top. Ovary very hirsute, with very numerous ovules to each placenta; stigma peltate. Capsule ¼ to ¾in. long, the hard almost woody

valves rough outside. Seeds numerous, red or brown.—DC. Prod. i. 346; Bot. Reg. t. 186; F. v. M. Pl. Vict. i. 224; *P. fulvum*, Rudge in Trans. Linn. Soc. x. 298. t. 20; DC. l.c.; Sweet, Fl. Austral. t. 25; *P. tomentosum*, Bonpl. Jard. Malm. 56 t. 24; Sweet, Fl. Austral. t. 38; DC. l.c.; *P. hirsutum*, Link, according to Putterl. Syn. Pittosp. 9.

Hab.: Moreton Bay, Brisbane River, and other southern parts.

6. **P. venulosum** (leaves beautifully veined), *F. v. M. Fragm.* vi. 186. A small tree of about 30ft.; the ultimate branchlets almost verticillate, slightly ferruginous-tomentose. Leaves lanceolate or ovate-lanceolate, acuminate, almost whorled, 2 to 4in. long, ¾ to 1½in. broad, the reticulations close and deep-coloured, margins thickened, tapering to the petioles. Flowers in a terminal corymbose panicle. Capsule globose-pyriform, 2 rarely 3-valved, slightly compressed, 8 to 10 lines long, yellow inside. Seeds 1 to 1¼ line long.

Hab.: Ranges of the tropical coast, as Rockingham Bay, &c.

7. **P. ferrugineum** (shoots rusty-tomentose), *Ait. Hort. Kew. ed.* 2, ii. 27; *Benth. Fl. Austr.* i. 112. A tree, flowering sometimes as a shrub, but attaining a height of 50 to 60ft., the young shoots thickly clothed with a loose rusty tomentum which soon wears off. Leaves from obovate or ovate, and obtuse or scarcely acuminate, to oblong or almost lanceolate, acuminate, and 3 to 4in. long, quite entire, narrowed into a petiole of ¼ to ¾in., rusty-tomentose on both sides when very young, but glabrous above, or on both sides when full grown. Peduncles terminal, usually clustered several together above the last leaves, each one bearing a cluster or umbel of rather small flowers, but sometimes the common peduncle grows out and the inflorescence becomes a thyrsoid or pyramidal panicle, not a corymb, as in *P. melanospermum*. Sepals lanceolate or lanceolate-subulate. Petals narrow, about 3 lines long, spreading only above the middle. Ovary villous, with 12 to 16 ovules to each placenta. Capsule sessile, nearly globular, scarcely 4 lines broad, ripening usually 3 or 4 black seeds.—DC. Prod. i. 346; Bot. Mag. t. 2075; *P. tinifolium (linifolium* by an error of the press), A. Cunn. in Ann. Nat. Hist. ser. 1, iv. 109; *P. ovatifolium*, F. v. M. Fragm. ii. 78.

Hab.: Common in the tropical parts.
Wood light grey, close in grain and tough.—*Bailey's Cat. Ql. Woods No.* 10.
Extends over the Malayan peninsula and adjoining islands and the Philippines. The Australian specimens have rather larger flowers and narrower-pointed sepals than the common Malayan form; but in this respect the Malacca specimens are very variable, some of them precisely resembling some of the Australian ones.

8. **P. Wingii** (after the editor of Southern Science Record), *F. v. M. in So. Sc. Rec.*, March, 1885. Leaves of almost herbaceous texture, on very short stalks, ovate or elongate-lanceolate, acuminate, hardly or slightly recurved at the margin, beneath prominently penninerved and as well as the branchlets brownish silky-tomentose; corymb umbelliform, solitary short-stalked or almost sessile; sepals velvet-hairy, narrow-lanceolate, gradually pointed; corolla about one-third longer than the calyx, its tube widened upwards, shorter than the bluntish and not much-spreading lobes; anthers fully half as long as the filaments, many times longer than broad; ovary brownish-silky; capsules not large, rather turgid, almost globular or somewhat depressed, velvet-hairy; valves 2, hard; funicles thick and very short. Seeds several, from garnet colour turning brown-black, somewhat viscid.

Hab.: Bellenden Ker and other high ranges of the tropical parts of the colony.

9. **P. rubiginosum** (foliage reddish-coloured), *A. Cunn. in Ann. Nat. Hist. ser.* 1, iv. 108; *Benth. Fl. Austr.* i. 112. A sparingly branched shrub; the branchlets, petioles, and inflorescence densely clothed with a rust-coloured tomentum, consisting of much more spreading hairs than in *P. ferrugineum*.

Leaves almost whorled, oblong-lanceolate, acutely acuminate, 5 to 10in. long, entire or slightly sinuate-toothed, narrowed at the base, but almost sessile, herbaceous, glabrous above, softly pubescent underneath. Peduncles stout. Sepals with scattered hairs, the petals 3 times as long as the sepals, cohering for about two-thirds of their length into an almost cylindrical tube, thin summits pointed and much recurved; filaments twice as long as the anthers. Capsule large, rugose, ovate-cordate, deep yellow.

Hab.: Common on the ranges about Bellenden Ker.
Sometimes infested with the fungus *Sphærella rubiginosa.*—Cke.

10. **P. phillyræoides** (Phillyrea-like), *DC.*; *Benth. Fl. Austr.* i. 112. A small graceful tree or slender shrub, quite glabrous in all its parts. Leaves usually oblong or linear-lanceolate with a small hooked point, 2 to 4in. long, quite entire, narrowed into a petiole, thick coriaceous and indistinctly veined, but in some forms short and broadly oblong, in others long and narrow. Pedicels axillary, solitary or in sessile or shortly pedunculate clusters or umbels, or the uppermost forming a terminal cluster. Flowers yellow, usually about 4 lines long, often diœcious, the females rather larger and fewer together than the males. Sepals short and very obtuse. Petals united to the middle or still higher, spreading at the top. Ovary pubescent, almost completely 2-celled, with 6 to 8 ovules in each cell. Fruit ovate or round-cordate, much compressed, quite smooth, varying from 4 to 9 lines in length, but usually about ½in. Seeds few, dark or orange-red.—Putterl. in Pl. Preiss. i. 192; F. v. M. Pl. Vict. i. 72; *P. angustifolium,* Lodd. Bot. Cab. t. 1859; *P. longifolium* and *P. Roëanum,* Putterl. Syn. Pittosp. 15, 16; *P. ligustrifolium,* A. Cunn. in Putterl. l.c. 16, and in Ann. Nat. Hist. ser. 1, iv. 110; Putterl. in Pl. Preiss. i. 190; *P. oleæfolium,* A. Cunn. in Putterl. Syn. Pittosp. 17; *P. acacioides,* A. Cunn. in Ann. Nat. Hist. ser. 1, iv. 109; *P. salicinum,* Lindl. in Mitch. Trop. Austr. 97; *P. lanceolatum,* A. Cunn. in Mitch. l.c. 272 and 291.

Hab.: Common on the inland downs.
Wood of a light colour, close-grained, and very hard.—*Bailey's Cat. Ql. Woods No.* 11.

2. HYMENOSPORUM, F. v. M.

(Seeds winged.)

Petals connivent or cohering in a tube to above the middle. Anthers ovate-oblong. Ovary incompletely 2-celled; style short. Capsule ovate, compressed, with thick coriaceous valves. Seeds numerous, horizontally imbricated, flat, reniform, surrounded by a membranous wing.—A shrub or tree, with the habit of *Pittosporum,* from which it only differs in its large flowers and in its seeds.

The genus is limited to a single species, endemic in Australia.

1. **H. flavum** (yellow flowers), *F. v. M. Fragm.* ii. 77; *Benth. Fl. Aust.* i. 114. A handsome evergreen shrub or tree, glabrous, except a loose pubescence on the inflorescence, and sometimes on the under side of the leaves. Leaves ovate-oblong or oblanceolate, acuminate, entire, from 3 to 5 or even 6in. long, narrowed into a petiole of ½in. or more, the upper ones often almost verticillate. Panicle terminal, loose, corymbose, often 6 to 8in. diameter, with small linear or lanceolate bracts. Flowers large, yellow. Sepals oblong-lanceolate, 3 to 4 lines long. Petals silky-tomentose outside, the erect base or broad claws nearly 1in., the spreading lamina nearly ½in. long. Ovary linear, silky-tomentose, with numerous ovules. Capsule stipitate, much flattened, 1in. or more long and nearly as broad. Seeds, including the wing, fully 4 lines broad.—*Pittosporum flavum,* Hook. Bot. Mag. t. 4799.

Hab.: Wide Bay district, Brisbane River, Ipswich, and many other localities in the south of the colony.
Wood whitish, close-grained, and tough.—*Bailey's Cat. Ql. Woods No.* 12.

XII. PITTOSPOREÆ.

3. BURSARIA, Cav.
(Capsules pouch-like.)

Petals narrow, spreading from near the base. Anthers ovoid. Ovary incompletely 2-celled; style short. Capsule shortly stipitate, flat, broadly orbicular, opening round the edge, with thinly coriaceous flat valves. Seeds 1 or 2 in each cell, flat, reniform, not winged.—Rigid, much branched shrubs or trees, often thorny. Leaves small, entire. Flowers small, in terminal panicles. Sepals very fugacious.

Leaves ½ to 1in. long, glabrous . . . . 1. *B. spinosa.*
Leaves 2 to 3in. long, hoary tomentose . . . . 2. *B. incana.*
Leaves 2½in. long, lanceolate, almost membranous . . . . 3. *B. tenuifolia.*

1. **B. spinosa** (spiny) *Cav. Ic.* iv. 30, *t.* 350; *Benth. Fl. Austr.* i. 114. A shrub or small tree, glabrous, and when young very bushy, the smaller branches often reduced to short subulate thorns. Leaves very variable, most frequently clustered, obovate, oblong or cuneate, obtuse, truncate or notched, ½ to 1in. long, narrowed at the base, and sometimes shortly petiolate, green on both sides; in luxuriant specimens they vary to oblong-lanceolate, 1 to 2in. long; in a few others they have occasionally a few coarse teeth at the top. Flowers white, usually very numerous, in a broad, pyramidal, terminal panicle, arranged along its branches in short racemes, on pedicels of 1 to 3 lines; occasionally the panicles are reduced to short racemes or to 1 or 2 terminal flowers. Bracts minute and very fugacious. Sepals small, also falling off long before the petals open. Petals narrow, about 2 lines long. Capsule 3 to 4 lines or, in the var. *incana*, sometimes 5 lines broad.—DC. Prod. i. 347; Bot. Mag. t. 1767; Hook. f. Fl. Tasm. i. 39; F. v. M. Pl. Vict. i. 74; *Itea spinosa*, Andr. Bot. Rep. t. 814.

Hab.: Common in southern Queensland.

2. **B. incana** (young shoots and foliage hoary), *Lindl. in Mitch. Trop. Austr.* 224. A small erect tree, the shoots, inflorescence, and under side of the leaves hoary-white, with a dense soft or close thin tomentum. Leaves 2 or 3in. long, oblong, obtuse, sometimes toothed at the end. Flowers and capsules of *B. spinosa*, but larger.

Hab : Mostly upon the ranges of the southern parts of the colony.
Wood white or light coloured, suitable for engraving, &c.—*Bailey's Cat. Ql. Woods No 13.*

3. **B. tenuifolia** (thin-leaved), *Bail.* A tall shrub or small tree, the branchlets more or less corrugated and bearing prominent lenticels. Leaves lanceolate, about 2¼in. long, the apex obtuse, tapering to a rather slender petiole 2 or 3 lines long, smooth and rather glossy with the erecto-patent parallel nerves prominent above, the under side covered with a thin tomentum, the nerves and principal veins showing as dark lines, margins entire. Panicles elongated upon peduncles of from only a few lines to 2in. Flowers pedicellate, bracts narrow-lanceolate, an upper one often very narrow, ferruginous. Sepals lanceolate, ciliate. Petals recurved, oblong-linear, 2¾ lines long, marked with 3 lines; filaments about as long as the petals, subulate. Ovary glabrescent.

Hab.: Barron River, *E. Cowley*; Shaw Island, *Lord Lamington*; Northcote, *R. C. Burton*; Herberton, *J. F. Bailey.* This species has more membranous leaves, more slender branchlets and panicles than any of the other species of the genus.

4. MARIANTHUS, Hueg.
(Dedicated to the Virgin Mary.)

(Calopetalum, *Harv.*; Oncosporum, *Putterl.*; and Rhytidosporum, *F. v. M.*)

Petals connivent at the base or above the middle, spreading at the top. Anthers oblong or ovate, shorter than the filaments. Ovary sessile or shortly stipitate, usually completely 2-celled, glabrous, except very rarely in *M. laxiflorus*. Capsule ovoid or oblong, turgid or slightly compressed, membranous or slightly coriaceous, the valves sometimes splitting septicidally. Seeds ovoid, reniform or globular.—Undershrubs, with procumbent, flexuose, or more frequently twining branches. Leaves entire, toothed, or the lower ones occasionally lobed. Flowers blue, white, and reddish, in terminal compact panicles, usually corymbose or almost umbellate, rarely solitary or apparently axillary from the extreme shortness of the flowering branch.

The genus is limited to Australia. It differs from *Billardiera* solely in the capsular not baccate fruit. The petals are in general more spreading than in *Billardiera*.—*Benth.*

1. **M. procumbens** (procumbent), *Benth. Fl. Austr.* i. 117. A low, prostrate or suberect, much-branched shrub, the branches sometimes flexuose and nearly 1ft. long, but usually much shorter, glabrous or slightly pubescent. Leaves crowded and sessile, in the northern varieties usually linear or linear-cuneate, pointed, entire or rarely toothed at the top, 4 to 6 lines long, rigid, with recurved margins; in the southern forms usually shorter, more cuneate or even obovate or ovate, and often toothed. Flowers small, white or tinged with red, solitary or 2 or 3 together, terminal or appearing axillary from the shortness of the flowering shoots, the pedicels 1 to 2 lines long and always shorter than the leaves at the time of flowering, rather longer and recurved when in fruit. Sepals lanceolate-linear, very pointed. Petals about 3 lines long or smaller, spreading from below the middle. Filaments dilated to the middle. Ovules 6 to 8 in each cell of the ovary. Style short. Capsule truncate, 3 lines broad and not quite so long. Seeds usually 3 or 4 in each cell, ovoid-reniform, transverse and laterally-attached, deeply wrinkled.—*Pittosporum procumbens* and *P. nanum*, Hook. Comp. Bot. Mag. i. 275; *Bursaria procumbens*, Putterl. Syn. Pittosp. 20; Hook. f. Fl. Tasm. i. 39; *B. diosmoides*, Putterl. l.c. (from the description I have not seen Sieber's n. 554); *B. Stuartiana*, Klatt, in Linnæa, xxviii. 568; *Rhytidosporum procumbens*, F. v. M. 1st Gen. Rep. 10; Pl. Vict. i. 75; *Campylanthera ericoides* Lindl. in Mitch. Three Exped. ii. 277.

Hab.: Southern parts of the colony.

5. CITRIOBATUS, A. Cunn.

(Derived from local name, "Orangethorn.")

(Ixiosporum, *F. v. M.*)

Petals connivent or connate to above the middle, in a cylindrical tube spreading at the top. Anthers oblong, shorter than the filaments. Ovary 1-celled, with 2 to 5 parietal placentas; style short. Fruit coriaceous or hard, globular, indehiscent. Seeds few or many, nearly globular, often enveloped in a viscous fluid.—Rigid much-branched shrubs, armed with short thorns or abortive branches. Leaves small, entire or toothed. Flowers small, sessile and usually solitary, surrounded by small sepal-like bracts.

The genus is limited to Australia.

Leaves cuneate, sessile, 4 to 6 lines long. Placentas 2, with 8 to 10 ovules each. Fruit 2 to 5 lines diameter, with few seeds . . . . 1. *C. multiflorus*.
Leaves obovate petiolate, 6 lines long. Placentas 5, with very numerous ovules. Fruit 1in. diameter or larger, with numerous seeds . . . . 2. *C. pauciflorus*.
Leaves entire, lanceolate, 1½in. long. Fruit 5 lines diameter . . . . . 3. *C. lancifolius*.

1. **C. multiflorus** (flowers numerous), *A. Cunn. in Loud. Hort. Brit. (name only) and in. Putterl. Syn. Pittosp.* 4; *Benth. Fl. Austr.* i. 121. ("Kary," Rockhampton, *Thozet.*) A straggling or prostrate very much branched shrub, with slender branches, rough with a minute pubescence, and bearing numerous subulate thorns or abortive branches. Leaves sessile, ovate, orbicular, obovate, or broadly cuneate, usually 4 to 6 lines long, entire or with a few small pointed or prickly teeth, rather thin, green and glabrous on both sides. Flowers about 2

lines long, always solitary in the axils, and not very numerous on the bush, notwithstanding the specific name. Ovary pubescent, with 2 parietal placentas and 8 to 12 ovules to each. Berry 2 to 5 lines diameter, containing from 2 to above a dozen seeds, which are not viscid.

Hab.: Southern parts of the colony.
Wood close in grain and very tough, colour light.—*Bailey's Cat. Ql. Woods No.* 13a.

2. **C. pauciflorus** (few flowers), *A. Cunn. in Loud. Hort. Brit. Suppl.* 585 *(name only); Benth. Fl. Austr.* i. 122. Habit of *C. multiflorus*, but stouter and more rigid, the branches similarly rough, with a minute pubescence, and thorny. Leaves from obovate to cuneate-oblong, rarely orbicular, mostly entire and obtuse, but occasionally mucronate or truncate and 3-toothed, rarely exceeding ½in. in length, often petiolate and more rigid than in *C. multiflorus*. Flowers larger than in that species, the petals 4 to 5 lines long, united into a complete tube to two-thirds of their length. Ovary pubescent, with 5 parietal placentas, covered with innumerable minute ovules. Style longer than in *C. multiflorus*. Fruit attaining 1 to 1½in. diameter, with a thick coriaceous pericarp. Seeds numerous, in a viscid pulp.—*Leiosporus spinescens*, F. v. M. Fragm. Phyt. Austr. ii. 76.

Hab.: Coast scrubs of central parts of the colony.
Wood close-grained, of a light uniform colour, and hard.—*Bailey's Cat. Ql. Woods No* 14.

8. **C. lancifolius** (lance-leaved), *Bail.* A small tree, bark whitish, branchlets slender, and the smaller ones often terminating in sharp spines. Leaves alternate, lanceolate, membranous, about 1½in. long, ½in. broad, on very short slender petioles, the veins very oblique, looping far within the margin, delicately reticulate and dotted with guttate oil-cells, margins entire. Flowers axillary or lateral, solitary or in pairs, on very short peduncles, bracts minute. Sepals 5, linear, recurved. Petals 5, linear, more or less imbricate, cohering in a tube of nearly their whole length. Stamens 5, shorter than the petals and opposite them; filaments flattened but tapering towards the anther, which is sagittate, free with 2 cell-slits the length of the anther. Style glabrous, stigma truncate or very shortly and obtusely lobed; ovary silky-hairy, seems to be 1-celled, superior. Fruit a berry, nearly globose, about 5 lines in diameter, with a thin coriaceous pericarp. Seeds 9 in the fruits opened, enveloped in a viscous fluid, somewhat reniform, flattened, dark-brown.

Hab.: Killarney, on border of scrub, in flower, *J. F. Bailey*; Warwick, fruit specimen, *C. J. Gwyther*.

6. BILLARDIERA, Sm.
(After J. J. Labillardière.)

Petals connivent or cohering in a tube to above the middle, spreading at the top. Anthers oblong or ovate, shorter than the filaments. Ovary sessile, or nearly so, completely or rarely imperfectly 2-celled, glabrous or pubescent. Fruit succulent or fleshy and indehiscent, ovoid or oblong. Seeds ovoid reniform or globular, often enveloped in a viscid pulp.—Undershrubs, with the branches usually twining. Leaves entire or sinuate. Flowers greenish-yellow, purple, or rarely blue, either solitary or clustered and pendulous, or in terminal cymes and erect.

The genus is limited to Australia, and differs from *Marianthus* only in the baccate, not capsular, fruit.

1. **B. scandens** (climbing), *Sm. Bot. Nov. Holl.* i. t. 1; *Benth. Fl. Austr.* i. 123. Stems twining, often to a considerable extent, or short and flexuose, nearly glabrous or more or less silky or velvety-pubescent, or hairy. Leaves from ovate-lanceolate to lanceolate or linear, obtuse or with a recurved point, usually 1 to 2in. long, entire or often with undulate margin, usually narrowed into a

short petiole. Flowers from greenish or pale yellow to violet or purple, pendulous on slender terminal pedicels varying from a line or two to above ½in., solitary or very rarely 2 together. Sepals lanceolate or lanceolate-subulate. Petals spreading from above the middle, so as to form a narrow-campanulate corolla, 8 to 10 lines or rarely 1in. long. Ovary glabrous or pubescent, 2-celled, with a very short style and broad hollow stigma. Berries cylindrical or ovoid-oblong, 2-celled, glabrous or downy. Seeds numerous, in a close double row in each cell and embedded in pulp.—DC. Prod. i. 345; Bot. Mag. t. 801; Sweet, Fl. Austral. t. 54; F. v. M. Pl. Vict. i. 79; *B. latifolia*, Putterl. Nov. Stirp. Dec. 47, but not of Klatt, Linnæa, xxviii. 570; *B. grandiflora*, Putterl l.c. 48 (all the above referring to specimens with pubescent ovaries and fruits); *B. mutabilis*, Salisb. Parad. Lond. t. 48; Bot. Mag. t. 1313; DC. Prod. i. 345; Hook. f. Fl. Tasm. i. 37 (with glabrous ovaries and fruits); *B. angustifolia*, DC. Prod. i. 345; *B. canariensis*, Wendl. Hort. Herrenh. t. 15.

Hab.: Sandy coast lands in the south of the colony.

7. CHEIRANTHERA, A. Cunn.

(Supposed resemblance of anthers to fingers of the hand.)

Petals spreading from nearly the base, obovate-oblong. Anthers longer than the filaments, all turned towards one side, opening by two pores at the top. Ovary 2-celled with a subulate style. Capsule oblong, hard, opening loculicidally in 2 valves, the valves also splitting septicidally. Seeds nearly globular.— Branches flexuose or twining. Leaves narrow. Flowers in terminal corymbs or cymes, or drooping from terminal solitary pedicels.

The genus is limited to Australia.

1. C. linearis (linear leaves), *A. Cunn. in Bot. Reg. under t.* 1719; *Benth. Fl. Austr.* i. 127. A low glabrous shrub or undershrub, with erect twiggy branches of 6in. to 1ft., or rarely longer. Leaves linear, acute or rather obtuse, ¾ to 1½in. or rarely 2in. long, entire or minutely toothed, flat, and ½ to 1 line broad, or the margins incurved, so as to be almost terete, with smaller leaves often clustered in the axils. Flowers blue and showy. Sepals lanceolate, 2 to 2½ lines long. Petals 8 to 10 lines. Filaments short. Anthers rather longer, but not reaching to the middle, and often not one-third of the length of the petals. Capsule very like those of *Marianthus pictus* and *lineatus* (two W. Aust. plants), oblong, much flattened, hard, but dehiscent when ripe.—Hook. Ic. Pl. t. 47; Fl. des Serres viii. t. 856; F. v. M. Fragm. i. 76; *C. cyanea*, Brongn. Voy. Coq. t. 77.

Hab.: Stanthorpe (in flower in Aug.)

Order XIII. TREMANDREÆ.

Flowers regular. Sepals 4 or 5, very rarely 3, free, valvate in the bud. Petals as many, hypogynous, spreading, induplicate-valvate in the bud. Stamens twice as many, hypogynous, free; filaments short; anthers oblong or linear, 2 or 4-celled, opening in a single terminal pore. Torus small or rarely expanded into a disk between the petals and stamens. Ovary sessile or nearly so, usually 2-celled; style filiform, deciduous, entire or minutely 2-lobed. Ovules solitary in each cell, or 2, one above the other, or rarely an additional small collateral one, pendulous, anatropous, with a ventral raphe. Capsule usually flattened, 2-celled, opening loculicidally at the edges. Seeds pendulous, the raphe usually expanded at the chalazal extremity into a twisted or strophiola-like appendage, rarely

XIII. TREMANDREÆ.

wanting; the testa crustaceous, glabrous or hairy; albumen fleshy or almost cartilaginous. Embryo small, straight, with a superior radicle.—Shrubs usually heath-like, glabrous or glandular-hairy, with small alternate opposite or verticillate leaves, rarely with a stellate tomentum and larger leaves. Flowers solitary, on axillary pedicels, usually red or purple. In many species, as in *Pittosporeæ* and *Polygaleæ*, a flower may here and there be found with a 3-merous ovary and fruit.

The Order is strictly confined to Australia, and although showing some affinity with *Cheiranthera* in *Pittosporeæ*, as well as with *Polygaleæ* proper, it is yet very different from either; the connection with *Lasiopetaleæ*, insisted upon by Steetz, appears to rest almost entirely on the valvate calyx, and on an occasional resemblance in habit, which is, however, partaken in by *Bauera* and several other genera of Australian heath-like shrubs, which have little else in common.—*Benth.*

1. TETRATHECA, Sm.

(From *tetra*, four, and *theka*, a box. Anthers 4-celled.)

Stamens apparently in a simple series, the anthers continuous with the filaments, 2-celled or 4-celled with 2 of the cells in front of the 2 others, more or less contracted into a tube at the top. Disk none. Capsule opening only at the edges. Seeds with an appendage at the chalazal end, usually contorted and glabrous or glandular-hairy. Leaves alternate, verticillate or scattered, heath-like and entire, or flat and toothed, or reduced to minute scales.

1. **T. thymifolia** (thyme-leaved), *Sm. Exot. Bot.* i. 41 t. 22. Intermediate between *T. ciliata* and *T. ericifolia*, it has usually the tall habit of the former, but is much more pubescent or hirsute. Leaves almost all verticillate in threes or fours, ovate-elliptical or lanceolate, the margins more or less recurved or revolute. Flowers of *T. ciliata*, except that the sepals are usually ovate-lanceolate, more acute or acuminate than in either of the two allied species, and seldom reflexed. Ovary glabrous, or more frequently pubescent, with 2 superposed ovules in each cell, and occasionally a third collateral one. Capsule broad, 2 to 4 lines long. Seed hairy.

Hab.: Sandy coast lands of the southern parts of the colony. Flowering about from Feb. to May.

Order XIV. POLYGALEÆ.

Flowers hermaphrodite, irregular. Sepals 5, free, much imbricate, the 2 inner ones usually larger and petal-like. Petals 3 or 5, rarely all free, most frequently 2 or 4 in pairs united at the base with the lower concave or helmet-shaped petal or keel and often with the staminal tube. Stamens 8, rarely 5 or 4, usually united to above the middle in a sheath open on the upper side. Anthers erect, 1 or 2-celled, usually opening by a single terminal or oblique pore. Torus small, or rarely expanded into a disk within the stamens. Ovary free, 2-celled or rarely 1-celled, or in a few flowers 3 to 5-celled. Style simple, usually curved at the top, with a variously shaped entire or 2-lobed stigma. Ovules usually solitary in each cell, pendulous, anatropous with a ventral raphe. Seeds pendulous, the crustaceous testa often hairy, and bearing a caruncle at the hilum or at the opposite end. Albumen fleshy or rarely deficient. Embryo straight, with flat, convex, or rarely thick and fleshy cotyledons.—Herbs, undershrubs, or small shrubs, rarely tall shrubs, climbers or trees (one tree in Queensland, *Xanthophyllum*), glabrous or hairy, but without stellate hairs. Leaves usually alternate and entire, without stipules, very rarely opposite. Flowers solitary or in spikes or racemes, rarely paniculate, the pedicels usually articulate at the base, with a subtending bract and 2 bracteoles.

XIV. POLYGALEÆ.

A considerable Order, widely dispersed over nearly the whole globe. Of the four Australian genera, one is the largest and most extensively diffused of the whole Order, here represented by a very few species of an Asiatic or African type; two others are Asiatic, extending to Australia; the third is endemic. - *Benth.*

Sepals nearly equal. Anthers 4 or 5. Flowers minute, in terminal spikes. Inner sepals larger and petal-like. Anthers 3. . . . . 1. SALOMONIA.
Capsule ovate or orbicular, scarcely contracted at the base. Seeds not comose.
  Lateral petals united with the carina (which is always crested in the Australian species) . . . . 2. POLYGALA.
  Lateral petals adnate to the staminal column, but distinct from the carina (which is not crested) . . . . 3. COMESPERMA.
Capsule cuneate, very narrow at the base. Seed hairs forming a long coma 3. COMESPERMA.
Sepals nearly equal. Petals 5. Stamens 8, free, 1 or imperfectly 2-celled. Ovules numerous. Fruit globose, indehiscent . . . . . . 4. XANTHOPHYLLUM.

1. SALOMONIA, Lour.
(After the Hebrew king Solomon.)

Sepals nearly equal, the 2 innermost rather larger. Petals 3, united in a single corolla open on the upper side, the keel not crested. Stamens united nearly to the top into a sheath open on the upper side, and adhering to the corolla at the base; anthers 4 or 5. Ovary 2-celled. Capsule thin, flat, obcordate or transversely oblong, usually ciliate, opening loculicidally at the edges. Seeds orbicular, with a minute or without any caruncle.—Small slender herbs, either annual or parasitical on roots. Leaves alternate, sometimes reduced to minute scales. Flowers very small, in terminal spikes.

The few species known are all natives of tropical Asia, the most common one extending into Australia; but none have yet been found in Africa.

1. S. oblongifolia (oblong-leaved), *DC. Prod.* i. 334; *Benth. Fl. Austr.* i. 138. A slender glabrous annual, erect and simple, or slightly branched at the base, 3 to 5, or rarely 6in. high. Leaves sessile, the larger ones oblong, 3 to 4 lines long, and scarcely above 1 broad, the lower ones small and ovate. Flowers pink, scarcely a line long, in terminal leafless racemes or loose spikes of about an inch or rarely longer. Capsule about 1 line broad, but not so long, flattened, didymous, bordered with a fringe of hairs or slender teeth.—Deless. Ic. Sel. iii. t. 19; *S. obovata*, Wight, Illustr. t. 22.

Hab.: All along the Queensland coast.

2. POLYGALA, Linn.
(From supposed effect of increasing the secretion of milk.)

Sepals unequal, the 2 innermost, or wings, large and petal-like. Petals 3, united in a single corolla open on the upper side, the keel bearing a crest-like appendage on the back near the top, or rarely (in species not Australian) 3-lobed. Stamens 8, united to above the middle in a sheath open on the upper side, and adnate to the petals at the base. Ovary 2-celled. Style various. Capsule thin or rarely coriaceous, flattened, obovate, ovate, or orbicular, usually notched at the top, opening loculicidally at the edges. Seeds ovate or oblong, hairy or glabrous, but the hairs not lengthened into a coma, with or without a caruncle at the hilum.—Herbs, undershrubs, or shrubs. Leaves usually alternate or whorled. Racemes or spikes terminal or lateral, rarely axillary.

A very large genus, abundant in tropical countries, and generally also in temperate regions, except in Australia, where it is, with one exception, limited to the tropical districts, and in New Zealand, where it is entirely absent. Of the 7 Australian species, 3 are widely spread over tropical Asia, and the 4 others, although endemic, are nearly connected also with corresponding Asiatic ones.—*Benth.*

Perennial. Style with 2 stigmatic lobes, one above the other. Seeds
obovate, shortly villous . . . . 1. *P. japonica.*
Annuals. Seeds oblong villous, the hairs much longer at the end furthest
from the hilum.
  Racemes long, terminal. Inner sepals petaloid, obtuse. Crest fringed.
    Stigma simple, terminal, capitate. . . . . 2. *P. leptalea.*
  Racemes terminal and extra-axillary. Pedicels curved . . . . 3. *P. persicariæfolia.*
  Racemes lateral. Inner sepals herbaceous, mucronate, usually falcate.
    Crest fringed. Style with 1 large hooked or reflexed stigmatic lobe.
  Racemes shorter than the leaves, or if longer, very dense. Leaves
    from obovate to linear.
    Capsules broadly winged and ciliate . . . . 4. *P. rhinanthoides.*
    Capsules wingless and glabrous or nearly so . . . 5. *P. arvensis.*

1. **P. japonica** (Japanese), Houtt. Syst. 8, t. 62, f. 1, *according to DC. Prod.* i. 324; *Benth. Fl. Austr.* i. 189. Rootstock perennial, often woody with age, emitting numerous rather slender leafy stems, decumbent or erect, rarely more than 6in. long, more or less pubescent. Leaves nearly sessile, the lower ones ovate, obtuse and small, the upper ones elliptical or lanceolate, acute, $\frac{1}{2}$ to $\frac{3}{4}$ or rarely 1in. long, of a rather firm consistence, glabrous and almost shining, distinctly veined. Racemes lateral, sometimes of 2 or 3 flowers only, and shorter than the leaves, sometimes 6 to 8-flowered and longer. Bracts small and deciduous, but less so than in most species. Outer sepals narrow-lanceolate; inner ones ovate, obtuse, 2 to 3 lines long and not oblique. Keel-petal crested. Ovary glabrous. Style thickened, incurved, with 2 unequal stigmatic lobes, the upper one arching over the lower short one. Capsule about 3 lines long and broad, including the rather broad wing. Seeds obovate, slightly pubescent, with a 3-lobed caruncle.—*P. veronicea*, F. v. M. Pl. Vict. i. 184.

  Hab.: Dawson, Brisbane, and Condamine Rivers.

  Also in the hilly regions of tropical Asia and northward to Japan. I can, indeed, find no difference between the Australian and the Japanese specimens, except that the flowers in the latter are rather larger; but several Khasia specimens are precisely like the Australian ones. *P. elegans*, Wall., from East India and China, differs slightly in the racemes most frequently terminal with numerous flowers.—*Benth.*

2. **P. leptalea** (weak plant), *DC. Prod.* i. 325; *Benth. Fl. Austr.* i. 189. An erect, glabrous, slender annual, simple or slightly branched, usually 1 to 1$\frac{1}{2}$ft. high. Leaves few, linear, the longer ones about 1in., the uppermost much smaller, and the lower ones sometimes shortly oblong. Flowers small, numerous, pendulous, in a 1-sided terminal raceme, on pedicels which rarely attain 1 line. Outer sepals narrow-oblong, obtuse, the lowest rather larger and concave; inner sepals nearly twice as large, petal-like, broadly oblong, obtuse, 2 to 2$\frac{1}{4}$ lines long. Keel-petal crested. Style scarcely thickened, much curved, inflexed at the summit with an entire capitate stigma. Capsule broadly oblong, rather shorter than the inner sepals, with a narrow transparent wing. Seeds hirsute with reflexed hairs, the caruncle very small.—*P. oligophylla*, DC. Prod. i. 325.

  Hab.: Endeavour and Gilbert Rivers; also Rockingham Bay.
  Frequent in northern and eastern India.

3. **P. persicariæfolia** (Persicaria-leaved), *DC. Hook. in Fl. of Brit. Ind.* i. 202. An erect or ascending slightly pubescent very much branched slender herb, 6 to 16in. high. Leaves linear or elliptic-lanceolate, flaccid, hardly petiolate. Racemes terminal and extra-axillary, slender, 1 to 2in. long, lax-flowered. Pedicels slender, curved. Bracts small, subulate. Wings broad-obovate, rather longer than the elliptic notched ciliate capsule. Seeds villous. Strophiole small, galeate.

  Hab.: Tropical parts of the colony.

Polygala.] XIV. POLYGALEÆ. 79

4. P. rhinanthoides (Rhinanthus-like), *Soland. in Herb. R. Br.; Benth. Fl. Austr.* i. 140. An erect branching slightly pubescent annual, from an inch or two to above a foot high. Leaves oblong-linear, or rarely obovate-oblong, obtuse or rarely acute, ¾ to 1½in. long, glabrous or ciliate, narrowed into a short petiole. Racemes lateral, short, rather dense, 6 to 10-flowered. Outer sepals lanceolate, with a fine point; inner sepals broadly ovate, oblique, mucronate, ciliate, 2 to 3 lines long. Keel-petal crested. Ovary broad, ciliate. Style slightly thickened, much curved, entire, with a broad almost petaloid decurved stigma, bearded underneath. Capsule 4 lines long and broad, including a broad wing, pubescent and ciliate. Seeds oblong, hirsute with reflexed hairs, the caruncle deeply 3-lobed.

Hab.: Endeavour River, *R. Brown.*

5. P. arvensis (field), *Willd. Spec. Pl.* iii. 876; *Benth. Fl. Austr.* i. 140. A procumbent or rarely erect annual, branching at the base only, sometimes not exceeding a couple of inches when in full fruit, sometimes the prostrate or ascending branches extending to 6 or 8in. or even more, and usually pubescent. Leaves from obovate to oblong or linear, ¼ to ¾in. long or rarely more. Flowers few, in short sessile racemes, usually lateral, often shorter than the leaves, and rarely lengthening to an inch. Outer sepals very small and narrow; inner sepals ovate-falcate, acute or mucronate, 2 to 3 lines long, herbaceous and glabrous or slightly pubescent. Corolla about as long, the lateral petals rather large, the crest of the keel fringed. Ovary glabrous. Style scarcely thickened, with an almost petaloid uncinate-decurved stigma, glabrous and glandular underneath. Capsule rather broad, glabrous or slightly pubescent, not winged. Seeds very hairy.—*DC. Prod.* i. 326.

Hab.: Rockhampton to the Endeavour River.

A very common East Indian weed, variable in foliage and stature; the following forms appearing sometimes constant enough to be considered as distinct species:—

Var. *obovata*. Leaves all obovate, giving the plant the aspect of a young *Euphorbia helioscopia*. Cavern Island, Carpentaria, *R. Brown.*

Var. *squarrosa*. Leaves narrow. Flowers small and numerous, in oblong racemes, mostly terminal, the inner sepals narrow and falcate. *P. squarrosa*, Soland. ms. Endeavour River, *R. Brown;* Upper Victoria River, *F. v. Mueller.*

3. COMESPERMA, Labill.
(From hairyness of seeds.)

Sepals unequal, the 2 innermost, or wings, large and petal-like. Petals 3, the keel not crested, the two lateral ones separately attached to the staminal column, and either overlapped by the keel or outside it at the top. Stamens 8, united to above the middle in a sheath, open on the upper side and adnate to the petals at the base. Ovary 2-celled. Style incurved, obliquely stigmatic and more or less 2-lobed at the top. Capsule coriaceous or almost membranous, usually cuneate and much narrowed at the base, rarely nearly orbicular, opening loculicidally at the edges. Seeds ovate or oblong, pendulous, pubescent or hairy, the hairs lengthening into a coma whenever the capsule is narrowed at the base, without any caruncle at the hilum, but the raphe often expanded into a caruncular appendage at the opposite end.—Herbs, undershrubs or shrubs, erect or twining. Leaves alternate, usually small. Racemes terminal.

A strictly Australian genus, with which was formerly united the Brazilian *Bredemeyera* (*Catocoma*, Benth.); but, besides the difference in habit, the latter has a more or less fleshy capsule, and the seeds have a long coma proceeding from the hilum; whilst in *Comesperma* the coma, when present, consists of the hairs of the testa, which always extend to the base of the capsule, although the seed is often not half so long. In two species the capsule is that of a *Polygala*, and the seeds have no coma; but in those the insertion of the lateral petals, very different from that of *Polygala* and approaching that of *Monnina*, is strongly marked. In *P. volubilis* (which was chiefly taken into account in verifying the characters for our "Genera

80   XIV. POLYGALEÆ.   [*Comesperma.*

Plantarum ") the arrangement of the petals is nearer to that of *Polygala,* but there the carpological characters are very decided. Besides that, the genus *Comesperma* is so natural a one that it is never liable to be confounded with any of those allied to it in structure. The precise arrangement of the petals in the smaller-flowered species, very difficult to ascertain in dried specimens, requires verification from the living plant.—*Benth.*

Capsule sessile. Seeds filling the cells, without a coma. Stems leafless.
  (Sect. **Prosthemosperma,** *F. v. M.*)
  Capsule orbicular. Flowers in a short terminal raceme . . . . 1. *C. sphærocarpum.*
  Capsule narrowed into a stipes, containing the long coma of the seeds, which only occupy the broad part of the cells.
    Outer sepals all free, much shorter than the wings.
      Branches twining or very short and almost leafless.
        Leaves few, mostly obtuse. Capsule not winged. Flowers blue or white. Pedicels glabrous . . . . 2. *C. volubile.*
      Stems erect, leafy.
        Leaves flat, ovate or oblong.
          Pubescent.
            Leaves small, broadly ovate, mucronate, crowded. Flowers 1 to $1\frac{1}{4}$ line . . . . 3. *C. secundum.*
          Glabrous.
            Leaves oblong, somewhat acute, pale on under side . . . 4. *C. præcelsum.*
            Leaves obtuse, green . . . . 5. *C. retusum.*
            Leaves mucronate, very glaucous . . . . 6. *C. sylvestre.*
        Leaves linear with revolute margins. Keel-petal not horned . . 7. *C. ericinum.*
    Outer sepals all free, nearly as long as the wings. (Sect. **Isocalyx,** *Steetz.*)
      Stems very slender, almost leafless . . . . 8. *C. defoliatum.*

1. **C. sphærocarpum** (capsule almost round), *Steetz, in Pl. Preiss.* ii. 814; *Benth. Fl. Austr.* i. 148. Rootstock woody but not thick, with slender, broom-like, or flexuose stems, sometimes perhaps slightly twining, $\frac{1}{2}$ to $1\frac{1}{2}$ft. long, glabrous and slightly sulcate. Leaves reduced to minute distant scales, or the lower ones rarely 2 lines long, and linear. Flowers 3 to 6, in a short loose terminal raceme, on pedicels of 1 to 2 lines, the bracts very minute and deciduous. Outer sepals oblong, rather acute, almost scarious, about half the length of the inner ones, which are broadly obovate, blue and petal-like, 2 to nearly 3 lines long. Corolla and style of *C. scoparium.* Capsule nearly orbicular, about 2 lines diameter, slightly cuneate at the base or at length quite obtuse, glabrous. Seeds ovate, shortly pubescent, with a short membranous hairy appendage at the lower or chalazal end.

  Hab.: Ranges about Brisbane.

2. **C. volubile** (twining habit), *Labill. Pl. Nov. Holl.* ii. 24 *t.* 163; *Benth. Fl. Austr.* i. 144. A glabrous twiner, with numerous branches, sometimes extending to a considerable length, rarely short and flexuose, or almost erect. Leaves few, the lower ones oblong-linear or lanceolate, sometimes above an inch long and narrowed into a petiole, the upper ones linear or rarely obovate, small and distant. Racemes axillary or terminal, loose, 1 or rarely 2in. long, sometimes 2 or 3 together. Flowers blue or rarely white, on pedicels of 1 to 2 lines. Outer sepals very broad, obtuse, about 1 line long; inner sepals fully 3 lines long, nearly orbicular, distinctly clawed. Keel-petal with 2 oblong lateral lobes turned inwards in æstivation and overlapped, at least at the top, by the 2 large, obovate lateral petals. Style dilated upwards but not winged. Capsule 4 to nearly 5 lines long, rounded, truncate and often slightly acuminate at the top, nearly $1\frac{1}{4}$ line broad, and gradually narrowed into a rather broad stipes. Seeds oblong, the long hairs forming the coma much fewer on the sides than on the edges.—DC. Prod. i. 334; Hook. f. Fl. Tasm. i. 31; F. v. M. Pl. Vict. i. 191; *C. tortuosum,* Steetz, in Pl. Preiss. ii. 308; *C. gracile,* Paxt. Mag. v. 145, with a fig.

  Hab.: Common on the damp sandy land of the southern coast.
  Var. *alba* only differs in the flower being white.

3. **C. secundum** (one-sided), *Banks in DC. Prod.* i. 334; *Benth. Fl. Austr.* i. 145. A low much-branched rigid shrub, with the habit of some *Epacrideæ*, the branches softly pubescent. Leaves crowded, spreading, ovate, mucronate, 2 to 3 lines long, rigidly coriaceous, rough with minute tubercular hairs. Flowers very small and numerous, in slender one-sided racemes of 1 to 2in., on very short pedicels. Outer sepals short, very broad and obtuse; inner sepals nearly three times as long, although scarcely exceeding 1 line, apparently pink. Keel-petal very broad, overlapping the narrow lateral ones. Style not winged. Capsule fully ½in. long, truncate, 3-toothed, and scarcely 1 line broad at the top, tapering into a slender stipes twice as long as the oblong part. Seed elongated, without any appendage, the long coma apparently very deciduous, but not seen quite ripe.

Hab.: Endeavour River and Cape Flinders.

4. **C. præcelsum** (tallest of the genus), *F. v. M. Fragm.* xi. 2. A tall shrub, said to attain the height of 12ft., the leafy branchlets scabrous-puberulent. Leaves glabrous, crowded, oblong or broadly-linear and somewhat acute, ½ to ¾in. long, under side pale. Flowers in short few or many-flowered corymbs. Bracts deltoid-lanceolate. Sepals, outer ones free, rotund-deltoid; inner ones pale, 3 lines long. Capsule oblong-cuneate, 4 or 5 lines long, 1¼ line broad at the top, where it is emarginate. Seeds oblique ellipsoid, 1 to 1¼ line long, velvety, the margins hairy but without appendage at the chalazal end.

Hab.: Ranges about Rockingham Bay.

5. **C. retusum** (form of capsule), *Labill. Pl. Nov. Holl.* ii. 22 t. 160; *Benth. Fl. Austr.* i. 145. Glabrous, erect, shrubby and much-branched, often several feet high, the branches mostly erect and not sulcate. Leaves oblong, obtuse, rarely above ½in. long, flat but rather thick, the midrib not prominent. Racemes short and dense, usually several in a terminal, leafy, flat corymb or pyramidal panicle. Outer sepals ovate, obtuse, about 1 line long; inner sepals nearly 3 lines. Petals rather shorter, the keel not horned. Capsule usually about 5 lines long, emarginate, with rounded lobes, and about 1½ line broad at the top, narrowed into a stipes much longer than the broad part. Seeds comose, without any membranous appendage.—DC. Prod. i. 334; Hook. f. Fl. Tasm. i. 32; F. v. M. Pl. Vict. i. 190.

Hab.: Sandy coast lands of the southern parts of the colony.

6. **C. sylvestre** (found in forests), *Lindl. in Mitch. Trop. Austr.* 342; *Benth. Fl. Austr.* i. 146. A glabrous and erect shrub of several feet, resembling *C. retusum*, with which F. v. Mueller proposes to unite it, but much more glaucous. Leaves larger, often ¾in. long and sometimes 3 lines broad, mucronate or pungent, often concave above. Flowers rather larger, with broader outer sepals. Capsule about ½in. long.—F. v. M. Fragm. i. 49.

Hab.: Open forest near Mounts Faraday and Pluto: sandy forest tableland on the Sutor River.

7. **C. ericinum** (heath-like), *DC. Prod.* i. 334; *Benth. Fl. Austr.* i. 146. Glabrous or minutely pubescent, usually erect, with rigid branches 1 to 2 or even 3ft. high, woody at the base. Leaves linear, erect or spreading, crowded or rather distant, obtuse or acute, rarely above ½in. long and usually shorter, the margins recurved or more frequently quite revolute. Racemes usually several and short in a leafy panicle, but longer and less dense than in *C. retusum*, rarely slender, and lengthening out to 3 or 4in. Outer sepals all free, ovate or ovate-lanceolate, ¾ to 1 line long; inner sepals about 3 lines. Keel-petal not horned. Capsule 3 to 4 lines long, truncate, with rounded angles or entirely rounded at the top, narrowed into a stipes usually longer than the broad part. Seeds oblong, comose, with a very small membrane at the lower or chalazal end.—Hook. f. Fl.

G

Tasm. i. 32; F. v. M. Pl. Vict. i. 190; *C. coridifolium*, A. Cunn. in Field. N. S. Wales, 337; *C. latifolium*, Steetz, in Pl. Preiss. ii. 295; *C. acutifolium*, Steetz, l.c. 296; *C. linariæfolium*, A. Cunn. in Steetz, l.c. 297.

Hab.: Moreton Bay, Glasshouses and Burnett Ranges.

Var. *patentifolium*. Leaves very spreading, often pungent, very broad at the base.—Burnett ranges in the interior of N. S. Wales, *F. v. Mueller. C. patentifolium*, F. v. M. Fragm. i. 48. (See F. v. M. Pl. Vict. i. 190.)

Var. *oblongatum*, R. Br. Leaves oblong-linear, obtuse and mucronate, longer and with less revolute margins than usual.—East coast, *R. Brown*.

8. **C. defoliatum** (few leaves), *F. v. M. Pl. Vict.* i. 189; *Benth. Fl. Austr.* i. 148. Allied in habit to *C. nudiusculum* with the flowers of *C. calymega*. Rhizome woody, with rigid and rush-like but slender and sometimes almost filiform stems, 1 to 2ft. high and glabrous. Leaves very few and distant, small, narrow-linear or sometimes all reduced to small linear scales. Racemes slender, 1 to 2in. long. Flowers rather larger than in *C. calymega*. Outer sepals all free, oblong, nearly as long as the inner ones. Capsule 3 or 4 lines long, contracted into a long narrow stipes. Seeds comose, without any terminal appendage.—*C. nudiusculum*, Steetz, in Pl. Preiss. ii. 308, not DC.

Hab.: Islands of Moreton Bay.

4. XANTHOPHYLLUM, Roxb.

(Leaves of some species yellowish.)

Sepals 5, nearly equal. Petals 5 or 4, nearly equal, the inferior keel-shaped, not crested. Stamens 8, distinct, 2 hypogynous, 6 attached to the base of the petals. Ovary stipitate, 1-celled; style curved; ovules various in number and insertion. Fruit 1-celled, indehiscent, 1-seeded. Seeds exalbuminous extrophiolate.—Timber trees. Leaves large, coriaceous, generally of a yellowish-green.

Abundant in the Malay Archipelago and Malacca, a few also in continental India. Only the one here mentioned in Australia.

1. **X. Macintyrii** (after D. Macintyre), *F. v. M. Fragm.* v. 8 *and* 57. Usually forming a small erect tree about 30 or 40ft. high, with a whitish bark on the trunk, reddish upon the branchlets. Leaves alternate, ovate or lanceolate, entire, 2 to 4in. long, ¾ to 2in. broad, glabrous and shiny on both sides, with often 1 or 2 glands on the under side, obtuse, penninerved, and reticulate veined; petioles short, eglandulose. Racemes axillary or terminal, the peduncle rhachis and pedicels shortly pubescent. Bracteoles 3, almost ovate, about 3 or 4 lines long, soon falling. Outside of sepals puberulous, about 1½ line long, ovate-orbicular. Petals 3 or 4 lines long, oblong-cuneate, lanuginoso-pubescent at the base. Stamens 8, filaments linear-setaceous 1½ to 2¼ lines long, the lower part broad and ciliolate; anther yellow. Style scarcely 2 lines long, glabrous in the upper part. Stigma 2-lobed. Disk annular, glabrous. Ovary stipitate, hoary-tomentose. Fruit globose.—*Macintyria octandra*, F. v. M. Fragm. v. 8.

Hab.: Common in the tropical scrubs.

Wood of a greyish colour, soft and easy to work, very light; useful for cigar boxes, lining, &c.

The fruit is sometimes infested with the fungus *Glæosporium carpophyllum*, Mass.

Order XV. FRANKENIACEÆ.

Flowers regular, hermaphrodite. Calyx tubular, persistent, with 4, 5, or rarely 6 lobes, valvate in the bud, and as many prominent angles and furrows. Petals as many, hypogynous, imbricate in the bud, free, the claws with an adnate plate or appendage on the inner face, the lamina spreading. Stamens usually 6, sometimes 4 or 5 or indefinite, hypogynous, free or shortly united in a ring at the base, filaments filiform or flattened; anthers 2-celled, versatile. Ovary free, sessile, 1-celled, with 3, rarely 2 or 4 parietal placentas, or very rarely a single one. Style filiform, with as many branches as placentas, the stigmas capitate or oblique. Ovules several, or rarely solitary, to each placenta, attached to rather long ascending funicles, amphitropous or nearly anatropous, with an inferior micropyle. Seeds ovoid or oblong, testa crustaceous, the hilum almost terminal. Embryo straight, in a mealy albumen, the radicle next the hilum, shorter than or as long as the cotyledons.—Low herbs or undershrubs, much branched and jointed at the nodes. Leaves opposite, small, without stipules, often clustered in the axils. Flowers usually pink or purple, sessile in the forks of the branches, forming a more or less dense, terminal, leafy cyme, sometimes contracted into a globular head.

The Order consists of a single genus, closely allied to the small group of *Diantheæ*, amongst *Caryophylleæ*, but distinguished by the parietal placentation of the ovary, and by the terminal hilum in the seed. The species are chiefly maritime, and generally distributed over the temperate regions of the globe, more especially of the northern hemisphere, less abundant within the tropics.—*Benth.*

1. FRANKENIA, Linn.

(After John Franken.)

Characters and distribution those of the Order.

1. **F. pauciflora** (few flowers), *DC. Prod.* i. 350; *Benth. Fl. Austr.* i. 151. Shrubby and procumbent or almost erect at the base, with ascending, erect, or divaricate dichotomous branches, nearly glabrous or hoary with a short down or scaly pubescence, often very low and spreading, sometimes above a foot high, attaining even 3ft. according to F. v. Mueller. Leaves opposite or the upper ones in whorls of 4, oblong or linear, obtuse or rarely almost acute, the margins usually revolute so as only to show a dorsal furrow, when very narrow above 3 lines long, but usually much shorter, the very short sheathing petioles ciliate on the edge, with smaller leaves often clustered in the axils. Flowers closely sessile in the last forks, forming a more or less dense terminal leafy cyme and sometimes unilaterally arranged along its branches owing to the abortion of one branch of each fork. Calyx 3 to 4 lines, or rarely only 2½ lines long. Petals with their claws cohering in an angular tube, the longitudinal appendage not very prominent, the lamina obovate, entire or crenulate. Stamens 5 or 6, with their filaments slightly dilated and usually cohering. Placentas 3 or rarely 2, with 2 to 4 ovules to each.—*Bot. Mag.* t. 2896; *Hook. f. Fl. Tasm.* i. 40; *F. scabra*, Lindl. in Mitch. Trop. Austr. 305.

Var. *serpyllifolia*. Pubescent or hirsute. Leaves, especially the lower ones short, from oblong to broadly ovate, the margins often much less recurved than in the typical *F. pauciflora*.—*F. serpyllifolia*, Lindl. in Mitch. Trop. Austr. 305.—Nive River, *Mitchell*; Murchison River, *Drummond*. Allied to this variety is the plant from Port Jackson, which De Candolle, Prod. i. 349, referred with doubt to the *F. pulverulenta*, Linn. The specimens in the herbarium of the Paris Museum have much the aspect of the latter species (very prostrate, with small broad flat leaves, more petiolate than is usual in *F. pauciflora*), yet I think they may prove to be only one of its numerous varieties, very near to the *serpyllifolia*.—*Benth.*

Var. *thymoides*. More woody, erect, and much-branched, with the habit of *Thymus vulgaris*, hoary all over, with a minute scaly indumentum. Leaves oblong, very obtuse, much revolute, 1 to nearly 2 lines long. Flowers rather small, the appendage of the petal-claws very prominent. Ovules 4 to 6 to each placenta.—Mount Goningbear, *Victorian expedition*.—*F. fruticulosa*, DC. Prod. i. 350, appears to connect this variety with the more common forms.—*Benth*.

Hab.: At the Georgina this variety is thickly, but loosely, incrusted with salt, even when found growing within a stone's throw of fresh water.

Order XVI. **CARYOPHYLLEÆ.**

Flowers regular, usually hermaphrodite. Sepals 4 or 5, persistent, free or united in a toothed calyx, imbricate in the bud. Petals either as many as the sepals, hypogynous or slightly perigynous, entire or lobed, imbricate and frequently contorted in the bud, or rarely minute and scale-like or none. Stamens 8—10 or fewer, inserted with the petals. Filaments filiform. Anthers 2-celled. Torus small or in a few *Sileneæ* lengthened into a gynophore, or in some *Alsineæ* forming a small disk, shortly adnate to the base of the calyx, or short glands between the stamens. Ovary free, 1-celled or partially divided especially at the base into 2 to 5 cells. Styles 2 to 5, linear and stigmatic along the inside from the base or towards the top, free or more or less united into 1 branching style. Ovules 2 or more, often numerous, attached to a short or columnar placenta in the centre of the ovary, amphitropous and usually curved. Capsule membranous or crustaceous, very rarely succulent, opening at the top in as many or twice as many teeth or valves as there are styles, very rarely indehiscent. Seeds several, rarely solitary by abortion, with a membranous or crustaceous testa. Albumen mealy. Embryo curved round the albumen, or rarely straight or nearly so, and excentrical, with the radical inferior, or when the embryo is circular turned upwards.— Herbs, very rarely shrubby at the base, usually thickened and jointed at the nodes. Leaves opposite and entire, usually connected by a transverse line or short sheath at the base. Stipules none, or small and scarious. Inflorescence centrifugal, usually forming a terminal leafy cyme, rarely paniculate or racemose, or the the pedicels all axillary.

A large Order, especially abundant in the extratropical regions of the northern hemisphere, rather less so in the high mountain-ranges of tropical America and Asia, and in the more temperate regions of the southern hemisphere, very rare in hot tropical countries. Of the Australian genera none are endemic. One, *Polycarpæa*, is chiefly tropical and almost limited to the Old World; another, *Drymaria*, is also chiefly tropical, but almost entirely American; a third, *Colobanthus*, is chiefly extratropical and limited to the southern hemisphere; a fourth, *Stellaria*, has almost as wide a range as the Order itself; the remaining genera and species, whether indigenous or introduced, are all European or East-Mediterranean.—*Benth*.

Tribe I. **Sileneæ.**—*Sepals united in a 4 or 5-toothed calyx. Petals and stamens hypogynous, often raised on a stalk-like torus. Styles distinct from the base. Stipules 0.*

Calyx broadly or obscurely 5-nerved. Styles 2 . . . . 1. Gypsophila.
Calyx obscurely veined. Styles 2 . . . . 2. Saponaria.
Calyx 10-nerved. Styles 3 . . . . 3. Silene.
Calyx 10-nerved. Styles 5 . . . . 4. Lychnis.

Tribe II. **Alsineæ.**—*Sepals free or only united by the disk at their base. Petals and stamens hypogynous or slightly perigynous, the torus not elongated. Styles distinct from the base. Stipules none, or rarely small and scarious.*

Petals usually 2-cleft.
  Capsule cylindrical or conical, opening equally in twice as many teeth as styles. Styles 5, opposite the sepals, or rarely 4 or 3 . . . . 5. Cerastium.
  Capsule globular or ovoid, opening in as many 2-cleft valves as styles. Styles 3, or if 5 alternate with the sepals . . . . 6. Stellaria.
Petals entire or none.
  Sepals 5. Styles usually 3. Capsule globular or ovoid.
    No stipules. Petals none . . . . 6. Stellaria.
    Stipules small and scarious. Petals pink . . . . 8. Spergularia.
    Stipules small and scarious. Leaves clustered so as to appear verticillate 7. Spergula.

XVI. CARYOPHYLLEÆ.

TRIBE III. **Polycarpeæ.**—*Sepals of* Alsineæ. *Petals usually very small or none. Stamens 5 or fewer, hypogynous or slightly perigynous. Style single at the base, with 2 or 3 branches or minute teeth. Stipules scarious or very minute.*

Petals lobed. Style very short. Stipules minute . . . . 9. DRYMARIA.
Petals entire. Style short. Stipules scarious . . . . 10. POLYCARPON.
Petals entire or notched. Style elongated. Stipules and sepals scarious . . 11. POLYCARPÆA.

1. GYPSOPHILA, Linn.
(From its preference for chalky soils.)

Calyx campanulate or turbinate-tubular, 5-toothed or 5-lobed, broadly 5-nerved, membranous between the nerves. Petals 5, with a narrow claw, and without any scale. Torus small. Stamens 10. Styles usually 2. Capsule globular or ovoid, opening to the middle or lower down in 4 valves. Seeds nearly reniform; embryo curved round the albumen.—Herbs, mostly glaucous, sometimes glandular or hirsute. Flowers usually small, numerous, and paniculate, or solitary in the forks of the stem.

A genus limited to the extratropical regions of the northern hemisphere in the Old World with the exception of the following species. It is chiefly distinguished from *Saponaria* by the calyx.—*Benth.*

1. **G. tubulosa** (tubular), *Boiss. Diagn. Pl. Or.* i. 11; *Benth. Fl. Austr.* i. 155. A slender erect dichotomous annual, often not above 2 or 3in., but sometimes 8 to 10in. high, more or less viscid-pubescent, and often slightly hirsute. Leaves linear-subulate, rarely attaining ½in. and often much shorter. Pedicels in the forks, or sometimes appearing axillary from 1 branch only being developed, 4 to 8 lines long, erect or spreading. Calyx erect, 1½ line long, narrower than in most *Gypsophilas*, with 5 prominent nerves, the teeth short and obtuse. Petals red, narrow-oblong, a little longer than the calyx. Capsule ovoid-oblong, rather exceeding the calyx. Seeds black, elegantly pitted under a lens.—F. v. M. Pl. Vict. i. 206; *Dichoglottis tubulosa*, Jaub. and Spach, Ill. Pl. Or. i. 14 t. 6; *D. australis*, Schlecht. Linnæa, xx. 681.

Hab.: Stanthorpe and a few other localities in the south.

A native of the East Mediterranean region of Europe and Asia, possibly introduced into Australia and New Zealand, where it is also found; yet from the localities where it was so early collected by R. Brown, and its general diffusion over extratropical Australia, it is difficult to conceive how a plant unknown in those parts of Europe whence the early colonists proceeded should have so promptly established itself. It is allied to the more common *G. muralis*, which, however, has not been detected in Australia, and is always quite distinct, especially in the form of the calyx, which is that of a true *Gypsophila*, whilst *G. tubulosa* is in this respect almost intermediate between that genus and *Saponaria*.—*Benth.*

*2. SAPONARIA, Linn.
(Bruised leaves when agitated in water produce a lather, like soap.)

Calyx more or less tubular, ovoid or oblong, 5-toothed, nerves obscure. Petals 5, clawed; limb entire or notched, with or without a basal scale. Stamens 10. Disk small, or produced into a gynophore. Ovary 1-celled, or imperfectly 2 or 3-celled. Styles 2, rarely 3; ovules many. Capsule ovoid or oblong, rarely subglobose, 4-toothed. Seeds reniform or subglobose, hilum marginal; embryo annular.—Annual or perennial herbs. Leaves flat. Flowers in dichotomous cymes.

Chiefly Mediterranean and West Asiatic.

1. **S. vaccaria** (from *vaccarius*, a cow-herb), *Linn.* Cowherb. A tall, robust, simple or sparingly branched, perfectly glabrous annual, 12 to 24in. high. Leaves radical, oblong, 1 to 3in. long, 3 to 9 lines broad, stem ones

sessile and linear-oblong. Cymes corymbose, many-flowered. Pedicels slender. Calyx ¼in. long, teeth triangular, margins scarious, with 5 broad green nerves, ventricose in fruit. Petals rosy, short, erose, obovate. Capsule included, broad ovoid. Seeds large, globose, black, granulate.

Hab.: A weed in cultivation paddocks, often introduced with wheat and other seeds.

*3. SILENE, Linn.
(Gummy secretion of leaves supposed like saliva.)

Calyx 10-nerved, rarely many-nerved, 5-toothed or 5-lobed. Petals 5, with a narrow claw, and usually with a double scale. Stamens 10. Torus usually elongated. Styles usually 3. Capsule opening in 6 or rarely 3 teeth or short valves. Seeds laterally attached; embryo curved round the albumen.—Herbs. Flowers solitary or cymose, often forming unilateral spikes or an oblong thyrsus or panicle.

A very large genus, chiefly abundant in Europe, N. Africa, and temperate Asia; with a few N. American and S. African species, and only introduced into Australia.

1. **S. gallica** (French), *Linn.; DC. Prod.* i. 871; *Benth. Fl. Austr.* i. 155. Catchfly. A hairy, slightly viscid, much branched annual, 6in. to nearly 1ft. high, erect or decumbent at the base. Lower leaves small and obovate, upper ones narrow and pointed. Flowers small, nearly sessile, generally all turned to one side, forming a simple or forked terminal spike, with a linear bract at the base of each flower. Calyx very hairy, with 5 slender teeth, at first tubular, afterwards ovoid and much contracted at the top. Petals very small, entire or notched, pale red or white, or in one variety with a dark spot.—*S. anglica, lusitanica, cerastoides* and *quinquevulnera,* Linn.; Reichb. Ic. Fl. Germ. vi. t. 272, 273.

Hab.: A plant probably of South European origin, now common in sandy, gravelly, and waste places, especially near the sea, in most parts of the world, as in this colony.

*4. LYCHNIS, Linn.
(From the Greek for lamp, alluding to brilliancy of flowers of some species.)

Calyx inflated, 5-toothed, 10-nerved. Petals 5, claws narrow, lamina 2-fid or laciniated. Stamens 10. Ovary 1-celled, with numerous ovules; styles 5, rarely 4. Capsule toothed at the top. Seeds tuberculate or smooth.—Herbs, with the habit of *Silene.*

Natives of the Arctic and temperate northern regions and of the Andes of South America.

1. **L. githago** (from the seed resembling a black aromatic grain, Gith or Git, used by the Romans in cookery), *Lam.* Corn cockle. A tall, erect annual, clothed with long, whitish, appressed hairs. Leaves long, narrow. Flowers on long leafless peduncles, rather large and red, remarkable for the long green linear lobes of the calyx projecting much beyond the petals; the latter broad, undivided, without scales. Stamens 10. Styles 5. Capsule opening in 5 teeth.

Hab.: A weed of cultivation belonging to Europe or the East Mediterranean.

5. CERASTIUM, Linn.
(Capsules horn-shaped.)

Sepals 5, rarely 4. Petals as many, usually notched or 2-cleft. Stamens 10 or fewer. Styles 5 or 4, opposite the sepals, or rarely 3. Capsule cylindrical or conical, often incurved, opening at the top in twice as many teeth as styles, all equal. Seeds more or less reniform.—Herbs, usually pubescent or hirsute. Leaves rarely subulate. Cymes terminal, dichotomous, leafy, or the floral leaves reduced to small or scarious bracts. Seeds usually pitted or muricate.

A considerable genus, distributed chiefly over the temperate regions of the northern hemisphere, more especially in the Old World, rare within the tropics except in mountain regions.

1. **C. vulgatum** (common), *Linn.; DC. Prod.* i. 415; *Benth. Fl. Austr.* i. 156. Mouse-ear chickweed. A coarsely pubescent usually more or less viscid annual, branching at the base, sometimes dwarf, erect, and much branched, at others loosely ascending to 1ft. or even 2ft., occasionally forming at the end of the season dense matted tufts, which may live through the winter and give it the appearance of a perennial. Radical leaves small and petiolate; stem leaves sessile, from broadly ovate to narrow oblong. Sepals 2 to 2½ lines long, green and pubescent, but with more or less conspicuous scarious margins. Petals seldom exceeding the calyx, and often much shorter, sometimes very minute or even none. Stamens often reduced to 5 or fewer. Capsule cylindrical, often curved and projecting beyond the calyx.—Reichb. Ic. Fl. Germ. v. t. 228, 229; *C. viscosum*, Linn.; DC. l.c. 416.

Hab.: The southern parts of the colony.

6. STELLARIA, Linn.
(Star-like flowers.)

Sepals 5, rarely 4. Petals as many, usually 2-cleft, rarely wanting. Stamens 10 or fewer. Styles 3, rarely 2 or 4, or very rarely 5, and then alternate with the sepals. Capsule globular, ovoid or oblong, opening to below the middle in twice as many valves as styles, or in an equal number of 2-cleft valves.—Herbs, usually diffuse, tufted or ascending, glabrous or pubescent. Leaves rarely subulate. Flowers solitary, or in loose leafless or leafy cymes. Seeds usually pitted or muricate.

A considerable genus, spread over nearly the whole globe, although within the tropics confined to mountain districts.

Petals longer than or nearly as long as the sepals.
  Leaves mostly sessile, linear or lanceolate. Pedicels axillary. Perennials.
    Leaves rigid and pungent, mostly linear-lanceolate, often recurved . . 1. *S. pungens*.
    Leaves linear, slender . . . . 2. *S. glauca*.
  Leaves mostly petiolate, ovate or ovate-lanceolate. Pedicels axillary.
    Perennial without any pubescent line . . . . 3. *S. flaccida*.
  Leaves sessile or petiolate, broadly ovate. Pedicels in the forks. Annual,
    with a pubescent line down each internode . . . . 4. *S. media*.

1. **S. pungens** (pungent), *Brongn. Voy. Coq.* t. 78; *Benth. Fl. Austr.* i. 157. Perennial and very much branched, decumbent or ascending amongst bushes, often to 3 or 4ft., with angular branches, smooth and shining, glabrous, or hirsute with loose scattered hairs. Leaves lanceolate to linear, rigid and pungent, mostly 3 to 4 lines long, and never exceeding ½in., often spreading or recurved, all sessile or scarcely narrowed at the base, the lower ones sometimes small and crowded. Pedicels axillary, very variable in length, but usually considerably exceeding the leaves. Sepals rigid, pungent, about 3 lines long, the outer ones prominently 3-nerved. Petals about as long or rather longer, deeply cleft.— Hook. f. Fl. Tasm. i. 44; F. v. M. Pl. Vict. i. 209; *S. squarrosa*, Hook. Journ. Bot. i. 250.

Hab.: Stanthorpe, towards the border of New South Wales.

2. **S. glauca** (grey-green colour of foliage), *With.; DC. Prod.* i. 397; *Benth. Fl. Austr.* i. 158. Perennial, usually glabrous, smooth and shining, with slender ascending or erect branches, often 1 to 2ft. high, but sometimes low and intricate. Leaves linear, acute, ¾ to 1½in. long, or the upper ones short. Pedicels axillary

or terminal, slender but rigid, longer than the leaves. Sepals very acute, 3-nerved, about 8 lines long when in flower. Petals about as long, or rather longer, deeply cleft. Capsule ovate, much shorter than the calyx, which usually lengthens after flowering.—Reichb. Ic. Fl. Germ. v. t. 228; Hook. f. Fl. Tasm. i. 44; F. v. M. Pl. Vict. i. 210; *S. angustifolia*, Hook. Journ. Bot. i. 250.

Hab.: Many localities in the southern parts of the colony.

3. **S. flaccida** (flaccid), *Hook. Comp. Bot. Mag.* i. 275; *Benth. Fl. Austr.* i. 158. Apparently perennial, with weak and decumbent very intricate branches, often extending to several feet, glabrous and shining, or with loose spreading scattered hairs, especially about the nodes. Leaves ovate to lanceolate, very acute, thin and flaccid, often undulate on the margin, narrowed and ciliate at the base, rarely exceeding ½in. without the petiole, which is long in the lower leaves, short or none in the upper ones. Pedicels all axillary, and usually 1 to 1½in. long. Sepals 2 to 2½ lines long, broadly lanceolate, acute, with a scarious border, usually 3-nerved, but the lateral nerves often very faint, often ciliate. Petals rather longer, deeply cleft. Capsule ovoid, usually exceeding the calyx.—*S. media*, var., Hook. f. Fl. Tasm. i. 43; F. v. M. Pl. Vict. i. 211.

Hab.: Southern parts of the colony.

*4. **S. media** (mediate), *Linn. DC. Prod.* i. 396; *Benth. Fl. Austr.* i. 159. A weak, much-branched annual, glabrous with the exception of a pubescent line down one side of each internode, and a few long hairs on the petioles, and sometimes on the sepals. Leaves ovate, shortly pointed, the lowest on long petioles, short and broad, and sometimes cordate, the upper ones on shorter petioles or quite sessile, ⅓ to ¾in. long, thin and flaccid. Pedicels slender, often drooping, in the forks of the branches, the upper ones usually forming a rather dense leafy cyme, very rarely one of the lowest axillary from the abortion of one fork. Sepals about 2 lines long, obtuse or rather acute, thin but green, with scarcely prominent nerves, and usually pubescent. Petals about as long, deeply cleft. Capsule scarcely longer than the calyx.—Reichb. Ic. Fl. Germ. v. t. 222.

Originating, probably, in the temperate regions of the northern hemisphere in the Old World, this plant is now a common weed in cultivated places, especially in gardens, as well as in waste places almost all over the globe, and as such is found in all of the Australian colonies.—*Benth.*

*7. **SPERGULA**, Linn.

(Because it scatters its seed.)

Sepals 5. Petals 5, entire. Stamens 10, rarely 5. Ovary 1-celled; ovules many. Styles 5. Capsule 5-valved, valves entire and laterally compressed, margins acute or winged.—Herbs, with dichotomous or fasciculate branches and small scarious stipules. Flowers pedicellate.

1. **S. arvensis** (field), *Linn.; Benth. Fl. Austr.* i. 161. A slender annual, branching at the base into several erect or ascending stems, 6 to 12in. high, glabrous or slightly pubescent. Leaves nearly subulate, 1 to 2in. long, in opposite clusters and spreading so as to appear verticillate. Stipules scarious, very minute, sometimes very difficult to see. Flowers small, white, on long pedicels, in terminal forked cymes. Sepals 5. Petals 5, undivided, generally rather shorter than the calyx. Stamens 10, or occasionally 5 or fewer. Styles 5, alternate with the sepals. Capsule deeply 5-valved. Seeds slightly flattened, with or without a scarious border.

Hab.: This common weed of Europe and temperate Asia has been met with in several localities in the southern parts of the colony.

XVI. CARYOPHYLLEÆ.

8. SPERGULARIA, Pers.
(Altered from *Spergula.*)
(Lepigonum, *Fries.*)

Sepals 5. Petals 5, entire or rarely none. Stamens 10 or fewer. Styles 3. Capsule 3-valved.—Herbs, usually diffuse. Leaves linear or filiform, often clustered in the axils so as to appear verticillate. Stipules small, scarious. Flowers pedicellate, pink or white, in the forks of the stem or in terminal cymes or 1-sided racemes. Seeds with or without a scarious border.

A small genus, widely dispersed over the temperate or subtropical regions of the globe, chiefly in maritime or saline localities, or heathy places, differing from *Arenaria* almost solely in the presence of stipules. The Australian species is the same as the common northern one.—*Benth.*

1. **S. rubra** (red), *Pers. Syn.* i. 504 (as a subgenus of *Arenaria*); *Benth. Fl. Austr.* i. 161. An annual, biennial or rarely perennial, glabrous or with a short viscid pubescence in the upper parts, with numerous stems branching from the base and forming spreading or prostrate tufts 3 or 4in., or when luxuriant 6in. long. Leaves narrow-linear, the scarious stipules at the base short but conspicuous. Flowers very variable in size, usually pink, on short pedicels, in forked cymes, usually leafy at the base. Petals shorter or rather longer than the sepals. Seeds more or less flattened, often surrounded by a narrow scarious border or wing.—A. Gray, Gen. Ill. t. 108; Hook. f. Fl. Tasm. i. 41; F. v. M. Pl. Vict. i. 207; *Arenaria rubra* and *A. media*, Linn.; DC. Prod. i. 401; *Lepigonum rubrum*, etc., Fries., Nov. Fl. Suec. Mant. iii. 32; *L. brevifolium*, Bartl. in Pl. Preiss. i. 243; *L. anceps* and *L. laxiflorum*, Bartl. l.c. 244 (of these last I have only seen authentic specimens of *L. anceps*); *Spergularia rupestris*, Fenzl. in Hueg. Enum. 9; Schlecht. in Linnæa, xx. 682.

Hab.: Southern parts of the colony.

Widely spread over Europe, temperate Asia, and North America, and some parts of South America, chiefly in maritime countries or in sandy heathy places more inland. There are two, often rather marked varieties; one chiefly occurring inland has slender leaves, small flowers, and short capsules, with the seeds less frequently bordered than in the larger variety, which has a sometimes perennial stock, thicker somewhat fleshy leaves, and larger flowers. Both forms occur in Australia and pass into each other as they do in Europe; the larger and more succulent ones are, however, the most common in Australia.—*Benth.*

9. DRYMARIA, Willd.
(From the plants being found in forests.)

Sepals 5, herbaceous or scarious at the edge. Petals 5, 2—6-cleft. Stamens 5 or fewer, slightly perigynous. Style 3-cleft. Capsule 3-valved. Seeds laterally attached; embryo curved round the albumen.—Herbs, usually diffuse, rarely erect, with dichotomous branches. Leaves flat, broad or narrow. Stipules very small, sometimes very fugacious or wanting. Flowers pedicellate, usually small, either solitary in the forks or in little axillary or terminal cymes. Petals usually shorter than the calyx.

The genus comprises a considerable number of American species.

1. **D. diandra** (two stamens), *Blume, Bijdr. tot de Fl. van Nederl. Indie* 68; F. v. M. *Papuan Plants* 86. Leaves glabrous, rhomboid or cordate-orbicular, conspicuously stalked. Stipules fringy-cleft; cymes paniculate, with elongated glandular-powdery peduncles. Flowers small; sepals only slightly scarious, their middle nerve forming a narrow pulverulent keel; petals deeply cleft into 2 segments, stamens usually 2, style almost none, stigmas 2. Fruit valveless or imperfectly 2-valved. Seeds large, 1 rarely 2, closely filling the cavity of the pericarp, black, opaque, glandular-scabrous.—F. v. M. Papuan Plants 86.

Hab.: Tropical parts of the colony.

XVI. CARYOPHYLLEÆ.

10. POLYCARPON, Linn.
(Plant loaded with seed.)

Sepals 5, keeled, scarious on the margin. Petals 5, small, entire or notched. Stamens 3 to 5. Style short, 3-cleft. Capsule 3-valved. Seeds laterally attached near the base; embryo excentrical, curved or nearly straight, the cotyledons incumbent or oblique.—Herbs, either diffuse or dichotomously branched, glabrous or pubescent. Leaves flat, usually ovate or oblong, often apparently, but not really, in whorls of 4. Stipules scarious. Flowers small, numerous, in terminal cymes, with scarious bracts.

A genus of very few species, dispersed over the temperate and tropical regions of the globe. The Australian species is identical with the commonest northern one.—*Benth.*

1. **P. tetraphyllum** (four-leaved), *Linn. f.; DC. Prod.* iii. 376; *Benth. Fl. Austr.* i. 162. A glabrous, much branched, spreading or prostrate annual, seldom more than 3 or 4in. long. Leaves obovate or oblong, really opposite, but placed as they usually are under the forks two pairs are so close together as to assume the appearance of a whorl of 4. Flowers very small and numerous, in loose terminal cymes. Sepals barely 1 line long. Petals much shorter and very thin. Stamens usually 3.—*F. v. M. Pl. Vict.* i. 205.

Hab.: Common in the southern parts.

Very common in sandy situations, chiefly not far from the sea, in Europe, temperate Asia, the greater part of Africa, and in many parts of North and South America; but unknown in tropical or subtropical Asia.

11. POLYCARPÆA, Lour.
(From the abundance of seed.)
(Aylmeria, Mart.)

Sepals 5, either entirely scarious, or herbaceous in the centre and scarious on the margin, but not keeled. Petals 5, entire or toothed. Stamens 5, hypogynous or slightly perigynous, free or united with the petals in a ring or tube. Style elongated, 3-furrowed, 3-toothed, or shortly 3-lobed at the top. Capsule 3-valved. Seeds obovoid or flattened; embryo curved or nearly straight; cotyledons usually (perhaps always) accumbent.—Annual or perennial herbs, erect or diffuse. Leaves narrow-linear or rarely ovate, often clustered in the axils so as to appear verticillate. Stipules scarious. Flowers usually numerous, in terminal cymes, sometimes loose and paniculate, sometimes dense and capitate, often remarkable for the white, pink or purple scarious sepals and bracts.

The genus is dispersed over the tropical and subtropical regions of the Old World, one—the commonest species—extending also into tropical America. The 6 Queensland species are all, except one, tropical; one is the abovementioned common one, the 5 others are endemic.—*Benth.* in part.

SECT. I. **Planchonia**, J. Gay.—*Petals and stamens united in a cup or tube, without staminodia.*
Stems hard and almost woody at the base, the radical leaves soon disappearing. Leaves all narrow. Flowers 3 to 4 lines.
  Stems tall, pubescent. Corolla-tube shorter than the free part. Stamens the length of the petals. Capsule short, obtuse . . . . 1. *P. longiflora.*
  Stems slightly pubescent. Capsule fusiform . . . . 2. *P. Burtoni.*
  Stems short, glabrous. Corolla-tube longer than the free part. Stamens much longer than the petals. Capsule oblong, tapering at the top . . . 3. *P. spirostyles.*
Stems herbaceous, several from a rosette of oblong or obovate radical leaves.
  Stem-leaves narrow. Flowers 1½ to 3 lines . . . . 4. *P. synandra.*

SECT. II. **Polycarpia**.—*Petals and stamens free or united in a ring at the base, without staminodia.*
Stems simple or hard and woody at the base. Radical leaves soon disappearing.
  Flowers 1½ line. Petals rounded and very obtuse. Capsule much shorter than the sepals . . . . 5. *P. corymbosa.*
  Flowers less than 1 line. Petals oval-oblong, acute, or toothed at the top. Capsule rather shorter or longer than the sepals . . . . 6. *P. breviflora.*

1. **P. longiflora** (long-flowered), *F. v. M. in Rep. Babb. Exped.* 8; *Benth. Fl. Austr.* i. 164. Pubescent, erect and rigid, 1 to 2ft. high, divided at the base into several erect branches. Leaves narrow-linear, acute or ending in a hair-like point, rigid, silky-hairy, often above ½in. long, with smaller ones clustered in their axils; the upper ones small and distant. Flowers large, brown red or purple, shortly pedicellate in dense terminal corymbose cymes or heads. Sepals fully 8 lines long, scarious, with a prominent midrib, the inner ones narrower, more acute and more deeply coloured than the outer. Petals hypogynous, united with the stamens in a campanulate tube not 1 line long, their free parts considerably longer and shortly bifid at the point. Filaments about as long as the petals. Ovary almost sessile. Style long and subulate. Capsule short ovoid, obtuse.

Hab.: North-western parts of the colony.

Var. *leucantha*. Leaves larger, broader, and less rigid. Sepals completely scarious and white, without any prominent midrib.

2. **P. Burtoni** (after R. C. Burton), *Bail. Proc. Roy. Soc. Ql.* Stems several, 9 to 12in. high, erect from a hard woody base, slightly pubescent. Leaves, those at base of stem linear-spathulate, about 1in. long, those of the stem almost filiform, about 1¼in. long, with bristle-like points. Flowers in terminal dense corymbs. Sepals narrow-lanceolate, 3 to 5 lines long, scarious, pinkish, with a midrib of a deep purple. Petals united with the stamens in a tube about 2½ lines long, purple, the free parts about the same length, and more or less lobed; filaments very slender, and with the free parts of the petals reflexed after flowering. Style shortly lobed at the end. Capsule fusiform. Seeds numerous.

Hab.: Northern parts, inland.

3. **P. spirostyles** (twisted styles), *F. v. M. in Rep. Babb. Exp.* 8; *Benth. Fl. Austr.* i. 165. Glabrous and often very glaucous, woody at the base, with numerous rigid opposite or dichotomous branches, our specimens not exceeding 6in. Leaves very narrow-linear, the margins revolute so as to be almost terete and filiform, rarely exceeding ½in., often clustered. Stipules small, with subulate points. Flowers large, on very short pedicels, either few in the upper forks or forming at length a broad corymbose cyme. Sepals 3 to 4 lines long, acute, white and scarious with a prominent midrib, the outer ones shorter and broader than the inner. Petals and stamens perigynous, united in a tube of fully 2 lines, with the slender filaments projecting considerably beyond the free oblong tops of the petals. Ovary shortly stipitate, tapering into a long spirally twisted deciduous style. Capsule stipitate, oblong, tapering at the top, nearly as long as the sepals. Seeds numerous, very small.

Hab.: Gilbert River, Herberton, Northcote.

Mr. S. B. J. Skertchly states that at Herberton this plant is intimately associated with copper deposits, and Mr. J. Brownlie Henderson, Government Analyst, found distinct traces of copper in plants brought from that district (*Report on Mines of Watsonville, &c., Geol. Surv. Q.* 1897). Mr. Skertchly informs me that the plant is a sure indication of copper deposits and is now frequently used by practical miners as a guide to that mineral.

4. **P. synandra** (anthers cohering), *F. v. M. in Rep. Babb. Exp.* 8; *Benth. Fl. Austr.* i. 165. A glabrous annual, with a rosette of petiolate spathulate or oblong radical leaves. Stems several, erect or decumbent, not above 6in. high, with dichotomous or clustered branches. Leaves narrow-linear, with recurved or revolute margins, the longer ones above ⅜in., but mostly shorter, and not much clustered. Stipules small, with fine points. Flowers rather larger than in *P. corymbosa*, in small rather loose corymbose cymes, all more or less pedicellate, the floral leaves all reduced to scarious bracts. Sepals about 2 lines or nearly 3 lines long in the capitate variety, white and scarious with a prominent midrib

often purple. Petals united with the stamens in a tube of about 1 line, their free part shorter and entire, sometimes very short, the filaments about the same length. Ovary sessile, with a subulate style. Capsule oblong, tapering at the top, with few seeds.

Hab.: North-western parts of the colony.
Var. *gracilis*. More slender. Sepals about 1½ line long. Petals rather broad, notched.

5. **P. corymbosa** (corymbose), *Lam. Illustr. n.* 2798; *Benth. Fl. Austr.* i. 166. Minutely pubescent or rarely almost glabrous, with erect, rather slender, but stiff branches, ¼ to 1 or even 1½ft. high. Leaves from narrow-linear to almost subulate, rarely linear-lanceolate, flat or with revolute margins, the longer ones ¼ to 1in., with small ones clustered in their axils, the upper ones much smaller and often few and distant. Stipules tapering to a fine point. Flowers numerous, in dense terminal corymbose cymes, sometimes all forming one dense mass on the top of an otherwise simple stem, sometimes the cymes numerous and loosely paniculate. Floral leaves all reduced to scarious bracts. Sepals about 1¼ line long, white and scarious, without any prominent midrib, but tapering to a fine point. Petals quite free, not ½ line long, broadly ovate, very obtuse and rather firm. Stamens often shorter. Style very short. Capsule ovoid or oblong, much shorter than the sepals.—DC. Prod. iii. 874; Wight. Ic. Pl. Ind. Or. t. 712.

Hab.: Port Curtis.
The species is common in tropical Asia and Africa and is found also in Brazil and Guiana.

6. **P. breviflora** (flowers short), *F. v. M. in Rep. Babb. Exp.* 9; *Benth. Fl. Austr.* i. 156. Glabrous or pubescent, and very nearly allied to *P. corymbosa*, but more slender and divaricately branched, and at once known by its very much smaller flowers. Sepals scarcely 1 line long, broader and less acuminate than in *P. corymbosa*, petals much narrower, not so obtuse and usually denticulate at the top; stamens much more perigynous; capsule longer in proportion, occasionally even exceeding the sepals.

Hab.: Various localities throughout the colony.

Order XVII. PORTULACEÆ.

Flowers regular, hermaphrodite. Sepals fewer than petals, usually 2, free or rarely adnate to the ovary at the base, usually broad, imbricate in the bud. Petals 4 or 5, rarely more, hypogynous or rarely perigynous, imbricate in the bud. Stamens inserted with the petals and often adhering to their base, of the same number or fewer and opposite to them or indefinite; anthers 2-celled. Ovary free or rarely half-inferior, 1-celled. Style more or less deeply divided into 3 or rarely 2 or more than 3 branches, stigmatic along the inner side. Ovules 2 or more, amphitropous, with an inferior micropyle, attached to funicles erect from the base of the cavity, and free or united in a central column, or in as many clusters as style-branches. Seeds several or solitary by abortion, usually more or less reniform, with a lateral hilum; testa crustaceous, sometimes with a caruncle at the hilum. Embryo more or less curved round the mealy albumen, or rarely nearly straight with very little albumen.—Herbs, rarely shrubby at the base, usually glabrous and succulent or clothed with long hairs. Leaves alternate or opposite, entire. Stipules scarious or split into hairs or none. Flowers terminal and solitary, or in racemes, cymes or panicles, or rarely axillary. Petals usually very fugacious or withering in a mass.

A small Order, chiefly American, with a few species dispersed over other parts of the world, especially S. Africa and Australia. The Queensland genera are none of them endemic, 1 of them being chiefly American. Of the other two, 1 is generally distributed over the globe, the

XVII. PORTULACEÆ.

other a naturalised plant. The chief characters, derived from the ovary and seeds, are those of *Caryophylleæ*, from which *Portulaceæ* differ in habit, in the number and position of the stamens, and especially in their calyx.—*Benth.* in part.
Ovary half-inferior. Petals and stamens perigynous . . . . . 1. PORTULACA.
Ovary superior. Petals and stamens hypogynous.
  Petals free.
    Sepals very often deciduous. Stamens 5 or many. Seeds strophiolate . 2. TALINUM.
    Stamens indefinite, often numerous, rarely and irregularly reduced to 5 . 3. CALANDRINIA.

1. PORTULACA, Linn.

(From the English name of Purslane.)

Sepals 2, united at the base in a tube adnate to the ovary, the free part deciduous. Petals 4 to 6, perigynous. Stamens indefinite, often numerous, sometimes 6 to 8, inserted with the petals. Ovary half-inferior, with several ovules. Style deeply 2 to 8-cleft. Capsule membranous, half-inferior, the free part circumsciss at maturity. Seeds reniform, shining, often granulate.—Herbs, more or less succulent. Leaves alternate or opposite, often clustered in the axils, the floral ones usually forming an involucre round the flowers. Stipules scarious, or more frequently reduced to a tuft of hairs, sometimes very minute or none. Flowers terminal, sessile, or pedicellate.

The species are mostly American, with a very few tropical Australian, Asiatic, or African ones, 2 of them widely dispersed over cultivated or sandy places in various parts of the globe. One of these is included among the Australian ones, of which the remainder are all endemic.—*Benth.*

Leaves mostly alternate.
  Stipular hairs minute or none.
    Leaves oblong-cuneate. Root slender. Capsule closely sessile . . . 1. *P. oleracea.*
    Leaves linear-terete. Root usually tuberous. Capsule narrowed into a
      short stipes . . . . 2. *P. napiformis.*
  Stipular hairs numerous and conspicuous.
    Leaves thick and short . . . . 3. *P. australis.*
    Leaves linear-terete, almost filiform. . . . . 4. *P. filifolia.*
Leaves all opposite.
  Stipular hairs short, but conspicuous. Flowers usually 3, within the floral
    leaves, and shortly pedicellate. Style-lobes subulate . . . . 5. *P. digyna.*
  No stipular hairs. Flowers solitary and sessile, within 4 bract-like floral
    leaves. Style-lobes flat and transparent.
      Leaves lanceolate or linear . . . . 6. *P. oligosperma.*
      Leaves orbicular . . . . 7. *P. bicolor.*
      Leaves cordate-orbicular . . . . 8. *P. Armitii.*

1. **P. oleracea** (from being used as a pot-herb),. *Linn.; DC. Prod.* iii. 353; *Benth. Fl. Austr.* i. 169. "Thukouro," Cloncurry, *Palmer.* Pig weed. A low, prostrate, or spreading annual, seldom exceeding 6in., somewhat succulent, and quite glabrous. Leaves mostly alternate, cuneate-oblong, obtuse, very rarely exceeding ½in., usually narrowed into a short petiole, the stipular hairs very minute, and sometimes quite disappearing. Flowers terminal and sessile, between 2 or more floral leaves, rarely solitary, usually several together in little heads which are either single or several in a dichotomous cyme. Sepals not much more than 2 lines long. Petals 5, scarcely longer than the calyx, slightly united at the base, yellow and very fugacious. Stamens 10 to 12 or rarely fewer. Style short, with 5 linear stigmatic lobes. Capsule sessile. Seeds minutely tuberculate, the funicles often united at the base into 5 clusters.—A. Gray, Gen. Ill. t. 99; F. v. M. in Rep. Babb. Exped. 10.

Hab.: Common.
The stalks are roasted in the ashes, which softens them, then eaten; also eaten raw. The plant is gathered in heaps, and after drying a little time the seeds fall off and are gathered with mussel-shells, ground between two stones and roasted.—*Palmer.*
Var. *grandiflora.* Sepals more obtuse, 3 to 4 lines long. Georgina River.
The species is common in maritime or sandy localities in most tropical countries, extending into the warm parts of the temperate regions, both of the northern and southern hemispheres.

2. **P. napiformis** (turnip-like root), *F. v. M. Herb.*: *Benth. Fl. Austr.* i. 169. "Karedilla," Cloncurry, *Roth*. Glabrous, with decumbent or erect stems of 6in. to near 1ft., the tap-root thickening into an oblong tuber. Leaves alternate, linear, succulent, apparently terete, $\frac{1}{2}$ to 1in. long. Stipular hairs exceedingly minute. Flowers smaller than in *P. oleracea*, usually 3 together, between 2 to 4 involucral leaves, but not quite sessile. Stamens about 16. Style rather long, 4-cleft at the top. Capsule small, contracted into a short stipes. Seeds smaller than in *P. oleracea*, black and shining, finely granulated.

Hab.: Leichhardt district.

The species is allied to the East Indian *P. tuberosa*, Roxb. but the flowers and fruits are much smaller, not so closely sessile, and there are not the long stipular and involucral hairs of that species.—*Benth.*

3. **P. australis** (Australian), *Endl. Atakta*, 7, t. 6; *Benth. Fl. Austr.* i. 169. Apparently decumbent and much branched, the stipular and involucral hairs copious, but otherwise glabrous. Leaves alternate, oblong, elliptical, thick, under $\frac{1}{2}$in. long. Flowers yellow, 1 or 2 together, sessile between 2 to 4 involucral leaves. Stamens numerous (Rockhampton specimens 20). Style elongated, 5 or 6-cleft (Rockhampton specimens 4 or 5). Seeds shining, granulate, the funicles united into as many clusters as styles.

Hab.: Leichhardt district and Gulf of Carpentaria.

It is not improbable that both this species and *P. filifolia* may prove to be forms of the tropical African *P. foliosa*.—*Benth.*

4. **P. filifolia** (thread-like leaves), *F. v. M. Fragm.* i. 169; *Benth. Fl. Austr.* i. 169. Annual, with erect or decumbent stems of $\frac{1}{2}$ to 1ft., the stipular and involucral hairs long and copious, but otherwise glabrous. The roots sometimes thick, but never tuberous. Leaves alternate, linear-terete, almost filiform, $\frac{1}{2}$ to 1in. long. Flowers rather large, yellow, 1 to 3 together, sessile between 2 to 4 involucral leaves. Sepals 2 to $2\frac{1}{2}$ lines, and petals twice as long. Stamens numerous. Style elongated, usually 4-cleft. Seeds shining, granulate, the funicles united in as many clusters as styles.

Hab.: In the interior common.

This may be a variety of *P. australis*, and only appears to differ from the tropical African *P. foliosa* in its more slender leaves, and from *P. tuberosa*, Roxb., in the roots not tuberous and in the large flowers.—*Benth.*

5. **P. digyna** (two-branched style), *F. v. M. Fragm.* i. 170; *Benth. Fl. Austr.* i. 170. A procumbent, glabrous annual of a few inches, with dichotomous or opposite branches. Leaves all opposite, ovate obovate or nearly orbicular, 2 to 3 lines long, very shortly petiolate. Stipular hairs very short. Flowers pink, very small, pedicellate, 1 to 3 together, between 2 or 4 involucral leaves, forming dichotomous leafy cymes. Sepals not 2 lines long. Petals 4, rather longer. Stamens about 10. Style long, with 2 long linear stigmatic branches. Ovules about 6, the funicles forming 2 clusters. Capsule elongate-conical, covered in the upper part with oblong papillæ. Seeds 1, 2, or 3, black, smooth, and shining.

Hab.: Northern interior and Stanthorpe.

6. **P. oligosperma** (few-seeded), *F. v. M. Fragm.* i. 170; *Benth. Fl. Austr.* i. 170. A little slender annual of 2 or scarcely 3in. with numerous opposite branches. Leaves all opposite, oblong, narrow-lanceolate or linear and semi-terete, 3 to 4 lines long. Stipular hairs none or quite microscopic. Flowers very small, pink, terminal, solitary and closely sessile within 2 or 4 involucral leaves, which do not exceed the calyx-tube, so that the flower appears pedicellate, with 4 calyx-like bracts at the summit of the pedicel. Sepals scarcely 1 line

long, and the petals apparently not longer. Stamens about 6, the anthers very transparent. Style divided into 2 to 4 lanceolate, transparent, and very delicate lobes. Seeds few, black, granulate.

Hab.: Cape River.
The Sturt's Creek specimens have smaller and rather broader leaves, and in the flower I examined the lobes of the style were broader than in those from Victoria River, but both are probably forms of one species, nearly allied to the East Indian *P. quadrifida*, but at once known by the absence of stipular hairs.—*Benth.*

7. **P. bicolor** (two-coloured), *F. v. M. Fragm.* i. 171; *Benth. Fl. Austr.* i. 170. A minute, prostrate, tuberous-rooted plant, with opposite branches, rarely above 1½in. long. Leaves all opposite, broadly ovate or orbicular, scarcely exceeding 2 lines. Flowers as in *P. oligosperma*, minute, solitary, terminal, and closely sessile between 4 bract-like floral leaves (appearing pedicellate, with 4 calyx-like bracts at the summit of the pedicel). Sepals not 1 line long. Petals minute, yellow. Stamens about 6. Style with 4 (or sometimes 2 ?) lanceolate, transparent, very delicate lobes. Capsule short, broad. Seeds several, small, black, granulate.

Hab.: Keppel Bay, *R. Br.*; Bustard Head, *Jas. Keys.*

8. **P. Armitii** (after W. E. Armit), *F. v. M. Fragm.* x. 97. Plant about 2 or 3in. high. Leaves 1½ to 3 lines long, opposite, cordate-orbicular, on very short petioles. Stipular hairs none. Pedicels 1½ line or less. Flowers solitary, deciduous, part of the calyx, 3 to 5 lines long. Petals purple, 3 lines long. Anthers oblong-oval. Style-branches shortly exserted and dilated. Operculum 1 to 1½ line long. Seeds numerous, turgid, shining.

Hab.: Robertson River.

*2. **TALINUM**, Adans.

(Said to be the name given to the plant by the negros of Senegal, who eat it as a salad.)

Sepals 2, herbaceous, ovate, deciduous or rarely subpersistent. Petals 5, hypogynous. Stamens indefinite, 5 to numerous, adhering to the base of the petals. Ovary free, ovules numerous. Style 3-fid or 3-sulcate. Capsule globose or ovoid, chartaceous, 3-valved. Seeds subglobose or compressed, strophiolate.—Succulent herbs or shrubs. Leaves flat, alternate or subopposite, no stipules. Flowers racemose or paniculated.

Found in warm countries throughout the world.

1. **T. patens** (spreading), *Willd.* A succulent perennial. Stems almost simple, 1 to 2ft. high, leafy to the middle, where the panicle begins. Leaves opposite or alternate, 2 to 6in. long, 1 to 2½in. broad, tapering much towards the base. The upper part of the plant composed of a panicle bearing dichotomous cymes of pink flowers. Pedicels filiform; sepals roundish, deciduous; petals small, obovate; stamens 15 to 20; style-branches divergent; capsule globose. Seeds minutely granulose.

Hab.: A S. American plant often found as a stray from garden culture near towns.

3. **CALANDRINIA**, H. B. and K.
(After J. L. Calandrini.)

Sepals 2, persistent or rarely deciduous. Petals 5 or more, or rarely fewer, hypogynous. Stamens indefinite, numerous or few, free or united in a ring at the base, or adhering to the petals. Ovary free, with several ovules, rarely reduced to 1 or 2. Styles 3 or rarely 4, free or united in a single style, 3 or 4-cleft, or furrowed at the top. Capsule globose, ovoid or oblong, opening in

3 or 4 valves, or almost indehiscent. Seeds reniform-globular or flattened, not strophiolate, shining or granulate. Embryo curved round the albumen.—Herbs, rarely half-shrubby at the base, glabrous or hirsute. Leaves alternate or in radical tufts, more or less fleshy. Stipules none. Flowers either solitary pedunculate and axillary, or arranged in terminal racemes or heads. Petals usually very fugacious.

A large genus, which besides numerous tropical, subtropical, or southern American species, only contains the Australian ones here described, which are all endemic. Formerly confounded with *Talinum*, it has been well distinguished from that genus chiefly by the absence of any strophiola or caruncle to the seeds, and differs from *Claytonia* in the stamens always indefinite, even when reduced to a number about the same as or fewer than that of the petals.—*Benth.*

Stamens numerous (20 to 100).
Scapes leafless, 1-flowered. Leaves radical, narrow-linear . . . . . 1. *C. uniflora.*
Stems more or less leafy, several-flowered.
  Perennial. Petals very broad. Anthers linear oblong. Styles united at the base . . . . 2. *C. balonensis.*
  Annuals. Petals oval-oblong. Anthers short. Styles free to the base.
    Styles and capsular valves 3 . . . . 3. *C. polyandra.*
    Style scarcely any, capsular valves 4 . . . . 4. *C. pleopetala.*
Stamens few. Capsule ovoid or oblong, very readily dehiscent.
Stamens mostly 8 to 10. Seeds pitted. Sepals broad and very obtuse. Leaves oblong or shortly linear.
  Stems short, ascending or diffuse . . . . 5. *C. pusilla.*
  Stems twining . . . . 6. *C. volubilis.*
Stamens mostly 3 to 5. Seeds very smooth and shining.
  Bracts very small. Sepals under 2 lines and often under 1 line, acute. Leaves oblong or linear-oblong, thick. Racemes loose. Pedicels at length 3 to 5 lines, reflexed . . . . 7. *C. calyptrata.*
  Sepals obtuse . . . . 8. *C. pumila.*
Stamens few. Capsule globular, or shortly ovoid, very smooth and shining, and scarcely dehiscent.
Leaves linear-terete. Stamens about 15. Anthers oblong. Capsular valves separating at the base . . . . 9. *C. spergularina.*
Leaves linear-terete. Stamens few. Seeds striate. Capsule cylindric, conical . . . . 10. *C. ptychosperma.*
Leaves short and broad. Capsule 3-valved, ovate . . . . 11. *C. pogonophora.*

1. **C. uniflora** (one-flowered), *F. v. M. in Trans. Phil. Inst. Vict.* iii. 41, and *Fragm.* i. 177; *Benth. Fl. Austr.* i. 172. Rootstock simple, cylindrical, erect, bearing a dense tuft of narrow-linear leaves of 2 to 4in. Scapes numerous from amongst the leaves, 8 to 10in. high, 1-flowered and leafless, except 1 or 2 minute scales. Flowers rather large. Sepals broad and thin, 3 to 4 lines long. Petals usually 6 or 7. Stamens very numerous, the inner ones much longer than the outer; anthers oblong. Styles 4, erect, shortly plumose and stigmatic along their whole length. Capsule about as long as the sepals, 4-valved. Seeds numerous, black and shining.

Hab.: Gilbert and Norman Rivers.

The species is nearly allied to two Chilian ones, *C. rupestris*, Barn., and *C. graminifolia*, Philippi.—*Benth.*

2. **C. balonensis** (from Balonne River), *Lindl. in Mitch. Trop. Austr.* 148; *Benth. Fl. Austr.* i. 172. Apparently perennial, erect, branching, 6in. to 1ft. high or rather more. Leaves thick and fleshy, the lower ones oblong-spathulate or obovate, 1in. long or less, the upper ones linear or lanceolate, often above 2in. Flowers large, purple, in loose terminal racemes, on pedicels of about 1in. Bracts scarious, acuminate, mostly opposite, but only one of each pair has a flower in its axil. Sepals very broad and obtuse, herbaceous, obscurely veined, with a scarious margin. Petals very broadly obovate, fully ¾in. long. Stamens very numerous; anthers narrow-oblong. Style 3-lobed, the lobes thick and nearly twice as long as the entire base.

Hab.: Sandy soil on the Balonne river, *Mitchell*.

3. **C. polyandra** (stamens numerous), *Benth. Fl. Austr.* i. 172. Annual, with decumbent or ascending branches of 6in. to 1ft. Leaves few, chiefly in the lower part of the stem, thick and fleshy, the lowest broadly linear or almost spathulate, the upper ones narrow-linear, occasionally almost opposite, mostly 1 to 1½in. long. Flowers of a red-purple, rather large, few together in a terminal raceme, the pedicels 1in. or more. Bracts small and scarious. Sepals very broad, rather obtuse, thin and slightly coloured, with scarcely prominent veins. Petals narrow-obovate, about ½in. long. Stamens very numerous, irregularly united at the base; anthers short. Style divided to the base into 3 linear stigmatic branches. Capsule ovoid or oblong, 3-valved. Seeds very numerous and small, black, minutely pitted.—*Talinum polyandrum*, Hook. Bot. Mag. t. 4833.

Hab.: In the interior.

4. **C. pleopetala** (petals numerous), *F. v. M. Fragm.* x. 70. A glabrous perennial, the radical leaves crowded, somewhat broad-linear, the stem ones short or wanting. Raceme of few or many flowers. Bracts very short, scarious. Pedicels spreading or refracted. Sepals persistent, ovate or orbiculate-cordate. Petals 8 or 9, 2 or 3 times larger than the calyx, cuneate-oblong. Stamens numerous; anthers ovate-rotund. Style scarcely any. Stigma 4-lobed, pubescent. Capsule cylindrical-oblong. Many-seeded 4-valved. Seeds brown, reniform-ovate, shining, smooth.

Hab.: Bowen Downs and Mueller's Range.

5. **C. pusilla** (small), *Lindl. in Mitch. Trop. Austr.* 360; *Benth. Fl. Austr.* i. 174. A small annual, the stems ascending from 1 to 3 or 4in. or rarely higher. Leaves radical or on the lower part of the stem, about ½ to 1in. long, much more succulent than in *C. calyptrata*, oblong or linear, mostly petiolate, but dilated and stem-clasping at the base. Racemes occupying a great part of the stems, but loose and few-flowered, with minute scarious bracts, except the lower ones, which are sometimes leafy. Flowers apparently pink, like those of *C. calyptrata*, except that the sepals are very broad and obtuse, coloured, with scarious margins, attaining 1½ line when in fruit. Petals 5 or 6, oblong, stamens 5 to 8; anthers small. Style divided to the base into 3 short, thick, stigmatic branches. Capsule narrow, longer than the calyx, opening in 3 valves. Seeds numerous, much smaller than in *C. calyptrata* and minutely pitted.

Hab.: On the Maranoa.

6. **C. volubilis** (twining) *Benth. Fl. Austr.* i. 174. Allied to *C. pusillæ*, and with that species considered by F. v. Mueller as a variety of *C. calyptrata*, but the seeds and flowers are different. Leaves crowded on a short, succulent, branching stock, linear-oblong, 1 to 1½in. long, narrowed below the middle, but dilated at the base. Flowering branches twining, almost leafless, except minute scarious bracts. Pedicels flexuose, 2 to 6 lines long. Sepals very obtuse, broad and succulent, 1¼ line when in flower, 2 lines when in fruit. Petals about as long, withering into a calyptra on the young fruit. Stamens 8 to 10, the filaments slightly dilated at the base, but scarcely united; anthers small. Style cleft almost to the base into 3 linear stigmatic branches. Capsule acuminate, twice as long as the sepals. Seeds strongly pitted.

Hab.: Stanthorpe.

7. **C. calyptrata** (withered petals form a calyptera or covering to the fruit), *Hook. f. in Hook. Ic. Pl.* t. 296; *Benth. Fl. Austr.* i. 174. A small annual, with petiolate linear-oblong or linear-spathulate radical leaves. Stems branching, prostrate or ascending, from 1 or 2 to 7 or 8in. long. Leaves few, smaller than the radical ones, varying from linear to almost obovate. Flowers very small, in a loose flexuose raceme, the pedicels 2 to 6 lines long, reflexed after flowering.

H

Bracts very small, the upper ones often scarious. Sepals acute, about 1 line long in flower, nearly 1½ when in fruit. Petals about as long, often persistent a long time after flowering, withered into a small calyptra on the top of the young fruit. Stamens about 5, with slender, free filaments; anthers ovate. Style very short, with 3 very short, oblong, stigmatic branches. Capsule rather longer than the calyx, 3-valved. Seeds numerous, small, very smooth and shining.—Hook. f. Fl. Tasm. i. 143; *Claytonia calyptrata*, F. v. M. Fragm. iii. 89.

Hab.: Darling Downs, &c.

8. **C. pumila** (dwarfish), *F. v. M. Fragm.* x. 68. A small, tufted, glabrous plant, with a thick succulent root. Leaves radical or nearly so, oblong or almost ovate, 3 to 4 lines long, but narrowed into a petiole twice that length. Racemes 1 to 4-flowered. Bracts small, scarious. Pedicels never very divergent. Sepals persistent, ovate or cordate-orbiculate. Petals 5, ovate. Stamens very few. Capsule almost globose, 3-valved, many-seeded. Seeds very minute, smooth, almost ovate, brown, shining.—*C. calyptrata*, var. *pumila*, in Flora Austr. i. 175.

Hab.: Balonne River.

9. **C. spergularina** (Spergularia-like), *F. v. M. Fragm.* i. 175; *Benth. Fl. Austr.* i. 176. A small annual, with a tuft of linear-terete leaves under 1in. long. Stems slender, decumbent, slightly branched, 2 to 4in. long or scarcely more. Leaves few, small, linear-terete. Flowers pink, very small, in a rather rigid often flexuose raceme on pedicels of 1 to 3 lines. Bracts very minute and scarious. Sepals acute, a little more than 1 line long in flower, 1½ line when in fruit. Petals 6, not twice as long as the calyx. Stamens about 15; anthers oblong, the cells adhering in the centre only. Style divided to the base into 3 linear stigmatic branches. Capsule small, the valves remaining coherent at the top, separating at the base, and falling off together. Seeds small, smooth, and shining.

Hab.: Cape York, Torres Straits, and Gulf of Carpentaria.

10. **C. ptychosperma** (referring to plait-like marking of seeds), *F. v. M. Fragm.* x. 70. Plant small, glabrous. Leaves, the radical ones crowded, 1 to 1½in. long and 1 line thick, acute; stem ones shorter. Racemes few-flowered. Bracts scarious, 1 to 1½ line long. Sepals persistent, roundly-ovate, acute, about 2 lines long. The dying petals forming a conical calyptra about 2 lines long. Stamens not numerous. Capsule 3 lines long, cylindric-conical. Seeds ½ line, neither rough nor reticulate but longitudinally striate.

Hab.: Bowen Downs.

11. **C. pogonophora** (beard-bearing), *F. v. M. Fragm.* x. 69. A small perennial. Leaves ¼ to ½in. long, lanceolate or rhomboid-ovate, sessile, bearded at the axil, crowded near the base of the stem. Racemes few-flowered, bracts minute-scarious. Pedicels 2 or 3 times as long as the calyx. Sepals 2 or 3 lines long, acute, deciduous. Corolla somewhat short. Capsule 3-valved, almost ovate, exocarp thin-cartilaginous, endocarp membranous. Seeds pale-red, pyramidal-trigonous, nearly ¼ line long, thinly reticulate or papillulose.

Hab.: Leichhardt district.

Order XVIII. ELATINEÆ.

Flowers regular, hermaphrodite. Sepals 2 to 5, free, imbricate in the bud. Petals as many, hypogynous, imbricate in the bud, occasionally wanting. Stamens as many or twice as many, hypogynous, free; anthers 2-celled. Torus small, without any disk. Ovary free, with as many cells as there are sepals; styles as many, free from the base, with terminal capitate stigmas. Ovules

XVIII. ELATINEÆ.

several in each cell, attached to the inner angle, anatropous. Capsule opening septicidally, the valves flat or concave, with the margins inflexed, leaving more or less of the dissepiments attached to the central column. Seeds straight or curved, testa crustaceous, usually wrinkled or ribbed, albumen none or very thin. Embryo filling the seed, cotyledons short, radicle next to the hilum.—Herbs or low undershrubs, aquatic, creeping or diffuse. Leaves opposite or rarely verticillate, entire or seriate. Stipules in pairs. Flowers small, axillary, solitary or in clusters or cymes.

A small Order, dispersed over nearly the whole globe, allied to *Hypericineæ* and *Caryophylleæ*, but differing from the former in habit, in the stipules, and in the perfectly isomerous flowers, from the latter chiefly in the ovary and fruit and want of albumen to the seeds; there is also considerable affinity, especially in habit, with *Lythrarieæ* and *Crassulaceæ*. The only two genera of the Order, both of them of wide geographical range, are represented in Australia.—*Benth.*

Sepals membranous, obtuse. Capsule membranous. Glabrous, aquatic or creeping herbs. Flowers 2 to 4-merous . . . . 1. ELATINE.
Sepals herbaceous in the middle or keeled, acute. Capsule almost crustaceous. Herbs or undershrubs. Flowers usually 5-merous, rarely 3 to 4-merous   2. BERGIA.

1. ELATINE, Linn.
(Leaves resembling the fir-tree.)

Flowers 3 or 4-merous, rarely 2-merous. Sepals membranous, obtuse, not keeled. Ovary globular. Capsule membranous, the dissepiments either disappearing or remaining attached to the central column.—Small glabrous herbs, either aquatic or creeping on mud. Leaves opposite or verticillate. Flowers usually solitary in the axils, and very small.

The genus is widely dispersed over the temperate and subtropical regions of the globe. The Australian species is considered by some as endemic, by others as identical with an American one.—*Benth.*

1. **E. americana** (American), *Arn. in. Edinb. Journ. Nat. Sc.* i. 431, var. *australiensis; Benth. Fl. Austr.* i. 178. A small, tender, glabrous annual, prostrate and creeping over mud in dense tufts, sometimes not 1in. in diameter, sometimes extending over a considerable surface. Leaves in the ordinary form ovate, obovate, or broadly oblong, 2 to 3 lines long, thin and of a bright green; but in some luxuriant specimens ovate-lanceolate or oblong, and exceeding ½in., almost always bordered by a few distant glands. Stipules very minute and deciduous, or rarely more persistent, and ¼ line long. Flowers very minute, sessile and solitary in one axil only of each pair of leaves, and in Australia almost always 3-merous. Sepals usually very minute and transparent, and the petals so very small and fugacious as to be rarely found in dried specimens, except in some western ones, where the petals are reddish and fully ½ line long. Stamens 3. Ovary depressed-globular, with 3 cells and 3 minute, punctiform, almost sessile stigmas. Capsule often 1 line in diameter, the dissepiments sometimes complete, sometimes obliterated at maturity. Seeds cylindrical, more or less curved or nearly straight, marked with longitudinal furrows and minute, transverse wrinkles.—Hook. f. Fl. Tasm. i. 47; *E. minima*, Fisch. and Mey. in Linnæa, x. 73; F. v. M. Pl. Vict. i. 195; *E. gratioloides*, A. Cunn. in Ann. Nat. Hist. iii. 26.

Hab.: Brisbane River and south Queensland generally.

2. BERGIA, Linn.
(After Dr. P. J. Bergius.)

Flowers 5-merous, or rarely 3—4-merous. Sepals herbaceous or keeled in the centre, acute, usually membranous and transparent on the edges. Ovary ovoid or globular. Capsule somewhat crustaceous, the valves sometimes induplicate on the edges and carrying off nearly the whole of the dissepiments, sometimes nearly flat, leaving more or less of the dissepiments attached to the axis.—Herbs or undershrubs, prostrate or much branched, often pubescent. Leaves opposite, entire or more frequently serrate. Flowers axillary, solitary or clustered in cymes, small, but usually larger than in *Elatine*.

The genus is widely distributed over the warmer regions of the globe.

Flowers small, clustered in the axils. Stamens of the same number as
  the petals and sepals.
    Stems pubescent . . . . 1. B. *ammannioides*.
Flowers solitary, pedicellate. Stamens twice the number of the sepals
  and petals.
    Stem woody, prostrate and tortuous. Pedicels short. Outer filaments
      much broader. Styles filiform . . . . 2. B. *perennis*.

1. **B. ammannioides** (like an Ammania), *Roth, Nov. Pl. Sp.* 219; *Benth. Fl. Austr.* i. 180. A rigid, much-branched annual, erect or decumbent, pubescent or hirsute, with spreading hairs, usually 6in. to 1ft. high. Leaves from oval-elliptical to oblong or lanceolate, the larger ones ½ to 1in., but mostly smaller, more or less serrate with mucronate or glandular teeth, narrowed at the base. Stipules lanceolate, serrate. Flowers very small, in dense axillary clusters, on very short filiform pedicels, usually 5-merous, but sometimes 4-merous or 3-merous. Sepals very narrow, acute, ciliate, about ½ line long. Petals narrow, very thin, about as long as the sepals. Stamens of the same number as the sepals and petals. Capsule rather shorter, the boat-shaped valves separating septicidally so as to leave the axis almost wholly without any remains of the dissepiments. Seeds very small, ovoid, nearly straight.—*Elatine ammannioides*, Wight, in Hook. Bot. Misc. iii. 93, t. 5; Wight, Ill. t. 25a; F. v. M. Fragm. ii. 147.

Hab.: Thursday Island and various other localities.
The species is common in East India and the warmer regions of Africa.

2. **B. perennis** (perennial), *F. v. M. Herb.; Benth. Fl. Austr.* i. 181. Stems prostrate, woody, tortuous, with very short leafy branches, glabrous or with a very few short hairs. Leaves from ovate to elliptical-oblong, mostly 3 to 4 lines long, rather rigid, glabrous and glaucous, often ciliate towards the base and narrowed into a short petiole. Stigmas lanceolate, ciliate. Flowers usually 5-merous, on solitary pedicels, rarely exceeding the length of the leaves. Sepals broadly-lanceolate, keeled, with scarious margins, nearly 2 lines long. Petals longer, rather narrow. Stamens usually 10, the 5 outer filaments dilated, especially below the middle. Styles filiform. Capsule rather shorter than the calyx, the valves leaving much of the dissepiments attached to the central column. Seeds oblong, curved, slightly furrowed and transversely wrinkled like those of *Elatine*.—*Elatine perennis*, F. v. M. Fragm. ii. 146.

Hab.: Banks of the rice swamps near Sturt's Creek, *F. v. Mueller*. The species is nearly allied to the S. African B. *anagalloides*, E. Mey, which is a perennial with the same styles and stamens, but its flowers are rather larger, on longer pedicels.—*Benth.*

Order XIX. HYPERICINEÆ.

Flowers regular, hermaphrodite. Sepals 5, rarely 4, imbricate in the bud. Petals as many, hypogynous, imbricate and usually contorted in the bud. Stamens indefinite, hypogynous, usually united or clustered into 3 or 5 bundles; anthers 2-celled. Ovary consisting of 3 to 5 carpels more or less united, either 1-celled with the placentas on the inflexed margins of the carpels or completely divided into cells by the union of the placentas in the axis. Styles as many as carpels, free or rarely united at the base, with terminal stigmas. Ovules usually several to each cell or placenta, anatropous. Fruit capsular or rarely fleshy and

XIX. HYPERICINEÆ.

indehiscent. Seeds straight or rarely curved, without albumen. Embryo straight or rarely curved, the radicle next the hilum.—Herbs, shrubs, or rarely trees. Leaves opposite or rarely verticillate, simple or entire or with glandular teeth. Stipules none. Flowers terminal or rarely axillary, solitary or in cymes or panicles. Leafy parts often marked with glandular, pellucid, or black dots.

The Order is dispersed over the greater portion of the globe, although represented in Australia by only one or two species, and those not endemic. It is closely allied to *Guttiferæ* and *Ternstræmiaceæ*.—*Benth.*

1. HYPERICUM, Linn.
(A name of Dioscorides.)

Sepals 5. Petals 5, not wholly inside. Capsule opening septicidally. Seeds not winged. Embryo oblong or cylindrical, with short cotyledons.—Herbs or shrubs. Leaves either small or thin, entire, or rarely minutely toothed. Flowers yellow or rarely white.

A large genus with nearly the same extensive geographical range as the Order.

Erect or ascending. Leaves usually subcordate . . . . 1. *H. gramineum.*
Procumbent. Leaves usually oblong or obovate . . . . 2. *H. japonicum.*

1. **H. gramineum** (often found among grass), *Forst.; DC. Prod.* i. 548; *Benth. Fl. Austr.* i. 182. A glabrous perennial, with erect or ascending angular stems, usually about 1ft. high, but sometimes nearly twice that height, or much shorter, slender, but rather rigid, branching at the base only or in the inflorescence. Leaves closely stem-clasping, ovate to oblong-lanceolate, obtuse, rarely exceeding ½in., entire, with numerous pellucid dots, the margins more or less revolute. Flowers 3 or more, in the forks or terminating the branches of a dichotomous cyme, with a pair of leafy bracts at the base of each fork; the pedicels erect and rigid, ¼ to ½in. long. Sepals lanceolate, acute, appressed, 2 to 3 or rarely 4 lines long. Petals entire, longer than the sepals. Stamens very variable in number, usually rather numerous and free. Styles 3, distinct. Capsule 1-celled, 3-valved, with narrow-linear placentas and numerous small seeds.—*DC. Prod.* i. 548; Labill. Sert. Austr. Caled. 53, t. 53; Hook. f. Fl. Tasm. i. 53; F. v. M. Pl. Vict. i. 198; *Ascyrum involutum*, Labill. Pl. Nov. Holl. ii. 32, t. 174; *Hypericum involutum*, Chois. in DC. Prod. i. 549; *H. pedicellare*, Endl. in Hueg. Enum. 12; *Brathys Billardieri* and *B. Forsteri*, Spach. in Ann. Sc. Nat. Ser. 2, v. 367.

Hab.: Frequent in all parts.

The species in the original form, above described, is common also to New Zealand and New Caledonia.

2. **H. japonicum** (Japanese), *Thunb. Fl. Jap.* 295, t. 81; *Benth. Fl. Austr.* i. 182. Very nearly allied to *H. gramineum*, and considered by F. v. Mueller as a variety only. It is much less rigid and usually very procumbent or diffuse, with ascending branches, terete or scarcely angled. Leaves smaller, flatter, and more obtuse, not so broad at the base. Flowers smaller, on shorter pedicels, the sepals less acute and the petals very seldom exceeding them.—DC. Prod. i. 548; Hook. f. Fl. Tasm. i. 53; *Ascyron humifusum*, Labill. Pl. Nov. Holl. ii. 33, t. 175; *H. pusillum*, Chois. in DC. Prod. i. 549; *Brathys humifusa*, Spach, in Ann. Sc. Nat. ser. 2, v. 367.

Hab.: Common in southern parts.

Order XX. **GUTTIFERÆ.**

Flowers regular, usually diœcious or polygamous. Sepals 2 to 6 or rarely more, much imbricate or in decussate pairs. Petals 2 to 6, rarely more, imbricate or contorted. Male flowers: Stamens usually indefinite, free or variously united; anthers adnate, innate, or sometimes immersed in the mass of filaments. Ovary none, or rudimentary, or more or less developed. Female or hermaphrodite flowers: Staminodia or stamens usually fewer and more free than in the males. Ovary 2 or more celled, rarely 1-celled, with 1 or more ovules in each cell, erect from the base or attached to the central angle. Stigmas as many as cells, radiating or united into one, sessile or raised on a simple or rarely branched style. Fruit usually fleshy or coriaceous, indehiscent or opening septicidally in as many valves as cells. Seeds thick, often arillate, without albumen. Embryo filling the seed, often apparently homogeneous, consisting either of a fleshy radicle, with minute or without any cotyledons, or of thick fleshy cotyledons, with a very short, usually inferior radicle.—Trees or shrubs, exuding a yellow resinous juice. Leaves opposite or rarely verticillate, thickly coriaceous and entire. Flowers terminal or axillary, solitary, clustered or in trichotomous cymes or panicles.

A tropical Order both in the New and in the Old World.

TRIBE I. **Garciniæ.**—*Ovary cells 1-ovuled; stigma sessile or subsessile, peltate, entire or with radiating lobes. Berry indehiscent. Embryo of a solid tigellus with minute cotyledons or none.*

Calyx of 4 or 5 sepals . . . . 1. GARCINIA.

TRIBE II. **Calophylleæ.**—*Ovary with 1, 2, or 4 erect ovules; style slender (rarely styles 2); stigma peltate or 4-fid or acute. Fruit fleshy, rarely dehiscent. Embryo of 2 fleshy free or consolidated cotyledons, with a small radicle.*

Ovary 1-celled, 1-ovuled; style 1, stigma peltate . . . . 2. CALOPHYLLUM.
Ovary 1-celled, 4-ovuled; style 1, stigma 4-fid . . . . 3. KAYEA.

1. GARCINIA, Linn.

(Name in honour of Laurence Garcin, M.D., a French botanist.)

Flowers polygamous or diœcious. Sepals 4, in opposite pairs. Petals 4 or 5. Male flowers: Stamens indefinite, free, tetradelphous or monadelphous; anthers erect or peltate, dehiscing longitudinally or circumscissile. Female or hermaphrodite flowers: Staminodia various, free or united; ovary 2 or many-celled; stigmas sessile, lobed, smooth or tuberculate; ovules solitary. Fruit a berry; embryo an undivided thick radicle (tigella*).—Glabrous trees, usually with a yellow juice. Leaves coriaceous or submembranous, opposite, or ternately verticillate. Flowers solitary, fascicled or subpaniculate, axillary or terminal.—Oliver in Fl. Trop. Africa.

Leaves narrow lanceolate, 2 to 3in. long . . . . 1. *G. Mestoni.*
Leaves lanceolate-ovate, 3 to 5in. long . . . . 2. *G. Warrenii.*
Leaves ovate-lanceolate, 3 to 4in. long. Fruit yellow, oval, 1¼in. long . . . 3. *G. Cherryi.*

1. **G. Mestoni** (after A. Meston), *Bail. Rep. Bell. Ker Exped.* 1889. Meston's mangosteen. An erect, slender, graceful tree of 20ft. or more, branches drooping. Leaves glossy dark-green, opposite, narrow-lanceolate, the points much elongated, 2 or 3in. long, somewhat wavy but with entire edges; petioles slender, ⅓in. or more long. Flowers (only a few very early buds seen, and these much injured by insects) probably small, either terminal or leaf-opposed, nearly sessile, with a few small bracts at the base. Sepals 4, small, imbricate. Petals white and seem to be hairy. Fruit depressed-globular, a

---

* A latinised word from the French *tigella*, diminutive of *tige*, a stem; the portion of the embryo between the radicle and the cotyledons.

pleasing green, 2in. diameter, but not fully grown, 8-celled. Seeds somewhat rugose. Stigmatic lobes 8, closely sessile on the fruit; the sepals closely appressed, persistent under the fruit, and probably not much enlarged.

Hab.: Bellenden Ker Range, at an altitude of 2000 feet.
Fruit of this tree were gathered in the ripe state by Messrs. Moston and Whelan on their first ascent of Bellenden Ker in 1889, and they describe the fruit as possessing a sharp, pleasant, acid flavour and very juicy, about 3in. in diameter.

2. **G. Warrenii** (after Dr. Warren), *F. v. M. Vict. Nat.* A glabrous tree of about 40ft., the branchlets robust, angular. Leaves 3 to 5in. long, of firm texture, mostly lanceolate-ovate, the primary lateral veins numerous, and somewhat prominent, particularly on the under side; petioles short. Flowers rather large, crowded into axillary clusters, the pedicels short and thick. Sepals almost semiorbicular, the inner only about $\frac{1}{8}$in. long, though exceeding the outer. Petals 4, pale, obovate or verging somewhat into an orbicular form, incurved, with broad base, sessile, seldom longer than 4 lines, in front slightly and irregularly denticulate, staminal mass of the male flowers divided almost to the base into 4 ovate lobes, about half as long as the petals, and to which they somewhat adhere. Anthers almost quadrivalvular, extremely numerous, densely covering the inner side of the lobes to near the base, pale, partly on very short filaments, partly sessile, their cells divergent, widely dehiscent; rudimentary style rather thick, angular, about $\frac{1}{8}$in. long, with a convex stigma. Female flowers and fruit not yet seen. The staminal arrangement resembles somewhat *G. cornea* and *G. merguensis*, and the leaves *G. neglecta*, Vieillard, and the venulation of them is much more prominent than in *G. subtilinervis.*—F. v. M. l.c.

Hab.: This second species of the genus was found by Stephen Johnson near the Coen River in 1891.

3. **G. Cherryi** (after F. J. Cherry), *Bail. n.sp.* A glabrous tree, about 30ft., with a somewhat thick bark, grey outside, the branchlets often dichotomous and rough from prominent lenticels. Leaves ovate-lanceolate, 3 or 4in. long and 1½ to 2in. broad above the middle, the lateral nerves distant, erecto-patent; the apex obtusely acuminate, tapering at the base to a petiole of about 6 or 9 lines. Flowers solitary, near the ends of the branchlets, on flattened peduncles from 6 to 9 lines long. (Flowers only seen in the bud state.) Buds globose. Sepals 5, imbricate. Petals 5, imbricate, larger but similar to the sepals. Stamens numerous. Lobes of stigma foliaceous. Fruit yellow, oval, 1¼in. long, slightly exceeding 1in. in diameter. Seeds 4, compressed, oblong, about 7 or 8 lines long, 4 to 4½ lines wide.

*Hab.: Coen, F. J. Cherry, who says "the fruit does not taste badly, and birds and insects are very fond of it."

2. CALOPHYLLUM, Linn.

(Name alluding to the beautiful leaves.)

Flowers polygamous. Sepals and petals together, 4 to 12, imbricate in 2 or 3 series. Stamens indefinite, free or nearly so; filaments shortly filiform; anthers ovate or oblong, 2-celled, opening longitudinally. Ovary 1-celled, with a single erect ovule; style elongated, with a peltate stigma. Drupe indehiscent, with a crustaceous endocarp. Seed erect, ovoid or globular, the testa thin, or thick and hard, or spongy and then often adhering to the endocarp.—Trees, with the leaves marked with numerous closely parallel transverse veins.

The genus is tropical, chiefly Asiatic, with a few American species.

Glabrous. Leaves oblong, or obovate-oblong, obtuse or emarginate . . . 1. *C. inophyllum.*
Young parts tomentose.
Leaves elliptic or linear-lanceolate, acuminate . . . . 2. *C. tomentosum.*
Leaves oblong, cuneate at the base. Fruit ribbed . . . . 3. *C. costatum.*
Leaves linear-oblong, apex blunt, base cuneate . . . . 4. *C. australianum.*

1. **C. inophyllum** (alluding to the thread-like veins of leaf), *Linn.; Benth. Fl. Austr.* i. 188. Alexandrian laurel; Tacamahac tree; Doomba tree of India. A middling-sized glabrous tree. Leaves petiolate, broadly oblong or obovate-oblong, rounded at the apex, about 6in. long, coriaceous and glossy on both sides, veins many, fine. Racemes in the upper axils often shorter than the leaves. Flowers ¾in. diameter, white, fragrant; buds nearly globular, sepals 4, the 2 inner ones more petal-like than the outer ones. Petals 4, longer than the calyx. Stamens numerous, more or less united at the base into 4 (or more?) bundles. Ovary globose, stipitate, style much exceeding the stamens, stigma peltate-lobed. Fruit globose, 1in. or more in diameter, smooth, yellowish.—Wight. l.c. t. 77.

Hab.: Rockingham Bay and other parts of the tropical coast.

The following analysis of the fruit is by Mr. K. T. Staiger, F.L.S.:—Shells, 62·5 per cent.; kernels, 37·5 per cent. Greenish-yellow oil, 43 per cent.; dry residue, 27. per cent.; moisture, 30 per cent. Ashes of whole kernels, 1·66 per cent.; ashes of exhausted residue, 6·15 per cent. Mr. Staiger finds the green oil on saponification gives a bright-yellow soap, the green pigment of the oil having changed into a bright yellow.

Wood of a reddish colour and pretty wavy figure, strong and durable; a useful wood for the joiner and cabinetmaker.—*Bailey's Cat. Ql. Woods No.* 16.

Rhede states that the resin is emetic and purgative. It is mostly used externally for plasters, like turpentine.

2. **C. tomentosum** (tomentose), *Wight. Ill.* i. 128; *It. t.* 110; *Hook. Fl. Brit. Ind.* i. 274. Keena or Poon spar tree. A tall straight tree, branches 4-angled, young parts tomentose. Leaves elliptic or linear-lanceolate, acuminate, margins wavy, 8 to 5in. long, 1¼ to 2in. broad, coriaceous, shining; veins many, close, slender, equally prominent on both sides; petiole ½ to ¾in. long, often tomentose. Racemes in the axils of the upper leaves or forming a terminal panicle, pubescent. Flowers upwards of ½in. diameter, pedicels long, slender. Sepals orbicular, outer ones smaller than the inner. Petals 4, ovate-oblong, larger than the sepals. Fruit ¾in. long, obliquely ovoid, pointed.

Hab.: Tropical coast scrubs.

A common tree in India and Ceylon.

This yields the Poon spars of commerce. It is used for bridgework in India, where the seeds are also said to give an oil. Yields a slightly astringent dark-coloured gum, soluble in water, which contains: Water, 18·5 per cent.; tannin, 4 per cent.; arabin, 77·5 per cent.—*Lauterer*.

Wood of a red colour, strong and durable; also a useful wood for the joiner and cabinet-maker.—*Bailey's Cat. Ql. Woods No.* 16a.

3. **C. costatum**, *Bail. n.sp.* A lofty tree, the branchlets not prominently angular, puberulent. Leaves oblong, tapering much towards the base, 2 to 2¾in. long, 1 to 1½in. broad, sometimes very shortly and broadly acuminate, margins somewhat wavy, lateral nerves numerous, oblique, midrib channelled above, prominent and more or less hairy on the under side. Petioles about ½in. long, flattened and puberulent. No flowers seen. Fruit picked off the ground under the trees, roundish-oval, pointed at each end, the largest measuring about 1¼in. long and 1in. diameter, epicarp thin, dark, and more or less prominently ribbed.

Hab.: Evelyn, *J. F. Bailey*, June, 1890. Figured in Q. Ag. Jl. vol. v.

4. **C. australianum**, *F. v. M.; J. Vesque's Guttiferæ in A. and C.; DC. Mono. Phane.* Branches acutely 4-angled, slender, ferruginous-tomentose. Leaves linear-oblong or lanceolate, petiolate, the apex obtuse, somewhat acute at the base, both sides shining; nerves somewhat ferruginous-tomentose, charta-ceous, 4 to 7in. long, 1 to 2in. broad. Petioles concave above, slightly pilose. Racemes axillary, short, bearing few flowers. (Flowers not seen.) Fruit globose, about 6 lines diameter. Epicarp thin, fragile, red with a pale-violet pruinose covering. Endocarp thin, crustaceous. Putamen ellipsoid, about 5 lines long, 3 or 4 lines broad.—J. Vesque l.c.

Hab.: Rockingham Bay, *J. Dallachy, F. v. M. l.c.*

XX. GUTTIFERÆ.

3. KAYEA, Wall.

(Named after Dr. R. Kaye Greville.)

Trees. Leaves opposite, veins rather distant, arched. Flowers hermaphrodite, either large and solitary or small and collected in terminal panicles. Sepals and petals 4 each, imbricate. Stamens numerous; filaments slender, free or connate at the base. Anthers small, subglobose, 2-celled, dehiscence vertical. Ovary 1-celled, style slender, stigma acutely 4-fid; ovules 4, erect. Fruit sub-drupacious, fleshy, indehiscent, 1 to 4-seeded. Seeds thick, testa thin and crustaceous.—Hooker's Flora of British India i. 276.

1. **K. Larnachiana** (after J. McD. Larnach), *F. v. M. Vict. Nat. Jan.* 1887. Supposed to be a tree about 20ft. in height, the bark of the branchlets somewhat cracked. Leaves on very short petioles, elliptic-lanceolate, in rather distant pairs, chartaceous; on the specimens seen from 5 to 7in. long and from 1½ to 2in. broad, nearly smooth, and scarcely shining on the upper surface, rounded at the base, the apex slightly pointed, very thinly penninerved, the faint reticulations immersed. Inflorescence in short terminal panicles or bundles without common peduncle; bracts obliterated or very fugitive; pedicels about the length of the calyx, bearing very minute deltoid bracteoles below the middle. Flower-buds globular, calyx glabrous, measuring hardly ¼in., thinly coriaceous, pellucid and imbricating at the edge, the sepals finally enlarged to an inch long, the two outer ones roundish, rough, developing a brownish film, the two inner ones more oval. Petals roundish, membranous, glabrous. Stamens numerous, slightly connate at the base. Filaments very thin, the summit pointed. Anthers almost orbicular, fixed above the base; the cells surrounding the short and broad connective, dehiscent along the margin. Style glabrous, subulate-filiform, short; stigmata minute, pointed. Fruit indehiscent, rather large, globular, somewhat pointed, 1-seeded, the pericarp coriaceous, the one seed filling the cavity, basifixed, sessile. Arillus none; testa chartaceous, smooth; embryo almost globular, carnulent.

Hab.: Mossman River.

The descriptive notes were elaborated by Baron Mueller from specimens with young flower buds and with over-ripe fruit.

This Australian species is evidently nearest allied to *K. racemosa*, but it has only faint nerves to the leaves, shorter petioles, and pluriseriate stamens; and perhaps the fruit of *K. racemosa*, when discovered, may show differences also.—*Vict. Nat., l.c.*

Order XXI. TERNSTRŒMIACEÆ.

Sepals 5, rarely 4 to 7, free or slightly connate, the innermost often larger. Petals 5, rarely 4 to 9, free or connate below, imbricate or contorted. Stamens numerous or definite, free or connate, usually adnate to the base of the deciduous corolla; anthers basifixed or versatile, dehiscing by slits or rarely by terminal pores. Ovary free or half inferior, sessile 3 to 5-celled, or many-celled; styles as many, free or connate, stigmas usually small; ovules 2 or many in each cell, rarely solitary, never orthotropous. Fruit baccate or capsular. Seeds few or numerous, placentas axile, albumen scanty or none, rarely copious; embryo straight or hippocrepiform, cotyledons various.—Shrubs, rarely climbing, or trees. Leaves alternate, simple, entire or often serrate, usually coriaceous, exstipulate. Flowers showy, seldom small, usually subtended by 2 sepal-like bracts, rarely diclinous, axillary, 1 or more together, rarely in lateral or terminal racemes or panicles.

Rare in temperate, abundant in tropical, Asia and America.

XXI. TERNSTRÆMIACEÆ.

1. SAURAUJA, Willd.

(After — Sauraujo, a Portuguese botanist.)

Sepals 5, strongly imbricate. Petals 5, usually connate at the base. Stamens numerous; anthers dehiscing by pores. Ovary 3 to 5-celled; styles as many, distinct or connate; ovules numerous. Fruit baccate, rarely dry and subdehiscent.—Trees or shrubs. Branches usually brown, with whitish tubercular dots, at first as well as the leaves more or less strigose, pilose, or scaly. Leaves approximate at the ends of the branches, usually serrate, with parallel veins diverging from the midrib. Inflorescence lateral, often from the axils of fallen leaves, cymose, subpaniculate, rarely few-flowered. Bracts usually small, remote from the calyx. Flowers usually hermaphrodite.

Met with in tropical and subtropical Asia and America.

1. **S. Andreana** (after E. André), *Oliver (inedited)*, F. v. M. in letter. A large spreading shrub, the branchlets, petioles, nerves on the under side of the leaves and inflorescence, more or less thickly covered with ferruginous strigose hairs. Leaves oblong-lanceolate attenuate-acuminate, 5 to 8½in. long, 2 to 3in. broad near the middle, the parallel nerves and cross veins prominent, margins setose-denticulate; petioles ½ to ¾in. long. Peduncles solitary, in the upper axils, from as long to twice as long as the petioles, bearing near the end from 1 to 3 flower-buds with a pair of bracts near them. Bracts narrow-linear, 4 or 5 lines long. Pedicels about 3 lines. Calyx densely-hairy, the sepals or calyx-lobes with a broad glabrous margin, 4 lines long. Petals white, oblong, sometimes twice as long as the sepals. Stamens numerous, filaments broad, frequently connate; anthers oblong opening in longitudinal slits. Styles 5 or fewer, connate near the base. Ovary glabrous, 5-celled. Fruit not seen quite ripe, oval, 5 or 6 lines long, seems to burst into valves near the top. Seeds very numerous, brown and very prominently reticulate.—*Dillenia Andreana*, F. v. M. Fragm. v. 175.

Hab.: Freshwater Creek near Cairns and creeks about Bellenden Ker, from which specimens I have drawn up the diagnosis here given. My specimens were identified as belonging to Oliver's species by Baron von Mueller in 1889.

Order XXII. MALVACEÆ.

Flowers regular, usually hermaphrodite or rarely partially diœcious or polygamous. Sepals 5, rarely 3 or 4, more or less united in a lobed or entire calyx, the lobes valvate or very rarely slightly imbricate. Petals 5, hypogynous, usually adnate at the base to the staminal column, contorted in the bud, rarely wanting. Stamens indefinite, hypogynous, more or less united at the base, the column divided into filaments at the top or bearing the filaments outside, below or up to the top. Anthers from globose to linear, often reniform or variously waved, 1-celled or spuriously divided into two cells by a thin and incomplete longitudinal septum. Torus small or conical and protruding into the centre of the ovary, not expanded into a disk. Ovary 2 or more-celled (very rarely reduced to a single carpel), entire or lobed, the carpels verticillate round the axis or (in genera not Australian) irregularly clustered. Style simple at the base, divided at the top into as many or twice as many branches or stigmas as there are cells, or rarely entire and clavate. Ovules 1 or more in each cell, ascending or horizontal, with a ventral or superior raphe, or reversed and pendulous, with the raphe dorsal. Fruit dry or rarely baccate, the carpels separating and indehiscent or 2-valved, or united in a loculicidally dehiscent capsule. Seeds with the testa usually crustaceous, without or with very little albumen; cotyledons usually folded and

Saurauja Andreana, Oliver.

XXII. MALVACEÆ.

often enclosing the curved or rarely straight radicle.—Herbs, shrubs, or soft-wooded trees, the hairs usually stellate. Leaves alternate, mostly toothed, lobed or divided, with palmate nerves or divisions, rarely digitately compound. Stipules free, usually subulate or small and deciduous, rarely leafy. Peduncles usually 1-flowered and articulate above the middle, rarely bearing a bract at the joint or several-flowered, all axillary or the upper ones forming a terminal raceme or panicle. Bracteoles either none or 3 or more, free or united, forming an involucre close to or adherent to the calyx. Flowers often large, usually purple, red, or yellow.

A large Order generally dispersed over all except the coldest regions of the globe, distinguished from *Sterculiaceæ* and *Tiliaceæ* by the 1-celled anthers, and from all others by the valvate calyx and monadelphous hypogynous stamens. Of the 15 following genera, 12 are more or less tropical, 7 being common to the warmer regions of both the New and the Old World; 4, *Malvastrum*, *Modiola*, *Pavonia*, and *Fugosia*, chiefly American, or American and African, but not Asiatic; and 1, *Thespesia*, African and Asiatic. *Lavatera* is a Mediterranean form, represented by one species in extratropical Australia, the remaining 2 are endemic or nearly so, *Plagianthus* being also represented in New Zealand and *Lagunaria* in Norfolk Island.—*Benth.* in part.

TRIBE I. **Malveæ.**—*Staminal column bearing filaments to the summit. Style-branches the same number as ovary-cells. Mature carpels separating more or less from the axis (imperfectly so in some Abutila).*
Ovules solitary in each cell, ascending with a ventral raphe.
  Style-branches lined with decurrent stigmas.
    Bracteoles 3 to 6, united at the base . . . . 1. LAVATERA.
    Bracteoles 3, distinct . . . . 2. MALVA.
  Stigmas terminal, capitate or truncate. Bracteoles 1 to 3 distinct, or none . 3. MALVASTRUM.
Ovules solitary in each cell, pendulous or horizontal with a dorsal raphe.
  Bracteoles none.
    Styles with decurrent stigmas. Flowers more or less diœcious . . . . . 4. PLAGIANTHUS.
    Stigmas terminal, capitate, or truncate . . . . 5. SIDA.
Ovules 2 or more in each cell. Bracteoles none. Stigmas terminal. Capsule
  5 to 20-celled, separating or cohering at least till the seed has shed . . . 6. ABUTILON.
  Bracteoles 3. Carpels with transverse septas inside . . . . 7. MODIOLA.

TRIBE II. **Ureneæ.**—*Staminal column truncate or 5-toothed at the summit, bearing the anthers or filaments on the outside. Style-branches twice the number of carpels. Carpels 1-seeded.*
Bracteoles 5, united at the base. Carpels muricate or glochidiate . . . . 8. URENA.
Bracteoles 5 or more, usually free. Carpels reticulate or smooth . . . . 9. PAVONIA.

TRIBE III. **Hibisceæ.**—*Staminal column truncate or 5-toothed at the summit, bearing the anthers or filaments on the outside, or rarely at the summit also. Style-branches or stigmas the same number as ovary-cells. Carpels united in a several-celled capsule, loculicidal or indehiscent.*
Style branched at the top or with radiating stigmas. Ovary 5-celled.
  Bracteoles 5 or more, free or united (sometimes very deciduous). Hairs or
    tomentum stellate . . . . 10. HIBISCUS.
  Bracteoles 3 (sometimes very deciduous). Tomentum of scurfy scales . . 11. LAGUNARIA.
Style undivided, with decurrent stigmas.
  Bracteoles 3 to 5, narrow, not cordate, sometimes very small.
    Ovary 3, 4, or rarely 5-celled. Capsule coriaceous, loculicidal . . . . 12. FUGOSIA.
    Ovary 5-celled. Capsule woody, sometimes indehiscent . . . . . 13. THESPESIA.
  Bracteoles 3, broad, cordate . . . . 14. GOSSYPIUM.

TRIBE IV. **Bombaceæ.**—*Staminal column in the Australian genera (only one genus represented in Queensland) divided at the top into numerous filaments, in other genera the filaments or anthers variously arranged. Style undivided, or with very short stigmatic lobes as many as ovary-cells. Carpels united in a loculicidal or indehiscent capsule.—A large tropical tribe, difficult to distinguish from arborescent Hibisceæ by a general character, although each genus has peculiarities not found among Hibisceæ.*
Calyx truncate in the bud, afterwards 3 to 5-cleft. Capsule 5-valved, densely
  woolly inside. Leaves digitate . . . . 15. BOMBAX.

1. LAVATERA, Linn.
(After the Lavaters of Zurich.)

Bracteoles united into a 3 to 6-cleft involucre. Calyx 5-lobed. Staminal column divided to the top into several filaments. Ovary-cells indefinite, 1-ovulate. Style-branches of the same number as cells, filiform, stigmatic along the inner side. Fruit-carpels in a depressed circle, indehiscent, verticillate round the torus or axis, which is usually prominent beyond them, either conical or variously dilated above them. Seed ascending.—Herbs, shrubs, or trees, tomentose or hirsute. Leaves angular or lobed. Flowers pedunculate, axillary or in a terminal raceme.

The greater number of species are from Western Europe or the Mediterranean region, one extending into central Asia; there are also two from the Canary Islands, besides the subjoined Australian species, which is endemic but nearly allied to one of the European ones.—*Benth.*

1. **L. plebeia** (plebeian) *Sims in Bot. May. t.* 2269; *Benth. Fl. Austr.* i. 185. A coarse, erect herb, becoming woody at the base and attaining the height of 5 to 10ft., more or less scabrous or softly tomentose with minute stellate hairs. Leaves on long petioles, orbicular-cordate, 5 or 7-lobed, the lower ones sometimes attaining 6in. diameter, the upper ones 1 to 2in.; the lobes short, broad, very obtuse and crenate, the central one of the upper leaves often longer than the others. Stipules narrow-lanceolate or triangular. Pedicels axillary, usually clustered, rarely solitary, sometimes very short and rarely exceeding 1in. Involucre deeply 3-lobed, the lobes ovate, obtuse, shorter than the 5-lobed calyx. Petals pale-rose or whitish, 1 to 1½in. long. Carpels of the fruit 6 to 15, in a close ring, with flat backs and sharp angles, the receptacle projecting from the central depression as a small conical point.—Hook. Fl. Tasm. i. 46; F. v. M. Pl. Vict. i. 166; *Malva Behriana*, Schlecht; *Lavatera Behriana*, Schlecht; *Malva Preissiana*, Miq.

Hab.: Southern parts of the colony.

In the early days of South Australia the aborigines of the Adelaide tribe largely used the young fleshy roots of this mallow for food, cooking in ovens sunk in the ground.—*F. M. B.*

*2. **MALVA**, Linn.

(Name from its emollient qualities.)

Bracteoles 3, distinct. Sepals 5, connate at the base. Petals emarginate, connate at the extreme base. Staminal-tube antheriferous to the top, without sterile teeth. Ovary many-celled; styles as many as the carpels, stigmas linear; ovules 1 in each cell. Ripe carpels 1-seeded, indehiscent, separating from a short conical torus. Seeds ascending.—Downy herbs. Leaves lobed. Flowers in axillary tufts.

The species of this genus are only found in the temperate regions of the Old World.

The following species are met with as cultivation weeds.
Plant erect, pubescent. Flowers in nearly sessile clusters . . . . . . 1. *M. verticillata.*
Plant erect, glabrous. Peduncle as long or longer than the flowers . . . 2. *M. sylvestris.*
Plant spreading, slightly pubescent, claw of petal bearded . . . . . . 3. *M. rotundifolia.*
Plant spreading, slightly pubescent, claw of petal glabrous . . . . . . 4. *M. parviflora.*

1. **M. verticillata** (whorled), *Linn.* Stem branched, 2 to 4ft. high. Leaves cordate, orbicular, 5 or 6-lobed, downy; petiole 6 or 7in. Flowers small, nearly sessile, densely crowded. Bracteoles linear. Sepals deltoid-lanceolate. Petals notched, slightly longer than the sepals. Carpels 10 to 12, enclosed within the accrescent calyx, netted on the sides, prominently ribbed at the back.

Hab.: Naturalised on waste places about townships.

2. **M. sylvestris** (forest plant), *Linn.* Annual, 1 to 3ft. high. Leaves cordate, rounded, lobed; petiole 4 or 5in. Peduncles about 1in. Bracteoles ovate, entire, shorter than the bell-shaped calyx. Corolla 1½in. diameter. Petals notched, claw-bearded. Carpels reticulate, downy or glabrous.

Hab.: Naturalised on waste places about townships.

3. **M. rotundifolia** (round-leaved), *Linn.* A much-branched, decumbent, or prostrate herb, sparingly villous. Leaves suborbicular, lobed, crenate; petiole 6 or 7 in. Peduncles 1½in. Bracteoles lanceolate half the length of the broadly-lanceolate sepals. Corolla 1in. diameter. Petals wedge-shaped, notched, twice the length of the sepals. Ripe carpels about 15, downy, flat or wrinkled.

Hab.: Naturalised on waste places about townships.

4. **M. parviflora** (small-flowered), *Linn.* A comparatively small spreading herb, slightly downy. Leaves roundish, obsoletely lobed. Peduncles short, spreading after flowering. Bracteoles linear. Sepals broad, acute. Petals notched, scarcely exceeding the sepals, claw glabrous. Carpels wrinkled, angular.

Hab.: Naturalised on waste places about townships.

8. MALVASTRUM, A. Gray.

(Altered from Malva.)

Bracteoles either none or 1 to 3, small and distinct. Calyx 5-lobed. Staminal column divided to the top into several filaments. Ovary-cells 5 or more, 1-ovulate; style-branches of the same number as the cells, filiform or club-shaped, with terminal small or capitate stigmas. Fruit carpels seceding from the short axis, indehiscent or slightly 2-valved, occasionally produced at the top into erect connivent beaks. Seeds ascending, reniform.—Herbs or undershrubs. Leaves entire or divided. Flowers red or yellow, shortly pedunculate or sessile, axillary or in terminal spikes.

A considerable genus, chiefly American, with a few South African species.

The genus, formerly confounded with *Malva* and *Sida*, is readily distinguished from the former by the styles, from the latter by the ascending ovules and seeds.

Tomentum stellate. Flowers mostly in a short terminal spike . . . . 1. *M. spicatum.*
Hairs appressed, parallel. Flowers mostly axillary. Calyx broad . . . 2. *M. tricuspidatum.*

1. **M. spicatum** (flowers in spikes), *A. Gray, Pl. Fendl.* 22, *and Bot. Amer. Expl. Exped.* i. 147; *Benth. Fl. Austr.* i. 187. An erect branching herb of 1 to 2ft., becoming almost woody at the base, scabrous or softly tomentose with stellate hairs. Leaves petiolate, ovate or ovate-lanceolate, acute or obtuse, 1 to 2in. long, irregularly serrate or crenate, very rarely obscurely 3-lobed. Flowers rather small, yellow, sessile in a dense terminal spike, rarely exceeding 1 to 1½in. in length, and often leafy at the base. Bracts narrow, shorter than the calyx, usually 2-lobed. Bracteoles 3, filiform, closely appressed to the calyx. Calyx softly pubescent, the lobes acuminate, and often bordered by long hairs. Petals about 4 to 5 lines long. Carpels 8 to 12, not close-pressed, angular on the edges, pubescent on the top, without points.—*Malva spicata*, Linn.; Cav. Diss. t. 20, f. 4; DC. Prod. i. 480; *M. ovata*, Cav. Diss. 81, t. 20, f. 2; *M. timoriensis*, DC. Prod. i. 430; *M. brachystachya*, F. v. M. in Linnæa, xxv. 878.

Hab.: Common in Queensland.

The species is common in tropical America, and has been found also in the Cape de Verd Islands and in Timor.

2. **M. tricuspidatum** (referring to the points on the carpel), *A. Gray, Pl. Wright, and Bot. Amer. Expl. Exped.* i. 148; *Benth. Fl. Austr.* i. 187. An erect branching herb, 2 to 3ft. high, hard and almost woody at the base, although sometimes annual, the branches sprinkled or covered with closely appressed hairs. Leaves on rather long petioles, from broadly ovate to lanceolate, 1 to 2in. long, irregularly toothed, hairy. Flowers yellow, almost sessile in the axils of the leaves, or clustered towards the ends of the branches. Calyx broadly 5-lobed,

with 3 small, narrow, external bracts. Carpels 8 to 12 or even more, closely packed in a depressed ring, each one reniform, with 3 minute unequal points on the upper edge, 1 at the inner angle, 2 dorsal.—*Malva tricuspidata.* Ait.; DC. Prod. i. 430; *Sida carpinoides,* DC. Prod. i. 460.

Hab.: Frequent in southern parts.

This species, probably of American origin, is much more widely scattered over the warmer regions of the Old World than the *M. spicatum.*

4. PLAGIANTHUS, Forst.

(Referring to the oblique petals.)

(Asterotrichon *and* Blepharanthemum, *Klotzsch;* Lawrencia, *Hook.;* Halothamnus, *F. v. M.*)

Bracteoles none or distant from the calyx. Calyx 5-toothed or 5-lobed. Staminal column divided at the top into several filaments. Ovary-cells 2 to 5, rarely 1 or indefinite, 1-ovulate. Style-branches as many as cells, filiform or club-shaped, stigmatic along the inner side, either the whole length or near the top. Fruit-carpels 1, 2, or more, seceding from the axis, indehiscent or irregularly breaking up. Seeds pendulous, with a dorsal raphe.—Shrubs or rarely herbs. Leaves entire or rarely lobed. Flowers usually small and white, more or less completely diœcious, axillary or terminal, usually clustered, rarely solitary or in short panicles.

The genus is confined to Australasia.

Sect. **Lawrencia** (Wrenciala, *A. Gray.*)—*Calyx with 5 prominent angles. Herbs or tortuous shrubs. Leaves thick or small, entire or toothed at the top, nearly glabrous or scurfy.*
A decumbent, much-branched herb, glabrous or slightly hoary . . . . 1. *P. glomeratus.*
A tortuous, branching shrub, covered with scurfy scales . . . . 2. *P. microphyllus.*

1. **P. glomeratus** (flowers clustered), *Benth. in Journ. Linn. Soc.* vi. 198; *Fl. Austr.* i. 190. A glabrous or slightly hoary, decumbent and much-branched herb, with ascending branches often above 1ft. high. Leaves cuneate-oblong, toothed at the end, resembling those of *P. spicatus,* but usually narrower and more gradually narrowed into the petiole. Flowers all axillary, usually 3 together and sessile, forming distant clusters along the leafy branches and never collected into a spike, the ends of the branches all barren. Flowers nearly those of *P. spicatus,* but smaller, and the stamens and styles much shorter.—*Lawrencia glomerata,* Hook. Ic. Pl. t. 417.

Hab.: Southern parts of the colony.

2. **P. microphyllus** (small leaves), *F. v. M. Fragm.* i. 29; *Benth. Fl. Austr.* i. 190; *Halothamnus microphyllus, F. v. M. Pl. Vict.* i. 159. A dwarf rigid shrub, clothed with scurfy scales, very tortuous and branchy, the smaller branches slender and often spinescent. Leaves from linear to oblong-cuneate, rarely exceeding ¼in. and usually much smaller, obtuse or 3-toothed at the end, more or less tapering at the base. Flowers small, sessile or nearly so, 1 to 3 together in the axils, not spicate. Calyx when in flower not above 1¼ line long. Carpel usually single, enclosed in the calyx and membranous.

Hab.: Southern border of the colony.

5. SIDA, Linn.

(A Greek name of a plant.)

Bracteoles none, or small and distant from the calyx. Calyx 5-toothed or 5-lobed. Staminal column divided at the top into several filaments. Ovary-cells 5 or more, verticillate, 1-ovulate. Style-branches as many as cells, filiform or slightly clavate, with terminal, capitate or truncate stigmas. Fruit-carpels either obtuse or with connivent points, seceding from the axis, indehiscent or opening shortly at the top in 2 valves. Seeds pendulous or horizontal, with a dorsal raphe.—Herbs or shrubs, usually clothed with a soft or whitish stellate tomentum. Stipules in all the Australian species except *S. Hookeriana*, subulate and deciduous. Flowers sessile or pedunculate, axillary or in terminal heads, spikes, or racemes, of various colours and sometimes large, but most frequently rather small, yellow, or whitish.

The genus, even as now limited to the exclusion of the *Abutilons*, is large and widely spread over the warmer regions of the globe, but most abundant in America. Of the Australian species three are common tropical weeds, the remainder all endemic.—*Benth.*

§ 1. *Calyx without prominent ribs or angles. Carpels strongly reticulate on the sides* (except *S. pleiantha*), *indehiscent or nearly so, never aristate. Perennials or shrubs. Leaves undivided.*
Flowers 1 or 2 together, on slender pedicels, articulate near the top.
  Calyx-lobes obtuse, not protruding beyond the broad part of the fruit.
    Carpels strongly wrinkled on the back. Fruit 2½ to 4 lines diameter   1. *S. corrugata.*
    Leaves roundish. Peduncles 4in. long. Calyx-lobes nearly deltoid.
      Carpels prickly on the back . . . .   2. *S. Spenceriana.*
    Leaves linear. Peduncles under 1in. long. Calyx-lobes deltoid.
      Carpels hairy . . . .   3. *S. argentea.*
    Carpels not or very slightly wrinkled. Fruit not exceeding 2 lines
      diameter. Leaves and flowers very small . . . .   4. *S. intricata.*
  Calyx-lobes acute or scarcely acuminate, remaining herbaceous, and not
    much enlarged after flowering.
    Leaves ovate or ovate-lanceolate, cordate at the base . . . . .   5. *S. macropoda.*
    Leaves lanceolate or oblong-lanceolate, not cordate . . . . . .   6. *S. virgata.*
  Calyx-lobes acuminate, with long, subulate, woolly points . . . .   7. *S. cryphiopetala.*
  Calyx-lobes enlarged and thinner or scarious after flowering. Leaves
    lanceolate or oblong. Carpels 6 to 8. Fruiting calyx about ½in.
    diameter, slightly spreading; lobes narrow, ovate-lanceolate . . .   8. *S. petrophila.*
Flowers clustered, several together. Pedicels short, not articulate.
  Flowers nearly sessile. Tomentum dense, or rarely scanty. Carpels
    reticulate on the side . . . .   9. *S. subspicata.*
  Flowers pedicellate. Tomentum thin or floccose. Carpels not reticulate 10. *S. pleiantha.*

§ 2. *Calyx 5-angled, prominently 10-ribbed. Carpels not reticulate on the sides, and opening in 2 short valves at the top. Herbs or undershrubs. Leaves undivided.*
Leaves ovate or narrow, whitish with a close tomentum on both sides.
  Carpels 5 . . . . 11. *S. spinosa.*
Leaves ovate or narrow, whitish with a close tomentum underneath.
  Carpels about 10 . . . . 12. *S. rhombifolia.*
Leaves broad, cordate (or rarely narrow). Tomentum soft, loose, or velvety. Carpels about 10 . . . . 13. *S. cordifolia.*

§ 3. *Calyx with 15 or 20 nerves prominent when in fruit. Carpels numerous. Styles free to the base. Leaves undivided.*
Calyx enlarging little after flowering, open at the top . . . . . . 14. *S. platycalyx.*
Fruiting calyx very large, membranous, quite closed over the fruit . . . 15. *S. inclusa.*

1. **S. corrugata** (fruit furrowed), *Lindl. in Mitch. Three Exped.* ii. 18; *Benth. Fl. Austr.* i. 192. Rootstock and often the base of the stem woody, the branches usually diffuse or procumbent and under 1ft. long, or in some varieties elongated, slender, and divaricate, attaining fully 2ft., more or less hoary as well as the leaves with stellate hairs or short pubescence. Leaves orbicular, ovate or lanceolate, crenate, mostly ½ to 1in. long, cordate or obtuse at the base, on petioles shorter than the laminæ, and sometimes very short. Pedicels axillary, 1 to 3 together, filiform or slender, rarely as long as the leaves, articulate below the top. Calyx tomentose, 2 to 2½ lines long, the lobes broad and obtuse, spreading under the fruit. Petals yellow, about twice the length of the calyx. Stamens 10 to 15. Fruit depressed-globular, varying from 2½ to near 5 lines diameter, tomentose or nearly glabrous, the obtuse often-raised centre marked with radiating furrows formed by the grooved connivent summits of the carpels, the circumference deeply wrinkled. Carpels 6 to 10, indehiscent, strongly reticulate on the sides. Seeds glabrous or slightly tomentose.—*F. v. M. Pl. Vict.* i. 168.

Hab.: On the Maranoa, *Mitchell*; in the interior, *Leichhardt*.

This plant assumes forms apparently so distinct that it is difficult to believe that some of them ought not to be considered as species. In attempting, however, to fix their limits, so many intermediate specimens have presented themselves that I feel compelled to follow F. v. Mueller in uniting them under one name. The following appear to be the most marked.—Benth. in Fl. Austr.

a. orbicularis. Stems short, diffuse, and tomentose. Leaves orbicular or broadly ovate, deeply and coarsely crenate, cordate at the base. Flowers and fruits rather large. *S. corrugata,* Lindl. l.c.; *S. interstans* and *S. spodochroma,* F. v. M. in Linnæa, xxv. 383. Chiefly in Victoria and N. S. Wales.

b. ovata. Stems usually more slender and elongated. Leaves mostly cordate-ovate, with small and regular crenatures, often softly tomentose. Petioles often short, and sometimes very short. Flowers and fruits rather small. *S. fibulifera,* Lindl. in Mitch. Three Exped. ii. 45; *S filiformis,* A. Cunn. in Mitch. Trop. Austr. 361.—N. Australia (including a var. with very short pedicels), Queensland, N. S. Wales, Victoria, and S. Australia. *S. pedunculata,* A. Cunn. ms., from Peel's Range, is a remarkable form, densely tomentose, with the lower leaves 2in. long, and the lower peduncles elongated, bearing a leafless raceme of several flowers, with rigid stipulary bracts; the inflorescence in the upper part quite normal. *S. nematopodo,* F. v. M. in Linnæa, xxv. 382, has smaller and less wrinkled fruits, although still much more so than in *S. intricata,* and the foliage is quite that of the present variety.

c. angustifolia. Stems slender, often nearly glabrous as well as the leaves. Leaves cordate-lanceolate, deeply toothed. Flowers and fruits small. Extends over the whole range of the species, and the only form hitherto found in W. Australia.—*S. humillima,* F. v. M. in Trans. Phil. Soc. Vict. i. 12, is a small hoary form, with larger leaves, approaching sometimes the first variety. Some specimens of A. Cunningham's from Dirk Hartog's Island have the leaves more densely white-tomentose.

d. trichopoda. Like the last, but the lanceolate or oblong-linear leaves are never cordate at the base, and the slender pedicels mostly exceed the leaves.—*S. trichopoda,* F. v. M. in Linnæa, xxv. 384. On nearly the whole range of the species, excepting W. Australia.

e. goniocarpa, F. v. M. Foliage of the last var., but the fruit larger, the angles of each carpel bordered by vertical wings, forming on the fruit as many very prominent angles as there are carpels. Nangavera in N. S. Wales, *Victorian Expedition.*

2. **S. Spenceriana** (after Mrs. F. Spencer), *F. v. M. in Wing's South. Sci. Rec., vol. I. (New Series), April,* 1885. Plant dwarf, covered with orbicular, silver-shining, densely-ciliate scales. Stipules linear-setaceous. Leaves from roundish to nearly ovate, irregularly denticulate, ¼ to 1½in. long, flat, on petioles of moderate length. Peduncles filiform, 1-flowered, about 4in. long, jointed near the summit. Fruit-bearing calyx not ½in. diameter, lobes nearly deltoid. Carpels numerous, broader than high, much-compressed, obliqae-ovate, short-pointed at the summit, prickly at the back, narrowly reticulate at the sides, tardily separating, Seeds slightly downy.—Baron v. Mueller in Wing's So. Sc. Rec., April, 1885.

Hab.: Thargomindah, Paroo River, and Yappunyah.

3. **S. argentea** (silvery), *Bail. Ql. Journ. Agri. vol. I. part 1, July* 1897. The stems, petioles, as well as most other parts of the plant closely clothed with silvery peltate, ciliate scales. Stems or the lower branches from a procumbent stem erect, slender, about 12in. high. Leaves rather distant, erecto-patent, narrow-linear, 1 to 2½in. long, 1 to 2 lines broad, slightly tapering towards the point, rounded at the base to a petiole of a few lines. Stipules subulate, nearly as long as the petioles. Peduncle axillary, solitary, filiform, about 8 or 9 lines, articulate above the middle. Calyx under 4 lines diameter, lobes deltoid, silky-hairy on the inside. Petals twice as long as the calyx, broadly-cuneate, almost roundly-lobed at the end, veined. Stamens under 10. Style-branches recurved. Carpels hairy, probably few, but only imperfect specimens to hand.

Hab.: Eulo, Paroo River, *J. F. Bailey,* Dec. 1896. The thick coating of the silvery scales gives to the thin stems the appearance of silver rods.

4. **S. intricata** (intricate), *F. v. M. in Trans. Phil. Soc. Vict.* i. 19, *and in Hook. Kew Journ.* viii. 9; *Benth. Fl. Austr.* i. 198. This form also is now reduced by F. v. Mueller (Pl. Vict. i. 168) to the *S. corrugata.* I am inclined

however to keep it distinct, as the characters appear on the dried specimens to be tolerably constant (Benth.). It is a small or slender, very much branched tomentose undershrub, resembling the var. *ovata* of *S. corrugata* in general characters, but with much smaller leaves and very much smaller flowers, on short slender pedicels, the fruits not above 2 lines diameter, consisting of 5 to 8 tomentose carpels, not furrowed at their points, and smooth or only very slightly wrinkled on the back.

Hab.: Various localities.

5. **S. macropoda** (long-footed), *F. v. M. Herb.; Benth. Fl. Austr.* i. 198. An erect, branching shrub, densely clothed with a stellate tomentum, thick and often yellowish on the branches, almost velvety on the leaves. Leaves ovate-cordate, obtuse, 1 to 2in. long, crenate, thick and soft, deeply wrinkled above, prominently veined underneath. Pedicels filiform, sometimes exceeding the leaves. Calyx-lobes acuminate or acute, closed over the fruit or spreading. Petals yellow, only shortly exceeding the calyx. Fruit 3 or 4 lines diameter, with the radiating striæ in the centre and the carpels wrinkled on the back as in *S. corrugata*, from which this species differs in stature, foliage, and the acute calyx-lobes.

Hab.: Various localities in the tropics, and Gulf of Carpentaria.

Var. (?) *cardiophylla*, F. v. M. Tomentum more dense, but closer; leaves shorter, and nearly orbicular; pedicels shorter.—Sturt's Creek, *F. v. Mueller*. This may possibly be a distinct species, but the specimens are not sufficiently advanced to determine. In other specimens in young bud only, these buds are sessile or nearly so; the pedicel probably grows out rapidly before the flower expands, and may sometimes remain very short. This will likely be met with in the colony.—*Benth.*

6. **S. virgata** (twiggy), *Hook. in Mitch. Trop. Austr.* 361; *Benth. Fl. Austr.* i. 194. This resembles at first sight, especially in the leaves, the *S. calyxhymenia* (an inland species, so far not been met with in Queensland), and in some respects some narrow-leaved forms of *S. corrugata*; but the calyx does not enlarge as in the former, and its lobes are not obtuse as in the latter, and the stellate tomentum is dense and soft, almost woolly, and often fulvous. It appears to be an erect shrub, with long twiggy branches. Leaves shortly petiolate, lanceolate, or oblong-linear, often exceeding 1in., obtuse at the base, denticulate, less tomentose above than underneath. Pedicels slender, but rarely as long as the leaves. Calyx very tomentose, not prominently ribbed, the acute lobes about as long as the cup. Petals yellow, twice as long as the calyx, varying from 3 to 4 lines. Fruit about 3 lines diameter, depressed, with the centre slightly projecting. Carpels 6 to 8 or rarely more, their radiating summits scarcely furrowed, wrinkled on the back, strongly reticulate on the sides.

Hab.: On the Maranoa, and other localities.

7. **S. cryphiopetala** (petals hidden in the calyx), *F. v. M. Fragm.* iii. 4; *Benth. Fl. Austr.* i. 194. A shrub, nearly allied to *S. virgata*, but the tomentum longer and denser, almost woolly or floccose. Leaves ovate-lanceolate or cordate, often 2in. long. Calyx densely woolly hirsute, the lobes attaining 3 or 4 lines, including their long soft hirsute filiform points, exceeding the petals in the specimens seen. Carpels 5 or more, wrinkled on the back, reticulate on the sides, their summits forming a strongly projecting centre to the fruit.

Hab.: Inland tropical parts.

8. **S. petrophila** (found on rocks), *F. v. M. in Linnæa*, xxv. 381; *Benth. Fl. Austr.* i. 194. A hoary tomentose erect shrub of 2 to 4ft., with the habit, foliage, and inflorescence of *S. calxhymenia*, but the flowers are not nearly so broad, the unexpanded bud rather ovoid than depressed-globular, the petals longer than the calyx, and the fruiting calyx not nearly so much enlarged, the

ovate-lanceolate lobes not exceeding 3 lines in length, not half so broad as in *S. calyxhymenia*, and of a much thicker consistence. Fruit depressed, tomentose, wrinkled on the circumference and furrowed between the carpels as in *S. calyxhymenia*, but the carpels are usually about 7.

Hab.: Inland.

9. **S. subspicata** (somewhat spicate), *F. v. M. Herb.; Benth. Fl. Austr.* i. 195. An erect shrub, sparingly tomentose and green, or densely tomentose like *S. virgata* and *S. macropoda*, but at once known by the inflorescence. Leaves from cordate-ovate to lanceolate, 1 to 2in. long, obtuse, crenate, cordate or rounded at the base, slightly wrinkled above, with the veins prominent underneath, scabrous, velvety or densely tomentose. Flowers small, nearly sessile, clustered or rarely solitary, the upper clusters forming often an irregular terminal spike, with few small floral leaves. Calyx not ribbed, the lobes acute, at least as long as the tube and closing over the fruit, but not covering it. Petals nearly twice as long. Stamens often under 10. Fruit nearly globular, but grooved between the carpels; carpels 5 or 6, tomentose, reticulate on the side, but not wrinkled on the back, and not acuminate.

Hab.: Common throughout the colony.

10. **S. pleiantha** (numerous flowers), *F. v. M. Herb.; Benth. Fl. Austr.* i. 195. A shrub or undershrub, with elongated branches, green or hoary with a loose stellate tomentum, sometimes floccose. Leaves petiolate, the smaller ones nearly orbicular, ½in. long, the larger ones ovate or ovate-lanceolate, 1 to 2in., toothed, rounded or scarcely cordate at the base. Flowers small, clustered several together, the pedicels 2 to 4 lines long, not articulate. Calyx broadly campanulate, when in flower about 1½ line long, with ovate-acute tomentose lobes, somewhat enlarged when in fruit, the lobes broad, herbaceous, glabrous, and connivent over the fruit, with projecting undulate sinuses. Stamens often not more than 10. Fruit depressed-orbicular, about 3 lines diameter, nearly glabrous, not wrinkled, but strongly grooved between the carpels. Carpels 7 to 10, not reticulate on the sides.

Hab.: Peak Downs, *F. v. Mueller.*

11. **S. spinosa** (see note below for derivation), *Linn.; DC. Prod.* i. 460; *Benth. Fl. Austr.* i. 196. An annual or sometimes perennial, and woody at the base, with the habit and inflorescence of the narrow-leaved forms of *S. rhombifolia*, but the whole plant, including both sides of the leaves, whitish with a minute tomentum, which is soft and more dense on the calyx. Leaves from ovate to lanceolate. Carpels almost always 5 only, more erect and less readily detached than in *S. rhombifolia*, often slightly reticulate, awnless or with short awns.— A. Gray, Gen. Ill. t. 123.

Hab.: Tropical parts.

The species is not uncommon in tropical Asia, more rare in America. It derives its name from the stipules in falling off often leaving a prominent tubercular base, more distinct in this than in any other species, although the character is even here not constant.

12. **S. rhombifolia** (name from form of leaf), *Linn.; DC. Prod.* i. 462; *Benth. Fl. Austr.* i. 196. A perennial or undershrub, very variable in stature, sometimes tall and erect with the larger leaves ovate and 3in. long, the Australian specimens more generally representing the more spreading forms, with rigid virgate minutely tomentose branches, and small narrow leaves, rarely exceeding 1in., varying from ovate-lanceolate to narrow-lanceolate, or from nearly obovate to oblong-cuneate, always shortly petiolate, toothed, nearly glabrous above and more or less whitened underneath with a short tomentum. Pedicels mostly longer than the petiole and sometimes as long as the leaf,

articulate about the middle. Flowers rather small, yellow. Calyx broad, glabrous or slightly hoary, prominently 10-ribbed at the base. Carpels about 10, with or without terminal erect-connivent awns, angled at the back, neither wrinkled nor reticulate, opening at the top into two very short valves.

Hab.: Abundant.

The species is one of the commonest tropical weeds, both in the New and the Old World, and includes *S. retusa*, Linn., *S. rhomboidea*, Roxb., *S. philippica*, and *S. compressa*, DC., and several other published forms.

Var. (?) *incana*. Leaves whitish on both sides as in *S. spinosa*, but carpels about 10, with long awns.—Nicholson River, *F. v. Mueller*; Comet River, *Leichhardt*; the specimens not complete.

13. S. cordifolia (leaves heart-shaped), *Linn.; DC. Prod.* i. 464; *Benth. Fl. Austr.* i. 196. A rather coarse, branching, erect, or rarely decumbent herb or undershrub, more or less clothed with a soft stellate tomentum or velvety hairs, the branches often also hirsute with spreading hairs. Leaves on rather long petioles, broadly cordate or almost orbicular or rarely ovate-lanceolate, 1 to 1½ or rarely 2in. long, usually soft and thick. Flowers small, yellow, on short axillary pedicels or clustered into short leafy racemes. Calyx 10-ribbed at the base, softly tomentose. Carpels about 10 or sometimes fewer, smooth or slightly wrinkled, opening at the top in 2 valves, and in the usual form terminating in rather long erect-connivent awns.

Hab.: Peak Downs and other inland parts.

The species is very abundant in almost all tropical countries, and includes *S. althæifolia*, Lam., and several other supposed species.

Var. (?) *mutica*. Carpels without the awns which generally distinguish the species. The leaves are very soft and velvety, but small and narrow; the specimens have, however, lost those of the primary branches.—Macarthur River, Gulf of Carpentaria.

14. S. platycalyx (broad calyx), *F. v. M. Herb.*; *Benth. Fl. Austr.* i. 197. Shrubby and densely clothed with a soft floccose or velvety stellate tomentum. Leaves ovate-cordate or nearly orbicular, obtuse, crenate, 1in. long or more, soft and thick. Pedicels as long as the leaves, soft, articulate above the middle. Calyx broadly campanulate, about 5 lines long, with a broadly obtuse base, the lobes erect or spreading, shorter than the tube, densely tomentose outside, each sepal marked with 3 prominent ribs, with another almost equally prominent at the junction of the sepals. Petals broad, shorter than the calyx. Stamens very numerous, the staminal tube almost truncate at the top. Carpels about 24, closely packed in a tomentose ring round the base of the styles, which are free almost to the base with small capitate stigmas. Fruit not seen.

Hab.: Tropics.

15. S. inclusa (enclosed), *Benth. Fl. Austr.* i. 197. A shrub, densely velvety tomentose or almost floccose. Leaves ovate or orbicular, often cordate, obtuse, crenate, mostly above 1in. long. Flowers not seen. Fruiting calyx on peduncles of about 1in., membranous and inflated, above 1in. diameter, tomentose, marked with numerous longitudinal veins or ribs, the short lobes connivent, so as completely to enclose the fruit. Carpels numerous, stellate-hirsute, echinate with rather soft hirsute spines, forming a depressed orbicular fruit of nearly 1in. diameter.

Hab.: Georgina.

This species and *S. platycalyx* are distinguished in the genus by their many-ribbed calyx; as the one is only known in fruit, and the other in flower, or scarcely past, the distinction between the two cannot be established with certainty, but *S. platycalyx* certainly shows no tendency to the singular enlargement of the calyx of *S. inclusa*.—*Benth.*

6. ABUTILON, Gaertn.

(Of Arabic origin, alluding to the yellow colour of the Mediterranean species.)

Bracteoles none. Calyx 5-lobed. Staminal column divided at the top into several filaments. Ovary-cells 5 or more, verticillate, each with 3 or more, rarely 2, ovules. Style-branches as many as cells, filiform or club-shaped, with terminal stigmas. Fruit-carpels united at the base or entirely seceding, rounded or angular or with diverging points (not connivent) at the top, opening in 2 valves, without internal appendages. Seeds nearly reniform, the upper ones usually ascending, the lower ones pendulous or horizontal.—Herbs or shrubs, rarely trees, usually clothed with a soft stellate tomentum. Leaves usually cordate, angular or lobed, rarely narrow; petioles usually long (except in *A. crispum*). Stipules in all the Australian species subulate and deciduous. Flowers in the Australian species axillary, yellow or rarely white, the pedicels articulate above the middle or near the top.

A large genus, distributed over the tropical and warm regions of the globe, chiefly American. The genus has frequently been united with *Sida*, but the characters derived from the diverging carpels with more than 1 ovule in each, as contrasted with the converging uniovulate carpels of *Sida*, are too constant and convenient to be neglected in groups so very numerous in species. The differential characters given to several of the following species from the tropical regions, or from the deserts of the interior, are as yet very unsatisfactory, owing to the imperfect state of many of the specimens, often mere fragments.—*Benth.*

§ 1. *Capsule truncate or concave at the top. Carpels (usually 2 or 3-seeded) angular-pointed or awned at the upper outer edge, persistent, or rarely at length deciduous leaving the filiform placenta attached to the axis.*

Carpels (usually 10 or fewer) not exceeding the calyx-lobes, the points erect, or rarely divergent. Stems usually (perhaps always) shrubby.
  Calyx-lobes shorter than the tube.
    Petals adnate high up the glabrous staminal tube. Calyx tubular, 1in. long . . . . 1. *A. tubulosum.*
    Petals shortly adnate to the pubescent base of the staminal tube. Calyx ⅓ to ⅖in., lobes acuminate or rather obtuse, spreading, much shorter than the tube.
      Petals above 1in. long . . . . 2. *A. leucopetalum.*
      Petals shortly exceeding the calyx . . . . 3. *A. Mitchelli.*
    Calyx about ⅓in., rather inflated, truncate, sinuate, or with very short obtuse lobes.
      Petals very small. Staminal column much longer than the calyx 4. *A. micropetalum.*
  Calyx-lobes longer than the tube or cup, acuminate.
    Calyx-lobes very concave and prominently keeled. Carpels about 10, scarcely acuminate . . . . 5. *A. otocarpum.*
    Calyx-ribs or angles scarcely prominent. Carpels 4 or 5, acuminate . 6. *A. subviscosum.*
Carpels usually exceeding the calyx-lobes, the points often divergent. Herbs usually tall, sometimes hard, almost woody at the base.
  Stems coarse and erect. Leaves broadly cordate.
    Capsule truncate. Carpels numerous, the points very short. Tomentum close and dense, usually without spreading hairs.
      Stipules small and subulate. Flowers mostly axillary . . . . 7. *A. indicum.*
      Stipules broadly semisagittate. Flowers in terminal leafless racemes or panicles . . . . 8. *A. auritum.*
    Capsule contracted and angular at the top. Carpels numerous, without points. Tomentum dense, mixed with long spreading hairs . . 9. *A. graveolens.*
  Stems rather slender. Leaves ovate or cordate-lanceolate. Capsule truncate, with short divergent points . . . . 10. *A. oxycarpum.*

§ 2. *Carpels (often 1-seeded by abortion) rounded or angled at the top, quite distinct, and seceding from the axis when fully ripe* (Gayoides, *Endl.*)

Carpels numerous (about 20), closely packed, very hirsute. Tall herbs, with large, broadly cordate leaves.
  Carpels angular at the top, leaving persistent filiform placentas . . . 9. *A. graveolens.*
  Carpels rounded at the top, completely deciduous . . . . 11. *A. muticum.*

Carpels rarely more than 10, glabrous or slightly tomentose, not scarious.
　Leaves mostly cordate orbicular.
　Densely velvety-tomentose (shrubby?) Petals shortly exceeding the
　　calyx . . . . 12. *A. Cunninghamii.*
　Low undershrub, shortly tomentose or pubescent, often with spreading
　　hairs. Petals fully twice as long as the calyx . . . . 13. *A. Fraseri.*

Distinct as the two sections are in some instances, they are closely connected by *A. graveolens* and some other intermediate species.

1. **A. tubulosum** (tubular), *Hook.; Walp. Ann.* ii. 158; *Benth. Fl. Austr.* i. 200. Tall and shrubby, clothed with a dense, soft, close or velvety tomentum. Leaves deeply cordate, ovate or lanceolate, almost acuminate, crenate, attaining 3 to 4in., very soft and velvety. Pedicels much shorter than the leaves. Buds acuminate, prominent-angled. Calyx tubular, about 1in. long, with 10 slightly prominent ribs, softly tomentose, the lobes acuminate, much shorter than the tube. Petals (yellow?) nearly ¾in. longer than the calyx, the claws adhering to nearly the middle of the glabrous staminal column. Capsule angular, about half the length of the calyx, softly villous; carpels 7 to 10, strongly acuminate on their outer edge, containing each usually 8 seeds.—*Sida tubulosa*, A. Cunn.; Hook. in Mitch. Trop. Austr. 390.

　Hab.: Open woods on the Mooni and Dawson Rivers.

　Var. (?) *breviflorum.* Petals shorter and broader, but glabrous and more adnate than in *A. leucopetalum;* the specimen, however, scarcely sufficient for accurate determination.—Dawson River, *Benth. in Fl. Austr.*

2. **A. leucopetalum** (petals white), *F. v. M. Herb.; Benth. Fl. Austr.* i. 200. A tall shrub, clothed with a soft velvety tomentum like *A. tubulosum*, but intermixed with long spreading hairs on the branches, and paler on the under side of the leaves. Leaves deeply cordate, from orbicular to nearly lanceolate, often shortly acuminate, irregularly crenate or almost lobed, mostly shorter than in *A. tubulosum.* Flowers large and white, on short pedicels. Calyx broadly tubular-campanulate, ½ to ¾in. long, 10-ribbed, scarcely acuminate in the bud, the lobes obtuse or shortly acuminate, shorter than the tube. Petals more than twice as long as the calyx, adnate only to the pubescent base of the staminal tube. Capsule as in *A. tubulosum*, but fully as long as the calyx-tube.—*Sida leucopetala*, F. v. M. Fragm. ii. 12.

　Hab.: Inland.

3. **A. Mitchelli** (after Sir T. Mitchell), *Benth. Fl. Austr.* i. 201. Apparently shrubby, clothed with a dense, soft, velvety tomentum mixed with long spreading hairs. Leaves deeply cordate, orbicular or broadly ovate, often shortly acuminate, 1½ to 2½in. long, crenate, very soft and thick. Pedicels shorter than the petioles. Calyx campanulate, 10-ribbed and somewhat 5-angled, 4 to 5 lines long, the acuminate spreading lobes shorter than the tube. Petals (yellow?) shortly exceeding the calyx, pubescent at the base. Ovary-cells and style-branches about 10. Fruit not seen.

　Hab.: Gullies in the ranges on the Maranoa. The plant has at first sight the aspect of *A. muticum*, but the calyx and ovary are quite different.

　Var. (?) *mollissima.* Tomentum very dense and soft, but without the long hairs of the other specimens. Stony Ridge, *Mitchell.*

4. **A. micropetalum** (small petals), *Benth. Fl. Austr.* i. 201. Shrubby, very densely and softly tomentose or velvety. Leaves deeply cordate, acuminate, 2 to 4in. long, crenate. Pedicels short, in the upper axils. Calyx loosely campanulate, almost inflated, very shortly sinuate-toothed or almost truncate, 4 to 5 lines long, tomentose, slightly 5-angled and 10-ribbed. Petals, in some flowers

at least, very small. Stamens very numerous, the slender column much longer than the calyx. Capsule as long as the calyx, truncate at the top ; carpels about 10 to 12, persistent angular, or scarcely pointed at the upper outer edge.—*Sida micropetala*, R. Br. Herb.

Hab.: Hills about Shoalwater Bay, *R. Brown.*

5. **A. otocarpum** (eared capsule), *F. v. M. in Trans. Phil. Soc. Vict.* 1855, 13, *and in Hook. Kew Journ.* viii. 10 ; *Benth. Fl. Austr.* i. 202. " Ballan-boor," Cloncurry, *Palmer.* A tall shrub, densely clothed with a soft velvety tomentum, the branches and petioles almost villous. Leaves deeply cordate, orbicular or broadly ovate, mostly $1\frac{1}{2}$ to $2\frac{1}{2}$in. long, rarely acuminate, crenate, very soft and thick. Pedicels much shorter than the leaves, often crowded at the ends of the branches. Calyx 4 to 6 lines long, very prominently 5-angled, deeply divided into very concave, almost boat-shaped, strongly keeled, acuminate lobes, making the calyx intruded at the base. Petals slightly exceeding the calyx. Capsule villous, shorter than the calyx-lobes, narrowed at the top, depressed in the centre; carpels about 10, rather obtuse or scarcely pointed on the upper outer edge. Seeds 3 or fewer.

Hab.: Stokes Range, on Gilbert River.

The natives peel the bark off, scrape it clean with mussel shells, and use it for making strong netting for game.—*Palmer.*

6. **A. subviscosum** (somewhat viscid), *Benth. Fl. Austr.* i. 202. Apparently shrubby, with much of the aspect of *A. indicum*, but the branches, petioles, and pedicels greener and clothed with a viscid stellate pubescence intermixed with longer hairs. Leaves broad, deeply cordate, abruptly acuminate, 3 to 4in. long, irregularly toothed, softly but sparingly pubescent above, tomentose and whitish underneath. Pedicels short. Calyx with slightly prominent angles, pubescent, deeply divided into acuminate lobes about $\frac{1}{4}$in. long. Petals exceeding the calyx, but imperfect in our specimens. Capsule shorter than the calyx-lobes, consisting of about 5 erect carpels, acuminate with rather long points.

Hab.: Subtropical regions of the interior.

7. **A. indicum** (Indian), *G. Don, Gen. Syst.* i. 504 ; *Benth. Fl. Austr.* i. 202. A tall biennial or perennial, clothed with a whitish tomentum, usually very close and short. Leaves cordate-orbicular, irregularly crenate, toothed or almost lobed, usually acuminate, attaining sometimes 5 to 6in., the upper ones much smaller. Pedicels shorter than the leaves. Calyx campanulate, 5 to 6 lines long, angular in the bud, the ribs scarcely prominent when in flower, deeply divided into acuminate lobes. Petals yellow, longer than the calyx. Capsule hairy, exceeding the calyx, truncate, and attaining sometimes 7 or 8 lines diameter at the top; carpels about 20, acute-angled or minutely acuminate at their upper outer edge, like all the preceding species not readily separating at maturity. Seeds 3 or fewer in each carpel.—*Sida indica*, Linn.; DC. Prod. i. 471 ; Wight, Ic. Pl. t. 12 ; *Sida asiatica*, Linn.; DC. Prod. i. 470 ; *Abutilon asiaticum*, G. Don, Gen. Syst. i. 503.

Hab.: Keppel Bay and Shoalwater Bay, Percy Island, Port Denison, Gulf of Carpentaria.

The species is widely spread over tropical Asia and Africa.

8. **A. auritum** (ear-form of stipule), *G. Don, Gen. Syst.* i. 500 ; *Benth. Fl. Austr.* i. 208. A tall herb or perhaps undershrub, softly clothed with a soft tomentum. Stipules broad, semisagittate, often 4 to 6 lines long, and persistent. Leaves deeply cordate, acuminate, denticulate, 2 to 4in. long, softly pubescent-tomentose above, white underneath. Flowers rather small, of a brown-reddish yellow; on very short pedicels, in almost leafless, terminal, branching racemes or

panicles, with a broad, whitish, deciduous, stipular bract under each pedicel. Calyx obtusely 5-angled, softly tomentose, deeply divided into broad acuminate lobes. Petals not twice as long. Stamens not very numerous. Capsule longer than the calyx, hirsute, truncate ; carpels numerous, with short divaricate points. —*Sida aurita*, Wall.; DC. Prod. i. 468 ; Bot. Mag. t. 2495.

Hab.: Keppel Bay, Percy Island.
The species is also found in Java and in the Philippine Islands.

9. **A. graveolens** (heavy-scented), *W. and Arn. Prod. Fl. Pen. Ind. Or.* i. 56 ; *Benth. Fl. Austr.* i. 204. A coarse annual or perhaps perennial, from 1 to 5ft. high, clothed with a viscid strong-scented tomentum, intermixed, especially on the branches and petioles, with long spreading hairs. Leaves broadly orbicular-cordate, resembling those of *A. Avicennæ* but softer. Flowers yellow, rather large, on pedicels about as long as the petioles. Calyx about 5 lines long, deeply divided into acuminate lobes, each with a prominent midrib. Petals twice as long. Capsule exceeding the calyx, 8 to 10 lines diameter, hirsute, contracted at the top so as to approach in form that of *A. muticum*, and the carpels are numerous and closely packed as in that species, but angular or very shortly pointed at the top and less deciduous, generally leaving the filiform placentas attached to the axis, the species thus connecting the true *Abutila* with the section *Gayoides*.— Hook. Comp. Bot. Mag. i. t. 2 ; *Sida graveolens*, Roxb.; DC. Prod. i. 473.

Hab.: Piper's Island and many parts along the tropical coast.
The species is widely spread over East India and tropical Africa. The petals have there usually a dark spot at the base which does not appear in our Australian specimens.

10. **A. oxycarpum** (carpels sharp-pointed), *F. v. M. Herb.; Benth. Fl. Austr.* i. 204. Herbaceous, diffuse or erect, attaining 2' or 3ft., clothed with a close tomentum or soft velvety pubescence, sometimes almost hirsute, the branches usually slender and divaricate. Leaves from cordate-ovate to ovate-lanceolate, crenate, obtuse or acuminate, 1 to 3in. long. Pedicels slender, often 2 together, 1 to 2in. long. Flowers small, yellow. Calyx deeply cleft, about 2 lines long. Petals not twice as long. Capsule closely tomentose or pubescent, about 4 lines long, truncate and somewhat dilated at the top ; carpels rarely above 10 and often much fewer, with short divaricate points at the outer angle, not separating till the seeds shed, and then leaving the filiform placentas attached to the axis. Seeds 2 or rarely 3.—*Sida oxycarpa*, F. v. M. Fragm. ii. 12.

Hab.: Keppel Bay, Brisbane River, Rockhampton, and other places.
There are two principal forms in our herbaria : 1, *acutatum*, softly tomentose, pubescent or almost hirsute ; leaves ovate-lanceolate, or lanceolate, acuminate ; the most common Brisbane and N. S. Wales form ; and 2, *incanum*, tomentum close and white ; leaves broadly cordate-ovate, obtuse or acuminate ; chiefly within the tropics and in the west. Both are readily recognized by the small calyx, usually not half so long as the capsule.

Var. (?) *malvæfolium*. Less tomentose, but hirsute with long spreading hairs. Leaves cordate-ovate, very obtuse, crenate, and more or less distinctly 3-lobed. Sepals almost as long as the carpels.—Mount Murchison in N. S. Wales, *Dallachy*. This may prove to be a distinct species.—*Benth. in Fl. Austr.*

11. **A. muticum** (without points to carpels), *G. Don, Gen. Syst.* i. 502 ; *Benth. Fl. Austr.* i. 204. Tall and erect, with the habit of *A. graveolens*, with which it is often confounded, but differs in the fruit. Tomentum dense and soft, but not usually mixed with spreading hairs. Leaves cordate-orbicular, often acuminate and irregularly toothed, 2 to 3in. diameter, thick and soft. Pedicels rarely exceeding the petioles. Calyx ½in. long, the lobes equal to or longer than the tube, the ribs not very prominent. Petals not twice as long, often with a dark base as in *A. graveolens*. Capsule longer than the calyx, depressed-globular

with a concave centre, 7 to 8 lines diameter, densely villous; carpels about 20, closely packed, rounded or very obtuse at the top and separating completely without leaving the persistent placentas of *A. graveolens.—Sida mutica*, Delil.; DC. Prod. i. 470.

Hab.: Keppel Bay, Percy Island, sources of the Burdekin and on the Dawson, and Rockhampton.

The specimens are not complete, but agree well with those from tropical Africa, where the species is common and generally referred to *A. asiaticum*, but is not *Sida asiatica* of Linnæus. *S. tomentosa*, Roxb., appears to be an E. Indian form of the same species, with the tomentum mixed with spreading hairs as in *A. graveolens*, from which it cannot always be distinguished without good fruit. It is this form which is represented as *Sida graveolens*, Bot. Mag. t. 4134. —*Benth. in Fl. Austr.*

12. **A. Cunninghamii** (after Allan Cunningham), *Benth. Fl. Austr.* i. 205. Allied to *A. Fraseri*, but apparently shrubby, much branched and densely clothed with soft, short, but velvety tomentum, without spreading hairs. Leaves cordate-orbicular, very obtuse, crenate, 1 to 2in. diameter, thick and soft. Flowers on rather long peduncles in the upper axis. Calyx 4 to 5 lines long, densely tomentose, deeply divided into broad acuminate lobes. Petals about $\frac{1}{2}$in. long. Carpels 10 or fewer, distinct and seceding completely from the axis, rounded at the top, densely but closely tomentose, and not scarious.

Hab.: Estuary of the Burdekin.

13. **A. Fraseri** (after C. Fraser), *Hook.; Walp. Ann.* ii. 158; *Benth. Fl. Austr.* i. 205. A low branching undershrub, rarely exceeding 1ft., shortly tomentose or pubescent, with longer hairs occasionally intermixed. Leaves cordate, from orbicular to ovate, crenate, often all under 1in. diameter, but sometimes 1$\frac{1}{2}$in. Pedicels rarely exceeding the petioles. Flowers rather large. Calyx 3 to 4 lines long, tomentose-pubescent and sometimes hirsute, divided to about the middle. Petals more than twice as long. Fruit usually exceeding the calyx, slightly tomentose or pubescent, 3 to 4 lines diameter, depressed in the centre; carpels 6 to 10, very disinct, and seceding completely from the axis, obtuse or almost pointed at the top, not scarious. Seeds 1 or 2 in each carpel, glabrous or minutely pubescent.—*Sida Fraseri*, Hook. in Mitch. Trop. Austr. 368.

Hab.: On the Maranoa, Sutton River and Broadsound, Comet River.

Var. *halophilum*. Leaves usually orbicular, very obtuse, often truncate or retuse, the carpels 5 or 6 lines long and very broad and obtuse.—*A. halophilum*, F. v. M. in Linnæa, xxv. 381.— N. S. Wales, S. Australia, and on Queensland border.

*7. MODIOLA, Mœnch.

(Carpels resembling the nave of a wheel.)

Calyx 5-cleft, with 3 bracteoles at the base. Carpels numerous, arranged circularly, 2-valved, spuriously 2-celled transversely by the inflexion of a valve-like process, 2-seeded. Radicle in the upper seed superior, in the lower seed inferior.—Prostrate and usually creeping herbs. Leaves divided. Peduncles axillary, 1-flowered.—A. Gray.

1. **M. multifida** (leaf much divided), *Mœnch*. Stems diffuse, more or less hirsute, often rooting at the joints. Leaves 1 to 2in. diameter, palmately 3 to 5-lobed; segments incised and toothed. Pedicels longer than the petioles. Bracteoles linear-lanceolate. Segments of the calyx ovate-lanceolate. Petals obovate, purplish-red, a little longer than the calyx. Stamens 15 to 18. Carpels 15 to 20, lunate, much compressed, hispid on the back, wrinkled on the sides towards the base. A rigid process rising from the back on the insides of the carpel extends to the axis, separating the upper from the lower seeds.—*Malva Caroliniana*, Linn.

Hab.: This American plant has become naturalised near many southern townships.

8. URENA, Linn.
(Its Malabar name.)

Bracteoles 5, united in a 5-cleft involucre, adnate to the calyx at the base. Calyx 5-toothed or 5-lobed. Staminal column bearing several filaments or almost sessile anthers outside, below the truncate or 5-toothed summit. Ovary-cells 5, 1-ovulate; style-branches 10, with terminal capitate stigmas. Fruit-carpels seceding from the axis, indehiscent, muricate, or covered with hooked bristles. Seeds ascending.—Rigid tall herbs or shrubs, more or less scabrous-tomentose. Leaves usually angled or lobed, at least the lower ones. Flowers sessile or on very short peduncles, often clustered, axillary or in terminal leafy racemes.

Besides the one or two species common in all tropical regions, the genus comprises two or three tropical Asiatic ones which appear distinct. As a genus, *Urena* scarcely differs from *Pavonia*.—Benth.

1. **U. lobata** (lobed), *Linn.; DC. Prod.* i. 441, var. *grandiflora; Benth. Fl. Austr.* i. 206. A hard erect herb or shrub of 2 to 4ft., covered on the stems and under side of the leaves with a whitish, close, often scabrous tomentum. Leaves petiolate, the lower ones nearly orbicular, the upper ones ovate or lanceolate, palmately 3 to 7-veined, irregularly toothed, angular, or broadly and shortly lobed, glabrous above or slightly scabrous-tomentose. Flowers sessile or nearly so. Involucre deeply-cleft into narrow-lanceolate lobes, in the single Australian specimen nearly ½in. long, and fully twice as long as the calyx, but often not longer than the calyx or shorter. Petals pink, about 1in. long in this specimen, but often much smaller. Carpels in our specimen shortly muricate.—Bot. Mag. t. 3043 (with short involucres).

Hab.: From Brisbane northward.

The species is widely spread over tropical America, Africa, and Asia, and is very variable in the shape of the leaf and proportions of the involucre, calyx, and petals, as well as in the carpels, more or less glochidiate or muricate; and most probably the *U. sinuata*, Linn., almost equally common, is only a variety with deeply-cut leaves.

2. **U. Armitiana** (W. E. Armit), *F. v. M. Fragm.* x. 78. An erect shrub, stellate-pillose. Leaves cordate-ovate, angular-denticulate, 1 to 1¼in. long, the lower ones larger and on longer petioles than the upper, under side pale, nerves glandular. Flowers racemose or paniculate. Pedicel 1 to 3 lines. Involucre-tube 1 to 2 lines long, lobes linear-lanceolate, 4 to 6 lines long. Calyx almost membranous, lobes puberulent at the margin, narrow-lanceolate. Petals rose-coloured, about 8 lines long. Staminal-tube glabrous; anthers 10, nearly sessile. Style-branches very short; stigmas barbillate. Carpels 1½ to 2½ lines long, not glochidiate. Seeds rugulose and puberulent.

Hab.: Etheridge River.

9. PAVONIA, Cav.
(After J. Pavon.)
(Greevesia, F. v. M.)

Bracteoles 5 or more, free or united at the base. Calyx 5-toothed or 5-lobed. Staminal column bearing several filaments on the outside, below the truncate or 5-toothed summit. Ovary-cells 5, 1-ovulate; style-branches 10, with terminal capitate stigmas. Fruit-carpels seceding from the axis, indehiscent or 2-valved at the top, with or without 1 or 3 awns or points, but not covered by the hooked bristles of *Urena*. Seeds ascending.—Herbs or shrubs, tomentose, hirsute, or glabrous. Leaves often angled or lobed. Flowers on axillary pedicels or in terminal heads or clusters.

A large genus, chiefly South American, with a few species scattered over the warmer regions of the Old World. The Australian species is the same as one of the South American ones.—*Benth.*

1. **P. hastata** (Halbert-headed form of leaf), *Cav. Diss.* 138, *t.* 47, *f.* 2; *Benth. Fl. Austr.* i. 207. A low spreading shrub, more or less hoary, with a minute close stellate tomentum. Stipules subulate. Leaves petiolate, from ovate-cordate to oblong-hastate, obtuse, 1 to 2in. long, coarsely crenate, scabrous above, hoary-tomentose underneath; when hastate, the lateral-lobes short and obtuse. Pedicels usually shorter than the leaves. Bracteoles 5, ovate, herbaceous, nearly as long as the calyx. Calyx tomentose, 2 to 3 lines long, divided to the middle into 5 ovate lobes. Petals in the perfect flowers twice as long as the calyx, of a reddish-purple with a dark centre, but in other flowers, equally fertile, they are very small and closed over the stamens, which are then reduced to 5, whilst they are much more numerous in the perfect flowers. Carpels obovoid, indehiscent, usually pubescent, strongly reticulate and with a slightly raised dorsal rib.—DC. Prod. i. 443; Reichb. Icon. Exot. t. 227; *Greevesia cleisocaly.r*, F. v. M. in Kew Journ. viii. 8 (founded on clandestine-flowered specimens).

Hab.: Moreton Bay, Brisbane River to Expedition Range.

Also a native of Monte Video in South America, where, as well as in Australia, it produces both kinds of flowers, although the clandestine ones appear never to have been observed until pointed out by F. v. Mueller.—*Benth.*

10. HIBISCUS, Linn.

(The ancient name of the Mallow.)

(Abelmoschus, *Medik.;* Paritium, *A. St. Hil.*)

Bracteoles several, rarely reduced to 5 or fewer, usually narrow, free or more or less united, sometimes very small. Calyx 5-lobed or 5-toothed. Staminal column bearing usually numerous filaments on the outside below the truncate or 5-toothed summit. Ovary 5-celled, with 3 or more ovules in each cell; style-branches 5, spreading, or rarely erect and subconnate or exceedingly short, with terminal dilated or capitate stigmas. Capsule membranous or coriaceous, loculicidally 5-valved, the endocarp not usually separating, and rarely produced into spurious dissepiments apparently doubling the number of cells. Seeds reniform or nearly globular, glabrous-pubescent or woolly.—Herbs, shrubs, or trees, hispid tomentose or glabrous, the hairs almost always stellate. Leaves various, often deeply divided. Stipules in the Australian species subulate or small and deciduous, except in *H. tiliaceus*. Flowers usually large, the petals almost always marked with a deeper colour at the base. Filaments usually short and numerous, crowded along the greater part of the elongated staminal column, rarely elongated, fewer and placed close round the top of the short column. Bracteoles usually persistent, but in a few species so deciduous as only to be seen on the very young buds.

A very large genus, widely dispersed over the tropical regions of the globe, a few extending into more temperate climates both in the northern and southern hemispheres. Of the Australian species four are generally distributed over E. India and Africa; of three others belonging to the section *Abelmoschus*, one is found in the Indian Peninsula, another is cultivated, if not wild, in the Indian Archipelago, the third is nearly allied to a corresponding E. Indian species, but in some respects distinct; an eighth species, of the section *Paritium*, is a common maritime tropical tree; the remaining 18 are all endemic.

§ 1. *Bracteoles free (sometimes very deciduous). Calyx 5-toothed, splitting open on one side and deciduous. Tall annuals.* (Abelmoschus, *Medik.*)

Glabrous or the inflorescence tomentose. Bracteoles small, falling off
 from the young bud. Flowers white . . . . 1. *H. ficulneus.*
Hispid. Bracteoles 8 to 12, linear, persistent. Flowers red . . . 2. *H. rhodopetalus.*
Glabrous or slightly setose. Bracteoles 5, broad-lanceolate, persistent.
 Flowers yellow . . . . 3. *H. Manihot.*
 Flowers white, with a reddish centre . . . . 4. *H. Notho-Manihot.*

§ 2. *Bracteoles free. Calyx shortly 5-lobed, inflated. Herb with deeply lobed leaves.* (Trionum, *Medik.*) . . . . 5. *H. trionum.*

§ 3. *Bracteoles free. Calyx deeply 5-lobed, the lobes 1 or 3-nerved, without thickened margins. Seeds bordered or covered by long woolly hairs. Low or slender shrubs or undershrubs.* (Bombicella, DC.)

Staminal tube short with long filaments round the summit . . . . . 6. *H. brachysiphonius.*
Staminal tube slender, the short filaments extending to the middle or lower.
  Plant densely and rigidly velvety-tomentose. Leaves ovate or lanceolate, mostly undivided. Bracteoles small . . . . 7. *H. microchlænus.*
  Plant close, rigid stellate hairs. Leaves (upper ones) narrow-lanceolate. Bracteoles 8 or 9, about 1 line long . . . . 8. *H. Burtonii.*

§ 4. *Bracteoles free. Calyx deeply 5-lobed, the lobes with a central nerve and thickened nerve-like margins. Seeds glabrous. Tall herbs or shrubs, often more or less armed with short prickles (except the last species).*

Herb, glabrous or with scattered hairs. Calyx ribs ciliate. Flowers white or pink . . . . 9. *H. radiatus.*
Tall shrubs, glabrous or with scattered hairs.
  Flowers axillary, without bracts under the pedicels.
    Flowers yellow. Calyx ciliate or setose . . . . 10. *H. divaricatus.*
    Flowers yellow, with dark centre. Bracteoles 12. Calyx deeply divided, velvety-hirsute . . . . 11. *H. Fitzgeraldi.*
    Flowers white. Calyx about 1in. long, lobes deltoid-lanceolate, immarginate . . . . 12. *H. Elsworthii.*
    Flowers white. Calyx densely tomentose . . . . 13. *H. heterophyllus.*
  Flowers in a terminal raceme, with a trifid bract under each pedicel. Calyx densely hirsute . . . . 14. *H. diversifolius.*
Tall shrub, densely velvety-tomentose or villous. Flowers large, pink. Calyx densely hirsute . . . . 15. *H. splendens.*
Tomentose or densely villous shrubs, without prickles. Calyx tomentose or villous.
  Flowers 1½ to 2in. long . . . . 16. *H. zonatus.*

§ 5. *Bracteoles free. Calyx deeply 5-lobed, the lobes 1 or 3-nerved, without thickened margins. Seeds glabrous or shortly pubescent.*

Low or slender shrubs or undershrubs, glabrous, scabrous-pubescent or bristly hispid.
  Leaves undivided.
    Scabrous-pubescent. Leaves ovate-lanceolate or oblong . . . . 17. *H. leptocladus.*
    Glandular viscid and rigidly setose. Leaves broad-cordate or orbicular . . . . 18. *H. setulosus.*
  Leaves deeply divided. Hirsute and densely setose. Calyx not ½in. Capsule glabrous . . . . 19. *H. geranoides.*
Small velvety-tomentose shrubs or undershrubs. Leaves shortly lobed.
  Bracteoles several, subulate . . . . 23. *H. Krichauffianus.*
    (See also 7, *H. microchlænus.*)
  Bracteoles 5, broadly ovate . . . . 22. *H. Normani.*
Tall coarse herbs or shrubs, densely tomentose and often setose.
  Bracteoles small, subulate. Capsule very prominently angled . . . 20. *H. vitifolius.*
  Bracteoles dilated above the middle. Capsule not angled . . . 21. *H. panduriformis.*

§ 6. *Bracteoles united at least at the base. Calyx 5-lobed.*

Tomentose shrubs or undershrubs. Leaves crenate or broadly and shortly lobed.
  Involucral teeth or lobes short or broad. Filaments long and few. Calyx-lobes obscurely nerved . . . . 24. *H. Sturtii.*
  Involucral bracts united at the base only. Filaments short and numerous. Calyx-lobes 1-nerved, with thickened margins . . . 16. *H. zonatus.*
  Calyx-lobes 5, ovate-lanceolate. Corolla 2-coloured, purple and yellow 25. *H. phyllochlanus.*
Glabrous tree. Leaves broad-cordate, entire . . . . 26. *H. tiliaceus.*

1. **H. ficulneus** (fig-leaved), *Linn.; DC. Prod.* i. 448; *Benth. Fl. Austr.* i. 209. "Cooreenyan," Cloncurry, *Palmer.* An erect annual of several feet, glabrous except a few scattered hairs on the leaves, and a velvety pubescence on the racemes and calyces. Leaves orbicular, 2 to 3in. diameter, the lower ones with 5 or 7 short broad lobes, the upper ones more deeply divided, with obovate or oblong lobes, all usually crenate. Flowers white, turning at length reddish, on short pedicels, in a terminal leafless raceme. Bracteoles few, small and so

deciduous as only to be seen on the very young buds. Calyx about ½in. long, shortly 5-toothed, splitting laterally and deciduous. Petals 1in. or rather more, glabrous. Capsule ovoid-oblong, acute, 5-angled, pubescent. Seeds hairy.—*Abelmoschus ficulneus*, W. et. Arn. Prod. i. 53; Wight, Ic. t. 154; *A. alborubens*, F. v. M. Fragm. i. 67.

Hab.: In basaltic tropical and subtropical plains, Fitzroy Plains, and Rockhampton.
The species is common in some parts of the E. Indian peninsula, and includes *H. strictus*, Roxb. Fl. Ind. iii. 206, and probably also *H. prostratus*, Roxb. l.c. 208. The plant figured by Reichenbach, Icon. Exot, t. 161, with persistent broad bracts, is a different species.—*Benth.*
Stem and root of young plant roasted in the ashes and eaten, like a potato.—*Palmer.*

2. **H. rhodopetalus** (red petals), *F. v. M. Herb.; Benth. Fl. Austr.* i. 209. An erect or decumbent coarse annual, of 1¼ to 3ft., more or less hirsute with long bristly hairs. Leaves (except the lowest) more or less deeply 5-lobed, the lobes of the lower ones short and broad, of the upper ones oblong or lanceolate, often 2 to 3in. long, more or less toothed, the lowest leaves often entire and cordate, and the uppermost lanceolate-hastate. Flowers large, red, on axillary pedicels longer than the petioles. Bracteoles 8 to 12, linear, distinct, persistent, usually shorter than the calyx. Calyx pubescent, 6 to 7 lines long, minutely 5-toothed, splitting laterally and deciduous. Petals 1½ to above 2in. long. Capsule oblong-ovoid, acute, 5-angled, longer than the bracteoles, very hispid.—*Abelmoschus rhodopetalus*, F. v. M. Fragm. ii. 112.

Hab.: Woody streams, Point Pearce, and Brisbane River.
This species is very nearly allied to the common East Indian *H. Abelmoschus*, Linn., differing chiefly, as observed by F. v. Mueller, in the colour of the flowers (red not yellow) and in smaller, more divided leaves.—*Benth.*

3. **H. Manihot** (Manihot), *Linn.; DC. Prod.* i. 448; *Benth. Fl. Austr.* i. 210. A tall herb, sprinkled with a few pungent bristly hairs, more copious on the peduncles, otherwise glabrous. Leaves deeply pinnate; lobes 5 to 9, lanceolate, the larger ones narrow, 4 to 5in. long, more or less toothed. Flowers large, yellow with a purple eye, on rather long pedicels in the axils of the upper reduced leaves. Bracteoles 5, herbaceous, broadly lanceolate, fully 1in. long, roughly pubescent, persistent long after the flower has fallen. Calyx shorter than the bracteoles, shortly 5-tooted, tomentose, deciduous. Petals fully 2½in. long. Capsule oblong, 1½ to 2in. long, 5-angled, hispid especially on the angles with stiff bristly hairs.—Bot. Mag. t. 3152; *Abelmoschus Manihot*, Walp. Rep. i. 311; *Hibiscus pentaphyllus*, Roxb. Fl. Ind. iii. 212.

Hab.: Shoalwater Bay and other tropical parts.
The species is frequently cultivated in eastern tropical Asia and in the islands of the Archipelago and the Pacific, but we have no certain record of it in a wild state.—*Benth.*

4. **H. Notho-Manihot** (a spurious form), *F. v. M. Fragm.* v. 57, ix. 130. A shrub, 15ft. high. Leaves palmately divided into from 5 to 9 lanceolate, acuminate, irregularly toothed lobes, the larger ones about 6in. long and ½ to 1in. broad; petioles 1 to 3½in. long, glabrous. Stipules 3 or 4 lines long, caducous, linear-subulate. Pedicels axillary; involucral bracts 5 or 6, lanceolate, acute, slightly pilose. Calyx silky inside, bilabiate, the apex of the lips with 2 or 3 short teeth or entire. Corolla white with a reddish centre, about 2in. long, staminal column about 1in.; anthers yellow. Capsule 1½ to 2½in. long, ovate, acuminate, 5-angled, hispid. Seeds numerous, ovate or renate-globose, puberulent, about 1½ line.

Hab.: Rockingham Bay, J. Dallachy (F. v. M.)

5. **H. trionum** (said to be derived from three divisions of the leaf), *Linn.; DC. Prod.* i. 453; *Benth. Fl. Austr.* i. 210. Bladder ketmia. An erect annual or perennial of short duration, usually 1 to 2ft. high, scabrous-pubescent or shortly hirsute. Leaves 2 to 3in. long,

deeply 3 or 5-lobed with oblong or lanceolate irregularly-toothed lobes. Flowers rather large, pale-yellow with a dark-purple centre, on axillary pedicels. Bracteoles 7 to 12, linear-setaceous. Calyx about ½in. long when in flower, twice that size in fruit, inflated, membranous with about 20 raised veins, glabrous or slightly hirsute, very shortly 5-lobed. Capsule ovoid-globose, hirsute, enclosed in the calyx. Seeds glabrous.—*Bot. Mag.* t. 209; *Reichb. Fl. Germ.* v. 181; *F. v. M. Fragm.* ii. 115; *H. Richardsoni*, Sweet; *Lindl. Bot. Reg.* t. 175; *H. trionioides*, G. Don, *Gen. Syst.* i. 483; *H. tridactylites*, Lindl. in *Mitch. Three Exped.* i. 85.

Hab.: Between the Burnett and Dawson Rivers.

A weed in the cultivation paddocks of the southern Downs.

Common throughout Africa and southern Asia, extending northwards to China and the Amur. Found also in New Zealand.—*Benth.*

6. **H. brachysiphonius** (short-tubed), *F. v. M. Fragm.* i. 67 and 243; *Benth. Fl. Austr.* i. 210. A low perennial or undershrub, with erect or decumbent stems, rarely above 1ft. long, slightly hirsute with short stiff stellate hairs. Lower leaves small, orbicular, undivided, crenate; upper ones divided into 3 obovate or oblong-cuneate coarsely crenate or lobed segments or deep lobes, mostly 1 to 1½in. long. Flowers rather small, pink, on axillary or terminal pedicels, sometimes very long. Bracteoles about 10, rather rigid, linear, shorter than the calyx. Calyx ciliate with a few stiff hairs, deeply divided into lanceolate 1-nerved lobes, not thickened at the margin. Petals about ½in. long. Staminal column short, bearing round the summit about 20 filaments much longer than in most species. Style-branches long, with large capitate stigmas. Capsule nearly globular, glabrous, 4 to 6 lines diameter. Seeds 4 to 6 in each cell, tomentose-villous.

Hab.: Mooni River, Peak Downs, Comet River, and other parts.

7. **H. microchlænus** (bracteoles minute), *F. v. M. Fragm.* ii. 116 (under *H. solanifolius*); *Benth. Fl. Austr.* i. 211. Apparently shrubby, densely clothed with a scabrous, rigid-velvety, or softer and almost floccose stellate tomentum. Leaves on rather short petioles, from ovate to oblong-lanceolate, 1 to 1¼in. long, obtuse, slightly toothed, thickly and rigidly tomentose. Flowers apparently pink or purple, on pedicels rather longer than the petioles. Bracteoles 7 to 9, sometimes very minute, sometimes half as long as the calyx. Calyx ½in. or rather more, densely scabrous-tomentose, deeply divided into lanceolate 1-nerved lobes. Petals 1 to 1½in. long, more or less stellate-tomentose outside where exposed in the bud. Capsule globular, glabrous or slightly hairy. Seeds more or less bordered or covered with long woolly hairs.—*H. brachychlænus*, F. v. M. *Fragm.* iii. 5.

Hab.: Cape River.

8. **H. Burtonii** (after R. C. Burton), *Bail. Bot. Britt.* ii. 7. A rather straggling small shrub, closely clothed with short rigid stellate hairs, which are somewhat longer on the leaves. Lower leaves not collected, those of the flowering shoots preserved, narrow-lanceolate or linear-oblong, 1 to 1¼in. long, rounded at the base, on petioles of 2 to 5 lines; margins slightly crenate or deeply and irregularly toothed. Flowers small, solitary, on slender axillary pedicels of about ¾in. long. Bracteoles 8 or 9, subulate, not over a line in length, covered with scabrous stellate hairs like the other parts of the plant. Calyx scarcely ½in. long, very deeply divided into narrow-lanceolate lobes, the midrib and also a parallel vein or rib near each margin prominent, scabrous, with stellate hairs outside, glabrous on the inside. Corolla probably lilac, the petals but slightly exceeding the calyx-lobes, hairy on the back, with rather stiff mostly simple hairs.

Staminal column about as long as the petals. Style-branches spreading; stigmas hispid, with white hairs. Capsule somewhat globose, not exceeding the calyx, more or less covered by short, bristle-like, usually simple hairs. Seeds bordered by silky laciniate scales.

Hab.: McKinley Ranges and Buckley River, *R. C. Burton.*

At first sight specimens of this plant remind one of *Solanum discolor*, and it is probable that shoots of it may have been mixed with those of *H. microchlæna* and the Western species *H. Pinonianus*, both of which Baron von Mueller at one time named *H. solanifolius*. *H. Burtonii* has much smaller flowers than either of the two species above named.

9. **H. radiatus** (radiated), *Cav. Diss.* 150, *t.* 54, *f.* 2; *Benth. Fl. Austr.* i. 212. An erect annual (or rarely perhaps perennial) of 2 to 3ft., glabrous or hispid in the lower part with a few rigid hairs, and often bearing also small conical prickles. Lower leaves broad and shortly lobed, upper ones deeply 3 to 5-lobed or the uppermost undivided, the lobes narrow, toothed and unequal, the central one often 2 to 3in. long. Flowers white or pink with a dark centre, on axillary pedicels usually very short, rarely attaining 1in. Bracteoles about 10, narrow-linear, often spreading or reflexed, and ciliate with a few rigid hairs. Calyx about ¾in. long, deeply divided into lanceolate acuminate lobes of a thin texture, but marked with a prominent midrib and thickened marginal nerves, more or less rigidly ciliate. Petals 1 to 1½in. long. Capsule globose, glabrous in the Australian specimens. Seeds few, glabrous.—DC. *Prod.* i. 449; *Bot. Mag.* t. 1911; F. v. M. *Fragm.* ii. 117.

Hab.: Palm Islands and Curtis Island.

The species extends over E. India and tropical Africa, but the extra-Australian specimens I have seen have always hirsute and less obtuse capsules. *H. Lindleyi*, Wall. Pl. As. Rar. i. 4, t. 4, is probably a purple-flowered variety. *H. cannabinus*, Linn., cultivated in Asia and Africa for its fibre, differs from *H. radiatus* only in the glands on the calyx.—*Benth.*

10. **H. divaricatus** (spreading), *Grah. in Edinb. Phil. Journ. July.*—Oct. 1830; *Benth. Fl. Austr.* i. 212. "Ngar-golly,". Cloncurry, *Palmer:* "Ithnee," Mitchell, *Palmer*. A tall erect glabrous shrub, with the foliage of some varieties of *H. heterophyllus* and the flowers of *H. radiatus*, the branches often beset with small conical prickles. Leaves on short petioles, entire or deeply 3-lobed, from round-cordate to ovate-lanceolate or oblong, often fully 4in. long, more or less toothed. Flowers large, yellow with a crimson eye, on short pedicels in the axils of the upper reduced leaves. Bracteoles 10 to 12, linear, rigid, ciliate. Calyx deeply divided into lanceolate lobes, with prominent midribs and margins as in *H. radiatus*, rigidly ciliate or rarely minutely tomentose. Petals 2 to 2½in. long. Capsule ovoid-globose, densely silky-hairy.—*Abelmoschus divaricatus*, Walp. Rep. i. 309; *Hibiscus magnificus*, F. v. M. *Fragm.* ii. 118.

Hab.: Shoalwater Bay, N.E. coast, Newcastle Range, Mackenzie and Dawson Rivers.

One of F. v. Mueller's specimens, with the calyx not ciliate but minutely tomentose, seems to connect this species with some forms of *H. heterophyllus.*—*Benth.*

Buds eaten raw, and the thick root is peeled and eaten raw.—*Palmer.*

11. **H. Fitzgeraldi** (after R. Fitzgerald), *F. v. M. Fragm.* viii. 242. A handsome tall shrub, the stem and branches almost glabrous or sparsely aculeate. Leaves glaucous green, ovate or rotundate, crenately dentate, shortly or not lobed, about 10in. long, on petioles of about 6in. Flowers large, yellow with a dark-purple centre, on axillary pedicels. Bracteoles about 12, linear-subulate, thinly tomentose. Calyx deeply divided, and more or less velvety-hirsute, with yellowish hairs. Staminal-column 2 or 3 times shorter than the petals; filaments purple, style shortly lobed. Ovary densely silky-hairy.

Hab.: Bowen River.

12. **H. Elsworthii** (after G. Elsworth), *F. v. M. Fragm.* viii. 241. Shrubby, the branches slightly pilose. Leaves 2 to 3in. long, ovate, acuminate, slightly cordate at the base, glabrous above, with a light-coloured thin tomentum

beneath, nearly chartaceous, margins crenulate. Petioles an inch or shorter. Stipules linear-subulate, soon falling off. Pedicels axillary, solitary. Bracteoles short, broad, subulate-linear, about 12, sparsely pilose as well as the calyx. Calyx about 1in. long, lobes deltoid-lanceolate, immarginate, inside velvety. Petals white, sparsely pilose. Staminal-column glabrous. Style-branches 3 or 4 lines long.

Hab.: Edgecombe Bay.

13. **H. heterophyllus** (leaves various), *Vent. Hort. Malm. t.* 103; *Benth. Fl. Austr.* i. 212. Native rosella; "Batham," N.Q., *Thozet*; "Yarra," Taromeo, *Shirley*. A tall shrub, small erect tree, glabrous except a stellate tomentum on the inflorescence and very young shoots, the branches often bearing small conical prickles. Leaves entire or deeply 3-lobed, linear, lanceolate or elliptical-oblong, often 5 to 6in. long, usually serrulate or crenulate, in some specimens white underneath. Flowers large, white with a purple centre, on short pedicels in the upper axils. Bracteoles about 10, linear, rigid, not ciliate. Calyx often above 1in. long, deeply divided into lanceolate lobes, densely covered with a stellate tomentum often concealing the venation, which, as in *H. radiatus*, consists of a midrib and the thickened margins of each lobe. Petals nearly 3in. long. Capsule ovoid-globular, acute, densely setose or silky-hairy. Seeds glabrous.—Bot. Reg. t. 29; DC. Prod. i. 450; *H. grandiflorus*, Salisb. Par. Lond. t. 22.

Hab.: Broadsound, Shoalwater Bay, Percy Isle, Port Curtis, Brisbane River, and Rockhampton.

Roots of young plants, young shoots, and leaves eaten without any preparation by natives of N.Q.—*Thozet*.

The northern specimens belong mostly to a broader-leaved form, distinguished by A. Cunningham under the name of *H. Margeriæ*.—*Benth.*

Wood of a pale yellow, open grain, smooth and tough; suitable probably for making musical instruments, being considered a good conductor of sound.—*Bailey's Cat. Ql. Woods No.* 17.

The gum contains true bassorin. Analysis: Bassorin 78%, water 22%.—*Lauterer.*

14. **H. diversifolius** (various-leaved), *Jacq.; DC. Prod.* i. 449; *Benth. Fl. Austr.* i. 218. A tall rigid herb or undershrub, sprinkled with a rigid pubescence, the branches and petioles more or less beset with small conical prickles. Leaves broadly cordate or nearly orbicular, irregularly toothed, angular or more or less 5-lobed. Flowers in a terminal raceme, on very short pedicels in the axils of small lanceolate or 3-fid floral leaves, often reduced, especially the upper ones, to small linear bracts. Bracteoles linear, and calyx with marginate lobes, as in *H. radiatus*, but the lobes are narrower, and usually densely hispid with rigid bristly hairs. Capsule acuminate, very hispid. Seeds glabrous.—Bot. Reg. t. 381; *H. Beckleri*, F. v. M. Fragm. ii. 117.

Hab.: Rockhampton, *Thozet* ?

The species is chiefly found in S. Africa, Mauritius, and Madagascar, but is also common in waste places in the Fiji and other S. Pacific islands. In E. India it appears to be in gardens only.—*Benth.*

15. **H. splendens** (showy flowers), *Fraser: Grah. in Edinb. Phil. Journ., Apr.—June,* 1830; *Benth. Fl. Austr.* i. 213. A tall shrub, of great beauty, attaining 12 to 20ft., densely clothed with a soft velvety tomentum, the branches and petioles armed with small scattered prickles or bristles. Leaves on long petioles, broadly ovate-cordate or palmately 3 or 5-lobed, often 6 or 7in. long, the lobes oblong-acuminate or lanceolate, often narrowed at the base. Stipules often 2 on each side. Flowers very large, rose-coloured, on pedicels about as long as the petioles. Bracteoles 10 to 15 or sometimes many more, linear-subulate, as long the calyx, densely hispid or softly villous. Calyx at least 1in. long, densely

tomentose or hispid, deeply divided into lanceolate lobes, with a dorsal and marginal nerve, as in *H. radiatus*. Petals 3in. long or more, glabrous. Capsule silky-hairy. Seeds glabrous.—Bot. Mag. t. 3025; Bot. Reg. t. 1629; *Abelmoschus splendens*, Walp. Rep. i. 309.

Hab.: Percy Island, N.E. coast, Rockhampton, Moreton Bay.

16. **H. zonatus** (zoned), *F. v. M. Fragm.* i. 221; *Benth. Fl. Austr.* i. 218. A shrub with a scabrous tomentum, sometimes short and close, sometimes dense and velvety, the rather slender branches occasionally hirsute or bristly. Leaves from orbicular-cordate to ovate, the larger ones attaining 3 or 4in., and shortly and broadly 3, 5, or 7-lobed, the upper ones entire or toothed and often narrow. Flowers rather large, pink, on very short pedicels in the upper axils. Bracteoles narrow and rigid, rarely exceeding half the length of the calyx, free or slightly united at the base. Calyx nearly $\frac{3}{4}$in. long, densely tomentose, deeply divided into lanceolate lobes, prominently 1-nerved and with thickened margins, as in the preceding species. Petals 1$\frac{1}{4}$ to 2in. long, nearly glabrous. Style-branches short, spreading. Capsule very hispid, nearly globular, shorter than the calyx. Seeds glabrous.

Hab.: Islands of the Gulf of Carpentaria.

17. **H. leptocladus** (slender-branched), *Benth. Fl. Austr.* i. 214. Apparently a low herb or undershrub, with slender branches, rough with short rigid stellate hairs. Leaves on rather long petioles, ovate-lanceolate, lanceolate or oblong, 1 to 2in. long, irregularly toothed, narrowed or rounded at the base, roughly pubescent on both sides with rigid stellate hairs. Flowers apparently pink, on rather long pedicels in the upper axils. Bracteoles about 7 to 9, linear-subulate, rarely exceeding half the length of the calyx. Calyx about $\frac{1}{2}$in. long, pubescent or hispid with stiff stellate hairs, deeply divided into lanceolate-acuminate 1 or 3-nerved lobes, without thickened margins. Petals 1 to 1$\frac{1}{2}$in. long, glabrous. Capsule nearly globular. Seeds 2 or 3 in each cell, glabrous.

Hab.: Islands of Carpentaria Bay, Daintree and Gilbert Rivers.

This species resembles in some respects *H. microchlænus*, but is much more slender and less tomentose, and both petals and seeds appear to be quite glabrous.—*Benth.*

18. **H. setulosus** (bristly), *F. v. M. Fragm.* i. 221; *Benth. Fl. Austr.* i. 214. A much-branched, viscid, strong-scented shrub of several feet, covered with resinous glands, the branches very hispid with long spreading bristles. Leaves broadly cordate or orbicular, mostly 1 to 1$\frac{1}{2}$in. long, toothed, more or less hirsute or pubescent with scattered rigid stellate hairs. Flowers rather large, pink with a dark centre, on axillary pedicels about as long as the petioles. Bracteoles linear, rigid, about as long as the calyx. Calyx about $\frac{3}{4}$in. long, pubescent and glandular like the leaves, deeply divided into lanceolate 3-nerved lobes. Petals about 1$\frac{1}{2}$in. long. Staminal-column conspicuously produced above the filaments and 5-toothed. Capsule globular, hispid, shorter than the calyx. Seeds glabrous or minutely scabrous.

Hab.: Gulf of Carpentaria.

19. **H. geranioides** (Geranium-like), *A. Cunn. Herb.; Benth. Fl. Austr.* i. 215. A low branching annual of 1 to 2ft., densely hispid with long rigid stellate hairs or bristles. Leaves deeply divided into 3 or 5 oblong-linear or cuneate segments, mostly about 1in. long, lobed or coarsely toothed, the lobes or teeth obtuse, hispid on both sides. Flowers small for the genus, on hispid pedicels often as long as the leaves. Bracteoles 8 to 10, linear-subulate, hispid. Calyx 4 to 5 lines long, hirsute, deeply divided into lanceolate-acuminate 3-nerved lobes. Petals about $\frac{3}{4}$ to 1in. long, dark at the base. Filaments short, along the upper part of the column. Stigmas capitate. Capsule small, globular, glabrous. Seeds glabrous.

Hab.: Islands of the Gulf of Carpentaria.

20. **H. vitifolius** (vine-leaved), *Linn.; DC. Prod.* i. 450; *Benth. Fl. Austr.* i. 215. A coarse, erect, divaricately-branched herb of several feet, in India usually shortly tomentose, more hispid in Africa, and in the Australian specimens still more beset with rigid hairs. Leaves broadly cordate, 2 to 3in. long and broad, usually broadly 3 or 5-lobed and toothed, very densely and softly villous-tomentose. Flowers rather large, pale-yellow with a purple centre, on short pedicels, the upper ones forming a short dense leafy raceme. Bracteoles 7 to 10, linear-subulate, shorter than the calyx. Calyx deeply divided into broadly lanceolate lobes, often enlarging after flowering. Capsule depressed globular, beaked in the centre, 5 to 8 lines diameter, hirsute with scattered hairs, the 5 acute angles raised into wings and transversely veined. Seeds glabrous.—F. v. M. Fragm. ii. 114.

Hab.: Keppel Bay, Percy Island, Dawson River, Palm Islands.

A very common species in E. India, extending into the warmer regions of Africa, and introduced into the W. Indies, readily known by its winged capsules.—*Benth.*

21. **H. panduriformis** (Fiddle-shaped), *Burm. Fl. Ind.* 151, t. 47, f. 2; *Benth. Fl. Austr.* i. 215. "Bee-allo," Mitchell River, *Palmer*. A tall, coarse herb or shrub, densely covered with a tomentum, usually thick and velvety on the upper side of the leaves, closer and whiter on the under side and on the petioles and branches, where it is often intermixed with long spreading bristly stellate hairs. Leaves broad-cordate, 3 or 4in. long and broad, or rarely narrow, usually 5-angled or broadly lobed and irregularly crenate. Flowers yellow, on very short pedicels in the axils of the upper reduced leaves, the side-branches often assuming the appearance of several-flowered peduncles. Bracteoles 6 to 8, linear or linear-spathulate, often as long as the calyx, more herbaceous than in most species and always dilated above the middle. Calyx 7 to 9 lines long, densely tomentose-hirsute, the lobes lanceolate, 1-nerved. Petals 1 to 2in. long, densely hirsute where exposed in the bud. Capsule ovoid-globular, very hispid. Seeds shortly pubescent or rarely glabrous.—*DC. Prod.* i. 455; *F. v. M. Fragm.* ii. 115; *H. tubulosus,* Cav. Diss. 161, t. 68, f. 2; DC. Prod. i. 447.

Hab.: Rockhampton.

The species is widely spread over tropical Asia and Africa. Burmann's figure represents a narrow-leaved form, not as yet found in Australia, and rare in India.—*Benth.*

Natives of Mitchell River make the bark, after being cleaned and twisted, into bags for carrying roots, game, &c.—*Palmer*.

22. **H. Normani** (after W. H. Norman), *F. v. M. Fragm.* iii. 4; *Benth. Fl. Austr.* i. 216. An undershrub, with apparently simple erect stems of about 1ft., densely velvety-tomentose. Leaves petiolate, from ovate to lanceolate, acute or obtuse, 2 to 3in. long, obscurely sinuate-toothed, tomentose on both sides, especially underneath. Peduncles 1½ to 2in. long. Involucre of 5 broadly ovate or rhomboidal leafy bracteoles, nearly as long as the calyx, distinct or scarcely united at the base. Calyx tomentose, about ½in. long, deeply divided into ovate-lanceolate 3-nerved lobes. Petals about twice as long or rather more, glabrous.

Hab.: Palm Island, Fitzroy Island.

23. **H. Krichauffianus** (after F. E. H. W. Krichauff), *F. v. M. Rep. Babb. Exped.* 7; *Benth. Fl. Austr.* i. 216. An undershrub, with the habit and foliage of some varieties of *H. Sturtii*, but the tomentum closer and whiter. Leaves ovate or ovate-lanceolate, obtuse, 1 to 1½in. long, irregularly and usually rather deeply crenate-toothed. Flowers rather larger than in most forms of *H. Sturtii*. Bracteoles linear-subulate, almost free, shorter than the calyx and sometimes very short. Calyx very tomentose. Petals 1 to 1½in. long. Seeds slightly pubescent.

Hab.: Cooper's Creek.

K

24. H. Sturtii (after Captain C. Sturt), *Hook. in Mitch. Trop. Austr.* 363; *Benth. Fl. Austr.* i. 216. A rather rigid, simple or branched undershrub, rarely exceeding 1ft., clothed with a whitish tomentum, either short and rather close, or dense and velvety or sometimes almost floccose. Leaves broadly cordate or ovate, rarely ovate-lanceolate, mostly 1 to 1½in. long, obtuse, irregularly crenate-toothed, usually rather thick and soft. Flowers few in the upper axils, rather small, white, pink, or yellow (at Rockhampton, *F. v. M.*) Involucre obconical or campanulate, with 7 or 8 teeth or short lobes, very variable in shape, but usually nearly as long as the calyx. Calyx very tomentose, the lobes shorter or rarely longer than the cup, thick and soft, obscurely 3-nerved. Petals varying from ¾ to fully 1½in. long. Staminal column slender, with scattered filaments as in most species, but the filaments not so numerous and longer than usual, showing an approach to those of *H. brachysiphonius*. Capsule globular, silky. Seeds glabrous or rarely woolly.—F. v. M. Fragm. ii. 13.

Hab.: Mackenzie, Burdekin, Suttor, and Dawson Rivers, Peak Downs, Fitzroy Island, Maranoa and Belyando Rivers.

This very variable species, remarkable for its cup-shaped short-lobed involucre, presents in our specimens the following principal forms:—

a. grandiflora. Involucre shorter than the calyx, with triangular or lanceolate, somewhat acute, erect teeth. Petals above 1in., and often 1½in. long.—Mount Goningbear, in N. S. Wales.

b. Muelleri. Involucre of the preceding variety with the small flowers of the following one.—Gathered by most collectors, as well as the following variety.

c. Sturtii. Involucre as long as the calyx, dilated, and spreading at the top, with short broad rounded lobes. Calyx 3 to 4 lines long, with rather short lobes. Petals rarely exceeding 1in., and often much smaller.—The most common N. S. Wales form.

d. campylochlamys, F. v. M. Both involucre and calyx more or less deeply divided into lanceolate acuminate lobes. Calyx otherwise rather longer than in the preceding varieties.—Victoria River and Sturt's Creek, *F. v. Mueller;* Dampier's Archipelago, *A. Cunningham.* In the latter specimens the seeds are woolly, but in the Victoria River plant they appear to be glabrous, as in the other varieties.

e. platychlamys. Very densely clothed with a somewhat rigid, velvety tomentum. Involucre very spreading, often above 1in. diameter, with broad lobes. Calyx exceeding ½in., with large ovate or ovate-lanceolate lobes.—Victoria River, *F. v. Mueller.*

25. H. phyllochlænus (referring to the leafy involucre), *F. v. M. Fragm.* ix. 128. Plant covered with a rusty or brown stellate tomentum. Leaves 1½ to 2in. long, lanceolate-oblong, crispate rugose, crenate-serrate, on petioles of from 2 to 6 lines about obtuse, apex somewhat acute. Peduncles solitary, 1-flowered, 1 to 1½in. long. Bracteoles 5, herbaceous, glabrescent, ovate-lanceolate. Calyx nearly 1in. long. Corolla 1½in., slightly ciliate, two-coloured—purple and yellow. Staminal column short, glabrous. Style-branches villous. Capsule silky-hispid. Seeds glabrous.

Hab.: Expedition Range.

26. H. tiliaceus (leaves Tilia-like), *Linn.; DC. Prod.* i. 454; *Benth. Fl. Austr.* i. 218. "Talwalpin," Moreton Bay, *Watkins.* A small tree. Leaves on long petioles, orbicular-cordate, shortly acuminate, entire or crenulate, white or hoary underneath with a close short tomentum, nearly glabrous above, 3 to 5in. diameter. Stipules large, broadly oblong, very deciduous. Flowers large, yellow with a dark crimson centre, on short peduncles in the upper axils or at the ends of the branches. Involucre campanulate, divided to about the middle into 10 to 12 lobes, about half the length of the calyx. Calyx nearly 1in. long, with lanceolate 1-nerved lobes. Petals 2 to 3in. long, slightly tomentose outside. Capsule nearly 1in. diameter, the valves bearing the dissepiments in their centre, and their thin margins turned inwards so as to make the capsule appear 10-celled.—*Paritium tiliaceum,* St. Hil. Fl. Bras. Mer. i. 256; Wight, Ic. Pl. t. 7.

Hab.: Islands of the Bay of Carpentaria, Rockhampton, Brisbane River.

A common seacoast tree in most tropical countries, particularly abundant in the islands of the Pacific.—*Benth.*

In Central America the fibre is known as "majagu," and in Bengal as "bola," and being little affected by moisture is therefore selected by surveyors for measuring lines. The Queensland blacks at one time used the roots and young growth of this tree for food. In the West Indies, in times of scarcity of bread-fruit, the mucilaginous bark is said to be sucked for food.

Wood close-grained, colour invisible green, beautifully marked, easy to work, and takes a good polish; supposed by some to resemble Pollard oak.—*Bailey's Cat. Ql. Woods, No. 19.*

11. LAGUNARIA, G. Don.

(After Andreas Laguna.)

Bracteoles 3 or 4, broad and united at the base, often very deciduous. Calyx very shortly 5-lobed. Staminal column bearing numerous filaments on the outside below the 5-crenate summit. Ovary 5-celled, with several ovules in each cell. Style clavate at the top, with 5 distinct ovate radiating stigmas. Capsule loculicidally 5-valved, the endocarp villous inside and separating from the pericarp. Seeds reniform, thick, glabrous.—A tree. Leaves entire, sprinkled or curved, with scurfy scales. Flowers large, axillary, on short thick pedicels.

The genus, scarcely perhaps sufficiently distinct from *Hibiscus*, is limited to a single species, represented, however, by two distinct varieties—one Australian, the other peculiar to Norfolk Island.—*Benth.*

1. **L. Patersoni** (after Colonel Paterson), *Don, Gen. Syst.* i. 485, var. *bracteata; Benth. Fl. Austr.* i. 218. A tree, the young parts and inflorescence more or less covered with minute scurfy scales, but otherwise glabrous. Leaves petiolate, oblong or broadly lanceolate, rarely ovate-oblong, 3 to 4 in. long, entire, somewhat coriaceous, white underneath when young, glabrous and pale-green on both sides when full grown, the scales of the under surface almost disappearing. Pedicels very short and angular. Bracteoles 3 to 5, very obtuse, united in a broad, shortly-lobed cup, usually persistent at the time of flowering in the Australian variety, but sometimes even these falling off early. Calyx 4 to 5 lines long. Petals narrow, above 1½ in. long, slightly tomentose outside.

Hab.: Port Denison, Port Cowper, Cumberland Islands.

The Norfolk Island form (*Hibiscus Patersonius*, Andr. Bot. Rep. t. 286; *H. Patersoni*, DC. Prod. i. 454; *Lagunæa Patersonia*, Bot. Mag. t. 769; *L. squamea*, Vent. Jard. Malm. t. 42) is much more scaly-tomentose, the leaves are broader and very white underneath, and the bracteoles fall off at so very early a stage that they have always been said to be entirely wanting. I had, on that account, at first considered the Australian plant as distinct, but I have since seen the bracts on very young buds of the Norfolk Island one, and observe them to be here and there very deciduous on Australian specimens, and the other characters, although as far as hitherto known constant, may not be sufficient to distinguish the two as more than varieties or races.—*Benth.*

Wood firm, close in grain, nearly white, and easy to work.—*Bailey's Cat. Ql. Woods, No. 20.*

12. FUGOSIA, Juss.

(After Bernard Cienfuegos.)

Bracteoles 3, distinct and narrow, or several united in a 3 to 6-toothed involucre. Calyx 5-lobed. Staminal column bearing numerous filaments on the outside, below the truncate or 5-toothed summit, or rarely quite to the top. Ovary 3 to 5-celled, with 3 or more ovules in each. Style thickened towards the top, grooved or divided into short, erect lobes, with decurrent stigmas. Capsule loculicidally 3 to 5-valved. Seeds obovoid-globular or slightly reniform, usually pubescent or woolly. Cotyledons much folded over the radicle.—Shrubs or undershrubs, with the habit of *Hibiscus*, but usually more glabrous. Leaves entire or lobed, rarely divided. Stipules small or subulate and deciduous. Flowers usually large, yellow or purple. Calyx often marked with black dots, but not the cotyledons.

The genus comprises several species from tropical and subtropical regions of America, and one from Africa, but none from Asia. It is very nearly allied on the one hand to *Hibiscus*, on the other to *Gossypium*, differing from the former chiefly in the style, from the latter in the bracteoles.—*Benth.*

1. **F. australis** (Australian), *Benth. Fl. Austr.* i. 220. An undershrub of several feet, hoary with a dense but very short tomentum. Leaves broadly or narrow-ovate, obtuse, 1½ to 2½in. long, entire or more or less sinuate or 3-lobed. Flowers rather large, pink, on very short pedicels, which are often clustered 2 or 3 together at the top of axillary peduncles, with a bract or small leaf under each. Bracteoles 3, linear, distinct. Calyx from ¼ to ¾in. long, tomentose and marked with black glandular dots, the lobes lanceolate or almost linear, varying very much in length. Petals 1½in. long, slightly tomentose outside. Capsule obovoid-oblong, shortly acuminate, tomentose, 3 or 4-valved. Seeds numerous, woolly.—*Gossypium australe*, F. v. M. Fragm. i. 46, and iii. 6.

Hab.: Georgina River to Gulf of Carpentaria.

In habit and foliage this much resembles the Brazilian *F. phlomidifolia*, St. Hil., which has, however, more numerous bracteoles and yellow flowers.—*Benth.*

13. THESPESIA, Corr.

(Name derived from its being planted near place of worship in India.)

Bracteoles 1 to 5, small or deciduous. Calyx truncate, minutely 5-toothed or rarely 5-lobed. Staminal column bearing numerous filaments on the outside, below or up to the summit. Ovary 5-celled, with few ovules in each cell. Style club-shaped at the top, 5-furrowed or obscurely divided into erect stigmatic lobes. Capsule hard, almost woody, indehiscent or loculicidally 5-valved. Seeds obovoid, glabrous or woolly. Cotyledons very much folded, enclosing the radicle, often black-dotted.—Trees or tall herbs. Leaves large, entire or angularly lobed. Flowers large, usually yellow.

A small genus, limited to tropical Asia, the Pacific isles, and eastern Africa, the Australian species being one which extends over the whole range. Closely allied to *Hibiscus*, *Fugosia*, and *Gossypium*, it differs from the former chiefly in the style, from the two latter generally either in the calyx or bracts, and from all in the more woody capsule.—*Benth.*

1. **T. populnea** (Poplar-leaved), *Corr.; DC. Prod.* i. 456; *Benth. Fl. Austr.* i. 221. Indian tulip-tree. A tree, with the young parts and under side of the leaves sprinkled with minute rust-coloured scales, otherwise glabrous. Leaves broad-cordate, acuminate, entire, 4 or 5in. long. Flowers reddish-yellow, rather large, on axillary pedicels usually shorter than the petioles. Bracteoles 1 to 3, lanceolate and deciduous, or sometimes wanting. Calyx very open, 6 to 8 lines diameter, truncate, with minute teeth. Petals broad, 1½ to 2in. long. Capsule fully 1½in. diameter, hard and woody, indehiscent or opening longitudinally when very dry.—Wight, Ic. t. 8.

Hab.: Islands of the Gulf of Carpentaria, Torres Straits, and N.E. coast.

Flowering in May and June.

The species is widely spread over the seacoasts of tropical Asia, extending from eastern Africa to the Pacific Islands. It is also introduced into the West Indies.—*Benth.*

14. GOSSYPIUM, Linn.

(From the Arabic name for softness.)

(Sturtia, R. Br.)

Bracteoles 3, large and cordate. Calyx much shorter, truncate or shortly 5-lobed. Staminal column bearing numerous filaments outside, below or up to the top. Ovary 5, rarely 4-celled, with several ovules in each cell. Style club-shaped at the top, furrowed, with decurrent stigmas. Capsule loculicidally 5, rarely 4-valved. Seeds angular or nearly globular, very woolly or nearly glabrous; cotyledons very much folded, enclosing the radicle.—Tall herbs, shrubs, or almost trees. Leaves 3 to 9-lobed, or rarely entire. Flowers large, yellow or purple. Bracteoles entire, toothed or cut, usually, as well as the calyx and cotyledons, marked with black dots.

The genus, besides the Australian species, which is endemic, comprises the cultivated Cotton, whose various forms, described as species, races, or varieties, are distributed either as indigenous or introduced plants over the warmer regions both of the New and the Old World, but not hitherto found in a wild state in Australia.—*Benth.*

1. **G. Sturtii** (after Captain C. Sturt), *F. v. M. Fragm.* iii. 6; *Benth. Fl. Austr.* i. 222. A shrub of several feet, glabrous and more or less marked with black dots. Leaves on rather long petioles, broadly ovate, entire, 1 to 2in. long, rather coriaceous and glaucous. Flowers large, purple with a dark centre, on short pedicels in the upper axils. Bracteoles cordate, entire, ¾ to 1in. long, many-nerved and black-dotted. Calyx not half so long, broad, truncate with minute or narrow-linear teeth, copiously black-dotted. Petals fully 2in. long. Capsule ovoid, shortly acuminate, much longer than the calyx, usually 4-celled, glabrous but copiously black-dotted. Seeds very sparingly and shortly woolly.—*Sturtia gossypioides*, R. Br. App. Sturt. Exped. 5.

Hab.: In the interior.

15. BOMBAX, Linn.
(From the Greek *bombyx*, raw silk.)
(Salmalia, *Schott.*)

Calyx cup-shaped, truncate, or splitting into 3 to 5 lobes. Staminal column divided into numerous filaments, of which the inner ones, or nearly all, are more or less connected in pairs and united at the base into 5 or more bundles. Ovary 5-celled, with several ovules in each cell; style club-shaped or shortly 5-lobed at the top. Capsule woody or coriaceous, opening loculicidally in 5 valves, the cells densely woolly inside. Seeds obovoid or globular, enveloped in the wool of the pericarp; albumen thin; cotyledons much folded round the radicle.—Trees. Leaves digitate, with leaflets usually entire. Peduncles 1-flowered, axillary or terminal. Flowers white or red.

The species are chiefly South American, with one from tropical Africa, and another from tropical Asia extending also into Australia.—*Benth.*

1. **B. malabaricum** (of Malabar), *DC. Prod.* i. 479; *Benth. Fl. Austr.* i. 223.. Silk-cotton tree. A large tree, the trunk covered with short conical prickles. Leaves on long petioles, deciduous; leaflets 5 to 7, petiolulate, elliptical-oblong, acuminate, 4 to 6in. long, coriaceous, entire, glabrous. Flowers large, red, on short pedicels, clustered towards the ends of the branches, which are then destitute of leaves. Calyx above 1in. long, thick and coriaceous, glabrous outside, silky-hairy inside, dividing into short, broad, obtuse lobes. Petals fully 3in. long, oblong, tomentose outside, nearly glabrous within. Staminal column short, filaments much longer, but shorter than the petals, 5 innermost forked at the top, each branch bearing an anther, about 10 intermediate ones simple, and the numerous outer ones shortly united in 5 clusters. Capsule large, oblong, and woody.—*Salmalia malabarica*, Schott, Meletem. 35; *Bombax heptaphylla*, Cav.; Roxb. Pl. Corom. iii. 43, t. 247; Wight, Ill. t. 29.

Hab.: This tree is of frequent occurrence on the borders of northern river scrubs, and there are probably two varieties if not species, but the specimens hitherto received do not allow of their determination.

In India the wood is not considered durable except under water. The cotton which surrounds the seeds is used for stuffing pillows, &c. Dr. Dymock says that according to Mahometan writers the young roots have restorative, astringent, and alterative properties.

Analysis of gum: Water 17·3, arabin 20·5, metarabin 16·4, impurities 12·3, tannin 33·5.—*Lauterer.*

Wood light, coarse-grained, and soft.—*Bailey's Cat. Ql. Woods No.* 22.

Order XXIII. STERCULIACEÆ.

Flowers regular, hermaphrodite or unisexual. Calyx usually persistent, more or less deeply divided into 5 or rarely 4 or 3 valvate lobes or segments, or rarely splitting irregularly, or the sepals entirely free. Petals either 5, hypogynous, free, or adhering to the staminal column, contorted-imbricate in the bud, or small and scale-like, or none. Stamens usually united into a ring, a cup, or tube, with 5 terminal teeth or lobes (staminodia) alternating with the petals, and one or more anthers sessile or stipitate (on distinct filaments) in each interval, the anthers 2-celled and opening outwards, in longitudinal slits, or exceptionally the anthers are numerous or the staminodia wanting, or the stamens 5, free and alternate with the sepals, or the anther-cells confluent or opening in terminal pores. Ovary free, 2 to 5-celled, with the carpels more or less united, rarely 10 or 12-celled, or reduced to a single carpel. Style entire, or divided into as many branches as there are cells, or rarely styles as many, nearly or quite free. Fruit various. Seeds sometimes hairy but not woolly, sometimes enveloped in pulp or strophiolate, the testa coriaceous, occasionally enclosed in an outer membranous integument; albumen fleshy or none; cotyledons usually foliaceous, flat or folded, the radicle shorter, next the hilum or rarely distant from it.—Herbs, shrubs, or trees, the tomentum or hairs stellate, rarely mixed with simple hairs. Leaves alternate or irregularly opposite, simple and pinnately or palmately nerved, entire toothed or lobed, or digitately compound. Stipules rarely wanting.

A large Order, chiefly tropical, dispersed over the New and the Old World, with some extra-tropical genera in S. Africa or Australia, and very few species without the tropics in the northern hemisphere. Of the 15 Queensland genera 10 are common to the tropical regions of the Old World or both of the Old and the New World, the remaining 5 are endemic in Australia, with the exception of single species of *Rulingia* and *Kerandrenia*, found in Madagascar.—Benth. (in part).

TRIBE I. **Sterculieæ.**—*Flowers unisexual or polygamous. Calyx often coloured. Petals none. Anthers 5 to 15, sessile or stipitate, surrounding the ovary at the top of a column or gynaphore. Fruit-carpels separate, sessile or stipitate.—Trees. Leaves simple or digitate.*

Anthers irregularly clustered. Seeds albuminous.
  Ovules 2 or more in each cell. Carpels follicular or opening along the
    inner edge . . . . 1. STERCULIA.
  Ovules single in each cell. Carpels winged, indehiscent . . . . . 2. TARRIETIA.
Anthers 5, in a ring. Ovules solitary. Carpels large, indehiscent. Albumen
  none . . . . 3. HERITIERA.

TRIBE II. **Helictereæ.**—*Flowers hermaphrodite. Petals 5, clawed, deciduous. Anthers on short filaments, surrounding or alternating with 5 teeth of the column or staminodia. Leaves simple.*

Anther-cells divaricate. Capsule membranous, inflated . . . . . 4. KLEINHOVIA.
Anther-cells divaricate or confluent into one. Fruit-carpels distinct, or
  spirally twisted . . . . 5. HELICTERES.
Anther-cells parallel. Fruit woody, 5-valved. Seeds winged . . . . . 6. PTEROSPERMUM.

TRIBE III. **Dombeyeæ.**—*Flowers hermaphrodite. Petals flat, persistent, longer than the calyx. Anthers in the only Queensland plant of the tribe 5.*

Stamens 5 (or in *Abroma* more), united at the base in a short cup or ring, or
  rarely free, with or without intervening staminodia, and surrounding the
  sessile ovary.
Stamens 5, united in a cup, with 5 intervening elongated flat staminodia 7. MELHANIA.

TRIBE IV. **Hermannieæ.**—*Flowers hermaphrodite. Petals marcescent, flat. Stamens 5. No staminodia.*

Stamens 5, united at the base without intervening staminodia.
  Ovary 5-celled . . . . 8. MELOCHIA.
  Ovary of one 1-celled carpel . . . . 9. WALTHERIA.

TRIBE V. **Buettnerieæ.**—*Flowers hermaphrodite. Petals with a short, broad, very concave base, and a sessile or stipitate lamina.*

Lamina of the petals stipitate, longer than the calyx. Staminodia 5, obcordate, with 2 to 4 stamens between each . . . . 10. ABROMA.

XXIII. STERCULIACEÆ.

Lamina of the petals short, sessile. Stamens 5.
   Staminodia single between each 2 stamens, lanceolate . . . . 11. RULINGIA.
   Staminodia 3 between each 2 stamens, all linear-spathulate, or the central
     one lanceolate, and the lateral ones subulate . . . . 12. COMMERSONIA.

TRIBE VI. **Lasiopetaleæ.**—*Flowers hermaphrodite. Petals small and scale-like or none.*
Anthers (linear-oblong) opening outwards in parallel slits.
   Calyx herbaceous, scarcely enlarged, and not coloured after flowering.
     Staminodia large. Carpels membranous, winged . . . . 13. SERINGIA.
   Calyx enlarged after flowering, thin and coloured. Staminodia single or
     none. Capsule or carpels membranous, rounded or rarely winged . . 14. KERAUDRENIA.
   Calyx strongly ribbed after flowering. Staminodia 3 between each 2
     stamens. Capsule hard or woody . . . . 15. HANNAFORDIA.

1. STERCULIA, Linn.

(Derived from the bad scent of the flowers of some species.)

(Brachychiton, Trichosiphon, *and* Pœcilodermis, *Schott ;* Delabechia, *Lindl.*)

   Flowers unisexual or polygamous. Calyx more or less deeply 5-cleft, rarely 4-cleft, usually coloured. Petals none. Staminal column adnate to the gynophore, bearing at the summit 15 or rarely 10 stamens, irregularly clustered in a head. Carpels of the ovary 5, distinct or nearly so, with 2 or more ovules in each. Styles united under the peltate or lobate stigma. Fruit-carpels distinct, spreading, either firm or woody, and scarcely opening along the inner edge, or thinner and opening as follicles, even long before they are ripe. Seeds 1 or more in each carpel, rarely winged; albumen adhering to the cotyledons, often splitting in two, assuming the aspect of fleshy cotyledons; real cotyledons flat or nearly so, and thin, the radicle next the hilum or at the opposite end, or intermediate.—Trees. Leaves undivided or lobed, or digitately compound. Flowers in panicles or rarely racemes, mostly axillary, sometimes very short; terminal flowers usually female; in these the staminal column is shorter and the anthers less perfect than in the males, surrounding the base of the ovary; in the males the ovary is often entirely abortive.

   A large genus, almost entirely tropical, and more abundant in Asia than in Africa or America, where however several species are found. The Australian ones are all endemic.

   The species of this genus were distributed by Schott into a number of genera, founded chiefly on the flowers and habit, afterwards reduced and rearranged by R. Brown, chiefly on carpological characters, without reference to habit or calyx. The majority of the Australian ones belong to the group distinguished by R. Brown chiefly by the seeds having a loose outer coating covered with hairs, which in some species are so adhesive that the seeds fall out in their inner coating only, leaving the outer coating adhering to the equally hairy endocarp, with the appearance of the cells of a beehive; and by the radicle next to the hilum. The seeds do not appear to cohere in all the species; in some they are hitherto unknown, and in flowers and habit *S. ramiflora* and *S. rupestris,* and *S. quadrifida* are more different from each other than from species belonging respectively to other groups. Among species not Australian, the position of the radicle unites two very heteromorphous ones under *Firmiana,* and would (as observed to me by M. Poinsot, of the Paris Herbarium) lead to separate *S. mexicana* from other digitate-leaved American species. I have, therefore, with Endlicher and others, considered Schott and Brown's genera as sections only.—*Benth.*

SECT. I. **Sterculia.**—*Radicle at the end remote from the hilum. Seeds and inside of the carpels glabrous.*
Leaves large, entire. Calyx-lobes 4, cohering at the tips . . . . . . 1. *S. quadrifida.*
Leaves large, entire. Calyx-lobes expanding . . . . 2. *S. laurifolia.*

SECT. 2. **Brachychiton.**—*Radicle next the hilum. Seeds and inside of the carpels usually villous, often cohering. Leaves entire or lobed (digitate only on some branches of* S. rupestris.) *Calyx-lobes spreading.*

Calyx-lobes (where known) with induplicate margins. Seeds (where known) scarcely cohering. Leaves tomentose or pubescent, at least underneath.
   Flowers large, sessile. (*Brachychiton,* Schott).

Leaves greyish-white, entire or 3-lobed, nearly orbicular in outline. Flowers in short axillary racemes . . . . 3. *S. Garrawayæ.*
Leaves green and softly tomentose or pubescent on both sides.
Leaves broad, entire, or obscurely 5 or 7-lobed. Calyx broadly campanulate . . . . 4. *S. ramiflora.*
The leafy parts and inflorescence clothed with loose short stellate pubescence (tomentum). Leaves entire or shortly lobed, orbicular-cordate. Inflorescence at the ends of the branchlets . . . . 5. *S. vitifolia.*
Leaves 3-lobed. Calyx tubular-campanulate . . . . 6. *S. Bidwilli.*
Leaves palmately 5 or 7-lobed . . . . 8. *S. lurida.*
Leaves white underneath.
Leaves angular or obscurely 5 or 7-lobed . . . . 7. *S. discolor.*
Calyx-lobes strictly valvate. Outer coating of the seeds usually remaining adherent to the endocarp. Leaves glabrous. Flowers in short panicles.
Calyx narrow, lobes lanceolate, shorter than the tube. Leaves palmately 5 or 7-lobed *(Trichosiphon,* Schott) . . . . 9. *S. Trichosiphon.*
Calyx broadly campanulate, deeply lobed *(Pœcilodermis,* Schott).
Leaves large, palmately 5 or 7-lobed. Flowers quite glabrous . . . . 10. *S. acerifolia.*
Leaves entire, ovate or cordate, or 3-lobed, acuminate. Flowers tomentose outside when young, glabrous inside. Follicles stipitate . . . 11. *S. diversifolia.*
Leaves cordate-acuminate, entire. Flowers tomentose outside, hirsute inside at the base. Follicles nearly sessile . . . . 12. *S. caudata.*
Leaves entire and lanceolate, or digitate. Flowers tomentose outside. Follicles long-stipitate . . . . 13. *S. rupestris.*

1. **S. quadrifida** (alluding to the four division of flower), *R. Br. in Benn. Pl. Jav. Rar.* 238; *Benth. Fl. Austr.* i. 227. "Ko-ral-ba," Cooktown, *Roth;* "Convavola," N.Q., *Thozet;* "Ku-man," Atherton, *Roth.* A tree of medium size, with a somewhat smooth grayish bark, glabrous, except the inflorescence. Leaves petiolate, ovate or cordate, obtuse or acuminate, mostly 3 to 5in. long. Racemes several, crowded within the uppermost leaves, 1 to 2in. long, clothed with a stellate tomentum. Bracts broad, acuminate, very deciduous. Pedicels 2 to 4 lines. Calyx about 4 lines long, tomentose, cleft to the middle, the lobes usually 4, lanceolate, connivent and cohering at the tips. Staminal column short. Follicles sessile, ovoid, 2 to 3in. long, hard and almost woody, minutely tomentose or glabrous. Seeds 2 to 4, ovoid, black, the radicle remote from the hilum.

Hab.: Islands of Torres Straits, delta of the Burdekin and Port Denison, Wide Bay, Moreton Bay.

The northern specimens have longer and more acute leaves, and rather smaller flowers on longer pedicels than the eastern ones.—*Benth.*

Wood of a light-grey, close-grained, light and easily worked. The bark yields a useful fibre; the seeds also are edible and of agreeable flavour.—*Bailey's Cat. Ql. Woods No.* 23.
Fibre from bark used for making kangaroo nets, &c.—*Roth.*

2. **S. laurifolia** (Laurel-leaved), *F. v. M. Fragm.* vi. 172. A tree of about 60ft. Leaves simple, entire, glabrous, 3 to 6in. long, 1½ to 2in. broad, shortly acuminate. Nerves and veins prominent, dotted on the upper side, and often bearded at the axils of the primary nerves on the under side. Petioles slender, somewhat terete, about 2in. Flowers numerous in panicles. Pedicels articulate about the middle. Calyx-lobes 5, yellow, valvate in the bud, when expanded about 3 lines long, narrow-lanceolate and velvety as well as the pedicels. Staminal column 1½ line high, almost ovate, the head containing about 15 yellow anthers scarcely ¼ line long. Column almost 1½ line high. Disk annular, tomentose. Female flowers and fruit unknown.

Hab.: Rockingham Bay, *J. Dallachy.*

3. **S. Garrawayæ** (after Mrs. J. Garraway), *Bail. Ql. Journ. of Agri., June,* 1899. "Morna," Palmer River, *Roth.* A tree attaining a height of 23ft., with a rough bark. Branchlets and leaves clothed with a thin, hoary, stellate pubescence. Leaves orbicular-cordate, entire or very bluntly 3 or 5-lobed, the middle lobe the smallest when only 3-lobed, the entire ones about 1½in. diameter; others

Sterculia Garrawayæ, Bail.

Sterculia vitifolia, Bail.

3in. long, 4in. broad. Nerves 5—7, palmate, prominent, as are the transverse veins; margins entire. Petioles slender, 1 to 3in. long, strongly striate. Flowers in short, few-flowered, axillary racemes. Calyx campanulate, about 9 lines long, dull red, cleft for about a-quarter down, lobes rounded, the margins induplicate; inside the tube, above the base, is a ring of broad, divided, tomentose reflexed scales. Staminal column densely clothed with rather large stellate hairs for half its length from the base, the filaments thence free and glabrescent to the head of anthers. No abortive ovary. Female flowers in appearance like the male; stipes of ovary with a dense ring of sessile sterile anthers at the top. Ovary consisting of 5 connate carpels; styles almost straight and free to the stigmas, densely stellate tomentose throughout. Stigmas recurved. Follicles on stipes of about 8 lines besides the 2-lines original stipes of the ovary, 2½in. long, ¾in. broad in the centre, rostrate at the end, sparsely stellate tomentose outside, densely villous inside, as well as the loose integument of the seeds. Seeds yellow, closely packed, about 12 or 13 in each follicle.

Hab.: Palmer River, *Mrs. J. Garraway*, who also forwarded some excellent sketches of the present and several other Palmer River plants. In flower about March. In many respects the present species approaches *S. ramiflora*, Benth. The leaves, however, are never angular or acuminate, and the flowers are pedicellate (not nearly sessile as given in the Fl. Austr. i. 217 for *S. ramiflora*).
Fruit eaten raw, Palmer River.—*Roth*.

4. **S. ramiflora** (flowers on the branches), *Benth. Fl. Austr.* i. 227. "An-gi-ur," Princess Charlotte Bay, *Roth*. A shrub or small tree, clothed with a soft stellate tomentum or pubescence, which rarely disappears on the upper surface of the older leaves. Leaves on long petioles, broadly ovate-cordate or nearly orbicular, mostly acuminate, entire, angular or obscurely 3 or 5-lobed, often attaining 5 or 6in. Flowers few, large, red, nearly sessile, and clustered in the axils of the upper leaves. Calyx broadly campanulate, 1 to 1½in. long, the lobes shorter than the tube, spreading, obtuse, 3-nerved in the centre, with broad induplicate margins; inside the tube at the base are 5 small, inflexed, and very villous double scales. Staminal column slender, hirsute at the base. Ovary pubescent; stigmas recurved. Follicles shortly stipitate, 3 to 4in. long, glabrous outside, villous inside, stipitate (according to R. Brown), with very numerous seeds; I have not seen them perfect.—*Brachychiton paradoxum*, Schott, Meletem. 34; *Brachychiton ramiflorum*, R. Br. in Benn. Pl. Jav. Rar. 234.

Hab.: Tropical parts.
Seeds roasted and eaten.—*Roth*.

5. **S. vitifolia** (Vine-leaved), *Bail.* A small tree, the branchlets, foliage, and inflorescence densely clothed with a loose, short, stellate, light-brown pubescence; branchlets rather slender, reddish-brown beneath the pubescence, somewhat terete, internodes often long. Leaves orbicular-cordate, 3 to 6in. diameter, entire or more or less 3-lobed, the lobes short and very obtuse, very rugose on the upper, prominently and closely reticulate on the under side, both surfaces clothed with a short, close, stellate pubescence; petioles rather slender, from 2 to nearly 5in. long. Flowers few, in pedunculate cymes at the ends of the branchlets; buds cylindrical-conic, about 8 lines long, 3 lines broad; lobes induplicate; expanded flower 8 to 9 lines diameter, seems to be of a purplish-red inside, densely stellate, hairy outside; inside of tube and lobes nearly glabrous. Staminal column glabrous under the head of anthers, hairy towards the base. Follicles rostrate-cymbiform, without the stipites 2½in. long, densely villous inside. Stipites rather long. Seeds about 6, villous.

Hab.: Fairview Telegraph Station; Laura, *T. Barclay-Millar*.

6. **S. Bidwilli** (after J. C. Bidwill), *Hook. Herb.: Benth. Fl. Austr.* i. 228. A shrub or tree, softly pubescent or tomentose in all its parts, closely allied to *S. ramiflora*, but differing in the leaves almost always deeply 3-lobed with

acuminate lobes, green, and softly villous on both sides, and especially in the calyx, which is narrow, tubular-campanulate, 1 to 1½in. long ; the red colour and induplicate lobes are the same as in *S. ramiflora.—Brachychiton Bidwilli*, Hook. Bot. Mag. t. 5133.

Hab.: Here and there in southern localities.

7. **S. discolor** (two-coloured), *F. v. M.; Benth. Fl. Austr.* i. 228. A tall tree, the young shoots tomentose. Leaves very broadly cordate, nearly orbicular, shortly acuminate, angular or very shortly and irregularly 5 or 7-lobed, glabrous above, white underneath with a very close tomentum, mostly 4 to 6in. diameter. Flowers (if correctly matched) like those of *S. ramiflora*, and similarly clustered. Calyx 1½ to 2in. long, broadly campanulate, tomentose inside and out, divided to the middle into broad lobes with induplicate margins. Follicles very shortly stipitate, 4 to 6in. long, acuminate ; densely rusty-tomentose outside.—*Brachychiton discolor*, F. v. M. Fragm. i. 1.

Hab.: Pine River and other southern localities.

Wood soft, coarse in grain, and light-coloured.—*Bailey's Cat. Ql. Woods No.* 24.

8. **S. lurida** (lurid), *F. v. M.; Benth. Fl. Austr.* i. 228. A tree. Leaves on long petioles, deeply 5 or 7-lobed, the lobes sinuate or even lobed as in *S. acerifolia*, and of the same size, but softly pubescent, especially underneath. Flowers like those of *S. discolor*, of a livid variegated colour. Calyx campanulate, 1½ to 2in. long, divided to the middle into broadly ovate lobes, with the margins thin and induplicate. Follicles (according to F. v. Mueller) shortly stipitate, large, tomentose, many-seeded.—*Brachychiton luridum*, F. v. M. Fragm. i. 1, and ii. 177.

Hab.: Southern scrubs.

9. **S. Trichosiphon** (staminal column hairy), *Benth. Fl. Austr.* i. 229. Broad-leaved bottle-tree ; "Ketey," N.Q., *Thozet*. A deciduous tree, quite glabrous, leafless when in flower. Leaves 4 to 8in. long and broad, more or less deeply cut into 5 or rarely 7 palmate lobes, sometimes broad and shortly acuminate, sometimes lanceolate with long points, and glabrous on both sides. Racemes short, mostly simple. Calyx narrow, tubular-campanulate, about ¾in. long, the lobes lanceolate, spreading, much shorter than the tube. Staminal column swollen and hairy in the middle. Stigma peltate. Follicles shortly stipitate, glabrous, oblong-triangular, 2 to 3in. long.—*Trichosiphon australe*, Schott, Melet. 84 ; *Brachychiton platanoides*, R. Br. in Benn. Pl. Jav. Rar. 234.

Hab.: Northumberland Island, Burdekin, Suttor and Dawson Rivers, Wide Bay, Leichhardt district.

Wood soft, spongy.—*Bailey's Cat. Ql. Woods No.* 25.

Roots of young plants and seeds eaten by natives of N.Q.—*Thozet*.

10. **S. acerifolia** (Maple-leaved), *A. Cunn. in Loud. Hort. Brit.* 392 *(partly)*; *Benth. Fl. Austr.* i. 229. Flame-tree. A large timber-tree, quite glabrous. Leaves on long petioles, deeply 5 or 7-lobed ; lobes oblong-lanceolate or almost rhomboid, occasionally deeply sinuate, the whole leaf often 8 or 10in. diameter, thin but shining and glabrous on both sides. Flowers of a rich red, in loose axillary racemes or small panicles of 2 to 3in. Calyx broadly campanulate, ¾in. long, quite glabrous, with short broad lobes, valvate in the bud. Ovary raised on a short column, quite glabrous, the carpels quite distinct, and the styles scarcely cohering at the broad radiating stigmas. Follicles large, on long stalks, quite glabrous.—*Brachychiton acerifolium*, F. v. M. Fragm. i. 1, and ii. 177.

Hab.: Common in the south, and here and there met with in the north.

Wood soft, light-coloured, and of light weight.—*Bailey's Cat. Ql. Woods No.* 26.

11. **S. diversifolia** (leaves various), *G. Don, Gen. Syst.* i. 516; *Benth. Fl. Austr.* i. 229. Kurrajong; "Dewtie," Taromeo, *Shirley;* "Kalan," Palmer River, *Roth*. A tree of from 20 to 60ft., quite glabrous except the flowers. Leaves on long petioles, glabrous and shining, either entire and from ovate to ovate-lanceolate, or more or less deeply 3 or rarely 5-lobed, the 2 lateral lobes sometimes very short, sometimes all lanceolate, 2 or 3in. long, the simple leaves or their lobes always ending in long points. Flowers in axillary panicles, rarely exceeding the leaves. Calyx very broadly campanulate, slightly tomentose when young, attaining when fully out 7 to 9 lines diameter, acutely lobed to the middle, of a yellowish-white and glabrous except the ciliate margins outside, reddish and glabrous within. Staminal column also glabrous. Ovary slightly tomentose. Follicles nearly ovoid, 1½ to 2 or even 3in. long, thick and glabrous, on stalks of 1 to 2in., the endocarp and outer coating of the seeds very shortly hirsute and cohering.—*Pœcilodermis populnea.* Schott, Melet. 33; *Brachychiton populneum*, R. Br. in Benn. Pl. Jav. Rar. 284; F. v. M. Pl. Vict. i. 156, and Suppl. 5.

Hab.: Dawson River, Rockhampton, and in the interior.

Wood soft, of coarse grain.—*Bailey's Cat. Ql. Woods No. 27.*

Twine made from bark by Palmer River natives.—*Roth.*

Specimens from this locality show entire leaves, linear-lanceolate to 6½in. long and 13 lines broad, with the same long thread-like points, coriaceous, with petioles 13 to 14 lines long. Fruit also larger than in normal form.

12. **S. caudata** (tailed), *Heward in Herb. Cunn.; Benth. Fl. Austr.* i. 280. "Kel-lan," Princess Charlotte Bay, *Roth.* A tree, quite glabrous except the flowers. Leaves ovate-cordate, entire, long-acuminate, mostly 3 or 4in. long, the veins more transverse than in any other species, some occasionally narrow-oblong or linear. Flowers rather small, in short axillary panicles, the rhachis and pedicels quite glabrous. Calyx broadly campanulate, deeply lobed, 6 to 7 lines diameter when fully out, very tomentose outside, pubescent inside, especially at the bottom, but without appendages. Staminal column slender in the males, short in the females, pubescent at the base. Ovary very tomentose. Follicles glabrous, ovoid, rather large and thick, almost sessile.—*Brachychiton diversifolium*, R. Br. in Benn. Pl. Jav. Rar. 284.

Hab.: Princess Charlotte Bay.—*W. E. Roth.*

Fibre used for dilly bags.—*Roth.*

13. **S. rupestris** (from growing in rocky places), *Benth. Fl. Austr.* i. 230. Narrow-leaved bottle-tree; "Binkey," N.Q., *Thozet*. A considerable tree, the trunk often swelling out to a large size, contracted at the top and bottom. Leaves quite glabrous, either quite entire, oblong-linear or lanceolate, 3 to 6in. long, or digitate, consisting of 5 to 9 linear-lanceolate sessile leaflets, often above 6in. long. Panicle tomentose, usually longer than the petioles. Calyx about 4 lines long, campanulate, deeply lobed, tomentose both inside and out. Staminal column short, hirsute at the base. Follicles ovoid, acuminate, about 1in. long, on stalks longer than themselves. Seeds, when deprived of the outer coating, which remains adherent to the endocarp, smooth and shining, marked with a large scar at the chalazal end, but the radicle in those I have opened always next to the true hilum.—*Delabechea rupestris*, Lindl. in Mitch. Trop. Austr. 155; *Brachychiton Delabechii*, F. v. M. Pl. Vict. i. 157.

Hab.: Isolated summits of the Grafton range; Wide Bay; Dawson, Mackenzie, and Burnett Rivers; Rockhampton, Peak Downs.

The digitate leaves grow on luxuriant barren branches.—*Benth.*

Wood soft and spongy.—*Bailey's Cat. Ql. Woods No. 28.*

Natives of N. Queensland eat the roots of the young plants and seeds, *Palmer, Thozet*. They also refresh themselves with the mucilaginous sweet substance afforded by the tree, and also make nets of its fibre, *Thozet.*

2. TARRIETIA, Blume.

(Java name for the original species.)

(Argyrodendron, *F. v. M.*)

Flowers unisexual. Calyx 5-cleft. Petals none. Staminal column short, adnate to the gynophore, bearing at the summit 10 to 15 anthers irregularly clustered in a head. Carpels of the ovary 3 to 5, nearly distinct, 1-ovulate, rarely 2-ovulate. Styles as many, shortly filiform, stigmatic on the inner edge. Fruit-carpels or samaras distinct, spreading, indehiscent, produced at the back into a wing. Seed oblong, albumen splitting in two, cotyledons flat.—Tall trees. Leaves digitately compound, glabrous or scurfy. Flowers small and numerous, in axillary or lateral panicles.

Besides the Australian species, which are endemic, there is another from the Indian Archipelago.

Leaflets 3 or 4, silvery or coppery on the under side . . . . 1. *T. Argyrodendron.*
Leaflets 3 to 9, glabrous . . . . 2. *T. actinophylla.*

1. **T. Argyrodendron** (leaves silvery beneath), *Benth.* "Boiong." All the varieties form tall straight-stemmed trees, with broad, flat abutments at the base; glabrous except minute scurfy scales on the young shoots, inflorescence, and under side of the leaves. Leaflets 3 or 4, or sometimes on young trees 5, petiolate, usually lanceolate, 3 or 4in. long, silvery on the under side. Petioles very variable in length, sometimes only a few lines, at other times from 1 to nearly 2in. Panicles dichotomous, the upper ones sometimes exceeding the leaves. Flowers very numerous. Calyx broadly campanulate, about 3 lines diameter. Carpels with a semi-orbicular or with a straight wing about 1in. long. —*Argyrodendron trifoliolatum*, F. v. M. Fragm. i. 2, ii. 177; Fl. Austr. i. 231.

Hab.: A common tree in the scrubs of southern Queensland.

The common or silvery-leaved "stavewood." Wood light-coloured, close-grained, tough and firm; may be used as a substitute for English beech.—*Bailey's Cat. Ql. Woods No.* 29.

Var. 1, *grandiflora*, Benth. Fl. Austr. Calyx 4 lines diameter. Stigmas short and broad.— Port Denison, *Benth., l.c.*

Var. 2, *angustifolia*, Bail. Cat. Ql. Woods, No. 29b, and No. 1 Occasional Papers on the Ql. Flora, 1886. This differs in foliage, the 3 leaflets being narrow, often under ¾in. wide, and 2 or 3in. long. Petioles of normal form, silvery or approaching coppery on the under side.— Endeavour River. Wood of a light-grey colour, close-grained, hard and tough, suitable for making tool handles.—*Bailey's Cat. Ql. Woods, No.* 29a.

Var. 3, *macrophylla*, Bail. Bot. Bull. ix. 5. A large tree. Leaflets 3, rarely 2, silvery with a slightly brownish tinge on the under side, oblong, often abruptly acuminate, 5 to 10½in. long, 1½ to 3½in. broad, on nearly terete, striate, petioles, 3 to over 6in. long. Flower-panicles rather large and loose, flowers small. Carpels not seen.—Wood light-coloured, prettily figured, tough and firm.—*Bailey's Cat. Ql. Woods No.* 29 *bis.*

Var. 4, *trifoliolata* (leaves of 3-leaflets) (*T. trifoliolata*, F. v. M. Fragm. ix. 43). A tall tree. Leaves with slender petioles 2 or 3in. long, usually bearing 3 lanceolate leaflets 3 or 4in. long, often dark from the numerous small coppery-coloured scales on the under side. Fruit usually coppery like the leaves, the wing nearly 2in. long. This tree is plentiful in many scrubs north and south. Wood similar to the foregoing kinds, but rather darker; known as "stavewood," *Bailey's Cat. Ql. Woods No.* 29a. Leaflets sometimes infested with the fungus *Dimerosporium Tarrietiæ*, Cke. and Mass.

Var. 5, *peralata* (wings of seed very large), Bail. Occ. Pap. on Ql. Pl. No. 1. Johnstone River Red Beech. "Peirir," Upper Barron River, *J. F. Bailey.* A large erect tree, the stem often exceeding 5ft. in diameter. Leaves trifoliolate, the petioles somewhat angular, mostly 2 or 3in. long. Leaflets lanceolate, from 4 to 7in. long and from 1 to 2in. broad towards the middle; the indumentum more silvery than usual in this species, but with the same numerous small brown scales. Inflorescence not sent by collectors. Samaroid carpels muricate, oval or globose, ⅞ to ⅞in. long by about ½in. diameter, furnished with an oblong, oblique, erect wing 2 to 4in. long by about 1¾in. broad, clothed with the same rusty stellate scales as the under side of the leaves. Wood resembling cedar in appearance, but harder; suited for indoor and cabinet work; soon decays if exposed to bad weather.—*Bailey's Cat. Ql. Woods No.* 29c.

2. **T. actinophylla** (leaves radiating), *Bail. Syn. Ql. Flora*, p. 37. A large tree, the young growth and inflorescence more or less covered with scurfy tomentum, otherwise glabrous. Petioles 3 to 9in. long, often curved upward at the end and bearing from 3 to 9 radiating oblong-lanceolate leaflets 3 to 9in. long, including the often rather elongated petiolule. Flowers in loose, broad panicles, 6 to 15in. long. Calyx densely tomentose; deeply lobed, campanulate, expanding to about 8 lines. Carpels 1 to 2in. long, including the wing, which is from $\frac{1}{2}$ to 1in. broad.

Hab.: Southern, and particularly mountain, scrubs.

Wood very tough, thought to resemble the English ash and to bend even better than that wood, therefore should be useful for chair-making and similar work.—*Bailey's Cat. Ql. Woods No.* 30

3. HERITIERA, Ait.
(After C. L. L. Heritier.)

Flowers unisexual. Calyx 5-toothed or 5-cleft. Petals none. Staminal column slender, bearing on the outside below the summit a ring of 5 anthers with parallel cells. Carpels of the ovary 5, nearly distinct, 1-ovulate; style short, with 5 rather thick stigmas. Fruit-carpels woody, indehiscent, keeled or almost winged on the back. Seeds without albumen, cotyledons very thick, the radicle next the hilum.—Trees. Leaves undivided, coriaceous, scurfy underneath, penninerved. Flowers small, in axillary panicles.

The genus consists of two tropical Asiatic seacoast trees, of which the one extending to Australia has the widest range.—*Benth.*

1. **H. littoralis** (found on the coast), *Ait.; DC. Prod.* i. 484; *Benth. Fl. Austr.* i. 231. A tree, attaining a considerable size. Leaves very shortly petiolate, oval or oblong, the larger ones fully 8 by 4in., but often much smaller, quite entire, coriaceous, glabrous above, silvery underneath with a close scaly tomentum. Flowers small, numerous, in loose tomentose panicles in the upper axils, much shorter than the leaves. Calyx about 2 lines long. Staminal column in the males, pistil in the females, much shorter than the calyx. Fruit-carpels sessile, ovoid, 2 to 3in. long, thick and almost woody, with a slightly projecting inner edge, and a strong, projecting, almost winged keel along the outer edge.

Hab.: In Queensland, as in India, this tree is found on the coast and in tidal forests. In Bengal it is known by the name of "Sundri," is considered durable, tough, and heavy, and used extensively in boat-building, buggy-shafts, and furniture.

The wood of the Queensland tree is firm, close-grained, and dark-coloured.—*Bailey's Cat. Ql. Woods No.* 31.

4. KLEINHOVIA, Linn.
(After M. Kleinhoff.)

Bracteoles small, ensiform. Sepals 5, deciduous. Petals 5, unequal, upper with longer claws, margins involute. Staminal column dilated above into a bell-shaped 5-fid cup, divisions each with 3 extrorse 2-celled anthers, cells divergent. Ovary inserted in the staminal cup, 5-lobed, 5-celled; style slender, stigma 5-fid. Capsule membranous, inflated, pyriform, loculicidally 5-valved. Seeds 1 or 2 in each cell, tubercled; albumen scanty or none; cotyledons convolute; radicle next the hilum.—A tree with palminerved, ovate, acuminate, quite entire leaves, and loose cymose inflorescence.

1. **K. hospita** (stranger), *Linn.* A tree with straight trunk, smooth bark, and spreading head. Leaves on petioles of about 1in. long, cordate, ovate, sub-acuminate, entire, palmately 3 to 5-ribbed, smooth on both surfaces, 6 to 12in.

long by 2 to 3in. broad. Stipules ensiform. Flowers purplish or rose-coloured. Petals 5, shorter than the lanceolate sepals. Seeds tubercled.—Rump. Amb. iii. 118.

Hab.: Near Port Douglas.

5. HELICTERES, Linn.

(From the Greek, alluding to the twisted carpels.)

(Methorium, *Schott.*)

Calyx tubular, 5-cleft at the top, often oblique. Petals 5, equal or the 2 upper ones broader, the claws elongated, and all or two of them often with a lateral appendage. Staminal column adnate to the gynophore, truncate at the top, or more frequently bearing 5 teeth or small lobes (staminodia), with 1 or two stipitate anthers between each, anther-cells divaricate, often confluent into one. Ovary nearly sessile on the top of the staminal column, 5-lobed, 5-celled, with several ovules in each cell. Styles 5, subulate, more or less connate, slightly thickened and stigmatic at the top. Fruit carpels distinct or separating, opening along their inner edge, straight or spirally twisted. Seeds with little albumen, cotyledons leafy, folded round the radicle.—Trees or shrubs, with stellate or branched tomentum. Leaves entire, serrate or obscurely lobed. Flowers axillary, solitary or clustered. Bracteoles none or distant from the calyx. Capsules usually tomentose, the clusters of tomentum often forming long woolly processes. The appendages on the claws of the petals appear to vary in different flowers of the same species.

A considerable genus, dispersed over the tropical regions both of the New and the Old World, but chiefly American. Of the Queensland species one is a common Asiatic one, the other is endemic. The frequently unilocular anthers closely connect the genus with *Malvaceæ*. The other characters are, however, more of *Sterculiaceæ*, and in some species the anthers are distinctly bilocular.—*Benth.* (in part).

Leaves hairy above, serrate. Fruit 1½in. long . . . . 1. *H. spicata.*
Leaves glabrous above, entire. Fruit size of a pea . . . . 2. *H. semiglabra.*

1. H. spicata (flowers in spikes), *Colebr.* (Masters) *in Hook. Fl. Brit. Ind.* A shrubby plant. Leaves 2 to 6in. long, 1 to 2in. broad, from ovate-oblong to lanceolate, acuminate, obliquely subcordate at the base, unequally serrate, on petioles of from 3 to 9 lines, stellate, hairy above, downy beneath; stipules setaceous, as long as the petiole. Peduncles shorter than the leaves, 3-flowered; pedicels shorter than the flowers. Calyx nearly ½in., campanulate curved, distending at the base, downy. Ripe carpels 1¼ to 1¾in.; stalks exserted from the persistent calyx, oblong-lanceolate, beaked, very shaggy.

Hab.: Inland, North Queensland.

2. H. semiglabra (almost glabrous), *F. v. M. Fragm.* iv. 43. A shrub of about 3ft. high. Leaves on short petioles, narrow-lanceolate, entire, glabrous on the upper, velvety with stellate hairs on the under side, from 2 to 4in. long, apex acute, base obtuse, Peduncles axillary, fasciculate, bearing 2 or a cyme of few flowers. Pedicels scarcely 3 lines long. Calyx shortish. Petals scarcely 2 lines longer than the calyx. Carpels erect, cohering in a globose woolly-tomentose fruit the size of a pea. Seeds glabrous, 2 to 4 in each.

Hab: Rockingham Bay, *Dallachy.*

Var. *procumbens.* Branches procumbent, ½ to 2ft. long; tomentum looser; leaves smaller and rounder, velvety-villous on the upper side; staminodia longer. Macadam range, *F. v. Mueller.*

Var. (?) *flagellaris.* Branches prostrate, 1 to 2ft. long; leaves nearly sessile, cordate or orbicular, 1 to 1½in. long; cymes on long slender peduncles. Port Essington, *Armstrong.*

6. PTEROSPERMUM, Schreb.
(Winged seed.)

Bracteoles 3, entire or laciniate, sometimes very deciduous or perhaps none. Calyx tubular, 5-cleft, deciduous. Petals 5, often very long, deciduous. Staminal column adnate to the gynophore, divided at the top into 5 linear-clavate staminodia, with 3 stipitate anthers between each ; anther-cells linear, parallel. Ovary sessile in the top of the column, 5-celled with several ovules in each cell. Style undivided, club-shaped, and 5-furrowed at the top. Capsule woody or coriaceous, ovoid or oblong, terete or angular, opening loculicidally in 5 valves. Seeds ascending, produced into a wing at the top; albumen little or none; cotyledons wrinkled or folded; radicle inferior, rather long.—Trees or shrubs, clothed with a stellate tomentum or scurfy scales. Leaves coriaceous, often oblique, entire, cuneate-toothed or angled at the upper end, penninerved or several-nerved at the base. Peduncles short, axillary, 1-flowered. Flowers often several inches long.

The genus is limited to East India and the Archipelago, the Australian species being probably the same as one of the Asiatic ones.—*Benth.*

1. **P. acerifolium** (Maple-leaved), *Willd.; W. and Arn. Prod.* 69 ? *Benth. Fl. Austr.* i. 238. I have seen a fragment only in very young bud, which agrees with this species in the very angular rusty-tomentose young calyx, and in the bracteoles divided into narrow-linear lobes and falling off at a very early stage. There are 3 leaves only, the largest is, as in *P. acerifolium*, coriaceous, broad at at the end, cordate at the base, nearly glabrous above, tomentose underneath, with about 11 prominent nerves radiating from the petiole; but it is much narrower than usual in that species, measuring 9 by 4in. The two others are as yet not half developed, but are broader in proportion, and although the specimen is insufficient for identification, it shows no character to separate it from *P. acerifolium.*—Wight, Ic. t. 631.

Hab.: Given as a Queensland tree, because a tree was at one time growing in the Brisbane Botanic Gardens, which was said to have been obtained on the southern border of the colony.

7. MELHANIA, Forsk.
(After Mount Melhan, in Arabia Felix.)

Bracteoles 3, persistent. Calyx divided almost to the base into 5 segments. Petals 5, persistent. Staminal cup very short, bearing 5 ligulate staminodia and 5 stipitate anthers alternating with them, the anther-cells parallel. Ovary sessile, 5-celled with 1 or more ovules in each cell. Style usually short, with 5 subulate branches, stigmatic along the inner side. Capsule opening loculicidally in 5 valves. Seeds with albumen; cotyledons folded, 2-cleft; radicle inferior.— Herbs, undershrubs, or small shrubs, softly tomentose. Leaves ovate or cordate, serrate-crenate. Peduncles axillary, 1 or few-flowered. Bracteoles often exceeding the calyx. Flowers yellow.

The genus extends over the tropical and subtropical regions of the Old World, but is most abundant in Africa. The Australian species are the same as the Indian ones. The habit is that of some *Malvaceæ.—Benth.* (in part).

Leaves scarcely toothed. Style elongate. Capsule shorter than calyx . . . 1. *M. incana*.
Leaves crenate-dentate. Style short. Capsule exceeding the calyx . . . 2. *M. abyssinica*.

1. **M. incana** (hoary), *Heyne ; W. und Arn. Prod.* 68 ; *Benth. Fl. Austr.* i. 234. A rather slender shrub of 1 or several ft., hoary or white except the upper side of the leaves, with a close or velvety tomentum. Leaves shortly petiolate, oblong or ovate-lanceolate, obtuse, scarcely toothed, 1 to 2 or even 3in. long, tomentose on both sides, or nearly glabrous above. Peduncles bearing 1, 2, or

rarely 3 or 4 flowers, the pedicels very short. Bracteoles narrow-linear or subulate, rather shorter than the calyx. Sepals lanceolate-subulate, tomentose, about 4 to 6 lines long. Petals rather longer, broad, yellow. Staminodia linear, often 3 lines long; anthers shorter, linear, on short filaments. Style elongated. Capsule tomentose, shorter than the calyx, with 2 or 3 seeds in each cell.—*M. oblongifolia*, F. v. M. Fragm. i. 69.

Hab.: Broadsound, Rockhampton and Burdekin Rivers, Port Curtis, Port Denison.

The species is also found in the East Indian peninsula, and a slight variety or closely allied species in tropical Africa.

2. **M. abyssinica** (Abyssinian), *A. Rich, Fl. Abyss.* i. 76, t. 18. Stock woody, tortuous, divided above into a number of crowded, subcæspitose, erect branches, the latter covered with greyish down. Leaf-stalks less than 1in. long, shorter than the subcordate, oval obtuse, crenate-dentate leaves, which are unicostate, downy on both services, paler beneath. Stipules hair-like. Peduncles axillary, equal to or exceeding the petioles, simple or bifurcate. Buds oblong, cylindrical. Flowers when expanded ¾in. across. Epicalyx 1-sided, of 3 linear-subulate bracteoles as long, or nearly so, as the ovate-lanceolate downy sepals. Petals yellow, convolute. Style short. Capsule roundish or slightly pointed, villose, slightly exceeding the persistent calyx. Seeds punctate, tuberculate.— *Mast.* in Oliv. Fl. Trop. Afr. i. 231; *Brotera ovata*, Cav. Ic. v. 20, t. 433; *B. Leprieurii*, Guill. et Perr., Fl. Seneg. i. 85; *Melhania ovata*, Boiss., Fl. Orient i. 841; *M. Leprieurii*, Webb, Fl. Nigrit. 110, t. 4, 5.

Hab.: Near Westwood Station, Central Railway, Dr. T. P. Lucas.

The small fragmentary specimen sent to me by Dr. Lucas for determination was certainly belonging to the above plant, but not sufficient to point out any distinctive form, neither can it as yet be stated with certainty whether it is indigenous or only an introduction.

8. MELOCHIA, Linn.

(Name in Arabic.)

(Riedleia, *Vent.*)

Calyx 5-lobed or 5-toothed, campanulate or inflated. Petals 5, spathulate or oblong. Stamens 5, united at the base, without any or with very minute tooth-like intervening staminodia; anther-cells parallel. Ovary sessile or shortly stipitate, 5-celled with 2 ovules in each cell, styles 5, free, or united at the base, often thickened at the stigmatic top. Capsule opening loculicidally in 5 or fewer valves, some of the cells occasionally abortive. Seeds usually solitary in each cell, ascending, with more or less of albumen; embryo straight, with flat cotyledons.—Herbs, shrubs, or rarely trees, the stellate tomentum occasionally mixed with spreading hairs. Leaves serrate. Flowers small, axillary or terminal, clustered or in cymes or panicles.

A large genus, dispersed over the warmer regions of the globe, the herbaceous and suffruticose species chiefly American. The two Australian species are both herbaceous; one belongs to the American series, the other is Asiatic.—*Benth.*

Capsule very angular, pyramidal, much longer than the calyx . . . . . 1. *M. pyramidata*.
Capsule small, globular . . . . 2. *M. corchorifolia*.

1. **M. pyramidata** (shape of capsule), *Linn.; DC. Prod.* i. 490; *Benth. Fl. Austr.* i. 234. Herbaceous, with a hard almost woody base, although sometimes annual only. Branches slender, divaricate, often 2 or 3ft. long, slightly pubescent in a decurrent line or all over. Leaves petiolate, lanceolate, or the lower ones ovate, the larger ones 1 to 2in. long, serrate, usually glabrous. Flowers small, purplish, 2 to 4 together in little almost sessile axillary umbels.

Calyx 10-ribbed. Petals about 2 lines long. Capsule 3 to 4 lines long, acuminate, the very prominent angles produced into short horizontal points, giving each valve a rhomboidal, and the whole capsule a pyramidal shape.—A. Gray, Gen. Ill. t. 144.

Hab.: Rockhampton, *Wallace*.
The species is very generally distributed over tropical America, and occurs also in E. Africa, the Mauritius, and the Pacific Islands.

2. **M. corchorifolia** (Corchorus-leaved), *Linn. Spec.* 944; *Benth. Fl. Austr.* i. 235. Herbaceous, with the habit of *M. pyramidata*, but usually more erect, glabrous or with slightly pubescent decurrent lines. Leaves petiolate, from broadly ovate to lanceolate, mostly 1 to 2in. long, serrate or crenate, glabrous. Flowers small, purplish, nearly sessile in clusters, usually several together in a broad, terminal, sessile cyme, rarely a few smaller clusters in the upper axils. Calyx 5-angled. Petals about 2 lines long. Capsule small, depressed-globular, with scarcely prominent angles, spinkled with a few hairs, the valves very rarely splitting septicidally.—*Riedleia corchorifolia*, DC. Prod. i. 491; W. and Arn. Prod. i. 66.

Hab.: The far northern parts of the colony.
The species is common in E. India, and includes *M. concatenata*, Linn., and *M. supina*, Linn., with all the synonyms referred to these plants respectively by Wight and Arnott (l.c.; under *Riedleia*). Some of the Australian specimens are much starved, with small, occasionally axillary, heads of flowers, apparently approaching *M. nodiflora*, Sw., another widespread tropical species, which however not only has all the flowers in axillary clusters, but the capsule is much more deeply furrowed, and usually septicidal as well as loculicidal, the carpels often entirely separating.—*Benth*.

9. WALTHERIA, Linn.
(After A. F. Walther.)

Calyx 5-lobed. Petals 5, spathulate, persistent. Stamens 5, united at the base, without intervening staminodia; anther-cells parallel. Ovary sessile, consisting of a single 1-locular, 2-ovulate carpel, style excentrical, thickened or fringed upwards. Capsule 2-valved, 1-seeded. Seed ascending, albumen fleshy; embryo straight, cotyledons flat.—Herbs, undershrubs, or rarely trees, the stellate tomentum usually mixed with spreading hairs. Leaves serrate. Stipules narrow. Flowers usually small, axillary or terminal in clusters, heads, cymes, or panicles.

The species are mostly American, two are African, and two from the Pacific islands. The Australian species is one which is very generally dispersed over the tropical regions of both the Old World and the New.—*Benth.*

1. **W. americana** (American), *Linn.; DC. Prod.* i. 492; *Benth. Fl. Austr.* i. 236. A perennial or undershrub, 1 to 2ft. or more high, densely tomentose or softly villous in every part. Leaves shortly petiolate, from ovate to oblong, 1 to 1½in. long, obtuse, toothed and plicately veined. Flowers small, yellow, in dense heads, almost sessile in the axils of the leaves, or the upper ones clustered in a short spike, or irregularly collected into dense cymes or leafy corymbs. Bracts narrow. Calyx 1¼ to 2 lines long. Petals nearly twice as long, narrow.—*W. indica*, Linn.; DC. Prod. i. 493.

Hab.: Cape Flinders, Port Denison, and other tropical parts.
The species is common within or near the tropics all round the globe.

10. ABROMA, Jacq.
(So called on account of it not being fit for food.)

Calyx 5-cleft. Petals 5, the claw dilated and concave at the base, the lamina stipitate, ovate, plane. Staminal cup with 5 obcordate lobes (staminodia) alternating with the petals, anthers 2 to 4 in each sinus, nearly sessile, with divaricate cells. Ovary sessile, 5-celled with several ovules in each cell; styles 5, short,

L

connivent. Capsule membranous, truncate, 5-angled, the angles winged and produced at the top into as many horn-like points, opening at the top loculicidally and septicidally. Seeds several, albuminous; embryo straight, with flat cotyledons.—Tall shrubs or small trees, with stellate pubescence. Leaves entire or palmately lobed. Peduncles leaf-opposed or terminal, few-flowered. Dissepiments of the capsule fringed at the inner edge with long hairs.

A genus of two or three species from tropical Asia, one of them the same as the Australian one.

1. **A. fastuosa** (disdainful), *R. Br.; DC. Prod.* i. 485; *Benth. Fl. Austr.* i. 286. A tall shrub, the branches softly pubescent, and bearing a few minute conical prickles. Leaves shortly petiolate, obliquely cordate-ovate, acuminate, 4 to 6in. long, undivided, slightly sinuate-toothed, nearly glabrous above, softly pubescent underneath. Peduncles very much shorter than the leaves, bearing a cluster of 3 to 5 shortly pedicellate flowers, one only usually fertile. Bracts linear, deciduous. Sepals narrow-lanceolate, about ½in. long. Petals rather exceeding them, the broadly ovate lamina supported above the concave base by a filiform stipes. Capsule hirsute with a few rigid hairs, or at length glabrous, 1½in. long, the wings of the angles nearly ½in. broad, besides the long incurved points of their upper angle. Seeds 10 to 12 in each cell.—Gærtn. Fr. i. t. 64; Salisb. Parad. Lond. t. 102.

Hab.: Tropical parts of the colony.

The species is widely distributed over the Eastern Archipelago.

This plant yields excellent strong fibre.

11. RULINGIA, R. Br.

(After J. P. Ruling.)

(Achilleopsis, *Turcz.*)

Calyx 5-lobed. Petals 5, broad and concave or convolute at the base, with a small, broad, or linear ligula at the top. Stamens shortly or scarcely connate at the base, 5 without anthers (staminodia), linear-lanceolate and petal-like, alternate with the petals and connivent or spreading; 5 short, opposite the petals, and perfect, the anther-cells parallel. Ovary sessile, 5-celled with 2 or rarely 3 ovules in each cell, styles connate, at least at the top, or rarely quite free. Capsule tomentose or beset with prickles or soft setæ, opening loculicidally in valves, or the carpels separating. Seeds 1 or 2 in each cell or carpel, ascending, usually strophiolate. Albumen fleshy; cotyledons flat.—Shrubs or undershrubs, with stellate tomentum or hairs. Leaves entire, toothed, or lobed. Stipules narrow, deciduous, the upper ones often laciniate. Flowers mostly white, small, in leaf-opposed or terminal, rarely axillary cymes. Petals shorter than the calyx. Strophiola of the seeds small, variable in shape in the same species.

The genus is confined to Australia, with the exception of one Madagascar species.—*Benth.*

A. *Leaves of the flowering branches or their lobes lanceolate or ovate-lanceolate, mostly above 1 and often 2 or 3in. long, entire or serrate, not undulate, crenate or crisped. Capsule loculicidal.*

Leaves or their lobes quite entire, softly hoary-tomentose . . . . 1. *R. salvifolia.*
Leaves or their lobes serrate, velvety or hirsute, at least underneath.
Capsule scarcely dehiscent, nearly glabrous, with rigid prickly setæ . . . 2. *R. pannosa.*
Capsule dehiscent, tomentose with soft pubescent setæ . . . . 3. *R. rugosa.*

1. **R. salvifolia** (Sage-leaved), *Benth. Fl. Austr.* i. 288. An apparently erect shrub, clothed with a soft but dense and close whitish tomentum. Leaves on very short petioles, lanceolate or lanceolate-linear, 2 to 4in. long, entire or deeply divided into 3 lanceolate lobes, the middle one the longest, all quite entire and softly tomentose on both sides, especially underneath. Cymes pedunculate,

but shorter than the leaves. Calyx spreading, about 3 lines diameter. Ligula of the petals linear, usually pubescent. Stamens very shortly united. Fruit not seen.—*Thomasia (?) salvifolia*, A. Cunn. Herb.; Steetz, in Pl. Preiss. ii. 333.

Hab.: Brisbane River, *A. Cunningham; Minto's Craig, Fraser*.

2. **R. pannosa** (alluding to the clothing), *R. Br. in Bot. Mag. t.* 2191; *Benth. Fl. Austr.* i. 228. A shrub of several feet, but flowering young so as to appear an undershrub, softly hirsute with velvety stellate hairs. Leaves on the full-grown plant shortly petiolate, ovate-lanceolate or lanceolate, mostly 2 to 3in. or sometimes longer, toothed, rounded or cordate at the base, scabrous-pubescent above, with impressed veins, densely velvety or hirsute underneath; on the younger plants they are broader and often 3 or 5-lobed. Cymes shortly pedunculate. Calyx tomentose, spreading to 3 or 4 lines diameter. Ligula of the petals linear, rather short. Staminodia pubescent, united with the perfect stamens higher up than in most species. Ovary glabrous, granulate. Capsule nearly glabrous, globular, hard and almost indehiscent, beset with rigid subulate bristles, glabrous except a stellate tuft at the tip.—Steetz, in Pl. Preiss. ii. 351; F. v. M. Pl. Vict. i. 150; *Commersonia dasyphylla*, Andr. Bot. Rep. t. 603; *Buettneria dasyphylla*, J. Gay, in DC. Prod. i. 486, and in Mem. Mus. Par. x. 200, t. 12; *B. pannosa*, DC. Prod. i. 486.

Hab.: Southern parts of Queensland.

3. **R. rugosa** (rugose), *Steetz; Benth. Fl. Austr.* i. 238. A shrub so closely resembling *R. pannosa* in indumentum and foliage that it is difficult to distinguish it without the fruit. Leaves usually narrower, more rugose, and almost bullate. Flowers in cymes, scarcely exceeding 2 lines in diameter when expanded. Ligula of the petals marked with 3 dark lines. Ovary tomentose. Capsule about 4 lines diameter without the setæ, not so hard as in *R. pannosa*, and readily dehiscent, beset with soft pubescent setæ, which is long in some specimens, short in others.

Hab.: Capalaba, *J. Shirley*.

12. COMMERSONIA, Forst.
(After M. Commerson.)

Calyx 5-lobed. Petals 5, broad and concave at the base, with a small broad or linear ligula at the top. Stamens united in a short cup at the base, 5 perfect with short filaments opposite the petals, alternating with staminodia in threes, the central one of each three lanceolate or spathulate, the latter ones linear or spathulate, attached at the base either to the central one or to the adjoining anther-bearing filament. Ovary sessile, 5-celled, with 2 to 6 ovules in each cell; styles distinct or united at least at the top. Capsule beset with soft pubescent setæ, opening loculicidally in 5 valves. Seeds usually 2 or 3, ascending, with a small strophiola; albumens fleshy; cotyledons flat.—Trees or shrubs, with stellate tomentum or hairs. Leaves toothed or lobed, often oblique. Flowers small, in terminal, leaf-opposed, or axillary cymes.

The species are all Australian, one is also widely dispersed over Eastern India, the Archipelago and Pacific Islands, the others are endemic.—*Benth.*

Tall shrubs or trees. Leaves mostly above 3in. long, acuminate. Ligula
of the petals linear or oblong.
Staminodia all linear-spathulate, elongated, the lateral ones attached to
the central . . . . 1. *C. Fraseri*.
Central staminodia lanceolate, lateral ones filiform.
Lateral staminodia attached to the central one. Ligula of the petals
oblong, rather short . . . . 2. *C. Leichhardtii*.
Lateral staminodia attached to the anther-bearing filaments. Ligula
of the petals long and linear . . . . 3. *C. echinata*.

1. **C. Fraseri** (after C. Fraser), *J. Gay; in Mem. Mus. Par.* x. 215, t. 15; *Benth. Fl. Austr.* i. 242. A tall shrub, with tomentose or hirsute branches. Leaves cordate-ovate, acuminate, 3 to 6in. long, irregularly toothed, often oblique at the base, glabrous or slightly pubescent above, white-tomentose or softly hirsute underneath, the lower ones in the young plants broad and 3 or 5-lobed. Cymes loosely dichotomous, many-flowered, but shorter than the leaves. Calyx tomentose, fully 3 lines diameter, the lobes acute. Petals with a very short broad concave base, the ligula oblong-spathulate, nearly as long as the calyx. Staminodia linear-spathulate, as long as the petals, the central one of each three rather broader and lanceolate at the base, the lateral ones filiform at the base and shortly adnate to the central one; anther-bearing filaments very short. Capsule large, densely beset with soft villous setæ.—Steetz, in Pl. Priess. ii. 359; F. v. M. Pl. Vict. i. 148.

Hab.: Southern Queensland.

2. **C. Leichhardtii** (after L. Leichhardt), *Benth. Fl. Austr.* i. 242. A medium-sized shrub, the branches densely velvety-tomentose or hispid. Leaves ovate-lanceolate or cordate, 2 to 3in. long (or specimens narrow-lanceolate-oblong), unequally toothed, rather harshly velvety-tomentose on both sides. Cymes nearly sessile, few-flowered. Calyx very tomentose, spreading to about 5 lines diameter; lobes broad and acute, sometimes reddish inside. Petals glabrous, violet, with an oblong ligula much shorter than the calyx. Central staminodium of each 3 lanceolate and fine-pointed, lateral ones filiform, attached to it near the base. Anther-bearing filaments very short. Ovary glabrous.

Hab.: Boyd and Cape Rivers and Rockingham Bay.

3. **C. echinata** (capsule prickly), *Forst.; DC. Prod.* i. 486; *Benth. Fl. Austr.* i. 243. Brown kurrajong; "Dim," Maroochie. A tall shrub or small tree, the young branches and inflorescence whitish-tomentose. Leaves ovate or cordate, acuminate, 3 to 6in. long or even more, irregularly toothed or nearly entire, often oblique at the base, glabrous or slightly-tomentose above, more densely whitish-tomentose underneath. Cymes pedunculate, many-flowered, but shorter than the leaves. Calyx tomentose, nearly 3 lines diameter, the lobes acute. Petals with a very short concave broad base, the ligula narrow-linear, nearly as long as the calyx. Central staminodium of each three lanceolate, pubescent, much shorter than the petals, lateral ones small, filiform, recurved, attached to the very short anther-bearing filaments. Anther-cells less divaricate than in the other species. Capsule often ½in. diameter, without the long, soft, villous setæ which cover it.

Hab.: Cape York, Endeavour River, Pine River, Upper Brisbane River, and other localities.

Wood soft, close-grained, white, and light. The bark yields a strong fibre, which was used by the aborigines for net-making and fishing lines.—*Bailey's Cat. Ql. Woods No. 32.*

18. SERINGIA, J. Gay.
(After M. Seringa.)

Calyx deeply 5-lobed, scarcely enlarged after flowering, and neither scarious nor coloured. Petals none. Stamens 5, alternate with the calyx-lobes, alternating with 5 subulate staminodia, and slightly united with them at the base; anther-cells parallel, opening by dorsal slits. Ovary 5-celled, with 2 or 3 ovules in each cell; styles cohering at the summit or nearly from the base. Fruit-carpels distinct, winged on the back, opening in 2 valves. Seeds strophiolate, albuminous, embryo straight, with flat cotyledons.—Shrub, with the habit nearly of a *Commersonia*. Flowers in dense, terminal, or leaf-opposed cymes. Bracteoles none.

The genus is now limited to a single Australian species.—*Benth.*

1. **S. platyphylla** (broad-leaved), *J. Gay, in Mem. Mus. Par.* vii. 443, *t.* 16, 17; *Benth. Fl. Austr.* i. 244. A tall shrub, with the habit nearly of *Commersonia Fraseri*, the young branches loosely whitish or rusty-tomentose. Leaves ovate to ovate-lanceolate, acuminate, coarsely toothed, 3 to 4 or even 5in. long, often oblique at the base, glabrous or sprinkled with minute stellate hairs, densely tomentose underneath. Cymes rather dense and many-flowered, but much shorter than the leaves. Calyx angular in the bud, attaining, when fully out, about 2 lines in length. Filaments and staminodia nearly similar, rather thick. Anthers oblong. Carpels about as long as the calyx, densely pubescent, the short broad vertical wing truncate at the top.—DC. Prod. i. 488; Steetz, in Pl. Preiss. ii. 849; *Lasiopetalum arborescens*, Ait. Hort. Kew. ed. 2 ii. 86.

Hab.: South Queensland.

14. KERAUDRENIA, J. Gay.

(After — Keraudren, a French nobleman.)

Calyx 5-lobed, enlarged and scarious or thin and coloured after flowering, the midrib of each sepal usually thickened without lateral ribs. Petals none, or minute and scale-like. Stamens 5, alternate with the sepals, free or shortly united at the base, with or without intervening staminodia, anther-cells parallel, opening by dorsal slits. Ovary 3 to 5-celled, with 3 or more ovules in each cell; styles cohering at the summit. Capsule membranous, villous or shortly setose, opening loculicidally, and usually separating into distinct carpels. Seeds strophiolate, albuminous; embryo straight or curved, with flat cotyledons.—Shrubs more or less stellate-tomentose. Leaves entire or sinuate-lobed. Stipules narrow, or small and deciduous. Cymes terminal or opposite the upper leaves, few-flowered. Bracteoles none.

Besides the Australian species, there is one other from Madagascar, which on a further examination proves more nearly allied to *K. lanceolata* than had appeared to us when preparing the "Genera Plantarum." The genus has the anthers of *Seringia* and *Hannafordia*, with the calyx nearly of *Thomasia*, and must include species in which, as in the Madagascar one, the carpels do not appear to separate, as well as those in which they are quite distinct.—*Benth.*

Bracts narrow. Carpels several-seeded, not always separating, the seeds
  nearly straight. Leaves mostly lanceolate, 1 to 3in.
Leaves quite glabrous and smooth above. Capsule scarcely septicidal.
  Leaves broad-lanceolate. Carpels angular, villous and setose . . . . 1. *K. lanceolata*.
  Leaves narrow-lanceolate or linear. Carpels rounded on the back, very
    villous, but not setose . . . . 2. *K. Hillii*.
  Leaves very rugose and pubescent above . . . . 3. *K. Hookeriana*.
  Leaves ovate-lanceolate, cordate at the base, 2 to 4in. long. Seeds black 4. *K. adenolasia*.

1. **K. lanceolata** (lanceolate), *Benth. Fl. Austr.* i. 245. A tall shrub, the young branches rusty-tomentose. Leaves shortly petiolate, oblong-lanceolate, 3 to 4in. long, rather thick, entire, glabrous above and smooth, or with the veins slightly impressed, white-tomentose underneath. Cymes short, few-flowered, very tomentose. Bracts narrow, deciduous. Calyx tomentose, spreading to 4 or 5 lines diameter, divided to about the middle, the midribs prominent and pubescent inside, the lobes of the fruiting calyx attaining 3 or 4 lines or more. Petals none. Filaments rather long, with slender staminodia intervening. Anthers linear. Ovary 5-celled, hirsute. Capsule truncate at the top, fully ½in. diameter, scarcely septicidal, but distinctly furrowed between the carpels, each carpel very angular on the edges, so as to make the capsule appear almost 10-winged, but it is so hispid and beset with short, soft, hirsute setæ as almost to disguise its form. Seeds several in each cell, obovoid; embryo straight.—*Seringia lanceolata*, Steetz, in Pl. Preiss. ii. 849.

Hab.: Port Bowen, *R. Brown, A. Cunningham*; also in Leichhardt's collection.
It is this species which is closely allied to one from Madagascar, which I had formerly referred to, *Thomasia*, on account of its capsule not separating into distinct carpels.—*Benth.*

2. **K. Hillii** (after Walter Hill), *F. v. M. Herb.*: *Benth. Fl. Austr.* i. 246. Very near to *K. lanceolata*, with the same inflorescence and flowers. Leaves much narrower, linear-lanceolate or linear, 1½ to 3in. long, coriaceous, glabrous without impressed veins above, white-tomentose, and often sprinkled with rusty stellate hairs underneath. Anther-bearing filaments scarcely dilated. Ovary of *K. lanceolata*. Capsule not so large, very hirsute, but without prominent setæ, furrowed between the carpels, which are rounded on the back, and not angular. Seeds of *K. lanceolata*.

Hab.: Southern Queensland, about the Brisbane River.

3. **K. Hookeriana** (after Dr. Hooker), *Walp. Ann.* ii. 164; *Benth. Fl. Austr.* i. 246. Branches rusty-tomentose or hirsute. Leaves mostly oblong-lanceolate, 1½ to 3in. long, entire, green, very rugose and velvety-pubescent above, densely white-tomentose underneath; the lower leaves or those of some branches often broader and shorter, almost ovate. Cymes or racemes 2 to 4-flowered, terminal or opposite the upper leaves, on very short peduncles. Bracts narrow, deciduous. Calyx divided nearly to the base into acute lobes, 3 or 4 lines long when in flower, 5 or 6 when in fruit. Petals small and scale-like or none. Filaments short, alternating with subulate staminodia. Anthers linear, much incurved. Ovary 5-celled, tomentose. Capsule very hirsute, 4 to 5 lines diameter, the carpels distinct and separating, each opening in 2 valves. Seeds several in each, cell obovoid; embryo straight.—*Seringia corollata*, Steetz, in Pl. Preiss. ii. 330; *Keraudrenia integrifolia*, Hook. in Mitch. Trop. Austr. 341, not Steud.; *K. Hookeri*, F. v. M. Fragm. i. 28, 242.

Hab.: Keppel Bay; Suttor, Burnett, Upper Pine, and Brisbane rivers; on the Maranoa and southward to Lindley's Range; Robinson's Range.

The petals are certainly present in those Carpentaria specimens which I have examined, and as certainly wanting in the flowers I opened of the more southern specimens, and the two are distinguished under different names in R. Brown's herbarium and notes, but I can discover no other character whatever.—*Benth.*

4. **K. adenolasia** (hairs glandular), *F. v. M. Fragm.* x. 96. Plant glandular-hirsute. Leaves herbaceous, ovate-lanceolate, cordate at the base, the margins irregularly crenate-denticulate, 2 to 4in. long, 8 to 20 lines broad, rugose. Stipules 2 or 3 lines long, not membranous, rhomboid or ovate-lanceolate. Peduncles bearing few flowers. Bracts small, linear-lanceolate. Calyx bluish, angular-plicate before expansion, ½in. broad when expanded; lobes acuminate. Petals very minute, rhomboid-orbicular. Anthers yellow, bent like a horseshoe, 2-lobed at the base. Filaments very short, turgid downwards. Staminodia 5, linear-setaceous. Capsule stellate-hirsute and setulose. Seeds black, 1 line long, with a minute strophiole.

Hab.: Robinson River, *W. E. Armit.*

15. HANNAFORDIA, F. v. Muell.

(After James Hannaford.)

Calyx 5-lobed, somewhat enlarged after flowering, with prominent raised ribs, 3 to each sepal, besides those connecting the sepals. Petals 5, lanceolate, slightly concave, shorter than the calyx. Stamens 5, opposite the petals; staminodia 3 or fewer between each 2 stamens, linear-subulate, all slightly connected in a ring at the base; anther-cells parallel, opening by dorsal slits. Ovary 3 or 4-celled, with 3 or 4 ovules in each cell. Style simple. Capsule hard, almost woody, opening loculicidally in 3 or 4 valves. Seeds strophiolate, albuminous; embryo straight, with flat cotyledons.—Shrub, with the habit of a *Thomasia*, but without stipules. Bracteoles 3, persistent.

The genus is limited to a few species. It has the anthers of *Keraudrenia* and *Seringia*, with the calyx nearly of *Guichenotia.—Benth.*

1. **H. Shanesii** (after P. A. O'Shanesy), *F. v. M. Fragm.* vi. 175. A shrub of about 2ft., with velvety-tomentose branches. Leaves from 1½ to 2½in. long, 5 to 10 lines broad, oblong, base cordate, repand-denticulate or almost entire, on petioles of 2 to 4 lines; glabrescent on the upper, velvety on the under side. Peduncles bearing 2 or few flowers. Pedicels about 2 lines long. Bracteoles 3, 1 line long. Calyx campanulate, 1in. long, striate, lobes 5, lanceolate, stellate-tomentose outside, glabrous and somewhat scarlet within. Petals dark purple, narrow-lanceolate, glabrous. Stamens 5 fertile, opposite the petals. Staminodia 3, subulate, 2 lines long. Anthers erect, extrorse, oblong. Ovary velvety, 4 or 5-celled. Capsule globose, about 5 or 7 lines, valves glabrous inside. Seeds 2 to 4 in each cell, black.—F. v. M. l.c. and x. 96.

Hab.: Leichhardt district, *O'Shanesy* (F. v. M. l.c.)

ORDER XXIV. **TILIACEÆ.**

Flowers regular, hermaphrodite or rarely unisexual. Sepals 5, rarely 3 or 4, free or more or less cohering, usually valvate. Petals as many or fewer or none, alternate with the sepals, inserted round the base of the torus. Stamens indefinite, rarely reduced to very few, inserted on the torus, which is often raised or disk-like. Filaments free or slightly united at the base. Anthers 2-celled, with parallel or rarely divaricate cells, opening in longitudinal slits or in terminal pores. Ovary free, sessile, 2 or more celled. Style simple and entire, or divided at the top into as many stigmatic teeth or lobes as there are cells. Ovules 1, 2, or more in each cell, erect, pendulous, or horizontal. Fruit capsular or indehiscent, with single or several-seeded cells, where several-seeded the cells often subdivided by spurious vertical or transverse partitions. Seeds without any arillus, the testa usually coriaceous or crustaceous. Albumen fleshy, rarely deficient. Embryo straight or rarely curved or slightly folded. Cotyledons leafy or rarely fleshy; the radicle next to the hilum, usually shorter than the cotyledons. —Trees, shrubs, or rarely herbs. Leaves alternate or very rarely opposite, simple, with pinnate or palmate nerves, entire, toothed, or rarely lobed. Stipules usually free and small or deciduous. Flowers axillary, terminal or leaf-opposed, usually in little cymes, often almost umbellate, either solitary and sessile or pedunculate, or arranged in panicles.

A large Order, chiefly tropical or subtropical, spread over both the New and the Old World, with one extratropical genus (*Tilia*) in the northern and another (*Aristotelia*) in the southern hemisphere. The Australian genera are none of them endemic, the extratropical *Aristotelia* is common to Chili and New Zealand. The others are all tropical and Asiatic, *Grewia* extending into Africa and *Corchorus* also partially into America, whilst *Triumfetta* belongs equally to the New and the Old World.—*Benth.*

SERIES A. HOLOPETALA.—Petals glabrous or rarely downy, coloured, thin, unguiculate, entire or nearly so, imbricate or twisted in the bud. Anthers globose or oblong, opening by slits.

TRIBE I. **Brownlowieæ.**—*Sepals combined below the cup. Anthers globose, cells ultimately confluent at the top.*
Anthers short, with confluent cells. Calyx irregularly 3 to 5-lobed. Petals entire. Capsule loculicidal, each valve 2-winged . . . . 1. BERRYA.

TRIBE II. **Grewieæ.**—*Sepals distinct. Petals glandular at the base. Stamens springing from the apex of a raised torus.*
Anthers short, with 2 parallel distinct cells opening longitudinally. Sepals distinct. Petals entire.
Drupe indehiscent, not echinate, entire or 2-lobed. Petals narrow, short, with a foveolate base. Trees or shrubs . . . . 2. GREWIA.
Fruit globular, echinate, indehiscent, or separating into 1-seeded cocci. Petals narrow, with a foveolate or pubescent base. Shrubs or herbs . . 3. TRIUMFETTA.

152   XXIV. TILIACEÆ.

TRIBE III. **Tilieæ.**—*Sepals distinct. Petals not glandular. Stamens springing from a contracted torus.*

Capsule 2 to 5-celled, with several seeds in each, opening in valves, usually long and smooth, rarely short and echinate. Petals usually obovate or broad, without a foveola. Shrubs or herbs . . . . 4. CORCHORUS.

SERIES B. HETEROPETALÆ.—Petals sepaloid, incised or none, induplicate or imbricate, not twisted. Anthers linear, opening by a terminal pore.

TRIBE IV. **Sloanieæ.**—*Anthers linear, dehiscent at the apex. Disk stamineiferous, plane or pulvinate. Sepals and petals inserted immediately around the stamens.*

Sepals 4, imbricate in 2 series. Capsule echinate, 4-valved . . . . . . 5. SLOANEA.

TRIBE V. **Elæocarpeæ.**—*Anthers linear, dehiscent at the apex. Petals around the base of a raised torus, from the inside of which the stamens arise.*

Sepals 4 or 5, valvate. Fruit a berry . . . . 6. ARISTOTELIA.
Sepals 4 or 5, valvate. Fruit a drupe . . . . 7. ELÆOCARPUS.

1. BERRYA, Roxb.
(After Dr. A. Berry.)

Calyx campanulate, irregularly 3 to 5-lobed. Petals 5, without any foveola at the base. Stamens numerous, free, without staminodia; anthers subglobose, the cells at length confluent into one. Torus not raised. Ovary (2? or) 3-celled, with 4 ovules in each cell; style subulate (2? or) 3-lobed (or the styles distinct?). Capsule nearly globular, opening loculicidally in 2 or 3 valves, each valve bearing 2 vertical, diverging, coriaceous wings. Seeds 1 or 2 in each cell, densely covered with rigid hairs; albumen fleshy; cotyledons leafy, flat.—Trees. Leaves entire, 5 or 7-nerved. Flowers small, white, the umbel-like cymes arranged in a terminal panicle.

The genus consists of a single species, common to tropical Australia and Asia.

1. **B. Ammonilla** (its name in Ceylon), *Roxb. Pl. Corom.* iii. 60, t. 264, var. *rotundifolia* (round-leaved); *Benth. Fl. Austr.* i. 268. A small tree, the young branches slightly tomentose. Leaves cordate-orbicular, very obtuse, 3 or 4in. diameter, rigidly membranous, glabrous when full-grown. Flowers of the Australian variety unknown, except from some fragments remaining about the fruits seen by R. Brown, in which he ascertained that the calyx was lobed and the stamens numerous. Capsule (always?) 2-celled, the wings broadly obovate, about ½in. long, sinuate-crenate on the margin. Seeds 1 or 2 in each cell.

Hab.: Cape York and Torres Straits Islands.

The shape of the fruit and its wings and the seeds are the same as in the Asiatic *B. Ammonilla*, Roxb., DC. Prod. i. 517, Wight, Ill. t. 34; but as that species has acuminate leaves and a 3-celled capsule, I had at first thought that this one might be distinct. I find, however, some Ceylon specimens with the same rounded leaves, and the Australian specimens are not sufficient to show whether the reduced number of carpels is more than accidental.—*Benth.*

2. GREWIA, Linn.
(After Dr. W. Grew.)

Sepals 5, distinct. Petals 5, with a foveola or thickened cavity at the base, usually shorter than the calyx, inserted round the base of the torus. Stamens indefinite, inserted on the raised torus. Ovary 2 to 4-celled, with 2 or more ovules in each cell; style subulate, minutely toothed or lobed. Drupe containing 1 to 4 pyrenes or nuts, entire or 2 or 4-lobed, the nuts either 1-seeded or 2 or more seeded, and then divided by transverse partitions between the seeds. Seeds ascending or horizontal, the albumen usually copious, the cotyledons flat.—Trees or shrubs, the hairs or tomentum stellate. Leaves entire or serrate, 3 to

7-nerved. Stipules narrow, deciduous. Flowers usually yellow, the umbel-like cymes axillary or terminal. In the Australian species (except *G. breviflora*) the ovary is 2-celled, but each cell is subdivided by a vertical, nearly complete partition, so as to appear 4-celled, with two or rarely more superposed ovules in each half-cell, each half-cell forming in the fruit a separate nut, with 1 or rarely more superposed seeds in each.

The genus is a large one, widely spread over the tropical and subtropical regions of the Old World.

Leaves glabrous or nearly so, 3-nerved at the base. Flowers hermaphrodite.
Sepals 7 to 9 lines. Petals small, the foveola very large. Torus elongated. Fruit depressed-globose, not lobed, ½in. diameter or more . . . 1. *G. orientalis.*
Sepals about 4 lines. Petals very small, the foveola large. Torus short, fruit small, 2-lobed (unless reduced to 1 carpel) . . . . 2. *G. multiflora.*
Leaves softly velvety-tomentose underneath, 3 or 5-nerved. Flowers hermaphrodite. Petals small, foveola large . . . . 3. *G. latifolia.*
Leaves white-tomentose underneath or scabrous, 3 or 5-nerved. Flowers polygamo-dioecious.
Leaves obovate-oblong to lanceolate. Foveolate base of the petals broader than the lamina . . . . 4. *G. polygama.*
Leaves small, ovate-obtuse. Stamens in the female flowers 1 or 2 apparently perfect, without staminodia. Buds not striate . . . . 5. *G. scabrella.*
Leaves large cordate-ovate. Buds of male flowers globose, in the female oblong. Corolla purple . . . . 6. *G. pleiostigma.*

1. **G. orientalis** (Oriental), *Linn.; W. and Arn. Prod.* 76; *Benth. Fl. Austr.* i. 270. A tall, rather weak shrub, glabrous, except a minute tomentum on the young shoots, or sparingly sprinkled on the under side of the leaves and more abundant on the inflorescence. Leaves shortly petiolate, from oval-oblong to oblong-lanceolate, acuminate, 3 to 4in. long, minutely crenulate, 3-nerved at the base. Peduncles 1 or 2-flowered, axillary or the upper ones forming a short terminal panicle. Sepals rusty-tomentose, 7 to 9 lines long. Petals not half so long, the foveolate base broader than and almost as long as the lamina, pubescent round the edge. Torus elongated. Stamens very numerous. Drupe depressed-globular, ½ to ¾in. diameter, flat-topped, slightly furrowed but not lobed, minutely tomentose with a few short straight hairs intermixed, containing usually 4 nuts, each with 2 or 3 horizontal, superposed seeds, separated by transverse partitions.

Hab.: Islands of the Gulf of Gulf of Carpentaria, N.E. coast, Northumberland Islands.

The species is not uncommon in Ceylon and a part of the Indian Peninsular.

Var. *latifolia.* Leaves ovate-cordate, crenate, fruit more densely pubescent. Port Denison, *Fitzalan.*

2. **G. multiflora** (numerous flowers), *Juss. in Ann. Mus. Par.* iv. 89, t. 47, f. 1; *Benth. Fl. Austr.* i. 270. A shrub or tree, with rather slender branches, glabrous or sprinkled with a few appressed simple or stellate hairs. Leaves from ovate-acuminate to elliptical-oblong or almost lanceolate, 3 or 4in. long or sometimes more, serrate, 3-nerved at the base. Peduncles axillary, usually 2 or 3 together, 2 to 5-flowered. Sepals lanceolate, about 4 lines long, minutely tomentose. Petals very short, the broad foveolate base villous round the edge, not longer than the short torus, the lamina still smaller. Stamens numerous. Ovary hirsute, with 2 superposed ovules in each half-cell. Drupe small, sprinkled with a few rigid hairs, deeply 2-lobed or entire by the abortion of one carpel, with 2 nuts in each carpel, each containing a single seed.—*DC. Prod.* i. 508.

Hab.: Percy Islands, *A. Cunningham.*

The species was originally described from Philippine Island specimens; our Australian ones agree well with Jussieu's figure, as well as with Cuming's specimens, n. 461, 701, and 1515. The common East Indian *G. sepiaria*, Roxb., as well as *G. prunifolia*, A. Gray, Bot. Amer. Explor. Exp. i. 77, said to be a common shrub on the leeward coast of the Fiji Islands, appear from our specimens to be the same species, which we have also from Java and Singapore, although not included in Miquel's Flora. It is, however, frequently confounded with *G. lævigata*, Vahl., which differs in longer flowers, a more raised torus, and several other points.—*Benth.*

3. **G. latifolia** (broad-leaved), *F. r. M. Herb.; Benth. Fl. Austr.* i. 271. A shrub or tree, the branches stellate-tomentose. Leaves petiolate, broadly cordate, ovate, 3 or 4in. long, irregularly serrate, scabrous-pubescent above and wrinkled, softy tomentose or hirsute underneath. Peduncles 2 or 3 together, 2 to 5-flowered, of unequal length, but scarcely exceeding the petioles. Sepals softly villous, 4 to 5 lines long, acute. Petals about one-third as long, the broad foveolate base as long as the small lamina. Torus considerably elevated. Stamens numerous. Ovary hirsute, 2-celled, with 2 superposed ovules in each half-cell. Fruit depressed-globular, 5 or 6 lines diameter, hirsute when young, at length shining and nearly glabrous, 2-lobed, each lobe containing 2 1-seeded nuts and slightly furrowed between them.—*G. Richardiana*, Hook. in Mitch. Trop. Austr. 388 ; not Walp.

Hab.: Islands off the N. coast, Bustard Bay. Brisbane River, Moreton Island, Peak Downs, and St. George's Bridge, on the Balonne.

The foliage is nearly that of *G. asiatica*, Linn., with the fruit of *G. polygama*, Roxb., and the flowers different from both. In some flowers I have seen the style divided some way below the dilated fringed stigmas.—*Benth.*

4. **G. polygama** (male, female, and hermaphrodite flowers on the same plant), *Roxb. Fl. Ind.* ii. 588 ; *Benth. Fl. Austr.* i. 271. " Kooline," Cloncurry, *Palmer;* "Karoom," Rockhampton, *Thozet;* "Ouraie," Cleveland Bay, *Thozet* and *Morrill;* "Pam-mo," Butcher's Hill, *Roth;* "Kou-nung," Middle Morehead River, *Roth.* An erect shrub, the branches tomentose or softly hirsute. Leaves almost sessile, from obovate-oblong to oblong-elliptical or almost lanceolate, 2 to 3in. long, serrate, wrinkled and softly pubescent or scarcely scabrous above, velvety-tomentose underneath. Flowers diœcious, 3 or 4 together on very short peduncles. Sepals about 4 lines long, silky-tomentose outside. Petals about one-third as long, the oblong lamina twice as long as the broad foveolate base. Male flowers : Stamens about 20, on the very hirsute torus, with a very rudimentary pistil or none at all. Female flowers: Stamens very short, with small anthers. Ovary very hirsute, with 2 superposed ovules in each half-cell. Style short, with broad, spreading, fringed, stigmatic lobes. Drupe depressed-globular, 5 or 6 lines diameter, hirsute when young, at length smooth and shining, 2-lobed, each lobe containing 2 1-seeded nuts and slightly furrowed between them.

Hab.: Islands of the Gulf of Carpentaria, Swears Island, Cape York and Port Molle, Bay of Inlets, Keppel Bay, Percy Islands, Rockhampton, Port Denison.

The species spreads over a great part of East India.

Fruit eaten by natives.

5. **G. scabrella** (rough), *Benth. Fl. Austr.* i. 272. A shrub with the habit of *G. orbifolia*, but the tomentum rather more sparing. Leaves broadly ovate, but not so rounded as in that species nor quite so rigid, 1 to 1½in. long. Flowers in small sessile clusters, apparently diœcious, the males not seen. Female flowers : Sepals softly tomentose, 2 to 2½ lines long, the buds not striate, as in *G. orbifolia*. Petals nearly as long as the sepals, glabrous, with a small foveola at the base, less distinct than in most species. Stamens 1 or sometimes 2 or 3, apparently perfect, without staminodia. Ovary oblong, villous, with 2 superposed ovules in each half-cell. Style very short, with broad, fringed, spreading, stigmatic lobes.

Hab.: Mackenzie and Dawson Rivers, *F. v. Mueller.*

6. **G. pleiostigma** (stigmas numerous), *F. r. M. Fragm.* viii. 4. An erect diœcious tree attaining the height of 50ft., the trunk seldom exceeding a diameter of 10in. Branches spreading horizontally ; bark smooth, very fibrous ; branchlets sparsely tomentose. Leaves cordate-ovate, often measuring 9in. in length and 5in. in breadth ; deeply and somewhat obliquely cordate at the base, and bordered

with small sharp teeth; quintuplinerved, penninerved, and transverse reticulations; under side thinly tomentose, upper with sparsely scattered stellate hairs. Petioles from 1 to 1½in. long. Stipules very soon deciduous, deltoid or lanceolate-ovate, about 2 lines long, entire. Flowers in axillary cymes, common peduncle 1 to 2in., each cyme often 6in. diameter. Flowers in twos or threes at the ends of the branchlets of the cyme, shortly pedicellate and surrounded by a more or less complete whorl of short, thick bracteoles. Buds before opening globose, or oblong. Sepals 5, mealy on the outside, hoary, frosted inside, about 3 lines long, lanceolate, valvate. Corolla of 5 petals, which are imbricate at the top and scarcely exceeding 2 lines long, rich-purple, glabrous except for a ring of hairs. The female flowers oblong in the bud, ovary globose, 4 or 5-celled, hirsute with white hairs and surrounded with staminodia. Stigma 3, once or twice forked. Male flowers on separate trees; buds globose. Stamens numerous, free, surrounding a small, silky, abortive ovary, crowned by a 3 or 4-branched style.

Hab.: Mulgrave River.

The above description is from these specimens, collected in 1889, and referred to in Report on the Botany of the Bellenden Ker Expedition, with the following remarks upon the wood:— "The wood of this tree, on account of its elasticity and toughness, may in a few years be in demand for the manufacture of oars, shafts, and for other purposes where strength and elasticity are required, for which other species of this genus are found valuable in India." Baron Mueller named and first described the species from specimens collected at Rockingham Bay by J. Dallachy.

3. TRIUMFETTA, Linn.
(After G. B. Triumfetti.)

Sepals 5, distinct, usually concave, or with a dorsal point or appendage at the top. Petals 5, thickened and globular, or foveolate at the base, inserted round the base of the torus, rarely wanting. Stamens indefinite, or rarely reduced to 5 or 10, free, inserted on the raised torus; anther-cells opening longitudinally. Ovary 2 to 5-celled, with 2 collateral ovules in each cell; style filiform, stigma minutely 2 to 5-toothed. Fruit globular or nearly so, echinate or bristly, indehiscent or (in species not Australian) separating into cocci. Seeds in each coccus or cell solitary, or, if 2, separated by vertical dissepiments, pendulous, albuminous; embryo straight; cotyledons flat, leafy.—Herbs, undershrubs, or shrubs, with the hairs or tomentum stellate. Leaves serrate, entire, or 3 or 5-lobed. Flowers yellow, in little pedunculate or almost sessile cymes or clusters, either leaf-opposed or lateral, rarely strictly axillary. Petals usually narrow and not exceeding the calyx, especially in the Old World species.

A considerable genus, widely spread over the tropical regions of both the New and the Old World. Of the Australian species, one, a maritime plant, extends to several of the South Pacific islands, the others are all endemic.

Ovary 3 to 5-celled. Fruit 3 to 8-celled, with 1 seed in each cell.
  Leaves round-cordate, entire or lobed. Fruit rather large, with 2 cells and seeds to each carpel.
    Stems prostrate. Leaves mostly lobed. Sepals 4 to 5 lines with minute pointed appendages . . . . 1. *T. procumbens*.
    Shrubby. Leaves roundish, hairy on both sides, with crisp margins. Sepals narrow, with a minute terminal appendage at the back . . 2. *T. Winneckeana*.
  Shrub, sometimes tall. Leaves polymorphous, often rhomboidal. Sepals oblong-apiculate . . . . 3. *T. rhomboidea*.
  Undershrub of a few feet, lower leaves lobed, upper ones ovate-lanceolate. Fruit 5-valved . . . . 4. *T. pilosa*.
  Fruit beset with long bristles, very dark coloured 4-celled . . . . 5. *T. nigricans*.

1. T. procumbens (procumbent), *Forst.; DC. Prod.* i. 508; *Benth. Fl. Austr.* i. 278. Stems procumbent or prostrate and rooting at the joints, often attaining several feet, the branches shortly ascending, tomentose. Leaves petiolate, broadly ovate-cordate or orbicular, obtuse, 1 to 2in. long, entire, crenate, or more or less deeply divided into 3 or 5 lobes, nearly glabrous above, more tomentose underneath. Peduncles short, few-flowered. Sepals 4 or 5 lines long, with small pointed appendages. Ovary hirsute and papillose, 3 or 4-celled, each cell again divided into 2. Fruit globular, about ½in. diameter, glabrous or villous, covered with hard conical prickles; endocarp hard, divided into 6 or 8 one-seeded cells.—Guillem. in Ann. Sc. Nat. Par. ser. 2, vii. 365; Hook and Arn. Bot. Beech. 60.

Hab.: Maritime sands, Cape York, Northumberland Islands, Fitzroy Island, Frankland Islands, Howick Islands.

The species is found in several islands of the Eastern Archipelago, and the Pacific, where the leaves are usually entire or not very deeply 3-lobed; Cunningham's specimens agree very well with these, in all the others (generally far advanced) the leaves are deeply 3 or 5-lobed, with glabrous fruits.—*Benth.*

2. **T. Winneckeana** (after M. Winnecke), *F. r. M. in Append. to Mr. Winnecke's Explo. Diary*, 1883. Leaves roundish or verging into an oval form, denticulated and somewhat crisp at the margins, velvet-hairy on both sides. Sepals narrow, dorsally terminated by a minute conical appendage. Petals downy towards the base. Stamens numerous. Ovary 3-celled. Fruit large, on a slender stalklet, almost globular, indehiscent, thinly tomentose, copiously beset by long, spreading, bristle-like, hooked prickles, the latter nearly glabrous at the summit.

Hab.: Inland southern border.

Allied to *T. leptacantha*, but the fruits are much larger and not glabrous.—*F. i. Mueller, l.c.*

3. **T. rhomboidea** (rhomboid), *Jacq., DC. Prod.* i. 507. Chinese Burr. A small shrub, glabrous or pubescent. Leaves polymorphous, ovate-rhomboid or cordate, 3 to 7-nerved, the apex acute or somewhat 3-lobed, serrate, variable in amount and quality of pubescence. Flowers about ¼in. diameter, yellow, in small dense cymes. Pedicels short. Flower-buds oblong, club-shaped, apiculate. Sepals oblong, apiculate. Petals oblong, ciliate at the base. Stamens 8 to 15. Capsule globose, the size of a small pea, albido-tomentose between the spines; spines hooked, glabrous or ciliated, 3 to 5-valved.—Masters in Fl. of Brit. Ind. and Trop. Afr.

Hab.: Cairns, Townsville, Cooktown, &c., where it has probably been introduced by the Chinese.

This pest, in one or other of its many forms, is found in India, Ceylon (ascending to 4000ft. in the Himalaya), Malay Islands, China, Tropical Africa, and West Indies.

4. **T. pilosa** (hairy), *Roth.; F. r. M. Fragm.* iv. 28. Plant herbaceous, bristly, bristles bulbous at the base. Upper leaves 3 to 4in. long, 2½in. broad, ovate or ovate-lanceolate, lower ones 3-lobed, stellate-hairy on both sides; petioles hairy, about ¾in. long. Stipules subulate-aristate, shorter than the petiole. Peduncles shorter than the petioles. Flowers yellow, ⅜in. diameter. Sepals linear, apiculate. Petals oblong-spathulate, scarcely shorter than the sepals, ciliate at the base. Stamens about 10 or more. Fruit globose, tomentose, covered with long hooked spines which are glabrous along the upper, hispid along the lower edge, 5-celled, cells 2-seeded.

Hab.: Mount Elliott, *E. Fitzalan.*

5. **T. nigricans** (blackish), *Bail.* Plant erect, 2 to 3ft. in height, clothed in most parts with short stellate hairs all round the stem and branches, but frequently more dense on one side than on the other; very dense on the back of the leaves. Branches nearly terete, stipules rather persistent, narrow, 4 lines long. Leaves ovate-lanceolate, palmately 3 to 5-nerved, 2¼ to 3½in. long, 1¼ to 2¼in. broad, coarsely serrate. Petioles slender, 1 to 1½in. long. Flowers yellow, solitary, or few in a pedunculate umbel. Bracts filiform, 2 lines long. Pedicels

about as long as the bracts. Buds narrow-oblong, crowned by the spreading sepal appendages. Sepals linear, 4 lines long, without the thread-like appendages. Petals spathulate, shorter than the sepals. Stamens 15 to 20, filaments glabrous. Style sulcate, glabrous. Ovary setose, 4-celled, 2 ovules in each cell. Fruit about 4 lines diameter, dark-coloured, 4-celled, 2 seeds in each cell; outside covered with slightly hairy hooked setæ about 4 lines long, the sharp hook at the ends often of lighter colour, glabrous between the setæ except for a few stellate hairs. Seeds oval, rough.

Hab.: Herberton and Tully River, *J. F. Bailey*. This plant is likely to become a pest, and should be included in "Noxious weeds to be destroyed." The fruit very dark or blackish on the specimens received.

4. CORCHORUS, Linn.
(From its supposed medicinal properties.)

Sepals 5, rarely 4. Petals as many, without any cavity at the base. Stamens indefinite, rarely few, inserted on a torus scarcely raised, but occasionally expanded in a disk round their base; anther-cells opening longitudinally. Ovary 2 to 5-celled, with several ovules in each cell; style short, simple. Capsule either long without prickles, or short or globular and more or less warted, muricate or echinate, opening loculicidally in 2 to 5 valves, with several seeds in each cell, rarely separated by transverse partitions. Seeds pendulous or horizontal, albuminous; embryo usually curved, with leafy cotyledons.—Herbs, undershrubs, or shrubs, with simple or stellate hairs. Leaves serrate. Peduncles very short, lateral or leaf-opposed, bearing 1 or several flowers. Bracts small. Flowers usually small, yellow.

A considerable genus, of which a few species appear to be limited to tropical America or to Australia, the remainder generally dispersed over various tropical regions in the Old as well as the New World. The fruit in this genus is often indispensable for determining the species.—*Benth.*

Annuals (or biennials), glabrous or loosely pubescent.
  Capsule globular or ovoid, very obtuse. Capsule slightly warted, 2 or 3 celled . . . . 1. *C. hygrophilus.*
  Capsule (½ to ¾in. long) rather thick, angular or winged.
    Capsule acute or acuminate, angular but not winged. Stamens numerous . . . . 2. *C. Cunninghamii.*
    Capsule 3-winged, truncate at the top, with 3 diverging points. Stamens under 20. Flowers very small . . . . 3. *C. acutangulus.*
  Capsule linear, not winged.
    Capsule under ½in., 2 or 3-celled. Leaves without setæ. Flowers very small. Stamens few. Pubescent plants. Capsule 2-celled, reflexed, very hirsute, rather acute . . . . 4. *C. pumilio.*
Undershrubs or shrubs more or less tomentose or hirsute.
  Fruiting pedicels recurved. Capsule linear, curved or twisted, more or less torulose, 2 or 3-celled.
    Low diffuse shrubs or undershrubs. Capsule few-seeded.
      Sepals under 2 lines. Stamens about 10. Capsule 3 or 4 lines long, very hispid, slightly curved . . . . 4. *C. pumilio.*
      Sepals 3 to 4 lines. Stamens numerous. Capsule tomentose, slender but not twisted . . . . 5. *C. tomentellus.*
    Erect or decumbent shrubs.
      Tomentum scabrous or almost villous. Sepals 2 or 3 lines. Petals narrow . . . . 6. *C. sidoides.*
      Stems purplish, decumbent. Pods 3 or 4in. long, erect. Valves 3 or 4, scabrous, ending in a short straight point . . . . 7. *C. trilocularis.*

1. **C. hygrophilus** (found near water), *A. Cunn. Herb.; Benth. Fl. Austr.* i. 276. A tall, erect, glabrous herb, apparently annual. Leaves petiolate, ovate or ovate-lanceolate, acuminate, 3 to 5in. long, acutely and irregularly toothed. Cymes several-flowered, reflexed, shortly pedunculate, but rarely equalling the

petioles. Flowers small, the buds obovoid, contracted at the base. Petals the length of the calyx. Stamens numerous, on a raised torus. Capsule globular or ovoid-oblong, very obtuse, 2 to 4 lines long, more or less tuberculate, 2 or 3-celled. Seeds 8 or more in 2 rows in each cell, without transverse partitions.

Hab.: Cleveland Bay, *A. Cunningham.*

2. **C. Cunninghamii** (after A. Cunningham), *F. v. M. Fragm.* iii. 8; *Benth. Fl. Austr.* i. 276. A tall, erect, glabrous herb, annual, or sometimes perhaps perennial. Leaves petiolate, from cordate-ovate to lanceolate, acuminate, 2 to 4in. long, coarsely serrate, without setæ. Peduncles short, bearing a cyme of 3 to 7 or 8 flowers, on rather long pedicels. Buds obovoid, narrowed at the base. Stamens numerous, on a raised torus. Ovary narrowed at the the top. Capsule narrow-oblong, acute, ½ to ¾in. long, slightly 3 or 4-angled, 3 or 4-celled, with numerous seeds in each cell.

Hab.: Dawson and Burnett rivers and Moreton Bay, *F. v. Mueller ;* Brisbane River, *Fraser*

3. **C. acutangulus** (alluding to angles of fruit), *Lam.; W. and Arn. Prod.* 73 ; *Benth. Fl. Austr.* i. 277. An annual, sometimes very small, but attaining 2ft., decumbent or erect, slightly pubescent and often sprinkled with a few rigid hairs. Leaves petiolate, ovate, serrulate, without setæ. Flowers 1 to 3, nearly sessile, and very small. Sepals little more than 1 line long. Stamens 15 to 20. Capsule straight, ½ to ¾in. long, rather thick, prominently 3-angled, or with 3 longitudinal wings, truncate at the top, with 3 spreading points or teeth, 3-celled. Seeds numerous. Very rarely the capsule has 4 cells and as many wings and teeth.—Wight, Ic. t. 789.

Hab.: Cape York. The species is common in tropical Asia and Africa, and occurs also—perhaps introduced—in some parts of S. America.—*Benth.*

4. **C. pumilio** (small), *R. Br. Herb.; Benth. Fl. Austr.* i. 277. A small, rigid, much-branched herb or undershrub, not much more than ¼ft. high, hirsute with spreading stellate hairs, the slender branches appearing almost woody at the base, although the plant flowers the first year. Leaves petiolate, ovate or oblong, obtuse, rarely above ½in. long, crenate, rugose and plicate, sprinkled with rigid stellate hairs. Flowers very small, in sessile clusters. Buds narrow-oblong. Sepals very narrow, acute, hirsute, 1 to 1½ line long. Petals narrow. Stamens about 10. Ovary very hirsute. Capsules reflexed, linear, 3 to 4 lines long, slightly curved, rather acute, very hirsute, 2-celled, with few oblong seeds.

Hab.: Islands of the Gulf of Carpentaria, *R. Brown.*

5. **C. tomentellus** (tomentose), *F. v. M. Fragm.* iii. 10; *Benth. Fl. Austr.* i. 278. A low, diffuse, stellate-tomentose shrub or undershrub. Leaves petiolate, from ovate to ovate-oblong, obtuse, ½ to 1in. long, crenate, slightly plicate and rugose, rather loosely stellate-tomentose, especially underneath. Flowers pedicellate, in nearly sessile clusters, much larger than in *C. vermicularis*. Buds obovoid. Sepals 3 to 4 lines long. Stamens numerous, the torus expanded into a prominent disk round their base. Capsule very slender, tomentose, ½ to ¾in. long, 3-valved, with few distant seeds, but scarcely contracted between them.

Hab.: Mackenzie River, *F. Mueller.* It is possible that this may prove a form of the very variable *C. sidoides,* but besides the difference in habit and foliage, the flowers appear to be larger and the disk much more developed.—*Benth.*

6. **C. sidoides** (Sida-like), *F. v. M. Fragm.* iii. 9; *Benth. Fl. Austr.* i. 278. An erect shrub of several feet, the branches densely but rather loosely tomentose. Leaves shortly petiolate, from oval-oblong to oblong-lanceolate, obtuse, 1 to 2in. long, rather thick, crenate, plicate and rugose, or on luxuriant specimens longer

and thinner, scabrous tomentose above, more densely tomentose underneath. Flowers in nearly sessile clusters. Calyx tomentose-villous, 2 to 3 lines long, the buds often tipped by the tooth-like points of the sepals. Petals narrow, in some flowers very small. Stamens numerous, on a small torus. Capsule slender, ¾ to near 2in. long, tomentose or villous, more or less torulose, 2 or 3-celled. Seeds oblong, often distant in each cell, although rather numerous on the whole.

Hab.: Islands of the Gulf of Carpentaria.

7. **C. trilocularis** (capsule 3-celled), *Linn.: Masters in Fl. Trop. Africa*, i. 262. Annual or perhaps perennial, with numerous erect or decumbent purplish, smooth or pilose, branching stems. Leaves elliptic, oblong or oblong-lanceolate, 1 to 3in. long, ½ to 1in. wide, crenate-serrate, either with or without basal lobes. Petioles very short, pilose. Stipules setaceous. Pedicels 2 to 3-flowered. Petals spathulate, bright yellow. Pods 2 to 3in. long, erect, straight or curved, slender, 3 to 4-angled, 3 to 4-valved ; valves scabrous, deeply pitted on the inner surface, and ending in a short straight point. Seeds numerous.

Hab.: Rockhampton, Burdekin River, and Rookingham Bay, *F. v. M. in Fragm.* viii. 5.

t

5. **SLOANEA,** Linn.

(After Sir Hans Sloane, principal founder of the British Museum.)

(Including *Echinocarpus australis*, Benth.)

Sepals or calyx-lobes 4 or 5, valvate or imbricate in 2 rows. Petals none or 1—4, imbricate or subimbricate, entire or dentate. Stamens numerous, free, covering the broad, thick, pitted disk from the petals to the ovary ; anthers linear, the cells placed back to back, and opening from the top in a slit extending more or less down the sides. Ovary 3 or 4-celled, with several ovules in each cell ; style subulate. Capsule thickly coriaceous or woody, densely echinate or covered with setæ, 3 or 4-celled or 1-celled by abortion, opening in 3 or 4 valves. Seeds several, or solitary and pendulous, ovoid ; testa hard ; albumen fleshy ; cotyledons broad, flat.—Trees. Leaves entire or sinuate-toothed, with pinnate veins. Peduncles axillary, 1-flowered, solitary or clustered, or forming terminal panicles.

The Australian species are endemic. The authors of the "Genera Plantarum" seemed to consider that the distinction between the genera *Sloanea* and *Echinocarpus* was scarcely enough to keep them apart, and Baron Mueller having described 3 out of the 4 Australian species under *Sloanea*, I have thought it better to follow him in this instance.

Leaves obovate-oblong, 6in. or more long, coriaceous, sinuate-toothed.
Petals glabrous. Capsule 3—4-valved . . . . 1. *S. australis.*
Leaves lanceolate-ovate, 3 to 5in. long, chartaceous, nearly or quite entire.
Petals velvety. Capsule 2-valved . . . . 2. *S. Langii.*
Leaves obovate, 2 to 6in. long, thin, coriaceous, crenulate.
No petals. Capsule pyriform-ovate, 2-valved . . . . 3. *S. Macbrydei.*
Leaves lanceolate to ovate-lanceolate, 3 to 5in. long, coriaceous, remotely toothed.
No petals. Capsule 2-valved, about ¾in. long . . . . 4. *S. Woollsii.*

1. **S. australis** (Australian), *F. v. M. Fragm.* iv. 91 ; (Echinocarpus, *Benth. Fl. Austr.* i. 279). A large tree, bark thin, scaly ; branchlets furfuraceous, adult foliage glabrous. Leaves obovate-oblong, 6 to 12in. long, shortly acuminate, more or less sinuate-toothed, much narrowed towards the base, but obtuse or slightly cordate at the petiole, coriaceous. Inflorescence furfuraceous-pubescent, forming a terminal raceme to the branchlets shorter than the last leaves. Pedicels 1 to 2in. long. Sepals ovate-oblong, 4 or 5, about 4 lines long. Petals glabrous, white, 4 or 5, oval, about 7 lines long, 3 or 4 lines broad in the middle.

Filaments short. Anthers scarcely pointed. Style acute, much longer than the stamens, the exserted portion deeply furrowed. Ovary 4-celled. Capsule opening in 4 hard almost woody valves, or sometimes 8-valved, about ½in. long, external setæ short and exceedingly densely crowded.

Hab.: Maroochie, Rockingham Bay, *F. v. M.;* Herberton scrubs, *J. F. Bailey.*
Wood pinkish, close-grained, light, suitable for lining boards.—*Bailey's Cat. Ql. Woods No.* 32a.

2. **S. Langii** (after Dr. G. Lang), *F. v. M. Fragm.* v. 28. A small tree, with a white, smooth bark; branchlets almost glabrous. Leaves 3 to 5in. long, lanceolate-ovate, chartaceous, glabrous, quite entire or indistinctly remotely denticulate, penninerved and copiously reticulate. Stipules long, persistent, canaliculate, almost silky, 3 lines long, denticulate. Flowers light-yellow, solitary or in corymbs. Pedicels 1 to 2in. long, very slightly velvety. Sepals valvate, white, lanceolate or orbicular-ovate, 4 to 6 lines long, the outside slightly the inside distinctly velvety. Petals white, ovate, thinner than the sepals, and a little imbricate and thinner on both sides. Stamens numerous, very short; anthers oblong-linear, about 2 lines long; style subulate, 1½ line long. Ovary 8-celled. Capsule mostly 2-valved, hispid.

Hab.: Mount Elliott, *E. Fitzalan* (F. v. M.) : Herberton scrubs and Tully River, *J. F. Bailey.*

3. **S. Macbrydei** (after J. Macbryde), *F. v. M. Fragm.* vi. 170. A large tree; wood yellow; branchlets at length glabrous. Leaves obovate, 2 to 6in. long, 1 to 2in. broad, chartaceous or thinly coriaceous, crenulate or repandly-denticulate, glabrous, base cuneate, lateral nerves distant with transverse and reticulate veins between. Flowers fragrant, in corymbose racemes, on elongated pedicels. Sepals 4, white, fleshy-coriaceous, valvate, ovate, about 4 lines long. Petals none. Anthers 60 to 70, tetragonous-linear, puberulous, 1½ line long, connectivum acute, style short. Ovary ovate, 2-celled. Capsule nearly 1in. long, slightly compressed, pyriform-ovate, 2-celled, woody, awned outside with subulate setæ 1 to 1½ line long.

Hab.: Rockingham Bay, *J. Dallachy* (F. v. M.)

4. **S. Woollsii** (after Rev. Dr. W. Woolls), *F. v. M. Fragm.* vi. 171. A tall tree with a deeply sulcate bark; branchlets slightly velvety. Leaves lanceolate or ovate-lanceolate, 2 to 5in. long, 1½in. broad, dark green and bordered by somewhat remote blunt teeth, on petioles about 1in. long. Flowers in racemose corymbs; pedicels 1 to 1½in. long. Sepals 4 or 5, rhomboid-orbicular, slightly imbricate, hoary-velvety on both sides. Petals none. Stamens numerous; filaments very short or scarcely any. Anthers almost linear. Capsule about ¾in. long, 2-valved, woody, covered with somewhat soft prickles on the outside, glabrous inside. Seeds pendulous, mostly solitary, ovate or ellipsoid, turgid.

Hab.: Mount Mistake and Bunya Mountains.
Wood of a light colour, close-grained and tough, useful for flooring and lining boards. When newly cut has somewhat the scent of celery.—*Bailey's Cat. Ql. Woods No.* 83.

6. ARISTOTELIA, L'Hér.
(After the philosopher.)
(Friesia, *DC.*)

Sepals 4 or 5, valvate. Petals as many, imbricate, 3-lobed, toothed or entire, inserted round the base of the thickened torus. Stamens indefinite, inserted on the torus, within a glandular ring; anthers linear, the cells placed back to back and opening from the top in short confluent slits. Ovary 2 to 4-celled, with 2 ovules in each cell; style subulate. Fruit a berry. Seeds few, ascending or pendulous; testa hard, often pulpy outside; albumen fleshy; embryo straight, with flat or undulate cotyledons.—Shrubs. Leaves mostly opposite or nearly so, entire or toothed. Flowers axillary or lateral, in racemes, or in the Australian species solitary or 2 or 3 together, often polygamous.

Besides the 2 Queensland species, which are endemic, the genus has 1 in Tasmania, 2 from New Zealand, and 1 from Chili.

Leaves ovate, acuminate, pubescent underneath. Berry globose . . . . 1. *A. australasica.*
Leaves oblong-lanceolate, pale underneath. Fruit ovate-acuminate, red, 6 to 9 lines long . . . . 2. *A. megalosperma.*

1. **A. australasica** (Australasian), *F. v. M. Fragm.* viii. 2. A slender shrub of several feet, with a few soft hairs on the young branches, petioles, and principal veins on the under side of the leaves, otherwise glabrous. Leaves opposite, on slender petioles, ovate-lanceolate, acuminate, pale on the under side, 2 to 3in. long, serrate, 3-nerved at the base. Peduncles about 2in. long, usually axillary, bearing 1 to 3 flowers. Pedicels about 9 lines long, with 2 narrow bracteoles, about ¾ line long at the base, which are soon deciduous. Sepals 5, ovate-oblong, about 2 lines long, membranous, pilose, with woolly-ciliate margins. Petals about 3 lines long, glabrous, tender, obovate-cuneate, the apex shortly 3-lobed. Stamens 12 to 16; filaments short. Anthers pointless, narrow-oblong. Ovary somewhat glabrous. Style scarcely 1 line long, filiform. Stigma very minute. Berry globular, about 4 lines diameter, nearly dry.

Hab.: Southern parts of the colony; North Coast line of railway.

2. **A. megalosperma** (long-seeded), *F. v. M. Fragm.* ix. 81. A small tree, the branchlets at first with appressed hairs. Leaves oblong-lanceolate, 2 to 4in. long, long-acuminate, pale on the under side. Peduncles very short or wanting; pedicels from 6 to 12 lines long, solitary or in pairs. Sepals lanceolate-linear, 3 lines long, puberulous inside. Petals 5, almost oblong, imbricate, shortly crenulate at the top, silky at the base. Stamens 12; anthers 1 line long, pointless, oblong-linear, hispidulous at the apex. Style undivided, 1½ line long, subulate. Stigma very small. Ovary glabrous. Fruit ovate, acuminate, red, ¼ to ¾in. long, 1 to 2-celled, 1 seed in each cell. Pericarp coriaceous; seeds about 3 lines long; testa brown, very thin.

Hab.: Rockingham Bay, *J. Dallachy* (F. v. M.)

7. ELÆOCARPUS, Linn.

(Supposed resemblance of the fruit of some species to the Olive.)

(Monocera, *Jack.*)

Sepals 4 or 5, usually valvate. Petals as many, fringed, lobed, or rarely entire, inserted round the base of the torus, induplicate-valvate, and embracing some of the outer stamens in the bud. Stamens indefinite, inserted on the torus, within a glandular ring; anthers oblong or linear, opening at the top in 2 valves (that is, the cells placed back to back and opening in short, terminal, confluent slits). Ovary 2 to 5-celled, with 2 or more ovules in each cell; style subulate. Fruit a drupe, with a hard often bony putamen, 2 to 5-celled or 1-celled by abortion. Seeds solitary in each cell, pendulous (or rarely erect?); testa hard; albumen fleshy; cotyledons broad, flat or undulate.—Trees. Leaves alternate or rarely opposite, entire or serrate. Flowers in axillary racemes, sometimes polygamous.

A large tropical Asiatic genus, extending to the Pacific Islands, New Caledonia, and New Zealand. The Australian species are all endemic.—*Benth.*

Young growth rusty-pubescent, the leaves becoming glabrous, ovate, obovate, 2 to 4in. long, 1 to 2in. broad.
Sepals 4. Petals 4, 7 lines long, top 3-lobed, silky outside. Drupe velvety, nearly globular, 1½in. long, putamen smooth . . . . 1. *E. Bancroftii.*

Leaves oblong, 1½ to 3in. long, 1½ to 1¾in. broad. Drupe bright-blue,
 ovoid, 6 to 7 lines long, 4 or 5 lines diameter, putamen tuberculate . . 2. *E. arnhemicus.*
Leaves oval-elliptical, 2 to 4in. long. Petals divided into about 7 linear
 obtuse lobes. Drupe globular-ovoid, blue, putamen rugose or tuberculate 3. *E. obovatus.*
Leaves oblong-lanceolate, 3 to 4in. long. Petals divided into 10 to 12 acute
 lobes, some united in pairs. Drupe globular-ovoid, blue, putamen
 rugose . . . . 4. *E. cyaneus.*
Leaves oblong, lanceolate, often with glaucous patches, 4 to 8in. long, 1 to
 1½in. broad. Drupe ovoid, about ½in. long, putamen rugose . . . 5. *E. Kirtonii.*
Leaves coriaceous, oblong-lanceolate, 3 to 5in. long, 1 to 1½in. broad.
 Drupe oval, ¾in. long, brown, with sometimes a bluish tinge . . . 6. *E. eumundi.*
Leaves oblong-lanceolate, 5 to 6in. long. Petals divided into about 5 deeply-
 fringed lobes, silky on the margins near the base. Drupe globular,
 blue, often more than 1in. diameter, tuberculate . . . . . 7. *E. grandis.*
Leaves coriaceous, 2 to 3in. long, 10 to 15 lines broad, lanceolate. Petals
 nearly entire, silky outside. Drupe elliptical, about 8 lines diameter 8. *E. foveolatus.*
Leaves 2 to 4in. long, 1 to 1½in. broad. Petals silvery-silky outside. Drupes
 blue, ½in. diameter . . . . 9. *E. ruminatus.*
Leaves chartaceous, broad-lanceolate, 4 to 6in. long, 1 to 1½in. broad.
 Petals finely fringed . . . . 10. *E. Grahami.*
Leaves oval-lanceolate, 2½ to 3½in. long, thin-coriaceous. Petals minutely
 denticulate . . . . 11. *E. sericopetalus.*

1. **E. Bancroftii** (after Dr. T. L. Bancroft), *F. r. M. et Bail. in Proc. Roy. Soc. Ql.* vol. ii. 142. Ebony-heart, of Cairns. A large handsome tree, often attaining more than 100ft. in height, with a diameter of more than 2ft. of stem. Bark scaly, of a brownish colour, about ½in. in thickness. resembling the American *Lignum-vitæ*, and might serve for the same purposes. The young growth and inflorescence clothed with a short rusty pubescence, the older leaves glabrous. Leaves clustered at the ends of the branches, ovate, obovate, or lanceolate, 2 to 4in. long, 1 to 2in. broad; the smaller ones at times very obtuse, tapering into a petiole of from 1 to nearly 2in. long; the midrib and few distant primary veins prominent, the small reticulations also often distinct on both sides. Flowers in lateral or axillary umbel-like racemes, bearing at the summit 3 to 5 rather large flowers, with often 1 or 2 lower down the stalk; pedicels about 5 lines long. Sepals 4, valvate, rigidly coriaceous, oblong or lanceolate, about 5 lines long and 2 lines broad, densely clothed on the inside by rather long silky hairs, the outside rusty. Petals 4, broadly cuneate, about 7 lines long and 6 lines broad at the top, which is wavy and lobed with 3 short obtuse lobes; the outside of the petals slightly silky. Stamens numerous, over 25; filaments flexuose, inserted on an annular lobed disk; anthers linear, about 2 lines long, the terminal 2-valved opening prominent, giving a lobed appearance to the apex. Ovary 4-celled; ovules generally 4 in each cell, hairy. Style subulate, hairy at the base. Stigma small. Drupe velvety, ovoid or nearly globular, about 1¼in. diameter; in drying, when fully ripe the thin epicarp readily separating from the mealy rather thick sarcocarp, this also separating freely from the hard smooth-pitted putamen. Putamen thick, but the 4 cells prominently marked, 3 of which are abortive; thus the fruit contains but a solitary seed, which resembles the kernel of a peach-stone, has an agreeable flavour, and is eaten by the settlers.

Hab.: Johnstone River and other tropical hillside scrubs.

Wood hard and durable, resembling the American *Lignum-vitæa*, and might serve for the same purposes.—*Bailey's Cat. Ql. Woods No.* 33c.

2. **E. arnhemicus** (from Arnheim's Land), *F. r. M. Rep. Intercol. Exhib.* 1867; *E. obovatus,* var. (?) *foveolatus, Benth. Fl. Austr.* i. 281. A small tree, height about 80ft., diameter of trunk about 8in., with a whitish-grey smooth bark; wood white, with a closely interlocked grain. The bark of the smaller branches or branchlets dark-brown and closely dotted with lenticels. Leaves oblong or broadly and obtusely ovate, 1½ to 3in. long, 1½ to 1¾in. broad, obscurely

crenate; the primary veins with glandular pits in their axils. Racemes solitary or in pairs, about 1in. long (no flowers sent with Mr. Jacobson's specimens). Drupe bright-blue, ovoid, 6 or 7 lines long, 4 or 5 lines diameter; sarcocarp of an agreeable acid flavour, putamen very prominently tuberculate; 1-seeded.

Hab.: Near Musgrave Electric Telegraph Station, Cape York Peninsular, *Geo. Jacobson*.
Wood of a light color with close interlocked grain and prettily marked.—*Bailey's Cat. Ql. Woods No.* 33d.

3. **E. obovatus** (obovate leaves), *G. Don; Benth. Fl. Austr.* i. 281. "Woolah," Moreton Bay, *Watkins*. A tree attaining 60ft., glabrous in all its parts. Leaves from oval-elliptical to obovate-oblong or almost lanceolate, obtuse or obtusely acuminate, 2 to 4in. long, irregularly sinuate-crenate, narrowed at the base, thinly coriaceous, the smaller veins much less numerous and less conspicuous than in *E. cyaneus*. Racemes solitary or clustered, many-flowered, but shorter than the leaves. Flowers small, white. Sepals acute, 1½ line long. Petals rather longer, divided to about the middle into about 7 linear obtuse lobes. Anthers short, obtuse or scarcely pointed. Ovary glabrous, 2-celled, with 4 ovules in each cell. Drupe globular or ovoid, often blue, the putamen rugose or tuberculate —F. v. M. Fragm. ii. 80; *E. parviflorus*, A. Rich. Sert. Astrol. 67, t. 24; *E. pauciflorus*, Walp. Rep. i. 364 (a mistake in the name and a wrong station).

Hab.: Many parts of South Queensland.
Wood light-coloured, close-grained, firm, and easy to work.—*Bailey's Cat. Ql. Woods No.* 34.

4. **E. cyaneus** (blue), *Ait. Epit. Hort. Kew; Benth. Fl. Austr.* i. 281. A tree, usually small, glabrous in all its parts. Leaves elliptical-oblong or oblong-lanceolate, acuminate, 3 or 4in. long or more when luxuriant, more or less serrate, acute at the base, coriaceous and very conspicuously reticulate. Racemes loose, shorter than the leaves. Sepals acute, 3 to 4 lines long, glabrous. Petals as long or rather longer, divided into 10 to 12 acute lobes, here and there united in pairs. Stamens numerous, within the undulate glandular disk. Anthers linear, the upper valve with a short point. Ovary glabrous, 2-celled, with 8 to 10 ovules in each cell. Drupe usually 1-seeded, globular or ovoid, blue outside, the putamen 4 to 6 lines long, hard and rugose.—DC. Prod. i. 519; Bot. Mag. t. 1737; F. v. M. Pl. Vict. i. 152; *E. reticulatus*, Sm. in Rees' Cycl. xii.; Bot. Reg. t. 657.

Hab.: All parts of South Queensland. The flowers usually white but in some inland localities rose-coloured.
Wood close-grained and light-coloured.—*Bailey's Cat. Ql. Woods No.* 35.

5. **E. Kirtonii** (after — Kirton), *F. v. M. (inedit.)* White beech of Bunya Mountains. A tall tree, often over 100ft. high, producing a fine timber; glabrous except the young shoots, which are more or less silky and often covered here and there with glaucous patches. Leaves oblong, lanceolate, acuminate, 4 to 8in. long, 1 to 1½in. broad, more or less cuneate at the base to a petiole of under 1in., sharply serrate and prominently reticulate. Inflorescence not seen. Drupe ovoid, about ½in. long, usually 2-seeded; putamen rugose.

Hab.: Forests of the Bunya Mountains. The trees are also abundant on Mount Mistake. Thus the tree seems only to be met with in high mountain scrubs.
Wood light-brown, fine-grained, suitable for furniture; thought to somewhat resemble English sycamore.—*Bailey's Cat. Ql. Woods No.* 33a.

6. **E. eumundi** (found at Eumundi), *Bail. Proc. Roy. Soc. of Ql., April*, 1894. A tree of considerable size and erect growth. Leaves more coriaceous than most other Australian species, mostly oblong-lanceolate, 3 to 5in. long and 1 to 1½in. broad near the middle, on somewhat slender petioles of 1½ to 2in. in length; the margins entire or with distant rather prominent blunt teeth in the

upper part; apex often elongated, but blunt. The young growth, petioles, and midrib more or less clothed with appressed, short gray hairs, which are also sometimes found sparsely scattered over the lamina on the under surface. Inflorescence lateral on the two-year-old wood. Racemes seldom exceeding 2in. in length, pedicels about ⅛in. Flowers not seen. Drupe (not quite ripe) oval, ⅜in. long; pericarp juicy, sharply acid; putamen deeply pitted, containing 1 or 2 seeds. The fruit structure reminds of the Indian species *E. oblonga*.

Hab.: Eumundi, *Field Naturalists*, March, 1894.

7. **E. grandis** (one of the largest trees in scrubs, hence the name), *F. v. M. Fragm.* ii. 81; *Benth. Fl. Austr.* i. 281. Quandong; "Moorqun," Upper Barron, *J. F. Bailey*: "Caloon," Nerang, *Schneider*. A tall tree, glabrous except the young shoots, slightly silky-hairy. Leaves on short petioles, oblong or lanceolate, obtuse or scarcely acuminate, 4 to 6in. long, crenulate, narrowed at the base, scarcely coriaceous, the smaller veins not prominent. Flowers large, in short dense racemes. Sepals fully ½in., including their long subulate points. Petals longer, divided into about 5 deeply fringed lobes, silky-pubescent on the margin towards the base. Stamens very numerous; anthers linear, the upper valve pointed and ending in 1 or 2 short fine setæ. Ovary silky-tomentose, 5-celled, with about 4 ovules in each cell. Drupe blue, globular, 1in. diameter; putamen hard and rugose.

Hab.: A scrub tree of North and South Queensland.

The large flowers, pubescent petals, and pointed anthers, refer this species to the section *Monocera*, usually considered as a distinct genus, but the group is neither natural nor accurately defined.—*Benth.*

Wood of a light colour, grain close, considered a useful timber where toughness is required.—*Bailey's Cat. Ql. Woods No.* 86.

8. **E. foveolatus** (foveolate) *F. v. M. Fragm.* v. 157, vi. 172, viii. 2. Branchlets silky-tomentose or glabrescent with age. Leaves coriaceous, 2 to 3in. long, 10 to 15 lines broad, almost entire or repand-crenulate, lanceolate, point obtuse, rusty-tomentose on the under side and foveolate in the axils of the principal nerves; reticulation copious, the upper side glabrous. Petioles 6 to 18 lines long. Racemes about 2in. long. Pedicels 5 to 8 lines long, rather stout, curved at the end, with a clothing of rusty hairs as well as the peduncles. Flower buds almost globose. Sepals fugaceous, 2 to 2½ lines long, silky outside. Petals valvate, almost entire, silky outside. Stamens 30 to 40. Anthers a line or a little more long. Style short, about a line long. Ovary ovate, 3-celled. Fruit elliptical-ovate, about 8 lines diameter, on a peduncle of about 6 lines.

Hab.: Mountain Ranges, Rockingham Bay, *J. Dallachy* (F. v. M.)

9. **E. ruminatus** (ruminate), *F. v. M. Fragm.* viii. 1, x. 4. A tree of about 60ft., the branchlets glabrescent. Leaves 2 to 4in. long, 1 to 1½in. broad, ovate-lanceolate, acuminate, repand-serrulate, under side pale, reticulation close, often foveolate at the axils of the principal nerves. Racemes 2 to 5in. long, hoary when young. Flower buds pyramidal-ovate. Petals silvery-silky outside. Stamens 20 to 25, anthers mucronulate, ovary silky. Fruit blue, about 6 lines diameter. Pericarp slightly spongy and acid. Putamen woody, usually 1-seeded.

Hab.: Dense scrubs, Rockingham Bay, *J. Dallachy* (F. v. M.)

10. **E. Grahami** (after Dr. George Graham), *F. v. M. Fragm.* x. 3. A tall tree, the branchlets silky tomentose. Leaves chartaceous, broad-lanceolate, 4 to 6in. long, 1 to 1½in. broad, point slender, crenate-serrulate, the principal nerves on the under side silky-pubescent, the reticulation a little prominent. Racemes 5in. long, numerous, slightly pilose. Pedicels 4 to 8 lines long, capillary. Bracts and bracteoles subulate, about 1 line long. Flower buds

slender-conical. Sepals 5, linear-lanceolate, almost 3 lines long. Petals narrow-oblong, fringe very fine. Stamens 15 to 17, puberulous. Anthers barbellate-mucronulate, about 1 line long. Style setaceous, 3 to 4 lines long, glabrous. Disk annular, glabrous. Ovary 2-celled.

Hab.: Daintree River, *E. Fitzalan* (F. v. M.)

11. **E. sericopetalus** (silky petals), *F. v. M. Fragm.* vi. 171. A tree 40ft. or more high. Leaves oval-lanceolate, 2½ to 3½in. long, thin-coriaceous, finely crenulate cuneate and entire at the base, usually shortly and obtusely acuminate at the point, glabrous on each face, nerves slender and patent, not foveolate, the upper reticulation somewhat prominent. Racemes with the short peduncle 1½ to 2½in. long, thinly hoary-silky. Flower buds globose, nodding. Pedicels 3 to 4 lines long. Sepals rigidly valvate, lanceolate, 2 lines long, inserted with the petals below the disk. Petals white, oblong, membranous, somewhat acute, imbricate above, almost valvate below, quite entire except the minutely denticulate apex, outside hoary-silky. Stamens 40 to 50, very shortly pubescent. Anthers pointless, linear-tetragonal, about ⅔ line long. Style about ⅔ line long, subulate. Disk flattened, annulate, very slightly crenulate.

Hab.: Mountains about Rockingham Bay, *J. Dallachy* (F. v. M.)

ORDER XXV. **LINEÆ.**

Flowers regular, hermaphrodite. Sepals 5, rarely 4, free or united at the base, imbricate or rarely almost valvate. Petals as many, hypogynous or rarely slightly perigynous, imbricate, usually contorted. Stamens as many as petals or twice or rarely thrice as many, united into a ring or short tube at the base; anthers 2-celled, with parallel cells opening longitudinally. Glands 5, adnate to or embedded in the outside of the staminal tube or rarely wanting. Disk none (besides the staminal tube). Ovary free, entire, 3 to 6-celled. Ovules 2 or rarely 1 in each cell, pendulous, anatropous, with a ventral raphe. Styles 3 to 5, distinct or more or less united, with terminal usually capitate stigmas. Fruit either a capsule, separating into cocci, usually dehiscent, or a drupe, with as many pyrenes as carpels, or more frequently reduced by abortion to 1. Seeds 1 or 2 in each coccus or pyrene; testa membranous or almost coriaceous; albumen fleshy, abundant or thin or entirely wanting. Embryo usually straight, with flat, ovate cotyledons; radicle superior.—Herbs, shrubs, or rarely trees, glabrous or rarely hirsute or tomentose. Leaves alternate or very rarely opposite, simple and entire or slightly serrate. Stipules lateral or within the petiole, sometimes minute or wanting.

An Order formerly almost limited to the genus *Linum*, but lately extended to include several small Orders or genera, chiefly tropical, from both the New and the Old World. The two Australian genera are the only two large ones, both of them widely dispersed, one chiefly in temperate regions, the other within the tropics.—*Benth.*

TRIBE I. **Eulineæ.**—*Petals contorted, fugacious. Perfect stamens, as many as petals. Capsule septicidally dehiscent. Herbs, rarely shrubs.*

Calyx glabrous or pubescent. Styles 5. Capsule 5, apparently 10-celled, with 1 seed in each cell . . . . 1. LINUM.
Calyx glabrous. Styles 3 or 4. Capsule 3 or 4-celled . . . . 2. REINWARDTIA.

TRIBE II. **Hugonieæ.**—*Petals contorted, fugacious. Perfect stamens 2 or 3 times as many as the petals. Fruit a drupe. Usually scandent shrubs, with hooked woody tendrils.*

Sepals subacute, tomentose, ebracteate . . . . 3. HUGONIA.

TRIBE III. **Erythroxyleæ.**—*Petals imbricate, rarely contorted, with a scale on the inner face, at length deciduous. Perfect stamens twice as many as the petals. Fruit a drupe. Shrubs or trees.*

Pedicels axillary. Petals with a double scale. Drupe 1-seeded . . 4. ERYTHROXYLON.

XXV. LINEÆ.

1. LINUM, Linn.
(From Linon, the old Greek name.)

Sepals 5. Petals 5, contorted, without appendages. Stamens 5, perfect; staminodia as many, alternating with the stamens, minute, tooth-like or hair-like, or sometimes scarcely conspicuous. Glands 5, small, scarcely prominent on the staminal tube, opposite the petals. Ovary 5-celled, with 2 collateral ovules in each cell. Capsule dividing into 5 cocci, with 2 seeds in each separated by an imperfect partition, or into 10 1-seeded cocci when the partition is more complete. Albumen thin.—Herbs. Leaves narrow, entire. Stipules none or minute and gland-like.

A large genus, widely distributed over the temperate or warmer extratropical regions of the globe, with a few tropical American species. The Australian species are endemic, but very closely allied to some of the commonest blue-flowered species of the northern hemisphere.—*Benth.*

Sepals acute or acuminate. Styles free . . . . 1. *L. usitatissimum.*
Sepals acute or acuminate. Styles united to above the middle . . . 2. *L. marginale.*
Sepals very obtuse . . . . 3. *L. suædæfolium.*
Flowers small, yellow . . . . 4. *L. gallicum.*

*1. **L. usitatissimum** (most useful), *Linn.* An erect annual 2 to 4ft. high, stem cylindrical, simple or corymbosely branched above. Leaves linear or lanceolate, narrow, sub-3-nerved, without stipular glands. Flowers blue, 1in. diameter, in broad cymes. Sepals ovate-acuminate, margins white, 3-nerved, eglandular, margins ciliate or not. Styles quite free, stigmas linear-clavate. Capsule scarcely exceeding the sepals.

Hab.: Europe. A stray from cultivation.

2. **L. marginale** (alluding to scarious border of sepals), *A. Cunn.; Planch. in Hook. Lond. Journ.* vii. 169; *Benth. Fl. Austr.* i. 283. A glabrous herb, forming a thick perennial rootstock, but also sometimes apparently annual, with erect or ascending slender stems of 1 to 2ft., corymbosely branched above the middle. Leaves linear or linear-lanceolate, acute or the lowest almost obtuse, often all under ½in., but the upper ones sometimes 1in. long. Stipular glands wanting. Flowers blue, on erect pedicels, forming a loose, irregular, terminal corymb. Sepals ovate or ovate-lanceolate, acute or cuspidate, 2 to 3 lines long with a strong midrib, the margins thin and often with a narrow scarious border. Petals from a little longer to twice as long. Styles united to above the middle. Capsule dividing into 10 1-seeded cocci.—Hook. f. Fl. Tasm. i. 46; F. v. M. Pl. Vict. i. 178; *L. angustifolium,* DC. Prod. i. 426 (as to the New Holland locality); Bartl. in Pl. Preiss. i. 161.

Hab.: Darling Downs.

3. **L. suædæfolium** (Suæda-leaved), *Planch. in Hook. Lond. Journ.* vii. 168; *Benth. Fl. Austr.* i. 288. Apparently an annual, with numerous short erect stems. Leaves crowded, linear, obtuse, 3 or 4 lines long, without stipular glands. Flowers and fruit of the small varieties of *L. marginale,* except that the sepals are very obtuse, those of the lower flowers almost dilated at the top.

Hab.: Balonne River, *Mitchell (Herb. Lindl.)* The specimen is very imperfect. It is probably a variety of *L. marginale,* with which some specimens in F. v. Mueller's Herbarium with less pointed sepals than usual would seem to connect it.—*Benth.*

*4. **L. gallicum** (French), *Linn.* Plant glabrous, usually producing many erect, slender stems. Leaves linear-lanceolate. Flowers small, yellow, in a terminal corym. Sepals ciliate at the base, subulate at the apex. Petals blunt, twice as long as the calyx.

Hab.: Mediterranean region. Now and again has been met with about Brisbane.

*2. REINWARDTIA, Dumort.
(After K. G. K. Reinwardt.)

Sepals 5, quite entire, lanceolate, acuminate. Petals 5, contorted, fugacious, much longer than the sepals. Stamens 5, hypogynous, connate below, alternating with as many interposed subulate staminodes. Glands 2 or 3, adnate to the staminal ring. Ovary 3 to 5-celled, 2-ovulate (the cells falsely 2-celled); styles 3 to 4, filiform, free or connate below, stigmas subcapitate; ovules 1 in each cell. Capsule globose, splitting into 6 to 8 cocci. Seeds reniform.—Undershrubs. Leaves alternate, quite entire or crenate-serrate; stipules minute, subulate, caducous. Flowers yellow, in axillary and terminal cymose fascicles, rarely solitary.

1. **R. trigyna** (three-styled), *Planch. in Hook. Fl. Brit. Ind.* An undershrub, 2 to 3ft. high, spreading by suckers (surculigerus). Leaves 1 to 3in. long, elliptic-obovate, usually rounded and mucronate at the tip. Flowers often 1in. diameter. Styles 3, free or connate at the base. Capsule shorter than the sepals.—*Linum trigynum*, DC.

Hab.: This Indian plant is frequently met with, as a stray from garden culture, near the principal towns.

3. HUGONIA, Linn.
(After Dr. A. J. Hugo.)

Sepals 5. Petals 5, contorted, fugacious. Stamens 10, hypogynous, with glandular swellings on the basal ring between the filaments, which are connate below. Ovary 5-celled; styles 5, filiform; stigmas capitate; ovules 2, collateral in each cell. Drupe globose. Seeds compressed, albuminous; embryo straight or slightly curved, cotyledons flat.—Climbing shrubs, often tomentose. Leaves alternate, serrate, stipulate. Inflorescence various. Flowers yellow, lower peduncles converted into spiral hooks.

Besides our species the genus is met with in tropical Asia and Africa.

1. **H. Jenkinsii** (after W. S. Jenkins), *F. v. M. Fragm.* v. 7. A showy climbing shrub, the branchlets slightly angular, dark but often with a glaucous covering. Stipules very short, unequal, setaceous. Leaves on somewhat short petioles, coriaceous, ovate to narrow-lanceolate, glabrous, repand-crenulate, attaining a length of about 7in. and a breadth of over 2in., glossy on both faces. Lateral nerves somewhat distant, the reticulation copious. The cieri or tendrils alternate or subopposite, stout, sulcate-striate, usually situated at the base of the present year's growth. Inflorescence racemose panicles in the upper axils. Bracts and bracteoles lanceolate-subulate. Pedicels short. The two outer sepals herbaceous, the others with membranous margins. Petals yellow, 3 to 4 lines long, obovate-cuneate, truncate or emarginate, on very short claws, contorted in the bud, soon deciduous. Stamens glabrous, some connate to near the middle. Anthers with oblique-ellipsoid cells. Styles capillary, 2 or 3 lines long, free, glabrous. Fruit globose-ovate, yellowish when fresh. Embryo green.

Hab.: Rockingham Bay, *Dallachy* (F. v. M.); Daintree River, *E. Fitzalan* (F. v. M.); Mourilyan Harbour, *W. Mumford*.

4. ERYTHROXYLON, Linn.
(From the red colour of wood of some species.)

Sepals 5, rarely 6, united into a lobed calyx, or free. Petals as many, with a 2-lobed appendage inside below the lamina. Stamens 10, rarely 12, the basal tube short, without glands, or more or less thickened into 10 glands, the filaments attached inside just below the crenulate top. Ovary 3 rarely 4-celled, with 1 or rarely 2 ovules in each cell. Drupe usually 1-seeded. Albumen copious, or thin, or none.—Trees or shrubs. Leaves entire. Stipules united into one within the the petiole, deciduous, or persistent especially on the leafless base of the young shoots. Flowers small, whitish, solitary or clustered in the axil of leaves or of leafless stipules.

A large tropical genus, abundant in S. America, less so in Africa and Asia. The two Australian species are perhaps endemic, but there is so much general similarity in the species of this genus, and their characters so vague and variable, that it is exceedingly difficult to determine their limits.—*Benth.*

Leaves oblong or narrow-elliptical, 1in. long or less, or the smaller ones cuneate-obovate, the veins few . . . . 1. *E. australe.*
Leaves obovate or ovate-elliptical, 1½ to 2½in. long, or the smaller ones rarely 1in., the veins numerous and finely reticulated . . . . 2. *E. ellipticum.*

1. **E. australe** (Australian), *F. v. M. in Trans. Vict. Inst.* iii. 22; *Benth. Fl. Austr.* i. 284. A small tree or glabrous shrub, with slender divaricate branches. Leaves elliptical-oblong, or the smaller ones cuneate or almost obovate, in some specimens all under ½in. long, in more luxuriant ones about 1in., the pinnate veins fewer and less reticulate than in many other species. Stipules small and deciduous. Pedicels solitary or rarely clustered, short or rarely attaining 3 lines, with minute bracteoles at their base. Flowers very small. Calyx not 1 line long, divided to below the middle, the lobes almost or quite valvate. Inner appendage of the petals with 2 very short crested lobes. Styles free or shortly cohering at the base. Drupe oblong, 3 to 3½ lines long, 3-celled, but with only 1 seed. Albumen thin; radicle slender, shorter than the ovate cotyledons.

Hab.: Brigalow scrub on the Burdekin, Suttor, and Dawson Rivers, *F. v. Mueller*; Comet River, *Leichhardt*; Rockhampton and Fitzroy River, *Thozet*; and many other localities in the south of the colony.

The late Mr. Staiger finds that the leaves do not contain cocaine, but they contain coca-tannic acid, and also a yellow dye-stuff which may prove of value.

Wood red in colour, close in grain, and prettily marked.—*Bailey's Cat. Ql. Woods No.* 37.

2. **E. ellipticum** (elliptic), *R. Br. Herb.; Benth. Fl. Austr.* i. 284 (*in part*), This species forms a tree about 35ft. high, with a trunk diameter of 12 or more inches. The young branches flattened. Leaves obovate or ovate-elliptical, very obtuse, 1½ to 2½in. long or the smaller ones rarely only 1in., on petioles of about 1 line, rather thin, the primary nerves not much more prominent than the reticulation, with very numerous and finely reticulated veins. Stipules usually about 2 lines long, and always longer than the petioles, deciduous. Flowers in axillary clusters of 3 to 6, the pedicels 2 to 4 lines long, with minute bracts at their base. Calyx about 1 line long, divided nearly to the base into lanceolate acute lobes, very slightly imbricate or almost valvate. Petals more than twice the length of the calyx, boat-shaped, very deciduous, the 2-lobed inner appendages very prominent and crumpled. Styles quite free, recurved. Drupe oblong, with a reddish, sweet pericarp, 3 to 4 lines long, 3-celled, 2 outer cells empty, the centre cell 1-seeded.

Hab.: Near Telegraph Station, Walsh River, *T. Barclay-Millar.* Previously only known from Dr. Robt. Brown's specimens gathered about 1802 on the mainland off Groote Eylandt.

The timber is excellent, resembles tulip-wood, is close in grain, and very durable.

Bark rough-corky, becoming hard with age.

Wood of a dark-brown colour, with light stripes, nicely marked, close-grained and firm; easily worked, and should be valuable for cabinet work.—*Bailey's Cat. Ql. Woods No.* 38.

Order XXVI. **MALPIGHIACEÆ.**

Flowers usually hermaphrodite. Calyx 5-cleft, the segments imbricate or rarely valvate, all, or more frequently 4 only (or rarely 3 or none of them), bearing 2 glands outside. Petals 5, usually equal, concave, toothed or notched, on slender claws. Disk scarcely prominent. Stamens usually 10, all perfect, or some of them deformed or without anthers, or sometimes wanting, the filaments usually united at the base; anthers 2-celled. Ovary usually 3-celled, or the 3 carpels distinct, with 1 ovule in each, ascending from a pendulous ventral funicle. Styles distinct, or united, or one only developed, with small terminal stigmas. Fruit-carpels 3 or fewer, either united in a berry, drupe, or hard capsule, or more frequently forming separate indehiscent nuts or winged samarae. Seeds without albumen, the testa usually membranous and double. Embryo straight or curved; cotyledons thin or fleshy, often unequal; radicle short, superior.—Trees, shrubs, or rarely undershrubs, frequently climbing. Hairs usually closely appressed and fixed by the centre. Leaves mostly opposite, with glands at the top of the petiole, and often on the margin underneath. Stipules usually small, deciduous, or none. Flowers usually yellow, red, or white, in racemes either simple and terminal or collected in corymbs or umbels, the pedicels articulate on the common peduncle.

A large tropical and subtropical Order, abundant in S. America, much less so in Africa and Asia. The only two Australian species belong to small genera spread over the Eastern Archipelago and S. Pacific Islands. Both genera are exceptional as being deprived of the calycine glands so general in the Order.—*Benth.*

TRIBE I. **Banisterieæ.**
Carpels with 1 vertical, large, oblong or incurved wing. Flowers in irregular
corymbs. Styles 3 . . . . 1. RYSSOPTERYS.

TRIBE II. **Hireæ.**
Carpels with several (7 or more) small linear, stellately spreading wings.
Flowers in simple racemes. Styles 1 or 2, unequal . . . . 2. TRISTELLATEIA.

1. RYSSOPTERYS, Blume.

(From the tubercles which cover the wings of the fruit.)

Calyx without glands. Petals scarcely clawed. Stamens all perfect, the filaments thickened at the base; anthers without appendages. Ovary 3-lobed, 3-celled, villous; styles 3, slender, with capitate stigmas. Samaras 1 to 3, expanded at the summit into a wing, of which the upper margin is thickened, tuberculate on the sides below the wing. Seeds oblong, with a slightly curved embryo.—Woody climbers. Leaves opposite. Inflorescence terminal or apparently axillary from the reduction of the flowering branches, compound, irregularly corymbose. Peduncles bracteate at the base, with 2 bracteoles at the articulation of the pedicels.

A small genus, dispersed over the Eastern Archipelago, one of the species extending into Australia.—*Benth.*

1. **R. timorensis** (also of Timor), *Blume; A. Juss. Malpigh.* 188; *Benth. Fl. Austr.* i. 285. A tall climber, the young shoots hoary-pubescent. Leaves on rather long petioles, broadly cordate-ovate or orbicular, obtuse or rather acute, 3 to 5in. long, somewhat coriaceous, glabrous above when full grown, hoary-pubescent underneath, with 1 or 2 prominent glands at the top of the petiole, those on the margin of the leaf very small. Flowers on pedicels of 2 or 3 lines, in short racemes arranged in irregular corymbs. Bracts and bracteoles very small. Fruit carpels or samaras pubescent, the lateral tubercles very prominent, the wing broadly semicircular, about ¾in. long and 5 or 6 lines broad.—Deless. Ic. Sel. iii. t. 85.

Hab.: Cape Cleveland, *A. Cunningham;* Fitzroy River, *Thozet.* The specimens are in fruit only, but agree perfectly with those we have in the same state from Timor. Some other species from the Archipelago are closely allied, but differ chiefly in the longer and narrower wing of the samaras.—*Benth.*

2. TRISTELLATEIA, Thouars.

(Referring to the star-like wings of the 3 carpels.)

Calyx without any or with very minute glands. Petals distinctly clawed. Stamens all perfect, filaments rigid, truncate, and articulate at the top; anthers acute. Ovary 3-lobed, style single or 2, or very rarely 3 unequal ones, the others reduced to small papillæ. Fruit-carpels 3, each one bearing about 7 small linear stellately spreading wings. Seeds obovoid; testa membranous, cotyledons fleshy, hooked.—Woody climbers. Leaves opposite or whorled, the petioles bearing 1 or 2 glands at the top, and minute stipules at the base. Flowers yellow, in terminal or lateral racemes.

A small genus ranging over Madagascar and the Indian Archipelago, one species from the latter region extending into Australia.—*Benth.*

1. **T. australasica** (Australasian), *A. Rich. Sert. Astrol.* 38, *t.* 15; *Benth. Fl. Austr.* i. 286. A tall, glabrous climber. Leaves opposite, on rather short petioles, ovate, acute, 2 to 4in. long, membranous, the glands of the petiole usually single and sometimes wanting. Racemes terminal, loose, 4 to 6in. long, Pedicels opposite, $\frac{1}{2}$ to 1in. long, articulate, with 2 minute bracteoles below the middle. Petals 3 or 4 lines long, spreading, the lamina ovate-cordate, the claw slender. Filaments much thickened below the middle, and very shortly united. Fruit (only seen in Archipelago specimens) quite glabrous, the wings of the carpels unequal, the longest often 8 lines long.

Hab.: Brown's River, *M'Gillivray.*

The species is found in various islands of the Indian Archipelago. The specimens described under the name of *Platynema laurifolium* by Wight and Arnott, in Jameson's Journal, and inserted in their "Prodromus," p. 107, as of doubtful Ceylonese origin, proved afterwards to have been from Singapore.—*Benth.*

Order XXVII. ZYGOPHYLLEÆ.

Flowers usually hermaphrodite and regular. Sepals 5 or 4, very rarely 6, free or connate at the base, imbricate or rarely valvate in the bud. Petals as many, free, imbricate or contorted, rarely valvate or wanting. Disk convex or depressed, rarely annular or undeveloped. Stamens usually the same or twice the number of the petals, the filaments most frequently with a scale or wings at or below the middle; anthers 2-celled, opening longitudinally. Ovary sessile or shortly stalked, often angular, with as many cells as petals or sepals, rarely more or fewer; style simple, with a simple or rarely lobed stigma. Ovules 2 or more in each cell, rarely solitary, pendulous or ascending, with a ventral raphe. Fruit sometimes drupaceous, never baccate, more usually separating into indehiscent or 2-valved cocci, the endocarp occasionally separating. Seeds solitary or rarely several, pendulous; testa membranous, crustaceous, or thick and mucilaginous when wetted; albumen usually thin. Embryo as long as the seed, green, straight, or rarely curved; cotyledons oblong or linear, radicle short, superior.—Shrubs, undershrubs, or herbs, the branches usually divaricate and articulate at the nodes. Leaves opposite, or rarely alternate by the abortion of 1 of each pair, 2-foliate or pinnate, rarely simple, the leaflets usually entire. Stipules in pairs. Peduncles axillary, 1-flowered, or rarely branching into cymes. Flowers mostly white, yellow, or red.

A small Order, nearly allied on the one hand to *Malpighiaceæ*, on the other to *Geraniaceæ* and *Rutaceæ*, dispersed chiefly over the subtropical regions of both the Old and New World, and most abundant in dry desert or saline regions. The three Australian genera are all common to Africa and Asia, and one of them extends also to Europe and America.—*Benth.*

Seeds exalbuminous.
  Leaves pinnate. Petals 5, flat. Fruit of 5 hard, indehiscent, usually prickly
    or tuberculate cocci . . . . 1. TRIBULUS.
  Leaves simple. Petals 5, concave. Fruit a drupe with a hard 1-seeded nut 2. NITRARIA.
Seeds albuminous
  Leaves with 2 leaflets or lobes. Petals 4 or 5, flat. Fruit a 4 or 5-angled or
    winged capsule . . . . 3. ZYGOPHYLLUM.

1. TRIBULUS, Linn.

(From *tribo*, to tear ; referring to the prickly fruits.)

(*Tribulopis*, R. Br.)

Sepals 5, rarely 6. Petals as many, flat. Disk annular, 10-lobed or sinuate, with a gland at the base of each of the inner stamens, alternating with the petals. Stamens twice as many as petals, the filaments filiform, without appendages. Ovary of 5 or sometimes more cells, with 1 or 2 to 5 superposed ovules in each cell. Fruit separating into as many cocci as carpels, hard, indehiscent, and each usually bearing 2 or more prickles or tubercles.—Herbs, usually prostrate or divaricate and hairy. Leaves abruptly pinnate, opposite, with one of each pair smaller than the other, or sometimes abortive or all alternate. Stipules small, lanceolate, or falcate. Pedicels solitary in the axil of the smaller leaf of each pair, or opposed to the leaf when alternate. Flowers white or yellow.

The genus is dispersed over the greater part of the tropical and warm regions of the globe, extending into Europe and N. America. Of the Queensland species, one is abundant in Asia, Africa, and S. Europe, another is most common in tropical America, less so in Asia and Africa, and the others are all endemic.—*Benth.*

Leaves, at least the upper ones, opposite. Glands of the disk not very
  prominent. Ovules 2 or more in each cell. (*Tribulus* proper).
  Cocci rounded at the back, without angular or winged edges.
    Cocci with 2 or 4 prickles, rarely minute or deficient.
      Leaves almost all opposite. Ovules 3 or 4 in each cell.
        Annual. Flowers small. Petals about ½in. . . . 1. *T. terrestris*.
        Perennial. Flowers large. Petals about ¾in. . . . 2. *T. cistoides*.
      Flowers large.
        Cocci 2 maturing, each 7 lines diameter, silky . . . . 3. *T. occidentalis*.
        Cocci covered with numerous nearly equal prickles . . . . 4. *T. hystrix*.
    Leaves (except *T. minutus*) all alternate. Glands of the disk prominent.
      Ovules solitary. Fruit pyramidal, the cocci with 2 or 4 tubercles or
      small prickles below the middle. (*Tribulopis*, R. Br.)
        Leaflets 2 pairs, the lowest much smaller. Perfect stamens usually 5 5. *T. pentandrus*.
        Leaflets about 3 pairs, ovate, or lanceolate, the lowest distant from the
          stem. Anthers 10, nearly similar. Flowers small . . . . 6. *T. Solandri*.
        Leaflets 4 to 6 pairs, linear. Anthers 10, similar. Flowers large . . . 7. *T. angustifolius*.
        Leaflets 2 or 3 pairs, linear. Flowers small, 5 perfect. Anthers 5, sterile. 8. *T. leptophyllus*.
        Leaflets 3 to 6 pairs, small, ovate or lanceolate. Leaves mostly opposite.
          Anthers 10, similar. Flowers very small , ... 9. *T. minutus*.

1. **T. terrestris** (from its prostrate habit), *Linn.; DC. Prod.* i. 703 ; *Benth. Fl. Austr.* i. 288. Cat's-head, or Catrops. A prostrate annual or biennial, more or less hirsute or silky-hairy, especially the young shoots, the stems extending often to 1 or 2ft. Leaves opposite, unequal ; leaflets of the larger one usually 5 to 7 pairs, obliquely oblong, 3 to 5 lines long. Pedicels shorter than the opposite larger leaf. Flowers small, the sepals rarely attaining 2 lines and often much less, the petals rather longer, but very rarely nearly twice as long. Anthers 10, all small and perfect. Ovules 3 or 4 in each cell. Cocci 5, hard, 2 to 3 lines long, glabrous or hairy, rounded on the back, with 2 marginal, divaricate, horizontal, subulate or conical prickles about half-way up, and often 2 smaller reflexed ones lower down, the rest of the surface usually tuberculate or shortly

muricate. Seeds 2 to 4 in each coccus, horizontal and separated by transverse partitions.—Reichb. Ic. Fl. Germ. v. t. 161; F. v. M. Pl. Vict. i. 99; *T. lanuginosus*, Linn.; DC. Prod. i. 704; Wight, Ic. t. 98; *T. acanthococcus*, F. v. M. in Trans. Phil. Soc. Vict. i. 9.

Hab.. Common in many parts of the colony both north and south.

The species is a common weed in S. Europe, temperate Africa, and S. Asia.—*Benth.*

2. **T. cistoides** (flowers like the Rock Rose), *Linn.; DC. Prod.* i. 708; *Benth. Fl. Austr.* i. 288. A perennial, forming at length a thick rootstock. Branches procumbent or ascending, attaining 1 to 2ft. Indumentum more silky than in *T. terrestris*. Larger leaf of each pair with frequently 7 or 8 pairs of leaflets. Flowers large, on longer peduncles than in *T. terrestris;* the sepals 3 or 4 lines long, very acute, silky-hairy; the petals obovate, at least ¾in. long. Anthers usually (perhaps not always) oblong or linear. Fruit like that of *T. terrestris* or rather larger, with 2 or very rarely 4 prickles to each coccus.—A. Gray, Ill. Gen. N. Am. t. 145.

Hab.: Gulf of Carpentaria, Northumberland Island, *R. Brown;* Port Curtis, Port Molle, *M'Gillivray;* Lord Howick's group, *F. v. Mueller;* Port Denison, *Fitzalan.*

The species is frequent in the West Indies and many parts of tropical America, and in the Pacific Islands, rare in tropical Asia and Africa.—*Benth.*

3. **T. occidentalis** (Western. First found in Western Australia), *R. Br., in App. Sturt Exped.* A diffuse or prostrate plant, the branches densely tomentose-hirsute or woolly. The upper leaves opposite, the larger one of each pair with 8 or 9 pairs of leaflets, silky-hairy, the base oblique, the apex pointed, about 5 lines long and 2 lines wide. Pedicels slender, 1¼ to 1¾in. long. Sepals narrow, 6 lines long. Petals of a deep-yellow, exceeding an inch in length, cuneate, 6 lines broad at the upper end. Stamens 10, long as the sepals; anthers all perfect, oblong; ovary covered with long barbellate bristles; style glabrous, together with the ovary equalling in length that of the stamens. Cocci usually but 2 coming to maturity, each of which are about 7 lines long and 5 lines broad, and clothed by a dense covering of soft white silky hairs, and numerous long hairy-subulate spines, thus the extreme diameter of fruit, including spine, will often measure over 1½in. The fruit examined not fully ripe.

Hab.: Diamantina, *Dr. Thos. L. Bancroft.*

The first specimens of this plant, Dr. Robt. Brown tells us, were gathered on the west coast of Australia, or on some of its islands, by the naturalists of the "Beagle." App. l.c. Mr. Bentham, Fl. Austr. i. 289, says: "In J. McDouall Stuart's collection is a fragmentary specimen from Fink River, with a much larger flower, which may possibly be a variety of *T. hystrix*, but is indeterminable without fruit." Both these notices agree with the Diamantina plant, and differ, in my opinion, sufficiently from *T. hystrix* to bear Dr. Brown's name as above.

4. **T. hystrix** (fruitlets porcupine-like), *R. Br. in App. Sturt Exped.* 6; *Benth. Fl. Austr.* i. 289. A diffuse or prostrate perennial or undershrub, the branches densely tomentose-hirsute or woolly. Lower leaves (at least in some specimens) alternate, upper ones opposite, the larger one of each pair with 6 to 8 or even more pairs of leaflets, rather broad and softly silky-hairy. Flowers smaller than in *T. cistoides*, but much larger than in *T. terrestris*, the petals generally about ½in. long. Ovary very hirsute, with 8 or 4 ovules in each cell. Cocci very villous, covered all over with hairy prickles, either subulate from the base or more or less thickened and conical.

Hab.: Towards the Gulf of Carpentaria.

5. **T. pentandrus** (five stamens), *Benth. Fl. Austr.* i. 290. A slender, branching, prostrate annual, often attaining 1ft. in length, more or less hairy. Leaves all alternate, with 2 pairs of oblong-lanceolate leaflets, the terminal ones 4 to 8 lines long, the lower pair much smaller, usually not half the size. Flowers

small, on slender pedicels. Petals oblong. Stamens usually 5, with globose or ovoid perfect anthers, and 5 small with imperfect capitate anthers, or entirely wanting. Ovules solitary in each cell of the ovary. Fruit pyramidal, 1 to 1½ line long, with 2 small tubercles at the base of each coccus.—*Tribulopis pentandra*, R. Br. in App. Sturt Exped. 7; F. v. M. Fragm. i. 48; *Tribulus Brownii*, F. v. M. Pl. of Vict. i. 99.

Hab.: Islands of the Gulf of Carpentaria, *R. Br.* (Benth. l.c.)

6. **T. Solandri** (after Dr. Solander), *F. v. M. Pl. Vict.* i. 99 (partly); *Benth. Fl. Austr.* i. 290. An annual, with prostrate or ascending stems, pubescent or nearly glabrous. Leaves alternate; leaflets usually 3 pairs, rarely 2 pairs, obliquely ovate or oblong-falcate, 3 to 6 lines long, the lowest pair distant from the stem and nearly of the size of the others, all glabrous except the ciliate margins or slightly hairy, those of the upper leaves sometimes narrower and lanceolate. Flowers small. Stamens usually all perfect, with small anthers. Fruit pyramidal, about 3 lines long, glabrous or slightly tomentose, with 2 pairs of prominent reflexed tubercles below the middle of each coccus.—*Tribulopis Solandri*, R. Br. in App. Sturt Exp. 7.

Hab.: Gilbert River, *F. v. Mueller;* Endeavour River, *Banks;* Lizard Island, *M'Gillivray*.

7. **T. angustifolius** (narrow-leaved), *Benth. Fl. Austr.* i. 290. Annual or sometimes forming a perennial rootstock, with procumbent, ascending, or erect stems, glabrous or silky-pubescent. Leaves all alternate; leaflets 4 or 5 pairs or sometimes more, linear, attaining 1in. in length, more or less silky-pubescent. Flowers large, the petals usually exceeding ½in. Stamens all perfect, with small anthers. Fruit 3 lines long, beside the rigid persistent style, which is about as long, with 2 minute tubercles at the base of each coccus.—*Tribulopis angustifolia*, R. Br. App. Sturt Exped. 7; *Tribulopis Solandri*, F. v. M. Fragm. i. 47; *Tribulus Solandri*, var. *angustifolia*, F. v. M. Pl. Vict. i. 99.

Hab.: Islands of the Gulf of Carpentaria, *R. Br.* (Benth. l.c.)

8. **T. leptophyllus** (slender-leaved), *Bail. Bot. Bull.* 3. A procumbent silky-pubescent annual extending 2 or more feet. Leaves all alternate; leaflets 2 or 3 pairs, linear, the end ones the longest, often attaining over 1in. length. Flowers small, yellow, on filiform pedicels, often as long or nearly as long as the leaves. Petals under 3 lines long. Stamens 5, perfect anthers, and 5 smaller ones with imperfect capitate anthers, ovary silky. Fruit tomentose, mixed with long hairs, 3 lines long, pyramidal. Style rigid, persistent, about half as long as the fruit; coccus tubercles basal.

Hab.: Walsh River, *T. Barclay-Millar*.

This new species is very closely allied to *T. angustifolius*, Benth., differing from that species in the less number of leaflets, in its small yellow flowers, and in only half the stamens having perfect anthers.

9. **T. minutus** (small), *Leichh.; Benth. Fl. Austr.* i. 291. Pubescent, apparently prostrate, and more slender than any other species. Leaves mostly opposite, those of each pair unequal or one occasionally abortive, the larger one of 3 to 5 pairs of obovate or oblong leaflets, about 2 or rarely 3 lines long. Flowers very small. Stamens 10, with the anthers all similar. Glands prominent. Ovules solitary (or sometimes 2 ?) in each cell. Fruit nearly of *T. Solandri*, but smaller; each carpel bearing a pair of small, reflexed, conical spines about the middle, and a pair of minute tubercles lower down.

Hab.: ? *Leichhardt Expedition*.

This species Bentham considered to connect the two groups, having the opposite leaves of *Tribulus* proper, with the fruit of *Tribulopis*.

XXVII. ZYGOPHYLLEÆ.

2. NITRARIA, Linn.

(First found near nitre-works in Siberia.)

Calyx small, 5-lobed. Petals 5, concave with inflexed points, induplicate-valvate in the bud. Disk not prominent. Stamens 15, rarely 10 to 14, the filaments free, without appendages. Ovary sessile, 2 to 6-celled, terminating in a short thick style, with 2 to 6 adnate stigmas; ovules solitary in each cell, ascending from pendulous funiculi, which are more or less adnate to their inner face. Fruit a drupe, with a berry-like sarcocarp; putamen ovoid-acute, hard, marked outside with irregular depressions, and opening at the top in 6 short, pointed valves, of which 3 inner ones smaller. Seeds solitary, pendulous, without albumen.—Rigid shrubs, often thorny. Leaves alternate or clustered, undivided, succulent. Stipules small. Flowers small, white, in once or twice-forked scorpioid cymes.

The genus, besides the widely-spread Australian species, comprises one other from Northern Africa. The raphe of the seed is described as dorsal by Spach, but we have always found it ventral in the ovary, although the seed sometimes hangs obliquely.—*Benth.*

1. **N. Schœberi** (after — Schœber), *Linn.; DC. Prod.* iii. 456; *Benth. Fl. Austr.* i. 291. A rigid, spreading shrub, attaining 3 to 6ft., glabrous or hoary with a very minute down, the smaller branches occasionally spinescent. Leaves from cuneate-oblong to lanceolate or linear, the lower ones obtuse and often 1in. long, those of the smaller branches smaller and more acute, all entire, thick and fleshy. Cymes usually shortly pedunculate, the flowers sessile or shortly pedicellate along the scorpioid branches. Petals about 1½ line long. Ovary 3-celled. Drupe varying from ovoid-globular to ovoid-oblong, the putamen from ¼ to more than ½in. long, the depressions in the lower part round or oblong, the upper part marked with 6 furrows, along which the valves ultimately open. Only 1 seed or very rarely 2 come to maturity.—Andr. Bot. Rep. t. 529; *N. Billardieri,* DC. Prod. iii. 456; F. v. M. Pl. Vict. i. 92, t. Suppl. 7; *N. Olivieri,* Jaub. and Spach. Ill. Pl. Or. iii. 143, t. 295; *Zygophyllum australasicum,* Miq. in Pl. Preiss. i. 164.

Hab.: Towards the South Australian border, on saline land, *F. v. M.* probably.
The species is spread over the hot, more or less saline, tracts of Western Asia and northern Africa. A careful examination leaves no doubt of the identity so often suggested of the Australian and northern plants.—*Benth.*

8. ZYGOPHYLLUM, Linn.

(From the leaves being in pairs.)

Sepals 4 or 5. Petals as many, flat, contracted into a short claw. Disk concave, angular or cup-shaped. Stamens twice as many as petals, inserted at the base of the disk; filaments filiform, with an adnate scale or wing-like appendage at the base, which however is wanting in some of the Australian species. Ovary sessile, 4 or 5-angled, narrowed at the top into an angular style, 4 or 5-celled, with 2 or more superposed ovules in each cell. Fruit capsular, with 4 or 5 angles or vertical wings, indehiscent or separating into cocci or opening loculicidally, the endocarp sometimes separating. Seeds 1 or more in each cell, pendulous; albumen scanty.—Shrubs or undershrubs, often prostrate. Leaves opposite, with 2 distinct leaflets or rarely 2-lobed, frequently fleshy. Stipules small. Peduncles 1-flowered, axillary, solitary, or rarely 2 together. Flowers white or yellow.

A considerable and widely-spread genus, though confined, with one exception, to the Old World, and chiefly numerous in the desert or saline regions of central and western Asia. North and South Africa. The Australian species are all endemic.—*Benth.*

Filaments winged at the base. Capsule angular, loculicidal.
  Capsule broad and truncate at the top, the angles usually produced into
    short appendages. Flowers mostly 5-merous . . . . . 1. *Z. apiculatum.*
  Capsule equally rounded at the top and the base.
    Capsule 4 to 8 lines long, the cells 2 to 4-seeded. Wings of the
      filaments toothed. Flowers usually 4-merous . . . . 2. *Z. glaucescens.*
    Capsule 2 to 3 lines long, the cells 1-seeded. Wings of the filaments
      small and entire. Flowers usually 5-merous . . . 3. *Z. iodocarpum.*
  Capsule oblong, the angles produced at the top into erect appendages . 4. *Z. prismatothecum.*
Filaments subulate, not winged.
  Capsule angular, loculicidal, broad and truncate at the top, narrow at
    the base . . . . 5. *Z. Billardieri.*
  Capsule indehiscent, the angles produced into broad membranous wings 6. *Z. fruticulosum.*
Varieties with leaves 2-lobed instead of 2-foliolate occur in *Z. iodocarpum, Z. prismatothecum, Z. Billardieri,* and *Z. fruticulosum*; with lobed or crenate leaflets in *Z. glaucescens* and *Z. iodocarpum*; and forms or states with minute flowers in several of the species.

1. **Z. apiculatum** (capsule apiculated), *F. v. M. in Linnæa* xxv. 373, *and Pl. Vict.* i. 101. A diffuse, glabrous undershrub. Leaflets 2, obliquely obovate or rarely oblong, ½ to 1in. long, on a short common petiole. Flowers usually 5-merous. Filaments with rather broad wings, adnate to above the middle and toothed at the top. Capsule about 4 lines long, opening loculicidally, broader and truncate at the top, the angles very obtuse, and produced at the upper outer corner into a short obtuse appendage. Seeds usually solitary in each cell.—*Roepera latifolia,* Hook. f. Fl. Tasm. i. 60; *Zygophyllum terminale,* Turcz. in Bull. Mosc. 1858, i. 437.

Hab.: Darling Downs; Mackenzie and Dawson Rivers, *F. v. M.* (a very small-flowered variety).

2. **Z. glaucescens** (glaucous), *F. v. M. Pl. Vict.* i. 228. Herbaceous, diffuse or erect and glabrous. Leaves of 2 broad leaflets as in *Z. apiculatum,* the petiole occasionally winged at the base. Flowers usually 4-merous. Filaments with toothed wings as in *Z. apiculatum.* Capsule usually above ½in. long, opening loculicidally, the angles equally rounded at the top and the base. Seeds 2 or 3 or sometimes 4 or 5 in each cell.—*Z. glaucum,* F. v. M. in Trans. Vict. Inst. i. 29; and Pl. Vict. i. 102; not of Sonder.

Hab.: Range about Toowoomba.
Var. *lobulatum,* Benth. A small annual. Leaflets irregularly 2 or 3-lobed or deeply crenate. Flowers and fruit precisely as in the ordinary form.—*Z. crenatum,* F. v. M. in Linnæa, xxv. 374, and Pl. Vict. i. 103, t. 6. On the Lachlan and Murray Rivers, and in the interior of S. Australia and Queensland, *F. v. M.*

3. **Z. iodocarpum** (violet fruited), *F. v. M. in Linnæa,* xxv. 372, *and Pl. Vict.* i. 105. A small, much-branched, diffuse annual. Leaflets oblong-cuneate or almost linear, very obtuse, rarely ½in. long, the petiole often 2-winged, especially towards the top. Flowers very small, usually 5-merous, the petals not 2 lines long. Filaments dilated at the base into short, narrow, entire wings, entirely adnate or very shortly free. Capsule 2 or rarely 3 lines long, loculicidal, the angles equally rounded at the top and the base. Seeds solitary in each cell.

Hab.: Queensland, *F. v. M.*

4. **Z. prismatothecum** (prism-shaped fruit), *F. v. M. in Linnæa,* xxv. 375. A much-branched, small annual. Leaves rather thick, the leaflets, in the few specimens seen, short and confluent with the more or less dilated petiole, so as to form a single two-lobed leaf. Flowers, which I have not seen, small and 4-merous according to F. v. Mueller, the filaments dilated at the base and toothed or entire. Capsules nearly sessile, oblong, 4-angular, about 4 lines long, of equal breadth at the base and the top, where the angles terminate in small erect leafy appendages. Seeds solitary in each cell.

Hab.: Border-land adjoining S. Australia, *F. v. M.*
The very few specimens seen have all the foliage of the 2-lobed varieties of *Z. iodocarpum, Z. Billardieri,* and *Z. fruticulosum,* but as in those species there is probably also a variety with normally 2-foliolate leaves.—*Benth.*

5. **Z. Billardieri** (after J. J. Labillardiere), *DC. Prod.* i. 705; *Benth. Fl. Austr.* i. 298. Herbaceous, prostrate or diffuse and much-branched. Leaves oblong, cuneate or linear, rarely obovate, ¼ to 1in. long, the petioles not usually winged. Flowers usually 4-merous, the size of those of *Z. apiculatum*. Sepals narrow, very acute. Petals about 3 lines long. Filaments subulate or slightly flattened, but not winged. Capsule 3 to 5 lines long, loculicidal, broad and truncate at the top, narrowed to the base, the angles acute or shortly pointed or scarcely rounded at the upper outer corner. Seeds 1 or rarely 2 in each cell.—Hook. f. Fl. Tasm. i. 60; F. v. M. Pl. Vict. i. 104; *Ræpera Billardieri*, A. Juss. in Mem. Mus. Par. xii. 454 (by inference); *Z. ammophilum*. F. v. M. in Linnæa, xxv. 376, in adnot.

Hab.: Warrego.

6. **Z. fruticulosum** (shrubby), *DC. Prod.* i. 705; *Benth. Fl. Austr.* i. 294. A low diffuse or divaricately branched shrub. Leaflets obliquely oblong or lanceolate, rarely ovate. Flowers 4-merous, the size of those of *Z. apiculatum*. Filaments subulate, without wings. Capsule ½in. long, indehiscent, or at length separating septicidally into cocci opening inside, the angles expanded into broad membranous wings, rounded at both ends and not splitting. Seeds solitary in each cell.—F. v. M. Pl. Vict. i. 105; *Ræpera fabagifolia*, A. Juss. in Mem. Mus. Par. xii. 525, t. 15; Deless. Ic. Sel. iii. t. 42; Miq. in Pl. Preiss. i. 164.

Hab.: Balonne River, *F. v. M.*

Var. *bilobum*. Leaflets narrow, continuous with the petiole, as in *Z. prismatothecum*.—*Ræpera aurantiaca*, Lindl. in Mitch. Three Exped. ii. 70; *Z. aurantiacum*, F. v. M. in Linnæa, xxv. 376 (note)..

Order XXVIII. GERANIACEÆ.

Flowers usually hermaphrodite, regular or irregular. Sepals 5, or rarely fewer, free, or rarely connate at the base, imbricate or (in genera not Australian) valvate in the bud. Petals as many or rarely wanting, hypogynous or slightly perigynous, variously imbricate in the bud. Torus more or less expanded into a disk, often bearing 5 glands alternate with the petals, and usually protruding into a short axis in the centre of the ovary. Stamens usually twice the number of the petals, 5 of them occasionally without anthers, or rudimentary, or in irregular flowers, 3 or more without anthers or wanting; filaments either free and filiform, or dilated or connate at the base; anthers with 2 parallel cells. Ovary usually 3 to 5-lobed, with as many cells, the carpels adnate to the axis up to the insertion of the ovules, and often produced above that into a beak bearing the style or stigmas; stigmas as many as cells, either raised on the style or sessile on the carpels, radiating from a connate base or rarely entirely connate. Ovules either 1 in each cell or 2 inserted nearly at the same point, 1 ascending, the other pendulous, or several in 1 or 2 rows. Fruit either a lobed capsule, the lobes 1-seeded, separating from the axis with the seed, and elastically rolled upwards along the beak, leaving the placentiferous portion attached to the axis, or the lobes several-seeded, remaining attached to the axis, but opening loculicidally, or in genera not Australian the fruit is a berry or separates into indehiscent cocci. Seeds pendulous or ascending; testa thin or rarely crustaceous; albumen usually scanty or none. Embryo straight or curved, radicle short and straight or long and curved or forked over the cotyledons.—Herbs or shrubs, or rarely, in genera not Australian, trees. Leaves opposite or alternate, toothed, lobed, or divided, very rarely quite entire. Stipules usually 2. Peduncles axillary, 1 or 2-flowered, or bearing an umbel of several flowers, very rarely a cyme or raceme.

The Order is chiefly dispersed over the temperate regions of the northern hemisphere, very abundant in Southern Africa, with a few extratropical South American and tropical species. Of the four Australian genera, two are common in the northern hemisphere, a third, although

XXVIII. GERANIACEÆ.

chiefly American, is represented in Australia by species of an extratropical European as well as American type, and the fourth is almost entirely South African. The Order is very closely allied to *Zygophylleæ.—Benth.*

TRIBE I. **Geranieæ.**—*Flowers regular. Sepals imbricate. Glands alternate with the petals. Antheriferous stamens as many or double or treble the number of petals.* Capsule beaked, the lobes 1-seeded, and elastically rolled upwards along the beak. Leaves toothed, lobed, or divided.

Anthers usually 10. Tails of the carpels glabrous inside . . . . 1. GERANIUM.
Anthers 5. Tails of the carpels bearded inside . . . . 2. ERODIUM.

TRIBE II. **Pelargonieæ.**—*No glands. Stamens declinate.* Flowers irregular, the upper sepal furnished with a linear tube or spur adnate to the pedicel. Anthers 5, 6, or 7 . . . . 3. PELARGONIUM.

TRIBE III. **Oxalideæ.**—*Flowers regular. Sepals imbricate. Glands none. Stigmas capitate.* Capsule opening loculicidally, the valves adhering to the axis. Leaves with 3 or many leaflets . . . . 4. OXALIS.

1. GERANIUM, Linn.

(From supposed resemblance of capsule to the head and beak of a crane.)

Flowers regular. Sepals 5. Petals 5. Glands 5, alternating with the petals. Stamens 10, all usually bearing anthers. Ovary 5-lobed, beaked, the beak terminating in the style, with 5 short stigmatic lobes. Ovules 2 in each cell. Capsule-lobes 1-seeded, separating from the placenta-bearing axis, enclosing the seed, and curled upwards on a long awn detached from the beak, and glabrous inside. Radicle of the embryo turned back on the folded or convolute cotyledons. Herbs, rarely undershrubs. Leaves opposite or alternate, toothed, lobed, or divided, the lobes or segments palmate, or rarely (in species not Australian) pinnate. Peduncles axillary or in the forks, 1 or 2-flowered.

A large genus, widely distributed over nearly the whole globe, but more abundant in the northern hemisphere, and rare within the tropics. The Queensland species is also in New Zealand and S. America, and extends up the whole length of that continent to the N.W., and in a slight variety also over most temperate parts of the northern hemisphere, but does not occur in S. Africa.—*Benth.*

1. **G. dissectum** (referring to the cut leaves), *Linn.; DC. Prod.* i. 648, var. *australe; Benth. Fl. Austr.* i. 296. Usually perennial, forming at length a thick rootstock, descending into a taproot. Stems diffuse, procumbent or shortly erect, more or less hairy with spreading or reflexed hairs, or hoary with a short pubescence. Leaves on long petioles, nearly orbicular in their circumscription, deeply divided into 5 or 7 segments, each one again more or less cut into 3 or more lobes, varying from broadly cuneate-oblong to linear, and usually pubescent or hairy, especially underneath. Peduncles 2-flowered, or rarely 1 or 3-flowered. Sepals 3-nerved, obtuse, acute, or very shortly mucronate; usually 2 or 3 lines long. Petals cuneate-obovate, entire or slightly notched, from rather longer than the sepals to twice as long. Anthers all perfect. Lobes of the capsule sprinkled with hairs, not wrinkled. Seeds covered with minute reticulations or rarely smooth.—Hook. f. Fl. N. Zeal. i. 39, and Fl. Tasm. i. 56; F. v. M. Pl. Vict. i. 173; *G. pilosum,* Forst.; DC. Prod. i. 642; Nees, in Pl. Preiss. i. 162; *G. parviflorum,* Willd.; DC. Prod. i. 642; *G. philonothum,* DC. Prod. i. 689 (from the character given); *G. potentilloides,* L'Hér., DC. Prod. i. 689; Hook. f. Fl. N. Zeal. i. 40; Fl. Tas. i. 57; *G. australe,* Nees, in Pl. Preiss. i. 162.

Hab.: Southern parts.

The original form of *G. dissectum,* as generally diffused over the temperate regions of the northern hemisphere, in the Old World, is an annual, with the petals very rarely exceeding the sepals, and the seeds very prominently reticulate. In the eastern United States of N. America, under the name of *G. carolinianum,* Linn., it is also annual or biennial, but has the petals often rather larger and the reticulations of the seeds are finer and less prominent. West of the Rocky Mountains the stock often appears to be perennial, and then it is undistinguishable from some Australian forms. The commonest Australian form frequently sent from extratropical S. America, and extends all along the mountainous regions of that continent to Mexico and the Rocky Mountains, often apparently together with and passing into the northern annual variety. The Australian plant again, both in that country and in New Zealand, is very variable, and may be generally subdivided into two principal races, although I have, after repeated trials, found it impossible to distribute our numerous specimens quite satisfactorily into the two groups, viz.:—

a. pilosum. Root thick. Stems erect, ascending or procumbent, usually hirsute. Seeds strongly reticulate. Common on downs country, the rootstock greatly relished by sheep.

b. potentilloides. Root and stock less thickened. Stems more slender and prostrate, less hairy, and usually only slightly hoary with more appressed pubescence. Seeds more finely reticulate, or rarely almost smooth. To this variety belongs generally the *G. potentilloides* of authors, and *G. australe,* Nees. It appears to be rather the more common form in the East, whilst the var. *pilosum* is more frequent in the West. But both are found throughout extratropical Australia.—*Benth.* Common on coast side of range.

2. ERODIUM, L'Hér.

(Fruit supposed to resemble the head and beak of the heron.)

Flowers regular or nearly so. Sepals 5. Petals 5. Glands 5, alternating with the petals. Stamens 5 bearing anthers, opposite the sepals, and 5 staminodia, usually scale-like, alternating with them. Ovary 5-lobed, beaked, the beak terminating in the style, with 5 short stigmatic lobes. Ovules 2 in each cell. Capsule-lobes 1-seeded, separating from the placenta-bearing axis, enclosing the seed and curled upwards on a long elastic awn, which separates from the beak and is usually twisted and bearded inside with long hairs. Radicle of the embryo turned back on the folded or convolute cotyledons.—Herbs or rarely undershrubs. Leaves unequally opposite or alternate, pinnately or rarely ternately lobed or divided. Peduncles axillary, bearing an umbel of several flowers, or rarely 1-flowered.

The species are numerous in Europe, North Africa, and temperate Asia, 2 or 3 are natives of S. Africa, and 2 or 3 more are now widely dispersed as weeds over many parts of the globe. Two of these are in Australia, one of them perhaps indigenous, but the common Australian species is endemic.—*Benth.*

Leaves of 3 lobed or divided segments, the middle one the largest . . . . 1. *E. cygnorum.*
Leaves pinnate with deeply-lobed narrow segments . . . . 2. *E. cicutarium.*

1. **E. cygnorum** (swan-like), *Nees, in Pl. Preiss.* i. 162; *Benth. Fl. Austr.* i. 297. An annual or biennial, with the habit of the coarser forms of *E. cicutarium,* sometimes slightly pubescent, sometimes very hispid, with the hairs of the stem spreading or reflexed. Leaves deeply 3-lobed or divided to the base into 3 lobes or segments, usually obovate or cuneate, and more or less deeply toothed or again 3-lobed, the central lobe larger, broader, and more lobed than the lateral ones. Flowers blue, usually 2 to 5 in the umbel. Sepals pointed. Petals obovate, scarcely exceeding the calyx or shorter. Filaments broad at the base, with subulate points; staminodia scale-like, often toothed. Capsule-lobes glabrous, hairy or hispid; beak usually above 2in. long.—F. v. M. Pl. Vict. i. 172.

Hab.: Many parts in the south; Peak Downs, *F. v. M.;* Maranoa River, *Mitchell.*

2. **E. cicutarium** (Hemlock-leaved), *L'Hér.; DC. Prod.* i. 646; *Benth. Fl. Austr.* i. 298. Usually an annual, but often forming a dense tuft, with a thick taproot, which may last over a second year, always more or less covered with spreading hairs, which are sometimes viscid. Stems sometimes exceedingly short, but lengthening out to near 1ft. Leaves mostly radical, pinnate, the segments distinct and deeply pinnatifid, with narrow, more or less cut lobes. Peduncles erect, bearing an umbel of from 2 or 3 to 10 or 12 small purple or pink

flowers. Sepals pointed, about the length of the obovate entire petals. Filaments and staminodia lanceolate-subulate. Lobes of the capsule slightly hairy, the beak ½ to 1¼in. long.—Nees, in Pl. Preiss. i. 161; Reichb. Ic. Fl. Germ. v. t. 189.

Hab.: Southern Downs country.

A very common weed in Europe and temperate Asia, and found in many other parts of the world, in many cases introduced, as in several or perhaps all of the Australian localities, but too widely spread now to be omitted from the Flora, even if it be not rarely indigenous.—*Benth.*

3. PELARGONIUM, L'Hér.

(Fruit supposed to resemble the head and beak of a stork.)

Flowers irregular. Sepals 5, shortly united at the base and produced into a tube or spur, adnate to the pedicel. Petals 5 or fewer, the 2 upper ones different from the others (usually larger), and inserted on the sides of or behind the spur. Disk without glands. Stamens usually 10, hypogynous, shortly united, 5 to 7 or rarely only 2 or 3 bearing anthers, the remainder without anthers or rudimentary. Ovary and fruit of *Erodium*. Cotyledons flat or folded.—Herbs, undershrubs, or shrubs. Leaves opposite or rarely alternate, entire, toothed, lobed, or variously divided. Peduncles usually axillary, bearing an umbel of several flowers.

A very large genus, but which, with the exception of 3 N. African or Levant species and the 2 Australian ones, is confined to S. Africa.—*Benth.*

1. **P. australe** (Australian), *Willd.; DC. Prod.* i. 654; *Benth. Fl. Austr.* i. 298. Herbaceous, often flowering the first year, but forming a perennial rootstock, either horizontal and almost creeping, or short and thick. Leafy stems decumbent or erect, sometimes short, but usually attaining 1ft. or more, generally pubescent or hirsute with spreading hairs. Leaves reniform-cordate, or very rarely broadly ovate-cordate, crenate, or very shortly lobed, very obtuse, rarely 2in. diameter, and usually much smaller, softly pubescent or hirsute. Stipules broad. Peduncles usually longer than the leaves, but not so long as in *P. Rodneyanum*, and sometimes very short. Flowers small, in an umbel, sometimes very dense, almost reduced to a head, sometimes loose with pedicels of ¼in. or more. Sepals acute, 2 to 3 lines long, usually very hairy, the decurrent tube rarely so long, and sometimes very short. Petals from a little longer than the sepals to about half as long again. Capsule-lobes pubescent, the beak from ½ to ¾in. long, the awns of the lobes bearded inside as in *Erodium*. Seeds smooth.—Sweet, Geran. t. 68; Hook. f. Fl. Tasm. i. 57; F. v. M. Pl. Vict. i. 170; *P. glomeratum*, Jacq.; DC. Prod. i. 659; *P. inodorum*, Willd.; DC. l.c.; Sweet, Geran. t. 56; *P. littorale*, Hueg. Bot. Arch. t. 5; *P. crinitum*, Nees, in Pl. Preiss. i. 163; *P. stenanthum*, Turcz. in Bull. Mosc. 1858, i. 149; *P. Drummondi*, Turcz. l.c. 421 (a robust form with large flowers).

Hab.: Stanthorpe, on rocks.

4. OXALIS, Linn.

(From *oxys*, sharp, the plants being very acid.)

Flowers regular. Sepals 5. Petals 5. Disk without glands. Stamens 10, free or united at the base, all bearing anthers. Ovary 5-lobed, 5-celled, without any beak or with a very short one; styles 5, with terminal stigmas, capitate or lobed; ovules 1, 2, or several in each cell. Capsule opening loculicidally, the valves persistent on the axis. Seeds with an outer fleshy coating, opening elastically, with the appearance of an arillus; testa crustaceous; albumen fleshy;

embryo straight.—Herbs. Leaves alternate or radical, compound; leaflets 3, digitate, or, in species not Australian, 3 or more and pinnate. Stipules scale-like or none. Peduncles axillary or radical, 1-flowered or bearing an umbel of several flowers.

A large genus, especially abundant in South America and extratropical South Africa, with a very few species widely dispersed over the temperate or tropical regions of the globe. Of the two Queensland species, one is common to New Zealand and Antarctic America, and perhaps not different from a common northern one.—*Benth.*

Stemless. Flowers lilac. Peduncles radical, many-flowered . . . . . 1. *O. corymbosa.*
Stem elongated. Flowers small, yellow. Peduncles axillary, 1 or more-flowered . . . . 2. *O. corniculata.*
Stems slender. Leaves abruptly pinnate. Peduncles terminal . . . . . 3. *O. sessilis.*

1. **O. corymbosa** (corymbose), *DC.* Stemless, with a rather deep transparent, clavate tuber, at the top of which are formed numerous bulbils, with brown membranous coats. Peduncles and petioles 6 to 12 lines long, more or less pilose. Leaflets 3, sessile, 1 to 2in. broad, deeply obcordate at the apex, cuneate at the base, pubescent beneath, and more or less marked, especially near the margins, with glandular yellow dots similar to those on the back of the sepals. Flowers in each corymb from 10 to 20. Sepals lanceolate, 2 or 3 lines long, rather blunt, with 2 linear dots on the back near the apex. Petals red, about 3 times as long as the sepals, oblong. Capsule oblong, seeds numerous, downy. —*O. bipunctata,* Grah.; *O. Martiana,* Zucc.

Hab.: This South American species has become quite a pest in Queensland gardens. According to Baker, Fl. Mauritius, it has also become naturalised at Mauritius.

2. **O. corniculata** (small horn bearing), *Linn.; DC. Prod.* i. 692; *Benth. Fl. Austr.* i. 301. Yellow-wood Sorrel. A decumbent, prostrate or ascending, much-branched, delicate perennial or sometimes annual, more or less pubescent, of a pale green, from a few inches to a foot long. Stipular scales small, adnate to the petiole. Leaves alternate; leaflets 3, broadly obcordate, usually 3 or 4 lines long, but sometimes half that size. Peduncles axillary, about the length of the petioles, bearing an umbel of several small yellow flowers, rarely reduced to 1 or 2, on reflexed pedicels. Capsule column-like, often above $\frac{1}{2}$in. long, with several seeds in each cell, rarely short and few-seeded.—Reichb. Ic. Fl. Germ. v. t. 199; Wight, Ic. t. 18; Hook f. Fl. Tasm. i. 59; F. v. M. Pl. Vict. i. 177; *O. microphylla,* Poir.; DC. Prod. i. 692; *O. perennans,* Haw.; DC. l.c. 691 (from the character given); *O. Preissiana* and *O. cognata,* Steud. in Pl. Preiss. i. 160.

Hab.: Islands of the coast as well as on the mainland, Keppel Bay, *R. Brown;* Percy Island, *A. Cunningham* and others; and in the interior as far north as the Burdekin, *F. v. M., Mitchell;* throughout the colony.

3. **O. sessilis** (sessile), *Hamilt.* Stems 1 to 6in. high, slender. Leaves 1 to near 2in. long, crowded with from 5 to 7 pairs of ciliate leaflets, the lowest smallest, orbicular-ovate, the middle ones truncate at the base, the terminal with a contracted oblique base and arched midrib, the lateral nerves prominent, horizontal, wavy; petiole puberulous. Pedicels sessile; sepals exceeding the pedicels, about 2 lines long, striate, with scarious margins. Corolla purple or red, or appears so in the dry specimen. Capsule slightly exceeding the calyx. Seeds tuberculate.—Hook. Fl. Brit. Ind. i. 437 (in part); *O. Petersii,* Klotz; *Biophytum Apodiscias,* Turcz.

Hab.: Musgrave Electric Telegraph Station, *T. Barclay-Millar.*
Leaves endowed with gentle movements under the influence of light, darkness, or shocks, like those of *Mimosa pudica, &c.*

Order XXIX. RUTACEÆ.

Flowers regular and hermaphrodite, or very rarely unisexual. Calyx usually small, 4 or 5-lobed, or divided into as many distinct imbricate sepals, rarely large, or with fewer or more numerous or valvate lobes. Petals of the same number as sepals, free or rarely cohering, hypogynous or slightly perigynous, imbricate or valvate in the bud. Stamens usually free, either equal in number to the petals and alternate with them, or double the number, or rarely more numerous, when twice as many as petals the sepaline ones (those opposite the sepals) usually longer than the others. Anthers usually versatile, with 2 parallel cells opening longitudinally, the connective occasionally tipped by a gland or projecting appendage. Torus usually more or less thickened into an entire crenate or lobed disk, within the stamens, under or round the ovary. Gynœcium of 4 or 5, rarely more or fewer carpels, more or less united into a single lobed or entire ovary, or rarely quite distinct, with one cell to each carpel. Styles as many as carpels, either free at the base but united upwards, or united from the base; stigma terminal, entire or lobed. Ovules usually 2 in each cell, superposed or rarely collateral or solitary, or more than 2; the micropyle superior. Fruit separating into 2-valved or rarely indehiscent cocci, or the carpels united in an indehiscent berry or drupe, or rarely in a loculicidally dehiscent capsule, the endocarp frequently separating from the pericarp. Seeds usually solitary in each cell; testa crustaceous and often shining, or rarely coriaceous or membranaceous; albumen fleshy or none. Embryo straight or curved, large in proportion to the seed; cotyledons flat or rarely folded; radicle superior.—Trees or shrubs, very rarely herbs, marked with glandular pellucid dots on the leaves and other thin herbaceous parts. Indumentum usually stellate, if any. Leaves opposite or alternate, simple or compound, entire or rarely toothed or lobed. Stipules none. Flowers axillary or terminal, solitary, clustered, cymose, or paniculate, very rarely racemose and seldom if ever spicate.

A large Order, ranging over the hotter and temperate regions of the whole World, but chiefly abundant within the tropics, in South Africa and in Australia. Among the Australian genera, the large tribe of *Boronieæ* is entirely endemic, with the exception of one New Zealand and one New Caledonian species. The monotypic genera, *Bosistoa*, *Medicosma*, and *Pentaceras*, and the small genus *Geijera*, are also endemic. *Melicope* extends to the Pacific Islands, and the remaining genera range over tropical Asia, three of them extending into Africa. *Zanthoxylum* alone, a wide-spread tropical genus, is common to America and Australia, and even here the Australian species belong to the exclusively Australasian section *Blackburnia*.—*Benth.*

Difficult as it is to distinguish *Rutaceæ* by well-marked floral or carpological characters from *Geraniaceæ*, *Zygophylleæ*, or *Simarubeæ*, they are so readily known by their dotted exstipulate leaves, that the ambiguous genera are remarkably few. They have usually been distributed into 3 or 4 Orders, *Rutaceæ* (including or not *Diosmeæ*), *Zanthoxyleæ*, and *Aurantieæ*, upon characters which break down upon a close scrutiny; the *Toddalieæ* being much nearer to the *Aurantieæ* than to the *Zanthoxyleæ* proper, which again have only vague differences to distinguish them from *Boronieæ*. We therefore, in our "Genera Plantarum," proposed the union of the whole into 1 Order, divided into 2 series, according as the ovary is lobed or entire, and subdivided into 7 tribes, of which 4 only are Australian.—*Benth.*

Tribe I. **Boronieæ.**—*Shrubs, very rarely arborescent. Leaves simple, 3-foliolate or rarely pinnate, with opposite small leaflets. Ovary lobed. Fruit separating into distinct, 2-valved cocci. Endocarp separating elastically. Seeds albuminous. Embryo usually terete.*

(The tribe differs from the S. African *Diosmeæ* chiefly in the presence of albumen.—*Benth.*)

Leaves opposite (except in one *Zieria*) simple or compound.
  Petals 4, united or connivent in a cylindrical or campanulate corolla.
    Leaves petiolate, simple . . . . 8. Correa.
  Petals 4, free, spreading.
    Stamens 4, inserted on 4 prominent glands or lobes of the disk . . . . 1. Zieria.
    Stamens 8. Disk without prominent glands (excepting *B. tetrandra*) . 2. Boronia.
Leaves alternate, simple.
  Flowers distinct or in sessile, erect heads.
    Petals free. Stamens twice as many, monadelphous.
      Stamens all perfect . . . . 6. Philotheca.

Petals free. Stamens twice as many, free.
  Calyx inconspicuous or none. Petals induplicate-valvate, tomentose
    outside . . . . 7. ASTEROLASIA.
  Calyx distinct but shorter than the petals.
    Petals broad, much imbricate, not scurfy, without inflexed tips.
      Filaments hairy.
        Anthers minutely or not at all apiculate . . . . 4. ERIOSTEMON.
        Anthers tipped with long, horn-like, hairy appendages . . . . 3. CROWEA.
    Petals valvate or slightly imbricate, with inflexed valvate tips, glabrous
      or scaly.
      Ovary of 5, rarely fewer carpels, the styles attached below the
        middle . . . . 5. PHEBALIUM.
      Petals free. Stamens of the same number, free . . . . 16. GEIJERA.

TRIBE II. **Zanthoxyleæ.**—*Trees or shrubs. Leaves pinnate or 3-foliolate with opposite leaflets, or 1-foliolate (truly simple in Geijera), the leaflets usually large. Ovary lobed. Fruit separating into distinct 2-valved cocci. Endocarp persistent, or separating elastically. Seeds with or without albumen. Cotyledons usually flattened and broader than the radicle.*

Stamens twice as many as petals.
  Leaves all or mostly opposite. Cocci dehiscent.
    Leaves pinnate. Petals valvate or slightly imbricate. Seeds without
      albumen . . . . 9. BOSISTOA.
    Leaves 3-foliolate. Petals valvate or slightly imbricate, with inflexed
      tips . . . . 10. MELICOPE.
    Leaves 1-foliolate. Petals large, broadly imbricate, not inflexed . . . 12. MEDICOSMA.
  Leaves alternate, pinnate. Petals valvate. Cocci winged, indehiscent . 17. PENTACERAS.
Stamens the same number as petals. Cocci dehiscent.
  Leaves all or mostly opposite, usually 3-foliolate . . . . 11. EVODIA.
  Leaves simple. Stamens 8. Filaments almost ovate, 4-petaloid, sterile . 13. BROMBYA.
  Leaves simple, or 2 or 3-foliolate. Stamens 10, free, all fertile . . . . 14. PAGETIA.
  Leaves alternate, simple . . . . 16. GEIJERA.
  Leaves alternate, pinnate . . . . 15. ZANTHOXYLUM.
  (See also *Flindersia* among *Meliaceæ*.)
  Leaves simple. Petals deciduous. Cocci 5 to 9 . . . . 18. PLEIOCOCCA.

TRIBE III. **Toddalieæ.**—*Trees or shrubs, with the habit of Zanthoxyleæ. Ovary not lobed. Fruit several-celled, indehiscent, or rarely loculicidally dehiscent. Seeds albuminous (in the Australian genus).*

Leaves 1-foliolate. Stamens twice as many as petals . . . . 19. ACRONYCHIA.
Leaves simple. Stamens 10 . . . . 20. HALFORDIA.

TRIBE IV. **Aurantieæ.**—*Trees or shrubs. Leaves pinnate, with usually alternate leaflets, or 1-foliolate or simple. Stamens twice as many as petals or more. Ovary not lobed. Fruit indehiscent. Seeds without albumen.*

Leaves all or mostly pinnate. No thorns.
  Flowers in terminal, flat, corymbose panicles. Filaments subulate
    Petals valvate or nearly so. Cotyledons much folded. Flowers small . 22. MICROMELUM.
    Petals imbricate, erect. Cotyledons flat. Flowers large . . . . 23. MURRAYA.
  Flowers in oblong, pyramidal, or loose axillary or terminal panicles. Fila-
    ments dilated at the base or middle.
    Ovules solitary. Leaflets few . . . . 21. GLYCOSMIS.
    Ovules 2 in each cell. Leaflets numerous . . . . 24. CLAUSENA.
Leaves all simple or 1-foliolate, coriaceous. Thorns axillary.
  Ovary 5 or fewer-celled, with 1 or 2 ovules in each cell . . . . . . 25. ATALANTIA.
  Ovary 6 or more-celled, with 4 or more ovules in each cell . . . . . 26. CITRUS.

1. ZIERIA, Sm.

(After M. Zier.)

Calyx 4-cleft. Petals 4, imbricate or almost valvate in the bud, spreading. Disk with 4 distinct gland-like lobes, alternating with the petals. Stamens 4, inserted on the outside of the glands of the disk. Carpels 4, distinct or nearly so; styles nearly terminal, short and united at least at the top; stigma capitate, 4-furrowed or shortly 4-lobed. Ovules 2 in each carpel, superposed. Cocci 4, 2-valved, the endocarp cartilaginous and separating elastically. Seeds solitary, or rarely 2 in each coccus, oblong; testa crustaceous.—Shrubs or rarely small trees, glabrous, hirsute or tomentose. Leaves usually opposite, with 3 leaflets, rarely alternate or simple. Flowers white, usually small, axillary, in small trichotomous cymes or rarely solitary.

The species are all endemic in Australia, and F. v. Mueller considers them as forming a section only of *Boronia*; but the characters and habit appear to me sufficiently distinct to justify the maintenance of so old-established and generally adopted a genus.—*Benth.*

Anthers distinctly apiculate. Plant glabrous or slightly pubescent.
    Leaflets with revolute margins. Cymes pedunculate.
    Branchlets angular, glabrous. Leaflets ½ to 1in. on a distinct common
      petiole . . . . 1. *Z. lævigata.*
    Branchlets terete, pubescent. Leaflets under ½in., sessile, appearing
      verticillate . . . . 2. *Z. aspalathoides.*
Anthers minutely apiculate. Plant pubescent or hirsute, rarely tomentose.
    Flowers 1 to 3, small. Calyx-segments very narrow, nearly as long as
      the petals . . . . 3. *Z. pilosa.*
Anthers not apiculate. Calyx-lobes short.
    Flowers 1 to 3, on short axillary pedicels. Leaves densely pubescent or
      tomentose.
    Leaflets 3, small, obovate or obcordate. Flowers very small . . . . 4. *Z. obcordata.*
    Flowers in pedunculate cymes or heads, with leafy bracts. Leaves
      densely tomentose or villous.
    Leaves all 3-foliolate. Cymes not capitate . . . . 5. *Z. cytisoides.*
    Flowers in loose pedunculate cymes, with small bracts.
    Glabrous or slightly pubescent.
      Leaflets flat, lanceolate. Petals distinctly imbricate . . . . 6. *Z. Smithii.*
      Leaflets narrow-linear. Flowers small, the petals almost valvate . 7. *Z. granulata.*

1. **Z. lævigata** (smooth-leaved), *Sm.; DC. Prod.* i. 723; *Benth. Fl. Austr.* i. 304. A glabrous, erect shrub, the branchlets angular. Leaflets 3, on a common petiole of 1 to 3 lines, linear, pointed, ½ to 1in. long, the margins closely revolute. Cymes few-flowered, mostly about as long as the leaves. Calyx-lobes short and broad. Petals fully 3 times as long as the calyx, broad, imbricate, slightly tomentose outside. Connective of the anthers distinct, produced beyond the cells into a short point or appendage. Style very short. Cocci and seeds of *Z. Smithii.*—Deless. Ic. Sel. iii. t. 49; Paxt. Mag. Bot. ix. 77, with a fig.; *Boronia lævigata*, F. v. M. Fragm. i. 101; *Z. revoluta*, A. Cunn. in Field, N. S. Wales, 330.

Hab.: Sandstone rocks near Mount Pluto, *Mitchell*: Mount Lindsay, *Fraser.*

Var. *laxiflora.* Leaflets longer (1 to 1½in.), on a longer common petiole. Flowers much smaller, in a looser cyme. Petals not twice as long as the calyx.—Stradbroke Island, *Fraser;* Moreton Island, *F. v. Mueller.*

2. **Z. aspalathoides** (Aspalathus-like), *A. Cunn. Herb.; Benth. Fl. Austr.* i. 305. A heath-like shrub, the branches terete and pubescent, but usually with a decurrent glabrous line. Leaflets 3, sessile or with the common petiole so exceedingly short that they appear verticillate, lanceolate or linear, rarely above 3 lines long, or when very luxuriant 4 or 5 lines, the margins revolute, glabrous or slightly pubescent. Cymes usually 3-flowered, rather longer than the leaves. Calyx-lobes broad, obtuse or acute. Petals about 2 or 3 times as long. Anthers tipped with a small obtuse appendage.--*Boronia lævigata*, F. v. M. Pl. Vict. i. 111 (in part).

Hab.: Near Mount Playfair, *F. v. M.*

3. **Z. pilosa** (pilose), *Rudge, in Trans. Linn. Soc.* x. 293, t. 17; *Benth. Fl. Austr.* i. 305. A shrub or undershrub, the branches terete and densely pubescent or hirsute. Leaflets 3, with a short common petiole, linear, oblong or lanceolate, obtuse, ½ to ⅔in. or rarely 1in. long, the margins recurved or revolute, slightly pubescent or glabrous above, more or less hirsute or tomentose underneath.

Flowers small, solitary and nearly sessile or 2 or 3 together on short pedicels. Calyx hirsute, with linear-subulate or narrow-lanceolate lobes, nearly as long as the petals and always much narrower than in any other species. Anthers minutely apiculate. Cocci hirsute, broader than in most species.—DC. Prod. i. 728; *Z. pauciflora*, Sm. in Rees. Cycl. xxxix.; DC. l.c.; *Z. hirsuta*, DC. l.c.; Deless. Ic. Sel. iii. t. 50; *Boronia hirsuta*, F. v. M. Fragm. i. 101.

Hab.: Herberton, *Rev. J. E. Tenison-Woods*, and *J. F. Bailey* a form of the above species.

4. **Z. obcordata** (form of leaflets), *A. Cunn. in Field, N. S. Wales*, 330; *Benth. Fl. Austr.* i. 305. A shrub of low growth, with elongated diffuse branches, terete and softly hirsute. Leaflets 3, with a very short common petiole, obovate or obcordate, 2 to 4 lines or rarely ½in. long, softly pubescent or tomentose above, more hirsute or velvety and whitish underneath, the margins recurved or revolute. Flowers 1 to 3 in the axils, very small, on short slender pedicels, the petals not above 1 line and the calyx about half as long with broad and obtuse segments. Anthers not apiculate. Cocci small, glabrous.—*Boronia minutiflora*, F. v. M. Fragm. i. 100.

Hab.: Glasshouse Mountains.

5. **Z. cytisoides** (resembling a Cytisus), *Sm.; DC. Prod.* i. 728 ; *Benth. Fl. Austr.* i. 306. A much-branched shrub, hoary all over with a soft close or more or less velvety tomentum. Leaflets 3, with a common petiole of 1 to 3 lines, obovate-oblong, about ½ or rarely ¾in. long, obtuse or minutely pointed, the margins revolute, narrowed at the base. Cymes dense but few-flowered, rarely much exceeding the leaves. Bracts leafy, as long as the pedicels or often nearly as long as the flowers. Calyx rather short, with broad acute segments. Petals rarely twice as long, much imbricate in the bud. Anthers not apiculate.

Hab.: Southern localities inland.

6. **Z. Smithii** (after Sir James Smith, founder of the Linnean Society), *Andr. Bot. Rep. t.* 606 (1810) ; *Benth. Fl. Austr.* i. 306. A tall shrub or small tree, glabrous or slightly pubescent with a very minute usually stellate down, the branches terete or compressed, occasionally covered with glandular tubercles. Leaflets 3, with a distinct common petiole, lanceolate or the larger ones oblong, elliptical, acute or rarely obtuse, 1 to 2in. long in the original form, flat or the margins slightly recurved. Flowers usually about 3 lines diameter, in axillary 2—3-chotomous cymes, shorter than the leaves. Calyx-lobes broad and short. Petals fully 3 times as long as the calyx, tomentose outside. Anthers obtuse, not apiculate. Cocci about 2 lines long, glabrous, usually glandular-tuberculate. Seeds shining, finely reticulate-striate.—Bot. Mag. t. 1395; Bonpl. Jard. Malm. 62, t. 24 ; *Z. lanceolata*, R. Br.; DC. Prod. i. 728; Hook. f. Fl. Tasm. i. 65 ; *Boronia arborescens*, F. v. M. Fragm. i. 100, and Pl. Vict. i. 111.

Hab.: Brisbane River, *A. Cunningham*; Stradbroke Island, *Fraser*.

Var. *parvifolia*. Leaflets rarely exceeding 1in.; cymes often as long.—Sandy Bay and Cape Hervey, *R. Brown*; New England, *Stuart*.

Var. *macrophylla*. More arborescent; leaflets often 3in. long; flowers larger than in the ordinary form; seeds broader and less reticulate.—*Z. arborescens*, Sims; Hook. Journ. Bot. i. 256 ; *Z. macrophylla*, Bonpl.; Deless. Ic. Sel. iii. t. 48; Bot. Mag. t. 4451. To this variety belong the Tasmanian and many of the Victorian specimens.

The stamens in this and other *Zierias* are figured in Delessert's "Icones," by some mistake, as attached inside instead of outside the glands or lobes of the disk. The name of *Z. lanceolata* was adopted by Smith (in Rees' Cycl. xxxix.), on the consideration that the synonym quoted in the Bot. Mag. was a sufficient publication ; Andrews' name, had, however, been published a year previous to the plate in Bot. Mag.—*Benth.*

7. **Z. granulata** (granulated), *C. Moore in Herb. Hook.; Benth. Fl. Austr.* i. 307. A tall shrub or small tree, glabrous or very minutely pubescent, and densely covered with glandular tubercles as in some varieties of *Z. Smithii*, with

which F. v. Mueller proposes to unite it. It differs chiefly in the narrow-linear leaflets, 1 to 2in. long, the margins revolute and whitish underneath, and in the very small flowers, with the petals almost strictly valvate. Cocci glabrous.—*Boronia granulata*, F. v. M. Fragm. i. 101.

Hab.: Belmont, near Brisbane.

2. BORONIA, Sm.
(After F. Borone.)

Calyx 4-cleft. Petals 4, either much imbricate or valvate in the bud, spreading. Disk thick, entire or (in one species only) with 4 gland-like lobes. Stamens 8, inserted outside the disk; anthers either all similar and perfect or 4 different from the others and imperfect. Carpels of the ovary 4, distinct or nearly so; styles terminal, united; stigma entire or 4-lobed. Ovules 2 in each carpel, superposed or rarely collateral. Cocci usually 4, 2-valved, the endocarp cartilaginous and separating elastically. Seeds solitary or rarely 2 in each coccus, oblong; testa crustaceous.—Shrubs, undershrubs, or rarely annuals, glabrous, pubescent or hirsute, rarely tomentose. Leaves opposite, simple, pinnate with a terminal leaflet, or once or twice ternately compound, the rhachis usually articulate at each pair of leaflets and often dilated between them. Peduncles axillary or terminal, either 1-flowered and jointed with a pair of minute bracts at the joint, or bearing an umbel or dichotomous cyme of several flowers with small bracts at the base of the pedicels. Flowers red, white, purple or blue. Calyx-segments or sepals usually valvate when the petals are valvate and sometimes also when they are imbricate, but in the latter case the sepals are usually also imbricate at the base. In some species the anthers and stigma are different in different individuals of the same variety. In most of the species the filaments of the sepaline stamens (those alternating with the petals) are longer and more distinctly clavate or capitate and glandular at the top than the petaline ones. Anthers usually very shortly stipitate, rather below the obtuse summit of the filament.

The species are all limited to Australia.

SERIES I. **Valvatæ.**—*Petals strictly valvate. Sepals usually valvate.*
Sepals longer than the petals. Leaves mostly or all pinnate . . . 1. *B. artemisiæfolia*.
Sepals much smaller than the petals.
  Inflorescence entirely axillary.
    Peduncles 1-flowered.
      Leaflets (usually 5 or 7) obovate or cuneate, glabrous, complicate.
      Flowers tomentose, rather large . . . . 2. *B. eriantha*.
      Leaflets 7 to 13 or more, small, linear or oblong, the margins revolute. Sepals lanceolate, subulate-acuminate . . . . . . 3. *B. alulata*.
      Leaflets 3, rarely 5, the margins recurved or revolute, tomentose or hoary underneath . . . . 4. *B. ledifolia*.
      Leaves simple.
        Leaves linear or linear-lanceolate. Flowers about 4 lines . . . 4. *B. ledifolia*.
        Leaves oblong-lanceolate. Flowers about 2 lines . . . 5. *B. lanceolata*.
    Peduncles bearing an umbel of several flowers.
      Leaves simple, lanceolate, tomentose underneath. Flowers small . 5. *B. lanceolata*.
      Leaves pinnate, glabrous. Flowers in umbels, 1—5. Filaments ciliate . . . . 6. *B. Bowmani*.

SECTION II. **Pinnatæ.**—*Petals imbricate. Anthers nearly uniform. Leaves pinnate. Peduncles mostly axillary.*

Peduncles mostly 3 or several-flowered. (Eastern species.)
  Glabrous. Leaflets small, thick, obovate . . . . 7. *B. microphylla*.
  Glabrous. Leaflets linear or oblong in distant pairs . . . . 8. *B. pinnata*.

SERIES III. **Variabiles.**—*Petals imbricate. Anthers nearly uniform. Leaves simple or 3-foliolate, or the terminal leaflet or all three again 3-foliolate. Flowers axillary, red or pink.*

Terminal leaflets or all three dentate, or again 3 or 5-foliolate. Erect or spreading shrub. Peduncles usually 3 to 5-flowered . . . . 10. *B. anemonifolia*.

Leaves mostly 3-foliolate.
Common petiole distinct.
  Leaflets flat, linear oblong or obovate. Anthers apiculate. Pedicels
  1-flowered . . . . 9. *B. polygalifolia.*
  Leaflets linear-terete, mucronate. Anthers not apiculate. Pedicels 1
  to 3-flowered . . . . 11. *B. falcifolia.*
  Leaves linear or lanceolate, acute, or the lower ones rarely cuneate.
  Low undershrub. Flowers all axillary. Sepals short . . . . . 9. *B. polygalifolia.*

SERIES IV. **Terminales.**—*Petals imbricate. Anthers nearly uniform. Leaves all simple (except in* B. filifolia, inornata, *and* oxyantha). *Flowers mostly or all terminal, sessile or on short 1-flowered peduncles.*

Terminal flowers solitary, or rarely 2 or 3, sessile or shortly pedicellate.
  Leaves linear or lanceolate, rarely oblong-cuneate, flat.
  Small undershrub. Filaments nearly glabrous. Anthers not apiculate 12. *B. parviflora.*

1. **B. artemisiæfolia** (Artemisia-leaved) *F. v. M. Fragm.* i. 66; *Benth. Fl. Austr.* i. 311. A shrub, clothed all over with a soft, hoary, close or velvety tomentum. Leaves all, or nearly all, pinnate. Leaflets 7 to 11 or more, crowded on a short common petiole, linear, obtuse, rarely exceeding ½in. and often much shorter, the margins closely revolute. Peduncles axillary, solitary, short, 1-flowered. Sepals lanceolate, tomentose, valvate, attaining 3 to 4 lines. Petals lanceolate, valvate and tomentose like the sepals, but smaller and enclosed in them in the bud. Filaments slightly hirsute, clavate and glandular at the top. Anthers scarcely apiculate. Ovary pubescent. Seeds smooth but scarcely shining.

Hab.: Islands of Gulf of Carpentaria, *R. Brown.*

2. **B. eriantha** (alluding to the woolly flowers), *Lindl. in Mitch. Trop. Austr.* 298; *Benth. Fl. Austr.* i. 313. A glabrous shrub, the branches angular. Leaves pinnate; leaflets 3 to 9, obovate or oblong-cuneate, obtuse or with a recurved point, rarely above 3 lines long, rather thick, and often folded upwards lengthwise, the margins never recurved. Peduncles axillary, short, 1 or rarely 2-flowered. Sepals ovate, acute, glabrous outside, minutely tomentose inside. Petals more than twice as long, attaining 3 or 4 lines, rather narrow, valvate, hoary-tomentose outside, with a prominent midrib. Filaments usually ciliate; anthers apiculate.

Hab.: Near Mt. Pluto, *Mitchell.*

With the aspect of *B. microphylla* this has the floral characters of *B. ledifolia*, with which F. v. Mueller proposes to unite it, but besides a totally different habit, the leaflets are thick, equally green on both sides, with the margins flat or folded upwards, not recurved with a pale or hoary-tomentose under surface as in *B. ledifolia.—Benth.*

3. **B. alulata** (small-winged), *Soland. in Herb. Banks; Benth. Fl. Austr.* i. 313. Apparently a divaricate or diffuse shrub, the young branches glandular-tomentose. Leaves pinnate; leaflets 7 to 13 or even more, oblong or linear, rarely almost ovate, obtuse, 2 to 3 lines long, the margins revolute, glabrous above when full-grown, hoary-tomentose underneath. Peduncles very short, axillary, 1-flowered. Sepals lanceolate, subulate-acuminate, from half to nearly as long as the petals. Petals about 3 lines long, mucronate, valvate in the bud but rather broad, glabrous outside with a prominent midrib, slightly tomentose inside. Filaments clavate and glandular upwards.

Hab.: Endeavour River, *Banks* and *Solander, R. Brown (Hb. Brit. Mus. and R. Br.)*

4. **B. ledifolia** (Ledum-leaved), *J. Gay; DC. Prod.* i. 722; *Benth. Fl. Austr.* i. 314. An erect shrub, the young branches glandular-tomentose. Leaves simple, 3-foliolate, or rarely pinnately 5 or even 7-foliolate; leaflets linear, oblong-linear, lanceolate or rarely broadly oblong, when single often above 1in. long, when several rarely above ½in., the margins recurved or revolute, glabrous

above when full-grown, hoary or rusty underneath with a minute tomentum. Peduncles axillary, 1-flowered, shorter than the leaves. Sepals broad, obtuse but valvate. Petals twice as long or more, attaining 4 or 5 lines, valvate in the bud, minutely tomentose outside, with a prominent midrib. Filaments short, as in several allied species, slightly ciliate or glabrous, clavate and glandular upwards; anthers more or less apiculate. Ovules usually, as in some allied species, almost or quite collateral. Style clavate, with a slightly furrowed stigma. Seeds smooth but not shining.—Reichb. Icon. Exot. t. 74; *Laxiopetalum ledifolium*, Vent. Jard. Malm. under n. 59; *Eriostemon paradoxum*, Sm. in Rees, Cycl. xiii.; *Boronia (?) paradoxa*, DC. Prod. i. 722.

Hab.: Burnett River, *F. v. M.*; Moreton Bay and Islands, *A. Cunningham, Fraser, etc.*

Var. *rosmarinifolia*. Leaves rigid, usually narrow, small, and all simple. Peduncles very short.—*B. rosmarinifolia*, A. Cunn. in Hueg. Enum. 16. To this form belong especially most of the Moreton Bay and other south Queensland coast specimens.

5. **B. lanceolata** (lanceolate), *F. v. M. Fragm.* i. 66; *Benth. Fl. Austr.* i. 314. A tall shrub with tomentose branches. Leaves simple, petiolate, oblong-lanceolate, obtuse or mucronulate, 1 to 2in. long, flat or the margins recurved, glabrous above, tomentose underneath. Peduncles very short, bearing an umbel of 3 to 5 small flowers, rarely reduced to a single flower. Sepals small, ovate, with a subulate point, sometimes very short, sometimes nearly as long as the petals. Petals broad, attaining about 2 lines in length, valvate in the bud, tomentose outside with a prominent midrib. Filaments glabrous, thickened and glandular at the top; anthers scarcely apiculate. Seeds smooth but not shining.

Hab.: Cave Creek, Dalby, *W. E. Armit* (F. v. M.)

6. **B. Bowmani** (after E. Bowman), *F. v. M. Fragm.* iv. 185; *B. platyr-rhachis*, F. v. M. Fragm. vii. 37. A slender-branched glabrous or nearly glabrous shrub with pinnate foliage. Leaflets usually 5, broad-linear, entire on a more or less winged rhachis. Peduncle axillary, short or very short, 1 to 5-flowered in the umbel. Sepals valvate, lanceolate. Petals lanceolate, about 2 lines long, greenish. Stamens with ciliate filaments, apiculate. Style very short, glabrous. Stigma minute. Cocci 2 lines long. Seeds black, hardly exceeding 1½ line long.

Hab.: Cape River, *Bowman* (F. v. M.); Percy and Gilbert Rivers, *R. Daintree* (F. v. M.)

7. **B. microphylla** (small-leaved), *Sieb. in Spreng. Syst. Cur. Post* 148; *Benth. Fl. Austr.* i. 318. A low stunted shrub, glabrous but often very glandular. Leaves pinnate: leaflets 5 to 11, obovate or oblong-cuneate, obtuse or acute, rarely above 3 lines long, and usually about 2 lines, thick and rigid. Peduncles in the upper axils, 1 to 3-flowered. Flowers of *B. pinnata*, or rather smaller, the anthers often conspicuously apiculate. Stigma slightly enlarged. Seeds in our specimens shining and reticulate.—Reichb. Icon. Exot. t. 72.

Hab.: Stanthorpe.

8. **B. pinnata** (leaves pinnate), *Sm. Tracts*, 290, t. 4; *Benth. Fl. Austr.* i. 318. A glabrous shrub, attaining several feet, but sometimes dwarf or diffuse, the small branches more or less angular. Leaves pinnate; leaflets 5 to 9 or rarely more, linear or oblong-lanceolate, acute, rigid, the pairs rather distant and the common petiole often dilated between them. Flowers rather large, usually 3 or more together, in loose axillary or subterminal corymbose cymes. Sepals small, acute. Petals attaining 3 to 5 lines, imbricate, glabrous or minutely tomentose inside, usually mucronate. Filaments woolly-hairy, especially towards the thickened summit; anthers very minutely or not at all apiculate. Style short. Seeds smooth and shining.—DC. Prod. i. 721; Andr. Bot. Rep. i. 58; Vent. Jard. Malm. t. 88; Bot. Mag. t. 1763; F. v. M. Pl. Vict. i. 115; *B. floribunda*, Sieb. in Spreng. Syst. Cur. Post. 148; Reichb. Icon. Exot. t. 71.

Hab : Islands of Moreton Bay and similar localities in the south of Queensland.

9. **B. polygalifolia** (Polygala-leaved), *Sm. Tracts* 297, *t.* 7; *Benth. Fl. Austr.* i. 8. Usually a low glabrous undershrub with a thick rhizome as in *B. parviflora*, or a small shrub, rarely stouter and 1 to 2ft. high. Leaves either simple with lanceolate or linear-lanceolate acute leaflets, mostly under ½in., but sometimes nearly 1in. long, or 3-foliolate with small acute leaflets, on a short common petiole. Pedicels axillary, solitary, and 1-flowered. Sepals short. Petals 2 or 3 times as long, imbricate, pink, and glabrous. Filaments hairy and glandular towards the top. Anthers conspicuously apiculate, the appendage erect or recurved. Seeds opaque and usually minutely tuberculate.—DC. Prod. i. 722; F. v. M. Pl. Vict. i. 114; *B. hyssopifolia*, Sieb. in Spreng. Syst. Cur. Post. 148; Hook. f. Fl. Tasm. i. 66; *B. tetrathecoides*, DC. Prod. i. 722; Hook. Comp. Bot. Mag. i. 277.

Hab.: Stradbroke Island and many other localities in southern Queensland.

Var. *robusta*. Leaves 3-foliolate, as in var. *trifoliolata*, but the stems are stout and the plant more shrubby, attaining 2ft. or more.—Moreton Island, *F. v. M.*

10. **B. anemonifolia** (Anemone-leaved), *A. Cunn. in Field, N.S. Wales*, 380; *Benth. Fl. Austr.* i. 321. A shrub of 2 or 3ft., glabrous or pubescent, and often glaucous. Leaves either simply 3-foliolate with the leaflets 3-toothed, or all 3 leaflets or the terminal one only again 3-foliolate or pinnately 5-foliolate, or sometimes some of them a third time divided, and all usually thick, linear-cuneate or, if entire, acutely linear. Flowers in axillary cymes of 3, 5, or even more, very rarely reduced to single flowers. Stamens and fruit of *B. polygalifolia*.

Hab.: Newcastle and Burnett Rivers, *F. v. M.*; near Lindley's Range, *Mitchell*.

Var. *anethifolia*. Leaves still more compound, often bipinnate, and leaflets narrower and more acute than in other varieties. Flowers 3 or more in the cyme.—*B. anethifolia*, A. Cunn.; Endl. in Hueg. Enum. 16; Lindl. Bot. Reg. 1841, under n. 47; *B. bipinnata*, Lindl. in Mitch. Trop. Austr. 225.—The common form in the interior of Queensland and N. S. Wales.—*Benth.* Fraser's Island, *Miss Lovell*.

11. **B. falcifolia** (curved leaves), *A. Cunn.; Lindl. in Bot. Reg.* 1841, *under n.* 47; *Benth. Fl. Austr.* i. 322. A glabrous, erect, heath-like shrub, with virgate branches. Leaves rather crowded, 3-foliolate; leaflets linear-terete, mucronate, mostly ¼ to ½in. long, on a common petiole rather shorter than themselves. Pedicels 1 to 3-flowered, in the upper axils. Bracts linear-subulate. Sepals lanceolate, subulate-pointed. Petals rather longer than the sepals, attaining 3 to 4 lines, acute, imbricate, glabrous. Filaments clavate and glandular upwards; anthers not apiculate. Stigma in some specimens capitate, in others not thicker than the style.—*B. paleifolia*, Endl. in Hueg. Enum. 16 (through a misreading of Cunningham's label).

Hab.: Islands of Moreton Bay *A. Cunningham, F. v. Mueller*, and others; Wide Bay, *Bidwill*.

12. **B. parviflora** (flowers small), *Sm. Tracts*, 295, *t.* 6; *Benth. Fl. Austr.* i. 324. A small, glabrous undershrub, forming a thick woody rhizome with numerous prostrate, ascending, or erect branching stems, usually under 6in. but sometimes nearly 1ft. long. Leaves all simple, from oblong to linear-lanceolate, rather acute, rarely ½in. long. Flowers small, terminal, solitary or few in a leafy terminal cyme, on short thickened pedicels, one or two rarely axillary by the abortion of the flowering branch. Sepals acute, 1½ to 2 lines long. Petals not much exceeding them, imbricate, glabrous. Filaments glabrous or slightly hairy and glandular towards the top; anthers very minutely or not at all apiculate. Ovary glabrous; style short and thick. Cocci small. Seeds smooth and shining.— DC. Prod. i. 721; F. v. M. Pl. Vict. i. 118; *B. pilonema*, Labill. Pl. Nov. Holl. i. 98, t. 126; DC. Prod. i. 722; Hook. f. Fl. Tasm. i. 66.

Hab.: Islands of Moreton Bay.

Some specimens much resemble at first sight some of the smaller forms of *B. polygalifolia*, but a careful examination of the inflorescence will always suffice to distinguish them, independently of the seeds.—*Benth*.

3. CROWEA, Sm.

(After James Crow, a British botanist.)

Calyx 5-cleft. Petals 5, imbricate in the bud. Disk annular. Stamens 10, shorter than the petals; filaments flattened, ciliate or woolly; anthers linear, hirsute, tipped with long hirsute appendages. Ovary 5-lobed; styles inserted above the middle of the carpels, immediately united into one filiform style with a small or globular stigma. Ovules 2, superposed or almost collateral. Cocci 2-valved, rounded or truncate at the top, the endocarp cartilaginous and separating elastically.—Glabrous shrubs or undershrubs. Leaves alternate, simple. Flowers rather large, red, purple or green, glabrous, solitary, axillary or terminal.

The genus is confined to Australia. It is united by F. v. Mueller with *Eriostemon*, from which it differs chiefly in the long hairy appendages of the anthers.—*Benth.*

1. **C. saligna** (Willow-leaved), *Andr. Bot. Rep. t.* 79; *Benth. Fl. Austr.* i. 329. Shrubby and erect, the branches prominently angular. Leaves mostly lanceolate, narrowed at each end, acute or obtuse, 1 to 2in. long, of a much thinner consistence than those of *Eriostemon salicifolius*, which this species sometimes resembles, in some specimens passing into a broadly oblong or elliptical-ovate shape, in others almost linear, like those of *C. exalata*. Flowers red, on axillary pedicels shorter than the leaves, thickened upwards, with 2 very minute bracts at their base. Sepals short and broad. Petals 7 to 9 lines long. Appendage of the anthers longer than the cells themselves. Style very short, with a large globular stigma. Cocci short, united to near the top. Seeds reticulate, somewhat shining.—Vent. Jard. Malm. t. 7; Bot. Mag. t. 989; DC. Prod. i. 720; *C. latifolia*, Lodd. in G. Don, Gen. Syst. i. 792; *Eriostemon Crowei* (partly), F. v. M. Pl. Vict. i. 119.

Hab.: Southern localities.

C. latifolia, Paxt. Mag. Bot. xiv. 222, with a fig., is one of the commonest forms of this species, In some specimens from Manly Beach, *Woolls (Herb. Muell.)*, the leaves are nearly twice as broad. In others from between Richmond River and Raymond Terrace, *A. Ralston (Herb. Muell.)*, they are linear, elongated, mostly rounded or truncate at the top. Again, in numerous specimens collected by *R. Brown* on the Hawkesbury River, they are linear, but smaller and more crowded, approaching those of *C. exalata*: but in all the pedicels are axillary and leafless.—*Benth.*

4. ERIOSTEMON, Sm.

(Alluding to the woolly stamens.)

Calyx 5-cleft or rarely 4-cleft. Petals 5, rarely 4, imbricate. Disk usually more or less thickened. Stamens 10, rarely 8, shorter than the petals; filaments hairy, attenuate or rarely obtuse at the top; anthers usually tipped with a very small point or appendage. Carpels 5, rarely 4 or fewer, distinct from the base (or in one species united to the middle), usually produced into a short appendage above the cells; styles inserted below the middle and immediately united into one; stigma small. Ovules 2 in each cell, superposed. Cocci 2-valved, usually more or less beaked at the top or at the outer angle; the endocarp cartilaginous and separating elastically. Seeds solitary.—Shrubs, either glabrous or slightly pubescent, without scurfy scales. Leaves alternate, simple, entire, the glands often large and prominent. Inflorescence axillary or terminal; peduncles bearing a single flower, or an umbel of few white, pink, or rarely blue flowers. Calyx small, with short broad lobes or sepals.

Besides the Australian species, which are all endemic, the genus comprises one from New Caledonia. F. v. Mueller proposes to extend its limits so as to include *Phebalium, Microcybe, Geleznovia, Crowea, Philotheca, Drummondita,* and *Asterolasia,* which are all no doubt nearly enough related to it to be equally well regarded as sections or as substantive genera; but as the majority of them have been long established and universally adopted, and are distinguished by characters easily recognised, their union into one vast genus seems to me to be scarcely justified.—*Benth.*

Inflorescence axillary.
  Filaments clavate and glandular at the top.
    Leaves linear or lanceolate, thick, obscurely 1-nerved. Bracts on the
      pedicel several, imbricate . . . . 1. *E. salicifolius.*
    Leaves oblong, finely 3-nerved. Bracts on the pedicel 1 to 3, distant   2. *E. Banksii.*
  Filaments subulate at the top, usually flattened below.
    Flowers 5-merous.
      Leaves oblong or lanceolate, 1 to 3 or 4in. long, flat, 1-nerved.
        Pedicels rigid, usually several-flowered. Carpels free from the
          base, rostrate when ripe . . . . 3. *E. myoporoides.*
      Leaves linear, or linear-spathulate, mucronate, with recurved
        margins and a prominent midrib . . . . 4. *E. hispidulus.*
      Leaves narrow-linear, convex underneath or terete,
        Filaments flat . . . . 5. *E. scaber.*
    Inflorescence terminal, appearing sometimes lateral by the elongation of
      the side shoots.
    Flowers solitary or rarely 2 or 3 together.
      Leaves small, flat or with recurved margins.
        Leaves not above 2 lines long, thick, warted or crenate with large
          prominent glands . . . . 6. *E. difformis.*
        Leaves flat, oblong or linear, 3 to 4 lines, crenate, with a prominent
          midrib . . . . 6. *E. difformis*, var.
            [*Smithianus.*
      Leaves flat, linear-cuneate, 2 to 4 lines, slightly crenate, nerveless . 7. *E. parvifolius.*

1. **E. salicifolius** (Willow-leaved), *Sm.; DC. Prod.* i. 720; *Benth. Fl. Austr.* i. 331. An erect shrub, the branches rigid and often angular, glabrous or minutely hoary. Leaves linear or linear-lanceolate, mostly 1 to 2in. long, rather thick and rigid, glabrous when full-grown, obscurely 1-nerved. Peduncles axillary, short and 1-flowered, with a few broad scale-like imbricate bracts at the base, hoary with a minute tomentum as well as the calyx and petals. Sepals short, orbicular, rigid. Petals pink, attaining about ½in. Filaments flattened, densely fringed with woolly hairs, clavate and glandular at the top, bearing the anthers on a short stipes as in *Boronia;* anthers tipped with a very short, broad, recurved appendage. Ovary glabrous; style slightly pubescent below the middle. Cocci truncate at the top, but not beaked, transversely wrinkled. Seeds smooth and shining.—Rudge, in Trans. Linn. Soc. xi. t. 26; Deless. Ic. Sel. iii. t. 46; Bot. Mag. t. 2854; *E. lanceolatus,* Gærtn. f. Fr. iii. 154, t. 210; *Crowea scabra,* Grah. in Edinb. Phil. Journ. 1827, 174.

Hab.: Queensland, *F. c. M.* (locality not given). .
The names or numbers of this and *Crowea saligna,* 295, interchanged in many herbaria, and *Fl. Mixt. n.* 536, and others.—*Benth.*
The synonym often quoted of *E. australasia,* Sm., is an error. Smith mentions no species in Trans. Linn. Soc. iv. 221, but in describing the genus gives the station Australasia, which has been mistaken for a specific name.—*Benth.*

2. **E. Banksii** (after Sir Joseph Banks), *A. Cunn.; Endl. in Hueg. Enum.* 15; *Benth. Fl. Austr.* i. 332. A large shrub, the young branches angular and loosely hairy. Leaves from obovate-oblong to oblong-lanceolate, often oblique, obtuse, 1 to 1½in. long, contracted into a very short petiole, thinly coriaceous, finely veined and obscurely 3-nerved, glabrous or slightly hairy. Peduncles very short, axillary, 1 or rarely 2-flowered, usually with 2 or 3 scale-like distant bracts. Sepals small, ciliate. Petals attaining about 3 lines, hoary outside, with a prominent midrib. Filaments slightly flattened, woolly outside, clavate and glandular at the top as in *E. salicifolius;* anthers not apiculate. Ovary glabrous, style pubescent. Carpels of the fruit 4 or 5 lines long, truncate, very shortly beaked.

Hab.: Sandy shores of the Endeavour River, *Banks* and *Solander,* R. Brown, A. Cunningham. The leaves have very much the aspect of the phyllodia of some *Acacias.*—*Benth.*

XXIX. RUTACEAE.

3. **E. myoporoides** (Myoporum-like), *DC. Prod.* i. 720; *Benth. Fl. Austr.* i. 338. A stout, usually tall, glabrous shrub, with the habit of a *Myoporum*, the glandular tubercles sometimes very prominent, sometimes almost inconspicuous. Leaves sessile, from obovate-oblong to lanceolate or linear-lanceolate, obtuse or rarely acute, always mucronate, 1 to 3 or rarely above 4in. long, rather firm and sometimes coriaceous, flat with the midrib prominent underneath. Peduncles shorter than the leaves, usually bearing an umbel of 3 to 9 flowers, very rarely reduced to 1 or 2, especially on the smaller-leaved branches. Flowers white or pink, rather large, the petals attaining about 4 lines. Filaments flat, more or less ciliate, attenuate at the top. Ovary glabrous. Cocci beaked.—Bot. Mag. t. 3180; Deless. Ic. Sel. iii. t. 47; F. v. M. Pl. Vict. i. 122; *E. cuspidatus*, A. Cunn. in Field, N. S. Wales, 331; *E. neriifolius*, Sieb. in Spreng. Syst. Cur. Post. 164; *E. lancifolius*, F. v. M. in Trans. Vict. Inst. i. 32.

Hab.: Glasshouse Mountains.

Var. *minor*. Leaves rarely much above 1in. long. peduncles mostly 1 or 2-flowered.—*E. intermedius*, Hook. Bot. Mag. t. 4439. I cannot, however, see in them any near approach to *E. buxifolius*.—*Benth*. Hab.: Stanthorpe.

4. **E. hispidulus** (stiff hairs), *Sieb. in Spreng. Syst. Cur. Post.* 164; *Benth. Fl. Austr.* i. 333. Shrubby, with elongated branches, more or less pubescent. Leaves sessile, linear or linear-spathulate, mucronate with a straight or recurved point, ½ to 1in. long, the margins revolute, usually pubescent especially underneath, rarely glabrous, often tuberculate with prominent glands. Peduncles axillary, shorter than the leaves, 1 or rarely 2-flowered, the pedicel thickened under the flower. Petals attaining 3 or 4 lines. Stamens, style, and fruit of *E. buxifolius*.

*Hab.: Stanthorpe.

F. v. Mueller considers this as a variety of *E. buxifolius*. The foliage appears to me, however, to be constantly distinct.—*Benth*.

5. **E. scaber** (rough), *Part. Mag. Bot.* xiii. 127, *with a figure*; *Benth. Fl. Austr.* i. 334. A shrub, with the general aspect of *E. hispidulus*, but with glabrous or very minutely pubescent branches. Leaves sessile, narrow-linear, acute and mucronulate, under 1in. long, thick and very convex underneath, flat or channelled above and often almost terete, the margins never revolute, more or less tuberculate with prominent glands. Inflorescence and flowers of *E. obovalis*. Carpels much compressed, prominently rostrate.

Hab.: Glasshouse Mountains.

This is considered by F. v. Mueller as a variety of *E. buxifolius*. It appears to me to be nearer to *E. obovalis*, and differs from both chiefly in foliage.—*Benth*.

6. **E. difformis** (leaves of irregular form), *A. Cunn.; Endl. in Hueg. Enum.* 15; *Benth. Fl Austr.* i. 335. A much-branched, compact shrub, glabrous or the younger branches minutely pubescent. Leaves in the normal form, small, numerous, obovate, oblong, or almost rhomboidal, very obtuse, rarely above 2 lines long, usually tuberculate or as it were crenate, with 2 or 3 very large prominent glands, thick and convex, the margins often recurved, glabrous on both sides. Flowers small, terminal, solitary or 2 or 3 together, on very short pedicels. Calyx very small. Petals 2 to nearly 3 lines long, usually pubescent outside. Filaments flattened, densely ciliate; anthers shortly apiculate. Ovary villous; style short. Cocci very shortly beaked.—F. v. M. Pl. Vict. i. 123; *E. rhombeus*, Lindl. in Mitch. Trop. Austr. 293.

Hab.: Mantua Downs, *Mitchell*; between Mackenzie and Dawson Rivers, *F. v. Mueller*; near Warwick, *Beckler*; near Broadsound, *Herb. F. v. M.*

Var. (?) *Smithianus*. Quite glabrous. Leaves flat, thin, oblong or linear, glandular crenate, 3 to 4 lines long, with a conspicuous midrib. Petals usually glabrous.—*E. Smithianus*, Hill, in Herb. Muell.—*Benth*. Hab.: Wide Bay, *W. Hill*; near Brisbane.

7. **E. parvifolius** (small-leaved), *R. Br. Herb.; Benth. Fl. Austr.* i. 335. A low, erect, compact, much-branched, glabrous shrub. Leaves crowded, linear-cuneate, obtuse, 3 to 4 lines long, slightly glandular-crenate, flat, coriaceous, without any conspicuous midrib. Flowers small, terminal, solitary, shortly pedicellate, glabrous. Sepals small. Petals 2 to 2¼ lines long. Filaments flattened, ciliate; anthers minutely apiculate. Cocci short, truncate, obscurely beaked. Seeds minutely tuberculate.

Hab.: Shoalwater Bay, *R. Brown* (Herb. R. Br.)

5. PHEBALIUM, A. Juss.

(One of the ancient names of the Myrtle.)

Calyx small, 5-cleft or 5-toothed. Petals 5, valvate or laterally imbricate, but always with valvate inflexed tips. Disk narrow or angular. Stamens 10, shorter or longer than the petals; filaments glabrous or rarely slightly ciliate, filiform or rarely flat, subulate at the top; anthers tipped with a small gland or not at all apiculate. Carpels 5, rarely 4 or fewer, distinct from the base or nearly so, usually produced into a short or long appendage above the cells; styles inserted below the middle and immediately united into one; stigma small; ovules 2 in each cell, superposed. Cocci 2-valved, usually more or less beaked at the top or the outer angle; the endocarp cartilaginous and separating elastically. Seeds usually solitary.—Shrubs, either glabrous or slightly stellate-pubescent or clothed with scurfy scales, very rarely hirsute. Leaves alternate, simple, entire or slightly toothed, the glands often large and prominent. Inflorescence axillary or terminal, peduncles rarely 1-flowered, usually forming an umbel-like short raceme, rarely reduced to a compact head. Flowers small, white or yellow, very rarely and exceptionally 4-merous or 6-merous.

Besides the Australian species, which are all endemic, the genus comprises one from New Zealand, nearly allied to, but apparently distinct from, one of the Australian ones. F. v. Mueller unites the genus with *Eriostemon*, but the æstivation of the corolla, besides the habit and a number of smaller characters, appear to me sufficient to warrant the maintaining it as distinct. Practically, the section *Leionema* may be at once distinguished from *Eriostemon* by the strictly valvate corolla, and *Phebalium* proper by the scurfy scales always present at least on the flower and ovary.—*Benth.*

Sect. I. **Leionema**, F. v. M.—*Glabrous or pubescent plants without scurfy scales. Petals strictly valvate, glabrous.*

Flowers terminal. Stamens usually exserted.
   Leaves obtuse, ½ to ¾in., thinly coriaceous . . . . 1. *P. elatius.*
   Leaves small, obovate or orbicular, rigid but not thick, flat or concave 2. *P. rotundifolium.*

Sect. 2. **Euphebalium.**—*The whole plant or at least the inflorescence and calyx, and often the petals and ovary, more or less covered with scurfy peltate scales, often fringed at the edge, those of the ovary often closely imbricate in one mass. Petals laterally imbricate or rarely almost valvate in the bud, with inflexed valvate tips.*

Umbels terminal.
   Leaves linear-cuneate, truncate or emarginate . . . . 3. *P. glandulosum.*
   Leaves oblong or linear, rounded or obtuse at the top, ½ to 1½in. long . 4. *P. squamulosum.*
   Leaves oblong, lanceolate or linear, obtuse or acute, from ⅜in. to 3, 4, or 5in. long, glabrous above, silvery underneath . . . . 5. *P. Nottii.*
Umbels terminal and lateral, loose. Leaves oblong or lanceolate or linear, 1 to 2in. long or more.
   Leaves silvery-white underneath. Petals distinctly imbricate, not scaly 6. *P. Billardieri.*

1. **P. elatius** (tall), *Benth. Fl. Austr.* i. 340. A tall shrub, glabrous or the branches very minutely pubescent, and usually tuberculate with prominent glands. Leaves linear-cuneate or oblong, obtuse, ½ to ¾in. long, entire or crenulate, thinly coriaceous, smooth and shining, narrowed into a very short petiole. Peduncles 2 or more-flowered, terminal or in the uppermost axils, form-

ing short terminal leafy corymbs or ovate panicles. Calyx very small. Petals valvate, not 2 lines long. Stamens exserted; filaments subulate, glabrous; anthers small. Ovary glabrous, on a raised almost stalk-like disk. Cocci obliquely obovate, very minutely beaked, about 2 lines long. Seed dark brown, shining.—*Eriostemon elatior*, F. v. M. Fragm. i. 181.

Hab.: Stanthorpe.

The species is very closely allied to the New Zealand *P. nudum*, Hook., differing chiefly in much smaller flowers, the calyx-lobes less prominent, the inflorescence not so flat-topped, etc.—*Benth*.

2. **P. rotundifolium** (leaves round), *Benth. Fl. Austr.* i. 341. An erect much-branched shrub, the young branches minutely pubescent. Leaves crowded, almost imbricate, small, obovate or orbicular, obtuse or minutely mucronate, mostly 2 to 3 lines long, flat or concave, coriaceous, glabrous, very shortly petiolate or almost sessile. Flowers several, in a terminal sessile umbel, almost contracted into a head in our specimens, which are not fully out. Sepals small. Petals valvate, glabrous. Filaments filiform, glabrous. Ovary glabrous, on a very short disk, the terminal appendages of the carpels very short.—*Eriostemon rotundifolius*, A. Cunn., Endl. in Hueg. Enum. 15.

Hab.: Stanthorpe.

3. **P. glandulosum** (plant glandular), *Hook. in Mitch. Trop. Austr.* 199; *Benth. Fl. Austr.* i. 342. Very closely allied to some of the smaller much-branched forms of *P. squamulosum*, with the same scurfy indumentum, inflorescence, and flowers, and recently united with that species by F. v. Mueller. (Pl. Vict. i. 130). It appears however to me to differ sufficiently in the leaves, which are narrowly linear-cuneate, emarginate or almost 2-lobed at the end, with revolute or recurved margins varying from 2 or 3 lines to ½in. in length. In the ordinary form also the branches and leaves are covered with large glandular tubercles.—*P. sediflorum*, F. v. M. in Trans. Vict. Inst. i. 80; *Eriostemon sediflorus*, F. v. M. Fragm. i. 102.

Hab.: On the Upper Maranoa, *Mitchell;* southern coast lands.

4. **P. squamulosum** (scaly), *Vent. Jard. Malm. t.* 102; *Benth. Fl. Austr.* i. 342. An erect shrub, varying in height, but never arborescent, the young branches brown with scurfy scales. Leaves shortly petiolate, oblong or linear, obtuse but often mucronulate, ½ to 1½in. long, somewhat coriaceous, the margins flat or slightly recurved, smooth above or slightly glandular-tuberculate, covered underneath with scurfy peltate scales. Flowers yellow, in terminal sessile, simple or compound umbels or corymbs, not exceeding the last leaves, the pedicels, calyx, and petals covered with comparatively large scurfy scales. Calyx very short, truncate, with minute or short and broad teeth. Petals barely 2 lines long, slightly imbricate with inflexed valvate tips. Stamens exserted (1 or 2 occasionally wanting); filaments glabrous; anthers tipped by a small gland. Ovary densely covered with white or brown scurfy ciliate scales. Cocci small, broad, obscurely beaked. Seeds scarcely shining.—DC. Prod. i. 720; A. Juss. in Mem. Soc. Hist. Nat. Par. ii. 132; *P. elæagnifolium*, A. Juss. l.c. 132, t. 11; *P. aureum*, A. Cunn. in Field, N. S. Wales, 331, with a figure (the specimens not so stunted as represented in the plate); *Eriostemon lepidotus*, Spreng. Syst. ii. 322; F. v. M. Fragm. i. 104, and Pl. Vict. i. 130.

Hab.: Stanthorpe, Main Range, Peak Downs, Rockhampton, etc.

5. **P. Nottii** (after Dr. Nott), *F. v. M. Fragm.* vi. 22. A shrub of about 10ft. in height, branches scaly. Leaves for the most part between ¾ and 1½in. long, and 2 to 3 lines broad, cuneate or ovate-oblong; obtuse, retuse, flat, a little repand; densely scaly on the under, and minutely scabrous on the upper side, on conspicuous petioles. Flowers lepidote, at the end of the branchlets, solitary or

o

in twos or threes on short pedicels of from 1 to 3 lines. Bracts linear-canaliculate, caducous. Calyx almost campanulate, about 2½ lines long, 6 or 7-cleft. Petals 6 or 7, lanceolate, 3 or 4 lines long, pale purple inside. Stamens 12 to 14. Anthers oblong, scarcely 1 line long. Style long as the stamens, and like the filaments glabrous, stigma small, peltate. Ovary densely lepidote, 6 or 7-merous. Cocci rhomboid-ovate, 2 lines long.

Hab.: Queensland, *F. v. M.*

5. **P. Billardiere** (after Dr. J. J. Labillardiere), *A. Juss.; Benth. Fl. Austr.* i. 344. An erect shrub or small tree, the branches angular and clothed with small brown scurfy scales. Leaves oblong, lanceolate or linear, obtuse or acute, rarely under ½in. and often 3in. long, or in very luxuriant specimens 4 or 5in. long, entire, coriaceous, flat or with recurved margins, glabrous above, silvery-white underneath with minute scales. Flowers in axillary corymbs, shortly pedunculate, but always shorter than the leaves; peduncles and pedicels thick and scaly. Calyx small, lobed. Petals about 2 lines long, glabrous, slightly imbricate, with inflexed valvate tips. Stamens exserted; filaments often hairy in the lower portion. Ovary glabrous. Cocci small, broad, with a very short beak. Seeds shining.—*Eriostemon squameus*, Labill. Pl. Nov. Holl. iii. t. 141; *P. retusum*, Hook. Journ. Bot. i. 254, and Ic. Pl. t. 57; *P. elatum*, A. Cunn. in Fields N. S. Wales 331; *P. elæagnoides*, Sieb. Pl. Exs.

Hab.: Fraser's Island, *H. Tryon.*

6. PHILOTHECA, Rudge.

(Referring to the tube formed by the lower part of the stamens.)

Calyx 5-cleft. Petals 5, imbricate in the bud. Disk slightly lobed. Stamens 10, shorter than the petals; filaments united into a glabrous tube at the base, free upwards, and very hairy; anthers oblong, all perfect, minutely apiculate. Carpels 5, nearly distinct from the base; styles inserted below the middle, and immediately united in a single style, hirsute in the middle; stigma small. Ovules 2 in each carpel, superposed. Cocci truncate, 2-valved, the endocarp cartilaginous and separating elastically.—Erect, heath-like shrubs, glabrous or nearly so. Leaves crowded, alternate, narrow-linear. Flowers terminal, nearly sessile, solitary, or 2 or 3 together.

A genus entirely Australian, differing from *Eriostemon* only in the monadelphous stamens.—*Benth.*

Leaves obtuse, mostly under 3 lines long . . . . 1. *P. australis.*
Leaves 3-angled, semi-terete, acute, 4 to 7 lines long . . . . 2. *P. calida.*

1. **P. australis** (Australian), *Rudge, in Trans. Linn. Soc.* xi. 298, t. 21; *Benth. Fl. Austr.* i. 348. Glabrous or sprinkled with a minute pubescence. Leaves numerous, linear, obtuse, rarely exceeding 3 lines, rather thick, flat or channelled above, very convex underneath, or almost terete. Flowers usually solitary, but sometimes 2 or 3 together. Sepals small, broadly triangular. Petals 3 or 4 lines long, broadly lanceolate, minutely hoary-pubescent on both sides, except a broad glabrous central line outside. Stamens rather shorter than the petals. Cocci shortly beaked.—*Eriostemon salsolifolius*, Sm. in Rees, Cycl. xiii.

Hab.: Near Mount Faraday, *Mitchell.*

Var. *parviflora.* Leaves more ciliate. Flowers much smaller; the petals scarcely 2½ lines long.—*P. ciliata*, Hook. in Mitch. Trop. Austr. 347.

2. **P. calida** (the most northern species), *F. v. M. Fragm.* vii. 21 and 38. Branches and foliage glabrous. Leaves trigono-semiterete, acute, crowded, 4 to 7 lines long, and about ½ line thick. Sepals oval, imbricate, 2 lines long, yellow,

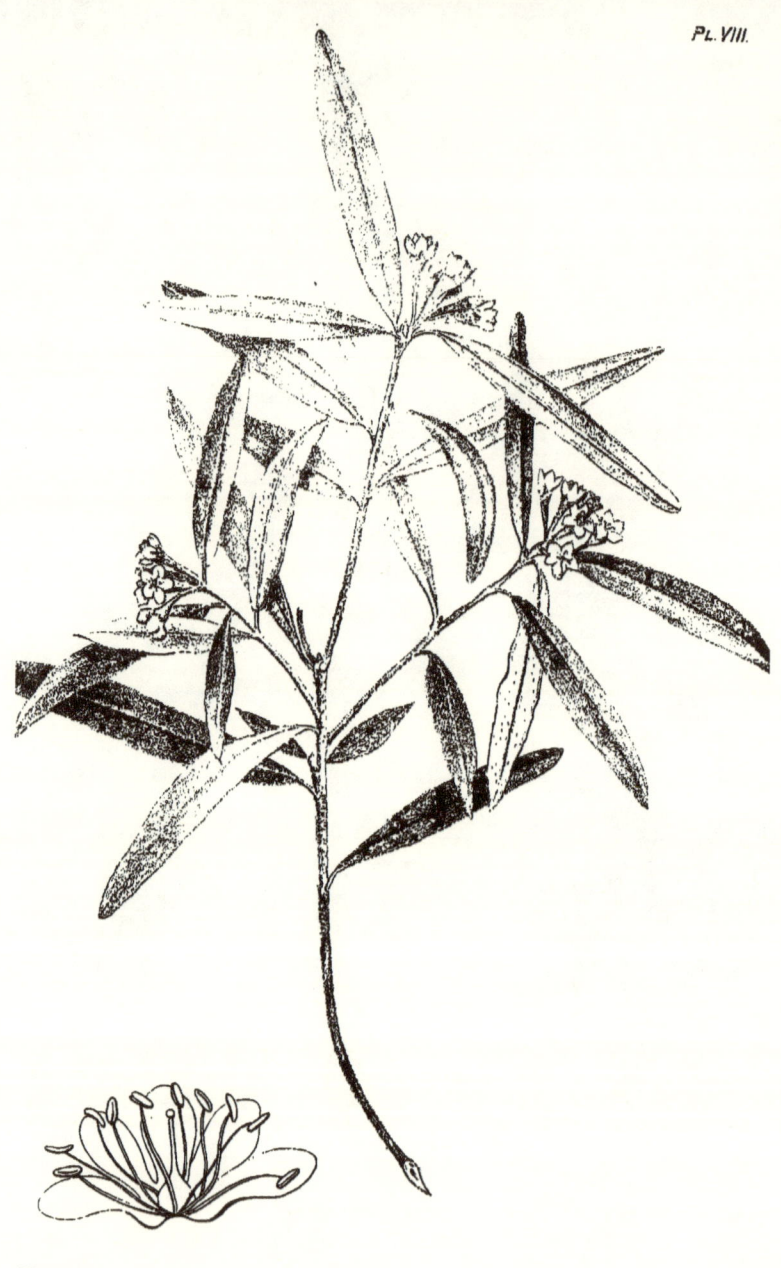

Asterolasia woombye; Bail.

slightly ciliate. Petals imbricate, sometimes ½in. long, lanceolate-oblong. The staminal tube with a ring of hairs a little above the base, free filaments villous. Fertile anthers yellow, narrow oblong-linear, about 1 line long, slightly bearded at the base. Style capillaceous, glabrous, nearly 1in. long. Stigma depressed-globose, ⅛ line broad. Carpels 5, erect, sessile, blunt, glabrous.

Hab.: Gilbert River, sandstone tablelands, and Cave Creek, *F. v. M. l.c.*

7. ASTEROLASIA, F. v. M.
(Stellate pubescence.)

Calyx often very minute or obsolete. Petals 5, tomentose outside, valvate and usually induplicate in the bud. Disk none. Stamens 10 or more, free, filaments filiform, glabrous or very slightly ciliate, anthers not apiculate. Carpels 2 to 5, united to the middle, or nearly to the top, into a single shortly-lobed or truncate ovary of 2 to 5 cells. Style inserted between the lobes, filiform, with usually a large reflexed peltate or deeply-lobed stigma. Cocci tardily separating, truncate, and often beaked, 2-valved; endocarp cartilaginous, separating elastically.—Shrubs or undershrubs, more or less stellate-tomentose, or the tomentum united into scurfy scales. Leaves alternate, simple. Flowers sessile or pedicellate, axillary or terminal, solitary or a few together.

The genus is limited to Australia.

1. **A. woombye** (found at Woombye), *Bail.* A tall shrub, clothed in most parts with an elæagnoid indumentum. Leaves membranous, upper face glabrous, deep-green, the under side grey, with close, silvery, stellate scales, and scattered brown ones, linear, almost linear-lanceolate, 1¼ to 2¾in. long, 3 or 4 lines broad, apex obtuse; petioles 2 or 3 lines, margins slightly repand, alternate, the two last ones close under the flower-head, nearly opposite. Flowers 6 to 10 or more in heads or clusters at the ends of the branchlets. Primary peduncle very short or wanting, sometimes is seen a secondary peduncle bearing three flowers. Pedicels 3 or 4 lines long. Calyx-teeth triangular, the points sometimes elongated and recurved, ½ to ¾ line long. Petals much imbricate in the lower part, valvate at the top, white with brown stellate scales on the back, the face glabrous, about 2 lines long. Stamens 10, filaments white, longer than the petals, filiform, glabrous or roughened by a few minute glands. Anthers oblong, yellow, ¾ line long. Style thicker and shorter than the filaments, glabrous. Stigma scarcely lobed. Ovary very scaly, showing 5 cocci in the flowers examined.

Hab.: Woombye, North Coast Railway, *W. French.*

8. CORREA, Sm.
(After Correa de Serra.)
(Didymeria, *Lindl.*)

Calyx cup-shaped, truncate and 4 or 8-toothed, or 4-lobed. Petals 4, valvate, connate in a cylindrical or campanulate tube, sometimes separating as the flower expands, spreading at the top. Disk shortly lobed. Stamens 8, free; anthers without appendages. Ovary of 4 carpels nearly distinct from the base; styles inserted above the middle, and immediately united into one filiform style, with a small often shortly 4-lobed stigma; ovules 2 in each carpel, superposed. Cocci 4, truncate, 2-valved, the endocarp cartilaginous and separating elastically.—Shrubs or rarely small trees, stellate-tomentose or rarely glabrous. Leaves opposite, petiolate, simple. Flowers rather large and showy, red, yellow, white or green, usually pendulous, solitary or 2 or 3 together, axillary or terminal. Petals usually mealy-tomentose outside.

The genus is limited to Australia.

1. **C. speciosa** (showy), *Ait. Espit. Hort. Kew* 866; *Benth. Fl. Austr.* i. 354.
A shrub, variable in size and habit, usually rigid and low, and rarely exceeding 6 to 8ft., the stellate tomentum very variable, usually loose and abundant on the branches or sometimes on the whole plant, dense and soft on the under side of the leaves, disappearing on the upper surface or sometimes on the whole plant, except the peduncles and flowers. Leaves very shortly petiolate, from broadly ovate or cordate to narrow-oblong or lanceolate, obtuse or retuse, usually from ¾ to 1½in. long, rarely all under 1in., or the larger ones attaining 2in. Flowers red, varying to white or yellowish-green, terminal, shortly pedicellate and pendulous, or a few rarely erect, solitary or 2 or 3 together. Calyx hoary or rusty-tomentose, truncate, with 4 minute teeth. Petals hoary-tomentose outside, united the greater part of their length into a cylindrical or slightly campanulate corolla of ¾ to 1½in., with 4 spreading lobes. Stamens exserted, the filaments of those opposite the petals more or less dilated below the middle.—DC. Prod. i. 719; F. v. M. Pl. Vict. i. 136.

Hab.: Stanthorpe district.

9. BOSISTOA, F. v. M.
(After J. Bosisto.)

Flowers hermaphrodite? Calyx small, 5-toothed. Petals 5, valvate or slightly imbricate, with inflexed tips. Disk thick. Stamens 10. Ovary of 5 distinct carpels; styles almost terminal, united upwards, but soon separating; ovules 2 in each carpel, superposed. Cocci distinct, large, coriaceous, 2-valved; endocarp cartilaginous, separating. Seeds solitary; testa membranous; albumen none; cotyledons thick and fleshy, radicle small.—A tree. Leaves opposite, pinnate. Panicles terminal.

The genus is limited to Australia, and allied in some respects to *Melicope* and *Evodia*, but very different in habit as well as in the seeds, which have the structure of *Pilocarpus* and some other American genera.—*Benth.*

1. **B. sapindiformis** (Sapindum-like), *F. v. M. Herb.; Benth. Fl. Austr.* i. 859. "Towra," Nerang, *Schneider*. A tree with the habit of a *Cupania*, the young shoots, petioles and inflorescence minutely pubescent. Leaves pinnate; leaflets 3 to 11, opposite in pairs, the terminal odd one occasionally wanting, oblong-lanceolate, 4 to 8in. long, more or less serrate-toothed, especially above the middle, narrowed at the base and attaining a width of 4in., on a short petiolule or nearly sessile. Panicles terminal, trichotomous, shorter than the leaves. Buds globular. Calyx small, very shortly and unequally toothed. Petals about 2 lines long. Filaments dilated at the base, attenuated upwards, glabrous; anthers large. Carpels very hirsute, on a raised disk. Styles short. Cocci broadly and very obliquely ovate, about 1in. long, hard, almost woody, tomentose and rugose outside.—*Evodia pentacocca*, F. v. M. Fragm. iii. 41.

Hab.: Nerang Creek, Rockhampton, Mt. Dryander, and other parts.
Wood close in the grain, yellow; very liable to split in drying.—*Bailey's Cat. Ql. Woods No.* 39.
Very hard timber used for handspikes and levers.—*Schneider.*

10. MELICOPE, Forst.
(Referring to the glands of the flower being notched.)

Flowers more or less unisexual. Sepals 4. Petals 4, valvate, or slightly imbricate, with inflexed tips. Disk thick, entire or lobed. Stamens 8. Ovary of 4 nearly distinct carpels; styles inserted above the middle, united immediately or at the summit into one, with a capitate 4-lobed stigma; ovules 2 in each carpel, superposed or collateral. Cocci distinct, spreading, 2-valved;

endocarp cartilaginous or horny, separating. Seeds usually solitary; testa crustaceous, shining; albumen fleshy, embryo straight or slightly curved, with oblong or ovate cotyledons.—Trees or shrubs. Leaves opposite, 3-foliolate, or (in species not Australian) 1-foliolate or simple. Flowers rather small, in terminal or axillary cymes or panicles.

Besides the Australian species, which are endemic, there are 2 from New Zealand and a few from the Pacific Islands. F. v. Mueller proposes to unite *Melicope* with *Evodia*, but the double number of stamens is a more constant character than many others distinguishing the received genera of *Zanthoxyleæ*.—*Benth.*

Leaves of a single leaflet. Petiole short. Petiolule scarcely any. Peduncles axillary, often very short, bearing often many flowers, on pedicels of about 6 lines	1. *M. Fareana*.
Leaves frequently of 5 leaflets, 2 rather distant pairs and a terminal one. Panicles terminal. Filaments glabrous. Cocci transversely wrinkled	2. *M. neurococca*.
Leaves of 3 leaflets, rarely reduced to the terminal one. Panicles terminal or in the upper axils. Filaments ciliate. Cocci divaricate	3. *M. erythrococca*.
Leaves of 3 leaflets, terminal one often 5in. long. Panicle terminal, about half as long as the petiole. Flowers few. Pedicels long as flowers	4. *M. Broadbentiana*.
Leaves of 3 leaflets, from 5 to 10in\ long. Petioles shorter than leaflets. Panicles axillary. Pedicels short. Petals minutely pubescent outside. Filaments ciliate	5. *M. australasica*.
Leaves of 3 leaflets, emarginate, subcoriaceous. Sepals minute. Petals thick, with incurved tips. Filaments broad, nearly glabrous. Disk yellow. Ovary glabrous. Style hairy on the lower half	6. *M. chooreechillum*.
Leaves of 3 or reduced to a single leaflet, 3 to 7in. long, pubescent. Peduncles axillary. Flowers shortly pedicellate. Sepals orbicular. Petals thick, linear. Filaments and style hairy. Ovary glabrous	7. *M. pubescens*.

1. **M. Fareana** (after M. Fare), *F. v. M. Fragm.* ix. 101. A small tree. Leaves unifoliolate, glabrous, ovate-lanceolate, chartaceous, 3 to 6in. long, 1 to 2in. broad, obtuse or acute, nerves very slender, widely spreading oil dots crowded. Petiole 1in. or less long; petiolule scarcely any. Peduncles short, sometimes very short, bearing few or many flowers. Pedicels 3 to 6 lines long, thickened upwards, and as the calyx glabrous. Sepals 4, rarely 5, about 2 lines long, lanceolate. Petals 4, rarely 5. Slightly imbricate, the apices minutely inflexed. Stamens 8, a little shorter than the petals, all fertile, filaments eglandulose, coherent at the base by their dense woolly clothing, the upper part glabrous. Anthers dorsally fixed, oblong-ovate, dehiscence introrse. Disk annular. Carpels 4, rarely 5. Style capillary, about 2 lines long.

Hab: Rockingham Bay, Barron River, and other parts in the tropics.

2. **M. neurococca** (referring to the nerves of cocci), *Benth. Fl. Austr.* i. 360. A small tree, the young branches, petioles, and peduncles pubescent with simple spreading hairs. Leaves of each pair generally unequal, the larger one with a common petiole of 2in. or more, the other with a much shorter petiole; leaflets 3, ovate-lanceolate or lanceolate, acuminate, mostly 3 to 4in. long, glabrous above, sprinkled with a few hairs underneath. Panicles terminal, trichotomous, corymbose. Sepals small, orbicular, concave, ciliate. Petals about 2 lines long, glabrous, valvate or nearly so. Filaments glabrous, dilated to the middle. Ovary hirsute, the carpels almost distinct from the base. Styles inserted below the summit. Cocci distinct, nearly erect, broad, about 3 lines long, the valves coriaceous and transversely wrinkled.—*Evodia neurococca*, F. v. M. Fragm. i. 28 and ii. 103.

Hab.: Brisbane River, *W. Hill* and *F. v. Mueller*; Wide Bay and Archer's Creek, used by the natives to make their spades, *Leichhardt*.

Wood very hard, and close-grained, of a uniform light-yellow colour.—*Bailey's Cat. Ql. Woods No.* 40.

3. **M. erythrococca** (cocci of a reddish colour), *Benth. Fl. Austr.* i. 360. "Thal-ango-thera," Forest Hill, *Macartney*. A lofty tree with a smooth whitish bark, quite glabrous. Leaves glabrous, at least on the old trees, opposite, sub-opposite or alternate. Leaflets 3 or rarely 1 only, oblong-lanceolate, obtuse, 1¼ to 3in. long, coriaceous, entire or obscurely crenulate, on a common petiole of ¾ to 1½in. Panicles terminal or in the upper axils, loose, scarcely longer than the leaves. Sepals small, triangular, slightly ciliate. Petals 1¼ line long, slightly imbricate, valvate at the tips, minutely pubescent outside. Disk obscurely lobed. Filaments dilated and ciliate to above the middle. Ovary slightly hirsute, the carpels almost distinct. Styles inserted above the middle. Cocci 4 or very rarely 5, very spreading, ovate, about 2 lines long, wrinkled, of a reddish colour.—*Evodia erythrococca*, F. v. M. Fragm. i. 28.

Hab.: Wide Bay, *C. Moore*; Moreton Bay and Brisbane River, *W. Hill*; Mackay, *W. Macartney*. The bark possesses a most peculiar acrid pungency, and promotes a great flow of saliva.

4. **M. Broadbentiana** (after K. Broadbent), *Bail*. A slender erect shrub, glabrous except the very young growth, branchlets 4-angular. Leaves in nearly equal pairs, sometimes one slightly shorter than the other; petiolules slender, 2 to 3½in. long; terminal leaflet often 5½in. long and 2½in. broad, ovate-lanceolate, often long-acuminate, equal sided to a petiolule nearly 1in. long, lateral ones smaller, sessile or nearly so, and very unequal-sided, the lamina on the upper side terminating some distance above the base, all the leaflets very thin; primary veins few, parallel, about 5 on each side of midrib with fainter intermediate one; veinlets few. Panicle terminal, trichotomous, not more than half the length of the petiole. Flowers few, small; pedicels long as the flower; sepals 4, very obtuse. Petals 4, white, minutely dotted, ovate-oblong, under 2 lines long. Stamens 8 (4 long, 4 short) the long ones alternating with the petals, the filaments thickened downwards, the lower half hairy. Disk broadly lobed. Ovary glabrous, 4-lobed. Styles glabrous, free at base, erect, inserted slightly below the apex at the internal angle of the carpels. Fruit not obtained.

Hab.: Palm Camp (4000ft. altitude), Bellenden Ker, *Bellenden Ker Exped.* 1889.

5. **M. australasica** (Australasian), *F. v. M.; Benth. Fl. Austr.* i. 360. A handsome tree, glabrous in all its parts. Leaves digitately 3-foliolate, the common petiole several times shorter than the leaflets; leaflets oblong-elliptical, or rarely obovate-oblong, obtuse or shortly acuminate, 6 to 10in. long, somewhat coriaceous, entire. Panicles axillary, trichotomous, loose and many-flowered, but much shorter than the leaves. Pedicels short. Sepals ovate. Petals narrow, about 4 lines long, of a firm consistence, reflexed above the middle, minutely pubescent outside. Filaments slightly dilated, ciliate and rigid, especially the larger ones, subulate upwards; anthers small. Disk inconspicuous. Carpels nearly glabrous, but tapering into strictly terminal short pubescent styles united at the summit. Cocci erect, distinct, angular, acuminate, not 2 lines long. Seeds shining.—*Evodia octandra*, F. v. M. Fragm. ii. 102.

Hab.: Pine River, *W. Hill* (F. v. M. Fragm. ix. 102), where the trees are said to attain 50ft. in height and have a white, variegated bark; filaments somewhat broad and smooth.

6. **M. chooreechillum** (native name of Mount Bartle Frere), *Bail. Rep. Bot. Bellenden Ker Exped*. A large shrub or small tree, glabrous, branches thick. Leaves opposite, of unequal size in each pair. Petioles 1 to 2½in. long, sulcate on the upper side. Leaflets 3, obovate, tapering from a broad emarginate end to the top of the petiole, 1½ to 3in. long, ¾ to 1½in. broad, of a thick cartilaginous or coriaceous texture; veins faint in the fresh leaf, but prominent in the dried specimens, the primary ones anastomosing far within the margin, the under surface closely covered with small dots (oil-dots). Flowers in short trichotomous

panicles, length of the petioles, pedicellate. Sepals 4, obtuse, minute. Petals 4, about 4 lines long, linear, white, thick, the apex incurved, Stamens 8; filaments broad at the base, very nearly glabrous. Anthers oblong-rotund. Disk a yellow ring. Ovary glabrous, 4-celled. Style hairy on the lower half. Stigma shortly 4-lobed.

Hab.: Summit of Bartle Frere, *Bellenden Ker Exped.* 1899.

7. **M. pubescens** (pubescent), *Bail. Bot. Bull. No. 3.* A small tree, with light-coloured bark, the branchlets somewhat flattened and usually opposite, the whole leafy part of the plant and inflorescence softly pubescent. Leaves opposite, 3-foliolate, often on lateral shoots reduced to a single leaflet or pair of leaflets on a petiole of ½in. or less. The pairs of leaves sometimes, but not always, of unequal size; leaflets sessile, lanceolate, glabrous except the veins on the upper surface, often sharply acuminate, the lateral ones unequal-sided at the base, 3 to 7in. long, 1 to 2¼in. broad; veins prominent on both sides, the primary ones looping some distance from the margin, margins entire; oil dots minute, not numerous. Flowers in lateral and axillary trichotomous cymes; peduncles shorter than the petioles: flowers on short pedicels, calyx-lobes 4, nearly orbicular, about 1 line. Petals 4, valvate, recurved when the flower is fully opened, thick linear, with inflexed tips, about 3 or 4 lines long, disk entire, glabrous. Stamens 8, those opposite the petals shorter than the other four, filaments much dilated and ciliate in the lower half. Style terminal, rather thick, long as the stamens, hairy in the lower half. Stigma small, slightly lobed. Ovary glabrous, 4-celled, 2 ovules in each cell. Fruit not yet collected.

Hab. Yandina and top of Blackall Range, March, 1891 (in full flower), *Field Naturalists.*

11. EVODIA, Forst.
(Sweet smell of foliage.)

Flowers more or less unisexual. Sepals 4 or 5, imbricate. Petals 4 or 5, valvate or very slightly imbricate. Disk sinuate. Stamens 4 or 5; filaments subulate or slightly dilated. Ovary of 4 or 5 carpels, usually distinct and style-like in the male flowers, more or less united in the females, styles attached below the middle, more or less united with a 4 or 5-lobed stigma. Ovules 2 in each carpel, collateral or superposed. Fruit separating more or less completely into coriaceous 2-valved cocci, the endocarp separating elastically. Seeds with a crustaceous testa, usually smooth and shining; albumen fleshy; embryo straight with ovate cotyledons.—Unarmed trees or shrubs. Leaves opposite, usually digitately 3-foliolate or pinnate, rarely 1-foliolate or simple; leaflets entire, often large. Cymes or panicles axillary or rarely terminal. Flowers small.

A considerable genus, spread over tropical Asia and the islands of the Pacific and of the Madagascar group; all but one of the Australian species are endemic. The genus differs from *Melicope* chiefly in the stamens equal to, not double, the number of petals, from *Zanthoxylum* by the leaves all or mostly opposite, generally by the more valvate petals and more united styles, besides minor characters offering occasional exceptions.—*Benth.*

Trees.
Petioles 1 to 2in. long. Leaflets 1½ to 3in. long, lanceolate, central one somewhat petiolulate. Cymes axillary or lateral; flowers crowded. Sepals small, orbicular. Petals about 2 lines long, glabrescent. Filaments ciliate, slightly dilated . . . . 1. *E. micrococca.*
Petioles 5 to 6in. long, minutely pubescent. Leaflets sessile, 6 to 10in. long, thin-chartaceous, shortly acuminate. Cymes many-flowered. Petals deciduous. Carpels glabrous on the outside, puberulent inside 2. *E. xanthoxyloides.*
Petioles 2 to 3in. long, semi-cylindrical. Leaflets 4 to 10in. long, thin coriaceous. Cymes umbellulate, many-flowered. Sepals orbicular. Petals reddish outside, densely bearded. Filaments glabrous. Style short, glabrous. Ovary densely tomentose . . . . 3. *E. Bonwickii.*

200   XXIX. RUTACEÆ.   [*Evodia.*

Petioles 1½ to 5¼in. long, margins slightly winged in the upper part,
often terete in the lower portion. Leaflets 3½ to 10in. long, sessile,
thin-herbaceous. Panicle in the upper axils 3 to 7in. long, lower
branches distant. Petals scarcely ¾ line long. Stamens shorter.
Style pubescent. Cocci rugose . . . .   4. *E. alata.*
Petioles 1 to 3in. long. Leaflets 2½ to 5in. long, shortly petiolulate.
Cymes lateral; flowers dense, peduncles short. Calyx puberulent.
Petals pink, 3 lines long. Filaments 4 lines long. Style 3 or 4 lines
long . . . .   5. *E. accedens.*
Petioles 2 to 3in. long. Leaflets oblong-acuminate, 5 to 6in. long,
coriaceous, glabrous, except the young growth. Panicles or cymes
near the ends of the branchlets. Sepals minute. Petals 1 line long,
base or claw woolly. Filaments hairy at the base . . . .   6. *E. littoralis.*
Shrubs.
Petioles 4 to 5in. long, somewhat terete. Leaflets 5 to 6in. long,
chartaceous, petiolulate. Panicles terminal. Peduncles long, thinly
pubescent. Flowers minute, shortly pedicellate. Petals scarcely 1
line long. Filaments slightly silky . . . .   7. *E. vitiflora.*

1. **E. micrococca** (small-fruited), *F. v. M. Fragm.* i. 144, *and* ii. 180; *Benth. Fl. Austr.* i. 361. A tree often of considerable size, quite glabrous. Leaves digitately 3-foliolate with long petioles; leaflets obovate-oblong, obtuse, mostly 1½ to 3in. long, entire, narrowed at the base, the central one almost petiolulate. Flowers in dense cymes or trichotomous panicles on short lateral peduncles below the young shoots. Sepals 4, orbicular, small. Petals 4, about 2 lines long, glabrous, slightly imbricate, with inflexed valvate tips. Filaments slightly dilated, ciliate, the attenuate tips folded inwards in the bud, exserted in the open flower. Cocci not 2 lines long, not separating so completely as in the *Melicopes*, rugose-glandular outside. Seeds black and shining.

Hab.: Many parts of southern Queensland.
Wood of a light-yellow colour, close in the grain, and tough.—*Bailey's Cat. Ql. Woods No.* 41.

2. **E. xanthoxyloides** (Xanthoxylon-like), *F. v. M. Fragm.* iv. 155. A tree, the bark smooth. Leaves trifoliolate. Petioles minutely puberulent, 5 to 6in. long. Leaflets 6 to 10in. long, thin-chartaceous, sessile, ovate, shortly acuminate, glabrous above, the nerves on the under side puberulent, margins entire. Cymes many-flowered. Peduncle and pedicels of fruiting specimens glabrescent, 2 to 4 lines long. Calyx 1 line or shorter, repandly 4-lobed. Petals deciduous (not seen). Cocci subovate, turgid, very obtuse, 3 to 4 lines long, outside glabrous, inside with an appressed pubescence. Seeds globose-ovate, scarcely 2 lines long, dark-brown, shining.

Hab.: Rockingham Bay, *J. Dallachy* (F. v. M., l.c.)
Wood hard, heavy-scented, yellowish (F. v. M., l.c.)

3. **E. Bonwickii** (after J. Bonwick), *F. v. M. Fragm.* v. 56. A large tree. Leaves trifoliolate, glabrous. Petioles 2 to 3in. long, semi-cylindric. Leaflets 7 to 10in. long, 3½ to 4½in. broad, obovate-lanceolate, sessile, the under side pale. Cymes umbellate, many-flowered. Primary peduncles about 1in. long, secondary shorter, thinly pubescent. Pedicels almost silky. Sepals imbricate, orbicular, nearly 1 line long. Petals valvate, about 1½ line long, glabrous and reddish outside, the inside densely bearded. Filaments subulate, about ⅓ line long, glabrous. Anthers purple, oblong, glabrous, ¾ line long. Disk inconspicuous. Indumentum of ovary yellow. Style ⅓ line long, glabrous. Stigma minute, undivided. Cocci 4, almost globose, slightly compressed, about 2 lines long. Seeds black, shining.

Hab.: Rockingham Bay, *J. Dallachy* (F. v. M., l.c.)

4. **E. alata** (winged), *F. v. M. Fragm.* vii. 142. A tree, seldom tall, with a short velvety pubescence. Leaves trifoliate. Petioles 1½ to 5½in. long, the upper portion more or less winged at the edges, the lower portion nearly terete. Leaflets

sessile, 8½ to 10in. long and often over 8in. broad, quite entire or slightly repand, thin-herbaceous. Panicles in the upper axils 8 to 7in. long, the lower branches rather distant. Flowers minute, greenish, clustered or here and there solitary. Sepals deltoid. Petals rhomboid-ovate, scarcely ⅔ line long. Stamens shorter than the petals, some sterile. Style short, pubescent. Stigma deeply 4-lobed. Cocci rotundly-obtuse, deeply bivalved. Epicarp rugose. Seeds globose, about 1 line diameter.

Hab.: Many parts of the tropics, and sometimes somewhat south.

5. **E. accedens** (near to another species), *Blume; F. v. M. Fragm.* ix, 102. " Bunnec-walwal," Moreton Bay; " Boogoobi," Herberton, *J. F. Bailey*. An erect tree 70 to 80ft. high, thinly-pubescent or glabrous. Bark light-coloured, somewhat thick and corky. Leaves trifoliolate. Petioles 1 to 8in. long. Leaflets 2½ to 5in. long, ovate, shortly acuminate, chartaceous, pale on the under side, shortly petiolulate. Cymes lateral, the flowers crowded, pink, turning bluish as they die away. Peduncles short. Pedicels about as long as the flowers. Calyx-lobes about 1 line long. Petals 2 to 3 lines long, slightly imbricate. Filaments glabrous, filiform, 4 lines long. Anthers oblong. Disk lobes semi-orbicular. Style 3 to 4 lines long, shortly pubescent. Stigma minute, capitate, 4-lobed. Ovary velvety. Cocci 4 or less by abortion, slightly compressed, globose-ovate, about 3 lines long. Seeds dark reddish-brown, ovate-globose, 1¼ line long.—*E. Elleryana*, F. v. M. Fragm. v. 4, 56, 179, and vii. 22.

Hab.: Not uncommon in damp scrubs in both southern and northern Queensland.
Wood very white, light and soft, furnishing a good substitute for that of the European lime-tree.—*Bailey's Cat. Ql. Woods No.* 42.
Leaves sometimes infested with the fungus *Phyllosticta Evodiæ*, Cke.

6. **E. littoralis** (of the coast), *Endl. Bot. Bull.* xiv. 7. An erect tree, glabrous except for a slight pubescence on the young growth and inflorescence, about 50ft., with a light-coloured bark, the branchlets very stout, terete, but bluntly 4-angled at the ends where the leaves and flowers are borne. Leaves somewhat crowded towards the end of the branchlets, digitately 3-foliolate, on petioles 2 to 2½in. long; leaflets oblong, lateral ones oblique at the base, the middle one sometimes tapering into a petiolule, all obtusely acuminate, 5 or 6in. long, and 1½ to 2in. broad, texture thick, deep-green, and somewhat glossy. Flowers small, white, in trichotomous panicles, the flowers borne in clusters of 3, 4, or more at the ends of the branchlets of the panicle. Sepals 4 minute. Petals 4, oblong, 1 line long, with reflexed tips, the base or very short claw woolly at the sides, stamens 4; filaments subulate, slightly hairy at base. Anthers ovate-oblong, light-coloured. Disk prominent, annular. Ovary deeply 4-lobed, glabrous; style short, thick. Fruit as yet not obtained.

Hab.: Eumundi, *Field Nat. Excursion*, Nov., 1895.
The above is, I believe, identical with the Norfolk Island tree described by Steph. Endlicher, and I think this the first time the tree has been met with in Australia since collected by Allan Cunningham near the Brisbane River in 1829. One of these specimens now in my possession has but imperfect flowers, and the leaflets are somewhat larger than those gathered at Eumundi.

7. **E. vitiflora** (vine-flowered), *F. v. M. Fragm.* vii. 144. A shrub with glabrous trifoliolate leaves. Petioles 4 to 5in. long, nearly terete, slightly canaliculate. Leaflets ovate-lanceolate, acuminate, 5 to 6in. long, 2 to 2½in. broad on narrow petiolules of about ½ or 1in. long. Panicles terminal, very slightly pubescent. Peduncles about 2in. long, flowers minute in small umbels on the branches. Pedicels about 1 line long. Calyx-lobes deltoid. Petals scarcely 1 line long, valvate, slightly silky. Stamens shorter than the corolla. Anthers ovate, about ⅓ line long. Ovary tomentose.

Hab.: Rockingham Bay, *J. Dallachy* (F. v. M., l.c.); near *E. aromatica*, according to F. v. M., l.c.

12. MEDICOSMA, Hook. f.

(Having the odour of lemons.)

Sepals 4, broad, imbricate. Petals 4, broad, much imbricate in the bud, the tips erect or recurved. Disk lobed. Stamens 8, filaments dilated, almost cohering by their woolly margins; anthers oblong. Ovary slightly 4-lobed, 4-celled. Style almost terminal, filiform, with a small 4-lobed stigma; ovules 2 in each cell, collateral. Fruit separating into distinct, 2-valved cocci; endocarp separating elastically. Seeds with a crustaceous shining testa, albumen fleshy; embryo straight with broad cotyledons.—A tree. Leaves mostly opposite, 1-foliolate. Flowers large, in axillary panicles.

The genus is limited to a single species endemic in Australia. F. v. Mueller proposes to include it as well as *Melicope* (with which it agrees in the double number of stamens) under *Evodia*, but the habit, that of *Acronychia*, and the large, much-imbricate petals, appear to be a sufficient distinction, unless nearly the whole of *Zanthoxyleæ* be united into one genus.—*Benth.*

1. **M. Cunninghamii** (after A. Cunningham), *Hook. f. in Benth. and Hook. Gen. Pl.* 297; *Benth. Fl. Austr.* i. 362. A small tree, glabrous, or the young shoots and inflorescence minutely pubescent. Leaves mostly opposite, consisting of a single leaflet obscurely articulate on a short petiole, oblong-elliptical or rarely obovate-oblong, obtuse or acuminate, 3 to 6in. long. Panicles axillary, 3-chotomous, with few large flowers. Sepals orbicular, 2 to 3 lines long, with a prominent midrib. Petals nearly ¾in. long, broadly ovate, minutely tomentose outside, with a prominent midrib. Disk thick and glabrous. Ovary hirsute; style slender. Cocci about 3 lines long, quite distinct, scarcely coriaceous, hirsute. Seeds black.—*Acronychia Cunninghamii*, Hook. Bot. Mag. t. 3994; F. v. M. Fragm. i. 27; *Evodia Cunninghamii*, F. v. M. Fragm. iii. 2.

Hab.: Brisbane River, Moreton Bay, *A. Cunningham*; common in the southern scrubs.

The subsucculent cocci, originally described in our "Genera Plantarum," are shown by subsequently received specimens to have been diseased.—*Benth.*

Wood of a light-yellow colour, close in the grain. A good cabinetmaker's wood.—*Bailey's Cat. Ql. Woods No.* 43.

13. BROMBYA, F. v. M.

(After Rev. Dr. J. E. Bromby.)

Calyx 4-lobed. Petals 4, deltoid or ovate-cordate, sessile, valvate. Disk quadrisinuate. Stamens 8; filaments almost ovate, those opposite the petals sterile with very small anthers, those opposite sepals fertile, the point subulate. Anthers introrse, 2-celled, dehiscing longitudinally. Style short. Stigma minute, 4-lobed. Ovary of 4 carpels, coherent inside. Ovules 2, superposed.—F. v. M. Fragm. v. 4.

1. **B. platynema** (broad filaments), *F. v. M. Fragm.* v. 4 *and* 56. A glabrous shrub or tree. Leaves simple, opposite, ovate, shortly acuminate, 2 to 4½in. long, 1½ to 2in. broad, base cuneate to a petiole of ⅓ to 1in. long. Panicles about as long as the leaves, the branches divergent. Pedicels about 1 line long. Calyx lobes deltoid, scarcely ⅓ line long. Petals white, about 1 line. Filaments ciliate, sterile ones spathulate-ovate, fertile ones acute, ovate. Fertile anthers cordate. Disk and style glabrous. Cocci 4, almost ovate, 2 lines long, 2-valved, 1 (rarely 2) seeded. Seeds almost oval, black, 1 line long.

Hab.: Ranges about Rockingham Bay, *J. Dallachy* (F. v. M., l.c.)

14. PAGETIA, F. v. M.

(After Dr. J. Paget.)

Flowers hermaphrodite. Calyx 5-parted, lobes semi-ovate. Petals 5, almost valvate in the bud. Stamens 10, free, all fertile. Filaments linear-subulate. Anthers ovate-cordate, dorsally attached, cells longitudinal. Hypogynous disk annular. Ovary 5-celled, 5-sulcate, or the 5 carpels confluent. Styles 5, short, twisted into 1, or reduced to 1 in the southern species. Stigmas minute. Ovules fasciculate. Ripe carpels bivalved. Seeds 1 or 2, matured, affixed laterally. Strophiole cordiform, membranous. Albumen none. Cotyledons ovate, planoconvex, not convolute, green, base emarginate. Radicle very short, cylindric, superior.—Trees, with opposite, simple, or here and there bi-tri-foliolate leaves and terminal trichotomous panicles of white flowers.—F. v. M. Frag. v. 178-215.

Leaves mostly 1-foliolate.
   Leaflets 3 to 4in. long, 2 to 4in. broad, broadly ovate. Styles 5 . . . . 1. *P. medicinalis*.
   Leaflets 4 to 6in. long, 1½ to 2½in. broad, narrow-oblong. Style single. . 2. *P. monostylis*.

1. **P. medicinalis** (medicinal), *F. v. Fragm.* v. 178-215, vi. 167, ix. 103. A tall tree, with opposite glabrous branchlets and leaves. Bark smooth, whitish. Leaves 3 to 6in. long, 2 to 4in. broad, on petioles of only a few lines, entire, shortly and obtusely acuminate, the under side pale, copiously veined, rarely all trifoliolate. Panicles about as long as the leaves. Flowers numerous in crowded cymes, very thinly puberulent. Calyx scarcely ⅔ line long, persistent. Petals deciduous, ovate, sessile, scarcely 2 lines long. Stamens opposite the sepals, longer than those opposite the petals. Anthers pale-yellow, blunt, scarcely ¼ line long, with an introrse dehiscence. Style scarcely 1 line long, glabrous. Ovary depressed-globose, thinly pubescent. Disk glabrous. Carpels ovate or rhomboid-globose, slightly compressed, 5 to 6 lines long. Seeds brown, glabrous. Strophiole 2 to 2½ lines, cordiform.—F. v. M., l.c.

Hab.: Near Rockhampton, *Messrs. Thozet* and *O'Shanesy* (F. v. M.); Crocodile Creek, *Bowman* (F. v. M.); Mount Buzzard, *J. Dallachy* (F. v. M.); Wide Bay, *Leichhardt* (F. v. M.)

2. **P. monostylis** (single-styled), *Bail. Bot. Bull.* xiii. 7. An erect, glabrous tree of about 60ft. in height, with a rather smooth, whitish bark; branchlets usually ternate, flattened, green, and cane-like, smooth except for the numerous lenticels; internodes long. Leaves opposite, mostly 1-foliolate with very short petioles, oblong, 4 to 6in. long, 1½ to 2½in. broad, base usually cuneate, apex often abruptly acuminate, blunt; the first leaves of the young growth are represented by linear, membranous bud-scales about 1in. long, 2 lines broad, and very deciduous; the pair of leaves under the inflorescence are 2 or 3-foliolate, the lateral leaflets often oblique at the base and nearly or quite sessile, lanceolate, about 3in. long. Flowers white, in terminal trichotomous, corymbose panicles, peduncles flattened, pedicels hairy. Bracts minute, hairy. Sepals about ½ line long, tomentose. Petals tomentose, 2 or 3 lines long. Stamens 10; filaments flattened, glabrous. Disk a glabrous, thick, slightly-lobed ring. Ovary hairy, of 5 pustulate lobes. Style glabrous, shorter than the stamens; stigma capitate, globose, slightly sulcate. No ripe fruit obtained.

Hab.: Eumundi, *Field Nat. Excursion*, Nov. 1895.

This graceful tree differs from *P. medicinalis*, F. v. M., in having narrow-oblong not broadly ovate leaves; the oil-dots are more prominent also in this fresh species. The flowers also have but one style. Further distinction will probably be found in the ripe fruit. The foliage has been distilled and a fair quantity of oil obtained, which has not yet been tested for medicinal virtues said to be contained in the leaves of the northern tree.

XXIX. RUTACEÆ.

15. ZANTHOXYLUM, Linn.

(Wood of a yellow colour.)

(Blackburnia, *Forst.*)

Flowers more or less unisexual. Calyx 3, 4 or 5-lobed. Petals 3, 4 or 5, imbricate or rarely valvate or wanting. Disk small or obsolete. Stamens in the males 3, 4 or 5, the ovary rudimentary or conical, or of 8, 4 or 5 distinct stylelike carpels. Female flowers without stamens or with scale-like staminodia. Ovary of 1 to 5 distinct carpels. Styles nearly terminal, distinct or united upwards; ovules 2 in each carpel, usually collateral. Fruit of 1 to 5 distinct cocci, dry or drupaceous, usually 2-valved; the endocarp separating or adherent. Seeds with a hard or crustaceous shining testa; albumen fleshy; embryo straight or curved, with broad flat cotyledons.—Shrubs or trees, often armed with scattered prickles, and sometimes climbing. Leaves alternate, usually pinnate. Flowers small, in axillary or terminal cymes or panicles.

A large genus, dispersed over the tropical and subtropical regions of the whole world. Of the following species, two are endemic in Australia, the third is also in Norfolk Island. All three belong to the section *Blacburnia*, characterised chiefly by solitary carpels, which are rare in the rest of the genus.

Tree, the trunk and branches prickly and here and there prickles upon
  the petioles, rhachis and costules.
  Leaflets 9 to 13, oblong-elliptical, acuminate, 1½ to 4in. long . . . 1. *Z. brachyacanthum*.
Tree, the trunk and branches often prickly.
  Leaflets 5 to 9, oblong-lanceolate, 3 to 6in. long, 1 to 2in. broad . . 2. *Z. veneficum*.
Tall climber, all parts prickly.
  Leaflets 5 to 9, broadly ovate, 3 to 5in. long, the margins with bristly
    teeth . . . . 3. *Z. torvum*.
  Leaflets scarcely oblique, not coriaceous. Panicles terminal. Flowers
    very numerous, under 1½ line . . . . 4. *Z. parviflora*.

1. **Z. brachyacanthum** (alluding to the short prickles), *F. v. M. Pl. Vict.* i. 108; *Benth. Fl. Austr.* i. 368. Satin-wood. A small tree, the trunk and branches covered with short conical prickles. Leaves pinnate, the common petiole 6 to 10in. long. Leaflets usually 9 to 13, opposite in pairs, with or without a terminal odd one, petiolulate, from ovate to oblong-elliptical, shortly acuminate, 2 to 3in. or more long, equal or oblique at the base, coriaceous and shining. Panicles axillary and terminal, much shorter than the leaves, irregularly 2—3-chotomous. Flowers white, on very short pedicels, the males nearly 3 lines long, the females shorter. Sepals 4, small and broad. Petals obtuse, much imbricate. Ovary rudimentary in the male flowers, in the females consisting of a single carpel with a large oblique stigma, nearly sessile or on a very short style, terminal but excentrical. Fruit opening wide to the middle in 2 valves. Seed black.—Benth, l.c.

Hab.: Common on the southern ranges; also met with at a few places in the tropics, according to F. v. M.

Wood a glossy yellow, superior to the wood used in Europe under the name of "satinwood."—*Bailey's Cat. Ql. Woods No.* 45.

2. **Z. veneficum** (poisonous), *Bail. 1st Suppl. Synop. Ql. Flora* ii., *Bot. Bull.* vii. A medium-sized tree, glabrous, the branches and sometimes the stem prickly. Leaves (including the petiole) from 6 to 12in. long. Leaflets 5 to 9, opposite in pairs, with or without a terminal oddone, petiolulate, oblonglanceolate, shortly acuminate, unequally sided at the base, 3 to 6in. long, 1 to 2in. broad, the primary veins prominent, almost transversely spreading, prickles none or scarcely any on the rhachis, petiolule 1 or costules. Panicles terminal and in the upper axils, trichotomous. Flowers small, in threes or fours at the ends of the branchlets of the panicle. Pedicels short. Buds globose. Sepals 4, scarcely over ½ line in diameter. Petals 4, ovate, spreading, 2¼ lines long,

imbricate in the bud. Filaments 4, flattened, longer than the petals, arching over the ovary. Anthers connivent, somewhat cordate, rather large. Ovary with 8 prominent wing-like angles. Stigma sessile.

Hab.: Johnstone River, *Dr. Thos. L. Bancroft*, who reports the wood and bark to be poisonous; Barron River, *E. Cowley*.
Wood yellow, of close grain, and easy to work.—*Bailey's Cat. Ql. Woods No.* 44a.
The bark possesses a poisonous principle as toxic as strychnine, to whose physiological action it has some resemblance. - *T. L. Bancroft*.

3. **Z. torvum** (unapproachable), *F. v. M. Fragm.* vii. 140. A tall scrub climber, the stems often several inches thick and armed with similar prickles to the other species. The branchlets, petioles, and peduncles clothed with a very short cano-fuscus pubescence and prickles. Leaves impari-pinnate. Leaflets 5 to 9, subcoriaceous, 3 to 5in. long, ovate, very shortly petiolulate, oblique, rounded at the base, somewhat long and obtusely acuminate, the margins with distinct bristle-like teeth, both sides glabrous and glossy. Panicles about 4in. long, shortly branched. Pedicels about as long as the flowers. Bracteoles at the base of the pedicels, very minute, linear-subulate. Petals imbricate, lanceolate-ovate, 1 line long, base truncate, yellow. Filaments of the male flower setaceous, almost 2 lines long. Anthers cordate-ovate, dorsally attached. Ovary of female flowers 4. Styles very short, coherent. Stigmas united into 1, peltate, 4—8, maturing 2—1, scarcely 3 lines, compressed, almost lenticular, obtuse. Seeds 1½ line, ovate-globular, glossy black, solitary.

Hab.: Many of the northern range scrubs.
Wood yellow and close-grained.—*Bailey's Cat. Ql. Woods No.* 44.

4. **Z. parviflorum** (small-flowered), *Benth. Fl. Austr.* i. 868. A small tree, glabrous and unarmed, or with very few minute distant prickles. Leaves pinnate, with a common petiole of 4 to 6in., angular but not winged ; leaflets usually 9 to 11, opposite in pairs, the terminal odd one occasionally wanting, ovate-lanceolate, acuminate, rarely above 2in. long, entire or slightly denticulate, usually oblique, the upper edge most rounded at the base, membranous or at length scarcely coriaceous. Panicles terminal, 3-chotomous, broad, with numerous small 4-merous flowers. Sepals small, triangular. Petals scarcely 1½ lines long, slightly imbricate. Stamens in the males 4, about as long as the petals. Ovary rudimentary, of 1 or 2 carpels. Female flowers not seen. Cocci solitary, 3 to 4 lines long, coriaceous, rugose outside, opening broadly to below the middle in 2 valves, endocarp persistent. Seeds with a hard bony testa enveloped in a thin black shining epiderm.

Hab.: Islands of the Gulf of Carpentaria.

16. GEIJERA, Schott.

(After J. D. Geijer.)

Flowers hermaphrodite. Sepals 4 or 5. Petals 4 or 5, valvate or imbricate. Disk thick and fleshy. Stamens 4 or 5 ; filaments subulate. Ovary depressed, partly immersed in the disk, 4 or 5-lobed ; styles terminal, immediately united into a single short style, with a capitate 4 or 5-lobed stigma. Fruit of 4 or 5 or sometimes fewer distinct, 2-valved cocci, the endocarp adherent or partially separating. Seeds with a hard or crustaceous shining testa ; albumen fleshy; embryo straight ; cotyledons broad.—Trees or shrubs. Leaves alternate, simple, not articulate on the petiole. Flowers small, in terminal panicles. Sepals small.

The genus is limited to Australia, and differs from *Zanthoxylum* chiefly in the simple leaves and hermaphrodite flowers.—*Benth.*

Panicles spreading. Petals imbricate. Carpels corrugated, oblong . . . . 1. *G. Helmsia.*
Panicles compact. Petals imbricate. Leaves broad. Carpels globose . . 2. *G. Muelleri.*
Panicles loose. Petals valvate.
   Leaves from ovate to lanceolate. Carpels globose . . . . . . 3. *G. salicifolia.*
   Leaves linear. Carpels globose. . . . . 4. *G. parviflora.*

1. **G. Helmsiæ** (after Mrs. Helms), *Bail.* A small glabrous tree with greyish bark; the branchlets wrinkled. Leaves smooth, ovate, tapering towards each end, 2 to 3½in. long, 1 to 2in. broad in the centre, obtuse or obtusely-acuminate, somewhat thick, coriaceous; petioles about ¾in. long. Glossy on both sides, but rather pale on the under side. Oil-dots small and numerous, not seen well without a lens; primary lateral nerves distant, fine, reticulate veins obscure. Panicles terminal and from the axils of the upper leaves, about 3in. long and 4in. broad. Flowers white and numerous, differing from *G. Muelleri* in nothing except in their somewhat larger size. Cocci 2 or 3 maturing, 4 or 4½ lines long, 3 lines broad, and deeply corrugated on the outside, cohering at the base. Endocarp freely separating, yellowish. Seeds pyriform, brown, with a large hilum, probably never so glossy as in other species.

Hab.: Childers, *Mrs. Helms.*

This new species differs from *G. Muelleri* somewhat in foliage, but principally in the cocci and seed. The cocci are very like those of *Melicope neurococca.*

2. **G. Muelleri** (after Baron von Mueller), *Benth. Fl. Austr.* i. 364. Nankeen dye-wood. A glabrous tree. Leaves ovate or obovate-oblong, 2 to 3in. long, narrowed into a rather long petiole, coriaceous, with a prominent midrib, the lateral veins slender and rather distant. Panicle compact, scarcely equalling the last leaves. Flowers rather larger than in the other species. Petals nearly 1½ line long, distinctly imbricate, obtuse, without inflexed tips. Cocci 2 to 3 lines long, distinctly but very shortly beaked, very spreading, but cohering at the base. Epicarp slightly tuberculous. Endocarp persistent. Seeds glossy, black.—*Coatesia paniculata*, F. v. M. Fragm. iii. 26 (in part).

Hab.: Cumberland Islands, *R. Brown;* Araucaria woods near Moreton Bay, *F. v. Mueller;* Curtis Island, *Henne.*

This species was generically distinguished by F. Mueller, on account of the imbricate æstivation of the petals, and a slight difference in the fruit, but the habit is that of the other species, and the genus is too closely allied to *Zanthoxylum*, which contains species with valvate as well as with imbricate æstivation, to admit of dividing it solely on that ground.—*Benth.*

Heart-wood dark, beautifully clouded, the rest of a light color, all hard and close-grained, suitable for cutting into veneers for cabinet work.—*Bailey's Cat. Ql. Woods* 45a.

3. **G. salicifolia** (Willow-leaved), *Schott Fragm. Rut.* t. 4; *Benth. Fl. Austr.* i. 364. A moderately-sized tree, glabrous or with a minute hoary pubescence on the inflorescence, and sometimes on the under side of the leaves. Leaves from ovate to ovate-lanceolate or rarely oblong-lanceolate, obtuse or acuminate, mostly 3 to 4in. long, entire, coriaceous, narrowed or rarely rounded at the base, with a rather long petiole. Panicles rather loose, broadly pyramidal, but much shorter than the last leaves, alternately branched, with numerous small white flowers. Petals about 1 line long, valvate. Cocci often reduced to 1 or 2, obovoid, not beaked, 2 to 3 lines long. Epicarp slightly tuberculous, the endocarp persistent or partially separating. Seeds glossy, black.—*G. latifolia*, Lindl. in Mitch. Trop. Austr. 236.

Hab.: Broadsound, *R. Brown;* Moreton Bay and Brisbane River, *A. Cunningham, F. v. Mueller,* and others; Brigalow scrub on the Burdekin, and near Warwick, *F. v. Mueller;* Wide Bay, *C. Moore;* Port Denison, *Fitzalan;* Rockhampton, *Thozet;* Mantua Downs, *Mitchell.*

Schott's figure represents a remarkably narrow-leaved form, which I have only seen in Brown's specimens, and in those from Warwick and Rockhampton. These, however, pass into the common broad-leaved form.—*Benth.*

Wood of a light-yellow (no dark heart-wood), hard, close-grained, of a somewhat greasy nature; suitable for engraving, skate-rollers, and hand-screws.—*Bailey's Cat. Ql. Woods No.* 46.

4. **G. parviflora** (small flowers), *Lindl. in Mitch. Trop. Austr.* 102; *Benth. Fl. Austr.* i. 364. "Wilga," southern border. A tall shrub or small tree, with slender, erect or pendulous branches, glabrous or the inflorescence and young parts slightly hoary. Leaves linear, acute or obtuse, 3 to 6in. long, and rarely

Geijera Helmsiæ, Bail.

above 8 lines broad, coriaceous, narrowed into a rather short petiole, the midrib prominent underneath. Flowers and fruit of *G. salicifolia,* or the flowers sometimes, but not always, rather smaller.—*G. pendula,* Lindl. in Mitch. Trop. Austr. 251. Possibly a variety only of *G. salicifolia,* Benth.

Hab.: Broadsound, *R. Brown;* Burdekin River, *F. v. Mueller;* Belyando River, *Mitchell.*

Wood hard, tough, of close grain, yellow, and when fresh of an agreeable fragrance.—*Bailey's Cat. Ql. Woods No.* 47.

17. PENTACERAS, Hook. f.
(Referring to arrangement of carpels.)

Sepals 5. Petals 5, valvate. Torus thick. Stamens 10; filaments subulate, glabrous. Ovary of 5 nearly distinct carpels, each with a glandular terminal appendage. Styles inserted below the middle and immediately united into one filiform style, with a small stigma; ovules 2 in each carpel, superposed. Fruit-carpels 5 or fewer, often solitary by abortion, indehiscent, expanded all round into a membranous wing, forming obovate or oval-oblong samaræ, the centre almost drupaceous, with a cartilaginous endocarp. Seeds usually solitary; testa thick; albumen not copious; embryo straight, with ovate cotyledons.—Trees. Leaves alternate, pinnate. Flowers numerous, small, paniculate.

The genus is limited to a single species, endemic in Australia. It differs from *Evodia* in its habit, alternate leaves, and in some measure in the ovary resembling that of several *Diosmeæ,* and from that and all other *Zanthoxyleæ* by the fruit, which, at first sight, is like that of an *Ailanthus;* but the dotted leaves and superposed ovules, which place it among *Rataceæ,* besides the inflorescence and other minor characters, amply distinguish *Pentaceras* from *Ailanthus—Benth.*

1. **P. australis** (Australian), *Hook. f. in Benth. and Hook. Gen. Pl.* 298; *Benth. Fl. Austr.* i. 865. A glabrous tree, small according to A. Cunningham, attaining 60ft. according to W. Hill. Leaves pinnate, with a common petiole of from 4 or 5in. to nearly 1ft.; leaflets usually 7 to 11, opposite in pairs, with a terminal odd one, ovate to lanceolate, obtuse or acuminate, 2 to 4in. long, entire or obscurely crenate, the lateral ones more or less oblique and decurrent on the petiolule on the lower side, like those of a *Clausena.* Panicles large, terminal, spreading, loose, with numerous white flowers, pedicellate along the ultimate branches. Petals about 1½ line long. Stamens nearly as long as the petals. Ovary glabrous. Ripe samaræ 1 to 1½in. or rather more in length, ½ to ⅜in. broad.—*Cookia australis,* F. v. M. Fragm. i. 25 and iii. 27; *Ailanthus punctata,* F. v. M. Fragm, iii. 42.

Hab.: Brisbane River, *A. Cunningham;* Moreton Bay district; M'Connell's Brush, *Leichhardt.*

Wood of a light-yellow, close-grained, and hard.—*Bailey's Cat. Ql. Woods No.* 48.

18. PLEIOCOCCA, F. v. M.
(Cocci numerous.)

Calyx 4-lobed, persistent. Petals deciduous. Filaments linear-subulate, ciliate. Anthers and styles not seen. Cocci or carpels 5 to 9, slightly attached to the axis for about a third of their length, the upper two-thirds free, tardily separating into 2 valves. Endocarp not readily separating. Seeds 1—2. Albumen very large. Embryo straight. Cotyledons nearly plane, oblong. Radicle slender.—A small tree with large opposite coriaceous leaves and axillary cymes of flowers. Carpels fleshy, epicarp sharply acid.—F. v. M. Fragm. ix. 117 (in part).

The genus seems intermediate between *Evodia* and *Acronychia.*

1. **P. Wilcoxiana** (after J. Wilcox), *F. v. M. Fragm.* ix. 117. So far as observed a small tree. Branchlets stout, deep green. Leaves of a single leaflet, on a petiole of about 1in., 6 to 8in. long, 2½ to 3½in. broad, oblong, obtuse or abruptly acuminate at the apex, and slightly cordate at the base, coriaceous and glossy, the primary nerves thread-like, but as well as the reticulate veinlets prominent, especially in the dry leaf. Flower cyme in the upper axils often reduced to a slender racemose panicle. Pedicels about 3 lines long, enlarging upwards. Fruit globular in outline, about ¾in. diameter, composed of from 5 to 9 white, fleshy, sharply acid and juicy carpels attached only by the inner angle to the axis, free above and connivent at the apex, the base of each carpel shortly prolonged below the attachment in the form of a short blunt spur. Seeds 1 or 2 in each carpel, flattish, reniform, tuberculous, blackish.

Hab.: Scrubs about Eumundi.

19. ACRONYCHIA, Forst.
(Referring to the hooked tips of petals.)
(Cyminosma, *Gærtn.*)

Flowers polygamous. Calyx 4-lobed. Petals 4, valvate. Torus thick. Stamens 8; filaments subulate. Ovary 4-celled; style terminal; stigma entire or obscurely 4-lobed, ovules 2 in each cell, superposed. Fruit 4-celled, usually succulent, with a coriaceous or hard endocarp, opening loculicidally, or drupaceous and indehiscent. Seeds usually solitary in each cell, with a crustaceous black testa; albumen fleshy; embryo straight; cotyledons oblong.—Trees or shrubs. Leaves opposite or alternate, 1-foliolate. Flowers white or yellowish, in axillary or rarely terminal small panicles or loose cymes.

The genus extends over tropical Asia and the islands of the S. Pacific, to New Caledonia and New Zealand. Of the Australian species, one is also found in New Caledonia, the six others are endemic.—*Benth.* (in part).

Petioles often long.
  Flowers minutely tomentose, in short oblong panicles. Petals ovate.
    Fruit 4-seeded . . . . 1. *A. Baueri.*
  Petioles 1in. long.
    Leaves 3 to 7in. long.
      Fruit oval, ½in. long, 1-seeded . . . . 2. *A. tetrandra.*
      Flowers glabrous, in axillary 3-chotomous cymes. Petals narrow.
        Leaves thin and scarcely coriaceous. Fruit 4-angled, depressed on the summit . . . . 3. *A. lævis.*
        Leaves very coriaceous. Fruit obcvoid-globular . . . . . 4. *A. imperforata.*
        Leaves 3-foliolate. Fruit yellow, 4-seeded . . . . . . 5. *A. melicopoides.*
    Fruit reddish, about ½in. diameter, juicy, acid . . . . 6. *A. Scortechinii.*
  Branchlets hirsute. Leaves often large.
    Fruit white, epicarp acid, large, irregularly corrugated . . . . . 7. *A. vestita.*

1. **A. Baueri** (after F. Bauer), *Schott Fragm. Rut.* t. 3; *Benth. Fl. Austr.* i. 366. A small or moderate-sized tree, glabrous or the young shoots or inflorescence minutely hoary-tomentose. Leaves opposite, of a single leaflet, on a rather long petiole, ovate, elliptical or obovate, obtuse or very shortly and obtusely acuminate, narrowed at the base, 3 to 4 or very rarely 5in. long, thinly coriaceous. Panicles axillary, oblong, the side branches and pedicels very short, sometimes reduced to a small spike. Flowers small, not numerous. Sepals very broad, short, ciliate. Petals ovate, valvate with inflexed tips, minutely pubescent outside, 1 to 1¼ line long. Filaments thin, dilated, and ciliate to above the middle. Ovary pubescent; style pubescent, short, with a rather large stigma. Fruit nearly globular or 4-angled, obtuse or shortly acuminate, ½in. diameter or rather smaller, not very succulent. Testa of the seeds hard and bony.—*A. Hillii,* F. v. M. Fragm. i. 26.

Hab.: Northumberland Islands and Richmond district, *R. Brown*; Moreton Bay and Brisbane River, *A. Cunningham, F. Mueller,* and others; Five Islands, *A. Cunningham.*

Wood of a uniform yellow, or sometimes darker towards the heart, rather hard and close-grained.—*Bailey's Cat. Ql. Woods No.* 49.

2. **A. tetrandra** (four stamens), *F. v. M. Fragm.* ix. 104. A shrub with very thinly pubescent branches. Leaves of a single leaflet, ovate-lanceolate, sometimes very shortly obtusely acuminate, 3 to 7in. long and about 2 to 3in. broad, chartaceous, glabrous, petioles about 1in. long. Panicles short and almost racemose, or on slender peduncles of about 3in. long, shortly branched towards the end, the small flowers velvety, on pedicels as long as themselves. Calyx 4-lobed, lobes deltoid. Petals ovate, acute, 1½ line long, valvate, silky outside, tomentose inside. Stamens shorter than the petals; filaments pilose, anthers cordate-rotundate, emarginate, yellow. Disk annular, glabrous. Style very short. Ovary velvety, 4-celled. Fruit oval, about ½in. long, 1-seeded.—*Evodia haplophylla*, F. v. M. Fragm. v. 179.

Hab.: On the ranges about Rockingham Bay, J. Dallachy (F. v. M., l.c.)

3. **A. lævis** (fruit smooth), *Forst. Char. Gen.* 58, t. 27; *Benth. Fl. Austr.* i. 366. A tree, sometimes large, glabrous except the stamens. Leaves irregularly opposite or alternate, of a single leaflet, obovate-oblong to oblong-elliptical, obtuse, 1½ to 3 or rarely nearly 4in. long, coriaceous when old. Cymes 2 or 3-chotomous, usually shortly pedunculate and few-flowered. Sepals very short, rounded, glabrous. Petals narrow, induplicate-valvate, with inflexed tips, 2 to 2½ lines long, glabrous. Filaments rather thick, dilated and ciliate towards the base, subulate and inflexed at the top. Ovary hirsute round the base of the style, otherwise glabrous; style rather long, the stigma not thickened, obscurely 4-lobed. Fruit succulent, with a crustaceous 4-celled endocarp, obtusely 4-angled, truncate at the top and depressed in the centre, ½in. diameter or rather smaller.—*Lawsonia Acronychia*, Linn. f.; Labill. Sert. Austr. Caled. 66, t. 65; *Cyminosma oblongifolium*, A. Cunn. in Bot. Mag. 3222; *Acronychia laurina*, F. v. M. Fragm. i. 27.

Var. *normalis*. Sometimes forming a rather large tree. Fruit hardly showing angles, leaves scarcely glossy, and smaller than in other forms. Hab.: Mountain scrubs and creek sides in southern and northern Queensland.
Wood close-grained, hard, and light-coloured.—*Bailey's Cat. Ql. Woods No.* 50.

Var. *purpurea*. A small tree. Fruit axillary, of a purplish or plum colour, leathery, with very prominent angles, very hollow; dark-green foliage. Hab.: Creek sides in southern Queensland.

Var. *leucocarpa*. A slender often tall tree. Fruit white, somewhat fleshy, with obtuse angles, borne in lateral cymes; leaves bright glossy-green, about 4in. long, and 1 to 1½in. wide. Hab.: Eumundi.

4. **A. imperforata** (referring to the minute oil-dots), *F. v. M. Fragm.* i. 26; *Benth. Fl. Austr.* i. 367. Usually a small tree, very nearly allied to *A. lævis*. Leaves of the same shape and size, but on much shorter petioles, and much more coriaceous, the minute pellucid dots only visible before a strong light. Inflorescence and flowers as in *A. lævis*, except that the peduncles are much shorter and the flowers rather larger. Filaments much ciliate. Fruit somewhat obovoid and obscurely or not at all angular, and not depressed at the top.

Hab.: N.E. coast, *R. Brown*; Brisbane River, *W. Hill*, *F. v. Mueller*; Herberton, *J. F. Bailey*.
Wood of a bright-yellow and hard.—*J. F. Bailey*.

5. **A. melicopoides** (Melicope-like), *F. v. M. Fragm.* v. 3. Tree attaining a height of from 30 to 50ft., with a smooth grey bark. The branchlets, petioles, and peduncles clothed with a very thin pubescence. Leaves trifoliolate. Petioles 1 to 3in. long, leaflets sessile, glabrous, lanceolate-ovate, 4 to 5in. long, thin-coriaceous. Cymes axillary and terminal, common peduncle about 1in. long, the secondary ones opposite, ½in. or shorter. Pedicels 1 to 1½ line. Bracts and bracteoles deltoid, minute. Sepals or calyx-lobes rounded, about 1 line long. Petals 4, yellow, about 3½ lines long, sessile, inflexed at the apex, linear-semilanceolate, glabrous. Filaments 2¾ to 3 lines long, flattened towards the

P

base, subulate towards the top, ciliate. Anthers subovate, dorsally fixed, introrse, ½ line or more long. Style somewhat long, tomentose in the lower portion, stigma small. Fruit yellow, about 5 lines long, 4-seeded, seeds black.—*Evodia acronychodes*, F. v. M. Fragm. iv. 117.

Hab.: Ranges, Rockingham Bay, *J. Dallachy* (F. v. M., l.c.); Herberton, *J. F. Bailey*.
Wood hard, of a light colour.—*J. F. Bailey*.

6. **A. Scortechinii** (after Rev. B. Scortechini), *Bail*. Scortechini's Crab or Logan Apple. A small tree, the branchlets often of a reddish colour. Leaves obtuse-lanceolate or obovate, subcoriaceous, 2 to 3in. long; petioles about ½in. long. Cymes axillary, few-flowered. Peduncle slender, about as long as the petioles; the 3 branches usually bearing 2 flowers at their extremities. Pedicels about 2½ lines. Sepals very short, broad, glabrous. Petals 4½ lines long, ciliate. Filaments dilated at the base, tapering upwards, margins densely ciliate except towards the top. Fruit globular, reddish, exceeding ½in. diameter, 4-celled, with a juicy epicarp. Seeds oval, testa brown, slightly rugose.

Hab.: Borders of scrubs, Logan River, *Rev. B. Scortechini*; Fraser's Island, *Miss Lovell*.

The fruit of this tree, which is of a sharp, pleasant acid taste and red colour, is useful for jam-making. The Logan River specimens I have referred to at times as a form of *A. acidula*, F. v. M., and the Fraser's Island ones as a form of *A. imperforata*, but now consider these two forms identical, and, although agreeing often somewhat in foliage and flowers, distinct from all others of the genus in fruit.

7. **A. vestita** (branches clothed with tomentum), *F. v. M. Fragm.* iv. 155, ix. 104. A small round-headed tree with a somewhat smooth light-coloured bark. Branchlets slightly or densely clothed with short hairs which extend more or less to the petioles and the principal nerves of the leaves on the under side. Petioles from 1 to 2in. long. Leaves 3 to 8in. or more long, and some 4in. or more broad, thin-coriaceous, ovate-oblong or ovate-lanceolate. Cymes of few flowers on rather long slender peduncles. Fruit white, the large ones about 1in. diameter, almost globose in outline, very irregularly corrugated, often forming a point at the apex and slightly tapering at the base, hard when dry but in a fresh state the epicarp fleshy with a sharply acid juice. Seeds oblique-oval, nearly black, 1½ to 2 lines long, more or less rugose.—*A. acidula*, F. v. M. Fragm. iv. 154.

Hab.: Borders of scrubs in the tropical parts of the colony.
Wood of a light colour, soft, and easy to work.—*Bailey's Cat. Ql. Woods No.* 51.

20. HALFORDIA, F. v. M.

(After Dr. G. B. Halford, Professor of Medicine, Melbourne University.)

Calyx 5-toothed. Petals 5, valvate in the bud. Stamens 10, free, all fertile. Filaments linear-subulate, ciliate. Anthers 2-celled, minutely apiculate. Style simple. Stigmas very minute. Fruit with a somewhat juicy epicarp, 3—5-celled. Seeds albuminous, pendulous, solitary in each cell. Embryo straight. Cotyledons leafy, narrow-oblong, radicle a little longer than broad.—Small trees with simple, alternate or subopposite leaves and terminal panicles of flowers.

Evergreen trees of southern and northern Queensland.

Fruit purple . . . . 1. *H. drupifera*.
Fruit red . . . . 2. *H. scleroxyla*.

1. **H. drupifera** (Plum-like), *F. v. M. Fragm.* v. 48, ix. 108. Small tree, branchlets angular. Leaves glabrous, lanceolate or ovate-lanceolate, with the short petiole 3 to 5in. long and 1 to 1½in. broad, somewhat thinly pellucidly-punctate, obtuse-acuminate, and decurrent upon the petiole. Paniculate cymes terminal with numerous shortly pedicellate white flowers. Pedicels with minute

bracteoles at the base. Calyx about 1 line long. Petals 2½ lines long, narrow-lanceolate. Filaments linear-subulate, ciliate, those opposite the petals shorter than those opposite the calyx-lobes. Anthers yellow, cordate, about ¼ line long, minutely apiculate. Style simple, very short, glabrous. Stigma very minute. Ovary glabrous, disk 10-ribbed. Drupe purple, about ½in. long, oval with a truncate base. Putamen bony. Seeds ellipsoid-cylindric, slightly angular, 2¼ lines long. Testa black, smooth, fragile. Albumen amygdaloid. Embryo about 2 lines long.—*Eriostemon Leichhardtii*, F. v. M. Fragm. v. 5.

Hab.: Fraser's Island.
Wood of a yellowish colour, close in grain, tough and durable.—*Bailey's Cal. Ql. Woods No. 52.*

2. **H. scleroxyla** (alluding to its hard wood), *F. v. M. Fragm.* vii. 142. "Ghittoe," Herberton, *J. F. Bailey*. A tree of about 60ft., but flowering as a shrub. Bark grey. Leaves coriaceous, obovate-lanceolate, tapering and decurrent upon the petiole, upper side glossy. Drupe a pretty red colour, globose-ovate, truncate or introrse at the base, 4 to 8 lines long, pericarp acidulous.

Hab.: Scrubs about Rockingham Bay, *J. Dallachy* (F. v. M., l.c.); Evelyn to Russell River, *J. F. Bailey*.
Wood when freshly cut yellowish, turning brownish with age; hard, tough, and very inflammable even in a green state.—*J. F. Bailey.*

21. GLYCOSMIS, Corr.

(Name from its fragrance.)

Calyx 5-cleft, the lobes broadly imbricate. Petals 5, imbricate in the bud. Stamens 10, filaments dilated at the base, anthers often tipped with a small gland. Ovary 3 to 5 or rarely 2-celled; style very short, thick and persistent, the stigma scarcely broader, ovules solitary in each cell. Berry succulent or almost dry, usually 1-seeded. Seeds with a membranous testa, without albumen; cotyledons fleshy.—Unarmed trees or shrubs. Leaves alternate, pinnate, with few alternate leaflets or 1-foliolate. Flowers small, in axillary or terminal panicles.

A genus of very few species, dispersed over tropical Asia and the Eastern Archipelago, the Australian one being the most widely spread over the whole region.—*Benth.*

1. **G. pentaphylla** (number of leaflets), *Corr.: Oliv. in Journ. Linn. Soc.* v. *Suppl.* 37; *Benth. Fl. Austr.* i. 367. A tall shrub or small tree, quite glabrous. Leaves occasionally 1-foliolate, on short petioles, but more generally pinnate, with 2 or 3 leaflets, from ovate-elliptical or ovate-lanceolate to oblong-lanceolate, obtuse or acuminate, 2 to 4 or rarely 5in. long. Panicles dense, shorter, or scarcely longer than the petiole of the pinnate leaves. Petals about 2 lines long. Ovary 5 or sometimes 4-celled, contracted into a very short, thick style. Berry globular, ½in. in diameter, or smaller.

Hab.: Northumberland Islands, *R. Brown*; islands of Torres Straits, *F. v. Mueller*; scrub near Rockhampton, *Thozet*.
The species has a very wide range in tropical Asia, and is very variable in the size of the leaves and the flowers, full details of which and of the consequently extended synonymy of the species will be found in Oliver's paper above quoted. The character given above has special reference to the Australian variety, which is almost identical with the Chinese and Eastern form, usually distinguished as *G. citrifolia*, Lindl.; Benth. in Fl. Hongk. 51, and figured as *Limonia parvifolia*, Hook. Bot. Mag. t. 2416.—*Benth.*

22. MICROMELUM, Blume.

(Small Apple; fruit small.)

Calyx 5-toothed or entire. Petals 5, valvate in the bud or nearly so. Stamens 10; filaments linear-subulate. Ovary 2 to 6 usually 5-celled, the dissepiments spirally twisted after the flowering; style deciduous with a small capitate stigma;

ovules 2 in each cell, superposed. Fruit a dry berry. Seeds usually 1 or 2; testa membranous; albumen none; cotyledons leafy, very much folded.—Unarmed trees. Leaves alternate, pinnate, with alternate oblique leaflets. Flowers small, in terminal corymbose panicles.

Besides the Australian species, which is widely dispersed over tropical Asia and the Eastern Archipelago, only 2 are known, from Penang or the Philippine Islands.—*Benth.*

1. **M. pubescens** (pubescent), *Blume; Oliv. in Journ. Linn. Soc.* v. *Suppl.* 40; *Benth. Fl. Austr.* i. 368. A small tree. Young branches and leaves more or less pubescent. Leaflets 9 to 15, or sometimes more, from ovate to broadly lanceolate, 1 to 3in. long, obtuse or shortly acuminate, oblique at the base, often becoming glabrous above, pubescent underneath. Corymbs nearly sessile above the last leaves, many-flowered. Calyx more or less 5-toothed. Petals about 2 lines long, more or less pubescent. Ovary usually hairy. Berry small, ovoid, glabrous or pubescent, red when fully ripe.

Hab.: Islands of the Gulf of Carpentaria, *R. Brown;* Albany and Cairncross Islands, and from the Burdekin to Moreton Bay, *F. v. Mueller;* Cape Upstart and Barnard Isles, *M'Gillivray;* Wide Bay, *Bidwill;* Rockhampton, *Thozet,* to the scrubs of the Brisbane River.

The various forms assumed by this species and the consequent synonymy are given in detail by Oliver in the above-quoted paper. The Australian specimens belong to the small-flowered variety, with rather broad leaflets, common in the S. Pacific islands, which I formerly described as *M. glabrescens,* in Hook. Lond. Journ. ii. 212.—*Benth.*

Wood of a light colour, and close-grained.—*Bailey's Cat. Ql. Woods No.* 53.

28. MURRAYA, Linn.
(After Professor Murray.)

Calyx 5-cleft. Petals 5, narrow, imbricate in the bud. Stamens 10, free; filaments subulate; anthers small. Ovary 2 to 5-celled. Style elongated, at length deciduous, stigma capitate. Ovules solitary or 2 in each cell, superposed or nearly collateral. Berry 1 or 2-seeded. Testa glabrous or woolly; albumen none; cotyledons equal, not folded.—Unarmed trees or shrubs. Leaves pinnate, leaflets alternate, usually oblique at the base. Flowers often rather large, in terminal corymbs, or few together in the upper axils.

The genus comprises a few species, dispersed over tropical Asia and the Eastern Archipelago; neither of the Australian ones are endemic.—*Benth.*

Ovary 2-celled. Flowers nearly ½in. long . . . . 1. *M. exotica.*
Ovary 5-celled. Flowers numerous, not 3 lines long . . . . 2. *M. crenulata.*

1. **M. exotica** (exotic), *Linn.; Oliv. in Journ. Linn. Soc.* v. *Suppl.* 28; *Benth. Fl. Austr.* i. 369. A shrub or small tree, glabrous, or the young branches and petioles pubescent. Leaflets usually 5 to 7, from ovate, cuneate-obovate, or almost rhomboidal to ovate-lanceolate, ¾ to 2in. long, coriaceous and shining when full-grown. Flowers white, very fragrant, in compact, terminal, sessile corymbs, or few together in the common varieties. Petals nearly ½in. long, erect at the base, spreading in the upper half. Ovary 2-celled. Berry globular or almost ovoid, usually 2-seeded.—*Wight, Ic.* t. 96.

Hab.: Islands of the Gulf of Carpentaria, *R. Brown;* scrub near Rockhampton, *Thozet.*

These specimens are past flower, and have only a few young fruits, which are more ovoid than they generally are in the species, but in other respects they appear to belong as well as Brown's to the few-flowered var. 2 of Oliver, or *M. paniculata,* Jack. The species is common from N.W. India to the New Hebrides.—*Benth.*

From my observations of this species at Tooloomba, the leaflets resemble those of the normal form, never being so long as in the form *paniculata,* and in habit this plant differs from both, it having a very straggling habit.

2. **M. crenulata** (leaflets crenulated), *Oliv. in Journ. Linn. Soc.* v. *Suppl.* 29? *Benth. Fl. Austr.* i. 369. A glabrous shrub or tree. Leaflets usually 7 to 11, very oblique, from oval-oblong to oblong-elliptical, obtuse or shortly acuminate, 2 to 3in. long, entire or obscurely crenulate. Flowers (in the Philippine specimens) in terminal corymbs, much more numerous and much smaller than those of *M. exotica*. Petals 2½ to nearly 3 lines long. Fruit depressed-globular, 5 or 6 lines diameter, 5-celled, but with 3 or 4 cells abortive. Seeds 1 or 2; cotyledons plano-convex, thick and fleshy.—*Glycosmis crenulata*, Turcz. in Bull. Mosc. 1858, i. 250.

Hab.: Eastern subtropical Australia, *Herb. F. v. M.*

These specimens are in fruit only, but the foliage, the inflorescence, and calyx are so precisely those of the Philippine Island ones that there is little doubt that they belong to the same species. The structure of the fruit is quite that of *Murraya*; the cotyledons of the seed very readily distinguish it from *Micromelum*, which in many respects has a similar habit and inflorescence.—*Benth.*

24. CLAUSENA, Burm.

(After P. Clauson, a Danish botanist.)

Calyx 4 or 5-cleft. Petals 4 or 5, broad, imbricate in the bud. Stamens 8 or 10; filaments dilated at the base or in the middle; anthers short. Ovary 4 or 5-celled, or rarely 2 or 3-celled; style deciduous, with an entire or lobed stigma; ovules 2 in each cell, collateral or superposed. Berry ovoid, oblong or globular. Seeds with a membranous testa; no albumen; cotyledons plano-convex.— Unarmed trees or shrubs. Leaves pinnate, with alternate, usually oblique leaflets. Flowers small, usually clustered in terminal or axillary panicles or racemes. Berries small.

The genus, although not large, comprises more species than any other one of the tribe *Aurantieæ*, and extends over tropical Asia and Africa; the only Australian species known is endemic.—*Benth.*

1. **C. brevistyla** (style short), *Oliv. in Journ. Linn. Soc.* v. *Suppl.* 31; *Benth. Fl. Austr.* i. 369. Apparently a shrub, glabrous, or the young branches and petioles slightly pubescent. Leaflets 10 to 15, very obliquely ovate or somewhat rhomboidal, shortly and obtusely acuminate and emarginate, mostly 2 to 4in. long, membranous, often obscurely sinuate-dentate, on petiolules of about 2 lines. Flowers 4-merous or 5-merous, in terminal, loose, oblong or pyramidal panicles. Petals about 2 lines long. Filaments thick and dilated at the base, arched. Ovary glabrous or nearly so, narrowed at the base, 4 or 5-celled. Style very short. Fruit not seen.

Hab.: Hope Islands, *M'Gillivray*.

The species is allied to *C. heptaphylla*, W. and Arn., from E. India, but the leaflets are much more oblique, the style much shorter, besides minor differences.—*Benth.*

25. ATALANTIA, Corr.

(Mythological Atalanta, the daughter of Schœneus.)

Calyx 3 to 5-cleft. Petals 3 to 5, imbricate in the bud. Stamens twice as many or rarely more, free or irregularly united at the base; anthers ovate or oblong. Ovary 2 to 5-celled; style deciduous, with a capitate stigma; ovules solitary or 2 in each cell, collateral or rarely superposed. Berry globular, with a thickened rind, 1 to 5-seeded. Seeds obovoid or oblong, testa membranous; albumen none; cotyledons flat or convex, more or less fleshy.—Shrubs or small trees, unarmed or thorny. Leaves simple, coriaceous. Flowers in axillary clusters or short racemes or small cymose panicles, occasionally solitary. Fruits usually larger than in the preceding genera.

The genus is dispersed over tropical Asia. The Australian species are both endemic; one, however, is in some measure doubtful, the flowers being unknown, and the other is slightly anomalous in character, though congener in essential points and habit. The genus, in the increased number of stamens of two species, and in the inflorescence, fruits, and seeds, connects the anomalous *Citrus* with the rest of the tribe.—*Benth.*

Leaves narrow. Spines straight or incurved. Pedicels clustered in the axils of
 the leaves . . . . 1. *A. glauca.*
Leaves ovate. Spines mostly recurved. Racemes short, axillary or terminal . 2. *A. recurva.*

1. **A. glauca** (glaucous), *Hook. f., in Benth. and Hook. Gen. Pl.* 305; *Benth. Fl. Austr.* i. 370. Native Kumquat. A rigid glaucous shrub of 2 or 3ft., often armed with straight or incurved axillary spines of ½in. or under, the young shoots whitish with a very minute pubescence. Leaves oblong-linear or slightly cuneate, very obtuse or emarginate, mostly 1 to 1½in. long, thick, rigid, veinless, narrowed into a short petiole; those on the barren shoots sometimes marked with a few coarse crenatures. Flowers usually 2 or 3 together in the axils, on pedicels of 1 to 2 lines. Sepals 3 or 4, short and broad. Petals 3 or more frequently 4, obovate or broadly oblong, 2 to 2¼ lines long, thin, concave, much imbricate. Stamens 8 to 12, or sometimes more, the filaments often slightly united at the base. Disk thick, annular. Ovary 4 or 5-celled, with 1 or occasionally 2 superposed ovules in each cell. Style rather thick. Berry globular, about ½in. diameter. Seeds 3 or 4, obovoid, slightly compressed; cotyledons slightly fleshy, but not thick.—*Triphasia glauca*, Lindl. in Mitch. Trop. Austr. 353; Oliv. in Journ. Linn. Soc. v. Suppl. 26.

Hab.: Broadsound, *R. Brown*; Maranoa River, *Mitchell*; Suttor and Burdekin Rivers, *F. v. Mueller*; Port Denison, *Fitzalan.*

The species, although anomalous in some respects, has the foliage and inflorescence of *Atalantia*, and is allied in several respects to *A. Hindsii*, Oliv., approaching like that species to *Citrus* in the increased number of stamens.—*Benth.*

Wood of a bright-yellow, with numerous brown streaks or veins, close-grained, and easily worked. Fruit useful for making preserves.—*Bailey's Cat. Ql. Woods No.* 54.

2. **A. (?) recurva** (recurved), *Benth. Fl. Austr.* i. 370. Glabrous, armed with axillary spines, very spreading or recurved. Leaves broadly ovate, obovate or elliptical, mostly very obtuse, 1½ to 2½in. long, coriaceous, on petioles of 1 to 3 lines. Racemes axillary, sometimes 2 together, ½ to 1in. long, or terminal and slightly branched. Pedicels very short. Calyx minute, 3 or rarely 4-lobed. Petals and stamens not seen. Berries globular, either 1-seeded and 3 or 4 lines diameter, or 2-seeded and larger.

Hab.: Islands of the Gulf of Carpentaria, *R. Brown* (Hb. R. Br.)

The flowers are wanting to determine absolutely the affinities of this species. R. Brown's specimens are, however, in very good fruit. Allan Cunningham's are in leaf only, with some remains of the inflorescence and calyx.—*Benth.*

26. CITRUS, Linn.
(From the Greek name Kitron.)

Calyx 3 to 5-lobed. Petals 4 to 8, thick, imbricate in the bud. Stamens indefinite, usually numerous, filaments flattened at the base and variously connate, anthers oblong. Disk large, cupular or annular. Ovary of 6 or more cells; style deciduous, with a capitate lobed stigma; ovules 4 to 8 in each cell, in 2 rows. Berry globular or oblong, with a thickened rind, several-celled, with thin dissepiments, the cells more or less filled with transverse pulpy cellules. Seeds with a

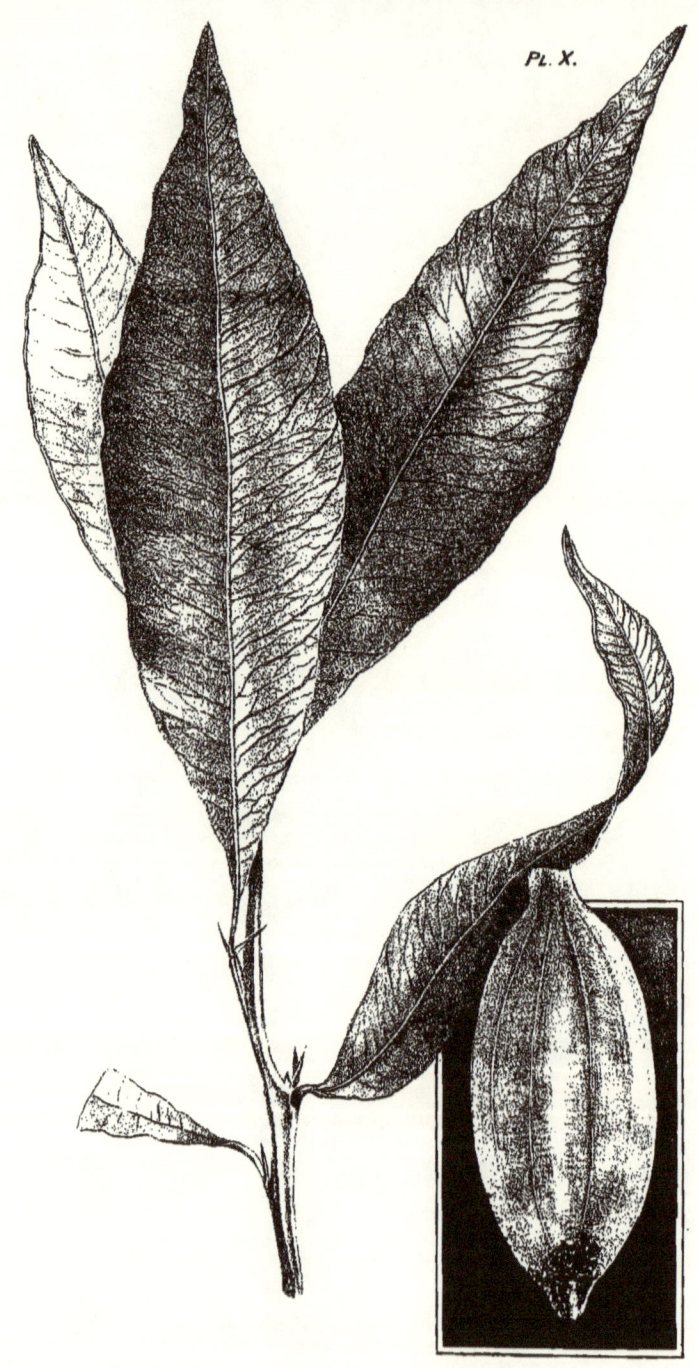

Citrus inodora, Bail.

coriaceous testa; albumen none; embryos often more than one; cotyledons fleshy, plano-convex.—Trees or shrubs, often armed with axillary spines. Leaves 1-foliolate, the petiole often winged. Flowers white, axillary, solitary, clustered or shortly paniculate.

The really wild species are few, chiefly from tropical Asia, but long culture in most hot countries has produced numerous permanent varieties. The Australian ones differ from the others in the short petiole not at all winged.—*Benth.*

Fruit globular. (Stamens about 10?) . . . . 1. *C. australis.*
Fruit oblong. Stamens about 20 . . . . 2. *C. australasica.*
Fruit oblong. Stamens above 30 . . . . 3. *C. inodora.*

1. **C. australis** (Australian), *Planch. in Hort. Donat.* 18 (*partly*); *Benth. Fl. Austr.* i. 371. Native Orange; "Dooja," Taromeo, *Shirley*. A tree of 80 to over 60ft., quite glabrous, with axillary straight thorns of about $\frac{1}{4}$in. Leaves ovate, obovate, or almost rhomboidal, 1 to 2in. long, obtuse or emarginate, the petiole not exceeding 3 lines and not winged, very small and linear on young plant. Flowers pinkish. Calyx teeth or sepals almost orbicular, about 1 line in diameter, sometimes only 4. Petals 2$\frac{1}{4}$ lines or more long, oblong. Stamens free, about 17, subulate. Ovary sulcate and including the style shorter than the stamens. Stigma large. Fruit globular, from 1 to 2$\frac{1}{2}$in. diameter, with a rough rind; cells 6 to 8, more or less pulpy, with usually 3 or 4 seeds in each.—*Limonia australis*, A. Cunn. in Sweet. Cat.

Hab.: Coast scrubs of southern Queensland.

Wood of a light-yellow, close-grained, hard and durable; useful for cabinet-work and might probably serve for engraving. The fruit useful for preserves.—*Bailey's Cat. Ql. Woods No. 55.*

2. **C. australasica** (Australian), *F. v. M. Fragm.* i. 26; *Benth. Fl. Austr.* i. 371. Finger Lime. A tall shrub or small tree, quite glabrous, with axillary straight slender spines of $\frac{1}{4}$in. or less. Leaves from obovate-oblong to oblong-cuneate or lanceolate, very obtuse and emarginate, 1 to 1$\frac{1}{2}$ or rarely 2in. long, coriaceous, the petiole usually very short, and not winged. Flowers solitary or rarely 2 together, on very short pedicels. Sepals 5, small, spreading, concave, minutely ciliate. Petals oblong, nearly 4 lines long. Stamens 20 to 25, free. Ovary in the flowers examined 6-celled. Style very short, with a thickened, obtuse, furrowed stigma. Ovules 4 in each cell. Fruit oblong, almost cylindrical, 2 or 3 times as long as broad, the largest seen about 4in. long, with usually 2 or 3 seeds in each cell.

Hab.: Mountain scrubs of southern Queensland.

Wood close-grained, of a yellowish colour. Fruit of each variety makes excellent preserve.—*Bailey's Cat. Ql. Woods No. 56.*

Var. *sanguinea*, Bail. Red-fruited Finger Lime. This differs from the ordinary form in colour of fruit only. This, however, being constant, it is better that the form should be known by a distinctive name. The fruit attains about 2 or 3in. in length and $\frac{3}{4}$in. diameter, is of a blood-red colour, thin-skinned, pulp sharply acid, and of a pink colour. Hab.: Tambourine Mountain, *J. Pindar.*

3. **C. inodora** (inodorous), *Bail. Rep. Bot. Bellenden Ker Exped.* 1889. North Queensland Lime. A tall shrub, the young branches flattened or more or less angular as in the common orange. Spines not very numerous, about $\frac{1}{4}$in. long, erect, 1 or 2 at the axils of the leaf. Leaves lanceolate or ovate-lanceolate, with distant sinuous crenations, 3 to 7in. long, 1$\frac{1}{2}$ to 2in. broad in the centre, the apex often elongated, and tapering towards the base to a very short petiole seldom exceeding $\frac{1}{4}$in. in length; substance coriaceous; midrib and primary veins more or less prominent, the latter rather close and anastomosing in an intra-marginal vein. Flowers axillary, nearly sessile, and, so far as observed, scentless. Calyx of 5 minute teeth. Petals 5, about 8 lines long, spreading. Stamens

free, over 80, of irregular length, shorter than the petals. Ovary, with style, scarcely exceeding 1 line in length. Fruit oblong, 2½in. long, 1¼in. diameter, showing 8 obtuse ribs, the divisions of the fruit; pulp a sharp agreeable acid. Seeds oval, somewhat flattened on one side, convex on the other, 3 lines long, 2 lines broad; testa white.

Hab.: Harvey's Creek, Russell River.
This new species of *Citrus* is well worthy of cultivation for its fruit, which is juicy, and of equal flavour with the West Indian Lime. In general appearance this tree somewhat resembles the orange, having the same angular branchlets.

Order XXX. SIMARUBEÆ.

Flowers regular, diœcious or polygamous, more rarely hermaphrodite. Calyx usually small, 3 to 5-lobed, or divided into as many distinct sepals. Petals 3 to 5, hypogynous or slightly perigynous, imbricate or valvate in the bud, rarely wanting. Stamens either equal in number to the petals, and alternating with them, or double the number; anthers usually versatile, with 2 parallel cells opening longitudinally. Disk annular, cupular, or elongated with the stamens, under or round the ovary, or rarely none. Gynœcium of 3 to 5, rarely more or fewer carpels, quite distinct, or more or less united into a single-lobed or rarely entire ovary, with one cell to each carpel. Styles as many as carpels, united from the base or by the stigmas only, or entirely distinct. Ovules solitary in each cell, or very rarely 2, the micropyle superior. Fruit-carpels either distinct, dry or drupaceous, usually indehiscent, or united in a single drupe or capsule. Seeds usually solitary in each carpel or cell, pendulous; testa membranous; albumen abundant, or little, or none. Embryo straight or curved; cotyledons flat or convex, rarely twisted; radicle superior.—Shrubs or trees, with a bitter bark. Indumentum of simple not stellate hairs. Leaves alternate or rarely opposite, pinnate or simple, usually without glandular dots. Stipules none, except in *Cadellia*. Flowers usually small, in axillary or rarely terminal panicles or racemes.

The Order consists of a considerable number of small genera, chiefly tropical, dispersed over the New as well as the Old World. Of the 6 Australian genera, 3 belong to tropical Asia, one of which extends also into Africa, 2 are endemic, and the sixth is on the seacoasts of all tropical countries. The Order as a whole is somewhat heterogeneous, and especially has no peculiar habit. In technical characters it is closely allied to *Rutaceæ*, from which it differs chiefly in the bitter bark, the want of pellucid dots to the leaves, and in the solitary ovules, but each of these characters has some exceptions.—*Benth.*

Tribe I. **Simarubeæ.**—*Ovary deeply divided, the carpels or lobes entirely distinct or connected by the styles or stigmas.*

Leaves pinnate.
  Stamens twice as many as petals. Fruit-carpels winged and samara-like . . 1. Ailanthus.
  Stamens equal in number to the petals. Fruit-carpels drupaceous . . . . 2. Brucea.
Leaves simple. Fruit drupaceous, variously winged . . . . 3. Samadera.
  Sepals nearly or quite as long as the petals. Styles free.
    Sepals spreading under the fruit. Leaves thin . . . . 4. Cadellia.
    Sepals connivent over the fruit. Leaves almost fleshy . . . . . . 5. Suriana.

Tribe II. **Picramnieæ.**—*Ovary 2 to 5-celled, entire or rarely shortly lobed.*

Leaves 3-foliolate . . . . 6. Harrisonia.

1. AILANTHUS, Desf.
(From *ailanto*, Tree of Heaven.)

Flowers polygamous. Calyx small, 5-lobed. Petals 5, valvate in the bud. Disk 10-lobed. Stamens 10, fewer or none in the female flowers; filaments without scales. Ovary 2 to 5-lobed; styles connate, with plumose stigmas;

ovules solitary in each cell. Fruit of 1 to 5, oblong, membranous samaræ, thickened in the centre round the seed. Seed flattened; testa membranous; albumen scanty; cotyledons leafy, nearly orbicular.—Trees. Leaves alternate, pinnate; leaflets oblique. Flowers small, in terminal panicles.

Besides the Australian species, which is endemic, the genus comprises 3 others, natives of the warmer regions of Asia, one of them much planted in various parts of the globe and in Queensland, and found as a stray from cultivation.

Panicle not much-branched . . . . 1. *A. imberbiflora*.
Panicle much-branched . . . . 2. *A. glandulosa*.

1. **A. imberbiflora** (beardless flowers), *F. v. M. Fragm.* iii. 42; *Benth. Fl. Austr.* i. 373. A tall tree, said to attain 100ft., quite glabrous in all its parts. Leaflets about 15 to 50, shortly petiolulate, apparently obliquely ovate-lanceolate and 2 or 3in. long, but much broken in the only specimen seen. Panicles not much branched. Male flowers on short pedicels, in little clusters along the upper part of the branches. Calyx very small. Petals about 1¼ line long, quite glabrous, valvate not induplicate, and the points scarcely inflexed. Stamens exserted. Female flowers not seen. Samaræ in our specimens attaining at least 2in. in lenth and ½in. in breadth.

Hab.: Brisbane River, Rockhampton; Mount Perry, *J. Keys*; Port Denison, *F. v. Mueller.*

Evidently, as suggested by F. v. Mueller, very nearly allied to the E. Indian *A. malabarica*, DC. Prod. ii. 89, Wight, Ic t. 1604, which indeed seems only to differ in a slight pubescence on the panicle and in rather large flowers and fruits.—*Benth.*

Wood yellow, porous, soft and light.—*Bailey's Cat. Ql. Woods No.* 57.

Var. *Macartneyi*, *Bail. Bot. Bull.* x. "Koorootha," Forest Hill, *Macartney*. A very large tree, exuding from the bark when wounded a copious flow of a colourless, transparent, resinous liquid. The young branchlets and inflorescence more or less covered with a yellowish mealy tomentum. Leaves slender, 6 to 9in. long including the rather long petiole, but probably much longer on young trees. Leaflets on the leaves near to the inflorescence 8 to 13, narrow-lanceolate, about 3in. long, membranous, very oblique, the under side pale, on petiolules of 1 to 3 lines. Panicles in the axils near the end of the branchlets (male), slender, about 5in. long, and with few short branches. Flowers singly on pedicels longer than the flowers. Calyx-lobes minute with ciliate margins. Petals glabrous, 1¼ line long, broad-lanceolate, valvate, with thickened margins, the tips slightly inflexed. Stamens scarcely exserted; filaments flexuose, 3 times longer than the rather large nearly globular anthers. The female flowers not seen, but from the panicles of fruit to hand it is probable that the female panicles are much longer than the male; one bearing nearly ripe samaræ measured over 9in. long, and had 2 or 3 rather long branches: the samaræ were usually in clusters of 3, about 2¼in. long and 8 lines broad, membranous, and delicately veined, and were borne upon pedicels of about 1in. or more. Hab.: Forest Hill, Mackay, *W. Macartney*, January, 1895. Wood of a light colour, soft and light.

Yields a large quantity of resin, which contains: Water, 7·8%; essential oil, 9·5%; alpha resin, 23·6%; beta resin, 55·4%; impurities, 3·8%. The essential oil of this resin agrees in its properties with that of *Ailanthus glandulosa*. It may be supposed, therefore, to act similarly to that; and, indeed, 2 drachms of the resin dissolved in lucca oil and given to a big dog produced strong purging. Very likely it acts as an anthelmintic too. While I worked with the hot resin, sublimating the oil or mixing it with fats, etc., I felt a nauseous sensation coming over me. It makes a good ointment for chronic ulcers, used pure or mixed with wax and lard. Fresh cuts and sores are not to be treated with this resin on account of the acrid oil. Technically, the resin might be adhibited for softening too hard and too quickly drying alcoholic varnishes, especially for tin and smooth metal surfaces, where a brittle varnish is liable to fall off after drying in the sun. With wax and turpentine it makes a good paste to fix paper labels permanently on tin boxes.—*Lauterer.*

Although stated to be quite glabrous, so far as my observations go *A. imberbiflora*, F. v. M., is always rusty-tomentose on the very young growth; there may, however, be several forms of the tree. The present seems to differ in the more copious flow of resin from its wounded stem, more slender branchlets, and the longer pedicels to its more distantly separated flowers. This form is very likely identical with the trees growing in the scrubs at Brookfield, Brisbane River. The form growing at Rockhampton and Mount Perry has the leaves much crowded at the end of the thick branchlets: the flowers also are more clustered, on shorter pedicels, and the stamens exserted.

2. ***A. glandulosa** (glandulous), *Desf.* Tree of Heaven. A tall fast-growing deciduous tree, the foliage when bruised emitting a disagreeable odour. Leaves 1 to 2ft. long. Leaflets 9 to 25, pubescent or subglabrous, divided very unequally on the rhachis, pale on the under side, often coarsely toothed near the base. Flowers greenish, small, in much-branched panicles. Filaments filiform, exserted. Several times exceeding the anthers, hispid at the base, the petals woolly-tomentose inside. Fruit of about 8 membranous linear-oblong samaras, about 1¼in. long, and about 4 lines broad, red. Seeds near the centre of the samara, about 3 lines long and a little over 1 line broad.

Hab.: Japan and China. Met with in a few localities in the southern inland parts of the colony as a stray from garden culture.

The leaves of this tree are said to be the favourite food of the silk moth " Bombyx cynthia."
The wood is yellowish-white, satiny, and well suited for cabinet-work.

2. BRUCEA, Mill.
(After J. Bruce, the traveller.)

Flowers polygamous. Calyx small, 4-cleft. Petals 4, minute, linear, imbricate in the bud, disk 4-lobed. Stamens 4. Ovary 4-lobed or of 4 distinct carpels, the styles free or connate at the base, the stigmas entire, spreading ; ovules solitary in each cell. Drupes 4, ovoid, scarcely fleshy, the putamen rugose. Seed with a membranous testa; albumen copious ; embryo straight, radicle superior.— Trees. Leaves alternate, pinnate; leaflets oblique. Flowers very small, in small cymes, in simple slender axillary spikes.

The genus comprises a very few species, spread over tropical Asia and Africa, extending into northern India. The Australian species is one of the commonest Asiatic ones.—*Benth.*

1. **B. sumatrana** (of Sumatra), *Roxb. Fl. Ind.* i. 449 ; *F. v. M. Fragm.* vi. 166; *Benth. Fl. Austr.* i. 378. A shrub or tree, the young branches and petioles softly tomentose. Leaves 1 to 1½ft. long or even more ; leaflets 5 to 11, ovate-lanceolate, acuminate, about 3in. long, coarsely toothed or quite entire, usually oblique at the base, softly pubescent or tomentose-villous, especially underneath. Flowers very small, purple, petals obovate, ¼ line long, anthers didymous, in little cymes or clusters along the peduncle, forming interrupted spikes or racemes of 6 to 10in. in the males, much shorter in the females. Drupes about 3 lines long.

Hab.: Arnhem's Bay, *R. Brown ;* Victoria River, *F. v. Mueller.* The latter specimen has the leaflets very densely and softly velvety on both sides; in R. Brown's specimens they are not more so than in the majority of Indian specimens.—*Herb. R. Br.* and *F. v. M.* Mount Elliot and Rockingham Bay, *F. v. M.*

The whole plant is bitter, and the seeds (called " Macassar Kernels") are much used in Java as a cure for dysentery. —*Trimen, Flora of Ceylon.*

3. SAMADERA, Gærtn.
(Derivation obscure.)

Calyx small, 3—5-partite, imbricate. Petals 3—5, much longer than the calyx, coriaceous, imbricate. Disk large, conical. Stamens 8-10, included in the corolla, with a small scale at the base. Carpels 4-5, distinct, free; styles free at the base, more or less united above, stigmas acute ; ovules solitary, pendulous. Fruit of 1-5 large, dry, compressed, 1-seeded drupes, each with a narrow unilateral wing. Large or small trees. Leaves simple. Flowers hermaphrodite, in peduncled axillary or terminal umbels.— Hook. Fl. Brit. Ind. i. 518.

Peduncles very short. The broad portion of filament hairy throughout . . . 1. *S. Bidwillii.*
Peduncles 2 or 3in. long. The broad portion of filament hairy only at the top 2. *S. Baileyana.*

1. **S. Bidwillii** (after J. C. Bidwill), *Oliver: Hook. Ic. Pl. t.* 2449. A small tree or tall shrub; branchlets minutely hoary-pubescent. Leaves alternate, sub-coriaceous, narrowing towards both ends, somewhat obtuse, sometimes emarginate, 3 to 4½in. long, 6 to 9 lines broad, lateral nerves parallel, rather obscure, confluent with an intromarginal one, upper side glossy-glabrous, the under side pale; petioles usually only a few lines long. Flowers 4 or 5-merous, usually few in axillary clusters or on a very short peduncle. Pedicels short, buds ovoid or subglobose. Sepals suborbular, scarcely ½ line long. Petals much imbricate, broad ovate-elliptical, slightly hairy on the back, 1¼ line or more long. Disk thick. Stamens about as long as the petals, filaments free, broadly dilated and abruptly constricted at the top, hirsute except the narrowed summit. Ovary hirsute, style glabrous, short. Stigma shortly 4 or 5-lobed. Fruit 4 or 5 lines long.—*Hyptiandra Bidwilli*, Hook. in Fl. Austr. i. 874.

Hab.: Wide Bay.

2. **S. Baileyana** (after F. M. Bailey), *Oliver: Hook. Ic. Pl. t.* 2450. An erect glabrous tree, about 20ft. Leaves coriaceous, oblanceolate, 5 to 9in. long, the larger ones 2in. broad in the centre, obtusely acuminate, narrowing to a petiole of about 8 lines; midrib and veins prominent; the lateral nerves looping some distance from the margin. Inflorescence terminal or in the upper leafless axils, the shoot ultimately growing out. Peduncles about 3in. long, forked at the end, the branches sometimes again shortly forked and bearing 2 or 3 pedicellate pink flowers, pedicels incurved. Calyx-lobes 5, coriaceous, about ½ line long, thick and obtuse. Petals 5, oblong, boat-shaped, about 5 lines long and 1 line broad, caducous. Stamens broad near the base, narrow above, glabrous except for a dense ring of hairs at the top of the expanded portion of the filament. Disk fleshy sulcate. Ovary glabrous or hairy at the base, 5-celled or 5 carpels. Style slender, stigma small, 5-lobed.—*Hyptiandra Bidwilli*, var. *grandiuscula*, Bail. and F. v. M. in Syn. Ql. Fl. iii Suppl. p. 12.

Hab.: Bellenden Ker, at an altitude of 4000ft. (Expedition 1889).

4. CADELLIA, F. v. M.
(After F. Cadell.)

Flowers hermaphrodite. Sepals usually 5, nearly as long as the petals, enlarged and stellately spreading under the fruit, imbricate in the bud. Petals 5, imbricate in the bud. Stamens 10, filaments filiform. Disk none. Carpels 1 to 5, free; styles distinct, inserted on the inner angle above or below the middle; stigma dilated or capitellate; ovules 2 in each carpel, collateral, pendulous or ascending. Fruit-carpels coriaceous, small, indehiscent or obscurely 2-valved. Seeds solitary, without albumen; testa membranous; embryo curved. Small trees. Leaves alternate, simple, with small, often deciduous stipules. Flowers in short loose axillary racemes.

The genus is limited to Australia. It only differs from *Suriana* in the arborescent habit and thinner spreading calyx.—*Benth.*

Carpels 5. Leaves mostly obtuse. Racemes very loose . . . . 1. *C. pentastylis.*
Carpels solitary. Leaves mostly acute or acuminate. Racemes short . . 2. *C. monostylis.*

1. **C. pentastylis** (styles five), *F. v. M. Fragm.* ii. 25, *t.* 12; *Benth. Fl. Austr.* i. 374. A tree of about 40ft., the smaller branches very slender and minutely pubescent. Leaves from obovate-oblong to elliptical or lanceolate, obtuse, about 1½ to 2in. long, entire, narrowed into a short petiole, occasionally bearing a gland on one side, glabrous, penninerved and reticulate, not dotted. Peduncles in the upper axils slender, bearing a short raceme of 2 to 4 flowers. Sepals nearly 3 lines long at the time of flowering, enlarged to 5 or 6 lines, and stellately spreading under the fruit. Petals white, slightly exceeding the sepals.

Carpels 5, the styles inserted above the middle. Ovules pendulous. Drupes about 1½ line long, nearly globular, with an inner angle, somewhat coriaceous, with a crustaceous endocarp. Embryo much curved or circinate like that of *Suriana*; cotyledons much broader than in that plant, variously folded according to F. v. Mueller, in the seed I opened flat, except following the general curvature of the embryo.—Benth.

Hab.: Near the border or N.S.W.

2. **C. monostylis** (style one), *Benth. Fl. Austr.* i. 375. A glabrous slender tree or shrub. Leaves petiolate, from ovate-lanceolate to elliptical-oblong, shortly acuminate, mostly 3 to 4in. long, narrowed at the base, membranous or thinly coriaceous. Racemes, in the few specimens seen, very short, slender, 2 to 4-flowered. Pedicels about 2 lines long, in the axils of minute bracts. Sepals nearly 2 lines long, shortly united at the base, membranous, persistent, and spreading after flowering. Petals rusty-tomentose, about twice as long as the sepals. Stamens 10, but in some of the flowers 1 or 2 are semi-abortive. Carpels in all the flowers seen solitary, with the style silky and quite basal as in *Suriana*. Ovules as in *C. pentastylis*, collateral, but horizontal or slightly ascending. Drupes oval, purple, about ½in. long.—*Guilfoylia monostylis*, F. v. M. Fragm. viii. 84, where he states that the tree attains the height of 100ft. I have never seen the Queensland form more than about 15ft. In later works the Baron placed the tree back in the old genus, as here given.

Hab.: Mt. Mistake (a tall shrub). Mueller says, Fragm. vi. and viii., "A tree of 100ft."

Wood of a yellowish colour, somewhat resembling some kinds of walnut and satinwood. It is of a pretty grain, and might be found useful for making toys; also for cabinet-work.—*Bailey's Cat. Ql. Woods No. 57a*.

5. **SURIANA**, Linn.
(After Dr. Surian, of Marseilles.)

Flowers hermaphrodite. Sepals 5, as long as the petals, persistent and closing over the fruit, imbricate in the bud. Petals 5, imbricate in the bud. Stamens 10, filaments filiform. Disk none. Carpels 5, free; styles distinct, filiform, inserted near the base of the carpels; stigmas capitellate; ovules 2 in each carpel, ascending. Fruit-carpels coriaceous, indehiscent. Seeds solitary, ascending, without albumen; testa membranous; embryo curved.—A maritime shrub. Leaves alternate, simple. Peduncles in the upper axils, 1 or few-flowered.

The genus is limited to a single species widely spread over the seacoasts of most tropical countries. It is in many respects anomalous in the structure of the flowers, but is certainly allied to *Cneorum* and *Castela*, and, with them, appears to be better placed among *Simarubeæ* than in any other Order to which it has been referred, although it is deprived of the bitter principle of the majority of *Simarubeæ*.—*Benth*.

1. **S. maritima** (coast), *Linn.; W. and Arn. Prod.* 861; *Benth. Fl. Austr.* i. 375. A rigid, much-branched shrub, more or less hoary or tomentose with simple, often capitate hairs. Leaves crowded, linear-spathulate, obtuse, 1 to 1½in. long, narrowed at the base, quite entire, rather thick, scarcely veined. Peduncles short in the upper axils, bearing 1 or very few flowers, often forming short leafy terminal corymbs. Sepals rather thick, acute or acuminate, 3 to 4 lines long, slightly enlarging and closing over the fruit. Petals yellow, scarcely as long as the sepals. Nuts or drupes about half as long as the calyx, minutely pubescent, with a thin epicarp and crustaceous endocarp. Embryo in the seeds examined as much curved as in *Cadellia*, but the cotyledons narrower.

Hab.: Islands off the N.E. coast, *R. Brown, F. v. Mueller*, and others.

XXX. SIMARUBEÆ.

6. HARRISONIA, R. Br.
(After — Harrison.)

Flowers hermaphrodite. Calyx small, 4 or 5-cleft. Petals 4 or 5, almost valvate. Disk hemispherical or cupular. Stamens 8 or 10, with a small 2-cleft scale at the base of the filaments. Ovary globular, entire or shortly lobed, 4 or 5-celled. Styles connate or distinct at the base; stigma furrowed. Ovules solitary in each cell, pendulous. Drupe small, globular, with 2 to 5 pyrenes or nuts. Seeds solitary, nearly globular; testa rather thick; albumen scanty; cotyledons folded towards the middle.—Trees, usually armed with prickles. Leaves alternate, compound. Flowers small, in pedunculate axillary cymes.

The genus comprises only two species, natives of the Indian Archipelago, one of them extending to Australia.

1. **H. Brownii** (after R. Brown), *A. Juss. in Mem. Mus. Par.* xii. 540, *t.* 28; *Benth. Fl. Austr.* i. 376. A shrub. Branches glabrous, often armed with short conical prickles, usually in pairs, one on each side of the leaf, but probably not really stipulary. Leaflets 3, ovate, acuminate, 1½ to 3in. long, the lateral ones petiolulate and oblique at the base, the terminal one narrowed at the base; all glabrous or sprinkled with a few hairs underneath. Flowers small, few together in axillary cymes, on slender peduncles, shorter than the leaves. Calyx and petals quite glabrous. Filaments hairy at the base. Drupe small, depressed, globular, furrowed between the nuts.

Hab.: Islands of the Gulf of Carpentaria, *R. Brown* (Herb. R. Br.) We have it also from Timor and from the Philippine Islands, and it probably extends over other intervening islands.

Order XXXI. OCHNACEÆ.

Sepals 4 or 5, free, imbricate, persistent. Petals 5, rarely 4 or 10, hypogynous, imbricate, longer than the sepals, deciduous. Disk enlarged after flowering, occasionally none. Stamens 4, 5, 8, 10 or indefinite, inserted on the disk; filaments persistent; anthers basifixed, sometimes deciduous, dehiscing longitudinally, or often opening by terminal pores. Ovary short, 2-celled, or elongated and 1 to 10-celled; placenta axile or parietal; style simple, subulate, acute, rarely divided at the extremity, stigma simple, terminal; ovules 1 to several in each cell or indefinite, ascending or rarely pendulous, raphe ventral, micropyle superior. Fruit indehiscent, drupaceous or baccate, compound, each drupe or pyrene 1 to 4-seeded; or capsular and 1 to 5-celled with septicidal dehiscence. Seeds solitary, few, or numerous; albumen fleshy or none; embryo straight or rarely curved, radicle superior or inferior.—Glabrous trees or shrubs, with watery juice. Leaves alternate, simple (very rarely pinnate), coriaceous; stipules 2. Inflorescence paniculate or occasionally umbellate (rarely flowers solitary), bracteate. Flowers hermaphrodite, conspicuous.

Inhabiting tropical regions of both hemispheres.

1. BRACKENRIDGEA, A. Gray.
(After W. D. Brackenridge).

Sepals 5, persistent, imbricate. Petals 5, equal with the calyx, deciduous, imbricate. Torus thick, conical, clavate. Stamens 10, inserted on the margin of the disk, filaments short; anthers dehiscing longitudinally. Ovary deeply divided into 5 parts or 5-celled; style basilary, columnar, 5-sulcate, stigma capitellate, 5-lobed; ovules solitary in each cell, sub-basilary, curved, hippocrepiform. Drupes 5, fleshy, seated on a broad disk. Seeds annular with a membranous testa; embryo annular, cotyledons linear.—Branching shrubs. Leaves alternate, petiolate, entire. Stipules jagged. Peduncles fasciculate, 1-flowered, axillary and terminal. Flowers somewhat small.

1. **B. australiana** (Australian), *F. v. M. Fragm.* v. 29 and 213. A large tree, quite glabrous. Leaves lanceolate or ovate-lanceolate, tapering to somewhat long points, petiole very short, chartaceous, 3 to 8in. long, 1 to 2in. broad, entire, shining on both sides, penninerved. Fascicles axillary or terminal, often few in pendulous clusters. Pedicels ¼in. or less, bracteoles minute. Sepals ovate or orbicular-ovate, about 3 lines long, somewhat thick. Petals membranous, cuneate-ovate; deciduous. Filaments very short; anthers curved-linear, connivent; yellow, scarcely over 1 line long, dehiscing by an apical pore. Style 1½ line long. Carpels 5 or fewer, round and moderately compressed, seated upon the pyramidal-hemispherical disk.—F. v. M., l.c.

Hab.: Rockingham Bay, *J. Dallachy* (F. v. M., l.c.)

Order XXXII. BURSERACEÆ.

Flowers regular, hermaphrodite or polygamous. Calyx usually small, 3 to 5-lobed or divided into as many distinct sepals. Petals 3 to 5, hypogynous or perigynous, imbricate or valvate in the bud. Stamens twice as many as petals, or rarely of the same number, inserted on or around the disk; anthers versatile, with 2 parallel cells opening longitudinally. Disk usually annular or cupular, often adnate to the base of the calyx. Ovary free, 2 to 5-celled, tapering into a single style, with an entire or lobed stigma. Ovules 2 in each cell or rarely solitary, usually pendulous, the micropyle superior. Fruit a drupe, either indehiscent or the epicarp opening in 2 valves, pyrenes 2 to 5, bony or chartaceous, distinct or united. Seeds solitary in each pyrene, pendulous; testa membranous; albumen none. Cotyledons usually membranous, folded or rarely thick and fleshy.—Shrubs or trees, often yielding a balsamic fluid. Leaves usually alternate, pinnate, or in genera not Australian 3-foliolate, without or rarely with stipules. Flowers small, in racemes or panicles.

The Order is spread over most tropical regions. The Australian genera are widely dispersed over tropical Asia, one is also in Africa, and the others in tropical America.—*Benth.* (in part).

Calyx 5-lobed, the disk lining the tube, with the stamens on the margin . . 1. GARUGA.
Calyx small, 4 to 6-partite. Stamens 8 to 12, inserted at the base of the annular disk. Flowers usually in panicles . . . . 2. BURSERA.
Calyx 3-lobed, the disk free, with the stamens outside or on the margin . . 3. CANARIUM.
Calyx 5-fid, valvate. Petals none. Stamens alternate with the calyx-lobes inserted at base of disk . . . . 4. GANOPHYLLUM.

1. GARUGA, Roxb.

Flowers polygamous. Calyx campanulate, 5-lobed, valvate. Petals 5, inserted above the middle of the calyx-tube, induplicate-valvate. Disk thin, lining the calyx-tube. Stamens 10, inserted with the petals. Ovary 4 or 5-celled; styles elongated; ovules 2 in each cell. Drupe indehiscent, with 5 or fewer bony nuts, rugose outside. Seeds solitary in each nut; cotyledons folded.—Trees. Leaves pinnate. Flowers rather large for the Order, in terminal panicles.

The genus is dispersed over tropical Asia and America; the Australian species extends at least o Timor, and is perhaps a variety of a common Asiatic one.—*Benth.*

1. **G. floribunda** (flowers numerous), *Dcne. Herb. Tim. Descr.* 149; *Benth. Fl. Austr.* i. 377. Branches thick, marked with the broad scars of the fallen leaves. Leaves crowded at the ends of the branches; leaflets 7 or 8 pairs, very

shortly petiolulate, very obliquely ovate-lanceolate, acuminate, 2 to 3in. long, crenate especially on the outer edge, glabrous when full grown, the common petiole 8in. to 1ft. long, slightly pubescent or at length glabrous. Panicles broad and dense, terminating leafless branches. Flowers numerous, much smaller than in the common Indian (*i. pinnata*, Roxb., arranged in cymes along the last ramifications, the pedicels and flowers hoary with a minute tomentum. Calyx about 2 lines long. Petals linear-oblong, twice as long as the calyx-lobes. Fruit not seen.

Hab.: Endeavour River.

It differs from some forms of *G. pinnata*, Roxb., in little besides the much smaller flowers in a more compound panicle.—*Benth.*

Wood tough, close-grained, firm, and easy to work; colour, grey.—*Bailey's Cat. Ql. Woods No. 57b.*

2. BURSERA, Linn.
(After Joachim Burser, a disciple of Caspar Bauhin.)

Flowers polygamous or hermaphrodite. Calyx small, 4 to 6-partite or toothed, imbricate. Petals 4 to 6, short, patent at length, reflexed, usually valvate. Disk annular crenate. Stamens 8 to 12, nearly equal, inserted at the base of the disk. Ovary free, ovoid or subglobose, 3 to 5-celled; style very short; stigma 3 to 5-lobed; ovules 2 in each cell. Drupe globose or ovoid, with 3 to 5 pyrenes. Balsamiferous trees. Leaves alternate, imparripinnate, or rarely 1-foliolate; panicle short-branched.

1. B. australasica (Australasian), *Bail. Bot. Bull.* v. 8, ix. 6; *Proc. Roy. Soc. Ql.* 1894. Carrot-wood of Eumundi.
So far as at present known a small glabrous tree; leaves alternate, pinnate; leaflets 3 to 5, most frequently 3, lanceolate, entire, coriaceous, 1½ to 2½in. long, obtusely acuminate; petiolules about 3 lines. When 3-foliolate the terminal pinna distant from the others, glossy on both sides, but the under paler than the upper; petiole about 1½in. long, angular. Panicles in the upper axils rather numerous near the end of the branches, and scarcely exceeding in length that of the petioles, of few branches. No flowers seen, but from their remains at the base of the fruit they appear to be very small; calyx 4-partite, the lobes under ¼ line long and somewhat triangular. Stamens 8, scarcely exceeding the calyx lobes; filaments very short; anthers lobed at the base, appearing sagittate. Drupe with purplish epicarp, oval, and showing more or less of 4 angles; endocarp hard, bony, with 4 very prominent ribs indicating the pyrenes or cells, only 2 seem to mature seeds.

Hab.: Eumundi, *J. F. Bailey* and *J. H. Simmonds.*

Wood of a grey colour, close-grained and firm, of a greasy nature; easily worked; does not shrink or warp much in drying. Useful for inside lining and other purposes.—*Bailey's Cat. Ql. Woods No. 58b.*

3. CANARIUM, Linn.
(From *Canari*, Malayan name.)

Flowers hermaphrodite or polygamous. Calyx campanulate, usually 3-lobed, valvate. Petals usually 3, valvate, or slightly imbricate in the bud. Disk annular, rather thick. Stamens twice as many as petals, inserted on the margin of or outside the disk. Ovary usually 3-celled; stigma sessile, capitate, 3-lobed; ovules 2 in each cell. Drupe ovoid or ellipsoid, often 3-angled, the putamen 1-celled by abortion. Seed solitary; testa membranous; cotyledons folded.— Trees, with large pinnate leaves. Flowers small, in axillary panicles.

The largest genus of the Order, dispersed over tropical Asia and especially the Indian Archipelago, with a few African species. The Australian ones are endemic.—*Benth.*

1. **C. australasicum** (Australasian), *F. v. M. Fragm.* iii. 15; *Benth. Fl. Austr.* i. 377. "Tchaln-ji," Bloomfield, *Roth;* "Kame," Batavia River, *Ward.*
Branches thick, marked with the broad scars of fallen leaves, the young ones minutely hoary. Leaflets 5 to 9, petiolulate, ovate or oval-oblong, or the lower ones nearly orbicular, very obtuse, or rarely shortly acuminate, 2 to 4in. long, glabrous, coriaceous, with parallel pinnate veins, and smaller reticulations conspicuous on both sides. Stipules linear-subulate, deciduous. Panicles raceme-like in the upper axils, shorter than the leaves, the cymes shortly pedunculate along the simple rhachis. Bracts and bracteoles small, deciduous. Flowering calyx 1 line long, tomentose. Petals about 2 lines, glabrous. Stamens 6, the filaments shortly united in a cup at the base. Drupes ellipsoid, the woody nut nearly 1in. long, smooth, usually 1-celled, rarely with 2 cells and seeds. Cotyledons much folded and crumpled.

Hab.: Many parts of tropical Queensland.
Wood of a grey colour, dark towards the centre; works easily, and would suit for lining-boards for houses.—*Bailey's Cat. Ql. Woods* No. 58b.
Gum used for cement by natives, *Roth.*

2. **C. Muelleri** (after Baron Mueller), *Bail. Cat. Ql. Pl.* The Queensland Elemi Tree. Tree of medium size, say 50 to 60ft., diameter of stem about 1ft., buds and quite young shoots rusty-tomentose, the smaller branches with prominent lenticels. Leaves pinnate, of from 5 to 7 leaflets; petioles slightly hoary or at length glabrous, angular, more or less flat on the upper side, 2 or 3in. long, slender, leaving a scar on the branch from whence they fall; leaflets glabrous, lanceolate, 3 or 4in. long, 1 to 1¼in. wide in the centre, apex blunt, the base nearly or quite equal-sided, and tapering to a petiolule of about 2 to 4 lines, primary veins rather distant, and smaller ones forming a more open reticulation than *C. australasicum.* Inflorescence in slender racemose panicles in the upper axils, nearly glabrous; buds in distant clusters, minute, nearly globose, about 1 line long; calyx with 3 minute teeth not half the length of the petals. Petals 3. Stamens 6. Filaments very short, attached to the edge of a very shallow cup.

Hab.: Johnstone River, *Dr. Thos. L. Bancroft.*
Although only shoots bearing the inflorescence in an early stage have been examined, this plant differs widely from the only other Australian species, as well as the other species of which I have descriptions. I therefore purpose giving to it specific rank.
Upon cutting a log of this tree Dr. Bancroft observed a flow of honey-like liquid of a turpentine-like odour, some of which was analysed by Mr. Mar, late Government Analyst, and Dr. Lauterer, who found it to contain a resin resembling elemi in its general chemical characteristics. Dr. Lauterer states that it may be used as a substitute for elemi, and that it is a very good healing agent for cuts, sores, and chronic ulcers.
Wood of a grey colour, in appearance like Queensland Beech, close-grained, easily worked, useful for flooring and lining-boards; also for joinery work.—*Bailey's Cat. Ql. Woods* No. 58b.

4. GANOPHYLLUM, Blume.
(Bright-leaved.)

Flowers polygamo-diæcious. Male: Calyx small, cup-like, 5-fid, valvate, no petals; stamens 5 to 7, inserted between the lobes of the disk, and alternate with the lobes of the calyx; filaments filiform, exserted; anthers oblong; disk annular, puberulous, 5-lobed; ovary rudimentary. Female: Stamens 5, very short, alternate with the calyx-lobes; anthers very small, ovate, sterile; style very short; stigma depresso-capitate, 2-lobed; ovary free, 2-celled. Drupe 1-celled, 1-seeded; rarely 2-celled, 2-seeded. Seeds turgid. Testa membranous. Albumen none. Cotyledons thick, fleshy, unequal, plicately incurved or involute. Radicle short.—Trees, the branchlets angular, foliage more or less covered with wax-like scales. Leaves alternate, imparipinnate; leaflets entire, falcate, many pairs. Flowers small, greenish, ebracteate, in axillary racemose panicles.—*B. et H.*, Gen. Pl., and *F. v. M.* Fragm. vii. 24.

1. **G. falcatum** (falcate), *Blume, Mus. Bot. Sugd.* i. 280; *F. v. M. Fragm.* vii. 24. A tree of about 80ft. Leaves pinnate, petiole ½ to 3in. long, rhachis about 1ft. long, bearing about 14 lanceolate-ovate leaflets, obtuse or shortly acuminate, 1½ to 3in. long, 1 to 1¾in. broad, entire. Panicle a few or many inches long. Calyx about 1 line long, teeth semilanceolate-deltoid, ciliate. Stamens 6 or 7; filaments 1 or 2 lines long, glabrous; anthers sulphur-coloured. Style in male flowers none, and in female flowers only rudimentary, obtuse and pilose, glabrous. Disk puberulous. Drupe 6 to 8 lines long. Seeds 4 or 5 lines long. Testa tawny-hoary. Embryo green. Cotyledons superior, strongly inflexed.

Hab.: Port Denison and Rockingham Bay, *J. Dallachy* and *E. Fitzalan*.

Order XXXIII. MELIACEÆ.

Flowers regular, usually hermaphrodite. Calyx small, 4 or 5-lobed, or divided into as many distinct sepals. Petals 4 or 5, rarely more, or 3 only, free or adnate to the staminal tube, imbricate or rarely valvate. Stamens as many, or more frequently twice as many, as petals; the filaments, in *Meliaceæ* proper, united in a tube; anthers sessile or shortly stipitate, within or at the summit of the tube; in *Cedreleæ*, filaments free. Disk various, often annular or tubular, free within the staminal tube. Ovary free, entire, 3 to 5-celled; style simple; stigma thick, disk-shaped or pyramidal. Ovules in each cell 2 or (in *Carapa* and the *Cedreleæ*) 4 or more, the micropyle superior. Fruit a capsule, berry, or rarely a drupe, indehiscent, or septicidally or loculicidally dehiscent. Seeds 1, rarely 2, or in *Cedreleæ* few in each cell, with a ventral hilum; albumen fleshy or none, embryo flat or nearly so, radicle superior.—Trees or shrubs, the wood often coloured and sometimes fragrant, the bark rarely bitter. Leaves alternate or very rarely opposite, simple, or more frequently pinnate, the petiole often continuing long to grow out and produce fresh leaflets; leaflets without dots, except in *Flindersia*. Flowers paniculate, often small.

The Order is found abundantly in the tropical or warm regions of Asia and America, more rarely in Africa. Of the 11 Australian genera, 4 are endemic, 3 are common to the tropical regions of both the New and the Old World, the remaining 4 are Asiatic, one of them extending also into Africa.—*Benth.* (in part).

Meliaceæ proper are at once known among the allied Orders by their staminal tube. *Cedreleæ*, with free stamens, are in that respect anomalous, and might technically be referred to some of the preceding Orders containing pinnate-leaved trees; but the habit, the large disk-like stigma, and some minor characters, have referred them with common consent to *Meliaceæ* as a tribe. *Flindersia*, however, with its pellucid-dotted leaves, is really as nearly connected with *Rutaceæ-Zanthoxyleæ* as with *Meliaceæ*, but retained among the latter on account of its fruit and seeds so nearly those of *Cedrela*.—*Benth.*

Tribe I. **Melieæ.**—*Stamens united in a tube. Ovules 2 in each cell. Seeds not winged, albuminous.*

Leaves simple. Petals very long and narrow . . . . 1. Turræa.
Leaves bipinnate . . . . 2. Melia.

Tribe II. **Trichilieæ.**—*Stamens united in a tube. Ovules 2, rarely 1, or (in Carapa) more than 2 in each cell. Seeds not winged, without albumen. Leaves pinnate.*

Disk tubular or cup-shaped, enclosing the ovary . . . . 3. Dysoxylon.
Disk annular, or undistinguishable from the thickened base of the ovary.
  Stamens equal in number to or not twice as many as petals. Flowers very
    small, globular . . . . 4. Aglaia.
  Stamens twice as many as petals.
    Staminal tube truncate or scarcely crenulate, the anthers included or
      scarcely protruding. Capsule hard.
    Ovules 1 (rarely 2 superposed) in each cell . . . . 5. Amoora.
    Ovules 2, parallel, attached to a pendulous placenta, which in the fruit is
      a thick arillus between the two seeds . . . . 6. Synoum.

Staminal tube crenulate. Petals and anthers 5 each . . . . 7. HEARNIA.
Staminal tube toothed, with the anthers protruding between the teeth.
Ovules solitary. Drupe globular, with a woody or stony putamen . . . 8. OWENIA.
Staminal tube truncate or crenate. Ovules more than 2 in each cell.
Leaflets reticulate . . . . 9. CARAPA.

TRIBE III. **Cedreleæ.**—*Stamens free. Ovules more than 2 in each cell. Seeds winged. Leaves pinnate or rarely simple.*

Petals erect. Disk thick. Capsule smooth. Leaves not dotted . . . . . . 10. CEDRELA.
Petals spreading. Disk broadly cupular. Capsule muricate. Leaves pellucid-dotted . . . . 11. FLINDERSIA.

1. TURRÆA, Linn.
(After George Turrea.)

Calyx 4 or 5-toothed or lobed. Petals 4 or 5, elongated, free. Staminal tube cylindrical, toothed at the summit, anthers 8 or 10, within the summit of the tube. Disk annular or none. Ovary 5, 10, or 20-celled; style filiform, with a disk-like stigma; ovules 2 in each cell, superposed. Capsule 5 or several-celled, opening loculicidally in as many coriaceous valves. Seeds oblong, with a broad ventral hilum, sometimes winged; albumen fleshy, cotyledons leaf-like.—Trees or shrubs. Leaves simple. Peduncles axillary, bearing few, white flowers.

The genus extends over tropical Asia and Africa; the Australian species is found also in the Indian Archipelago.—*Benth.*

1. **T. pubescens** (pubescent), *Hellen; Willd. Spec. Pl.* ii. 555; *Benth. Fl. Austr.* i. 379. A deciduous shrub or small tree. Leaves at the time of flowering small, from obovate and emarginate to ovate-lanceolate and acuminate, pubescent as well as the young shoots; when full-grown ovate, shortly acuminate, 2 to 3, or even 4in. long, somewhat coriaceous, quite glabrous or slightly pubescent underneath. Flowers white, sweet-scented, in axillary clusters or short racemes of 3 to 6. Petals narrow, linear-spathulate, 1 to 1¼in. long. Staminal tube rather shorter, with 10 short teeth, each one more or less divided into 2 to 4 lobes, or rarely entire. Style exserted. Fruit nearly globular, 5-celled, furrowed opposite the dissepiments, 3 to 4 lines diameter in some specimens, ½in. in others, opening loculicidally in 5 valves, leaving the greater part of the membranous dissepiments attached to the axis. Seeds not winged.—*T. Billardieri*, A. Juss. in Mem. Mus. Par. xix. 218; Benn. Pl. Jav. Rar. 181 (from the character given); *T. concinna*, Benn. Pl. Jav. Rar. 182.

Hab.: Not uncommon in scrubs north and south.

Wood close-grained, hard, the centre or heart-wood dark, the outer part of a somewhat bright yellow.—*Bailey's Cat. Ql. Woods No.* 59.

The species appears to be generally dispersed over the Indian Archipelago; the lobes of the teeth of the staminal tube, upon which the distinction of *T. pubescens*, *T. Billardieri*, and *T. concinna* is chiefly founded, are very variable, even on the same specimen.—*Benth.*

2. MELIA, Linn.
(The Greek name of the Manna Ash.)

Calyx 5 or 6-cleft. Petals 5 or 6, linear-spathulate, spreading. Staminal tube 10 or 12-toothed; anthers 10 or 12, within the summit. Disk annular. Ovary 3 to 6-celled; style slender, with a capitate lobed stigma; ovules 2 in each cell, superposed. Drupe succulent, with a bony 1 to 5-celled putamen. Seeds solitary in each cell; testa crustaceous; albumen fleshy, sometimes scanty or none, cotyledons leaf-like.—Trees. Leaves usually twice or thrice pinnate, with petiolulate toothed leaflets. Flowers paniculate.

The genus comprises but very few species, natives of tropical Asia, one of them generally planted in many parts of the globe. The Australian species is one of the Asiatic ones.—*Benth.*

1. **M. composita** (referring to the compound leaves), *Willd.; W. and Arn. Prod.* 117; *Benth. Fl. Austr.* i. 380. White Cedar. An elegant tree, the young leaves, shoots, and inflorescence sprinkled with a mealy stellate tomentum which disappears with age. Leaves twice or rarely thrice pinnate; leaflets petiolulate, opposite with a terminal odd one, ovate to almost lanceolate, acuminate, 1 to 2 in. long, entire, coarsely toothed or sometimes lobed. Panicles loose, shorter than the leaves, retaining the mealy tomentum late, especially on the calyx and petals. Sepals small, ovate. Petals 4 to 5 lines long. Staminal tube hirsute inside behind the anthers, the teeth alternately entire and 2-cleft; anthers glabrous or slightly hirsute. Ovary 5-celled. Drupe ovoid, $\frac{1}{2}$ to $\frac{3}{4}$ in. long.—*M. Australasica*, A. Juss. in Mem. Mus. Rar. xix, 257.

Hab.: Burdekin River, *F. v. Mueller*; Broadsound, *R. Brown*; Rockhampton, *Thozet*; Herberton, *J. F. Bailey*.

The Australian tree appears to me identical with the *M. composita* of East India and the Archipelago, and scarcely differs from the more common *M. Azedarach*, except in the more abundant mealy tomentum, especially on the inflorescence and flowers. The drupe is also usually larger and more ovoid.—*Benth.*

The gum is a good substitute for arabic gum. It contains: Arabin 79%, water 21%.—*Lauterer.* Wood easily worked, light-red, soft and light.—*Bailey's Cat. Ql. Woods No. 60.*

8. DYSOXYLON, Blume.

(Name derived from the disagreeable odour of the wood of *D. alliaceum.*)

(Hartighsea, *A. Juss.*)

Calyx small, 4 or 5-toothed, or divided into 4 or 5 sepals. Petals 4 or 5, free or adnate to the staminal tube, spreading at the top. Staminal tube truncate or 8 or 10-toothed; anthers 8 or 10, within the summit. Disk tubular, as long as or usually much longer than the ovary. Ovary 3 to 5-celled; style elongated; stigma disk-like; ovules 2 in each cell, or rarely solitary. Capsule globular or pear-shaped, 1 to 5-celled, opening loculicidally in 2 to 5 thickly coriaceous valves. Seeds with or rarely without an arillus, oblong, without a broad ventral hilum; testa coriaceous; albumen none; cotyledons large.—Trees, often fœtid. Leaves pinnate, leaflets opposite or alternate in the same species, entire, often oblique. Panicles axillary, loose, but often small. Flowers not very small.

A considerable genus, spread over tropical Asia and the Indian Archipelago, extending also to New Zealand. The Australian species are all endemic. The genus is readily known by the tubular disk enclosing the ovary within the staminal tube.—*Benth.*

Calyx gamosepalous, dentate or rarely partite. Staminal tube free.
 Shrub or small tree. Branchlets smooth, dark-red, with pale lenticels. Panicles much shorter than the leaves. Petals puberulent; apex subulate. Ovary 5-celled, cells 2-ovulate. Fruit 7 or 8 lines long, seed 1 in each cell . . . . 1. *D. arborescens.*
 Small tree, branchlets green. Panicles very shortly branched, almost racemose. Petals glabrous, apex sharp-acuminate. Fruit sub-globose, rufescent, glabrous, about 1in. long, 4-valvate . . . . 2. *D. latifolium.*
 Small tree. Leaves glabrous. Racemose panicles produced from the stem. Calyx large before expansion, entire, bursting to the middle into 2 undivided or slightly cleft lobes. Petals 4, white, elongate-oblong, silky outside, free. Fruit 1in. or more long, 4-celled, 1 seed in each cell with arillus . . . . 3. *D. Schiffneri.*
 Small tree. Branchlets glabrous, brownish. Panicles narrow, puberulent on the 2-years-old wood. Petals silky outside. Fruit subglobose, 4-celled, 4-seeded, about $\frac{3}{4}$in. diameter . . . . . . 4. *D. oppositifolium.*
 Small tree. Branchlets terete, glabrous. Panicles with short branches. Pedicels silky, with 2 bracts at the base. Petals 4, very slightly silky outside, about 6 lines long, cohering high up . . . 5. *D. Klanderi.*
Staminal tube more or less united with the petals.
 A large deciduous tree. Bark reddish. Young branchlets stout, rough, with prominent lenticels. Flowers in erect axillary spikes. Petals 4, slightly hairy outside near the base. Fruit pyriform, marked with white lenticels, 4-angular, 4-celled . . . . 6. *D. Pettigrewianum.*

Tree of medium size. Branchlets brownish, at first puberulent, and
bearing round lenticels. Panicles much shorter than the leaves, the
branches few-flowered. Petals glabrous. Fruit subglobose, 4-valved,
1 to 1¼in. long, pyriform, rubrescent glabrous . . . . . . 7. *D. Fraseranum.*
Tree, the wood having a strong odour of onions. Lenticels on the
branchlets pale and minute. Panicles on shoots of the present year.
Petals yellow, glabrous. Fruit globose, glabrous, 4 to 6-valved, 4 to
6-seeded, 7 to 8 lines diameter . . . . 8. *D. Lessertianum.*
A tall tree, the branchlets puberulent; the pedunculate panicles
slightly hirsute, pyramidal. Flowers dense. Petals glabrous,
laciniate, acute. Fruit 4-celled; arillus orange . . . . . . 9. *D. Muelleri.*
Tree, the young branches velvety, greenish. Panicles half the length
of the leaves. Petals puberulent, laciniate; apex sharply acuminate.
Ovary with 2-ovulate cells . . . . 10. *D. Becklerianum.*
Tree. Branchlets pilose, glabrescent. Panicles short. Petals 5,
yellowish. Ovary 4-celled, cells 2-ovulate . . . . 11. *D. Nernstii.*
Calyx polysepalous. Sepals imbricate. Staminal tube free.
A small tree, the branchlets clothed with a close yellow pubescence.
Panicles long, drooping, often from the axils of the lowest leaves.
Petals 5; apex acute. Fruit globose, 5-celled, nearly 1in. diameter 12. *D. rufum.*
A small tree. Young branchlets clothed with a ferruginous pubescence.
Fruiting panicles racemose, near the end of the branchlets. Fruit
pyriform or globose, the pericarp deeply wrinkled, clothed with
short orange-coloured hairs; 5-celled, 1 or 2 ovules in each cell . 13. *D. cerebriforme.*

1. **D. arborescens** (tree-like), *Miq.; F. v. M. Fragm.* ix. 184; *C. DC. Mono. Phan.* i. 489. A small tree, tall shrub, the branches glabrous, dark-red with pale lenticels. Petioles moderately long, impari-pinnate. Leaflets about 9 pairs, on short petiolules, elliptic-oblong, narrowed into an unequal-sided base, the apex somewhat obtusely cuspidate, glabrous on both sides; terminal one often 6in. long and nearly 2in. broad, the lateral ones smaller. Panicle glabrous, flowers pedicellate. Calyx minutely 5-dentate, teeth sharp, pubescent outside. Petals 5 to 7 mill. long, membranous, oblong, the apex subacute, puberulous outside. Staminal tube cylindrical, crenulate, glabrous. Disk crateriform, outside glabrous, inside beset with reversed hairs. Ovary hirsute, 5-celled, 2 ovules in each cell, collateral. Style pilose at the base. Fruit 15 mill. long, seed in cell solitary.—*Goniocheton arborescens,* Bl. Bijdr., p. 177; Miq. Fl. Ind. Bat. i. 540; *Trichilia arborescens,* Spreng. Syst.; *Hartighsea acuminata,* Miq. Fl. Supp.; *Disoxylum lampongum,* Miq. var. 2 l.c.

Hab.: North-east coast, locality (?)

2. **D. latifolium** (leaflets broad), *Benth. Fl. Austr.* i. 881; *F. v. M. Fragm.* ix. 61. Leaves glabrous; leaflets in our specimens 4 or 5, ovate or broadly oval-oblong, shortly acuminate, 3 to 4in. long, oblique at the base, somewhat coriaceous. Flowers in sessile or shortly pedunculate clusters, along a simple, axillary, nearly glabrous peduncle of 4 to 5in. Pedicels short, slightly pubescent. Calyx cupular, not 1 line long, with 4 very short broad teeth. Petals 4, pubescent outside, about 3 lines long, valvate in the bud, free from the staminal tube. Staminal tube truncate and shortly and irregularly 8-toothed. Disk broadly tubular, sprinkled with a few minute hairs. Ovary in the flowers examined 2-celled, with 2 ovules in each cell, pubescent, tapering into an elongated style; stigma disk-like. Fruit 1 or 2-celled, globose or pyriform-oval, 1 or 2, rarely 3 or 4-seeded.

Hab.: Islands off Rockingham Bay.

3. **D. Schiffneri** (after Dr. R. Schiffner), *F. v. M. (section Cleisocalyx)* in *Melb. Chemist,* 53 (1881). A tree about 80ft. in height, having a greyish-brown smooth bark and yellowish wood. Leaves and petioles almost glabrous. Leaflets ovate to lanceolate, opposite, thin-chartaceous, of few pairs, 2 to 5in. long, inequilateral at the base, oil dots minute. Racemes 2 or 3 together, 2 to 4in. long, flowers fragrant, silky. Pedicels long as the flowers. Calyx large, in

the bud oval, perfectly entire, without any rupture or sutural lines; afterwards torn to about the middle into 2 undivided or slightly cleft lobes. Petals 4, free, elongated-oblong, ½in. long, being about a third longer than the calyx, silky outside, white, upwards slightly imbricate, downwards valvate. Staminal column glabrous, broadly tubular, 7 or 8-toothed, teeth semi-lanceolate, about three times shorter than the tube, their connectives often minutely pointed. Disk cupshaped, free, slightly crenulated, glabrous. Style filiform, densely tomentose at the base as well as the ovary, stigma depressed-hemispherous; ovary 4-celled with 2 superposed ovules in each cell. Fruit globular, glabrescent, brown outside. Seed ripening solitary in each cell, turgid, almost longitudinally adnate. Testa thin, dark-brown. Albumen none. Cotyledons plano-convex, collateral. Radicle very short, terminal, almost concealed between the minute lobes of the cotyledons.—F. v. M. l.c.

Hab.: Mount Bellenden Ker Ranges, *Karsten* (F. v. M., l.c.); Harvey's Creek; Russell River scrubs.

Baron von Mueller notes that the present plant, from the exceptional structure of its calyx, might be raised to generic distinction under the sectional name used by him l.c., *Cleisocalyx*, on account of the fruit not splitting into valvular divisions; in which event *D. Klanderi* must be included, or both might be placed in Blume's genus, *Epicharis*, which would only seem to differ from *Dysoxylum* in the structure of the calyx.

4. **D. oppositifolium** (opposite-leaved), *F. v. M. Fragm.* v. 144, 177; *C.DC. Mono. Phan.* i. 501. A medium-sized tree with a brownish smooth bark. Leaves opposite, 2 to 8-foliolate. Leaflets alternate, almost glabrous, lanceolate-ovate or lanceolate, shortly acuminate, a little unequal at the base, about 4in. long and 2in. broad; rhachis sub-tetragonal. Panicles racemose, shorter than the leaves, axillary. Flowers subsessile. Calyx cupuliform, 4-dentate, puberulent outside. Petals 4, pale-yellow, 2 or 3 lines long, valvate, puberulent, semi-lanceolate, free. Staminal tube truncate, glabrous, very slightly crenulate. Anthers 8 ovate, mucronulate. Disk patelliform, pilose inside. Ovary densely hirsute, 4-celled, 1 to 2-ovulate, collateral, style glabrous. Capsule (sometimes 6-celled, *F. v. M.*) 4-celled, subglobose, 4-seeded, about 1in. thick. Seeds oblongelliptic, slightly 3-angled. Arillus fulvous. Cotyledons plano-convex. Albumen none. Radicle short.

Hab.: Rockingham Bay, *J. Dallachy* (F. v. M.); Endeavour River, *W. E. Armit.*

Wood reddish towards the centre, with a small, prettily-marked heartwood, and a large quantity of yellow wood towards the bark; grain close, easily worked, and fragrant. A useful wood for both joiner and cabinetmaker.—*Bailey's Cat. Ql. Woods No.* 63b.

5. **D. Klanderi** (after Dr. Klander), *F. v. M. Fragm.* v. 176, ix. 61, 97 *and* 184. A tree of medium size, with terete glabrous branches. Leaves glabrous, 4 to 8-foliolate. Leaflets opposite on long petiolules, lanceolate-ovate, equal sided, acuminate, 4 to 6in. long, 1½ to 2in. broad, both sides glossy, oil-dots linear. Panicles very long, often 2ft., pendulous, the branches very short and patent, racemose, or bearing at the end little corymbs of flowers. Bracteoles at base of pedicels, 2-opposite, canaliculate-subulate, and with the pedicels silky. Pedicels somewhat thick, 1½ to 2 lines long. Calyx scarcely 1 line long, deltoidesquadridentate. Petals about ½in. long, more or less coherent. Staminal tube at the apex 8-dentate. Anthers 8, sessile below the apex of the tube, oblong-linear. Disk cylindrical, half including the pistillum, minutely 8-dentate. Style about as long as the staminal tube. Stigma truncate. Ovary glabrous. Fruit globosepyriform, about 1in. in diameter, very thinly puberulous, usually 4-celled, 4-seeded. Seeds trigono-ovate. Cotyledons almost semi-ovate. Radicle very short, lateral. Albumen none.—*Schleichera ptychocarpa*, F. v. M. l.c.

Hab.: Rockingham Bay, *J. Dallachy* (F. v. M.); now known from many other localities in North Queensland.

6. **D. Pettigrewianum** (after Wm. Pettigrew), *Bail. Bot. Bull.* Scrub Ironbark or Cairns Satin-wood. A tall, deciduous tree, with large umbrageous head; stem erect, often flanged at the base; bark reddish. The young branches stout, containing a large proportion of pith; the bark rough, with lenticels and the strong decurrent lines from the bases of the petioles. Leaves alternate, puberulent; leaflets opposite, lanceolate, 5 or 6 pairs and a terminal one, which is the largest and about 9in. long and 4in. broad; all shortly petiolulate, more or less elongated at the apex and rounded at the base; primary veins prominent, nearly parallel, often nearly opposite, from 12 to 20 on each side of the midrib; petioles short and sharply angular. Flower spikes erect, in the axils of the leaves, about 3in. long and covered with a short grey pubescence. Calyx cupular, about 1¼ line long, with 4 short sharp teeth, which are very deciduous. Petals 4, linear, 4 lines long, glabrous or slightly hairy on the back near the apex. Staminal tube three-quarters length of the petals, crenulate, glabrous. Anthers 8, oblong. Disk tubular, about half as long as the staminal tube, sprinkled with a few minute hairs outside, densely hirsute inside, the mouth ciliate with longish hairs. Style and ovary hirsute, stigma large. Fruit pyriform, with 4 prominent acute angles, glabrous or nearly so, the outside marked by numerous white lenticels; when ripe about 1¼ in. long, 4-celled or less by the abortion of one or more cell.

Hab.: Scrubs at the base of the Bellenden Ker Range, *Bellenden Ker Exped.* 1889, and the Barron River, *E. Cowley.*

Wood hard, deep-coloured towards the centre of the stem; outer wood yellowish, close-grained, and firm. Useful for cabinet-work and joinery.—*Bailey's Cat. Ql. Woods No.* 61.

7. **D. Fraseranum** (after C. Fraser), *F. v. M. Fragm.* ix. 61; *Benth. Fl. Austr.* i. 381. A tree of 80 to 180ft., the young leaves and shoots slightly pubescent, glabrous when full-grown. Leaves pale-green, the terminal one sometimes wanting. Leaflets 5 to 9, oblong-lanceolate or elliptical, acuminate, 3 to 6in. long, narrowed and equal at the base, bearing occasionally tufts of hairs in the axils of the principal veins underneath. Panicles in the upper axils short, loose, divaricately branched, racemose slightly pubescent. Calyx cupular, about 1 line long, shortly and broadly 4-lobed. Petals 4, about 3 lines long, nearly glabrous, adnate to the staminal tube to about half their length. Staminal tube 8-toothed, glabrous outside. Disk broadly tubular, rather longer than the ovary. Ovary hirsute, 8-celled, with 2 ovules in each cell. Fruit 1 to 1½in. long, globose or pyriform, tardily dehiscing, seeds usually 4 trigono-ovate, ½ to ⅜ line long, arillus ample.—*Hartighsea Fraserana,* A. Juss. in Mem. Mus. Par. xix. 262, t. 15.

Hab.: Southern ranges.

In New South Wales said to furnish a beautiful red wood, of value for cabinetmaking, turning, and carving.

8. **D. Lessertianum** (after M. Lessert), *F. v. M. Fragm,* v. 145; *Benth. Fl. Austr.* i. 382. A tall tree with erect trunk, wood red, fragrant (*C. Moore*), quite glabrous, or the young shoots and panicles minutely pubescent. Leaflets 4 to 10, usually without any terminal odd one, elliptical or lanceolate, shortly and obtusely acuminate, 4 to 5in long. Panicles loose, extra-axillary, 3 to 4in. long. Calyx short, cupular, entire or irregularly crenulate. Petals 4 or 5, glabrous, more or less adherent to the staminal tube at their base, rarely at length free. Staminal tube glabrous, 8 or 10-toothed. Tubular disk broad, scarcely longer than the ovary. Ovary hirsute, 4 or 5-celled, with 1 ovule in each cell. Fruit hard, globose, glabrous, 4, 5, 6-valved, 4, 5, 6-seeded.—*Hartighsea Lessertiana,* A. Juss. in Mem. Mus. Par. xix. 264.

Hab.: Southern parts.

9. **D. Muelleri** (after Baron Mueller), *F. v. M. Fragm.* v. 143; *Benth. Fl. Austr.* i. 381. "Kedgy Kedgy," Nerang (*Schneider*). A tree of 80 to 90ft. or more, glabrous or nearly so, except the very young shoots and inflorescence.

Leaves 1 to 2ft. long; leaflets 11 to 21, from ovate to almost lanceolate, shortly acuminate, 3 to 6in. long, very oblique at the base, one side rounded, the other truncate and shorter, almost coriaceous. Panicles pyramidal, $\frac{3}{4}$ to 1ft. long, much-branched and many-flowered. Calyx cupular, $\frac{1}{2}$ to $\frac{3}{4}$ line long, pubescent, 4-lobed. Petals 4, nearly glabrous, about 5 lines long, adhering to the staminal tube to about two-thirds their length. Staminal tube truncate and minutely crenulate, hirsute outside. Disk narrow-tubular, nearly half as long as the staminal tube. Ovary hirsute, 4-celled, with 1 ovule in each cell. Fruit tardily splitting into 4 valves; arillus orange-coloured, enveloping the seed.

Hab.: Brisbane River, Moreton Bay, and other southern parts.

Wood of a red colour, prettily marked; a useful wood for the joiner and cabinetmaker.—*Bailey's Cat. Ql. Woods No. 61a.*

10. **D. Becklerianum** (after Dr. Beckler), *C.DC. Mono. Phan.* i. 509. The young branches velvety, green. Leaves abrupto-pinnate, on moderately long petioles, 4 or 5-jugate. Leaflets opposite, very shortly petiolulate, oblong-lanceolate, base equal-sided and acutely cuneate, the apex very shortly, somewhat obtusely cuspidate, membranous, opaque, faintly punctulate, nerves on under side hairy, about $3\frac{1}{2}$in. long, the upper ones a little longer. Petiole and rhachis densely villous, about 2in. long, terete. Panicle elongate, branches very short, puberulous, flowers subsessile. Calyx cupuliform, with 4 roundish lobes, puberulous outside. Petals 4, about $2\frac{1}{2}$ lines long, apex sharply acuminate. Disk cylindrical, glabrous, one-third the length of the petals. Ovary hirsute, cells 2-ovulate, ovules superposed.—*D. Lessertianum*, var. *pubescens*, Benth. Fl. Austr. i. 382.

Hab.: Southern portions of colony.

11. **D. Nernstii** (after J. Nernst), *F. v. M. Fragm.* v. 176. Branches pilose or glabrescent. Leaves alternate, glabrous, with few leaflets. Leaflets ovate, acuminate, on short petiolules almost equilateral at the base, thin-chartaceous, 2 to 4in. long. Panicle short or very shortly pedunculate, the secondary peduncles and pedicels very slightly pilose. Calyx acutely 5-dentate. Petals 5, yellowish, 3 or 4 lines long, quickly falling. Teeth of staminal tube alternately larger. Anthers ovate, $\frac{1}{3}$ line long, alternately short and tall, somewhat exserted. Disk $\frac{2}{3}$ line high. Style 2 lines long, the lower part and the ovary pilose. Stigma truncate-hemispherical. Ovary 4-celled, cells 2-ovulate.

Hab.: Scrubs about Rockingham Bay, J. Dallachy (F. v. M., l.c.)

12. **D. rufum** (red), *Benth. Fl. Austr.* i. 382. A slender tree of 30 to 70ft., with a smooth bark, the young branches, petioles, and under side of the leaves clothed with a soft often rust-coloured pubescence. Leaves alternate, $1\frac{1}{2}$ to 2ft. long; leaflets numerous, very shortly petiolulate, ovate-lanceolate or lanceolate, acuminate, 3 to 6in. long, very oblique at the base, glabrous on the upper side. Panicles axillary or lateral, not much branched, pubescent. Flowers sessile, very fragrant. Sepals 5, almost free, orbicular, imbricate, about 1 line long. Petals 5, pubescent, $\frac{1}{2}$in. long, adhering to the staminal tube to about the middle. Staminal tube truncate, with 10 retuse short lobes or teeth; anthers tipped with a short point. Disk broadly tubular, very hairy. Ovary hirsute, 5-celled, with 2 ovules in each cell. Fruit depressed-globular, 1in. diameter, densely hirsute with short rigid, almost golden hairs. Seeds arillate.—*Hartighsea rufa*, A. Rich, Sert. Astrol. 29, t. 11.

Hab.: Southern scrubs.

Wood a light-yellow, hard, and close-grained.—*Bailey's Cat. Ql. Woods No. 62,*

Var. (?) *glabrescens.* Leaves quite glabrous. Fruit tomentose, with very short golden hairs. Rockhampton, Port Denison, Elliott and Herbert Rivers. Wood straw-coloured, of coarse grain, easy to work.—*Bailey's Cat. Ql. Woods No. 62.*

18. **D. cerebriforme** (pericarp marked like the brain), *Bail. Bot. Bull.* xiv., *Pl.* 1 and 2. A small tree, leaves on the fruiting branchlets, 6 to 9in. long, more or less hirsute, impari-pinnate, leaflets 7 to 9, 1½ to 3¼in. long, 1¼in. at the widest part, those near the end of leaf lanceolate, the lowermost ones abruptly rounded at the base, points elongated, petiolules very short. (On a leaf sent separate from the fruit-bearing specimen the petiole and rhachis measured 1ft. long, and the leaflets were of the same form but measured 8½in. long and 3in. wide.) The young growth shows a ferruginous pubescence, but the older leaves are hirsute with white hairs, some of which are much longer than the others. Flowers not seen. Fruit on short racemes or panicles near the summit of the branchlets, pyriform-globose, 1¼in. long, 1¼in. in diameter, 5-celled, 5-seeded; in some fruits one or more of the cells 2-seeded; seeds attached near the base, testa orange-madder coloured, bluntly 3-sided, the arillus only covering a third of the seed. Pericarp prominently wrinkled, covered with a dense coating of short orange-coloured hairs besides the longer white ones with which other parts of the plant are sprinkled.

Hab.: Freshwater Creek. near Cairns, *L. J. Nugent.*

The only other Australian species that I know of with a deeply-wrinkled pericarp is one of which I gathered fruiting specimens on the Mulgrave River in 1889; this had the 5-celled fruit and the very oblique base to the leaflets of *D. rufum*, Benth. Thus I placed my specimens under that species, but now I am inclined to think it may prove a distinct species or variety.

4. AGLAIA, Lour.
(One of the Graces.)
(Milnea, *Roxb.*; Nemedra, *A. Juss.*)

Flowers polygamous. Calyx 4 or 5-toothed or cleft. Petals 4 or 5, short, connivent, imbricate in the bud. Staminal tube globular or urceolate, entire or shortly toothed; anthers as many as petals or rarely more, within the summit of the tube. Disk none, or not distinct from the base of the ovary. Ovary 2 or 3-celled, with a short, thick style and disk-like stigma; ovules 1 or 2 in each cell. Fruit coriaceous or almost succulent, indehiscent. Seeds 1 or 2, enveloped in a mealy pulp, without any arillus.—Trees, either glabrous or clothed with small scurfy scales or rarely with stellate tomentum. Leaves pinnate, with entire leaflets. Flowers very small, nearly globular, in axillary panicles.

The genus is dispersed over tropical Asia and the islands of the Indian Archipelago and the Pacific. The only Australian species is also a native of New Caledonia and New Guinea.—*Benth.*

1. **A. elæagnoidea** (Elæagnus-like), *Benth. Fl. Austr.* i. 388. A tree of 40 to 50ft. Wood yellow, bark smooth, grey. The young branches, inflorescence, and under side of the leaves covered with silky or rust-coloured scurfy scales, often fringed at the edges. Leaflets 3 or often 5, petiolulate, ovate-oblong, or the terminal one obovate, acuminate, rarely ovate-lanceolate, 2 to 3in. long or rarely more, coriaceous, glabrous above when full-grown. Flowers globular, about 1 line diameter, numerous in loose panicles which rarely exceed the leaves. Calyx shortly 5, rarely 4-lobed. Petals 5, rarely 4, much imbricate, sprinkled as well as the ovary with the scurfy scales that cover the calyx and inflorescence. Anthers usually 5, but in some flowers 6, 7, or even more, within the short urceolate tube, which is thickened into raised filaments below the anthers. Ovary 3-celled, with 1 (or sometimes 2 ?) ovules in each cell. Fruit globose or obovoid, ½ to 1in. long, covered with minute orange-yellow or rust-coloured scurfy scales. Seeds 1 or 2, enveloped in a mealy pulp.—*Nemedra elæagnoidea*, A. Juss. in Mem. Mus. Par. xix. 259, t. 14; *Aglaia odoratissima*, Benth. in Hook. Lond. Journ. ii. 213, but probably not of Blume.

Hab.: Islands of the Gulf of Carpentaria, *R. Brown* (specimens in fruit and flower); Entrance Island, Endeavour Straits, *Leichhardt*; Mt. Elliott and Rockingham Bay, *F. v. M.*

Found also in New Caledonia, the New Hebrides, and in New Guinea. The station, King George's Sound, given by A. de Jussieu on the authority of the Paris Herbarium, is evidently one of those errors of locality which occurs in many of the early collections of Australian plants deposited there. A. de Jussieu having found as many as 10 stamens, gives that as the typical number, although he observes at the same time that there are sometimes fewer. We, therefore, not having then any Australian specimens, failed to recognise his plant, and from the technical characters referred it in our " Genera Plantarum" to *Amoora*. Having since, however, examined Leichhardt's and R. Brown's Australian specimens, and also some flowers from A. de Jussieu's specimens, kindly transmitted to me by M. Brongniart, I have been able satisfactorily to identify the species, which, notwithstanding an occasional increase in the number of stamens, belongs undoubtedly to *Aglaia*, a very natural genus if extended so as to include *Milnea*. In the majority of specimens examined I find almost always 5 stamens, and only now and then 6. Out of three unexpanded flowers from A. de Jussieu's plant, I found 7 stamens in two of them, and only 5 in the third.—*Benth.*

5. AMOORA, Roxb.

(From the Bengalese name of *A. cuculata*.)

Flowers polygamous. Calyx 3 to 5-toothed or lobed. Petals 3 to 5, imbricate in the bud, free from the staminal tube. Staminal tube urceolate or nearly globular, truncate or crenate; anthers within the tube, twice as many as petals. Disk none besides the thickened base of the ovary. Ovary 3 to 5-celled or rarely 2-celled, with 1 or 2 superposed ovules in each cell; style short or long with a disk-like stigma. Capsule obovoid or globular, coriaceous or hard, opening loculicidally in 3 to 5 valves (or sometimes indehiscent?). Seeds solitary in each cell, enclosed in a fleshy arillus (or sometimes without an arillus?).—Trees. Leaves pinnate, with entire leaflets. Flowers small, but usually larger than in *Aglaia*.

The genus is spread over tropical Asia and the Indian Archipelago; the Australian species is endemic.—*Benth.*

1. **A. nitidula** (shining), *Benth. Fl. Austr.* i. 388. "Jimmie Jimmie," Herberton, *J. F. Bailey*. A tall tree, quite glabrous. Leaflets 2 or 4, opposite, without any terminal odd one, elliptical-oblong, 3 to 4in. long or sometimes more, obtuse or shortly and obtusely acuminate, somewhat coriaceous and shining, narrowed at the base, the common petiole often slightly dilated towards the end. Panicles axillary, loose, but shorter than the leaves. Calyx very short, with 5 short teeth or lobes. Petals 5, about 2 lines long, glabrous or minutely ciliate. Staminal tube broadly urceolate; anthers 10, the tips slightly protruding. Ovary 2 or 3-celled, with 1 ovule in each cell. Fruit obovoid, hard and almost woody, narrowed almost into a stipes at the base, 2 or 3-celled. Seeds nearly globular, laterally attached near the top, apparently without any arillus.

Hab.: Southern scrubs.

The species has much of the habit of some *Dysoxyla*, but the want of any free disk and the form of the staminal tube agree better with *Amoora*.—*Benth.*

Wood of a light colour, hard, tough, and close in grain.—*Bailey's Cat Ql. Woods No. 64.*

6. SYNOUM, A. Juss.

(The ovules united.)

Calyx 4 rarely 5-cleft. Petals 4, rarely 5, valvate or slightly imbricate in the bud. Staminal tube cylindrical, slightly crenulate; anthers twice as many as petals, within the summit of the tube. Disk continuous with the thickened base of the ovary. Ovary 3-celled; style short, with a disk-like stigma; ovules 2 in each cell, attached collaterally to a thickish placenta pendulous from the apex of the cavity. Capsule 3-celled, opening loculicidally in 3 valves, or reduced by abortion to 2 valves and cells. Seeds 2 in each cell, attached by a broad lateral hilum, and half embedded collaterally in a fleshy arillus formed by the enlarged placenta.—A tree. Leaves pinnate, with entire leaflets.

The genus is endemic in Australia.

Leaflets 5 to 9. Panicles very short. Flowers dense. Ovary 3-celled,
3-sulcate . . . . 1. *S. glandulosum.*
Leaflets 7. Panicles about the length of the leaves. Ovary 2-celled . . . 2. *S. Muelleri.*

1. **S. glandulosum** (glandulous), *A. Juss. in Mem. Mus. Par.* xix. 227, *t.* 15; *Benth. Fl. Austr.* i. 884. A moderate-sized tree, glabrous or the young leaves and shoots slightly silky-tomentose. Leaflets 5 to 9, elliptical-lanceolate, acuminate, mostly 2 to 3in. long, narrowed at the base, somewhat coriaceous, the lateral veins few and scarcely prominent. Flowers in short dense axillary panicles, rarely exceeding 1in. Sepals small, orbicular, spreading. Petals about 2¼ lines long. Staminal tube broad, slightly crenulate, glabrous or with a few hairs inside; anthers sometimes slightly protruding. Ovary villous. Capsule depressed-globular, glabrous, about ¾in. diameter, furrowed opposite the dissepiments so as to be almost 3-lobed.—*Trichilia glandulosa*, Sm. in Rees' Cycl. xxxvi.

Hab.: Southern scrubs.

It has the general habit of some *Dysoxyla*, but, besides the want of any free disk and the curious insertion of the ovules and seeds, it is easily recognised by its very short inflorescence.—*Benth.*

Wood of a red colour, close-grained, and easy to work; a useful wood for the cabinetmaker.—*Bailey's Cat. Ql. Woods No.* 68a (cut in mistake for *Dysoxylon Fraserianum*).

2. **S. Muelleri** (after Baron von Mueller), *C. DC., Mono. Phan.* i. 598. Branchlets glabrous. Leaves shortly petiolate, impari-pinnate, 3-jugate, about 6 or 7in. long; leaflets opposite, petiolulate, elliptic-lanceolate, equal and acute at the base, the apex obtuse cuspidate, glabrous on both sides, subcoriaceous, the superior leaflet attaining about 4in. long and nearly 2in. broad, the others smaller, secondary nerves fine. Panicle long as or longer than the leaf, the branches elongated, glabrous. Buds globose-ovate. Flowers on long pedicels. Calyx-teeth 5, ovate, glabrous. Petals 5, elliptic, acute, glabrous, staminal tube urceolate-cylindric, glabrous. Anthers 10, elliptic, glabrous. Ovary 2-celled, densely covered with yellow hairs, the style also hairy.—C. de Candolle l.c.

Hab.: Rockingham Bay, *F. v. M.*

7. HEARNIA, F. v. M.

(After William E. Hearn, LL.D.)

Calyx 5-dentate, 5-partite, or 5-sepalous, very short. Petals 5, distinct or connate at the base. Staminal tube a little shorter than the petals, free, apex crenulate or dentate. Anthers 5, rarely 6, at the apex of the teeth, connective incrasated often mucronulate. Disk none, or inconspicuous. Ovary 1 or 2-celled, often sterile, 1 to 2-ovulate, ovules collateral. Style none. Stigma minute, globose. Fruit indehiscent, baccate, septa frail, 1-celled, 1 or 2-seeded. Seeds almost ovate, exalbuminous, cotyledons superposed, thick, semi-ovate, radicle short included. Trees with alternate impari-pinnate foliage. Leaflets entire. Inflorescence axillary, paniculate. Flowers hermaphrodite (or sometimes polygamous diœcious?), minute, globose.—C. de Candolle, Mono. Phan. i. 628.

1. **H. sapindina** (Sapindus-like), *F. v. M. Fragm.* v. 55, *DC. l.c.* "Boodyarra," Herberton, *J. F. Bailey.* A small tree, with smooth bark and pale-coloured soft wood. Branchlets, petioles, and peduncles covered over with a very thin tomentum. Petioles 2 to 4in. long. Leaves 2 or 3, rarely 1, jugate. Leaflets chartaceous, the young ones stellate-puberulent on both sides, when dry glaucescent, opaque, pellucido-punctate (dots

often elongate), 3 to 11in. long, 1 to 5¼in. broad, ovate or ovate-elliptic or lanceolate-ovate, acute or shortly acuminate, the petiolules ½in., or sometimes less, rarely longer. The penninerved thin, and the reticulate veinlets very thin, base obtuse or rarely decurrent. Panicles axillary and terminal on very short peduncles. Flowers pedicellate. Calyx scarcely ½in. long, teeth ovate-acute. Petals 5, elliptic, distinct, glabrous, yellow, about 1 line long. Anthers 5, ovate, obtuse mucronulate. Fruit ½ to ¾in. diameter, yellowish. Seeds ovate.

Hab.: Many of the tropical scrubs, Rockingham Bay, Barron River, etc.
Wood of a grey colour, hard and tough.—*Bailey's Cat. Ql. Woods No.* 65.

8. OWENIA, F. v. M.
(After Professor Owen.)

Sepals 5, short, orbicular, much imbricate. Petals 5, imbricate in the bud. Staminal tube short or long, with 10 entire or 2-lobed teeth; anthers protruding between the teeth. Disk small, annular or not distinct from the ovary. Ovary 3 or 4-celled, or in one species 12-celled, with 1 ovule in each cell; style rather thick; stigma globular or conical, entire or lobed, on a disk-like expansion of the summit of the style. Drupe globular, the epicarp more or less succulent, putamen thick, woody or bony, rugose outside, 2 to 4-celled, or in one species 12-celled. Seeds solitary in each cell, the outer coating spongy, the hilum broad lateral; cotyledons oblong, thick.—Trees, with the juice often (perhaps always) milky, the young shoots often viscous or gummy. Leaves pinnate. Flowers small, in axillary panicles. Fruit rather acid, eaten by the aborigines.

The genus is endemic in Australia, and differs from all other known *Trichiliæ* in its globular drupaceous fruit.—*Benth.*

Leaflets numerous, lanceolate, acute.
  Leaflets 1-nerved. Panicles narrow. Flowers 2¼ lines long . . . . . . 1. *O. acidula.*
  Leaflets with the lateral veins conspicuous. Panicles divaricate. Flowers very numerous, about 1 line long . . . . 2. *O. vernicosa.*
Leaflets 2 to 10 pairs, obtuse, penninerved or reticulate.
  Leaflets oblong or broadly lanceolate, narrowed at the base, quite glabrous. Fruit 4-celled . . . . 3. *O. venasa.*
  Leaflets about 10 pairs. Fruit 2 rarely 1-seeded; pericarp pulpy-rimulose . . 4. *O. cepiadora.*
  Leaflets large, ovate or ovate-lanceolate, broad and sessile at the base, very prominently reticulate underneath . . . . 5. *O. reticulata*

1. **O. acidula** (fruit acid), *F. v. M. in Hook. Kew Journ.* ix. 304, and *Fragm.* iii. 14; *Benth. Fl. Austr.* i. 385. "Emu Apple," *Gruee;* "Bulloo," *J. F. Bailey;* "Dilly Boolen," St. George, *Wedd.* A small or moderate-sized tree, glabrous, with the young shoots glutinous. Leaves crowded at the ends of the often pendulous branches; leaflets from 9 to nearly 30, linear-lanceolate, acute or mucronate, 1 to 1½in. long, oblique, the midrib prominent underneath, but otherwise almost nerveless, the common petiole 3 to 6in. long. Panicles narrow, shorter than the leaves. Flowers nearly sessile, in clusters or on short branches of the panicle. Sepals about 1 line long. Petals about 2 lines. Teeth of the staminal tube subulate, but more or less connected by an undulate crenate or almost fringed membrane. Disk small, annular. Ovary 3-celled. Drupe ¾ to 1in. or rather more in diameter, said to resemble a russet apple, the epicarp pulpy, of a rich crimson; putamen very hard.

Hab.: A common tree of the western plains.
Wood reddish, close-grained, hard, but very easy to work; useful for cabinetmaking and turnery.—*Bailey's Cat. Ql. Woods No.* 66.

2. **O. vernicosa** (appearing varnished), *F. v. M. Fragm.* iii. 15; *Benth. Fl. Austr.* i. 385. Quite glabrous. Branches thick, marked with the broad scars of the fallen leaves, the young shoots glutinous. Leaves crowded at the ends of the branches, larger than in *O. acidula,* the common petiole slightly flattened;

leaflets 15 to nearly 30, lanceolate, acuminate, often above 2in. long, oblique, with a prominent midrib and transverse reticulations. Panicles 3 or 4 in. long, with divaricate branches and numerous flowers, much smaller than in *O. acidula*. Sepals about ½ line long, slightly ciliate. Petals little more than 1 line. Staminal tube short, with 10 subulate teeth. Fruit the size of that of *O. acidula*, the stony endocarp thicker and harder, usually 3-celled.

Hab.: Musgrave, Cape York Peninsula, *Geo. Jacobson.*
Var. (?) *pubescens.* Young shoots and inflorescence softly pubescent; flowers still smaller and more numerous.—Mouth of the Victoria River, *F. v. Mueller* (Herb F. v. M.)

3. **O. venosa** (veins prominent), *F. v. M. in Hook. Kew Journ.* ix. 804; *Benth. Fl. Austr.* i. 386. "Crow's Apple." A tall tree, quite glabrous, the young shoots slightly glutinous. Leaflets 6 or 8, obliquely oblong or ovate-lanceolate, obtuse or emarginate, 2 to 8 or rarely 4in. long, coriaceous, prominently penninerved, slightly reticulate underneath, the petiole angular or sometimes broadly winged. Panicles narrow, 3 to 5in. long, glabrous. Flowers not yet open in our specimen, but apparently like those of *O. acidula*, except that the staminal tube is exceedingly short, but possibly it may grow out as the bud advances. Sepals orbicular, about 1 line diameter. Fruit globose, 1 to 1½in. diameter.

Hab.: Between the Dawson and Burnett Rivers, *F. v. Mueller;* Rockhampton, *Thozet;* and many other localities of south Queensland.
Wood resembles that of the Emu Apple; very durable.—*Bailey's Cat. Ql. Woods No. 67.*

4. **O. cepiodora** (Onion-scented), *F. v. M. Fragm.* xi. 81. A tall tree, branchlets thick, prominently and copiously cicatrisate, neither glutinous nor lactescent. Leaves crowded near the end of the branches, about 10-jugate, impari-pinnate, leaflets 2½ to 5in. long, ¾ to 1¼in. broad, quite entire, often somewhat sharply acuminate, sometimes obtuseate, very seldom retuse, upper side deep the under pale green, oblique oblong-lanceolate; petiolules 2 to 3 lines long, narrow. Panicle about 7in. long, wide or at length divaricately branched. Pedicels about 1½ line long. Bracts minute, solitary, at the base of the pedicels, semi-lanceolate or linear-subulate. Sepals 5, orbicular, persistent, about 1 line long, imbricate, connate at the base. Petals deciduous, 1½ to 2 lines long, oblong-lanceolate, acute at the base, white, the margins slightly imbricate. Stamens 10; tube free, about 1 line long; teeth 10, very short, truncate or bifid at the end. Anthers scarcely 1½ line long, almost terminal, oblong-linear, somewhat acute, introrse, dehiscing longitudinally. Disk very short. Stigma globose or conical, ovate, red, style scarcely ⅓ line long, bifid, lobes cohering. Ovary 2-celled. Drupe spherical, 8 lines diameter, pericarp red outside, pulpy and white inside, putamen bony, smoothish. Seeds 2, rarely 1, erect, oval, basally attached; arillus none; albumen none; embryo upright or a little oblique; cotyledons dimidiato-ovate, amygdaline, greenish; radicle very short, almost globose.

Hab.: Southern coastal scrubs.

5. **O. reticulata** (reticulate), *F. v. M. in Hook. Kew Journ.* ix. 805; *Benth. Fl. Austr.* i. 386. A small tree, quite glabrous. Leaves often above 1ft. long, the common petiole angular or slightly dilated, terminating in a short point. Leaflets 4, 6, or 8, sessile, ovate or broadly ovate-lanceolate, obtuse, 4 to 8in. long, oblique at the base, coriaceous, smooth above, with very prominent pinnate veins and numerous raised reticulations underneath. Panicles loose, very divaricate, the branches often 6in. long or more. Flowers sessile, clustered. Sepals above 1 line long, orbicular. Petals twice as long. Staminal tube often divided to near the middle into 10 flat 2-lobed teeth or lobes. Ovary 2 or 3-celled. Fruit 1½in. diameter, the epicarp fleshy but not thick. Putamen hard and very rugose. —*O. xerocarpa,* F. v. M. Fragm. iii. 18.

Hab.: Islands of the Gulf of Carpentaria, *R. Brown, F. v. Mueller, Henne.*

XXXIII. MELIACEÆ.

9. CARAPA, Aubl.
(*Caraipe*, its name in South America).
(Xylocarpus, *Kœn.*)

Calyx small, 4 or 5-lobed. Petals 4 or 5, free, imbricate in the bud. Staminal tube urceolate, crenate or lobed; anthers 8 or 10, within the summit. Disk thick, surrounding the ovary. Ovary 4 to 5-celled, with 2 to 6 ovules in each cell; style short, with a large disk-like stigma. Capsule globular or ovoid, fleshy or woody, the dissepiments often disappearing. Seeds several in a compact mass round the remains of the central axis, large, thick, with a ventral hilum; testa spongy; cotyledons superposed, often united; radicle dorsal.—Maritime trees. Leaves pinnate with entire leaflets. Panicles axillary.

The species are few, ranging over the tropical seacoasts either of America and Africa or of Africa and Asia. The Australian one belongs to the latter category.—*Benth.*

1. **C. moluccensis** (also of the Moluccas), *Lam.*; *DC. Prod.* i. 626; *Benth. Fl. Austr.* i. 887. Cannon-ball tree. A tree, glabrous in all its parts. Leaflets 4, rarely 2 or 6, opposite, ovate, obtuse, shortly acuminate or rarely acute, 2 to 3 or rarely 4in. long, somewhat coriaceous, more reticulate than in any of the preceding genera. Panicles short, loose, and few-flowered, sometimes reduced to simple racemes or with few divaricate branches. Calyx small, irregularly lobed. Petals 4 or rarely 5, 2½ to 3 lines long. Staminal tube crenate or splitting into short lobes. Ovary very small, in the centre of a large thick depressed disk. Ovules 2, 3, or 4 in each cell, excessively minute. Fruit often 3 or 4in. diameter, irregularly globular. Seeds usually 4 to 6, large, irregularly shaped, closely packed; testa very thick, of a hard spongy consistence.—*Xylocarpus Granatum*, Kœn.; Willd. Spec. Pl. ii. 328.

Hab.: Common along the tropical coast.

Common on the seacoasts of tropical Asia, extending westward to E. Africa and eastward to the Moluccas. It varies considerably in the more compact or loose inflorescence, in the size of the flowers, and in the teeth of the staminal tube.—*Benth.*

Wood resembling Red Cedar, of close grain, prettily-marked; a good cabinet wood.—*Bailey's Cat. Ql. Woods No.* 68.

10. CEDRELA, Linn.
(From *Cedrus*, the Cedar-tree.)

Calyx small, 5-cleft. Petals 5, imbricate. Disk thick or raised. Stamens 4 to 6, inserted on the summit of the disk, alternating sometimes with as many staminodia, filaments subulate, anthers versatile. Ovary 5-celled, style filiform, with a disk-like stigma; ovules 8 to 12 in each cell, in 2 rows. Capsule membranous or coriaceous, 5-celled, opening in 5 valves, leaving the dissepiments attached to the persistent axis. Seeds flattened, winged; albumen scanty; cotyledons flat; radicle short, superior.—Tall trees, with coloured wood. Leaves pinnate. Flowers small, in large panicles.

The genus is spread over tropical America and Asia. The Australian species is a common Asiatic one.

1. **C. Toona** (Indian name), *Roxb. Pl. Corom.* iii. 83, t. 238; *Benth. Fl. Austr.* i. 887. Red Cedar. "Mamin" and "Mugurpul," Brisbane; "Boolboora," Forest Hill, *Macartney*; "Woota," Wide Bay; "Wanga," Herberton, *J. F. Bailey*. A tall, handsome tree, quite glabrous or the young shoots minutely pubescent. Leaves large, deciduous; leaflets 11 to 17, opposite or irregularly alternate, ovate-lanceolate, acuminate, 3 to 5in. long, oblique at the base, petiolulate, membranous. Panicles large, pyramidal, many-flowered, glabrous. Pedicels short. Sepals orbicular, ciliate, very small. Petals nearly 3 lines long.

Stamens 5, as long as the petals, inserted in cavities on the outside of the very thick pubescent disk. Ovary half-immersed in the disk. Capsule glabrous, oblong, 1 to 1½in. long.—*Wight, Ic. t. 161 ; C. australis, F. v. M. Fragm. i. 4.*

Hab.: Moreton Bay, *Herb. F. v. Mueller ;* Mackenzie's Station, *Leichhardt.*
The gum contains 45% of arabin, 36% of metarabin, and 19% of water, *Lauterer.* Mr. J. H. Maiden found a sample of gum from a New South Wales tree to contain : Arabin 68·8%, metarabin 6·8%, hygroscopic moisture 19·54%, and ash 5·16%.
Wood beautifully grained, of a red colour, easy to work, and very durable ; the common furniture wood, and in the north frequently used in house-building.—*Bailey's Cat. Ql. Woods No.* 69.

11. FLINDERSIA, R. Br.
(After Captain N. Flinders.)
(Oxylea, *A. Cunn.;* Strzeleckia, *F. v. M.*)

Calyx small, 5-lobed. Petals 5, imbricate in the bud, spreading. Disk broad, concave. Stamens 5, inserted on the outside of the disk, with as many or fewer staminodia alternating with them, sometimes wanting; filaments subulate; anthers versatile. Ovary 5-celled, 5-lobed; style short, thick, inserted between the lobes ; stigma capitate; ovules 4 to 6 in each cell. Capsule oblong, hard, tuberculate or muricate, opening septicidally in 5 boat-shaped valves or cocci, without any persistent axis. Seeds flat, winged, 2 or 3 on each side of a flat placenta, which almost divides each cell into two; albumen none ; cotyledons flat, radicle very short.—Trees. Leaves alternate or more frequently opposite, pinnate or rarely simple, marked with pellucid dots. Flowers in terminal panicles.

Besides Australia, this genus is found in New Guinea, Amboyna, and New Caledonia. The genus, although allied to *Cedrela* and therefore placed by common consent in *Meliaceæ,* is nevertheless, as observed by R. Brown, very closely connected with *Rutaceæ-Zanthoxyleæ,* and might be very well placed there next to *Geijera,* with which it is connected, especially through *F. maculosa.*—Benth.

Leaves alternate, pinnate, 1 to 3-jugate. Petals pubescent outside. Fruit 2 to 3in. long, the valves adherent at the base, not separating into free, boat-shaped pieces. Seeds usually winged at the top only . . . . 1. *F. australis.*
Leaves opposite, pinnate, 2 to 5-jugate. Petals glabrous or nearly so. Fruit 3 to 4in. long, separating when ripe into free, boat-shaped pieces. Seeds winged at both ends . . . . 2. *F. Oxleyana.*
Leaves opposite, pinnate, 2-jugate. Leaflets acute-acuminate at the apex. Petals glabrous. Ovary subglobose, 5-celled, hirsute. Ovules 4 in each cell ; 2-seriate . . . . 3. *F. Leichhardtii.*
Leaves opposite, pinnate. Leaflets 3 to 5, acute or obtuse at the apex. Petals glabrous or nearly so. Fruit 2 or 3in. long. Seeds winged at upper end only, or slightly also at the lower . . . . . 4. *F. Bennettiana.*
Leaves opposite, pinnate. Leaflets about 7. Petioles sharply angular. Fruit 3 or more inches long, oblong, prominently marked with musselshaped scars . . . . 5. *F. Chatawaiana.*
Leaves opposite, pinnate, 3 to 5-jugate. Petals glabrous inside, slightly pilose outside. Fruit about 2in. long, separating into free, boat-shaped pieces. Seeds winged at the top . . . . 6. *F. Ifflaiana.*
Leaves opposite or alternate, pinnate, deep-green. Leaflets 3 to 5, ovatelanceolate. Fruit oblong, muricate, the top often forming a 5-rayed star 7. *F. Mazlini.*
Leaves opposite, pinnate. Leaflets 3 to 9, or rather long petiolules. Petals red, glabrous inside, silky-pubescent on the back. Fruit acutely echinate ; cells 2-seeded . . . . 8. *F. Pimenteliana.*
Leaves opposite, pinnate. Leaflets 3 to 9, almost sessile. Petals glabrous inside. Fruit large, echinate. Seeds winged at both ends . . . 9. *F. Bourjotiana.*
Leaves opposite, pinnate, 2 to 6-jugate. Petals hairy on both sides. Fruit about 3in. long, separating into free, boat-shaped pieces containing 2 seeds each . . . . 10. *F. Brayleyana.*
Leaves opposite, pinnate, 4 to 6-jugate and sometimes terminal-pedicellate, one nearly or quite sessile. Petals hairy at the base on both sides. Fruit 4 or 5in. long, separating into free, boat-shaped pieces. Seeds winged at both ends . . . . 11. *F. Schottiana.*

Leaves opposite, pinnate, 6 or 7-jugate with sometimes a terminal one on a long petiolule, often denselypubescent like the branches. Petals woolly-hairy inside, and at the base outside. Fruit about 5½in. long, separating into free, boat-shaped pieces. Seeds winged at both ends . . . . . 12. *F. pubescens*.
Leaves opposite, pinnate. Leaflets 3 to 5, sessile, equilateral. Petals thickened in the centre, hairy on both sides. Fruit 1½ to 2in. long, separating into free valves . . . . 13. *F. collina*.
Leaves opposite, pinnate. Leaflets very irregular as to number, 3 to 6 or less, sessile, linear. Petals glabrous on both sides. Fruit about 1in. long, separating into free valves . . . . 14. *F. Strzeleckiana*.
Leaves opposite, simple, 1 to 2in. long, linear-oblong. Petals glabrous. Fruit about 1in. long . . . . 15. *F. maculosa*.

1. **F. australis** (Australian), *R. Br. in Flind. Voy.* ii. 595, *t.* 1; *Benth. Fl. Austr.* i. 888. Crow's Ash. A tree of moderate size, with a rugged bark. Branchlets striate, lepidote-puberulent. Leaves alternate, crowded at the end of short barren branches, glabrous; leaflets 3 to 6, broadly lanceolate or oblong-elliptical, obtuse or scarcely acuminate, 2 to 4in. long, scarcely oblique. Panicles much branched, terminating short branches without any leaves except a few scale-like bracts, sprinkled with a stellate tomentum. Flowers numerous. Calyx open, tomentose, with 5 short broad obtuse lobes, terminal or lateral. Petals about 2 lines long, tomentose outside, except a narrow border, slightly pubescent inside. Fruit almost woody, 2 or 3in. long, the valves not separating at the base. Seeds winged at the upper end only.

Hab.: Hill sides in many parts of southern Queensland.

Wood yellow, close-grained, very hard, and of great strength and durability; does not rust iron.—*Bailey's Cat. Ql. Woods No.* 70.

2. **F. Oxleyana** (after J. Oxley), *F. v. M. Fragm.* i. 65, iii. 25; *Benth. Fl. Austr.* i. 389. Yellow Wood. A tall, much-branched tree, attaining often 100ft. Leaves opposite, crowded under the panicles; leaflets 4 to 10, with or without a terminal odd one, broadly lanceolate, obtuse or shortly acuminate, 2 to 4in. long, oblique and almost falcate, narrowed into a distinct petiolule, glabrous or sprinkled underneath with minute stellate hairs, thinly coriaceous, rather sparingly glandular-dotted. Panicles loose and many-flowered, but shorter than the leaves. Sepals very small. Petals about 2 lines long, obovate-oblong, glab-rous or nearly so. Fruit woody, 3 to 4in. long, muricate. Seeds winged at both ends.—*Oxleya xanthoxyla*, A. Cunn. in Hook. Bot. Misc. i. 246, t. 54.

Hab.: Brisbane River, *Fraser, A. Cunningham, F. v. Mueller*.

Wood a bright-yellow, strong, and fibrous. Used in cabinet-work; not readily attacked by white ants; adapted for making handscrews and buggy shafts, &c.—*Bailey's Cat. Ql. Woods No.* 72.

3. **F. Leichhardtii,** *C. DC. Mono. Phan.* i. 781. Branchlets densely puberulent, greyish yellow, elenticellose. Leaves opposite, rather long petioles, impari-pinnate, 2-jugate; leaflets opposite, shortly petiolulate, sub-oblique oblong-elliptical, equal-sided and acute at the base, the apex acuminate, glabrous on both sides, membranous, subpellucid, pellucid-punctate, about 4in. long, about 1in. broad, veins thin. Panicles axillary, about as long as the leaves, thinly puberu-lent, peduncle about 1in. long. Flowers shortly pedicellate. Calyx of 5 round-ovate, ciliate sepals. Petals 5, glabrous, about 2 lines, membranous, elliptical, obtuse. Staminodia 5, filaments bearded at the top. Anthers scarcely ½-line long, reniform; apex obtuse. Disk membranous, entire, connate at the base with the filaments. Ovary subglobose, 5-celled, hirsute. Ovules in each cell 4, in 2 series. Style glabrous, shorter than the ovary.—From C. DC. Mono. Phan. i. 781.

Hab.: Moreton Bay, *Leichhardt* (in Herb. Mus. Par.), *C. DC. l.c.*

I have no knowledge of this tree, neither can I find it recorded in Baron Mueller's writings.

4. **F. Bennettiana** (after Dr. Geo. Bennett), *Benth. Fl. Austr.* i. 389; *F. v. M. Herb.* "Bogum-bogum." A large tree. Branchlets angular. Leaves opposite, the petioles somewhat flattened with thin, sharp edges, crowded under the panicles; leaflets 3 or 5, from ovate to ovate-lanceolate or oblong-elliptical, obtuse or scarcely acuminate, 2 to 3in. long in some specimens, 4 to 5in. and 2½in. broad in others, glabrous, very coriaceous, not oblique, and scarcely petiolulate, the common petiole angular. Panicles ample, sometimes short, sometimes exceeding the leaves, minutely stellate-pubescent. Petals about 2 lines long, rather broader than in *F. Oxleyana*, glabrous or nearly so. Fruit 2 or 3in. long, muricate. Seeds 2 in each cell, broadly winged at the upper end only, or some with a very small wing also at the lower end, but only seen in one capsule.—*F. australis*, F. v. M. Fragm. iii. 26, not of R. Brown.

Hab.: Wide Bay, *Bidwill;* Brisbane River, Moreton Bay, *A. Cunningham, Fraser, W. Hill.*

Wood hard, fine in grain, and light in colour.—*Bailey's Cat. Ql. Woods No. 74.*

5. **F. Chatawaiana** (after the Hon. J. V. Chataway, M.L.A.), *Bail.* "Narroo," Herberton district, and "Arrago," Tully River, *J. F. Bailey.* Red Beech or Cardwell Maple. A large tree with a trunk diameter of from 3ft. to over 4ft. Leaves opposite, petioles and rhachis more or less sharply angular; leaflets usually 7, oblong-falcate, obtuse or with a more or less acuminate obtuse point, sometimes very oblique at the base, 3 to 4½in. long, 1½ to 2½in. broad, under side somewhat pale; primary lateral nerves rather distant, 9 or 10 on each side of rhachis; petiolules slender, ½ to ¾in. long. Panicles rather large and spreading. No flowers available for examination. Fruit 3in. or more long, oblong, but tapering at each end, pentagonal, prominently marked with dark mussel-shaped scars, which gives to the fruit somewhat the appearance of a fir-cone. Seed winged all round, including wing about 2¾in. long, ½in. broad.

Hab.: Cardwell to Herberton, *J. F. Bailey.*

Wood of a lightish colour, grain cedar-like, often very prettily figured; used extensively in house and boat building.—*J. F. Bailey.*

6. **F. Ifflaiana** (after S. Iffla, M.D.), *F. v. M. Fragm.* x. 94. Cairns Hickory. A tall evergreen tree, the branchlets, petioles, and peduncles slightly puberulent. Leaves opposite on long petioles; leaflets thick-chartaceous, almost ovate, glabrous on both sides, 6 to 10, pari-pinnate, 2½ to 4½in. long, rhachis angular. Panicles terminal, peduncle somewhat short, branches numerous. Flowers very numerous. Calyx not 1 line long. Petals ovate, glabrous inside, outside thinly pilose, a little exceeding 1 line long. Stamens very short. Staminodia somewhat acute, as long as the fertile filaments; anthers pointless. Style very short, stigma almost sessile, peltate. Capsule about 2in. long, tuberculate with large and small rough tubercles, valves separating. Seed 4 or 5 lines, winged at the upper end only.

Hab.: Scrubs of the Barron River.

Wood of a yellowish colour, close-grained, very hard, of a greasy nature; useful for building purposes.—*Bailey's Cat. Ql. Woods No.* 74a.

7. **F. Mazlini** (after W. Mazlin), *Bail.* A large tree, stem diameter exceeding 2ft. Leaves glabrous, deep green, usually opposite, but here and there alternate; leaflets 3 or 5, ovate-lanceolate or oblong, with more or less acuminate blunt points, 2½ to 4in. long, 1½ to 1¾in. broad, thin-coriaceous, lateral-primary nerves thin, close, and rather dark-coloured. Petioles 1 to 2in. long, rhachis about the same length, petiolules 2 to 4 lines, all slender. Panicles of few branches, about as long as the leaves. No flowers available for describing. Fruit oblong, muricate, 2 to 3¼in. long, the protuberances very irregular as to

size, glossy, slightly tapering at the base, the valves protruding at the end of the fruit and forming a 5-rayed star. Seeds winged all round, often solitary on one side of the placenta and 2 or 3 on the other, the single one including wings 1¾in. long, when 2 or 3 about half that length.

Hab.: Evelyn, near Herberton, *J. F. Bailey.*
Wood light-coloured, but as yet not much used.—*J. F. Bailey.*

8. **F. Pimenteliana** (after J. M. de O. Pimentel), *F. v. M. Fragm.* ix. 132. An umbrageous tree of about 40ft., bark smooth. Branchlets, petioles, and peduncles slightly puberulent. Leaves pari or impari-pinnate, glabrous, opposite, 3 to 9-foliolate, the rhachis not winged. Leaflets on rather long petiolules, unequal-sided, lanceolate or falcate-oblong, or ovate, apex obtuse, long protracted, 3 to 4in. long, upper side deep green, under opique. Pellucid dots scattered. Panicles ample, on long peduncles, branches opposite. Calyx scarcely over ⅙ line long. Petals 1½ line long, deep red, glabrous inside, silky-pubescent on the back. The fertile filaments hispid about the apex; staminodia very short, somewhat acute. Anthers yellow; connective not excurrent. Style very short; stigma dilated. Fruit densely echinate; cells 2-seeded. Seeds winged at both ends. Young fruit only examined.

Hab.: Rockingham Bay, Mount Macallister, *J. Dallachy* (F. v. M., l.c.)

9. **F. Bourjotiana** (after Dr. A. Bourjot), *F. v. M. Fragm.* ix. 188. A large umbrageous tree; bark light-coloured; the branchlets, petioles, and panicles bearing small stellate hairs. Leaves scattered opposite, on long petioles, 3 to 9-foliolate. Leaflets chartaceous, almost sessile, opaque on both sides, epunctate, equilateral, ovate-lanceolate, the nerves and veinlets not prominent. Panicles divitiflorous, branches opposite except the upper ones, which are scattered. Calyx lobes semi-ovate or semi-orbicular. Petals not bearded on the inside. Fertile filaments pilose towards the top. Anthers pointless, staminodia broadish, a little acute. Style very short. Stigma peltate. Fruit large, echinate. Seeds winged at both ends.

Hab.: Ranges about Rockingham Bay, *J. Dallachy* (F. v. M., l.c.); also at other localities in the tropical parts of the colony.
Wood strong, durable, easily worked, light-coloured.—*Bailey's Cat. Ql. Woods No.* 73a.

10. **F. Brayleyana** (after E. W. Brayley), *F. v. M. Fragm.* v. 143, vi. 252. A large, umbrageous tree, bark of the trunk smooth, pale coloured. Branchlets glabrous, green, elenticellose. Leaves large, opposite on somewhat long petioles, pari-pinnate; rhachis not winged. Leaflets 2 to 6 petiolulate pairs, upper side subglancescent, almost ovate or oblique, ovate-lanceolate, oil-dots copious. About 6in. long and 2in. broad, veins indistinct. Panicles ample with opposite glabrous branches. Flowers fragrant, pedicellate. Calyx deeply 5-toothed, teeth glabrous, ovate. Petals 5, hairy on both sides, white, ovate, scarcely 1¼ line long. Stamens and staminodia 5 each. Anthers mucronulate. Ovary globose, 5-celled, style shorter than the ovary. Stigma orbicular, ovules in each cell 2, collateral. Fruit ellipsoid, 2½ to 3in. long, 1in. diameter, shortly tuberculose, carpels or cells 2-seeded, seeds winged at both ends.

Hab.: Several of the tropical scrubs.

11. **F. Schottiana** (after H. Schott), *F. v. M. Fragm.* iii. 25; *Benth. Fl. Austr.* i. 388. "Bunji Bunji," Herberton, *J. F. Bailey.* A tree of moderate size, or sometimes tall. Branchlets very thinly puberulent. Leaves opposite, crowded under the panicle; leaflets 8 to 12, with or without a terminal odd one, ovate-lanceolate, obtuse or acuminate, 4 to 5in. long, more or less falcate, sessile, with a broad very oblique base, somewhat

R

coriaceous, glabrous on both sides or softly pubescent underneath when young. Panicles terminal, ample and many-flowered, but not exceeding the leaves. Petals about 2 lines long, glabrous outside, sprinkled on both sides at the bottom as well as the anthers with a few hairs. Fruit echinate, 4 or 5in. long, separating into 5 boat-shaped valves. Seeds nearly 2in. long ; winged at each end.

Hab.: Wide Bay, *Bidwill;* Cumberland Islands, *Herb. F. v. Mueller;* Brisbane River, *A. Cunningham;* and many other southern and northern localities.

12. **F. pubescens** (pubescent), Bail. *Ql. Agri. Journ.* iii. part 3. A large umbrageous tree ; branches, especially those bearing leaves, closely covered with a velvety tomentum, stout, and strongly marked by the scars from the fallen leaves. Leaves opposite, impari-pinnate, or pari-pinnate from the abortion of terminal pinna, on stout petioles of about 5in., semiterete. Rhachis from 10 to 15in. long, and with the petioles clothed with a similar tomentum to the young branches. Leaflets 6 or 7 pairs and a terminal one, 5 to 9in. long, 1½ to 3in. broad, more or less rugose, lateral parallel nerves, 18 or 20 on either side of the midrib, glabrous, punctate and glossy above, more or less hairy on the under side, oblong-lanceolate or some oblong ; the lateral ones auriculate at the base, the terminal one cuneate at the base and on a petiolule of 1in., the lateral petiolules very short, the leaflets nearly sessile ; pellucid dots minute, only visible with the aid of a lens. Panicles terminal, widespread, and dense. Flowers pedicellate, numerous, white, and very fragrant. Calyx-lobes rotund, about ¼ line diameter, the margin laciniate-ciliate. Petals white, oblong-linear, about 2 lines long, 1 line broad, woolly-hairy on the face in the lower half. Filaments incurved, hairy at the top. Anthers greenish, ovate-apiculate. Disk orange, connate to the base of the filaments, undulately lobed. Stigma sessile, minute. Ovary hairy. Fruit echinate outside, about 5½in. long, dividing into 5 boat-shaped valves. Seeds about 2½in. long, winged at both ends.

This may be the tree alluded to in Fragm. v. 143 as *F. Schottiana,* var. *pubescens,* but even so it should, in my opinion, rank as a species.

Hab.: The above description refers to trees now (October, 1898) flowering on Wickham Terrace and other town reserves. The seed from which they were raised was obtained by Mr. Walter Hill, the late Colonial Botanist, from tropical Queensland. As a shade tree it would be difficult to find its superior. The first tree planted has borne fruit for the past few years, and Mr. Robt. McDowall (who has charge of the reserves) has carefully collected the seed, and raised plants which have been used for planting and distribution.

13. **F. collina** (found on hills), Bail. *Ql. Agr. Journ.* iii., part 3. A small tree, the bark falling off in rather larger patches than in allied species, but leaving the same pale-coloured patches upon the stem ; branchlets corrugated, and when young more or less covered with short ferruginous hairs. Leaves opposite, impari-pinnate, petioles about 1in. long, flattened, the edges thin but scarcely winged. Leaflets 3 to 5, obovate-cuneate, sessile, the terminal one sometimes 3in. long and ¾in. broad, the lateral ones smaller, equilateral ; apex obtuse, truncate, often emarginate, parallel lateral nerves erecto-patent, numerous, often very prominent on the upper side, which is very glossy, under side thinly hoary or pale-coloured. Rhachis slightly winged. Panicles terminal, about 2½in. long and nearly as broad (or in the Childers specimens 4 or 5in. long and broad), densely branched, usually on very short peduncles, more or less covered with a close stellate pubescence. Flower-buds globose, slightly 5-angled. Calyx small. Petals imbricate, ovate-oblong, thick in the centre, hairy on both sides. Filaments shorter than the petals, glabrous, rather thick and angular. Anthers ovate-cordate. Fruit echinate, oblong, 1½ to 2in. long, dividing tardily into 5 separate valves.

Hab.: Ranges southern parts of the colony.

Wood hard, close-grained, yellow, strong, and durable.—*Bailey's Cat. Ql. Woods No.* 730 (given as *F. Strzeleckiana,* var. *latifolia*).

14. F. Strzeleckiana (after P. E. de Strzelecki), *F. v. M. Fragm.* i. 65. Spotted Tree. A small tree, the bark falling off in thick, hard, scale-like pieces at irregular times, leaving light-coloured indented patches upon the stem, hence the name Spotted Tree; branchlets slender, more or less quadrangular, grey from a close covering of stellate scales. Leaves opposite, impari-pinnate, petioles about ½in. long, slightly winged. Leaflets 3 to 6, linear-oblong, usually less than 1in. long and 2 lines broad, the pairs usually distant, thick, veins obscure from the scaly indumentum. Rhachis narrowly winged. Panicles terminal, pyramidal, attaining 4in. in length. Branches opposite, short; peduncle about ½in. long. Flowers numerous, shortly pedicellate, buds globose. Calyx-lobes or sepals ovate, ciliate, carinate. Petals oblong, 1½ line long, membranous, glabrous on both sides. Filaments 5, glabrous, about half to three-quarters the length of the petals. Anthers glabrous, cordate. Disk prominent, glabrous, lobed, fleshy. Ovary densely hairy. Style short, corrugated; stigma brown, of moderate size. Fruit scarcely 1in. long, elliptical, almost glabrous, echinate-tuberculate, dividing into separate boat-shaped valves.

Hab.: Brigalow scrubs of the Leichhardt district and other inland parts.
Wood hard, close-grained, yellow, strong and durable.—*Bailey's Cat. Ql. Woods* No. 73b.
Dr. Lauterer found the gum to contain 83·5% of arabin and 16·5 of water, and states that for most purposes it is as good as gum arabic.

15. F. maculosa (stem spotted), *F. v. M. in Journ. Pharm. Soc. Vict.* ii. 44. Leopard Tree. A small tree, bark of trunk falling off in scale-like pieces, giving a spotted appearance to the stem, hence called "Leopard Wood" and "Spotted Tree." Branchlets slender, terete, bark wrinkled, when young clothed with minute silvery scales and stellate hairs. Leaves simple, opposite, linear oblong 1½ to 2in. long, 6 or 7 lines broad, obtuse at the apex, tapering at the base to a petiole of a few lines long, closely scaly on the under side, nearly glabrous on the upper side except the midrib; parallel lateral nerves erecto-patent, numerous, plainly visible on the upper side, invisible on the under side from the close indumentum. Panicles terminal, pyramidal, seldom much exceeding 1¼in., on short peduncles. Flowers globose in the bud, very shortly pedicellate. Calyx of 5 rotundate-ovate ciliate sepals. Petals 5, glabrous, about 2 lines long, imbricate, obovate, membranous, stamens glabrous. Anthers cordate-acuminate. Disk membranous, serrulate, glabrous, connate with the base of the filaments. Ovary hairy, globose, 5-celled. Ovules 2, collateral in each cell. Stigma almost or quite sessile, discoid, 5-angular. Fruit elliptical, oblong, 1in. or more long, densely muricate. Seeds winged at both ends.—*Elæodendron maculosum*, Lindl. in Mitch. Trop. Austr. 384.

Hab.: St. George, *Jos. Wedd.*
This small tree somewhat resembles *Geijera parviflora*, Lindl., but the foliage is lighter-coloured.
Wood bright-yellow, nicely marked, grain close, very hard; suitable for bearings of shafting. *Bailey's Cat. Ql. Woods* No. 73.
The gum, which contains 81·4% of arabin and 18·4 of water, may be used as a substitute for gum arabic.—*Lauterer*.

Order XXXIV. OLACINEÆ.

Flowers regular, hermaphrodite or rarely unisexual. Calyx small, 4 or 5, rarely 6-toothed, free or adnate to the disk (in *Cansjera* scarcely distinguishable from the corolla). Petals 4, 5, or rarely 6, free or united in a campanulate or tubular corolla, valvate in the bud (except *Villaresia*). Stamens as many or twice as many as petals or rarely fewer, adnate to the base of the petals, or free and hypogynous; anthers 2-celled, versatile, or rarely adnate. Disk free, or adnate to the ovary or to the calyx, or divided into scale-like glands. Ovary free or immersed in the disk, 1-celled or imperfectly 2 or 3-celled; style simple; stigma entire or lobed. Ovules 2, 3, or rarely 1, pendulous from a central placenta into the imperfect cells, or from the side or apex of the cavity. Fruit usually an indehiscent drupe, either superior or inferior by the growth over it of the disk and tube of the calyx. Seed solitary, pendulous, or sometimes, owing to the adnate nerve-like remains of the placenta, apparently erect; testa very thinly membranous; embryo very small in the apex of a fleshy albumen, or larger and axile; or, in a genus not Australian, occupying the whole seed without albumen; cotyledons flat or terete; radicle superior.—Trees, shrubs, or climbers. Leaves usually alternate, entire, penninerved, without stipules. Flowers few and axillary, or rarely in terminal panicles, usually small.

The Order is widely dispersed over the tropical and subtropical regions of the globe. The ten Australian genera are none of them endemic, one extending to New Zealand, four to tropical Asia, three to tropical Asia and Africa, one to tropical Asia and America, and one is common to Asia, Africa, and America. The Order is more nearly allied to *Loranthaceæ* among *Calyciflorœ*, and especially to *Santalaceæ* among *Monochlamydeæ*, than to any (except *Ilicineæ*) of the *Disciflorœ*, among which it is technically placed.—*Benth.*

TRIBE I. **Olacineæ.**—*Stamens twice as many as petals or fewer, or if the same number as petals opposite to them. Ovary often 2 or 3-celled at the base, 1-celled at least at the top; placenta central, with 2 or 3 pendulous ovules.*

Calyx not enlarged after flowering. Stamens twice as many as petals; anthers oblong or linear . . . . 1. XIMENIA.
Calyx enlarged and enclosing the fruit. Stamens 3; staminodia (in the Australian species) 5; anthers short . . . . 2. OLAX.

TRIBE II. **Opilieæ.**—*Stamens as many as petals and opposite to them. Ovary 1-celled, with 1 ovule.*

Perianth apparently simple, shortly 4-lobed. Stamens 4, included, alternating with 4 glands or scales . . . . 3. CANSJERA.
Calyx minute. Petals 5, free. Stamens 5, exserted, alternating with 5 scales . . . . 4. OPILIA.

TRIBE III. **Icacineæ.**—*Stamens as many as petals and alternate with them. Ovary 1-celled, with 1 or 2 pendulous ovules.*

Filaments glabrous or shortly pilose, thick, dilated above, hollowed in front to receive the anthers. Anthers pendulous. Ovules 2. Cymes axillary or lateral . . . . 5. GOMPHANDRA.
Petals glabrous within. Filaments glabrous. Anthers innate, sagittate at the base. Ovary oblique . . . . 6. APODYTES.
Petals strictly valvate. Ovule 1, the placenta not prominent. Flowers in a much-branched corymbose panicle . . . . 7. PENNANTIA.
Calyx deeply 5-parted, lobes imbricate. Flowers unisexual. Ovary 1-celled. Style acute. Albumen much-lobed . . . . 8. PHLEBOCALYMNA.
Petals slightly imbricate. Ovules 2, the placenta forming a half-dissepiment on one side of the cavity. Flowers in a narrow raceme-like panicle 9. VILLARESIA.
Genus of doubtful affinity. Milky-juiced climber. Sepals and petals imbricate. Fruit dry-winged . . . . 10. CARNIOPTERIS.

1. XIMENIA, Linn.
(After F. Ximenes.)

Calyx minutely 4 or 5-toothed, not enlarged after flowering. Petals 4 or 5, bearded inside, valvate in the bud. Stamens twice as many as petals, free; filaments filiform; anthers linear, erect. Ovary 3-celled at the base; stigma capitate; ovules 3, descending into the incomplete cells from a central placenta. Drupe ovoid or globular, with a thick sarcocarp. Seed spuriously erect; embryo minute.—Shrubs or trees, often thorny. Flowers white, rather large for the Order, in small axillary cymes or solitary.

The Australian species is spread over almost all tropical countries, the few other species are American or African.—*Benth.*

1. **X. americana** (American), *Linn.*, *DC. Prod.* i. 533; *Benth. Fl. Austr.* i. 391. "Gotoobah," Bellenden Ker. (*Mrs. Gribble*). A glabrous shrub, or sometimes a small tree, with spreading branches, often armed with axillary spines (abortive peduncles). Leaves petiolate, ovate, obtuse, or scarcely acute, 1 to 2in. long, entire, the veins inconspicuous, except the midrib. Peduncles short, bearing little cymes of 3 to 7 yellowish sweet-scented flowers, rarely reduced to a single one. Calyx minute, 4-toothed. Petals 4, recurved, 8 to 4 lines long, densely bearded inside with long white hairs. Stamens 8, filaments crocked, anthers large. Drupe yellow, attaining 1in. diameter or rather more.—*X. elliptica*, Forst.; Labill. Sert. Austr. Caled. 34, t. 37; *X. laurina*, Delile, in Ann. Sc. Nat. ser. 2; xx. 89; *X. exarmata*, F. v. M. in Trans. Phil. Inst. Vict. iii. 22.

The species is widely spread over the tropical regions of both the New and the Old World, varying in most places with or without thorns. The Pacific and New Caledonian *X. elliptica* has been distinguished from the common form as having a globular, not elliptical fruit; but some of Gardner's specimens from Brazil have certainly also the fruit globular. The Australian specimens, like the majority of those in our herbaria, are without fruit; they are unarmed, or have only small nascent spines in the axils of some of the young leaves.—*Benth.*

Hab.: Clermont, *H. Salmon*; Hammond Island, where it is known as Yellow Plum, *Mrs. Smyth*; Somerset, *F. L. Jardine.*

Wood close-grained, tough, hard, and light in colour. It works like English Box, and might be suitable for engraving.—*Bailey's Cat. Ql. Woods No.* 74d.

Fruit eaten by natives, *Mrs. Gribble.*

2. OLAX, Linn.

(Flowers furrowed or imbricated.)

(Spermaxyrum, *Labill.*)

Calyx small, cup-shaped, truncate, enlarged after flowering and enclosing the fruit. Petals 5 or 6, free, or slightly cohering, valvate in the bud. Stamens usually 3, alternate with the petals, the filaments adnate to the petals and connecting them in pairs; staminodia as many as petals and opposite to them, filiform or flat, entire or 2-cleft. Ovary free, 1-celled, or very shortly 3-celled at the base; stigma entire or slightly 3-lobed; ovules 3, pendulous from a central placenta. Drupe globular or oblong, enclosed in the enlarged calyx, but free from it, the sarcocarp thin. Seed spuriously erect; embryo very small in the apex of a fleshy albumen.—Trees, shrubs, or undershrubs, rarely half climbing, the Australian species all erect shrubs, with small alternate, entire, distichous leaves, the veins inconspicuous, except the midrib. Flowers axillary, solitary in the Australian species, several in short racemes or spikes in some others.

The genus is confined to the Old World, extending over tropical Asia and Africa. The Australian species are all endemic, and differ from all except the E. Indian *O. nana*, Wall., in their solitary axillary flowers and small leaves. They have all 5 petals, 3 stamens, and 5 staminodia.—*Benth.*

Staminodia 2-cleft to the middle. Leaves rather thin, narrow, retuse (Eastern species) . . . . 1. *O. retusa.*
Staminodia undivided.
Leaves narrow-oblong, mucronate. Staminodia linear, bearded at the base . 2. *O. stricta.*
Leaves reduced to minute scales. Flowers densely bearded inside. Staminodia linear . . . . 3. *O. aphylla.*

1. **O. retusa** (referring to the blunt end of the leaf), *F. v. M. Herb.* as a var. of *O. stricta*; *Benth. Fl. Austr.* i. 392. A glabrous shrub, with the slender virgate branches of *O. stricta*. Leaves linear-cuneate or narrow-oblong, truncate and emarginate, or almost 2-lobed, minutely mucronate, rarely exceeding ½in. and smaller on the lateral branches, rounded at the base. Pedicels very short. Flowers about 2 lines long. Filaments glabrous, dilated at the base; staminodia bearded below the middle, glabrous above and divided into 2 linear lobes. Fruit ovoid-oblong, not exceeding 3 lines in the specimens seen.

Hab.: Islands of Moreton Bay, Stanthorpe.

2. **O. stricta** (upright), *R. Br. Prod.* 358; *Benth. Fl. Austr.* i. 392. An erect, glabrous shrub, of 2 or 3ft., with slender virgate branches. Leaves narrow-oblong or linear, acute or obtuse, but always mucronate, ¼ to ½in. or rarely ¾in. long, flat, with a prominent midrib, narrowed or rarely rounded at the base. Pedicels scarcely 1 line long. Petals varying from 2 to 3 lines. Filaments flattened to very near the anthers, glabrous; staminodia linear, entire, more or less bearded below the middle. Fruit obovoid-oblong, often 4 lines long or rather more.

Hab.: Edges of lagoons, Moreton Island.

3. **O. aphylla** (leafless), *R. Br. Prod.* 358; *Benth. Fl. Austr.* i. 393. A shrub of several feet, with numerous wiry, virgate, slightly pubescent branches. Leaves all reduced to minute scales. Flowers very small, almost sessile in the axils of orbicular ciliate bracts rather longer than the calyx, towards the ends of the branches. Petals scarcely more than 1 line long, densely bearded inside about the middle. Staminodia linear and entire, or slightly spathulate and emarginate at the top. Fruit ovoid, about 2 lines long.

Hab.: Gulf of Carpentaria.

3. CANSJERA, Juss.

(From *Tsjerou-Cansjeram*, Malabar name.)

Perianth apparently simple, the calyx very minute and often not distinguishable, at the base of the tubular or urceolate 4-lobed corolla. Stamens 4, opposite to the petals or corolla-lobes, and more or less adherent at the base; filaments filiform; anthers small. Hypogynous scales (or lobes of the disk) 4, alternating with the stamens. Ovary small, fleshy; ovule 1, apparently erect or suspended from a short placenta in the centre of the minute cavity. Drupe with a thin sarcocarp. Seed erect; embryo small or sometimes elongated.—Weak or climbing shrubs. Leaves alternate, entire. Flowers small, in short axillary spikes.

Besides the Australian species, which is also in New Ireland, the genus comprises 2 or perhaps 3 from tropical Asia.—*Benth.*

1. **C. leptostachya** (slender spikes of flowers), *Benth. in Hook. Lond. Journ.* 281, *and Fl. Austr.* i. 394. A rambling shrub, never very tall, glabrous or the young shoots very minutely tomentose. Leaves ovate-lanceolate, long-acuminate, 2 to 3in. long, membranous, glabrous. Spikes 1 or 2 together in the axils, rarely exceeding ½in. Flowers in the young bud strigose-pubescent, sessile in the axils of narrow minute bracts which soon fall off, when fully open about 1 line long, nearly globular and glabrous, the lobes very short and spreading. Filaments slender, but shorter than the perianth. Hypogynous scales short, broad, entire or rarely 3-toothed. Fruit yellow, oval, 1in. long, taste not unpleasant.

Hab.: Cape York and islands off the N.E. coast, *A. Cunningham, M'Gillivray.*

The species is also in New Ireland. The flowers are about half the size of those of the common *C. Rheedii*, Gmel., and I have not succeeded in detaching the calyx from the corolla, as I have readily done in Malacca specimens of *C. Rheedii* or of an allied species.—*Benth.*

4. OPILIA, Roxb.

(Not explained by author.)

Calyx minute, 5 or rarely 4-toothed. Petals 5, rarely 4, hypogynous, valvate in the bud. Stamens as many, alternating with the petals, free; filaments filiform; anthers ovate. Disk of 5, rarely 4 scales, alternating with the stamens. Ovary 1-celled, tapering into a short thick truncate style; ovule solitary, suspended from a central filiform placenta very early adnate to it. Drupe with a thin sarcocarp and crustaceous endocarp. Seeds spuriously erect; embryo linear, short, or nearly as long as the albumen.—Shrubs or small trees, sometimes climbing. Leaves alternate, entire. Flowers in axillary racemes; pedicels 3 together in the axils of peltate bracts, which are imbricate at an early stage but fall off before the flowers expand.

A genus of 2 or perhaps 3 species, natives of tropical Asia and Africa, the Australian species one of the widest dispersed.—*Benth.*

1. **O. amentacea** (inflorescence in catkins), *Roxb. Pl. Coron.* ii. 31, t. 158; *Benth. Fl. Austr.* i. 394. A scrambling, half-climbing shrub or small weak tree, glabrous or the young leaves and shoots minutely tomentose-pubescent. Leaves petiolate, ovate, ovate-lanceolate, or almost oblong, acute or acuminate, 2 to 3 or even 4in. long, or rarely shorter and very obtuse, entire, thinly coriaceous, the veins usually prominent though fine. Racemes before flowering resembling little cylindrical cones of $\frac{1}{2}$in., the peltate imbricate but almost squarrose bracts alone visible, when in flower slender, about 1in. long, without bracts. Flowers very small, on filiform pedicels of about 1 line. Petals about $\frac{1}{2}$ line long, very deciduous. Drupe ovoid or globular, $\frac{1}{2}$ to $\frac{3}{4}$in. long, or about the size of a pigeon's egg. Pericarp white, flesh edible. Embryo linear, nearly as long as the albumen.— Wight Illustr. t. 40; *O. javanica.*

Hab.: Rockingham Bay.

5. GOMPHANDRA, Wall.
(Swollen stamens—anthers.)

Calyx minute, cup-shaped, 4 or 5-lobed. Corolla campanulate, 4 or 5-lobed, lobes acuminate, inflexed, rarely free, midrib prominent within. Stamens 5, hypogynous, alternate with the petals, filaments thick, dilated above, hairy at the back (in most species), hollowed in front to receive the anthers. Anthers pendulous from the filiform apex of the filament, 2-lobed, dehiscing lengthwise; pollen-grains triangular. Hypogynous disk thick, angular or none. Ovary sterile in the male, oblong in the female flower, 1-celled; style conic, stigma minute, or style crowned by a stigmatiferous disk; ovules 2, collateral, pendulous, funicle dilated into an "obturator." Fruit drupaceous, surmounted by the remains of the disk (stigma?), stone crustaceous. Seed pendulous, surrounded by the raphe, albumen fleshy, bipartite; embryo minute. Tree with alternate leaves, simple 1-nerved and petiolated. Flowers in axillary, terminal, or leaf-opposed cymes; dichlamydeous, hermaphrodite or polygamo-diœcious.—Hook. Fl. Brit. Ind. i. 585.

A genus probably of few species. Tropical Asia.

Leaves often 8in. long, 4in. broad near the rounded base . . . . 1. *G. australiana.*
Leaves about 5in. long, 2in. broad near the centre, base cuneate . . . . . 2. *G. polymorpha.*

1. **G. australiana** (Australian), *F. v. M. Fragm.* vi. 3 and 258 ix. 150. A tall tree, bark pale-coloured. Leaves glabrous, 6in. or more long and 2 or 4in. broad, broadly ovate, tapering from about the centre to an acuminate apex, very minutely pellucidly punctate, the under side paler than the upper; lateral nerves prominent, rather distant, texture chartaceous. Peduncles axillary, cymes rather few-flowered. Corolla tubular, about 2 lines long, toothed or lobed at the top, glabrous. Stamens 5, a little exceeding the corolla. Filaments somewhat thick, attenuated at the base. Anthers ovate, scarcely $\frac{1}{4}$ line long, introrse. Ovary glabrous. Fruit ellipsoid, 1in. long. Pericarp yellowish; endocarp cartilaginous-crustaceous, the outside striate with thick longitudinal nerves. Radicle superior, tenni-cylindric. Cotyledons rotundo-ovate.

Hab.: Scrubs of tropical parts of the colony.

2. **G. polymorpha** (many-formed), *Wight;* var. 6 *Bail. Bot. Bull.* viii.
A handsome tree about 60ft. in height, the inflorescence and young shoots puberulent, otherwise glabrous. Leaves alternate, 3 to 5in. long, ovate-lanceolate, with a more or less elongated blunt apex and cuneate base, petioles about 4 or 5 lines long, dark-green on the upper, pale on the under surface; the primary veins distant, only 3 or 4 on each side of the midrib, reticulate veinlets obscure, margins entire. Flowers in short axillary, dichotomous cymes. Calyx very short and cup-shaped, with almost entire edge. Corolla-tube about 2 lines long with minute teeth, the tips inflexed, from which proceeds down the inside of the corolla-tube a prominent rib. Stamens glabrous, 5, at length exserted and widely spreading, but incurved again near the anthers. Ovary glabrous, 4-angled, stigma sessile. Fruit not as yet been gathered.

I cannot separate this plant from the East Indian species, of which several (5) varieties are named and described in Hook. Fl. of Brit. Ind. i. 586. From the fragmentary specimens which I have of *G. australiana*, F. v. M., the present plant seems to differ considerably both in foliage and inflorescence.

Hab.: Scrubs of the Barron River, *E. Cowley.*

Wood of a light colour, nicely marked.

6. APODYTES, E. Meyer.

(Divested of flowers.)

Calyx minute, cup-shaped, 5-toothed. Petals 5, free, valvate. Stamens 5, alternate with the petals, filaments dilated; anthers long or short, oblong, basifixed, sagittate, 2-lobed, dehiscing longitudinally; pollen triangular. Ovary 1-celled, obliquely gibbous; style excentric, curved, stigma small; ovules 2, pendulous, superposed. Drupe obliquely ovoid, compressed, stone crustaceous. Seed pendulous; embryo small in the apex of fleshy albumen, cotyledons narrow. Trees or shrubs. Leaves alternate, petiolate simple, coriaceous, usually black in drying. Flowers small in terminal or axillary corymbose cymes.—Hook. Fl. Brit. Ind.

The species, besides the Queensland one, are natives of tropical Asia and Africa.

1. **A. brachystylis** (short styled), *F. v. M. Fragm.* ix. 149, *and 2nd Add. 3rd Supp. Syn. Ql. Flora at end of Cat. Ind. and Nat. Pl. of Ql.* p. 107. Tree of medium size. Leaves alternate, lanceolate or oblanceolate, 3 or 4in. long, 1 to 1¾in. broad, tapering to a petiole of 2 to 4 lines, both sides more or less glossy, primary veins few and distant, the small reticulation very close and wavy. Inflorescence terminal, axillary, or lateral, in cymes or racemes about 1in. long, clothed with short appressed light-brown hairs. Calyx cup-shaped, with 5 minute teeth. Petals 5, valvate, inflexed at the point. Stamens 5; filaments flat, broadening upwards, glabrous except at the top, where the anther is surrounded with a dense mass of scale-like hairs, like in *Lasianthera*. Style very short, slightly curved; stigma lateral; ovary large, glabrous. Fruit bluish-green and glossy, about 1¼in. long, 6 to 8 lines broad, flat, emarginate, with prominent midrib and 3 branching prominent veins on either side, on the outer side the midrib in the fresh fruit expands into a soft fleshy gibbosity of a dirty-white or light-brown colour, in the dried specimens this part dries to a rather loose roll of skin over the midrib or suture of the fruit. Seeds flat, lanceolate, three-quarters the length of the fruit.

Hab.: Rockingham Bay, *J. Dallachy* (F. v. M.); Johnstone River, *Dr. T. L. Bancroft;* Herberton District, *J. F. Bailey.*

This is in all probability the plant described by Baron Mueller, l.c., he only having flowering specimens. The flower I examined was only a single bud, in which the stamens were probably only staminodia, for I found no anthers. Thus the flowers may be polygamous.

Wood of a yellow colour, hard, and considered useful for many purposes.—*J. F. Bailey.*

XXXIV. OLACINEÆ.

7. PENNANTIA, Forst.
(After Thomas Pennant.)

Flowers diœcious or polygamous. Calyx minute. Petals 5, hypogynous, glabrous, valvate in the bud. Stamens 5, alternating with the petals; anthers oblong-sagittate. Ovary 1-celled; stigma nearly sessile, entire or 3-lobed; ovule solitary, suspended from the apex of the cavity. Drupe with a hard putamen, or almost baccate with a slightly coriaceous endocarp. Seed pendulous; embryo small within the apex of the fleshy albumen.—Trees. Leaves thinly coriaceous, entire or (in New Zealand species) coarsely toothed. Flowers in terminal corymbose panicles.

Besides the Australian species, which is endemic, there is one from Norfolk Island and another from New Zealand.

1. P. Cunninghamii (after A. Cunningham), *Miers, in Ann. Nat. Hist. ser.* 2. ix. 491, *and Contrib.* 80, *t.* 12; *Benth. Fl. Austr.* i. 395. A glabrous, sub-erect, tall shrub. Leaves ovate or broadly elliptical, acuminate, 4 to 6in. long, entire, coriaceous, and shining when old, narrowed into a petiole of ½in. or more. Flowers numerous, in broad rather dense panicles, either terminal or in the upper axils, the males only known. Calyx scarcely prominent. Petals nearly 1½ line long. Filaments bent in below the summit in the bud; anthers oblong, sagittate. Rudimentary ovary narrow, with 2 or three erect style-like lobes, and occasionally containing an imperfect pendulous ovule. Drupes or berries ovoid, about ½in. long, the endocarp scarcely hardened. Seed pendulous; testa thinly membranous; embryo much shorter than the albumen.

Hab.: Mt. Mistake. Specimens in early fruit only.

The ovaries described by Miers appear to me to have been imperfect, at least I find none but male flowers in the specimen he examined, nor in any others I have seen. It is probable that the female flowers, as in the New Zealand species, are smaller, and have therefore not attracted the notice of collectors.—*Benth.*

8. PHLEBOCALYMNA, Griff.
(Veiny involucre.)

Sepals 5, distinct, imbricate. Corolla tubular, limb 5-parted. Stamens 5, alternate with the petals, adherent to the tube of the corolla. Anthers ovoid-oblong, dorsifixed, 2-lobed, dehiscing longitudinally. Disk fleshy, hypogynous, 5-lobed, lobes opposite the petals. Ovary conic, 1-celled. Style subulate. Ovules 2, pendulous. Fruit oblong, with a crustaceous rind; seeds pendulous. Albumen coriaceous, lobulate, ruminate. Embryo minute.—Trees. Leaves alternate, petiolate, simple, coriaceous. Flowers polygamous; males in globose heads, borne on short spikes; females shortly pedicellate.

Besides the Queensland plant, the species are found in the Malay Peninsula and Islands.—*Hooker's Flora of British India.*

1. P. lobospora (seeds lobed), *F. v. M. Fragm.* ix. 151. A small glabrous tree. Leaves thin, coriaceous, lanceolate-ovate, pale beneath, sparsely or almost verticillately crowded, 2 to 4in. long, ⅔ to 1¼in. broad, thinly and very widely penninerved, distantly and minutely glandular denticulate. Inflorescence terminal; peduncles short or very short, somewhat thick at the ends. Pedicels to the fruit 2 lines or less. Fruit 8 to 10 lines long, almost truncate, a little protracted or apiculate, 1 or 2-seeded. Seeds deeply lobed. Albumen amygdaloid, testa membranous, shiny brown. Embryo ovate, scarcely ¼ line long.

Hab.: About Rockingham Bay, *J. Dallachy* (F v. M., l.c.)

XXXIV. OLACINEÆ.

9. VILLARESIA, Ruiz and Pav.
(After Mattias Villarez, a Spanish botanist.)
(Pleuropetalum, *Blume;* Chariessa, *Miq.*)

Flowers hermaphrodite or polygamous. Sepals 5, distinct, broad, imbricate. Petals 5, with the midrib prominent inside, imbricate or almost valvate in the bud. Stamens 5, alternating with the petals; anthers cordate. Ovary 1-celled, the cavity marked on one side with a raised ridge half dividing it; style short, thick; ovules 2, suspended from the summit of the raised ridge. Drupe ovoid or globular, the endocarp forming a prominent half-dissepiment which penetrates into a deep vertical furrow in the seed. Embryo small, in the apex of the albumen.—Lofty trees (or tall woody climbers?) Leaves alternate, coriaceous, entire or toothed. Flowers in small cymes, along the simple rhachis of a raceme-like panicle.

Besides the Australian species, of which one is and the other may be endemic, there is one (perhaps not really different) from the Indian Archipelago, one from the S. Pacific Islands, and several from S. America. The genus is exceptional in *Olacineæ* by the more or less imbricate petals. I have not seen the 2 cells to the ovary which Miers met with in one species, possibly in accidentally abnormal flowers.—*Benth.* (for the most part).

A large tree. Petals about 1 line long . . . . 1. *V. Moorei.*
Small tree or shrub. Petals 2 or 3 lines long . . . . 2. *V. Smythii.*

1. **V. Moorei** (after C. Moore), *F. v. M. Herb.; Benth. Fl. Austr.* i. 396. A lofty handsome tree, glabrous except the inflorescence. Leaves ovate-lanceolate or oblong, acuminate, 3 to 4in. long, entire, narrowed into a short petiole, coriaceous and shining, but not so thick as in the American species. Raceme-like panicles irregularly lateral or axillary, 2 to 4in. long, hoary with a minute pubescence. Cymes numerous, few-flowered, on short peduncles along the rhachis. Flowers almost sessile in the cymes, those seen all males. Petals 1 line long, very slightly imbricate. Drupes globular, the putamen hard, about ¾in. diameter, rugose outside, the half-dissepiment projecting quite to the centre of the cavity and there slightly thickened, forming a column, up the centre of which the placenta appears to pass, as if the endocarp had grown over it as in the New Zealand *Pennantia.* Seed quite enclosing the half-dissepiment, its transverse section being horseshoe-shaped.

Hab.: Bunya Mountains.

Wood of a light-grey colour, close-grained, and prettily marked; useful both for the cabinet-maker and joiner.—*Bailey's Cat. Ql. Woods No.* 74a.

The Javanese *V. suaveolens* (*Pleuropetalum suaveolens*, Blume) is unknown to me, but must, from the character given, be nearly allied to this species. *V. Samoensis* (*Pleuropetalum Samoense,* A. Gr.) which we have also from the Fiji Islands, appears to be quite distinct.—*Benth.*

2. **V. Smythii** (after R. B. Smyth), *F. v. M. Fragm.* v. 156, ix. 150. Sometimes attaining the size of a small tree. Branchlets thinly pubescent. Leaves 3 to 5in. long, 1½ to 2½in. broad, the nerves thinly pubescent on the under side, ovate, acuminate, membranous, very thinly reticulate veined, quite entire. Panicles racemose, the branches from 1 to 4in. long. Sepals broad-ovate, ⅔ line long. Petals white or slightly yellowish, oblong-cuneate, 2½ to 3 lines long. Filaments glabrous. Anthers cordate-ovate, ⅓ line long. Style setaceous, a little exceeding 1 line long, with the ovary also glabrous. Fruit unknown.

Hab.: Rockingham Bay, *J. Dallachy* (F. v. M., l.c.); Johnstone and Daintree Rivers, *W. Hill* (F. v. M., l.c.)

10. CARDIOPTERIS, Wall.
(Fruit heart-shaped.)

Calyx 4 or 5-parted; lobes imbricate, persistent, but not or only slightly accrescent. Corolla deciduous, between rotate and funnel-shaped, 4 or 5-lobed; stamens 4 or 5, inserted on the base of the tube of the corolla, alternate with its lobes;

filaments short, glabrous; anthers 2-celled, introrse, dehiscing longitudinally; pollen-grains 4-angular. Ovary free, surrounded at the base by a thick fleshy annular disk, oblong, compressed, 1-celled; ovule 1 (rarely 2), pendulous, naked, micropyle ultimately superior. Style 2-branched, one branch deciduous, curved, capitate at the apex; the other accrescent, ultimately deciduous, divided at the apex into 2 unequal, ovate, rather obtuse divisions. Fruit ovate-orbicular, emarginate or obcordate, compressed, very broadly winged, 1-celled, indehiscent. Seed solitary, linear, furrowed; embryo minute, in hard fleshy albumen.—A climbing herb with milky juice. Leaves alternate, long-petioled, simple or lobed, cordate, palminerved. Flowers ebracteate, in axillary racemose or paniculate cymes, bisexual, dichlamydeous.—Masters in Hook. Fl. Brit. Ind.

This genus is placed in Hooker's Flora of British India in *Olacineæ*, but as a genus of doubtful affinity.

1. **C. lobata** (lobed), *R. Br.* Stem terete, striate. Leaves 3 to 5in. by 3 to 4½in., glabrous, membranous, polymorphous, usually more or less angular and slightly lobed; base 7 to 9-nerved, cordate; lobes acute or acuminate, widely divergent; petiole 3 to 5in. Peduncles 2½ to 4in., solitary, axillary, dichotomous; pedicels puberulous, erect, ultimately spreading or recurved. Flowers rather crowded, secund, ebracteate. Calyx puberulous. Corolla slightly exceeding the calyx, whitish, deciduous. Fruit 1 to 1½in. by 1¼in. The plant varies much in the consistence and form of the leaves, occasionally even on the same specimen; hence, by some writers, several species have been proposed.—Hook. l.c.

Hab.: Barron River, *E. Cowley*, 1892.

The Australian form is that known as var. *moluccana* (*C. moluccana*, Blume, Rumph. iii. 207 f. 2). The plant has some resemblance to a yam, *Dioscorea*; and by some has been mistaken for a species of that genus.

Order XXXV. ILICINEÆ.

Flowers regular, hermaphrodite or unisexual. Calyx of 4 or 5, rarely 3 or more than 5 sepals, imbricate, usually persistent. Petals 4 or 5 or rarely more, hypogynous, imbricate in the bud, sometimes united in a lobed corolla. Stamens of the same number as petals, hypogynous, free or adhering to the corolla at the base; anthers 2-celled, opening inwards. Disk none, except the thickened base of the ovary. Ovary free, 3 to 5-celled, rarely many-celled; stigma broad or capitate, sessile or supported on a distinct style. Ovules 1 or 2 in each cell, pendulous, with a superior micropyle. Fruit a drupe, with as many one-seeded pyrenes as cells. Seeds pendulous; testa membranous; embryo very small in the apex of a fleshy albumen.—Trees or shrubs. Leaves alternate, simple, without stipules. Flowers small, in axillary umbels or cymes, rarely solitary or terminal. Fruits small.

The Order, limited to the large genus *Ilex*, and two small ones separated from it, is dispersed over the greater part of the world, but most abundant in America, very rare, however, in Africa, absent from New Zealand, and represented by one species only in Australia.—*Benth.*

1. ILEX, Linn.

(Latin name of Holly Oak.)

Calyx 4 or 5-lobed or parted. Corolla rotate. Petals free or connate at the base. Stamens 4 or 6, adhering to the base of the corolla in the male; sometimes hypogynous in the female. Ovary 2 to 12-celled. Style none or very short. Stigmas free or confluent on the top of the ovary. Drupe globose, very rarely ovoid, with 2 to 16 stones.

The genus contains about 150 species.

1. **I. peduncularis** (pedunculate), *F. v. M. Fragm.* vii. 105. A small glabrous tree, perhaps diœceous. Branchlets nearly terete. Leaves chartaceous, ovate or lanceolate-ovate, quite entire, short-acuminate, 2 to 4in. long, 1½ to 1⅔in. broad, here and there very minutely denticulate. Petiole 6 to 8 lines long. Peduncles often 2in. long or now and then shorter, bearing umbels or little corymbs. Pedicels 2 to 4 lines long. Flowers 5-merous. Calyx scarcely 1 line high. Lobes semi-rotund, very thinly ciliate. Corolla white; lobes 1½ to 2 lines long, broad-ovate. Filaments free, subulate-capillare, about 1 line long. Anthers yellow, dorsifixed. Style short.

Hab.: Rockingham Bay, *J. Dallachy* (F. v. M., l.c.)

Order XXXVI. **CELASTRINEÆ.**

Flowers regular, hermaphrodite or polygamous. Calyx small, persistent, 4 or 5-cleft, rarely 3 or 6-cleft. Petals as many as calyx-segments, spreading, imbricate or rarely valvate in the bud. Stamens as many as petals and alternate with them, inserted round the base or on the margin of the disk, or upon the disk itself; filaments usually short, incurved; anthers short, 2-celled, the cells in a few genera confluent into one. Disk usually conspicuous, more or less fleshy, flat or broadly cup-shaped, or thick and conical, nearly free, or adnate to the base of the calyx or confluent with the ovary. Ovary sessile on the disk, 2 to 5-celled, tapering to a short style with an entire or lobed stigma; ovules usually 2 in each cell, ascending with a ventral raphe, occasionally several, rarely 1 only, or pendulous with a dorsal raphe. Fruit a capsule, berry, drupe or samara, rarely divided into distinct carpels. Seeds usually enveloped in an arillus, sometimes winged; albumen fleshy or almost horny or none; embryo usually rather large, with flat cotyledons and a short radicle next to the hilum.—Trees or shrubs, occasionally thorny, or woody climbers. Leaves opposite or alternate, entire or toothed. Stipules minute and very deciduous or none. Flowers small, white or greenish, in axillary cymes or small racemes or in terminal panicles.

A considerable Order, dispersed over the greater part of the globe, more abundantly within the tropics than in temperate regions. Of the eleven Australian genera three only are endemic, the others are all Asiatic, one extends to Africa and S. Europe but is not American, one is also tropical American but not hitherto found in Africa, and two are both in America and Africa. The peculiar disk readily characterises the greater number of genera, where that is wanting the insertion of the ovules and inferior radicle are the chief points separating *Celastrineæ* from *Ilicineæ*; from *Rhamneæ*, with which the real affinity is much closer, the stamens alternating with the petals is a constant distinctive mark. The majority of *Celastrineæ* assume also when dry a peculiar pale-green colour, very rare in allied Orders.—*Benth.*

Tribe I. **Euonymeæ.**—*Leaves opposite (in the genus represented in Queensland). Fruit a capsule, dehiscent.*
Petals free, efoveolate. Ovules 2 in the axis of each cell . . . . 1. Euonymus.

Tribe II. **Celastreæ.**—*Stamens the same number as petals, inserted round the disk or on its margin. Seeds albuminous.*
Leaves alternate. Ovules 2 in each cell. Capsule loculicidal, coriaceous.
   Flowers in racemes or panicles. Stamens on the margin of the disk . . 2. Celastrus.
   Flowers in cymes. Stamens under the disk . . . . 3. Gymnosporia.
Leaves alternate. Ovules 3 or more in each cell.
   Flowers few, axillary or terminal. Peduncles filiform. Anthers sessile . 4. Hedraianthera.
Leaves alternate or opposite. Ovules many.
   Flowers in short axillary cymes. Filaments short. Anthers reniform . 5. Hypsophila.
Leaves alternate. Ovules 3 or more in each cell. Capsule loculicidal, woody or bony.
   Flowers in cymes. Stamens on the margins of the disk . . . . . 6. Denhamia.
Leaves alternate. Flowers 5-merous. Petals valvate. Fruit a small globose berry, 2—4-celled, 2 to 4-seeded . . . . 7. Caryospermum.
Leaves mostly opposite. Ovules 2 in each cell. Drupe indehiscent, 2 or 3-celled . . . . 8. Elæodendron.

XXXVI. CELASTRINEÆ.

TRIBE III. **Hippocrateæ.**—*Stamens 3, rarely 2, 4, or 5, inserted on the face of the disk. Filaments flattened, sometimes adnate to the ovary, recurved and causing the anthers when dehiscing to become extrorse. Seeds albuminous. Leaves opposite, except* in Siphonodon. Fruit flattened, dehiscent. Seeds winged. Scandent shrubs. Leaves opposite . . . . 9. HIPPOCRATEA.
Fruit a berry. Seeds not winged. Erect shrubs. Leaves opposite . . 10. SALACIA.
Fruit indehiscent. Seeds not winged. Leaves alternate . . . . . . 11. SIPHONODON.

1. EUONYMUS, Linn.
(Well named.)

Calyx 4 or 5-fid, spreading or recurved. Petals 4 or 5. Stamens 4 or 5, inserted on the disk; anthers broad, 2-celled. Disk large, fleshy, 4 or 5-lobed. Ovary sunk in the disk, 3 or 5-celled. Style short or none, stigma 3 or 5-lobed; ovules 2 in each cell, attached to the inner angle, ascending and suspended. Capsule 3 or 5-celled, 3 or 5-lobed, angled or winged, coriaceous, rarely echinate; cells 1 or 2-seeded, loculicidal. Seeds covered by the aril, albuminous.—Trees or shrubs, erect, rarely scandent, glabrous. Leaves petiolate, rarely subsessile, stipules caducous.

The species are natives of the mountainous regions of tropical Asia, the Malayan Archipelago, a few scattered over Europe and North America, and one in Australia (Queensland).

1. **E. australiana** (Australian), *F. v. M. Fragm.* iv. 118, vi. 202. A tall evergreen shrub with pale smooth bark, the branchlets compressed-angular, at length terete. Leaves opposite, 2 to 3in. long, 1 to 1¾in. broad, subovate, lanceolate, or orbicular-ovate, repand-crenulate, entire below the middle, shining on both sides, pale beneath, the apex obtuse or very shortly and obtusely produced. Cymes axillary, on bony peduncles, solitary, bearing many or few flowers, dichotomously branched. Flowers pedicellate. Bracts opposite, minute, almost deltoid, soon deciduous. Sepals unequal, 1 to 2 lines, glabrous, imbricate, fimbriate, the 2 outer ones almost semi-orbicular, 3 inner ones almost rotund. Petals ochroleucus, deciduous, orbicular, about 2 lines, fimbriate on short broad claws. Stamens glabrous, inserted on the disk; filaments setaceous, about 1 line long. Anthers scarcely 1 line long, almost renate, cells divaricate, distinct, extrorse. Style conico-subulate, ⅔ line long. Ovary 5-celled, cells 2-ovulate. Capsule 4 to 6 lines long, loculicidally 5-valved, cells 1 rarely 2-seeded, dark-brown, ovate, glossy, about 2 lines long. Arillus cupular, about 1½ line long. Albumen white, amygdaline-cartilaginous. Cotyledons pale-green, plane, ovate-orbicular. Radicle inferior, very short.

Hab.: Rockingham Bay, *J. Dallachy* (F. v. M., l.c.)

2. CELASTRUS, Linn.
(Name referring to the late time of ripening its fruit; *Kelas*, the latter season.)

Flowers polygamous. Calyx 5-cleft. Petals 5, spreading. Disk broad, concave. Stamens 5, inserted on the margin of the disk; filaments subulate, flattened at the base; anthers ovoid or oblong. Ovary not immersed in the disk, 2 to 4-celled; style usually short, the stigma lobed, spreading; ovules 2, collateral, erect, the funicle cup-shaped. Capsule globular, oblong, or obovoid, coriaceous, 2 to 4-celled, opening loculicidally. Seeds 1 or 2 in each cell, usually enveloped partially or wholly in a fleshy arillus, sometimes connecting the seeds in a mass, sometimes nearly or quite wanting; testa membranous or almost crustaceous; albumen fleshy; cotyledons leafy.—Trees or shrubs, often climbing, unarmed. Leaves alternate, petiolate, entire, or serrate. Stipules minute and deciduous, or none. Flowers small, in terminal or axillary oblong panicles or racemes. Pedicels articulate. Bracts very small.

254   XXXVI. CELASTRINEÆ.   [*Celastrus.*

The genus extends chiefly over tropical and eastern extratropical Asia, with 1 Mascarene and a few N. American species. The Australian species are all endemic, although one is nearly allied to a common Indian one.—*Benth.*

Tall climber. Panicles terminal. Ovary 3-celled . . . . 1. *C. australis.*
Trees or tall shrubs. Racemes or pedicels lateral or axillary. Ovary 2-celled.
  Leaves ovate or elliptical.
    Leaves quite entire, much narrowed into a long petiole. Flowers 4-merous . . . . 2. *C. dispermus.*
    Leaves entire or toothed, petiole short. Flowers 5-merous . . . . 3. *C. bilocularis.*
  Leaves linear or narrow-lanceolate, entire . . . . 4. *C. Cunninghamii.*

1. **C. australis** (Australian), *Harv. and Muell. in Trans. Phil. Soc. Vict.* i. 41; *Benth. Fl. Austr.* i. 398. A tall, woody, glabrous climber. Leaves from ovate-lanceolate to oblong-elliptical or lanceolate, acuminate, 2 to 4 in. long, entire or minutely and usually remotely serrate, narrowed into a petiole of 1 to 3 lines. Panicles terminal, or rarely in the upper axils, narrow, loose, rarely above 2 in. long. Flowers white. Calyx-lobes broad, rounded, ciliate. Petals twice as long, attaining a little more than 1 line, broadly ovate or orbicular. Disk almost free from the calyx. Ovary 3-celled; style short, with 3 spreading stigmatic lobes. Capsule nearly globular, rarely exceeding 3 lines diameter. Seeds enveloped in a fleshy arillus.—Reissek, in Linnæa, xxix. 265; F. v. M. Fragm. iii. 94.

Hab.: Many parts of southern Queensland.

The species differs slightly from the E. Indian *C. paniculatus,* Willd., in the narrower and more acuminate, not obovate leaves, usually more coriaceous, and in the rather smaller flowers and fruits.—*Benth.*

2. **C. dispermus** (two-seeded), *F. v. M., in Trans. Phil. Inst. Vict.* iii. 81, and *Fragm.* vi. 208; *Benth. Fl. Austr.* i. 399. A small glabrous tree. Leaves elliptical, obovate-oblong, or rarely broadly lanceolate, obtuse or slightly acuminate, 2 to 3 in. long, quite entire, much narrowed into a rather long petiole. Racemes axillary or lateral, not seen in flower, when in fruit 1 to 1½ in. long, the pedicels 1 to 2 lines. Flowers polygamous. Calyx lobes 4, deltoid, semi-orbicular, ½ line long. Petals 4, glabrous, membranous, ovate, truncate at the base, a little exceeding 1 line long. Stamens 4. Filaments in the male flowers twice, rarely three times, as long as the anthers, in the female flowers the same or a little longer. Anthers ⅓ line long, yellow, cordate, obtuse, dorsifixed. Style very short. Ovary free above. Capsule obovoid or obcordate, slightly compressed, 3 to 4 lines long, 2-celled and 2-valved, with usually 2 seeds, covered at the base, according to F. v. Mueller, with a thick arillus, but I find no remains of it on our specimens; very rarely the capsule is 3-angled and 3-celled.—Benth.

Hab.: Araucaria forests near Moreton Bay, *F. v. Mueller*; Port Denison, *Fitzalan*; Rockingham Bay and Mackay.

Wood of a light colour, close-grained, and prettily marked.—*Bailey's Cat. Ql. Woods No.* 75a.

3. **C. bilocularis** (two-celled), *F. v. M. in Trans. Phil. Inst. Vict.* iii. 31; *Benth. Fl. Austr.* i. 399. A small much-branched glabrous tree. Leaves ovate, oblong, or broadly lanceolate, obtuse or slightly acuminate, 1½ to 2½ or very rarely 3 in. long, entire sinuate or bordered by acute teeth, rounded or cuneate at the base, on a short petiole. Racemes axillary or lateral, rarely 1 in. long. Pedicels 1 to 2 lines. Calyx-lobes 5, broad and short. Petals 5, ovate, about 1 line long. Ovary 2-celled; style exceedingly short, with 2 broad short spreading stigmatic lobes. Capsule 2-valved, coriaceous, pear-shaped or nearly globular, under 3 lines diameter. Seeds enclosed in a thin arillus.

Hab.: Dawson and Burnett Rivers, *F. Mueller*; Brisbane and Logan Rivers, *Fraser* (all with entire or slightly toothed leaves); Warwick, *Beckler* (with sharply toothed leaves).—*Benth.*

Wood of a light-grey, close in grain, hard and tough.—*Bailey's Cat. Ql. Woods No.* 75.

4. **C. Cunninghamii** (after A. Cunningham), *F. v. M. in Trans. Phil. Inst. Vict.* iii. 30; *Benth. Fl. Austr.* i. 399. A tall shrub or small tree, quite glabrous and often somewhat glaucous. Leaves linear or narrow-lanceolate, mucronate, 2 to 3in. long in some specimens, all under 1in. in others, entire, rigid, the midrib alone prominent underneath. Flowers small, in short loose axillary or lateral racemes, occasionally growing out into leafy branches. Pedicels slender, 2 to 3 lines long. Calyx-lobes 5, orbicular, not ciliate. Petals broadly ovate, about 1 line long. Disk rather thick, but less so than in *Gymnosporia*. Ovary 2-celled, with a short style and 2 short spreading stigmatic lobes. Capsule globular or ovoid, 2 lines diameter, or rather more, 2-valved, 1 or 2-seeded. Seeds enclosed in a pulpy arillus.—*Catha Cunninghamii*, Hook. in Mitch. Trop. Austr. 387.

Hab.: Islands of the Gulf of Carpentaria, Broadsound, *R. Brown;* Moreton Bay (?), *A. Cunningham;* Rockhampton, *Thozet;* Warwick, *Beckler;* St. George's Bridge, *Mitchell.*
This and the three preceding species appear to have the erect habit but not the cymose inflorescence nor the thick disk of *Gymnosporia*, and the stamens always proceed from the margin of the disk.—*Benth.*
Wood of a pinkish colour, nicely marked; useful for cutting into veneers for cabinet-work.—*Bailey's Cat. Ql. Woods No.* 76.

3. GYMNOSPORIA, W. and Arn.
(Naked seeds.)

Calyx 4 or 5-cleft. Petals 4 or 5, spreading. Stamens 4 or 5, inserted under the disk; filaments subulate; anthers short. Disk broad, sinuate or lobed. Ovary attached by a broad base or partially immersed in the disk, 2 or 3-celled; style short; stigma 2 or 3-lobed; ovules 2 in each cell. Capsule obovoid or nearly globular, 2 or 3-celled, opening loculicidally. Seeds 1 or 2 in each cell, the arillus complete or imperfect, or sometimes wanting; testa coriaceous; albumen fleshy; cotyledons leafy.—Shrubs or small trees, the small branches often thorny. Leaves alternate, entire or serrate, without stipules. Flowers small, in dichotomous cymes, either axillary or on the old nodes.

The genus is widely diffused over the warmer regions of the Old World, one species being found as far north as Spain, and a few extending to the Pacific Islands. The Australian species is an Indian and African one.—*Benth.*

1. **G. montana** (mountain), *W. and Arn. Prod.* 159 (under *Celastrus*); *Benth. Fl. Austr.* i. 400. A tall glabrous shrub or small tree, the smaller branches occasionally terminating in stout thorns. Leaves obovate, very obtuse, 1½ to 2½ or rarely 3in. long, entirely or minutely crenulate, narrowed into a petiole of 2 or 3 lines, membranous or thinly coriaceous, of a pale-green. Cymes 2 or 3 together in the axils or on the old nodes, rarely above 1in. long, with slender dichotomous branches. Calyx-lobes 5, very short, broad, ciliate. Petals 5, obovate, about 1 line long. Ovary 3-celled; style very short, with 3 spreading stigmatic lobes. Capsule flat at the top, obtusely 3-angled, about 3 lines diameter in the Australian specimens, usually smaller in India. Arillus of the seeds cup-shaped. *Celastrus montanus*, Roxb.; W. and Arn. l.c., with all the synonyms quoted; Wight, Ic. Pl. t. 882.

Hab.: Cape York, *M'Gillivray.*
Common in the Indian Peninsula, and apparently the same as the tropical African *Celastrus senegalensis*, Lam.; I have seen no specimens from the Indian Archipelago. The Australian specimens are unarmed, but that is frequently the case with Indian ones, with which they agree in every respect except the larger capsules.—*Benth.*

4. HEDRAIANTHERA, F. v. M.
(Sessile anthers.)

Calyx 5-fid, persistent. Petals 5, membranous, inserted below the disk, naked, entire, ovate, sessile, deciduous imbricate in the bud. Stamens 5, opposite the calyx-lobes, alternate with the petals. Filaments none. Anthers sessile on the top of the glandular disk, 2-celled, cells divaricate, slits vertical, 2-valved. Disk annular, crenulate. Stigma orbicular, undivided, sessile. Ovary 5-celled, with many adscendent ovules in each cell. Fruit bony, almost globose, perfectly 5-celled. Seeds adscendent.—F. v. M. Fragm. v. 59.

1. **H. porphyropetala** (flowers rich purple), *F. v. M. Fragm.* v. 59. A glabrous shrub with alternate foliage. Leaves ovate-lanceolate, 2½ to 4in. long, 10 to 16 lines broad, slightly unequal-sided, quite entire or repandulous, the lateral nerves and veinlets thin, upper side deep green and shining, under side pale. Peduncle axillary and terminal, 2 or 3-flowered or corymbs of few flowers, thin-filiform, from a few lines to about 1in. long. Pedicels ¼in. or less long, the basal bracteoles lanceolate or linear-subulate, about 1 line long. Calyx ¾ line high, lobes semi-orbicular. Petals dark-purple, about 1½ line long, cohering a little at the base. Anthers didymous. Ovary attenuated to the stigma at the apex. Fruit 5-valved, base thin, cells dehiscent. Seeds often 3, maturing in a cell, superposed, angular-ovate, glabrous, 3 or 4 lines long. Testa brown. Arillus ochroleucus. Albumen copious. Embryo straight. Cotyledons elliptic. Radicle inferior.

Hab.: Rockingham Bay, *J. Dallachy* (F. v. M., l.c.)

5. **HYPSOPHILA**, F. v. M.

(A lover of heights; found at a high elevation.)

Sepals 5, roundish, much overlapping in bud, persistent; the 2 outer smaller than the 3 inner. Petals 5, orbicular or ovate, entire, smooth or hairy, deciduous. Stamens 5, opposite the sepals, very short, inserted towards the centre of the disk; filaments free, filiform; anthers almost reniform, the cells divergent, dehiscent by an external slit. Disk depressed, crenulated. Style extremely short; stigma orbicular, convex, with a central cavity. Ovary immersed in the disk, 3-celled; ovules many, axillary, ascendant. Fruit almost ellipsoid, somewhat pointed; pericarp coriaceous, tardily ruptured into 3 valves; septa membranous. Seeds several in each cell, almost ovate, somewhat oblique, long-persistent on elongated crisped funicles, and clasped at the base by an obconic-cupular arillus. Testa thinly crustaceous. Raphe longitudinal, prominent. Albumen carnulent. Embryo straight, somewhat shorter than the albumen. Cotyledons flat, about twice as long as the cylindric inferior radicle—A tree, with scattered or opposite coriaceous oval or lanceolate-elliptical entire leaves, with short axillary cymes of small flowers and with rather large particularly elongated fruits.—F. v. M., Vict. Nat., April 1887.

Leaves scattered. Petals smooth . . . . 1. *H. Halleyana.*
Leaves mostly opposite. Petals hairy . . . . 2. *H. oppositifolia.*

1. **H. Halleyana** (after the Rev. J. J. Halley, of Melbourne), *F. v. M., Vict. Nat.*, 1887. An evergreen glabrous tree attaining the height of about 40ft. Branchlets angular, often flexuous. Leaves flat, rather blunt, attaining 4in. in length and 1½in. in breadth, on short petioles; upper surface dark-green, lower much paler; not shining; nerves faint, veins concealed. Cymes much shorter than the leaves, at times reduced to 2 or 3 flowers. Bracts and bracteoles minute, lanceolar-deltoid, the latter distant from the calyx. Petals about twice as long as the inner sepals, dull-red according to the collector's notes, hardly attaining above ¼in. in length. Filaments not reaching beyond the disk, persistent; anther-cells ellipsoid, deciduous. Stigma not broader than the style. Fruit 1 to 1½in. long. Arillus of about ¼in. in length, the funicle sometimes quite as long. Seeds seldom exceeding ¼in. in length; testa brown, slightly wrinkled, subtle-dotted.—F. v. M., l.c.

This new genus is readily separable from *Euonymus* on account of the scattered leaves, the entire stigma, and the many-seeded not prominently angular fruit; from *Lophopetalum* in deeper and more unequally divided calyx, not appendiculated petals, and form of the anthers; from *Hedraianthera* in neither anthers nor stigma sessile, not hard pericarp, and the shape of the arillus; from *Denhamia* in the structure of the anthers, undivided stigma, not bony pericarp, and seeds much emerging from the arillus.—*F. v. M.*, l.c.

Hab.: In the highest regions of Mount Bellenden Ker, descending to about 4000ft., *W. Sayer F. v. M.*) The shrub is also abundant at Pimpama and on Tambourine Mountain; flowers minute, deep-red.

2. **H. oppositifolia** (opposite leaved), *F. v. M., Vict. Nat.*, May 1892. Leaves 1½ to 3in. long, on rather long petioles, mostly opposite, ovate-lanceolate, bluntly acuminate. Panicle many-flowered, cymose, terminal and axillary. Sepals connate towards the base. Petals ovate, beset outside with short appressed hairs. Filaments much incurved, considerably longer than the anthers, stoutish, dark-purplish, hispidulous. Style very short, stigma much broader, depressed; disk and ovary glabrous. Fruit not seen.

Hab.: Mount Bartle Frere, *Stephen Johnson* (F. v. M., l.c.)

6. **DENHAMIA**, Meisn.
(After Captain Denham, an African explorer.)
(Leucocarpon, *A. Rich.*)

Calyx 5-cleft. Petals 5. Stamens 5, inserted on the margin of the disk; filaments subulate; anthers ovate. Disk broadly cupular, rather thick. Ovary 1-celled, with 3, or rarely 4 or 5 parietal placentas, or completely divided into as many cells; style short, with as many stigmatic lobes as cells or placentas. Ovules 3 to 8 to each cell or placenta. Capsule ovoid or globular, opening in thick woody or bony valves, bearing the placentas or dissepiments in their centre. Seeds enclosed in a fleshy arillus; albumen fleshy; cotyledons flat.—Shrubs or small trees, glabrous and more or less glaucous. Leaves alternate, rigid, entire, or toothed. Flowers small, in few-flowered cymes or racemes.

The genus is exclusively Australian, and on account of the parietal placentation of two species has been by some referred to *Bixineæ*; but the disk, stamens, general habit, etc., are those peculiarly characteristic of *Celastrineæ.—Benth.*

Ovary 1-celled; placentas (4 to 8 ovulate) not meeting in the axis. Veins
of the leaves not very prominent.
Flowers racemose. Style distinct . . . . 1. *D. oleaster*.
Flowers in cymes or narrow panicles. Style very short, branched . . 2. *D. obscura*.
Ovary 3-celled, placentas (3 or 4-ovulate) united in the axis. Leaves
prominently veined . . . . 3. *D. pittosporoides*.
Placentas 4 to 6-ovulate. Leaves green on both sides . . . . . . 4. *D. viridissima*.

1. **D. oleaster** (Oleaster-like), *F. v. M. in Trans. Phil. Inst. Vict.* iii. 29; *Benth. Fl. Austr.* i. 401. A tall shrub with slender branches. Leaves lanceolate, acute, or rarely obtuse, 2 to 3in. long, entire or remotely toothed, narrowed into a very short petiole, coriaceous, the veins scarcely conspicuous. Flowers in short, simple, axillary or terminal racemes, the pedicels very rarely bearing 2 flowers. Calyx-segments broadly ovate or orbicular. Petals nearly 2 lines long. Disk thicker, and filaments longer than in the other two species. Ovary 1-celled, tapering into a style of at least ½ line, the stigmatic lobes very short. Placentas 3, with 4 to 6 ovules to each. Fruit not seen.—*Melicytus (?) oleaster*, Lindl. in Mitch. Trop. Austr. 388.

Hab.: St. George's Bridge, Balonne River, *Mitchell.*

2. **D. obscura** (obscure), *Meisn. in Walp. Rep.* i. 208; *Benth. Fl. Austr.* i. 401. A tall shrub or small tree, the young branches generally pendulous. Leaves mostly oblong-lanceolate, acuminate, 2 to 3in. long, entire, with often wavy margins, narrowed into a rather long petiole, coriaceous, finely but not

prominently veined; on barren branches the leaves are sometimes broadly ovate and bordered by coarse prickly teeth like those of a holly. Flowers in small pedunculate cymes in the upper axils, or forming a short oblong terminal panicle. Calyx-segments ovate. Petals rather broad, 1½ lines long. Ovary 1-celled, with 3 to 5 placentas; style very short, with 3 to 5 oblong-linear stigmatic branches. Ovules 4 to 8 to each placenta. Capsule ovoid or globular, attaining about 1in., of a pale-whitish hue when dry, the thick valves bearing slightly projecting placentas along their centre.—*Leucocarpon obscurum*, A. Rich. Sert. Astrol. 46, t. 18; *Denhamia xanthosperma*, F. v. M. Trans. Phil. Inst. iii. 28, and *D. heterophylla*, F. v. M. l.c. 29.

Hab.: Broadsound, *R. Brown*; Newcastle Range, between Gilbert and Burdekin Rivers, *F. v. Mueller*.

Wood of a light-yellow colour, very close in the grain, easily worked; useful for engraving and cabinet-work.—*Bailey's Cat. Ql. Woods No. 77a*.

3. **D. pittosporoides** (Pittosporum-like), *F. v. M. in Trans. Phil. Inst. Vict.* iii. 30; *Benth. Fl. Austr.* i. 402. A tree, the trunk, according to Thozet, beautifully striated. Leaves lanceolate or rarely ovate-lanceolate, obtuse, 2 to 3 or rarely 4in. long, obtusely serrate, narrowed into a petiole, coriaceous, with very prominent pinnate and reticulate veins, not so glaucous as in the other two species. Cymes pedunculate, few-flowered, on short leafless branches on the old wood or at the base of young leafy branches. Calyx-segments broadly orbicular. Petals ovate, about 1 line long, rather thick at the base. Ovary fleshy, completely 3-celled, with 3 or 4 ovules in each cell. Capsule globular, attaining in our specimens ½in. or rather more, but many of them opening when not half that size, the thick woody valves bearing the dissepiments on their centre.

Hab.: Wide Bay, sources of the Burnett River, Rockhampton, Warwick, Keppel Bay, Fitzroy River, Taylor's Range. In flower in September.

Wood of a uniform pale colour, resembling English Elder; suitable for engraving, pattern-making, and similar uses.—*Bailey's Cat. Ql. Woods No. 77*.

4. **D. viridissima** (very green), *Bail. and F. v. M.* A small tree with dense head of dark-green foliage. Leaves alternate or nearly opposite, often clustered at the ends of the branchlets, lanceolate, 3 or 4in. long, 1 to 1¾in. broad in the centre, tapering to a very short petiole, the apex more or less elongate, margins entire, green on both surfaces, and the reticulation more or less prominent. Flowers in terminal pedunculate cymes or racemes, but the tree only met with in early fruit. Capsules 3-valved, somewhat fusiform, prominently obtusely 3-angular, 1 to 1¼in. long, bursting luculicidally, bearing the seeds on the prominently projecting placentas. Seeds 4 to 6 to each placenta, half enclosed in a rather large fleshy arillus.

Hab.: Bellenden Ker *Expedition* 1889.

7. CARYOSPERMUM, Blume.

(From its nut-like seeds.)

Calyx cupular, 5-fid, open in the bud. Petals 5, 3-angular, reflexed, valvate, keeled inside. Stamens 5, inserted on the margin of the disk; filaments subulate. Anthers subglobose. Disk sinuate-lobed, somewhat thick. Ovary half sunk in the disk, not at all confluent with it, 3 to 4-celled. Stigma subsessile, obsoletely 3 to 4-lobed. Ovules solitary, erect, in each cell. Berry globose, small, 2 to 4-celled, 2 to 4-seeded. Seed erect, subrotund, exarillate; testa thick, crustaceous, fleshy outside.—Glabrous shrubs, leaves alternate, petiolate, subcoriaceous, ovate-oblong, acuminate, serrulate. Stipules minute, caducous. Cymes axillary, short; flowers small.

A genus of few species inhabiting, besides Queensland, Java and Amboyna. Habits and flowers of *Rhamnads*, but the stamens are alternate with the petals. - *Benth. and Hook. Gen. Pl.* i. 367.

C. arborescens (a tree-like shrub), *F. v. M. Fragm.* vi. 202, viii. 279. A tall glabrous shrub or small tree. Leaves chartaceous, ovate-lanceolate, obtuse, acuminate, denticulations remote and minute, 2 to 4in. long, 1 to 1½in. broad, under side pale, upper deep-green, lateral nerves looping some distance from the margin. Petiole narrow, about ⅓in. long. Stipules minute. Panicles axillary or lateral, sometimes very short. Pedicels about 1¼ line long. Flower-buds truncate-globose below, the upper part pyramidal. Calyx-lobes 5, semi-lanceolate, ½ line long, a little distant in the bud. Petals persistent, valvate, yellow. Stamens 5, alternate with the petals, very short. Anthers cordate-rotund, yellow, obtuse, dorsifixed. Style very short. Stigma simple. Ovary depressed, glabrous, 2-celled. Cells 2-ovulate. Fruit red, globose, 3 to 5 lines diameter, 1 or 2-seeded. Pulp macilaginous. Seeds basifixed, erect. Testa tenui-crustaceous. Albumen fleshy. Embryo 1½ line long. Cotyledons ovate-orbicular. Radicle tenui-cylindric.

Hab : Rockingham Bay, *J. Dallachy* (F. v. M., l.c.)

8. ELÆODENDRON, Jacq. f.

(From the resemblance of the fruit to the Olive.)

Flowers often polygamous. Calyx 4 or 5-cleft, rarely 3-cleft. Petals as many as calyx-segments, spreading. Disk thick. Stamens as many as petals, inserted under the edge of the disk ; filaments short ; anthers nearly globular. Ovary continuous with the disk, conical, 3-celled, rarely 2 or 4 or 5-celled ; style very short ; ovules 2 in each cell. Drupe succulent or nearly dry, the putamen hard, 1, 2, or 3-celled. Seeds usually solitary, without any arillus ; testa membranous or spongy ; albumen scanty or copious, cotyledons flat.—Shrubs or small trees, usually quite glabrous. Leaves opposite or alternate, entire or crenate. Flowers small, in dichotomous cymes, usually axillary or lateral, often clustered.

The species are numerous in East India and southern Africa, with a very few in tropical America ; none are known from tropical Africa. The two Australian ones are endemic.—*Benth.*

Ovary 2-celled. Drupe red. Veins of the leaves scarcely conspicuous
above . . . . 1. *E. australe.*
Ovary 3-celled. Drupe black. Veins of the leaves conspicuous on both
sides . . . . 2. *E. melanocarpum.*

1. **E. australe** (Australian), *Vent. Jard. Malm.* t. 117 ; *Benth. Fl. Austr.* i. 402. A glabrous, small or middle-sized tree. Leaves opposite, or here and there alternate, ovate, obovate, elliptical, or oblong-lanceolate, obtuse or obtusely acuminate, 2 to 4in. long, entire or broadly crenate, narrowed into a very short petiole, coriaceous, the reticulate veins slightly prominent underneath and scarcely conspicuous above. Flowers 4-merous, in slender cymes, much shorter than the leaves. Calyx-segments broadly ovate. Petals from a little more than 1 line to nearly 2 lines long, ovate, often broadly and shortly 3-lobed. Ovary confluent with the disk in a conical mass, 2-celled ; style either very short or attaining ¾ line. Drupe ovoid or globular, rarely above ½in. long, of a bright-red colour, which it often retains in the dried specimens. Putamen hard and woody, usually 1-seeded, but showing the traces of the abortive cell. Albumen copious.—F. v. M. Fragm. iii. 61 ; *Portenschlagia australis,* Tratt. Arch. t. 250.

Hab.: Wide Bay, Moreton Bay, Ipswich. In flower in September.
Var. *angustifolia.* Leaves lanceolate or narrow-oblong, entire or nearly so ; fruit more ellipsoid.—*Portenschlagia integrifolia,* Tratt. Arch. t. 284 ; *Elæodendron integrifolium,* G. Don, Gen. Syst. ii. 12.—Burnett, Dawson, and Pine Rivers, in Queensland, *F. v. Mueller ; Warwick, Beckler.*

According to F. v. Mueller, the fruit in *E. Australe* is occasionally 3-celled ; but this must be rarely the case, as I have never found more than 2 cells to the ovary in any of the numerous specimens I have examined. The above references to Trattinick's Archiv are quoted after G. Don ; I do not find the second volume of that work in any of our libraries.—*Benth.*

Wood of a pinkish colour, close in grain, and very tough, but warps a good deal in drying if cut up before perfectly seasoned ; useful for tool-handles.—*Bailey's Cat. Ql. Woods No.* 78.

2. **E. melanocarpum** (black-fruited), *F. v. M. Fragm.* iii. 62; *Benth. Fl. Austr.* i. 408. A glabrous tree. Leaves opposite, obovate or oval-elliptical, broadly crenate, scarcely to be distinguished from those of *E. australe*, except that the veins are more conspicuous on the upper as well as the lower side. Flowers smaller than in *E. australe*, the males more numerous, in slender cymes like those of the small-flowered Indian *Hippocrateas*, usually 3-merous. Female flowers in less-branched cymes and often 4-merous. Ovary 3-celled, but very imperfect in the flowers examined. Drupe ovoid or globular, shining-black, rather larger than in *E. australe*, the hard putamen always 3-celled, or showing the traces of a second or third cell when reduced to one. Albumen copious.

Hab.: Keppel Bay, Port Bowen, Rockingham Bay, Fitzroy and Lizard Islands; Port Denison, Rockhampton.
Wood tough, of a light colour and fine grain.—*Bailey's Cat. Ql. Woods No.* 79.

9. HIPPOCRATEA, Linn.
(After Hippocrates, the father of physic.)

Calyx small, 5-cleft. Petals 5, valvate or imbricate. Stamens usually 3, the filaments thick at the base, connivent round the ovary, recurved at the top; anthers at first divided into 2 or 4 cells, at length confluent into 1 transverse cell. Disk conical or broad. Ovary 3-celled, style short, stigma 3-lobed; ovules 2 or more in each cell. Fruit of 3 distinct, flat, coriaceous carpels, opening along the middle in 2 boat-shaped valves. Seed compressed, usually produced at the base into a wing adnate to the raphe; albumen none; embryo in the upper end of the seed; cotyledons flat, connate; radicle inferior.—Small trees or woody climbers. Leaves opposite, entire or serrate. Stipules very small and deciduous. Flowers in axillary cymes or panicles.

A large genus, widely distributed over tropical Asia, Africa, and America, the Australian species being one of the common Asiatic ones. It belongs to the section with comparatively large flowers and valvate petals. The other section common in India, including *H. indica*, with minute globular flowers and imbricate petals, has not yet been observed in Australia.—*Benth.*

1. **H. obtusifolia** (obtuse-leaved), *Roxb.*; *W. und Arn. Prod.* 104, var. *barbata*; *Benth. Fl. Austr.* i. 404. A tall, woody, glabrous climber. Leaves ovate, obovate, or oblong, obtuse or obtusely acuminate, 2 to 4in. long, entire, coriaceous, somewhat shining. Flowers in short, loose, axillary cymes, the upper ones forming sometimes large leafy terminal panicles. Petals fully 2 lines long, lanceolate, rather thick, valvate in the bud, and in the Australian specimens bearded inside above the middle, the disk and ovary also occasionally villous or pubescent. Ovules 6 to 10 in each cell of the ovary. Carpels about 2in. long, either broadly oblong and entire or broader and emarginate at the top.—*H. macrantha*, Korth. Verhand. Nat. Gesch. Bot. 187, t. 39; *H. barbata*, F. v. M. in Trans. Phil. Inst. Vict. iii. 23.

Hab.: Taylor's Range, Moreton Bay. Ripe in November.
The species is widely distributed over tropical Asia. The common Indian form, figured in Wight,'Ic. t. 968, has glabrous petals, but the variety with bearded petals as described by Korthals from Borneo, and of which we have specimens from Ceylon, is the same as the Australian one; and the amount of hairiness both on the petals and ovary appears to be variable. —*Benth.*

10. SALACIA, Linn.
(After Salacia, the wife of Neptune.)

Calyx small, 5-parted. Petals 5, imbricate. Stamens 3, rarely 2 or 4, continuous with the disk, recurved. Ovary conical, immersed in the disk, 3-celled. Style very short, stigma simple or 3-lobed, ovules 2 to 8 in each cell, in 1 to 2 series, inserted on the inner angle. Fruit a berry, 1 to 3-celled, sub-woody or fleshy. Seeds large, angular.—Scandent or sarmentose shrubs or small trees. Leaves opposite, petiolate, exstipulate. Flowers few or many, clustered in the axils of the leaves or extra-axillary, more rarely in cymes.

The species belong to the tropics of both hemispheres.—*M. A. Lawson in Hook. Fl. Brit. Ind.* i. 625.

1. **S. prinoides** (Prinos-like), *DC. Prod.* A small straggling tree or large climbing shrub. Branchlets somewhat terete. Leaves petiolate, very coriaceous, 1½ to 3in. long, ¾ to 1¼in. broad, oblong, obtusely acuminate, serrate, slightly repand. Flowers mostly axillary, 3 to 6 (1, 2, or 3, *F. v. M.*) from each tubercle, pedicels under ½in. Sepals puberulous, ciliate. Petals clawed, entire, about 1¼ line, broadly-ovate. (Anther cells divergent, *F. v. M.*) Fruit globose, size of a small cherry, smooth 1-celled, 1-seeded, black (2-seeded, *F. v. M.*)

Hab.: Recorded from Queensland by *F. v. M.* without locality.

11. SIPHONODON, Griff.
(Teeth united into a tube.)

Calyx 5-cleft. Petals 5, spreading. Disk not distinct from the base of the calyx. Stamens 5, connivent round the pistil, the filaments flattened. Ovary half immersed in the disk or base of the calyx, conical, the summit hollowed and stigmatic in the cavity round a central style-like column ; cells numerous, in 2 to 4 series; ovules solitary in each cell, alternately ascending and pendulous. Drupe globular, hard-fleshy, with numerous 1-seeded bony pyrenes superposed in rings of about 10 round the central axis. Testa of the seed membranous ; albumen almost horny ; cotyledons large, flat ; radicle short.—Glabrous tree. Leaves alternate, entire or crenate. Stipules minute, deciduous. Peduncles short, axillary, few-flowered.

Besides the Australian species, which are endemic, it comprises only one from the Indian Archipelago.

Leaves coriaceous, pale-coloured, oblong, obtuse, 3 to 4½in. long, 1¾in. broad, unequal-sided and tapering much towards the base. Fruit usually oval, about 1in. long . . . . 1. *S. australe.*
Branches drooping. Leaves falcate, about 5½in. long, 1in. broad, texture thin, pale-coloured. obtuse, tapering at the base. Fruit globose, about 2in. diameter . . . . 2. *S. pendulum.*
Leaves of thin texture, 4½ to 6in. long, 1 to 1¾in. broad, apex sharply acuminate, slightly cuneate at the base. Fruit globose-turbinate, 2in. diameter, smooth, more or less sunk at the apex, and often deeply 5-sulcate . . . . 3. *S. membranaceum.*

1. **S. australe** (Australian), *Benth. Fl. Austr.* i. 408. A tree of 60ft. or more. Leaves obovate or broadly oblong, obtuse, 2 to 3in. long, entire or slightly sinuate, coriaceous, drying of the pale colour so frequent in *Celastrineæ*. Peduncles very short. Calyx irregularly 5-partite. Lobes imbricate. Petals cuneate-orbicular, imbricate before expansion, white, deciduous, about 2½ lines. Stamens 5, alternate with the petals ; filaments broad, scarcely 1 line long. Anthers 2-celled; cells broadly divergent. Disk very short. Drupe globular, ¾ to 1in. diameter, the flesh hard and dry, with the stigmatic scar at the top, and the scar of the calyx at the base, as in *S. celastrineus*. Nuts numerous, appearing to have been arranged in two rows in each of 5 cells, irregularly ovoid, somewhat compressed, 3 to 4 lines long. Testa of the seed brown ; albumen not very thick; cotyledons broadly ovate.

Hab.: Scrubs of southern Queensland.
Wood white, very close in grain, firm, and easily worked ; an excellent wood for the cabinet-maker, and might serve for engraving.—*Bailey's Cat. Ql. Woods No.* 80.

Var. *Keysii.* Leaves coriaceous, pale-coloured, obovate, 2½in. long, 2in. broad, tapering and equal-sided at the base. Fruit nearly globular, 1¼in. long. Hab.: Mount Perry, *J. Keys*.

2. **S. pendulum** (weeping), *Bail. Bot. Bull.* iv. (1891). Weeping Ivorywood tree. "Aguridal," Palmer River, *Roth*. A tree with a thick, rough, corky outer bark on the trunk, the wood close-grained, when fresh yellowish; the branches dividing at their extremities into numerous long, slender, thong-like, drooping branchlets. Leaves usually falcate, about 5in. long and seldom exceeding 1in. in breadth, obtuse or at times minutely apiculate, the texture thin, almost membranous; primary veins distant and very oblique, the veinlets undulately anastomosing, but not prominent. Peduncles about 1in. long, bearing a few flowers at the end. (Flowers not forwarded). Fruit nearly globose, 2½ by 2in., on pedicels of about 1in., spuriously 5-celled. Nuts irregularly ovate, about 5 lines long and 3 or 4 lines broad, with a smooth but somewhat lacunose face, and when dry freer from the surrounding mealy substance than is the case in *S. australe*. Putamen hard and thick. Testa of seed brown.

Hab.: Musgrave Electric Telegraph Station, Cape York Peninsula, *Geo. Jacobson*, who says that the fruit is edible, and resembles in taste the common white guava (which it is not unlike in size and appearance).
Fruit eaten raw, Palmer River, *Roth*.
This new species differs from *S. australe* in the texture and form of leaves, the pendulous branches, the size of fruit, and in the figure of its wood. I think it may prove identical with one growing in the scrubs of Tringilburra Creek, of which I picked up fruit, but could not identify the tree from which they had fallen.

3. **S. membranaceum** (membranous), *Bail. Ql. Agri. Journ.* v. 388. A tree of 60 or more feet in height, the branchlets rather slender and often deeply striate. Leaves oblong-lanceolate, membranous in comparison with those of the other Queensland species, 4 to 6in. long, 1 to 1¾in. broad, the margins somewhat wavy; apex acuminate, slightly cuneate at the base; petioles short. Flowers small, few in axillary cymes. Fruit yellow, globose-turbinate, attaining a diameter of 2in.; smooth, more or less sunk at the apex, and often deeply 5-sulcate, soft and the inner substance mealy, but if kept a few months becoming quite hard as in the other kinds.

Hab.: Evelyn, Herberton district, *J. F. Bailey*.

Order XXXVII. STACKHOUSIEÆ.

Flowers regular, hermaphrodite. Calyx small, 5-lobed or 5-cleft. Petals 5, perigynous or almost hypogynous, with elongated claws, usually free at the base but united upwards in a tubular corolla, with spreading lobes, imbricate in the bud. Disk thin, lining the calyx-tube. Stamens 5, inserted on the margin of the disk; filaments free, slender; anthers oblong. Pollen grains smooth or echinate. Ovary free, 2 to 5-lobed, 2 to 5-celled; style single, with 2 to 5 lobes, stigmatic along the inner side. Ovules solitary in each cell, erect, anatropous. Fruit of 2 to 5 globular, angular, or winged indehiscent cocci, at length seceding from the axis. Seeds solitary, erect; testa membranous; albumen fleshy; embryo straight; cotyledons short; radicle inferior.—Herbs, usually forming a perennial stock, with erect, little branched, virgate stems, often assuming a yellowish colour, rarely dwarf and tufted. Leaves alternate, narrow, entire, often somewhat fleshy. Stipules none or very minute. Flowers in terminal spikes, rarely solitary, with 3 minute or linear bracts (1 bract and 2 bracteoles) at their base. Stamens included in the corolla-tube, of very unequal lengths. Pistil almost always 3-merous.

The Order is limited to two genera, almost endemic in Australia, one species extending to the Philippine Islands, and another represented by a closely allied species in New Zealand.—*Benth.* (in part).

Petals perigynous. Filaments slender. Pollen-grains sub-4-lobed, echinate . 1. STACKHOUSIA.
Petals almost hypogynous. Filaments very short. Pollen-grains ovate, very smooth . . . . 2. MACGREGORIA.

1. STACKHOUSIA, Sm.
(After J. Stackhouse.)
(Tripterococcus, *Endl.*; Plokiostigma, *Schuch.*)

Calyx-tube hemispherical, lobes imbricate. Petals erect, imbricate, often connate. Stamens alternate with the petals. Ovary lobed. Herbs with simple leaves.

Corolla-lobes oblong, obtuse. Stems elongated. Spikes terminal.
Cocci acutely angled or winged. Leaves obovate or obovate-oblong . . 1. *S. spathulata.*
Cocci obovoid or globular, reticulate. Leaves lanceolate, linear or filiform.
Spikes dense at the top, usually interrupted as the flowering advances.
Flowers 4 to 6 lines long.
Leaves flat, lanceolate or linear or rarely terete. Bracts small. . 2. *S. monogyna.*
Spikes filiform. Flowers distant, not 3 lines long. Leaves narrow, often very few . . . . 3. *S. muricata.*
Corolla-lobes acute or acuminate. Cocci obovoid or globular, reticulate.
Corolla 3 lines or less.
Leaves very narrow, subulate, pointed. Cocci tuberculous . . . . 4. *S. intermedia.*
Spikes long and slender. Flowers or clusters of flowers distant. Leaves oblong or linear. Sometimes few or very small . . . . . . 5. *S. viminea.*

1. S. spathulata (like a spatula), *Sieb. in Spreng. Syst. Cur. Post.* 124; *Benth. Fl. Austr.* i. 406. Glabrous, usually much branched at the base, with stout decumbent or ascending branches of about ¼ft., but sometimes lengthening to 1ft. or more. Leaves from obovate to oblong, usually very obtuse, rather thick, and ½ to ¾in. long, but in luxuriant stems lengthening out to 1in. or more and almost acute. Spikes dense, with the flowers almost of *S. monogyna*. Corolla-tube 3 to 4 lines long, lobes much shorter, oblong, obtuse. Cocci fully 2 lines long, with 3 prominent vertical acute angles or narrow wings.—F. v. M. Fragm. iii. 86; *S. maculata*, Sieb. in Hook. Journ. Bot. ii. 421; Hook. f. Fl. Tasm. i. 79 (the name originating in a clerical error in Sieber's label); *Tripterococcus spathulatus*, F. v. M. in Hook. Kew Journ. viii. 208; Schuch. in Linnæa, xxvi. 20; F. v. M. Fragm. iii. 86; *S. monogyna*, Labill. Pl. Nov. Holl. i. 77, t. 104 (as to the fruit).

Hab.: Sandy Cape, Hervey Bay, *R. Brown*: Moreton Island, *M'Gillivray, F. v. Mueller*; Logan River, *Rev. B. Scortechini*.

2. S. monogyna (single-styled), *Labill. Pl. Nov. Holl.* i. 77, t. 104 (partly); *Benth. Fl. Austr.* i. 406. Glabrous, with a perennial base and erect, simple or slightly branched, stout or slender stems, usually 1 to 1½ft., but sometimes twice that height. Leaves linear or lanceolate, acute or obtuse, crowded or few and distant, usually ½ to 1in. long, or when very luxuriant 2in. Racemes at first dense, but often lengthening out to 4 or 5in., the lower bracts sometimes leaf-like, passing into the very small lanceolate upper ones, and often all very small. Calyx-lobes narrow. Corolla-tube 3 to 4 lines long; lobes much shorter, oblong, obtuse. Cocci obovoid, prominently reticulate, not angled.—Lindl. Bot. Reg. t. 1917; Hook. f. Fl. Tasm. i. 79; *S. obtusa*, Lindl. in Bot. Reg., under n. 1917; *S. linariæfolia*, A. Cunn. in Field. N. S. Wales, 356; F. v. M. Fragm. iii. 87; *S. Gunnii*, Hook. f. Fl. Tasm. i. 79; Schlecht. Linnæa, xx. 642; *S. aspericocca*, Schuch. in Linnæa, xxvi. 12; *S. Muelleri*, Schuch. l.c. 16; *S. Gunniana*, Schlecht. in Schuch. l.c. 18.

Hab.: Keppel Bay, Broadsound, *R. Brown*: Port Curtis, *M'Gillivray*: Dawson and Bowen Rivers, *F. v. Mueller*.

Although Labillardière confounded this species with *S. spathulata*, and represented and described the fruit of the latter species, yet the common one, of which he described the flowering specimens, has been so universally known under his name that it would only increase the confusion to adopt a later name for that species. Among its numerous forms, the luxuriant specimens with more conical spikes which commonly pass for the true *S. monogyna*, and the smaller ones with fewer flowers and the young spike more obtuse, published by Lindley as *S. obtusa*, pass into each other by innumerable gradations. It is to the former that Schlechtendal gave the name of *H. Gunnii*, whilst Hooker's variety of that name is nearer to *H. obtusa*. A rather more distinct variety, with elongated slender stems, narrow and more distant leaves, sometimes very few and small, and rather smaller flowers, with smaller and smoother cocci, is amongst the more common Victorian and S. Australian forms, and is more especially the *S. linariæfolia*, A. Cunn., or *S. Muelleri*, Schuch. It has sometimes the almost terete leaves of *S. Huegelii*, from which it then differs in its very short bracts. The calyx in this variety is often strongly ribbed after flowering, but still more so in a slender northern variety, which has larger almost muricate cocci. A few Queensland specimens (Port Denison, *Fitzalan*), very slender, with small flowers in short dense spikes, seem almost to connect this with *S. muricata*. Indeed, different as are the extreme forms, the numerous specimens I have had before me show scarcely any definite limits between *S. monogyna*, *pubescens*, *Huegelii*, *flava*, *muricata*, and *viminea*.—Benth.

3. **S. muricata** (rough), *Lindl. in Bot. Reg. under n.* 1917; *Benth. Fl. Austr.* i. 408. Glabrous. Stems slender, simple or branched, often above 1½ft. long. Leaves narrow-linear, sometimes almost filiform, ¼ to 1¼in. long. Spikes long, very slender, with distant clusters of 2, 3, or more small flowers, usually under 3 lines and sometimes not 2 lines long. Calyx-lobes small, obtuse. Corolla-lobes narrow but obtuse, sometimes as long as the tube, sometimes not half so long. Cocci strongly reticulate, sometimes almost muricate.—Schuch. in Linnæa, xxvi. 25.

Hab.: Port Curtis and Dunk Island, *M'Gillivray*; Brigalow scrub in the interior, *Mitchell*; Peak Downs, *F. v. Mueller*.

This species, which we have also from the Philippine Islands, varies considerably and sometimes approaches *S. viminea*, but the leaves are never so broad, and the corolla-lobes obtuse. The Sturt's Creek specimens belong to a more branched and compact form, with very small flowers more frequently solitary, and the leaves few, small, and distant. Some smaller specimens, like those from the Philippine Islands, are less branched and perhaps sometimes annual.—Benth.

4. **S. intermedia** (intermediate), *Bail. Ql. Agri. Journ.* iii. 281. An erect annual. Stem striate or sulcate, only a few inches high; branches few, terminating in rather long slender spikes of minute flowers. Leaves very narrow, ¾ to 1¼in. long, margins revolute; apex shortly subulate, pointed. Flowers solitary or in twos or threes. Calyx-lobes about ¼ line with acute points. Corolla-tube scarcely longer than the calyx-lobes; lobes spreading, narrow, long as the tube. Cocci tuberculous.

Hab.: On Damp Rocks, Lizard Island.

The position of this species is between *S. muricata*, Lindl., and *S. viminea*, Sm. It was found in company with *Drosera indica*, Linn.; *Buchnera linearis*, R. Br.; and that curious rock-loving grass, *Diplachne loliiformis*, F. v. M.

5. **S. viminea** (shoots flexible), *Sm. in Rees' Cycl.* xxxiii.; *Benth. Fl. Austr.* i. 408. Glabrous. Stems erect or ascending, slender, often 1 to 1½ft. high. Leaves on the barren shoots often rather broad, oblong, obtuse, ¼ to 1in. long, narrowed at the base, on the flowering-stems fewer, often small and narrow-linear, and sometimes scarcely any. Spike slender, elongated, with distant clusters of small flowers, sometimes numerous in the clusters, sometimes solitary or nearly so. Calyx small, with acute lobes. Corolla rarely exceeding 3 lines and often not above 2 lines long, slender, with narrow acuminate or acute lobes. Cocci small, strongly reticulate or muricate.—Schuch. in Linnæa, xxvi. 22; *S. nuda*, Lindl. in Bot. Reg. under n. 1917; Schuch. l.c. 22; *S. monogyna*, Sieb. Pl. Exs.; *S. dorypetala*, Schuch. l.c. 24.

Hab.: Warwick, Stanthorpe, Islands of Moreton Bay.

Var. *elata*. Branches numerous and more erect, attaining 5ft. according to Maxwell, but several of Drummond's are under 1ft.; leaves all narrow; the whole plant drying more yellow than usual in the eastern variety, although some specimens of the latter are also yellow.—*S. elata*, F. v. M. Fragm. iii. 86. To this variety belong Maxwell's specimens above mentioned and Drummond's n. 92. A few Port Jackson ones can scarcely be distinguished from them.—Benth. Islands of Moreton Bay.

Bentham, in Flora Austr. i. 409, aptly remarks that the distinctions between this species and *S. muricata*, and the value of the character derived from the acute or obtuse corolla-lobes, require further investigation on the living plant.

2. MACGREGORIA, F. v. M.
(After John Macgregor, of Victoria.)

Sepals 5, narrow-lanceolate, imbricate in the bud, free, persistent. Petals 5, almost hypogynous, free, membranous, alternate with the sepals, imbricate in the bud, oblong or obovate, spathulate, deciduous. Stamens 5, alternating with the petals. Filaments free, very short. Anthers linear-oblong, straight, introrse, basifixed, 2-celled, dehiscing longitudinally. Pollen-grains oval, very smooth. Stigmata 5, very short, sessile, linear-subulate. Cocci 5, superior, obovate, indehiscent, 1-seeded. Columella very short, the upper membrane ample. Pericarpum very thick. Seeds erect from basil cavities at base. Albumen fleshy. Embryo pyriform, straight, scarcely shorter than the albumen. Cotyledons plano-convex. Radicle short, inferior. A glabrous annual with but few linear leaves and terminal racemes of flowers.—F. v. M. Fragm. viii. 160.

1. **M. racemigera** (flowers in racemes), *F. v. M. Fragm;* viii. 161; *Hook. Ic.* 1280. A small, erect, or ascending branched herb. Leaves quite entire, somewhat acute, slight narrowing to the base, $\frac{1}{4}$ to $\frac{3}{4}$ in. long, $\frac{1}{4}$ to $\frac{3}{4}$ line broad. Pedicels very slender, 1$\frac{1}{2}$ to 4 lines long. Sepals about 1 line long, with a very narrow membranous margin, joined together only just at the very base. Petals 8 to 5 lines long, narrowing to a rather long base. Anthers $\frac{1}{4}$ line long, light-yellow, terminated by a very small, oval, white appendage. Cocci scarcely $\frac{1}{4}$ line long, sprinkled with very short hooked hairs. Testa shining, smooth, and marked with very fine dots. Embryo yellowish.—F. v. M., l.c.

Hab.: Towards Lake Nash, *M. Costello.*

Order XXXVIII. RHAMNEÆ.

Flowers regular, hermaphrodite, or rarely polygamous. Calyx campanulate, urceolate, or cylindrical, the tube persistent and often adnate to the ovary or disk; lobes 4 or 5, valvate, usually with a raised longitudinal line inside, and deciduous. Petals 4 or 5, concave or hood-shaped, inserted at the base of the calyx-lobes, alternating with and rarely exceeding them, or none. Stamens 4 or 5, alternating with the calyx-lobes, inserted with the petals and opposite to them when present; filaments short, filiform; anthers small, often enclosed in the petals, rarely oblong or exserted. Disk rarely wanting, usually filling the calyx-tube or lining it, or annular round the ovary when inferior, rarely cup-shaped and free. Ovary sessile on the disk or immersed in it, or more or less inferior, 3-celled, or rarely 2 or 4-celled; style short, entire, or with as many lobes or branches as ovary-cells; stigmas terminal, capitate or club-shaped. Ovules solitary in each cell, erect, anatropous, with a dorsal or rarely lateral raphe. Fruit a drupe or capsule, the border of the adnate base of the calyx forming a ring at the base or round the fruit or at the summit; epicarp thin and dry or fleshy; endocarp separating into as many membranous coriaceous or hard cocci as cells, or woody or bony, divided into cells. Seeds solitary, erect, usually ovate and somewhat compressed, often arillate; testa coriaceous or crustaceous and shining or rarely membranous; albumen fleshy or almost horny, often scanty, rarely wanting; embryo usually straight, with flat rather thick cotyledons and a short inferior radicle.—Shrubs or trees, very rarely, in genera not Australian, herbs, erect or climbing. Leaves alternate or rarely opposite, undivided, entire, or toothed. Stipules usually present but very deciduous, rarely spinous and persistent. Flowers small, usually green or yellowish, in cymes or umbel-like clusters, either solitary or forming axillary or terminal compound cymes, racemes or panicles.

A considerable Order, ranging over the tropical and temperate regions of both the New and the Old World. Of the 12 Australian genera, 3 are widely spread tropical or northern genera, and 1 tropical Asiatic, all represented in Australia by single or very few species, a fifth is South

XXXVIII. RHAMNEÆ.

American, with one Australian and one New Zealand species, the remaining 7, several of them numerous in species, are endemic or nearly so, *Alphitonia* extending to the Pacific Islands, and *Pomaderris* to New Zealand. The Order is a well-marked one, the floral characters separating it very readily from all except *Ampelideæ*, from which it is distinguished by the habit, by the drupaceous or capsular, not baccate fruit, and by the seeds; but most of the genera, even the most natural ones, are difficult to characterise. The differences in their flowers and fruits are very trifling; they often pass into each other by the finest gradations, and habit, foliage, and inflorescence must often be relied upon for fixing generic limits.—*Benth.*

TRIBE I. **Ventilagineæ.**—*Scandent unarmed shrubs or small trees. Leaves alternate. Disk filling the calyx-tube. Ovary superior or half-superior. Fruit dry, 1-celled, 1-seeded, girt at the base or middle by the calyx-tube. Seeds exalbuminous.*

Leaves penninerved. Panicle branches elongated and raceme-like. Nut 1-seeded, produced into a long wing-like appendage . . . . 1. VENTILAGO.

TRIBE II. **Zizypheæ.**—*Shrubs or trees. Disk filling the calyx-tube. Fruit a dry or fleshy drupe, with a 1 to 3-celled stone, the base or middle girt by the calyx-tube.*

Leaves 3 or 5-nerved. Drupe succulent, the putamen woody or bony, 1 to 4-celled. Stipules usually spinescent . . . . 2. ZIZYPHUS.

TRIBE III. **Rhamneæ.**—*Shrubs or trees. Disk lining or filling the calyx-tube. Ovary superior or half-superior. Fruit dry or fleshy, of 3 (rarely 2 or 4) pyrenes or cocci.*

Calyx-tube very short. Petals clawed. Style shortly bifid. Ovary 2-celled. Disk margins free. Fruit juicy outside. Endocarp indehiscent, 1 rarely 2-celled . . . . 3. DALLACHYA.
Calyx deeply 5-cleft, lobes deciduous. Petals orbicular-rhomboid, rolled inwards. Filaments filiform. Disk slightly undulate at the margin. Fruit 3-lobed. Seeds roundish . . . . 4. SCHISTOCARPÆA.
Drupe with a thin epicarp, covering membranous or crustaceous cocci. Unarmed . . . . 5. COLUBRINA.
Panicle or cyme 2 or 3-chotomous. Endocarp separating into cocci.
    Ovary immersed in the disk. Epicarp thick. Leaves white or rusty underneath . . . . 6. ALPHITONIA.
    Ovary sessile on the disk. Epicarp thin. Leaves green on both sides . 7. EMMENOSPERMUM.
Calyx campanulate or tubular. Disk none, or annular, or lining the calyx-tube. Ovary partially or wholly inferior. Leaves alternate, usually small and entire (except a few *Pomaderrises*). Fruit under 2 lines diameter.
    Calyx-tube entirely adnate, or lined by the disk up to the lobes. Petals none, or concave, not enclosing the anthers, which are either oblong or on long filaments. Flowers usually pedicellate. Bracts very deciduous . . . . 8. POMADERRIS.
    Calyx-tube produced above the ovary and disk.
        Flowers sessile, or nearly so, in cymes, often contracted into heads surrounded by imbricate brown bracts . . . . 9. STENANTHEMUM.
        Flowers solitary or in leafy spikes, sometimes contracted into heads, or pedicellate, individually surrounded by brown bracts . . . . 10. CRYPTANDRA.

TRIBE IV. **Colletieæ.**—*Ovary free or half-immersed at the base of the calyx. Calyx-tube urceolate, campanulate, or cylindrical, produced above the ovary and annular disk. Stamens in the orifice of the calyx. Fruit coriaceous, containing 2 or 3 cocci, or drupaceous and the putamen 1 to 3-celled.—Small trees or shrubs, the branches opposite, often spinescent. Leaves opposite, sometimes small or none. Flowers fasciculate or solitary.*

Calyx campanulate or tubular, the tube produced above the ovary and annular disk. Spines and small leaves opposite . . . . 11. DISCARIA.

TRIBE V. **Gouanieæ.**—*Ovary inferior. Disk various. Fruit coriaceous, containing 3 or 4 cocci, very often 3-winged or 3-angled, crowned by the calyx-limb.—Shrubs, rarely herbs. Leaves alternate, often broad. Flowers racemose, cymose.*

Fruit 3-winged. Flowers in racemose spikes . . . . 12. GOUANIA.

1. VENTILAGO, Gærtn.

(From *ventilo*, to blow, and *ago*, to drive away; the winged fruit being easily carried by the wind.)

Calyx 5-lobed, spreading. Petals hood-shaped or none. Stamens 5, scarcely exceeding the petals when present. Disk flat or concave, filling the short calyx-tube. Ovary more or less immersed in the disk, 2-celled; style short, with 2 short erect stigmatic lobes. Nut globular at the base, produced into an oblong or linear coriaceous wing, 1-celled and 1-seeded, indehiscent. Seed globular; testa membranous; albumen none; cotyledons thick and fleshy.—Climbing shrubs or trees. Leaves alternate, penninerved. Flowers small, clustered along the branches of axillary or terminal panicles.

The genus is dispersed over the tropical regions of the Old World. The Australian species are endemic, differing from the others in habit and foliage as well as in the absence of petals.

A small tree. Branchlets glabrous. Leaves narrow-lanceolate, 2 to 5in. long, entire. . . . . 1. *V. viminalis.*
A tall climber. Branchlets and inflorescence hairy. Leaves 1½ to 2in. long, ovate-lanceolate, the margins denticulate-repandulate . . . . 2. *V. ecorollata.*

1. **V. viminalis** (branches flexible), *Hook. in Mitch. Trop. Austr.* 369; *Benth. Fl. Austr.* i. 411. "Thandorah," Cloncurry, *Palmer*. A small glabrous tree. Leaves narrow-lanceolate, 2 to 4 or even 5in. long, entire, narrowed into a petiole, coriaceous, the pinnate veins very oblique and sometimes almost parallel with the midrib, without the elegant transverse venation of the rest of the genus. Panicles not much branched or almost reduced to simple racemes, shorter than the leaves, solitary or clustered in the axils. Calyx about 1 line long. Petals none. Disk entirely adnate to the short broad calyx-tube. Ovary slightly immersed in the disk. Fruit glabrous, about 1in. long including the wing, the turbinate adnate base of the calyx not attaining above a quarter the length of the globular nut.

Hab.: High sandy ridges on the Maranoa, *Mitchell;* Gulf of Carpentaria, *F. v. Mueller;* Rockhampton, *O'Shanesy;* Barcoo and many other inland localities.

This is one of the woods recorded as being used by the aborigines for making boomerangs and fire-sticks.—*Palmer.*

2. **V. ecorollata** (without corolla), *F. v. M. in Wing's So. Sc. Rec.,* 1883. A tall, climbing plant, the branchlets hairy-pubescent. Leaves on short petioles, ovate-lanceolate, shining on both sides, 1½ to 2in. long, the margins denticulate-crenulate; texture chartaceous. Racemes axillary, about 1in. long, scarcely pedunculate, hairy. Pedicels fascicled, of equal length or shorter than the flowers. Calyx about 1 line long, yellow inside. Petals none. Filaments inserted on the margin of the disk. Anthers oblong or oval-dorsifixed. Disk entire, slightly adnate to the tube of the calyx. Style somewhat thick, compressed, very shortly 2-fid, the lower portion pilose. Fruit surrounded only at the base by the tube of the calyx, its appendage narrowly lanceolate-oblong, glabrous.—*Berchemia ecorollata,* F. v. M. Fragm. ix. 141.

Hab.: Rockingham Bay, *J. Dallachy,* and Endeavour River, *Persich* (F. v. M., l.c.).

2. ZIZYPHUS, Juss.

(From its Arabic name.)

Calyx 5-lobed, spreading. Petals hood-shaped or rarely none. Stamens 5, included in the petals or scarcely exceeding them, when present. Disk flat, filling the short calyx-tube. Ovary immersed in the disk, 2, rarely 3 or 4-celled; style shortly branched or styles distinct; stigmas small. Drupe ovoid or globular, putamen woody or bony, 1 to 4-celled, 1 to 4-seeded. Seeds with a smooth fragile testa; albumen none or scanty; cotyledons thick.—Trees or shrubs, usually armed with stipular prickles. Leaves alternate, 3 or 5-nerved, often distichous and very oblique. Flowers small, greenish, in axillary cymes. Fruit often edible.

The genus ranges over the tropical and subtropical regions of the New and the Old World. The two Queensland species are also common Asiatic ones.

Leaves green on both sides, softly pubescent or villous, or at length glabrous. Drupe small, 2-celled . . . . 1. *Z. Œnoplia.*
Leaves white or rusty underneath, with a close tomentum. Ovary and drupes 2-celled . . . . 2. *Z. jujuba.*

1. **Z. Œnoplia** (wine jujube), *Mill.; W. and Arn. Prod.* 168 (with the synonyms adduced, except *Z. Napeca*); *Benth. Fl. Austr.* i. 412. A shrub of several feet, with very divaricate branches, the young ones rusty-pubescent or villous. Stipular spines short, in pairs, one straight and deciduous, the other hooked or recurved and more persistent. Leaves very obliquely ovate, obtuse or slightly acuminate, 1 to 2in. long, entire or crenulate, 3 or 5-nerved, membranous, green on both sides, softly pubescent or villous, especially underneath, or sometimes glabrous when full-grown. Cymes small, compact, few-flowered, and almost sessile. Ovary 2-celled, style short, the stigma scarcely divided. Drupe globular, 2 or 3 lines diameter, 2-celled or 1-celled by abortion.—*Z. celtidifolia*, DC. Prod. ii. 20 (from the character given); Fenzl in Hueg. Enum. Pl. Ceyl. 20; *Z. rufula*, Miq. Fl. Ind. Bat. i. part 1, 643.

Hab.: Thursday and other islands of Torres Straits; Islands of the Gulf of Carpentaria, *R. Brown*. Common in East India and the Archipelago, but apparently not in Africa.

Of the two Linnæan *Rhamni* doubtfully referred here by Wight and Arnott, *R. Œnoplia* is quite correct; *R. Napeca*, however, is *Zizyphus lucida*, Moon; Thw. Enum. Pl. Ceyl. 74. The Linnæan herbarium has very good authentically named specimens of both.—*Benth*.

In India a decoction of the bark of the fresh root is said to promote the healing of flesh wounds. The fruit, which has a pleasant acid flavour, is also eaten. According to J. S. Gamble, the wood is used in India for saddle-trees, &c.; weight about 50lb. per cubic foot.

2. **Z. jujuba** (common jujube), *Lam.; W. and Arn. Prod.* 162 (with the synonyms adduced); *Benth. Fl. Austr.* i. 412. A tall shrub or small tree, with short stipular prickles, occasionally wanting. Leaves ovate or nearly orbicular, usually very obtuse, 1 to 3in. long, entire or toothed, 3-nerved, glabrous above, covered underneath, as well as the petioles and branches, with a close white or rusty tomentum. Cymes small, compact, and nearly sessile. Ovary 2-celled, tapering into a short 2-lobed style. Drupe globular, usually about ½ to nearly ¾in. diameter, 2-celled or 1-celled by abortion.

Hab.: Torres Straits, *Dubouzet*.

Very common, both wild and cultivated, throughout tropical Asia, extending also to tropical Africa.—*Benth*.
Fruit eaten by natives, *Thozet*.
Wood whitish, hard, and close-grained.—*Bailey's Cat. Ql. Woods No.* 82.
Common in India, where the fruit is prepared into pectoral lozenges called "Pate de Jujube," and the bark is employed in the Moluccas as a remedy for diarrhœa.

3. DALLACHYA, F. v. M.
(After John Dallachy.)

Calyx 5-fid, tube very short. Petals unguiculate. Stamens 5. Anthers ovate-cordate, introrse, pointless. Style shortly bifid. Disk cupular, margin free. Ovary 2-celled, exocarp juicy. Endocarp indehiscent, 1, rarely 2-celled, cartilaginous. Seeds ovate. Albumen none. Cotyledons collateral, plano-convex, green. Radicle very short.—An unarmed small tree. Leaves chartaceous, almost ovate, with distant penninerves. Flowers in axillary clusters.—F. v. M. Fragm. ix. 140.

1. **D. vitiensis** (also of Viti), *F. v. M. Fragm.* ix. 140. "Murtilam," North Queensland, *Thozet*. A small tree, flowering as a shrub, quite glabrous, the branches slender. Leaves ovate or oval-oblong, shortly acuminate, 2 to 3in. long, entire or serrate-crenate, green on both sides, thin and apparently deciduous. Flowers in axillary sessile clusters, yellow inside, on slender pedicels of 3 or 4 lines. Calyx about 2 lines long, the tube broadly hemispherical; the inner lobes triangular, rather thin. Petals involute, enclosing the stamens. Disk concave, broadly cup-shaped, slightly crenulate, the margins free. Style compressed, sometimes undivided. Ovary broadly sessile, 2-celled, tapering into the short style. Drupe ovate, black, 3 to 4 lines long. Endocarp indehiscent; cells 1,

rarely 2, woolly, 1-seeded. Testa membranous, blackish. Cotyledons dimidiate-ovate.—F. v. M., l.c., who also mentions having heard through A. Thozet of a pubescent variety; *Colubrina ritiensis*, Seem. Mission to Viti, p. 434; *Rhamnus ritiensis*, Benth. Fl. Austr. i. 418; Seem. Fl. Viti. p. 42.

Hab.: Coast from Rockhampton to Somerset.
Fruit eaten by natives, *Thozet*.

4. SCHISTOCARPÆA, F. v. M.
(Fruit divided.)

Calyx deeply 5-cleft, its lobes semi-lanceolate, deciduous. Petals orbicular-rhomboid, short-stipitate, longitudinally rolled inward. Stamens hardly longer than the petals, much concealed by them. Filaments filiform. Anthers almost ovate, basifixed, longitudinally dehiscent. Style very firm, longer than the stigmas. Disk slightly undulate at the margin. Ovary 3-celled, almost fully emerged. Fruit roundish, somewhat turgidly 3-lobed, by the persistent short calyx-tube surrounded only at the base; exocarp crustaceous, irregularly 3-valvular; endocarp receding, thinly pergamenous, each of the three portions splitting to near their base along the inner side, much ruptured and twisted on the outer side. Seeds roundish, very convex at the outer side, much flattened and somewhat 3-gonous at the inner side; testa chartaceous; albumen none; cotyledons outward very convex; radicle minute, ovate included.—Vict. Nat., March, 1891.

This genus must stand near *Colubrina*, to which it could be referred as a section; but the course of the primary venules of the leaves is different, the calyx-tube under the fruit is shorter and less completely adherent, the cotyledons are outward very turgid and the albumen is wanting. The last-mentioned characteristic this new plant of ours has in common with *Scutia* and *Dallachya*, but both have a fruit of different structure, ours approaching that of *Macrorhamnus*.—*F. v. M., l.c.*

1. S. Johnsoni (after S. Johnson), *F. v. M., Vict. Nat.*, March, 1891. A plant of laurinaceous aspect. Leaves on very short petioles, scattered, of firm texture, mostly lanceolate-ovate, acuminate, entire, glabrous, 3 to 7in. long, 1 to 2½in. broad, shining on both sides but paler beneath, their primary venules rather distant, costular-adscending; their ultimate venules reticularly joined. Stipules semi-lanceolate, fugacious. Panicles small or even diminutive, axillary and terminal, formed by cymose clusters of flowers, beset with short scattered hairs. Bracts very small, varying from almost lanceolate to nearly deltoid. Calyx about ⅙in. long. Petals somewhat shorter than the calyx, membranous, pale-yellowish, as well as the stamens glabrous. Style and ovary bearing short hairs. Ripe fruit measuring rather more than ⅓in., dark outside, glabrous; valves of the exocarp somewhat bifid from the summit; endocarp of each fruitlet after secession divaricately spreading. Seeds about 2 lines; testa greyish-brown, without lustre, irregularly reticulate-regular. Embryo almost amygdaline. —F. v. M. l.c.

Hab.: On Mount Bartle Frere. *Stephen Johnson* (F. v. M., l.c.)
Colubrina Travancorica, doubtfully admitted by Beddome into that genus, has some resemblance to our new species, but the leaves are almost opposite, bear some indument and are distinctly serrulated; and as the mature fruit remains unknown, the generic position continues also dubious.—*F. v. M., l.c.*

5. COLUBRINA, L. C. Rich.
(The twisted stems supposed to resemble snakes.)

Calyx 5-lobed, spreading. Petals hood-shaped. Stamens 5, included in the petals. Disk thick, filling the calyx-tube. Ovary immersed in the disk, 3 or rarely 4-celled, tapering into a 3, rarely 4-cleft style, with obtuse stigmas. Drupe nearly globular, obscurely lobed, the epicarp thin or succulent, the endocarp separating into 3, rarely 4 membranous or crustaceous cocci, opening inwards by a longitudinal slit. Seeds without any arillus; testa smooth, shining, coriaceous; albumen fleshy but thin; cotyledons flat or incurved, thin or rather thick.—Erect or half-climbing shrubs or trees. Leaves alternate, 3-nerved at the base or penninerved in species not Australian. Stipules small, deciduous. Flowers small, in axillary cymes or clusters.

The species are nearly all American, tropical or subtropical, with one from tropical Asia, extending also into Australia.

1. **C. asiatica** (Asiatic), *Brongn.; W. and Arn. Prod.* 166 (with the synonyms adduced); *Benth. Fl. Austr.* i. 413. A large shrub or small tree, unarmed, and quite glabrous, with long, slender, often flexuose branches. Leaves petiolate, ovate or broadly cordate, acuminate, 2 to 3in. long, crenate-serrate, 3-nerved and penninerved, smooth and shining, but scarcely coriaceous. Cymes shortly pedunculate, rarely exceeding the petioles. Flowers greenish, about 2 lines diameter. Fruit about 4 lines diameter, depressed at the top, furrowed opposite the dissepiments, the endocarp separating more or less perfectly into 3 or rarely 4 membranous cocci.

Hab.: Cape York, *M'Gillivray*; Cape Grafton and Rodd's Bay, *A. Cunningham*; Howick's Group, *F. v. Mueller*; Shoalwater Passage, *R. Brown*; Port Denison, Rockingham Bay, and Russell River, of later collectors.

The species is common in tropical Asia, extending to the Pacific Islands.

6. ALPHITONIA, Reissek.

(Seeds like pearl barley.)

Calyx 5-lobed, spreading. Petals involute. Stamens 5, included in the petals. Disk thick, filling the calyx-tube. Ovary immersed in the disk, 2 or rarely 3-celled, tapering into a shortly lobed style. Drupe globular or broadly ovoid, the epicarp of a dry, mealy, or somewhat corky substance; endocarp of 2 or 3 hard coriaceous nuts or cocci, opening inwards by a longitudinal slit. Seeds with a shining hard testa, completely enclosed in a membranous brown shining arillus, open at the top, but with the edges folded over; albumen cartilaginous or horny; cotyledons flat.—Tree. Leaves alternate, penninerved. Cymes dichotomous, many-flowered. Seeds often persisting on the torus after the pericarp has fallen off.

The genus is probably limited to a single species, ranging from Australia to the Pacific Islands.

1. **A. excelsa** (tall), *Reissek, in Endl. Gen.* 1098; *Benth. Fl. Austr.* i. 414. Red Ash. "Ane," Batavia River, *Ward*; "Meeamee," Moreton Bay, *Watkins*; "Baragara," Cairns, *Cowley*; "Dunanya," Taromeo, *Shirley*. A tall hard-wooded timber-tree, the young branches, petioles, and inflorescence hoary or rusty with a close tomentum. Leaves petiolate, varying from broadly ovate or almost orbicular and very obtuse to ovate or lanceolate and acute or acuminate, usually 3 to 6in. long, entire, coriaceous, glabrous or slightly hoary above, white, or rarely rust-coloured underneath with a close tomentum, the parallel pinnate veins very prominent. Flowers 2 to 3 lines diameter, in little umbel-like cymes, arranged in dichotomous cymes in the upper axils or in a terminal corymbose panicle. Calyx tomentose. Disk broad and nearly flat. Fruit 3 or 4 lines diameter, or sometimes rather larger.—*Colubrina excelsa*, Fenzl, in Hueg. Enum. 20.

Hab.: Islands of the Gulf of Carpentaria, Sweers Island, Curtis Island, Rockhampton, Port Denison, Brisbane River, Moreton Bay.

The Carpentaria island specimens belong to a variety with remarkably large obtuse leaves, the flowers rather larger than usual, and the tomentum somewhat rusty. To this belongs *Zizyphus pomaderroides*, Fenzl, in Hueg. Enum. 20, judging from R. Brown's specimens corresponding to

Bauer's. *Alphitonia zizyphoides*, A. Gray, Bot. Amer. Expel. Exped. i. 278, t. 20 (*Rhamnus zizyphoides*, Soland.), which extends from Borneo and New Caledonia to the Pacific Islands, does not appear to differ at all from some of the eastern Australian specimens; whilst *A. franguloides*, A. Gray, l.c. 280, is very like some of the more tomentose N. Australian specimens.—*Benth.*

Wood near the outside somewhat pinkish, the inner wood dark-brown, or parti-coloured throughout, close-grained, very tough, warps in drying, but probably a good wood for the cabinet-maker. The wood usually, by keeping, becomes deeper in colour.—*Bailey's Cat. Ql. Woods No. 84.*

The leaves are sometimes infested with the fungi *Glœosporium Alphitoniæ*, Cke. and Mass., and *Pestalozzia Guepini*, Desm.

7. EMMENOSPERMUM, F. v. M.

(Referring to the seeds remaining attached to the torus.)

Calyx 5-lobed, the tube campanulate. Petals hood-shaped, inserted with the stamens on the margin of the disk. Stamens 5, enclosed in the petals. Disk thin, lining the calyx-tube. Ovary inserted on the disk in the bottom of the calyx-tube, but not immersed, 2-celled or rarely 3-celled, tapering into a shortly-cleft style. Fruit almost capsular, with a very thin almost dry epicarp, the endocarp separating into 2 or rarely 3 cartilaginous almost crustaceous cocci, opening along the inner face in 2 valves. Seeds inserted on a turbinate or slightly cup-shaped funicle, without any arillus; testa hard and shining; albumen cartilaginous; cotyledons flat.—Trees. Leaves opposite or alternate, penninerved. Cymes or panicles trichotomous, many-flowered. Seeds often persisting on the torus after the pericarp has fallen off.

The genus is endemic in Australia. It is closely allied in technical characters to the S. African *Noltia*, but with a different habit.—*Benth.*

Leaves opposite or nearly so . . . . 1. *E. alphitonioides.*
Leaves alternate . . . . 2. *E. Cunninghamii.*

1. E. alphitonioides (Alphitonia-like), *F. v. M. Fragm.* iii. 68; *Benth. Fl. Austr.* i. 415. A tall hard-wooded timber tree, quite glabrous. Leaves opposite or nearly so, petiolate, ovate, acuminate, 2 to 3in. long, entire, coriaceous, shining above, green on both sides. Flowers numerous, in little dense umbel-like cymes, arranged in trichotomous cymes or corymbose panicles in the upper axils or terminal. Calyx-lobes almost petal-like, nearly 1 line long. Fruits apparently about 3 lines long, but either unripe or already open in our specimen. Seeds persistent, like those of *Alphitonia*, but without the peculiar arillus of that species.

Hab.: Brush of Brisbane River, *M'Arthur*; Peri Creek, *Leichhardt*; Rockingham Bay, *J. Dallachy*.

Wood yellow, with a small, pinkish centre, close-grained; useful for cabinet-work.—*Bailey's Cat. Ql. Woods No. 84a.*

2. E. (?) Cunninghamii (after A. Cunningham), *Benth. Fl. Austr.* i. 415. Leaves alternate, similar to those of *E. alphitonioides*, except that the petioles are longer. Flowers not seen. Umbel-like cymes apparently not numerous, in a terminal corymbose panicle. Fruits rather larger than in *E. alphitonioides*, 3 or 4-celled; epicarp scarcely any; cocci 2-valved. Seeds red and shining as in that species, but not persistent on the torus, and the funicle very small.

Hab.: This species has been recorded by Baron Mueller for Queensland.

8. POMADERRIS, Labill.

(From *poma*, a lid, and *derris*, a skin; membranous covering of seed vessels.)

Calyx-tube entirely adnate to the ovary, the limb divided to the base into 5 lobes, usually deciduous or reflexed. Petals either concave or nearly flat, not enclosing the anthers, or none. Stamens 5, the filaments long and usually suddenly inflected and attenuate near the top; anthers oblong or ovoid. Disk annular, surrounding the ovary at the base of the calyx-lobes, often scarcely conspicuous, and never very prominent. Ovary half-inferior or rarely almost entirely inferior. Style 3-cleft, or rarely almost entire. Capsule protruding above the border of the calyx-tube, septicidally 3-valved, the endocarp separating into 3 crustaceous or membranous cocci, opening by a broad operculum at the base of the inner face, or by the separation of the whole inner face, or rarely by a longitudinal slit. Seed inserted on a short, thickened turbinate or cup-shaped funiculus.—Shrubs, with the young branches and under side of the leaves white, hoary or rusty with a close stellate tomentum, often mixed with or concealed by longer, simple, soft, often silky hairs. Leaves alternate, penninerved. Stipules brown and scarious, usually very deciduous. Flowers pedicellate, in small, umbel-like cymes, usually forming terminal panicles or corymbs, or rarely solitary in the axils of the leaves. Bracts brown and scarious, but so deciduous as to be seldom visible at the time of flowering.

The genus is confined to Australia and New Zealand; the Australian species are all endemic and from the eastern and southern districts, with the exception of two which are also found in New Zealand.—*Benth.*

Flowers with petals.
  Calyx-tube turbinate, at least half as long as the lobes. Cocci opening by an operculum below the middle.
    Leaves mostly ovate-lanceolate, 2 to 3in. long. Panicles many-flowered.
      Leaves hoary or tomentose above, softly tomentose underneath. Calyx about 2 lines long, very villous . . . . 1. *P. lanigera.*
      Leaves glabrous or sparingly scabrous-pubescent above, densely ferruginous, tomentose underneath. Calyx 1 to 1½ lines long, softly hairy . . . . 2. *P. ferruginea.*
    Leaves ovate and obtuse or oblong-elliptical, often above 2in. long, glabrous above, white underneath. Panicles many-flowered. Calyx 1 to 1½ line, closely tomentose or hairy . . . . 3. *P. elliptica.*
    Leaves firm, rarely above 1in. long. Panicles small and compact. Calyx of *P. elliptica* . . . . 4. *P. phillyreoides.*
Flowers without petals. Cymes rather loose, numerous in much-branched panicles. Calyx softly hairy, with a turbinate tube. Leaves mostly obtuse, scabrous above, often crenulate and rugose . . . . . . 5. *P. prunifolia.*

1. **P. lanigera** (woolly), *Sims. Bot. Mag.* t. 1823; *Benth. Fl. Austr.* i. 416. An erect branching shrub, nearly allied to *P. elliptica*, with which it is united by F. v. Mueller, differing chiefly in the leaves softly though minutely tomentose on the upper side, and the larger more villous flowers. Leaves oblong or ovate-lanceolate, the under side as well as the young branches clothed with a soft velvety tomentum often rust-coloured. Panicles often larger and less corymbose than in *P. elliptica.* Calyx about 2 lines long, very densely and softly hairy, the turbinate tube about half as long as the lobes. Petals ovate, concave, on slender claws. Fruit as in *P. elliptica*, but larger and more hairy.—DC. Prod. ii. 33, excluding the var. 2; *Ceanothus laniger*, Andr. Bot. Rep. i. 569; *P. obscura*, Sieb. Pl. Exs.

Hab.: Stanthorpe, *I. A. Bernays, junr.*

2. **P. ferruginea** (ferruginous), *Sieb.; Fenzl, in Hueg. Enum.* 21; *Benth. Fl. Austr.* i. 417. Very near *P. elliptica*, and united with it by F. v. Mueller, having the leaves glabrous above, and the small flowers of that species, but the leaves are usually rather longer for their breadth and more acute, and the down of the under side is much more dense, velvety and usually ferruginous. The flowers are more numerous, the calyx more softly and densely hairy, and the petals usually narrower. The fruits are the same.—Hook. f. Fl. Tasm. i. 76;

P. lanigera, *var.* 2, DC. Prod. ii. 33; *P. viridirufa*, Sieb. Pl. Exs.; *Ceanothus Wendlandianus*, Rœm. and Schult. Syst. v. 299 (from the character given); *Pomaderris Wendlandiana*, G. Don, Gen. Syst. ii. 39.—Benth.

Var. *canescens*, Benth. Flora Austr. i. 417 (in part). Leaves 3 to 4in. long, white, and less ferruginous underneath. Intermediate almost between *P. ferruginea* and *P. elliptica*. Percy Island, *A. Cunningham* (Benth, l.c) Of this I have no authentic specimens, but from some received from Mr. F. C. Simmonds it would seem that the variety is growing on Taylor's Range. This plant forms a shrub 5 or 6ft. high. Branchlets softly velvety, with light or dark brownish longer hairs. Leaves ovate-lanceolate, 2½ to 4½in. long, 1 to 1½in. broad, rather obtuse, apiculate, on petioles of about ⅙in., margins entire, glabrous above, the under side densely clothed with a nearly white tomentum and longish scattered brownish hairs. Panicles corymbose, terminal and in the upper axils. Stipules hairy outside, glabrous and purplish-brown inside, ovate-lanceolate, 3 lines long, frequently toothed. Bracts scarious, almost orbicular, margins ciliate. Flowers minute, with pedicels not exceeding 2 lines. Calyx with long hairs outside, glabrous, and satin-white inside. Petals very fugaceous, shorter than the calyx-lobes, white, spathulate, toothed on the margins, serrulate at the end; filaments glabrous, anthers slightly exserted, style-branches with broad, sometimes almost didymous stigmas.

3. **P. elliptica** (elliptic), *Labill. Pl. Nov. Holl.* i. 61, *t.* 86; *Benth. Fl. Austr.* i. 417. A tall shrub or small tree, the young branches rusty with a very close stellate down, intermixed occasionally with a few longer hairs. Leaves petiolate, ovate, oblong or ovate-lanceolate, obtuse or rarely almost acute, usually 2 to 3in. long and ¾ to 1½in. broad, entire or the margins slightly waved, glabrous above and smooth or scarcely scabrous, white underneath with a very close tomentum, the prominent midrib and principal parallel veins often rust-coloured. Cymes numerous, in dichotomous panicles, usually more or less corymbose. Stipules lanceolate, brown and scarious as well as the broad concave bracts, but all falling off in a very early stage so as to be rarely seen at the time of flowering. Calyx about 1½ line long, white with a minute stellate tomentum, often intermixed with longer simple hairs, especially on the turbinate tube. Petals usually broadly cordate or nearly orbicular, concave, on slender claws, but often much narrower, sometimes deeply toothed and occasionally abortive. Style-branches short, with capitate stigmas. Capsule about 1½ line diameter, slightly hairy, the free part rather shorter than the adnate portion, the cocci opening in a round valve or operculum below the middle.—Bot. Mag. t. 1510; DC. Prod. ii. 33; Hook. f. Fl. Tasm. i. 76; F. v. M. Fragm. iii. 69.

Hab.: Near the tops of many hills in southern and tropical Queensland.

Two species are usually distinguished, *P. elliptica*, with broader more obtuse leaves and without any silky hairs mixed with the stellate tomentum of the calyx, and *P. discolor*, DC. Prod. ii. 33, Sweet, Fl. Austr. t. 41, with the calyx, at least the tube, more or less silky-hairy and the leaves often less obtuse. Labillardière's specimens belong to the former, but his description agrees better with the latter; and in many instances the two forms pass one into the other. Sieber's specimens, n. 208 *(P. malifolia,* Sieb.; *P. multiflora,* Fenzl, in Hueg. Enum. 21), are very broad-leaved, with the tomentose calyx of the first form; n. 213 *(P. discolor)* belongs to the second; n. 210 *(P. intermedia,* Sieb; DC. Prod. ii. 33) has the leaves narrower than usual and the indumentum of the calyx variable. *Ceanothus discolor,* Vent. Jard. Malm. t. 58, has the more acute leaves of the second form with the close tomentum of the first. *P. acuminata,* Link. Enum. Hort. Berol. 235, is probably established on the same garden-plant as Ventenat's.—*Benth.*

4. **P. phillyreoides** (Phyllyrea-like), *Sieb. in DC. Prod.* ii. 33; *Benth. Fl. Austr.* i. 418. A shrub, said to be of much smaller stature than *P. elliptica*. Down of the young branches sometimes very close and white or rusty, sometimes loose and more rusty, almost as in *P. ferruginea*. Leaves much smaller than in any of the preceding species, seldom attaining 1½in., and usually much shorter, oblong or oval, obtuse or acute, entire, of a firm consistence, glabrous or minutely hoary above, soft underneath with a white or rusty down. Flowers rather larger than in *P. elliptica*, but variable in size, the cymes compact, in small terminal panicles. Calyx softly silky-hairy, the turbinate tube shorter than the lobes.

T

Petals nearly of *P. elliptica*, but usually narrower. Styles more deeply cleft, the branches club-shaped at the top, with somewhat decurrent stigmas. Capsule of *P. elliptica*.—*P. andromedæfolia*, A. Cunn. in Field N.S. Wales, 351; Bot. Mag. t. 3219; *P. phillyreæfolia*, Fenzl, in Hueg. Enum. 22 (from the character given), Benth.

Hab.: Between Stanthorpe and the border.

Var. *nitidula*. Leaves more coriaceous, usually acute; tomentum closer, very white on the under side of the leaves. Stanthorpe; Mount Lindsay, *W. Hill*.

5. **P. prunifolia** (Plum-leaved), *A. Cunn.; Fenzl, in Hueg. Enum.* 22; *Benth. Fl. Austr.* i. 420. Stellate tomentum of the branches and under side of the leaves dense and white, or sometimes ferruginous. Leaves ovate or oblong, obtuse or mucronate, seldom above 1½in. long, wrinkled, and often scabrous above, with short simple or stellate hairs. Flowers small and numerous, in many-flowered compact cymes, arranged in thyrsoid terminal panicles as in *P. ligustrina*. Calyx obovoid, about 1 line long, the tube turbinate, the stellate tomentum usually concealed by long silky hairs. Petals none. Styles cleft nearly to the base. Capsule about 1 line diameter, hirsute, obtuse, only slightly protruding from the adnate tube of the calyx.—F. v. M. Fragm. iii. 75.

Hab.: Between Stanthorpe and the border.

9. STENANTHEMUM, Reissek.
(Flowers slender.)

Flowers sessile in heads, surrounded by small, persistent, imbricate brown bracts. Calyx-tube adherent at the base, free, slender, and often deciduous above the ovary and disk, 5-lobed at the top. Petals 5, hood-shaped, enclosing the anthers and inserted with the stamens at the top of the calyx-tube. Disk scarcely prominent, round the top of the ovary at the base of the calyx-tube. Ovary wholly inferior, 3-celled. Style entire or minutely 3-toothed. Capsule enclosed in the base of the calyx-tube, which is often contracted over it or deciduous; the endocarp separating into 3 membranous or crustaceous cocci opening in 2 valves. Seeds of *Pomaderris*.—Shrubs, with a habit of *Spyridium*. Flowers sessile, in heads, or in one species in a cyme, surrounded by small, persistent, imbricate brown bracts, and sometimes with 1 or 2 floral leaves, as in *Spyridium*.

The genus is confined to Australia. The floral characters are those of *Cryptandra*, with the inflorescence of *Spyridium*.—Benth.

1. **S. Scortechinii** (after Rev. B. Scortechini), *F. v. M., Austr. Chem. and Drug.*, 1884. An erect shrub of several feet, the branches pubescent with stellate hairs. Leaves lanceolate, acute, ½ to ¾in. long, 2 to 4 lines broad, shining above, grey beneath, the margins recurved. Flower-heads dense and somewhat compound, terminal, ½ to 1in. diameter; bracts scarious, dark-brown, almost oval, keeled. Calyx small, enveloped in white intricate hairs; lobes shorter than the tube, woolly; petals and stamens inserted near the summit of the calyx-tube. Style short, capsule glabrous.—F. v. M. l.c.

Hab.: Severn River.

10. CRYPTANDRA, Sm.
(From the anthers being hidden in the petal.)
(Wichurea, *Nees*.)

Calyx-tube adherent at the base, free, campanulate or tubular and persistent above the ovary and disk, 5-lobed at the top or to the middle. Petals 5, hood-shaped, enclosing the anthers and inserted with the stamens at the top of the calyx-tube. Disk annular, or often scarcely prominent round the top of the ovary, at the base of the calyx-tube. Ovary wholly inferior, or slightly prominent in the calyx-tube, 3-celled. Style entire or minutely 3-toothed. Capsule enclosed in the base of the persistent calyx-tube, but often partially free within it, the endocarp or the whole capsule separating into three crustaceous or rarely membranous cocci usually opening inwards in 2 valves. Seeds of *Pomaderris*.—Shrubs, mostly heath-like or thorny. Leaves small, narrow, often clustered, rarely ovate and flat, often nearly cylindrical, the under surface usually tomentose and whitish, but often concealed by the closely revolute margins. Flowers sessile or shortly pedicellate, mostly surrounded by persistent imbricate brown bracts, either distinct along the smaller branches or clustered in terminal spikes or heads intermixed with leaves, never in cymes.

A genus confined to Australia. Like the majority of *Rhamneæ*, it is chiefly distinguished by habit. The floral characters of the first section are nearly those of *Stenanthemum*, of the second scarcely distinct from *Discaria*.—*Benth.*

SECT. 1. **Cryptandra.**—*Disk usually pubescent, continuous with the summit of the ovary, either undistinguishable from it or forming a slightly prominent ring round it.*
Flowers pubescent or hairy, closely sessile in terminal or lateral heads.
  Brown bracts acuminate. Calyx tubular. Heads many-flowered. Calyx
  narrow. Ovary almost entirely inferior . . . . 1. *C. ericifolia.*
Flowers pubescent or hairy, sessile in spikes or short heads, or not crowded.
  Brown bracts obtuse, very much shorter than the calyx-tube.
  Calyx 1 line long or more, the tube longer than the lobes.
    Calyx narrow, glabrous outside at the base, tomentose above. Adnate
    base of the ovary longer than the free top . . . . 2. *C. spinescens.*
    Calyx broadly campanulate or urceolate, tomentose all over. Free
    part of the ovary longer than the adnate base . . . . 3. *C. amara.*

SECT. 2. **Wichurea.**—*Disk glabrous or villous, distinct from the ovary, usually annular.*
Calyx glabrous or very slightly tomentose.
Leaves linear, with revolute margins. Calyx campanulate, deeply lobed.
  Disk and ovary glabrous . . . . 4. *C. longistaminea.*

1. **C. ericifolia** (heath-like), *Sm. in Trans. Linn. Soc.* x 294, *t.* 18, *f.* 1; *Benth. Fl. Austr.* i. 438. Branches elongated and twiggy, with few smaller branchlets, always unarmed, more or less pubescent with simple appressed hairs. Leaves linear-terete, or with a slightly prominent midrib, 2 to 4 lines long, often clustered or crowded, glabrous or pubescent with simple appressed hairs. Flowers crowded in little terminal heads surrounded by leafy bracts, and each flower by several imbricate, acuminate, and ciliate brown bracts, often half as long as the calyx. Calyx narrow-campanulate, about 2 lines long, silky-hairy outside, the lobes short and spreading. Ovary very small, slightly projecting above the very short adnate part. Style pubescent at the base. Disk inconspicuous. Cocci opening in 2 valves.—*C. capitata,* Sieb. Pl. Exs.

Hab.: Stanthorpe.

2. **C. spinescens** (spinous), *Sieb. in DC. Prod.* ii. 38; *Benth. Fl. Austr.* i. 439. Nearly allied to *C. amara*, and with nearly the same foliage, but the branches are usually more twiggy and the spinous branchlets more densely crowded. Leaves usually linear or linear-oblong, 2 or rarely 3 lines long, but occasionally small and obovate. Flowers smaller than in *C. amara*, and more distinctly although very shortly pedicellate. Calyx 1½ to 2 lines long, narrow-campanulate, the adnate base glabrous and suddenly contracted into a little stipes about the length of the imbricate brown bracts, the free part white-tomentose outside. Ovary almost entirely inferior, the pubescent summit slightly prominent above the adnate part and obscurely grooved opposite the stamens, but without any distinct disk. Capsule oblong, 1½ to 2 lines long, almost included in the glabrous, elongated, adnate base of the calyx-tube, shortly free in the upper part. Cocci thinly crustaceous.—*C. pyramidalis,* R. Br., Brongn. in Ann. Sc. Nat. x. 373.

Hab.: Southern parts of the colony.

3. **C. amara** (bitter), *Sm. in Trans. Linn. Soc.* x. 295, *t.* 18, *f.* 2; *Benth. Fl. Austr.* i. 440. A rigid, wiry, decumbent or suberect, much-branched shrub, the young branches minutely hoary with a close stellate down, the smaller ones often ending in a fine thorn. Leaves solitary or clustered, linear or linear-oblong, usually 1 to 2 and rarely 3 lines long, obtuse or acute, rigid, glabrous or nearly so, the margins usually recurved. Flowers almost sessile, solitary within the bracts, but usually several together, forming short leafy spikes or racemes on the smaller branches. Calyx at the time of flowering 1 to 1½ line long, campanulate, white outside with a close minute down, very shortly adnate by its obtuse base, the lobes usually shorter than the tube, the brown imbricate bracts not exceeding the adnate base and very obtuse. Ovary densely pubescent, included in the tube, but adnate only below the middle, the disk not distinct. Fruiting calyx often 3 lines long, enclosing the capsule, which remains adherent at the base only or below the middle. Cocci crustaceous.—DC. Prod. ii. 88; F. v. M. Fragm. iii. 66; *C. Sieberi*, Fenzl, in Hueg. Enum. 23; Hook. f. Fl. Tasm. i. 74; *C. campanulata*, Schlecht. Linnæa, xx. 689; F. v. M. Fragm. iii. 67, partly; *C. nervata*, Reissek, in Linnæa, xxix. 291; *C. largiflora*, F. v. M., Reissek, in Linnæa, xxix. 292.

Hab.: Kent's Lagoons, *Leichhardt;* Mount Mitchell, *Beckler;* Cunningham's Gap, and other southern localities.

Independently of the diversity in the size of the flowers resulting from age, there appear to be two distinct varieties with large and small flowers, the calyx in the latter usually broader and more deeply lobed, both of them included among Sieber's specimens; the southern ones belong chiefly to small-flowered varieties. These have usually the free part of the ovary less prominent, but in Cunningham and Fraser's specimens from the interior the ovary and capsule are very prominent, whilst the calyx is small and much more loosely pubescent than usual. Some specimens are remarkable for their short, almost ovate leaves.—*Benth.*

4. **C. longistaminea** (long stamens), *F. v. M. Fragm.* iii. 64; *Benth. Fl. Austr.* i. 440. A much-branched unarmed shrub of 2 or 3ft., the smaller branches minutely hoary-tomentose. Leaves ovate or oblong, obtuse, 1 to 2 lines long, the margins recurved or revolute, glabrous above, minutely silky-tomentose underneath or almost glabrous. Flowers numerous, crowded on the smaller branches, but not quite sessile. Brown bracts imbricate round the base of the calyx-tube. Calyx about 2 lines long, minutely silky outside, divided below the middle into spreading lobes. Petals on slender claws, at first enclosing the stamens, but reflexed after the calyx opens, leaving the stamens erect and apparently exserted. Disk annular, glabrous or very minutely tomentose, quite distinct from the ovary. Ovary sessile or slightly immersed in the disk. Style very shortly 3-lobed. Fruit not seen.

Hab.: Condamine River, *C. H. Hartmann.*

11. DISCARIA, Hook.
(Broad disk.)
(Tetrapasma, *G. Don.*)

Calyx campanulate or tubular above the ovary, shortly 4 or 5-lobed. Petals hood-shaped, inserted with the stamens at the base of the calyx-lobes or none. Stamens 4 or 5, with short filaments, included in the petals when present. Disk annular in the base of the calyx-tube, the margin shortly free. Ovary more or less immersed in the disk, 3-lobed, 3-celled. Style slender, with a shortly 3-lobed stigma. Drupe or capsule coriaceous, 3-lobed, the endocarp separating into 3 2-valved crustaceous cocci. Seeds with a coriaceous testa; albumen fleshy; cotyledons orbicular.—Much-branched rigid shrubs, with opposite, often thorny branchlets. Leaves small, opposite, 1-nerved or penninerved. Stipules and bracts small. Flowers axillary.

The genus is chiefly S. American, extratropical or alpine, with one species endemic in Australia and another in New Zealand.—*Benth.*

1. **D. australis** (Australian), *Hook. Bot. Misc.* i. 157, t. 45; *Benth. Fl. Austr.* i. 445. A scrubby, much-branched, thorny shrub of 1 to 2ft., usually glabrous. Branches green, terete, the smaller ones reduced to stout spines of 1 to 1½in. Leaves often appearing clustered from the shortness of the shoots, oblong or cuneate, obtuse or emarginate, rarely exceeding ½in. Pedicels solitary or clustered in the axils of small leaves, which soon fall off from the very short branches, the flowers then appearing densely clustered under the spines. Flowers sweet-scented, white. Calyx-tube broadly campanulate above the disk, the limb spreading to about 2 lines diameter. Petals spathulate or oblong-cuneate, hood-shaped. Ovary deeply immersed in the disk, the short free part 3-lobed. Fruit 2 to 3 lines diameter.—Hook. f. Fl. Tasm. i. 69; Reissek. in Linnæa, xxix. 266; F. v. M. Fragm. iii. 88; *Colletia pubescens*, Brongn. in Ann. Sc. Nat. x. 366; *Tetrapasma juncea*, G. Don, Gen. Syst. ii. 40; *Colletia Cunninghamii*, Fenzl, in Hueg. Enum. 23.

Hab.: Eton Vale, Darling Downs.

12. GOUANIA, Linn.
(After A. Gouan.)

Flowers polygamous. Calyx-tube short, obconic, adhering to the ovary; limb 5-lobed. Petals 5, hood-shaped or flat, inserted below the margin of the disk. Stamens 5, included under the petals or exceeding them. Anthers dehiscing longitudinally. Disk glabrous or hairy, epigynous, filling the calyx-tube, 5-angled or lobed. Ovary immersed in the disk, 3-celled. Style 3-partite or lobed. Stigma small. Fruit coriaceous, quite inferior, crowned by the persistent calyx-lobes, 3-winged, 3-coccous, cocci woody, indehiscent, separating from the 6-partite axis. Wings large, rotundate. Seeds obovate, plano-convex. Testa horny, shining. Albumen thin. Cotyledons rotundate, broad. Radicle very short. Tall, climbing, cirrhose shrubs, glaucous or tomentose. Branches slender, elongate. Leaves alternate, petiolate, entire or dentate, 3-nerved from the base or penninerved. Stipules oblong or lanceolate, deciduous. Flowers small, in terminal or axillary racemes; rhachis often passing into tendrils.

The species are scattered through the tropics.

Leaves on short petioles. Fruit, including the wings, 4 lines diameter . . 1. *G. australiana*.
Leaves on longish petioles. Fruit very small, including the wings about 2 lines . . . . 2. *G. Hillii*.

1. **G. australiana** (Australian), *F. v. M. Fragm.* iv. 144, *and Bail. Rep. Bell. Ker Exp.*, 1889. A tall, velvety-tomentose climber. Leaves cordate or lanceolate-ovate, acuminate and sharply denticulate towards the apex, 2 to 4in. long, 1¼ to 3in. broad, penninerved, tomentose on both sides. Petioles somewhat short, the spiciferous panicles wide-spreading. Bracts narrow, semi-lanceolate, scarcely 1¼ line long. Flowers subsessile, densely hoary. Fruit 3-winged, 3-celled, including the wings about 4 lines diameter, tomentose, like the rest of the plant. Wings veined. Seeds glossy, light-brown, 1 line long, showing about 3 blunt angles.

Hab.: Rockingham Bay, *J. Dallachy* (F. v. M., l.c.); scrubs of the Mulgrave River.

2. **G. Hillii** (after Walter Hill), *F. v. M. Fragm.* viii. 168. The young branches and petioles fusco-tomentose. Leaves ovate or cordate-orbicular, entire except the somewhat acute apex, the upper side glabrous, under side moderately pilose, 2 to 3in. long, 1¼ to 2in. broad. Petioles somewhat long. Nerves conspicuous on the under side. Stipules caducous. Flowers in spiciform terminal panicles, about 6in. long. Fruit glabrous, including the wings about 2 lines broad.

Hab.: Daintree River, *W. Hill*, whom Baron Mueller says described the species as a small tree.

Order XXXIX. AMPELIDEÆ.

Flowers regular, hermaphrodite or unisexual. Calyx small, entire or 4 or 5-toothed. Petals 4 or 5, free or cohering, valvate in the bud. Stamens 4 or 5, opposite the petals, inserted on the outside of the disk at its base or between its lobes. Disk free or adnate to the ovary. Ovary usually immersed in or surrounded by the disk, more or less perfectly 2 to 6-celled; style short and conical or subulate, or none; stigma small, capitate or lobed. Ovules 2 in each cell where there are 2 cells, solitary where there are more cells, erect, anatropous, with a ventral raphe. Fruit a berry, the dissepiments frequently disappearing. Seeds 1 to 6; testa hard, the inner coating frequently penetrating into the fissures of the ruminate albumen. Embryo short, in the base of the albumen; cotyledons oval; radicle short, inferior.—Woody climbers or rarely erect shrubs or small trees. Branches often articulate. Leaves alternate or the lower ones opposite, simple or compound, the petiole usually articulate with the stem and expanded into a membranous stipule. Flowers small, in little umbels, cymes, racemes, or spikes, arranged in leaf-opposed, cymose, thyrsoid, or elongated panicles.

The Order, almost or quite limited to the two following genera, is widely dispersed over the tropical and warm regions of the globe, more abundant in the Old World than in America, and the smaller genus confined to the Old World. It is very nearly allied to *Celastrineæ*, and especially to *Rhamneæ*, from which it differs in habit, in the more developed petals, in the baccate fruit, and in the smallness of the embryo.—*Benth.*

Stamens free. Ovary 2-celled with 2 ovules in each cell. Woody climbers, with tendrils . . . . 1. Vitis.
Stamens and petals connate with the disk. Ovary 3 to 6-celled with 1 ovule in each cell. Erect, without tendrils . . . . 2. Leea.

1. VITIS, Linn.
(An ancient name of the Grape vine.)
(Cissus, Linn.)

Petals free or cohering at the tips, and falling off together. Stamens inserted round the base of the short, annular, or lobed disk. Ovary 2-celled (sometimes imperfectly so), with 2 ovules in each cell.—Woody climbers or rarely bushy shrubs, with leaf-opposed tendrils (abortive inflorescences). Leaves simple or compound, sometimes marked with pellucid dots. Panicles in the Australian species cymose or rarely reduced to solitary umbels. Petals very concave, almost hood-shaped, but without the dorsal appendages of some Asiatic species.

The genus comprises nearly the whole of the Order, extending over the whole of its geographical area. The Australian species appear tolerably constant in the division of their leaves, but that character is not to be absolutely relied on, for the trifoliolate, digitate, and pedate forms will occasionally pass one into the other.—*Benth.*

Leaves simple.
  Leaves ovate, penniveined, or 3-nerved at the base, rather fleshy.
    Leaves shortly acuminate, mostly toothed. Berries globular. Tall, woody climbers . . . . 1. *V. antarctica.*
    Leaves very obtuse, quite entire. Berries obovoid. Bushy tree . . 2. *V. oblonga.*
  Leaves broad-cordate, 5-nerved, membranous.
    Branches glaucous. Veinlets reticulate, not prominent. Flowers at least 1 line diameter . . . . 3. *V. cordata.*
    Not glaucous. Veinlets transverse. Flowers not ½ line diameter . . . 4. *V. adnata.*
Leaflets 3.
  Leaflets ovate, rather thick and firm, shining. Cymes nearly globular, on very short peduncles. Stigma very broad . . . . 5. *V. nitens.*
  Leaflets thin-coriaceous, ovate-acuminate, 2 to 4in. long. Panicle repeatedly trichotomous-puberulent . . . . 6. *V. brachypoda.*

Leaflets large, broadly ovate or cordate, membranous. Cymes loose, divaricate.
  Leaves glabrous, or nearly so. Flowers fully 1 line diameter, on stout pedicels . . . . 7. *V. saponaria.*
  Leaves hairy on both sides. Flowers about ½ line diameter, on filiform pedicels . . . . 8. *V. acris.*
Leaflets mostly under 2in., rather thick or almost fleshy, coarsely toothed. Cymes loose, divaricate . . . . 9. *V. trifolia.*
Leaflets cordate, the apex often with long narrow points, the rounded basal lobes often overlapping the petiolule, over 3in. long; young stems, petioles, and main nerves bristly . . . . 10. *V. strigosa.*
Leaflets 5 to 9, pedate.
  Leaflets lanceolate, sharply toothed. Petioles and peduncles often long and slender . . . . 11. *V. japonica.*
  Leaflets small, ovate, acuminate, deeply toothed. Disk very prominent 12. *V. clematidea.*
  Leaflets 2 to 3in. long, oblong or cuneate, minutely and remotely serrate or entire. Disk inconspicuous . . . . 13. *V. acetosa.*
Leaflets 3, rarely 5, digitate.
  Leaflets obtuse at the base, on a distinct slender petiolule, coriaceous, and very reticulate . . . . 14. *V. hypoglauca.*
  Leaflets narrowed into a very short petiolule or sessile.
    Leaflets very coriaceous. Berries ovoid . . . . 15. *V. sterculifolia.*
    Tendrils few or none. Leaflets obovate, reticulation prominent. Peduncles 2in. long . . . . 16. *V. penninervis.*
    Leaflets membranous. Berries globular. Leaflets linear-cuneate to oblong or obovate. Cymes loose . . . . 17. *V. opaca.*
  Not a climber. Stems prickly, erect; from a rootstock . . . . 18. *V. Gardineri.*

1. **V. antarctica** (most southern), *Benth. Fl. Austr.* i. 447. Young shoots more or less clothed with short rust-coloured hairs, rarely entirely glabrous. Leaves simple, petiolate, ovate or oblong, mostly acuminate and slightly cordate, 3 to 4in. long and 1½ to 2in. broad, entire, sinuate or irregularly toothed, rather firm or almost coriaceous, penniveined and obscurely 3-nerved, with glands on the under side in the axils of some of the principal veins. Cymes dense, broadly corymbose, shorter than the petioles. Flowers tomentose-pubescent, the buds nearly globular, under 1 line diameter. Petals 4, separately deciduous. Disk prominent, undulate, obscurely 4-lobed. Style shortly conical. Berry globular.—*Cissus antarctica*, Vent, Choix, t. 21; DC. Prod. ii. 629; Bot. Mag. t. 2488; *C. glandulosa*, Poir. Dict. Suppl. i. 105.

Hab.: Brisbane River, Moreton Bay, and other parts of southern and northern Queensland. The specific name, although inappropriate, is too generally sanctioned by use to be altered.—*Benth.*
Wood dark-brown, coarse-grained.—*Bailey's Cat. Ql. Woods No. 85.*
The fungus *Phyllosticta neurospilea*, Sacc. and Berk., sometimes attacks the leaves.

2. **V. oblonga** (oblong), *Benth. Fl. Austr.* i. 447. A small bushy tree (according to Henne's notes, but R. Brown's specimens have tendrils), quite glabrous or the young shoots minutely rusty-tomentose, the branches rigid and flexuose. Leaves petiolate, broadly oblong or ovate-oblong, very obtuse, 1½ to 2½in. long, quite entire, firm but thinner than in *C. antarctica*, very finely penniveined and obscurely 3-nerved, with 2 large glands underneath in the axils of the lateral nerves. Cymes bearing small umbels of thinly velvety flowers. Calyx truncate, scarcely ½ line high. Fruiting cymes on short peduncles, bearing few obovoid berries.

Hab.: North-east scrubs.
On some cymes of this species and also *V. antarctica* the berries are replaced by a monstrous growth of dichotomous branches covered with small, broad, leafy scales, forming dense globular tufts of 3 or 4in. diameter, like those often observed on some *Mæsas*.—*Benth.*

3. **V. cordata** (cordate), *Wall. Catal. n.* 6008 (partly); *Benth. Fl. Austr.* i. 447 (in part). Very glabrous and often somewhat glaucous in all its parts, the young stems succulent and disarticulating in the dried specimens. Leaves on

rather long petioles, broadly cordate, 2½ to nearly 4in. long and nearly as broad, entire except small, almost bristle-like distant teeth, 5-nerved, the smaller veins reticulate, very few or none, transverse, and faintly conspicuous. Flowers deep crimson, in corymbose trichotomous cymes, the buds about 1 line diameter. Calyx minute, truncate. Petals 4, thick cymbiform, with inflexed tips, often cohering at the top and falling off together. Stamens 4, filaments very short. Anthers large, golden. Hypogynous disk annular, prominent. Style subulate. Berries obovoid-globular.—Benth. Fl. Hongk. 54; *Cissus cordata*, Roxb. Fl. Ind. i. 407; *Vitis cardiophylla*, F. v. M. Fragm. ii. 78.

Hab.: Barnard Islands, *M'Gillivray*; Burdekin River, *F. v. Mueller*; Rockhampton, *Thozet*; Bundaberg, *Rev. B. Scortechini*. Common in the Archipelago and Eastern India, extending northward to Sikkim and Hongkong and New Guinea.

4. **V. adnata** (adnate), *Wall.; Wight and Arn. Prod.* 126 (with the synonyms adduced); *Benth. Fl. Austr.* i. 448. Young shoots and under side of the leaves more or less covered with a short tomentum, which sometimes disappears with age. Leaves petiolate, broadly cordate, almost orbicular, acuminate, 3 to 6in. or more diameter, bordered with small bristle-like teeth, 5-nerved and penniveined, the primary veins connected by transverse veinlets. Flowers scarcely ½ line diameter, numerous in corymbose cymes. Petals 4, cohering by the tips and falling off together. Style shortly subulate, at least in the fertile flowers. Fruit globular, small.—*Cissus adnata*, Roxb.; Wight, Ic. t. 144.

Hab.: Ranges, Barron River, *E. Cowley* and *L. J. Nugent*.

5. **V. nitens** (shining), *F. v. M. Fragm.* ii. 78; *Benth. Fl. Austr.* i. 448. Quite glabrous. Leaflets 3, ovate or oval-oblong, acuminate, mostly 3 to 4in. long, remotely toothed, narrowed at the base, the lateral ones scarcely oblique, on short petiolules, rather firm, smooth and shining above. Umbel-like cymes almost glabrous, dense and nearly globular, 2 or 3 together or solitary on a very short common peduncle, the pedicels very short. Flower-buds ovoid, rather more than 1 line long. Petals 4 or rarely 5, oblong, falling off separately. Disk inconspicuous. Style very short and thick, with a broad, flat, almost fringed, slightly 2-lobed stigma. Berry ovoid, blackish.

Hab.: Herbert, Dawson and Burnett Rivers, *F. v. Mueller*; Brisbane River, *Fraser*, *F. v. Mueller*.

Wood soft and spongy, of a brown colour and coarse grain.—*Bailey's Cat. Ql. Woods No.* 85a.

6. **V. brachypoda** (petiole short), *F. v. M. Fragm.* ix. 125. Branches almost terete. Petioles somewhat thick, corrugate, from a few lines to 1in. long. Leaflets 3, thin, coriaceous, ovate-acuminate, 2 to 4in. long, shining, rather pale on the under side. Lateral nerves distant, often foveolate in the axils. Reticulation prominent on both sides, contracted into very short petiolules. Peduncle about 1½in. long. Panicle repeatedly trichotomous, puberulent. Flowers pedicellate, 3-merous; buds obtuse. Calyx patelliform, not toothed, about ¾ line broad. Petals yellow, thinly puberulent outside, sometimes 5. Anthers cordate. Style and ovary glabrous. Stigma not at all dilated.

Hab.: Rockingham Bay, *J. Dallachy* (F. v. M., l.c.).

7. **V. saponaria** (stems used as substitute for soap in Fiji), *Seem. Syst. List. Vit. Pl.* 4; *Benth. Fl. Austr.* i. 448. Young leaves and shoots and inflorescence minutely hoary-tomentose. Leaflets 3, very broadly ovate, acuminate, entire or crenate, attaining 4 to 6in., thin and glabrous when full-grown, penniveined and more or less distinctly 5-nerved at the base, especially the lateral ones, with transverse veinlets, the central one rounded at the base, the lateral ones obliquely

cordate. Cymes loose, divaricate, many-flowered, on long peduncles. Flowers nearly globular, above 1 line diameter. Petals 4, usually falling off together. Disk broad. Style conical. Berry depressed-globular.

Hab.: Walsh River, *T. Barclay-Millar*; Torres Straits, *R. Brown*; Cape York and Piper's Island, *M Gillivray*. Also in the Fiji Islands, where, according to Seeman, the stems are used in washing linen.

A. Gray, in Bot. Amer. Expl. Exped. i. 272, had referred this plant with doubt to *Cissus geniculata*, Bl., and perhaps correctly so, for although Blume describes the central leaflet as oblong-lanceolate, yet he mentions a broad-leaved variety, but with more pubescent leaves. All are closely allied to the common E. Indian *V. pedata*, Wall., and may be a 3-foliolate variety of that very variable species.—*Benth.*

8. **V. acris** (acid), *F. v. M. Fragm.* ii. 75; *Benth. Fl. Austr.* i. 449. Branches and leaves softly pubescent or hairy. Leaflets 3, broadly ovate, acuminate, crenate, 3 to 4in. long, thin, hairy on both sides, penniveined with transverse veinlets, the lateral leaflets oblique, obscurely cordate, and more or less 5-nerved at the base, on petiolules of ¼ to ½in. Cymes loose and divaricate on long slender peduncles, the branches almost filiform and nearly glabrous. Flowers nearly globular, about ½ line diameter. Petals 4, apparently distinct. Disk very prominent. Style short, conical.

Hab.: Brisbane, Burnett, and Pine Rivers.
The foliage is that of *V. mollissima*, Wall., from the Archipelago, from which the species appears to differ chiefly in the very slender inflorescence and small flowers. These may, however, not be full-grown in the very few specimens seen.—*Benth.*

9. **V. trifolia** (3 leaflets), *Linn. Spec. Pl.* 298; *Benth. Fl. Austr.* i. 499. Softly hoary-pubescent all over, especially the young shoots, or sometimes nearly or quite glabrous, arising from a tuberous, underground stem. Leaflets 3, ovate-acuminate, obovate or rhomboid, usually 1 to 2in., rarely 3in. long, coarsely and irregularly toothed or crenate, softly herbaceous, usually thick and sometimes almost fleshy, the lateral ones very oblique, on short petiolules. Cymes many-flowered, divaricate, on long peduncles, hoary or pubescent. Flowers nearly globular, about 1 line diameter. Petals 4, distinct. Disk very prominent. Style in some specimens short with a broad peltate stigma, in others slender with a small stigma. Berry small, depressed-globular.—*Cissus carnosa*, Lam.; DC. Prod. i. 630; *C. cinerea*, Lam.; DC. l.c. 681; *C. crenata*, Vahl. DC. l.c.; *Vitis carnosa*, W. and Arn. Prod. 127; Wight, Ic. t. 171 (a broad-leaved form); *V. psoralifolia*, F. v. M. Fragm. ii. 75.

Hab.: Cape York, *M'Gillivray*, and many other parts of tropical Queensland.
The species is very common in East India and the Archipelago, and is probably described under several names besides those above quoted.—*Benth.*
Tubers said to be edible, *F. v. M.*
The roasted roots used for food, *Roth.*

10. **V. strigosa** (hairs stiff), *Bail.* A tall scrub-climber with rather stout angular branches prominently striated between the angles. Leaves 3-foliolate. Petiole about 3in. long; petiolules of lateral leaflets less than ½in. long, of the central leaflet about 1in. long, when young hispid. Leaflets cordate, often long, acuminate, broadly crenulate, 3 to 5in. long, 2 to 3in. broad above the base, lateral ones very unequal sided at the base and usually larger than the terminal one, the nerves all more or less bristly, particularly on the under side towards the base, the lowest pair of nerves starting from near the top of the petiole. No tendrils or flowers on the specimens examined. Fruit depressed-globular, 4-seeded. Seeds almost hemispherical.

Hab.: Ranges about the Barron River, *L. J. Nugent.*

11. **V. japonica** (Japanese), *Willd.*; *J. E. Planchon in. A. and C. De Candol. Mono. Phan.* v. 561. Whole plant nearly or quite glabrous, the branches and under side of the leaves more or less glaucous. Branches striate, sometimes

minutely puberulent. Leaves pedately 5 (often 7 in the Queensland examples), foliolate, the petioles and secondary ones often long and slender. Leaflets 5 to 7, the lateral ones on rather short, the terminal ones on rather long petiolules; lanceolate to obovate-oblong, acutely serrate. Tendrils bifid. Panicles on peduncles as long as the petioles, one branch only bearing flowers and fruit, the other a closely-curled tendril. Flowers minute; pedicellate. Bracts small. Calyx patelliform, membranous. Corolla tumid at the base. Petals oblong-ovate, 3-angular, pointless, not readily separating at the apex. Berry oval, about 4 lines long. Seeds 2 to 4-ovate, trigonal, rugose, the back polyhedric-angular, foveolate.

Hab.: Endeavour River, *Planchon, l.c.;* ranges about the Barron River, *E. Cowley* and *L. J. Nugent.*

12. **V. clematidea** (Clematis-like), *F. v. M. Fragm.* ii. 74; *Benth. Fl. Austr.* i. 449. "Mor-bir," Mt. Cook, *Roth.* Minutely tomentose, pubescent or glabrous, underground stem forming tubers about the size of walnuts. Branches angular-striate. Leaflets usually 5, pedate, petiolate, ovate, acuminate, coarsely toothed or lobed, usually 1 to 2in. long, narrowed at the base, herbaceous, rather thick and pubescent or thin and glabrous. Cymes divaricate, rather dense, on long peduncles, minutely hoary-tomentose. Pedicels short. Flowers globular, about 1 line diameter. Petals apparently separating. Disk very prominent, entire. Style filiform. Berries depressed-globular, small.

Hab.: Brisbane River, *Fraser;* Pine River, *J. W. Statter;* Mount Perry, *J. Keys;* ranges about Cairns, *L. J. Nugent.*

Natives eat the root after being beaten on stone and roasted, *Roth.*

13. **V. acetosa** (sharp flavour of fruit), *F. v. M. Herb.; Benth. Fl. Austr.* i. 449. Climbing over shrubs, never very high. Glabrous or the young shoots and inflorescence very slightly hoary-tomentose. Leaflets 5 to 7, pedate, petiolulate or the central one nearly sessile, oblong or obovate-cuneate, obtuse or rarely shortly acuminate, 2 to 3in. long or rarely longer, entire, or bordered by small teeth or minute distant serratures, narrowed at the base, herbaceous, but rather firm, pale underneath. Cymes pedunculate, dense, divaricate or almost thyrsoid, the flowers often shortly racemose along the branches, on short pedicels. Flowers purple-red, ovoid-globular, about 1 line long, glabrous. Petals separating. Disk indistinct. Style very shortly conical or scarcely any, with a truncate stigma. Berries ovoid-globose, about 5 lines long. Seeds transversely sulcate. —*Cissus acetosa,* F. v. M. Trans. Vict. Inst. iii. 24; *Ampelocissus acetosa,* Planch. in A. and C. De Candolles Mon. Phanero. v.

Hab.: Cape York and other localities on the north-east coast; Palmer River, *Dr. W. E. Roth* (specimens hoary, particularly on under side of leaf).

The specimens first described were, according to F. v. Mueller's notes, from tall herbaceous not climbing stems—the one I have named *V. Gardineri.*

Mrs. Frank L. Jardine says that the aborigines use this plant as a remedy in cases of bites from the death adder.

14. **V. hypoglauca** (under side of leaf pale), *F. v. M. Pl. Vict.* i. 94; *Benth. Fl. Austr.* i. 450. "Billangai," Barron River, *E. Cowley.* A tall climber. Young shoots rusty-tomentose or villous, adult specimens usually quite glabrous. Leaflets 5, digitate, obovate, oval or oblong-elliptical, shortly and often acutely acuminate, 2 to 3in. long, the lateral ones smaller than the central ones, entire or toothed towards the top, obtuse at the base, on rather long petiolules, coriaceous, penniveined and finely reticulate, pale or glaucous underneath. Cymes rather dense, shortly pedunculate. Flowers yellowish, glabrous, ovoid,

fully 1 line long. Petals separating or slightly cohering. Disk 4-lobed, but not very prominent. Style conical. Berry nearly globular, rather small.—*Cissus hypoglauca*, A. Gray, Bot. Amer. Expl. Exped. i. 272; *C. australasica*, F. v. M. in Trans. Phil. Soc. Vict. i. 8.

Hab.: Coastal scrubs from the border of New South Wales to Cairns.
Wood soft and spongy, of a grey colour.—*Bailey's Cat. Ql. Woods No.* 86.

15. **V. sterculifolia** (Sterculia-like), *F. v. M. Herb.; Benth. Fl. Austr.* i. 450. "Yaroong," Moreton Bay, *Watkins*. Fruiting specimens quite glabrous. Leaflets 5, digitate, elliptical-oblong or somewhat obovate, shortly and obtusely acuminate, 3 to 4in. long, entire, narrowed into a very short petiolule, coriaceous, or sometimes thin chartaceous, penniveined, the reticulate veinlets much less conspicuous than in *V. hypoglauca*, with glands or foveolæ in the axils of some of the primary veins underneath. Flowers 4-merous. Calyx crenulate. Petals about 1 line long. Style very slender and short. Stigma not dilated. Fruiting cymes on short peduncles. Berries ovoid, rather large.

Hab.: Islands of Moreton Bay and scrubs of the southern coast.
Wood light-brown, soft and spongy.—*Bailey's Cat. Ql. Woods No.* 86a.

16. **V. penninervis** (feather-veined), *F. v. M. Fragm.* vi. 177. A glabrous climber with somewhat terete branches, tendrils few or none. Leaves digitate, 3 to 5-foliolate. Petioles 1 to 2½in. long. Leaflets 2 to 3in. long, chartaceous, obovate, apex obtuse or somewhat acute, reticulation close, decurrent upon the petiolules. Peduncles about 2in. long. Flowers bisexual, pedicellate, almost in umbels, buds globose-ovate. Calyx truncate or very short and obtusely 4-lobed. Petals 4, brownish, free, about ¾ line long. Anthers minute cordate when expanded. Disk annular. Style short, thickened below. Ovary 2-celled. Stigma scarcely as broad as the style. Berry obovate, 2-celled, 2-seeded, ½in. or more long. Seeds 2 to 6 lines long, obtuse at both ends, rugulose, brown, shining.

Hab.: Rockingham Bay, *J. Dallachy* (F. v. M., l.c.); ranges about the Barron River, *L. J. Nugent*.

17. **V. opaca** (on account of its thick growth), *F. v. M. Herb.; Benth. Fl. Austr.* i. 450. Pepper Vine of Fraser's Island. "Wappo Wappo" and "Yaloone," Rockhampton, *Thozet*. Underground tuberous stems large, quite glabrous. Leaflets 5, rarely 3 or 4, digitate, from linear-cuneate to elliptical-oblong, obovate or narrow rhomboidal, obtuse or acuminate, mostly 1 to 2in. long, entire or slightly toothed, narrowed at the base into very short petiolules or almost sessile, rather firm but not coriaceous, smooth, obscurely penniveined, usually pale underneath. Cymes rather loose, but not large. Flowers glabrous, globular, about 1 line diameter. Petals yellow, 5 or rarely 4, separating. Disk prominent, entire or scarcely lobed. Style short, conical. Berries depressed-globular.—*Cissus opaca*, F. v. M. in Trans. Vict. Inst. iii. 23.

Hab.: Common throughout the colony.
Berries and yam both eaten without any preparation, *Thozet*.
Fruit used for jam-making, *Miss Lovell*.

18. **V. Gardineri** (after T. R. Gardiner), *Bail.* An herbaceous species producing from the rootstock annually a growth of long prickly stems, upon which the flowers appear before the leaves. Primary petioles about 1½in. long, the two lateral ones nearly as long, all tomentose and bearing from 2 to 5 palmately-arranged leaflets. The terminal leaflet between the two lateral petioles on a petiolule nearly as long as the secondary petioles, ovate-oblong, shortly acuminate, rounded and usually equal-sided at the base; leaflets on the lateral petioles narrowing to a somewhat elongated apex and at the base to short petiolules, the margins with more or less prominent glandular teeth, length varying from 1 to 2½in., some very unequal-sided at the base, upper side slightly tomentose, the

under side with a close felt-like tomentum. Peduncles longer than the petioles, bearing at the summit two branches, one sterile (a tendril), the other developing into a panicle several inches long. Flowers (only seen in the bud) pedicellate. Calyx truncate or crenulate. Petals free. Filaments glabrous; anther almost orbicular, scarcely half as long as the filament. Disk lobed, style very short. The bunches of ripe fruit attaining 1 to 2lb.; berries black, ovoid-globose. Seeds 4, the sides transversely corrugated.

Hab.: Walsh River, *T. R. Gardiner*, who says that the best fruit and largest bunches are found on plants growing upon limestone country. The specimens examined bore no tendrils except on the inflorescence.

2. LEEA, Linn.
(After J. Lee.)

Petals united in a campanulate corolla with 5 spreading or recurved lobes. Disk (resembling a staminal tube) cup-shaped, conical, or nearly globular, 5-lobed, enclosing the ovary. Stamens inserted in grooves outside the disk, the filaments incurved at the top, with the anthers inside the disk in the bud. Ovary enclosed in the disk, 3 to 6-celled, with 1 ovule in each cell.—Shrubs or small trees, without tendrils. Leaves once, twice, or thrice pinnate, with large entire or toothed penniveined leaflets. Panicles or cymes leaf-opposed, corymbose. Flowers usually larger than in *Vitis*.

The genus is dispersed over tropical Asia and Africa, the only Australian species being the most common among the Asiatic ones.—*Benth.*

1. **L. sambucina** (supposed to resemble an Elder tree); *Willd. Spec. Pl.* i. 1177; *Benth. Fl. Austr.* i. 451. "Kalet." A tall, glabrous, coarse shrub, the young branches occasionally furrowed. Leaves mostly twice or thrice pinnate; leaflets few in each pinna, from ovate to oblong-elliptical or lanceolate, acuminate, usually 3 to 6in. long and 1½ to 2in. broad, but sometimes twice as long, irregularly crenate, the primary arcuate pinnate veins and transverse veinlets very prominent underneath. Cymes large, divaricate, trichotomous, on short peduncles. Flowers about 2 lines long, on very short pedicels. Ovary 5-celled. Berries small, depressed-globular, usually ripening 4 to 6 seeds.—DC. Prod. i. 635; *L. staphylea*, Roxb., W. and Arn. Prod. 182, with the synonyms adduced; Wight. Ill. t. 58 and Ic. Pl. t. 78.

Hab.: Islands of Howick's group, *F. v. Mueller;* numerous localities on the N. E. coast.
The species is common in tropical Asia, and is, perhaps, the same as a common African one.—*Benth.*

Wood hard, close-grained, of a pinkish color, and nicely marked.—*Bailey's Cat. Ql. Woods* No. 87.

2. **L. Brunoniana** (after R. Brown), *C. B. Clarke in Brit. Journ. Bot.* xix. 105. Nearly glabrous, upper leaves 2 (or often 3), pinnate; leaflets elliptic, very shortly acuminate, primary nerves numerous, continued nearly to the margin, often setulose, corymbs glabrous.

Hab.: Australia, *R. Br.*, No. 5272; Port Darwin, *Schultz*, No. 627. Called *L. sambucina* by Benth., Fl. Austr. i. 451. But not merely the colour of the flowers, but the venation of the leaves totally differs from *L. sambucina*, Willd. The present species is like a very handsome, well-developed *L. rubra* or *L. setuligera.—Clarke, l.c.*

The notice of this genus is here given in full from the Flora Austr. as well as Clarke's (from the Journ. of Bot.) Mr. Bentham probably had only very imperfect specimens, thus I would be in favour of retaining Mr. Clarke's name. I have so far only met with one species. Baron Mueller, in the 2nd Syst. Census of Austr. Plants, records two species, viz., *L. Brunoniana*, Clarke, and *L. staphylea*, Roxb., this latter is only another name for *L. sambucina*, Willd. I cannot, however, find any reason given by him for recording these two species for Australia.

Order XL. SAPINDACEÆ.

Flowers usually polygamous. Sepals 4 or 5, free or united in a small toothed or lobed calyx, imbricate or rarely valvate in the bud. Petals as many as sepals, or 1 fewer, sometimes minute or wanting, frequently bearing a scale inside. Disk various, in some genera unilateral, rarely wanting. Stamens 8, rarely fewer or more, inserted round the ovary within the disk (except in a few genera not Australian), sometimes unilateral; anthers versatile or erect. Ovary entire or lobed, 1 to 4-celled, most frequently 3-celled. Style simple, with a single stigma, or more or less divided. Ovules 1, 2, or rarely more in each cell, ascending, or rarely horizontal, with the micropyle inferior. Fruit dry or succulent, dehiscent or indehiscent, entire or separating into cocci. Seeds with or without an arillus, without albumen (except in a few genera not Australian). Embryo usually thick, frequently folded or spiral, the cotyledons usually unequal, collateral or superposed; radicle short, turned downwards or reascending towards the hilum.—Trees, shrubs, or rarely almost herbaceous, often climbers (especially in genera not Australian). Leaves alternate (or in genera not Australian opposite), usually compound, pinnate with (or more frequently without) a terminal odd one, the leaflets often irregularly alternate, rarely decompound; 3-foliolate or simple. Flowers usually small.

Sapindaceæ are abundant within the tropics, both in the New and in the Old World, more rare in the temperate regions of the northern hemisphere, and those, chiefly of the genera *Æsculus, Acer*, and their allies, unrepresented, except by one, in Australia; there are very few also in southern extratropical Africa or America.—*Benth.* (in part).

The majority of *Sapindaceæ* are readily known by the disk outside, not inside, the stamens, and by the 8 stamens in a 5-merous flower, with a 3-merous gynœcium; but all these characters have exceptions, which renders the technical limitation of the Order difficult, although really doubtful genera are very few. The position of the micropyle appears to be constant, but often difficult to observe. The arboreous genera with pinnate leaves, often numerous in species, especially in tropical Asia, may require considerable modification as to their characters, and probably some reduction, when those proposed by Blume come to be better known as well as to flower as fruit.—*Benth.*

Tribe I. **Sapindeæ.**—*Stamens inserted inside the disk, sometimes unilateral. Seeds exalbuminous. Leaves exstipulate, alternate, or rarely opposite.*

Flowers irregular, either 1 petal fewer than the sepals, or the stamens or disk unilateral, and ovary excentrical.
　One ovule in each cell of the ovary.
　　Herbaceous or half-herbaceous climber with biternate leaflets. Capsule inflated, membranous . . . . 1. Cardiospermum.
　　Trees with pinnate leaves. Petals 1 fewer than sepals.
　　　Calyx valvately 5-lobed. Capsule loculicidally 3-valved . . . 2. Diploglottis.
　　　Petals 5. Disk complete, deeply 5-lobed. Ovary 2-celled, cells 1-ovulate. Fruit irregularly dehiscent . . . . 3. Castanospora.
　Shrubs or trees, with 1 or 3 digitate leaflets. Sepals 4, broadly imbricate. Petals 4 or none. Fruit of 1 or 2 indehiscent lobes . 4. Schmidelia.
Flowers regular. Disk annular or none. Stamens all round the ovary.
　One ovule in each cell of the ovary. Trees or tall shrubs. Leaves pinnate (except *Heterodendron* and sometimes in *Atalaya*).
　　Capsule loculicidally 3-valved.
　　　Sepals distinct, broadly imbricate . . . . 5. Cupania.
　　　Calyx small, toothed, or the lobes valvate or slightly imbricate . . 6. Ratonia.
　　Fruit separating into winged samaras . . . . 7. Atalaya.
　　Fruit divided into indehiscent or 2-valved lobes or irregularly loculicidal, the valves not separating from the axis.
　　　Leaves pinnate.
　　　　Sepals broadly imbricate in 2 rows. Petals usually exserted. Fruit lobes smooth, indehiscent . . . . 8. Sapindus.
　　　　Calyx-teeth or lobes valvate or slightly imbricate. Petals very small or none. Fruit-lobes smooth (in Australia), indehiscent or 2-valved . . . . 9. Nephelium.
　　　　Calyx-segments imbricate. Petals very small or none. Fruit-lobes tuberculate or muricate, indehiscent . . . . . . 10. Euphoria.

Leaves coriaceous, simple, entire or pinnatifid. Calyx entire or
minutely toothed . . . . 11. HETERODENDRON.
Two ovules in each cell of the ovary.
Trees with pinnate leaves. Petals 4 or 5.
Calyx deeply divided into imbricate segments. Disk inconspicuous . 12. HARPULLIA.
Calyx campanulate, shortly lobed. Disk broad .' . . . . . 13. AKANIA.

TRIBE II. **Acerineæ.**—*Flowers regular. Stamens inserted on the disk. Lobes of the fruit indehiscent. Seeds exalbuminous. Leaves opposite, exstipulate.*
A tree. Flowers in a hard, deeply-divided involucre . . . . 14. BLEPHAROCARYA.

TRIBE III. **Dodonæeæ.**—*Flowers regular. Stamens inserted outside the disk. Seeds exalbuminous. Leaves alternate, exstipulate.*
Shrubs or rarely small trees. Leaves simple or pinnate with small
leaflets. Calyx cup-shaped. Petals none. Disk inconspicuous.
Stamens in the male flowers 10 or fewer, usually 8 . . . . 15. DODONÆA.
Stamens in the male flowers more than 10 . . . . 16. DISTICHOSTEMON.

1. CARDIOSPERMUM, Linn.

(Named from the heart-shaped scar of seed.)

Flowers polygamous. Sepals 4, broadly imbricate, the 2 outer ones small. Petals 4, 2 larger with a large scale, 2 smaller with a crested scale. Disk one-sided, almost reduced to 2 prominent glands opposite the lower petals. Stamens 8, oblique. Ovary excentrical, 3-celled, with one ovule in each cell; style very short, with 3 stigmatic lobes. Capsule vesicular, membranous, more or less 3-cornered, 3-celled, opening loculicidally. Seeds globose, with a thick funicle or small aril; testa crustaceous; cotyledons large, transversely folded.—Herbs or undershrubs, mostly climbing. Leaves dissected. Flowers few, small, on long axillary peduncles, which usually bear a tendril under the panicle.

A small genus, chiefly American, of which 2 species are also spread over the Old World within the tropics, and a third is perhaps confined to the Old World. The Australian species is one of those most widely diffused in both worlds.—*Benth.*

1. **C. Halicacabum** (old generic name of the plant), *Linn.; DC. Prod.* i. 601; *Benth. Fl. Austr.* i. 453. Heart Pea, or Balloon-vine. A straggling or somewhat climbing annual or perhaps perennial, attaining several feet in length, glabrous or slightly pubescent. Leaf-segments usually twice ternate, ovate or ovate-lanceolate, coarsely toothed or lobed, the upper leaves smaller, narrower, and less divided. Peduncles 2 to 3in. long, bearing a double or treble short recurved tendril under the small panicle, which is often reduced to an umbel of few small white flowers. Capsules flat on the top, usually pubescent.—A. Gray, Gen. Ill. t. 181; Wight, Ic. t. 508.

Hab.: Common to many parts of the colony, both coastal and inland.
The species is common in most tropical regions. The Australian specimens belong either to the variety with fruits scarcely ¾in. diameter, often considered as a distinct species (*C. microcarpum*, H.B. and K.), or are intermediate between that and the typical form, with fruits above 1in. diameter.—*Benth.*

2. DIPLOGLOTTIS, Hook. f.

(Double-throated.)

Calyx deeply 5-lobed, valvate. Petals 4, the place of the fifth vacant, the inner scale divided into two. Disk one-sided, crescent-shaped. Stamens 8, ascending, unequal. Ovary 3-celled; style short, incurved; stigma entire or obscurely 3-lobed. Ovules solitary in each cell. Capsule nearly globular, thick, somewhat fleshy, loculicidally 3-valved. Seeds enclosed in a pulpy arillus.—A tree, with large pinnate leaves, more or less villous-tomentose. Flowers not very small, in large axillary panicles.

The genus is limited to a single species endemic in Australia.

1. **D. Cunninghamii** (after A. Cunningham), *Hook. f. in Benth. and Hook. Gen. Pl.* 895; *Benth. Fl. Austr.* i. 454. Native Tamarind. Often a large tree with a brownish smooth bark, the young branches, petioles and inflorescence densely clothed with a soft rust-coloured tomentum. Leaves very large, sometimes exceeding 2ft.; leaflets 8 to 12, opposite or irregularly alternate, oblong-elliptical to ovate-lanceolate, acute or obtuse, usually 6 to 8in., but sometimes above 1ft. long, glabrous above, pubescent underneath, with raised parallel pinnate veins. Flowers numerous, on pedicels of 1 to 2 lines, clustered along the branches of the ample panicle. Calyx about 1½ line long, rusty-tomentose. Petals about as long as the calyx, orbicular, thin, ciliate, the two inner scales not united, about as long as the petal itself, but thicker and very hairy. Stamens exserted in some specimens, shorter than the petals in others. Fruit about ½in. diameter, of 2 or 3 roundish lobes, each containing a round seed enclosed in an amber-coloured juicy arillus of an agreeable acid flavour, tomentose.—*Cupania Cunninghamii*, Hook. Bot. Mag. t. 4470.

Hab.: Southern Queensland in river or coastal scrubs, and on the Bunya and other mountain ranges.
Wood light-coloured except near the centre, close-grained, and very tough.—*Bailey's Cat. Ql. Woods No.* 88.
The arils used for jam-making.
The leaflets of this tree in the Brisbane River scrubs are at times much infested with the blight fungi *Uromyces diploglottidis*, C. and Mass., and *Phoma diploglottidis*, C. and Mass.

Var. *Muelleri* (*Cupania diphyllostegia*, F. v. M.) This differs from *D. Cunninghamii* in being a larger tree with smaller leaves of a thinner texture, and the shoots and leaves having a less dense covering of tomentum; panicle smaller, the flowers with 5 or 6 petals. The arils which enclose the seed of this, like those of the southern tree, make excellent jam. As a shade tree it is well worthy of being planted. Although belonging to the tropical scrubs, it thrives well in the neighbourhood of Brisbane, keeping its distinctive characteristics. Hab.: Tropical scrubs.

3. CASTANOSPORA, F. v. M.
(Fruit resembling a Chestnut.)

Flowers almost symmetrical. Calyx 5-partite, the parts almost valvate in the bud. Petals 5, equal the calyx or scarce above it, rhomboid-ovate, on very short claws, bearing 2 bearded scales on the face near the base. Disk deeply 5-lobed. Stamens 8, inserted inside the disk. Filaments several times longer than the anthers. Anthers minute, ovate-cordate. Style very short. Stigma minute, terminal, 2-lobed. Ovary 2 (or sometimes 3) celled, cells 1-ovulate. Capsule indehiscent (according to F. v. Mueller), bursting in all the examples I have seen in Queensland, smooth. Pericarp thick-crustaceous, or coriaceous. Septa very thin or none. Seed solitary, basifixed. Arillus absent. Hilum large. Albumen none. Cotyledons equal, erect, seceding, the outside much swelled but neither plicate nor twisted. Radicle very short, inferior. Evergreen trees of tropical Queensland. Leaves alternate, simply pinnate. No stipules. Leaflets large, entire. Flowers polygamous, in racemose panicles. Calyx minute. Seeds like chestnuts.—F. v. M. Fragm. ix. 92.

Capsule on a short stipes . . . . 1. *C. Alphandi*.
Capsule on a long stipes . . . . 2. *C. longistipitata*.

1. **C. Alphandi** (after M. Alphand), *F. v. M. Fragm.* ix. 92. A tree of 30 or 40ft., the branchlets and panicles with a thin covering of fuscous tomentum. Leaves 4 to 10-foliolate. Leaflets 3 to 6in. long, ¾ to 2in. broad, thin-coriaceous, lanceolate or ovate-lanceolate, usually obtuse, acuminate, quite entire, the under side indumentum very thin, fulvido-cinerascens, petiolules 5 or 6 lines long. Racemose branches of panicle divergent, a few or several inches long. Pedicels ½ to 1½ line long. Bracts and bracteoles minute triangular. Calyx 5-toothed. Petals 5, rhomboid-ovate, shortly clawed, ⅔ line long, the lower part bearing 2

bearded inflexed scales. Stamens 8, filaments about 1 line long, dilated and hairy below. Ovary subovate. Style short. Capsule compressed or depressed-globular, very shortly stipitate, about 1in. diameter. Seeds resembling a chestnut in colour.—*Ratonia Alphandi*, F. v. M. Fragm. iv. 158.

Hab.: Borders of creeks, tropical Queensland; Rockingham Bay; Johnstone, Mulgrave, the Barron and other river scrubs.

Wood of a yellowish color, hard, and close-grained.—*Bailey's Cat. Ql. Woods No.* 87a.

2. (?) **C. longistipitata** (fruit long-stalked), *Bail*. A tree of about 40ft. in height, the branchlets angular and puberulent. Leaves alternate; leaflets from 2 to 9 pairs with occasionally a terminal one, 6 to 9in. long, 1½ to 2½in. broad, oblong-lanceolate, nearly or at length quite glabrous, the primary nerves prominent, oblique, parallel, rather distant, the reticulate veins obscure. Petioles about 3½in. long, enlarged and shortly decurrent upon the stem; petiolules about ½in. long, puberulent as well as the petioles and rhachis. Panicles racemose, in the axils of the upper leaves, 8 or 9in. long. No flowers seen. Capsule nearly globular, 1¼in. or more in diameter, tapering slightly at the base to a stipes of 1in., the outside velvety puberulent, indehiscent, or at length opening into 2 valves. Seeds solitary in the fruits examined; testa loose, chartaceous.

Hab.: Scrubs of the Barron River, *E. Cowley*.

4. SCHMIDELIA, Linn.
(After C. C. Schmidel.)

Flowers polygamous. Sepals 4, broadly imbricate, the outer ones smaller. Petals 4, small, or rarely none. Disk one-sided, usually lobed or divided into 4 glands. Stamens 8, more or less one-sided. Ovary excentrical, 2 or rarely 3-celled; style 2 or 3-lobed; ovules solitary in each cell. Fruit of 1 or rarely 2 small ovoid or globular indehiscent, fleshy or almost dry berries. Seeds with a short arillus; embryo curved, cotyledons folded.—Shrubs or trees. Leaves with 1 or 3 leaflets. Flowers very small, in simple or loosely paniculate axillary racemes.

The species are numerous in tropical America, with several African ones, and a few in tropical Asia and the Indian Archipelago, one of the common Asiatic ones extending to Australia. The genus is one of the most easily recognised in the Order, by its foliage as well as by its small flowers and fruit.—*Benth*.

1. **S. serrata** (serrate), *DC. Prod*. i. 610; *Benth. Fl. Austr*. i. 455. A straggling shrub or small tree, the young leaves and shoots pubescent-tomentose, often glabrous when full-grown. Leaflets 3, ovate or obovate-oblong, obtuse or slightly acuminate, 2 to 4in. long, irregularly and coarsely toothed, or rarely quite entire, sessile or narrowed into a short petiolule, glabrous above, pale or pubescent underneath, often bearing hairy tufts in the axils of the principal veins. Racemes slender, simple or slightly branched. Flowers ⅓ to nearly 1 line diameter, on short pedicels, clustered along the pubescent rhachis. Petals cuneate, with a minute scale. Disk of 4 small lobes or glands. Stamens glabrous. Berries small, globular.—W. and Arn. Prod. 110; *Ornitrophe serrata*, Roxb. Pl. Corom. i. 44, t. 61; *S. timoriensis*, DC., Dcne. Herb. Timor. 115.

Hab.: Common along the tropical coast.

The latter specimens are nearly glabrous, with the leaflets more sessile and narrowed at the base, as described in *S. timoriensis*. Some of R. Brown's are similar; others are more pubescent, like the common form in India, where these characters are very variable; and, as suggested by W. and Arn., these plants may all be varieties only of *S. Cobbe*, Linn., which would thus have a very wide range over tropical Asia, including the Archipelago.—*Benth*.

5. CUPANIA, Linn.
(After F. F. Cupani.)

Flowers regular, polygamous. Sepals 4 or 5, imbricate in the bud. Petals either as many as sepals, small, with or without scales inside, or none. Disk usually annular. Stamens usually 8 to 10, inserted inside the disk; filaments short, rarely as long as the calyx. Ovary 2 or 3-celled, rarely 4-celled, with 1 ovule in each cell. Capsule obovoid or rarely globular, coriaceous or hard, 2 or 3, rarely 4-celled, often angled or lobed, opening loculicidally in as many valves as cells. Seeds usually more or less covered by an arillus; testa crustaceous or coriaceous; embryo curved; cotyledons plano-convex. — Trees or rarely tall shrubs. Leaves alternate, pinnate; leaflets alternate or opposite, with or without a terminal one. Flowers small, in small axillary or terminal panicles, sometimes almost reduced to simple racemes. Petals rarely as long as the sepals.

A large tropical genus, both in the New and the Old World, the precise limits of which are very difficult to fix, and are very differently viewed by different botanists. The Australian species are all endemic, as far as hitherto known.—*Benth.*

Leaflets 2 to 4, 2 to 3in. long, cuneate, broad at the apex, sessile, entire, glabrous. Capsule 2 or 3-lobed, about 1in. broad, velvety outside . . 1. *C. Wadsworthii.*
Leaflets 4 to 15, from very small to 2½in. long, cuneate, the end and upper margin sharply toothed, pubescent, sessile. Capsule ¼ to ⅜in. long, pubescent outside, silky-hairy inside . . . . 2. *C. Shirleyana.*
Leaflets 6 to 10, 2½ to 5in. long, elliptic-oblong, entire, petiolulate, entire, almost glabrous. Capsule sometimes exceeding 8 lines broad and acutely 3-lobed, tomentose outside . . . . 3. *C. anacardioides.*
Leaflets 3 to 6, 1½ to 2½in. long, oblong with sometimes a cuneate base, entire, petiolulate. Capsule 4 or 5 lines broad, angular . . . . . 3. Var.
Leaflets 13 or more, irregular in size, to 6in. long, ovate-lanceolate, dentate, petiolulate, hairy on the under side. Panicle branches long, thong-like, 1 to 2ft. long . . . . 4. *C. flagelliformis.*
Leaflets 4 to 11, 2¼ to 3in. long, ovate-lanceolate, petiolulate, entire, glabrous. Capsule 1in. broad, rufescent outside . . . . 5. *C. Robertsonii.*
Leaflets 7 to 14, 3 to 6in. long, oblong-lanceolate, petiolulate, serrate, glabrous except the nerves. Capsule when young globose, tomentose, 3-lobed, 6 to 8 lines broad . . . . 6. *C. serrata.*
Leaflets about 20, oblong-lanceolate, 6in. long, bordered by large curved teeth 7. *C. curvidentata.*
Leaflets 5 to 13, 3 to 6in. long, oblong-lanceolate, glabrous, petiolulate, repando-crenate with dimples in the axils of the principal nerves. Capsule about ⅔in., 3-angular-globose, woody, silky outside, woolly inside . . . . 8. *C. foveolata.*
Leaflets 8 to 10, 1½ to 3in. long, obovate-oblong, petiolulate, crenate, rusty-hairy underneath. Capsule 3-angled, velvety-tomentose and rugose, about ¾in. broad . . . . 9. *C. tomentella.*
Leaflets 13 to 21 or more, lanceolate, 1½ to 4in. long, oblique, pubescent underneath, petiolulate, the margins sharply serrate. Capsule globular, densely hirsute, ¼ to ⅜in. diameter . . . . 10. *C. pseudorhus.*
(Name only recorded by F. v. Mueller in Vict. Nat. iii. 170, and placed in 2nd Syst. Census of Austr. Plants between *C. pseudorhus* and *C. xylocarpa* as here given) . . . . 11. *C. pleurophylla.*
Leaflets 3 to 6, 2 to 3in. long, ovate, obovate, petiolulate, sometimes with a few teeth or entire, pubescent underneath with tufts of hairs in the axils of the principal nerves. Capsule about ⅜in. broad, nearly globular, 3-angular, woody, minutely tomentose outside . . . . . 12. *C. xylocarpa.*
Leaflets 3 to 6, 2 to 6in. long, lanceolate, petiolulate, crenate or entire, glabrous, foveolate in the axils of the principal nerves on the under side. Capsule nearly globular, about ⅜in. broad, woody, glabrous, or nearly so outside . . . . 13. *C. nervosa.*
Leaflets 2 to 4, 3 to 6in. long, ovate-oblong or ovate-lanceolate, entire or with a few teeth, glabrous on both sides. Ovary hirsute . . . . 14. *C. Bidwilli.*
Leaflets 2 to 6, 3 to 5in. long, lanceolate-ovate, entire. Ovary glabrous. The young fruit 3-angular . . . . 15. *C. Mortoniana.*
Leaflets 2 to 10, 4 to nearly 12in. long, lanceolate, entire, petiolulate. Capsule about 10 lines, subglobose, red, corrugated and glabrous . . 16. *C. erythrocarpa.*
Leaflets 2 to 7, distant upon the rhachis, 3 to 6in. long, oblong or ovate-lanceolate, petiolulate. Filaments with short, white, scale-like hairs . 17. *C. sericolignis.*

1. **C. Wadsworthii** (after R. Wadsworth), *F. v. M. Syst. Cens. Austr.* Pl. 1882. Usually a small slender tree, the branchlets bearing a thin appressed pubescence. Leaves 1 or 2-jugate, glabrous, on rather short petioles. Leaflets sessile, chartaceous, broad-cuneate, the truncate apex sometimes ending in 2 or 3 very broad short lobes, 1½ to 3in. long, ¾ to 1½in. or more broad, both sides deep-green and shining, lateral nerves distant. Rhachis not winged. Peduncles axillary, slender, bearing narrow panicles shorter than the leaves, bearing an appressed pubescence. Bracts cymbro-semi-lanceolate, scarcely over 1 line long. Pedicels short. Sepals persistent, 2 outer ones orbicular-ovate, slightly exceeding 1 line long; 3 inner ones subrotund, petaloid, about 2 lines long, all broadly imbricate in the bud, and silky-pubescent outside. Petals 5, sessile, about 1 line long, scales bearded. Stamens 8 or 9. The filaments pubescent in the lower part. Anthers oblong or oval, glabrous, about 1 line long, yellow. Pollen grains smooth, yellow, spherico 4-angular. Ovary central, rudimentary in the male flowers. Disk narrow, glabrous. Capsule 2 or 3-lobed, about 1in. broad, velvety outside. Seeds about ½in. long, broadly ovate, included in a thin orange-coloured arillus.—*Harpullia Wadsworth*, F. v. M. Fragm. iv. 1, Pl. xxvi., and Fragm. ix. 89.

Hab.: Rockhampton to Mount Elliott.
Wood close-grained, tough, with a somewhat pinkish tinge.--*Bailey's Cat. Ql. Woods* No. 88a.

2. **C. Shirleyana** (after J. Shirley, B.Sc.), *Bail. Sny. Ql. Fl. 2nd Suppl.* 15. "Kooraloo," Bundaberg, *Keys*. A large shrub or small tree, the branchlets rusty-pubescent. Leaves 2 to 5in. long; leaflets 7 to 15, the lowest the smallest, often not more than 3 lines in diameter, very unequal-sided at the base, and overlapping the stem like stipules, increasing in size upwards, the end ones attaining the length of 2in., cuneate in outline, the upper part sharply toothed, the end very obtuse or truncate, glossy above, more or less pubescent beneath; primary veins very oblique and prominent, as are also the smaller reticulations. Flowers small, nearly sessile, in slender racemes about as long as the leaves. Sepals orbicular, silky, much imbricate. Petals ovate, glabrous, nearly 1 line long, the inside scale divided into 2 hairy spathulate lobes as long as the petal; disk prominent. Stamens usually about 8, but very irregular as to number; scarcely exserted. Ovary hairy. Capsule ½ to ¾-in. long, shortly stipitate, pubescent outside and silky-hairy inside. Seeds dark-brown, somewhat pear-shaped, and more than half enveloped in a reddish arillus.

Hab.: Sankey's scrub, off the Logan Road, near Brisbane.
Wood of a yellow colour, close-grained, and tough.—*Bailey's Cat. Ql. Woods* No. 89.

3. **C. anacardioides** (Anacardium-like), *A. Rich. Sert. Astrol.* 33, t. 13; *Benth. Fl. Austr.* i. 458; *F. v. M. Fragm.* ix. 91. "Tuckeroo," Moreton Bay, *Watkins*. A slender tree with a pale smooth bark, quite glabrous or with a minute hoariness on the inflorescence. Leaflets 6 to 10, usually 8, from broadly ovate or obovate to elliptical-oblong, very obtuse, 2½ to 4in. long, rounded at the base, and shortly petiolulate, quite entire, coriaceous. Flowers rather large for the genus, in pedunculate cymes along the branches of loose panicles. Sepals orbicular, the inner ones 2 lines broad, slightly ciliate. Petals small, orbicular, with 2 very short obovate hirsute scales at the base. Stamens 10 or sometimes 8; filaments short, hirsute; anthers oblong. Ovary villous. Capsule glabrous, coriaceous, acutely and divaricately 3-lobed, 6 to 8 lines broad, very shortly attenuate at the base, tomentose inside, arillus red.

Hab.: Brisbane River, Moreton Bay, *Fraser*, *A. Cunningham*, *F. v. Mueller*; Burdekin River, *F. v. Mueller*; Rockhampton, *A. Thozet*.
Wood of a light-pinkish colour, close-grained, and tough; might serve for making handles for tools.—*Bailey's Cat. Ql. Woods* No. 91.

Var. *parvifolia*. A small tree, minutely hoary. Leaflets 3 to 6, opposite or subopposite, 1½ to 2½in. long, ½ to 1in. broad, oblong with sometimes a cuneate base, the apex emarginate; petiolules about 2 or 3 lines, margins entire. Panicles seldom more than 3 or 4in. long, and broad.

Capsule about 4 or 5 lines broad, shortly stipitate, angular, rugose outside, densely hairy inside with short rusty hairs. Seeds black, nearly enclosed in the arillus. Hab.: Main Range and several other localities in southern Queensland; also Mt. Perry, *J. Keys.* Wood light-coloured, close-grained, very tough.—*Bailey's Cat. Ql. Woods No.* 91a.

4. **C. flagelliformis** (from the thong-like branches of panicle), *Bail. Bot. Bull.* viii. 78. " Maraguigi," Barron River, *Cowley.* A shrub or small tree, the branches angular, dark-coloured except in a young state, when, like the young foliage, they are clothed by a ferruginous or grey tomentum. Leaves, the petiole and rhachis together, measuring from 1ft. to 16in. in length, the leaflets scattered, 18 or more, size very irregular, some attaining 6in. in length and a width of 1¾ in., the point often much elongated, the base shortly cuneate to a petiolule of about ¼in., margins dentate with large teeth, the primary nerves prominent and parallel, with strongly-marked reticulations between, all more or less hairy on the under side, upper surface of leaflet glabrous, and the reticulations not prominent. Panicles velvety, near the ends of the branches, of few (3 to 5) slender thong-like branches, some of which being about 2ft. in length; flowers in distant sessile clusters, expanded flower about 4 lines in diameter. Sepals orbicular, much imbricate, silky on the back, irregular as to size. Petals buff-coloured, broad-cuneate, undulately lobed at the top, much shorter than the sepals, and bearing at the base of each 2 incurved, hairy, scale-like appendages, much shorter than the petal. Stamens 8, included; filaments hairy; anthers slightly longer than the filaments, oblong, angular, hairy between the blunt angles. Ovary hairy, but probably abortive in the flowers examined. Capsules not yet obtained.

Hab.: Scrub about the Barron River, *E. Cowley.*

5. **C. Robertsonii** (after Dr. J. Robertson), *F. v. M. Fragm.* v. 146, ix. 94. A tree with a smooth bark attaining the height of about 40ft. Branchlets slightly angular. Leaflets 4 to 11, very shortly petiolulate, lanceolate or ovate, green on both sides, the upper shining, margins quite entire, attaining as much as 8in. in length and 3in. in breadth. Rhachis angular. Panicle rather large, with racemose branches. Pedicels as long as the calyx. Sepals much imbricate, the larger ones slightly over 1 line long. Petals ciliate, exappendiculate, about ½ line long, rhomboid, sometime the face pubescent. Disk entire, glabrous. Filaments 1⅓ line long, hairy on the lower part. Anthers oval, almost ½ line long. Ovary 3-angular, 3-celled, 3-ovulate. Style about 1 line long. Capsule obtuse-angular, about 1in. broad, rufescent outside, valves hard, coriaceous. Seeds about 5 lines long, ovate-globose. Hilum broad-orbicular.

Hab.: Rockingham Bay, *J. Dallachy* (F. v. M., l.c.)

6. **C. serrata** (serrate), *F. v. M. Fragm.* iii. 48, ix. 94; *Benth. Fl. Austr.* i. 458. A tree, but flowering when still shrubby, the young branches rusty with a close tomentum. Leaflets usually 6 to 10, sometimes many more, ovate-lanceolate or lanceolate, acute or acuminate, 3 to 6in. long, sharply and coarsely serrate, rounded at the base and nearly sessile, rigid but not thick, shining above, very prominently pinnately veined and reticulate underneath. Panicles in the upper axils, little branched or almost reduced to dense racemes of 2 to 3in., softly tomentose or pubescent. Flowers rather large, on very short pedicels. Sepals orbicular, the innermost fully 2 lines long. Petals much shorter, broad with a short 2-cleft scale at the base. Anthers 8, oblong, on very short filaments. Ovary in the males rudimentary, villous. Style very short, if not wanting. Stigma globose-ovate, capsule 3-celled, when young globose, tomentose, silky tomentose inside, thick and deeply 3-valved. Seeds ovate or rarely globose, black and glossy, about ½in. long, the arillus covering three-quarters of the seed.

Hab.: Pine River, Moreton Bay, *W. Hill*; Rockingham Bay, *J. Dallachy.*

7. **C. curvidentata** (referring to the curved teeth), *Bail. Ql. Agri. Journ.* v. 483. A small tree, not, so far as known, attaining a height of more than 20ft. Leaves about 1½ft. long, pinnate, with alternate leaflets. Petiole stout, almost terete, about 3in. long, petiolules 1 to 1½in. long, slender; leaflets about 20, oblong-lanceolate, to 6in. long and 2in. broad, coriaceous, bordered by rather large, blunt, incurved teeth, acuminate at the apex, cuneate and more or less unequal-sided at the base, upper side smooth, the veins showing but slightly, the numerous parallel nerves and reticulate transverse veins very prominent on the under side; the petiole and rhachis dark-coloured and closely dotted with small light-coloured lenticella. Panicles puberulous, erect, narrow, about 11in. long, with a few racemose branches below the middle. Flowers in small clusters, nearly or quite sessile; sepals 5, dark-purple, the colour concealed on the outer side by the hairy surface, nearly orbicular, about 2 lines long. Petals yellowish, wavy, about half as long as the sepals. The scales at their base very wavy and hairy. Stamens 8, the short filaments hairy at the base. Ovary ferruginous, hairy, 3-lobed; stigma coloured. No fruit to hand.

Hab.: Stony Creek, near Cairns, *L. J. Nugent.*

8. **C. foveolata** (foveolate), *F. v. M. Fragm.* ix. 95. A tree attaining about 40ft.; bark somewhat smooth, wood hard. Branchlets, petioles, and peduncles slightly puberulent. Leaves with from 5 to 13 leaflets, rhachis angular; the leaflets oblong-lanceolate, from 3 to 6in. long and ¾ to 1¼in. broad, thin, coriaceous, shortly decurrent upon the petiolule, the apex elongated but blunt, glabrous on both faces, the under of somewhat paler colour, repando-crenate, the principal veins on the under side with dimples in their axils. Branches of panicles angular. Sepals rather large, nearly glabrous; inner ones roundish, 1¼ to 2 lines long. Petals minute, bearing auriculate hairy scales at their base. Stamens 8. Filaments longer than the calyx, slightly tomentose. Anthers oval, ½ line long. Disk crenulate, silky-tomentose as well as the ovary. Capsule about ⅔in., 3-angular-globose, woody; the outside silky-tomentose, the inside woolly. Seeds nearly covered by the arillus.

Hab.: Various localities in tropical Queensland, *Carron, Dallachy,* and *W. Hill.*

9. **C. tomentella** (tomentose), *F. v. M. Herb.; Benth. Fl. Austr.* i. 458 in part. Possibly a variety of *C. serrata,* of which it has the flowers. Branches, petioles, and inflorescence softly tomentose, almost villous. Leaflets 5 to 8, oblong or obovate-oblong, obtuse, 2 to 3 in. long, minutely and remotely denticulate or nearly entire, on petiolules often 2 lines long, thinly coriaceous, glabrous above, softly tomentose underneath. Panicles not much branched. Bracts rather large, tomentose, deciduous. Flowers nearly sessile. Sepals orbicular, and petals small with a short scale as in *C. serrata.* Anthers oblong, slightly pubescent. Capsule 3-angled, thickly coriaceous, velvety-tomentose and rugose, ⅔in. broad.

Hab.: Moreton Bay, *W. Hill.*

10. **C. pseudorhus** (resembling a *Rhus*), *A. Rich. Sert. Astrol.* 84, *t.* 14; *Benth. Fl. Austr.* i. 459; *F. v. M. Fragm.* ix. 92. "Moorjung," Taromeo, *Shirley.* A spreading tree of moderate size, the young branches and petioles densely rusty-tomentose. Leaves crowded under the panicles; leaflets 13 to 21 or even more, lanceolate or ovate-lanceolate, acuminate, 1½ to 3in. long or rarely more, very oblique or almost falcate, nearly glabrous and shining above when full-grown, more or less tomentose or pubescent underneath. Panicles usually much-branched and rather dense, rarely exceeding the leaves, tomentose. Flowers rather small, on very short pedicels. Sepals ovate, less imbricate than in the preceding species, the longest scarcely exceeding 1 line. Petals red, orbicular,

rather exceeding the sepals, the inner scales hirsute, as long as the lamina. Stamens 8 to 10. Filaments puberulent. Anthers oblong. Ovary purple, villous. Capsule globular, slightly lobed, almost woody, densely hirsute with short velvety hairs, about ½in. diameter. Arillus small.

Hab.: Common from the Brisbane to the Barron River ranges.
Wood of a light colour, grain close, very tough; would be excellent for pick-handles.—*Bailey's Cat. Ql. Woods No. 93.*

11. **C. pleurophylla**, *F. v. M. Vict. Nat.* iii. 170 (1887.)

12. **C. xylocarpa** (woody carpel), *A. Cunn. Herb.; F. v. M. Trans. Vict. Inst.* iii. 27 ; *Benth. Fl. Austr.* i. 459. A moderate-sized tree, the young branches rusty-tomentose. Leaflets 3 to 6, rarely more or reduced to 2, ovate-obovate or elliptical-oblong, obtuse or scarcely acuminate, 2 to 3in. long or rarely more, slightly and irregularly sinuate-toothed or entire, glabrous and shining above, more or less pubescent underneath or rarely almost glabrous, with hairy tufts almost always conspicuous in the axils of the raised primary veins. Panicles short and little branched, often reduced to simple racemes, and rarely above 2in. long, shortly tomentose. Flowers small, the upper male ones sessile, the lower hermaphrodite and pedicellate. Sepals ovate, tomentose, under 1 line long, unequal and slightly imbricate. Petals very small, with a minute scale at the base. Stamens 8 to 10. Filaments oblong. Ovary tomentose, occasionally 4-merous. Capsule nearly globular, 3-angled, about ½in. broad, woody, glabrous or minutely tomentose outside, the valves villous inside. Arillus small.

Hab.: Burnett River, *F. v. Mueller;* Brisbane River, *A. Cunningham;* Logan River, *Fraser;* Curtis Island, *Henne.*

The foliage of this species often closely resembles that of *Nephelium tomentosum.—Benth.*

Wood very tough, of a light-yellow colour, the grain resembling that known in Europe as Lancewood: would be useful for making tool-handles.—*Bailey's Cat. Ql. Woods No. 94.*

13. **C. nervosa** (veins prominent), *F. v. M. in Trans. Vict. Inst.* iii. 27; *Benth. Fl. Austr.* i. 459. A moderate-sized tree, the young branches and inflorescence minutely hoary-tomentose, otherwise glabrous. Leaflets 3 to 6, rarely more or reduced to 2, lanceolate or rarely elliptical-oblong, mostly 3 to 6in. long, sinuate-toothed or entire, glabrous, with very rarely small tufts underneath in the axils of the raised primary veins. Racemes usually simple, axillary, 1 to 2in. long, the flowers all pedicellate and larger than in *C. xylocarpa.* Sepals narrow-ovate, slightly imbricate, above 1 line long. Petals very small, with a very short scale. Anthers oblong, hirsute at first, but soon glabrous. Capsule nearly globular, 3-angled, about ½in. broad, woody, glabrous or nearly so outside, the valves villous inside.

Hab.: Moreton Bay, *F. v. Mueller;* Rockhampton, *Thozet;* also in A. Cunningham's and Leichhardt's collections without the precise station *(Benth.)*

Cunningham's and Leichhardt's are the only specimens I have seen in flower, the others are in fruit only, and may possibly include some glabrous specimens of *C. xylocarpa,* to which this species is very nearly allied. It is also closely allied to, although not quite identical with, *C. falcata,* A. Gray, from the Fiji Islands.—*Benth.*

The outer wood light-coloured, that nearer the centre dark-brown, close-grained, very tough and strong.—*Bailey's Cat. Ql. Woods No. 94a.*

14. **C. Bidwilli** (after J. C. Bidwill), *Benth. Fl. Austr.* i. 460. A tree, the young shoots and inflorescence minutely tomentose. Leaves 2 to 4, ovate-oblong or ovate-lanceolate, obtuse or scarcely acuminate, 3 to 6in. long, entire or obscurely sinuate-toothed, glabrous on both sides, with few or no tufts in the axils of the raised primary veins underneath. Panicles terminal, much branched, but shorter than the leaves. Flowers small, all pedicellate. Sepals tomentose, narrow-ovate, slightly imbricate, about 1 line long. Petals rather shorter than

the calyx, oblong, concave, with 2 minute hirsute auricle-like scales at the base of the lamina. Stamens about 8. Filaments nearly as long as the calyx. Anthers oblong. Ovary hirsute. Fruit not seen.

Hab.: Wide Bay, *Bidwill.*
Although I have not seen the fruit, this species has all the appearance of a true *Cupania.* It has some general resemblance to a Philippine Island species, n. 1237 of Cuming, which is, I believe, as yet unpublished.—*Benth.*
C. nervosa and *C. Bidwilli* are considered as varieties only of *C. xylocarpa,* A. Cunn. by F. v. M.

15. **C. Mortoniana** (after L. Morton), *F. v. M. Fragm.* v. 177, ix. 94. A small glabrous tree, leaves 2 to 6-foliolate. Leaflets 3 to 5in. long, 1¼ to 2in. broad, chartaceous, lanceolate-ovate, acuminate, quite entire, both sides shining, very slenderly penninerved and copiously reticulate. Panicles a few or several inches long. Flowers scattered, somewhat large. Sepals 5, broadly imbricate, 1½ to 2¼ lines long, membranous, ovate-orbicular. Petals 5, white, nearly 2 lines long, deciduous, their scales much shorter and barbellate. Disk slightly crenulate. Stamens 8, the lower part of the filaments slightly hairy. Anthers ovate, introrse. Style scarcely 1 line long. Ovary glabrous, ovules erect, 1 in each cell. The young fruit 3-angular, contracted at the base.

Hab.: Mackay River, *J. Dallachy* (F. v. M., l.c.)

16. **C. erythrocarpa** (arillus scarlet), *F. v. M. Fragm.* v. 7, ix. 91. A tree of about 30ft., glabrous, with 10-foliolate leaves, the rhachis angular. Leaflets lanceolate, long-acuminate, quite entire, shining on both sides, decurrent upon the short petiolules, thin-coriaceous or somewhat thick chartaceous, 4 to 8in. or near 1ft. long, from 2 to near 4in. broad. Panicles several inches long. Calyx-lobes deltoid, scarcely 1 line long. Petals 5, sometimes exceeding the calyx, rhomboid-orbicular, glabrous outside, the inner scales oblong, lanuginose-barbate. Disk annular, repandulous. Stamens usually 8. Filaments 2 lines long or shorter, pubescent, glabrous towards the top. Anthers almost ovate, obtusate. Style short, glabrous. Stigmas very short. Capsule about 7 to 10 lines, subglobose or triangular, shortly narrowed at the base, red outside, corrugated and glabrous. Seeds about 5 lines long, black, ovate; arillus scarlet, small, perhaps sometimes wanting.

Hab.: Ranges about Rockingham Bay, *J. Dallachy* (F. v. M., l.c.)

17. **C. sericolignis** (Silkwood), *Bail. Bot. Bull. No.* v. Tree glabrous, said to produce a good timber; branchlets angular. Leaves alternate, pinnate, leaflets from 2 to 7, usually about 5, irregular both as to number and position upon the rhachis, oblong to ovate-lanceolate, obtuse or bluntly acuminate, the terminal one the largest, and attaining 6in. in length and 2in. in width, texture thin, sessile or on petiolules of 1 or 2 lines, the primary veins prominent on the under side, rather distant. Inflorescence (all male flowers on the specimens examined) lateral or axillary, often forming clusters of delicately slender racemes, about 1¾in. long at the nodes on the branches below the leaves. Flowers white, minute when expanded, about 1¼ line diameter. Sepals 5, imbricate, orbicular, unequal in size. Petals 5, larger than the sepals, oblong, with a tuft of ciliæ on either side near the base, otherwise glabrous. Disk annular, dark, glabrous. Stamens 8, subulate, of about equal length with that of the petals or rather shorter; the whole length of the short filament densely clothed with white soft scale-like hairs. Fruit not yet obtained.

Hab.: Mulgrave River, *Bellenden Ker Expedition;* scrub about the Barron River, *E. Cowley.*
The specimens which I obtained on a creek off the Mulgrave River had rather longer racemose panicles, and the leaflets were more sharply acuminate, with a paler under side, and were gathered off small-sized creek-side trees.
Wood soft, of a greyish colour.

XL. SAPINDACEÆ.   295

6. **RATONIA**, DC.

(From *Raton*, a name of a species in St. Domingo.)

(Arytera, *Blume.*)

Flowers regular, polygamous. Calyx small, cup-shaped, 4 or 5-toothed or lobed, open, valvate, or slightly imbricate in the bud. Petals 4 or 5, small, with or without scales inside or none. Disk usually annular. Stamens 7 to 10, inserted inside the disk. Filaments filiform, longer than the calyx. Ovary 2 or 3-celled, with 1 ovule in each cell. Capsule either 2-celled and compressed, or 3-celled and 3-angled or 3-lobed, loculicidally 2 or 3-valved, rarely almost indehiscent. Seeds more or less covered by an arillus; testa crustaceous; cotyledons thick, often curved or folded.—Trees. Leaves alternate, pinnate; leaflets alternate or opposite, usually without a terminal one. Flowers small, in terminal or axillary panicles. Petals rarely as long as the calyx.

A large tropical genus, with the same range as *Cupania*, but especially numerous in America. The Australian species are all endemic. It is closely allied to *Cupania*, with which it is usually joined, but the gamosepalous calyx and long filaments appear to give it at least as great a value as several other generally admitted genera of *Sapindaceæ*.

Leaflets 1 to 4, 2½ to 4in. long, ovate or lanceolate-ovate, slightly undulate; petiolules short. Capsule 7 to 10 lines long, stipitate, angular, hairy, 2 or 3-valved and 2-angular. Arillus light-orange . . . . 1. *R. lachnocarpa.*

Leaflets numerous, 12 to 16in. long, 4 to 5½in. broad, lanceolate-oblong, glabrous, entire; petiolules 6 to 8 lines long. Capsule glabrous, clavate, 3-ribbed, brownish, 1 or 2-seeded . . . . 2. *R. grandissima.*

Leaflets 3 to 8, 4 to 6in. long, ovate or ovate-lanceolate, entire, very coriaceous, glabrous except the young growth; petiolules ½in. or more. Capsule globular, pear-shaped, 4 lines diameter, on long stipes, 3-ribbed 3. *R. pyriformis.*

Leaflets 2 to 4, 2 to 4in. long, ovate or ovate-lanceolate, glabrous; petiolules ¼ to ½in. long. Capsule pear-shaped, 3-angled, about ½in. broad, the valves almost woody . . . . 4. *R. anodonta.*

Leaflets 3 to 6, 2 to 3in. long, oblong-lanceolate, glabrous except when young; petiolules 3 or 4 lines long. Capsule 3-angular, depressed at the top, ½in. broad, stipitate . . . . 5. *R. stipitata.*

Leaflets 4 to 7, 3 to 4in. long, obliquely ovate-lanceolate, entire, minutely pellucidly-dotted; petiolules about ½in. long. Capsule glabrous, 3-angled, flat at the top, 4 lines broad . . . . 6. *R. punctulata.*

Leaflets 4 to 6, 5 to 9in. long, ovate-lanceolate, entire, membranous, on short petiolules. Capsule 1in. diameter on a thick stipes, 3-celled, sharply angled at the sutures, yellow, glabrous. Seeds chestnut-like, bearing at the base a yellow strophiole 3 lines broad . . . . . . 7. *R. Nugentii.*

Leaflets 2 to 5, 2 to 8in. long, ovate to lanceolate, glabrous above, pilose below; petiolules 3 to 9 lines long. Capsule glabrous, including the stipes 8 to 12 lines long, clavate . . . . 8. *R. exangulata.*

Leaflets 4 or 5, 4 to 7in. long, lanceolate, membranous; petiolules slender. Capsule red, ¾in. long, globose-pyriform, triquetrous towards the base, 1-seeded . . . . 9. *R. Lessertiana.*

Leaflets 2 to 5, 2 to 5 in. long, broad-obovate, entire, glabrous; petiolules very short. Capsule brownish, trigonous, shortly stipitate. Arillus yellow or orange . . . . 10. *R. O'Shanesiana.*

Leaflets about 8, 2 to 6in. long, broad or lanceolate-ovate, discolor; margins repand-undulate, rarely crenate, almost glabrous. Capsule 6 to 8 lines long, not stipitate, thick, woody, 3-valved. Arillus bilobed, short, pale-yellow . . . . 11. *R. Dæmeliana.*

Leaflets 2 or 3 or 1, 2 to 4in. long, broad or ovate-lanceolate, entire, the costule on the under side pubescent. Capsule woody, glabrescent, acutely 3 or 4-angled, about ½in. broad, reddish outside, at length expanding horizontally . . . . 12. *R. Martyana.*

Leaflets 4 to 10, about 5in. long, almost glabrous on both sides, lanceolate-ovate, imperfectly crenulate. Capsule hard, acutely trigonous, shortly stipitate, scarcely ½in. long, glabrous outside . . . . 13. *R. Cordieri.*

Leaflets 3 to 8, 2 to 5in. long, oblong-elliptical or lanceolate, much narrowed towards the base, scarcely petiolulate. Capsule flattish, 2-celled, about ¾in. long, and broad at the top, reddish-yellow outside . . 14. *R. tenax.*

1. **R. lachnocarpa** (woolly fruit), *F. v. M. Fragm.* iv. 157, v. 6, ix. 91. A tree from 20 to 30ft. high, bark smooth, branchlets tomentose. Leaves 1, 2, or sometimes 3 or 4-foliolate. Leaflets often slightly undulate, 2½ to 4in. long, 1½ to 2in. broad, ovate or lanceolate-ovate, more or less acuminate, nerves on the under side tomentose-pubescent, wide-spreading, and as well as the reticulation prominent; petiolules short. Spikes 1 to 3in. long, tomentose, pedicels very short. Calyx deeply deltoid-dentate, scarcely one line long. Filaments shortly exserted, glabrous. Anthers very minute, almost rotund. Disk annular, crenulate. Styles loose, at length revolute. Capsule 7 to 10 lines long, stipitate, angular, hirsute, fuscescent, 3-valved or 2-valved and 2-angular, hairy inside. Seeds solitary, subglobose or often ovate. Arillus very thin, yellow or orange, 3 or 4 lines long.

Hab.: On the Rockingham Bay coast, *J. Dallachy* (F. v. M., l.c.)

2. **R. grandissima** (from its very showy appearance), *F. v. M. Fragm.* iv. 156, ix. 91. A small tree or tall shrub. Leaves plurifoliolate. Leaflets 12 to 16in. long, 4 to 5½in. broad, the lateral nerves widely spreading, the reticulation close, lanceolate-oblong, acuminate, the point short or very short, glabrous, entire; petiolules 6 to 8 lines long. Racemes 7 or 8in. long. Pedicels to the fruit 1 or 2 lines long, the calyx 1 line or less, with 5-fid deltoid teeth. Capsule glabrous, clavate-pyriform, 3-ribbed, brownish outside, hairy inside, 1 or 2-seeded. Arillus at length pale and thin.

Hab.: Rockingham Bay, *J. Dallachy* (F. v. M., l.c.)

3. **R. pyriformis** (pear-shaped), *Benth. Fl. Austr.* i. 461. A tree of considerable size, but flowering sometimes as a shrub, glabrous except a minute hoariness on the young shoots and panicles. Leaflets 3 to 6, ovate or ovate-lanceolate, shortly acuminate, 4 to 6in. long, entire, very coriaceous, on petiolules of ¼in. or more. Flowers very small, shortly pedicellate, singly or in little cymes of 2 or 3 along the raceme-like branches of the panicle. Calyx nearly 1 line diameter, shortly and broadly 5-lobed. Petals 5, scarcely exceeding the calyx-lobes, cuneate or spathulate, the inner scales lining and bordering the base of the lamina. Stamens in the male flower 8, much exserted, the filaments slightly hirsute, in the females few, with short filaments. Ovary stipitate, slightly hirsute, style filiform, with 3 diverging stigmatic lobes. Capsule globular-pear-shaped, about 4 lines diameter, narrowed into a long stipes, glabrous, with 3 raised ribs, appearing almost drupaceous and scarcely dehiscent. Seeds often reduced to 2 or 1, enclosed in the arillus; cotyledons much folded. Radicle somewhat broad and much compressed.—*Schmidelia pyriformis*, F. v. M. Fragm. i. 2.

Hab.: Brisbane River, Moreton Bay, and other localities in southern Queensland.
Wood of a light colour, firm and tough, suitable for axe-handles.—*Bailey's Cat. Ql. Woods No. 95.*

4. **R. anodonta** (toothless), *Benth. Fl. Austr.* i. 461. A tree of considerable size, flowering also as a shrub, quite glabrous. Leaflets 2, 3, or rarely 4, ovate or ovate-lanceolate, obtuse or obtusely acuminate, 2 to 4in. long, coriaceous, but not thick, very much reticulate, narrowed into a petiolule of ¼ to nearly ½in. Panicle glabrous, slender, not much branched. Calyx glabrous, about ¾ line diameter. Petals none. Filaments exserted, glabrous. Ovary stipitate, almost glabrous; style shortly subulate, with diverging stigmatic lobes. Capsule pear-shaped, somewhat 3-angled, nearly ½in. broad, the valves almost woody, densely villous inside. Stipes sometimes as long as the valves. Seeds often reduced to 2 or 1, enclosed in the arillus. Embryo much curved; cotyledons folded, but less so than in *R. pyriformis*.—*Schmidelia anodonta*, F. v. M. Fragm. i. 2; *Cupania anodonta*, F, v. M. Fragm. ii. 76.

Hab.: Brisbane River, Moreton Bay, *A. Cunningham, W. Hill*; Mackenzie River, *Leichhardt*; Rockingham Bay, *J. Dallachy* (F. v. M.)

5. **R. stipitata** (stipitate), *Benth. Fl. Austr.* i. 461. A moderate-sized tree, glabrous except a minute tomentum on the young branches and inflorescence. Leaflets 3 to 6, oblong-lanceolate, acute, 2 to 3in. long, narrowed into a petiolule of 3 or 4 lines, coriaceous, very rigid, shining above, the primary veins very prominent underneath. Panicles axillary and terminal, divaricately branched. Flowers not seen. Fruiting pedicels 2 to 3 lines long. Calyx persistent, very small, acutely 4 or 5-lobed. Capsule 3-angled, depressed at the top, ½in. broad, narrowed into a short but distinct stipes, valves thickly coriaceous, almost woody, glabrous and reddish inside. Seeds shining, in a thin arillus.—*Cupania stipata*, F. v. M. Fragm. ii. 75 and 175.

Hab.: Near border of N.S.W., locality ?

6. **R. punctulata** (referring to the pellucid dots of leaf), *F. v. M. Fragm.* ix. 91. A tall shrub, quite glabrous. Leaflets usually 4 to 7, on a long slender common petiole, very obliquely ovate-lanceolate, acuminate, 3 to 4in. long, quite entire, thinly coriaceous, smooth and shining, minutely pellucid-dotted, narrowed into a petiolule of ½in. or more. Flowers not seen. Fruiting panicles short, slender, clustered in the axils or at the ends of the branches. Pedicels short. Sepals often persistent or reflexed, orbicular, about 1 line long, glabrous. Capsule glabrous, 3-angled, flat at the top with the remains of the style forming a point in the centre, about 4 lines broad, contracted into a short obconical stipes, half opening in 3 coriaceous valves. Seeds not seen.—*Cupania*, F. v. M. Fragm. iii. 12, and Benth. Fl. Austr. i. 458.

Hab.: Cumberland Islands, *E. Fitzalan* (F. v. M., l.c.)

7. **R. Nugentii** (after L. J. Nugent), *Bail. Bot. Bull.* xiv. 9. "Chambin," Barron River, *J. F. Bailey.* A small tree, branchlets dark-coloured and angular. Leaves with from 4 to 6 distant leaflets; petioles 3 to 4in. long, angular and swollen where they join the stem; rhachis slightly longer than the petiole; leaflets ovate-lanceolate, 5 to 9in. long, 2 to 3¾in. broad; base cuneate; apex abruptly acuminate; texture membranous; margins entire; petiolules short, swollen where they join the rhachis. Panicles in the upper axils, slender on very short peduncles, 2 or 3 of the lower branches 4 or 5in. long, the others much shorter. Flowers pedicellate in scanty clusters. Bracts minute, hairy. Pedicels about 2 lines long, slender. Calyx 5-lobed, the lobes slightly hairy, and almost triangular, ¾ line. Petals 5, tomentose, twice the size of the calyx-lobes, very broad, truncate, emarginate or toothed at the end, and tapering to the base, marked on the back with 8 lines, scale on face large, very woolly, bearing near the end of the back prominent orange-coloured glands. Stamens about 8; filaments hairy, twice as long as the petals; anthers glabrous. Ovary hairy, 3-angular. Capsule 1in. diameter, on a thick stipes, 3-celled and sharply keeled at the sutures; pericarp thick, succulent, yellow, glabrous; seeds 1 in each cell; testa chestnut-coloured, bearing at the base a yellow cordate strophiole about 3½ lines broad.

Hab : Freshwater Creek, Cairns, *E. Cowley* and *L. J. Nugent.*

8. **R. exangulata** (referring to the ribs), *F. v. M.* iv. 156, ix. 91. A tall shrub or small tree, the branches, petioles and inflorescence clothed with a ferruginous or fuscescent indumentum. Leaves 2 to 5-foliolate. Leaflets mostly alternate, ovate, lanceolate-ovate or lanceolate, chartaceous, 2 to 8in. long, 1 to 3½in. broad, glabrous above, pale on the under side and pilose, nerves tomentose; petiolules 3 to 9 lines long. Panicles racemiform, from 1½ to several inches long. Bracts subulate-semilanceolate, about 1 line long. Calyx 5-parted, scarcely 1 line long. Lobes semi-lanceolate. Petals minute, barbate. Filaments exserted,

pubescent below. Anthers dorsifixed, almost ovate. Capsules including the stipes, 8 to 12 lines long, clavate-pyriform, rufescent, glabrous, not regularly valvate. Style rostellate. Seeds oval, nearly ¼in. long. Testa brown.

Hab.: Rockingham Bay, *J. Dallachy* (F. v. M., l.c.)

9. **R. Lessertiana** (after M. Lessert), *Benth. and Hook. f., Gen. Pl.* i. 400; *Bail. Bot. Bull.* ix. 7. A large shrub or small tree. The branchlets, leaf-petioles, rhachis, and inflorescence puberulent, but the dark-reddish colour of the bark plainly visible. Leaflets 4 or 5 (on the specimens examined), opposite or alternate, usually lanceolate, 4 to 7in. long, and from 1¼ to 2in. broad in the widest part; the apex obtuse, but often elongated; base cuneate to the short slender petiolule; texture thin; primary veins distant, and the reticulation delicate. Inflorescence a racemose panicle, slender and drooping, 4 to 8in. long; the branches very short, or one 1½ or 2in. long; flowers minute, only seen at base of fruit. Capsule red, ¾in. long, globose-pyriform, glabrous, stipitate, triquetrous towards the base; all examined 1-seeded.

Hab.: Daintree River, *E. Cowley*.

The above species, with which I believe our Australian to agree, and under which I place it, enjoys a wide range, being, according to Sir J. D. Hooker, met with in Tenasserim, South Andaman Islands, Malacca.

10. **R. O'Shanesiana** (after P. O. O'Shanesy), *F. v. M. Fragm.* ix. 96. A shrub or small tree with a whitish bark. Leaves 2 to 5-foliolate; rhachis slightly angular; petioles 1½ to 2in. long. Leaflets somewhat thick, chartaceous, 2 to 5in. long, 1 to 2in. broad, obovate, quite entire and glabrous, on very short petiolules. Panicles axillary or terminal, often several inches long. Calyx minute; lobes deltoid-semi-ovate. Petals ciliate, inflexed, 2-lobed. Stamens 8. Anthers oval, puberulent. Capsule deeply split, almost woody, obtuse, trigonous, shortly stipitate, rufescent or fuscesent outside, lanuginous inside, glabrous. Valves about ½in. long. Seeds broad, ovate, brown outside. Arillus yellow or orange, thick.

Hab.: Gracemere, Rockhampton, *O'Shanesy* (F. v. M., l.c.)

11. **R. Dæmeliana** (after E. Dæmel), *F. v. M. Fragm.* ix. 96. A small tree, the young branches thinly tomentose. Leaflets about 8, on short petiolules, broad or lanceolate-ovate, 2 to 6in. long, 1 to 2in. broad, discolour, the margins repandly-undulate, rarely crenate, almost glabrous. The upper side deep-green and glossy; veins very thin. Apex produced into a blunt point; rhachis angular. Panicle thinly-tomentose, rather long. Calyx minute, lobes semi-lanceolate, deltoid, ⅛ line or less long. Petals cuneate or orbicular-obcordate, about 1 line long, glabrous outside, the margins inside puberulent, the two scales semi-adnate, bearded. Filaments many times longer than the anthers, thinly pilose near the top. Disk glabrous, somewhat thick, completely annular. Capsule 6 to 8 lines long, not stipitate, thick, woody, 3-valved, obtuse 3-angular, densely silky tomentose inside. Arillus bilobed, pale-yellow, much shorter than the seed.

Hab.: Cape York, *Dæmel*.

12. **R. Martyana** (after M. Marty, a French botanist), *F. v. M. Fragm.* v. 6, ix. 94. A tree attaining the height of 40ft., with brownish smooth bark, the branchlets and petioles and peduncles ferruginous-tomentose. Leaves 2 or 3-foliolate, rarely reduced to 1; rhachis nearly terete. Leaflets somewhat thick, chartaceous, 2 to 4in. long, 16 to 20 lines broad, broad or ovate-lanceolate acuminate, quite entire, the veins and costula on the under side pubescent; petiolules shortly pubescent. Peduncles, pedicels, calyx, and ovary tomentose. Panicles about 6 or 7in. long. Bracts at the base of the pedicels, 1 to 1½ line long, subulate-semilanceolate, long persistent. Pedicels 2 or 3 lines long.

Calyx-lobes ovate-semilanceolate, about 1 line long. Petals orbicular with very short claws, scarcely 1 line long. Scales barbate-ciliate. Filaments tomentellous, 2 lines long. Anthers glabrous, almost ovate, ⅓ line long, emarginate at both ends. Style nearly 3 lines long, tomentellous, subulate. Disk annuliform, turgid. Capsule woody, glabrescent, acutely 3 or 4-angled, about ½in. broad, the valves almost semi-orbicular, compressed, and sharply angled on the back, reddish outside, at length expanding horizontally. Seeds ellipsoid-ovate, about 3 lines long, black or dark-brown. Arillus thin, yellow.

Hab.: Rockingham Bay, *J. Dallachy* (F. v. M., l.c.)

13. **R. Cordierii** (after the brothers Cordier), *F. v. M. Fragm.* ix. 93. A tree of about 30ft. high, the petioles and peduncles angular and with the branchlets tomentellous. Leaves 4 to 10-foliolate, the rhachis acutely angular. Leaflets somewhat thick, chartaceous, lanceolate-ovate and imperfectly repand crenulate, about 5in. long and 2in. broad, both sides almost glabrous and shining, the lateral nerves broadly spreading, the petiolules short or very short. Panicle rather large, with spike or racemose branches. Flowers in clusters. Pedicels very short, elongating slightly under the fruit. Calyx minute; lobes deltoid. Petals ovate-orbicular, scarcely clawed and scarcely 1 line long; scales barbate. Stamens 8. Filaments 1 to 2 lines long, glabrous at the top, lanuginous at the base. Anthers ovate, obtuse. Disk annular, glabrous, repandulous. Capsule hard, acutely trigonous, very shortly stipitate, scarcely ½in. long, glabrous outside, the inside margin of the valves densely tomentose. Seeds ellipsoid-ovate.

Hab.: Rockingham Bay, *J. Dallachy* (F. v. M., l.c.)

14. **R. tenax** (tough), *Benth. Fl. Austr.* i. 461. A moderate-sized tree, quite glabrous except the inflorescence. Leaflets usually 3, but varying from 2 to 8, from obovate to oblong-elliptical or lanceolate, obtuse, 1½ to 2 or rarely 5in. long, much narrowed at the base but scarcely petiolulate, thinly coriaceous, shining above, pale or sometimes slightly glaucous underneath, with foveoles in the axils of the nerves. Panicles small, little branched. Calyx a little above 1 line broad, 5-lobed. Petals small, broad, the scale inside very hairy. Stamens about 8, the exserted filaments woolly-hairy. Anthers ovate-globose, a little didymous, sometimes barbellate. Ovary stipitate, 2 or rarely 3-celled. Style rather short, with spreading stigmatic lobes. Capsule usually flattened, 2-celled, about ½in. broad, contracted into a very short stipes; valves thick, densely villous inside. Seeds apparently only half enveloped in the arillus, but much injured in the specimens examined.—*Cupania tenax*, A. Cunn. Herb.

Hab.: Brisbane River, *A. Cunningham, W. Hill, F. v. Mueller;* Port Curtis, *C. Moore;* Rockingham Bay, *J. Dallachy* (F. v. M.)

Wood light in colour, dark towards the centre, very tough and close grained.—*Bailey's Cat. Ql. Woods No.* 96.

7. ATALAYA, Blume.

(Indian name.)

Flowers regular, polygamous. Sepals 5, much imbricate in the bud. Petals 5, exceeding the sepals, with an inner scale or tuft of hairs. Disk annular. Stamens 8, inserted inside the disk. Ovary 3-celled, with 1 ovule in each cell. Style short, undivided. Fruit separating into 3 distinct carpels or samaræ, 1-celled, 1-seeded, and indehiscent at the base, terminating in a long wing. Seeds without any arillus, testa coriaceous; cotyledons thick, unequal.—Trees or shrubs. Leaves pinnate or rarely simple. Flowers usually larger than in *Cupania* and *Ratonia*, in axillary or terminal panicles.

The genus is endemic in Australia, with the exception of one species, which extends to Timor. The flowers are nearly those of *Sapindus*, with the fruit of *Thouinia* and *Acer.—Benth.*

Flowers and fruit more or less pubescent or tomentose.
Leaflets ovate or broadly oblong, the petiole not winged. Panicle
 pedunculate, many-flowered. Carpels divaricate . . . . 1. *A. multiflora.*
Leaflets narrow-oblong or linear, or leaves undivided, the petiole often
 winged. Carpels diverging.
 Plant glabrous, except the flowers . . . . 2. *A. hemiglauca.*
 Branches, young leaves, and panicles velvety-tomentose. Leaflets and
 petiole-wings much reticulate . . . . 3. *A. variifolia.*

1. **A. multiflora** (flowers numerous), *Benth. Fl. Austr.* i. 468. A tall shrub or small tree, glabrous except the inflorescence. Leaflets 2 to 6, ovate or oblong, very obtuse, 2 to 3in. long or rarely more, sometimes 2½in. broad, distinctly petiolulate, coriaceous and strongly reticulate. Panicle pedunculate above the last leaves, oblong or pyramidal, minutely tomentose-pubescent. Flowers very numerous, the small scale-like bracts more conspicuous than in the other species. Flowers of *A. hemiglauca.* Ovary slightly pubescent. Samaræ 1 to 1¼in. long, including the straight or falcate wing, very divaricate, pubescent or nearly glabrous.

Hab.: Cape York and Trinity Island, *M'Gillivray;* Brisbane River, *W. Hill, F. v. Mueller;* Rockhampton, *Thozet.*

2. **A. hemiglauca** (referring to light colour on under side of leaf), *Benth. Fl. Austr.* i. 468. White Wood, Cattle Bush. "Boorbal," flower "Boorbalmin," St. George, *Wedd.* A tall shrub or small tree, quite glabrous except the flowers, and more or less glaucous. Leaves usually pinnate; leaflets few, from narrow-oblong to linear, obtuse or scarcely acute, from 2 or 3 to 7 or 8in. long, often somewhat falcate, narrowed at the base, but rarely petiolulate, rigidly coriaceous, with numerous pinnate and reticulate veins and a somewhat thickened margin, the common petiole terete or nearly so; sometimes, however, the petiole becomes winged, or the leaves are quite simple, oblong, or linear, or the leaflets are decurrent on the petiole forming a large 2 or 3-lobed leaf, or rarely the simple leaf is ovate-lanceolate, and 8 to 10in. long. Panicles rather dense, the rhachis and branches glabrous or nearly so; pedicels 1 to 2 lines long. Sepals orbicular, nearly glabrous, 1½ or the inner ones nearly 2 lines long. Petals pubescent, oblong, 3 to 4 lines long, with a hirsute scale at the base. Filaments pubescent. Ovary densely silky-pubescent. Samaræ pubescent, with minute appressed hairs, 1 to 1½in. long including the wing, which is nearly as broad as long, the cavity hairy or nearly glabrous inside.—*Thouinia hemiglauca,* F. v. M. Fragm. i. 98.

Hab.: E. coast, *R. Brown;* Oxley's Station, *Leichhardt;* Rockhampton, *Thozet;* Brisbane River, *A. Cunningham, Fraser;* Mooni River, *Mitchell;* and Cooper's Creek, *Victorian Expedition* and others.

This small tree is often cut down for fodder in times of drought, hence known as Cattle Bush. Wood of a close grain, yellowish, and hard.—*Bailey's Cat. Ql. Woods No. 98.*

3. **A. variifolia** (leaves variable), *F. v. M. Herb.; Benth. Fl. Austr.* i. 468. A tall shrub or small tree, the young branches and panicles softly velvety-tomentose. Leaves or leaflets from oblong to linear, apparently as variable as in *A. hemiglauca,* but longer, often above 8in., very much more reticulate, the common petiole usually broadly winged, the wing also much reticulate. Panicle loose. Sepals silky pubescent, about 1¼ line long. Petals twice as long. Filaments hairy. Samaræ softly tomentose, 2in. long including the wing, which is fully twice as long as broad, the cavity pubescent inside.—*Thouinia variifolia,* F. v. M. Fragm. i. 46.

Hab.: Given as a Queensland plant by F. v. Mueller in Census of Austr. Pl.

8. SAPINDUS, Linn.
(Fruit used for soap.)

Flowers regular, polygamous. Sepals 4 or 5, much imbricate in the bud. Petals as many, usually exceeding the sepals, with 1 or 2 inner scales or without any. Disk annular. Stamens usually 8 to 10. Ovary 2 to 4-lobed, 2 to 4-celled,

Sapindus.] XL. SAPINDACEÆ. 801

with 1 ovule in each cell. Style with 2 to 4 stigmatic lobes. Fruit fleshy or coriaceous, divided into 2 to 4 globular or ovoid indehiscent lobes, not muricate. Seeds without any arillus; embryo straight or curved; cotyledons thick.—Trees or shrubs, rarely climbing. Leaves pinnate, rarely 1-foliolate. Flowers in terminal or axillary panicles.

The genus is widely dispersed over tropical regions, but less numerous in America than in Asia. The Australian species is, so far as known, endemic; but, like many others of the genus, it must remain in some measure doubtful until the fruit has been seen.—*Benth.*

1. **S. (?) australis** (Australian), *Benth. Fl. Austr.* i. 464. Young branches, petioles, and panicles pale or hoary with a very minute tomentum. Leaflets, in our specimens, 4 or 6, broadly ovate, obtuse, 3 to 5in. long, entire, often oblique, narrowed into a short petiolule, coriaceous, glabrous, much veined, of a pale, almost glaucous colour. Panicle loose, longer than the leaves. Flowers shortly pedicellate, in little loose cymes along the divaricate branches. Sepals in the male flowers, the only ones seen, hoary tomentose, rather above 1 line long. Petals nearly 2 lines long, oval-oblong, narrowed into a short claw, pubescent outside, with a single short broad scale inside fringed with long hairs. Stamens usually 8, as long as the petals. Filaments hairy.

Hab.: Cape York, *M'Gillivray.*

In the absence of female flowers and fruit, I have referred this plant to *Sapindus*, from its general resemblance in habit and male flowers to *S. emarginatus*, Roxb.—*Benth.*

Baron Mueller, in 2nd Census of Austr. Plants, places this species as *Atalaya ausrralis*, giving for the authority Radlkofer in Sitzungsaber der Akad zu Münch 327, 1878; his writings, however, throw no further light upon the subject, and I have not the work referred to.

9. NEPHELIUM; Linn.

(Ancient name of Burdock, supposed resemblance of fruit.)

Flowers regular, polygamous. Calyx small, cup-shaped, with 4 or 5 rarely 6 teeth or lobes, valvate or slightly imbricate in the bud. Petals none, or as many as calyx-lobes, small, with a 2-cleft scale or 2 scales inside. Disk annular. Stamens 6 to 10, inserted within the disk; filaments in the Australian species short, in others elongated. Ovary 2 or 3-celled, usually lobed, with 1 ovule in each cell. Style with 2 or 3 stigmatic lobes. Fruit usually deeply 2 or 3-lobed, or rarely entire, 2 or 3-celled, or reduced to a single carpel, the lobes indehiscent or 2-valved, or opening irregularly, muricate, or in the Australian species smooth. Seeds usually wholly or partially enclosed in an arillus; testa coriaceous; cotyledons thick.—Trees, with the habit of *Cupania*. Leaves abruptly pinnate; leaflets opposite or alternate, the primary parallel pinnate veins prominent underneath in all the Australian species except *N. microphyllum*. Flowers small, in axillary or terminal panicles.

The genus extends over tropical Asia, especially the Archipelago. The Australian species differ from the majority of the Asiatic ones in their smooth fruit and shorter filaments. The flowers are nearly those of *Ratonia*; but the fruit does not open in septiferous valves, even when, as in *N. connatum*, it is scarcely lobed. It is also very nearly allied to *Euphoria*, differing chiefly in the smaller gamosepalous calyx. The distinctions, however, between *Cupania, Ratonia, Nephelium, Euphoria,* and several others, are very slight.—*Benth.*

Leaflets 2 to 6, 2 to 4in. long, oblong-elliptical, narrowed to short petiolules, glabrous, somewhat glaucous, margins entire. Capsule 4 to 5 lines diameter, glabrous; lobes divaricate, compressed . . 1. *N. semiglaucum.*

Leaflets 2 to 6, 2½ to 4in. long, obovate to oblong, scarcely petiolulate, entire or obscurely sinuate, glabrous, somewhat glaucous or minutely tomentose underneath. Capsule 3-furrowed or 3-lobed, hoary . . 2. *N. connatum.*

Leaflets 2 to 6, 4 to 7in. long, lanceolate; petiolules rather long, grey on the under side. Capsules of 1 to 3 globose lobes, about 3 lines diameter, velvety . . . . 3. *N. semicinereum.*

Leaflets 2 to 6, ovate or ovate-lanceolate, irregularly sinuate-toothed, glabrous on both sides. Capsule usually 2-lobed, compressed-globular, united to the top . . . . 4. *N. subdentatum.*
Leaflets 4 to 8, 2 to 4in. long, oval-oblong, velvety-tomentose on the under side, at length glabrous above, the margins usually acutely toothed. Capsule softly tomentose-villous; lobes 2 or 3, compressed . . . . . 5. *N. tomentosum.*
Leaflets 2, 2½ to 8in. long, obovate-oblong, entire, coriaceous, glabrous, the under side glaucous, on a short petiolule. Capsule hoary-tomentose, mostly 3-lobed, lobes nearly globular . . . . 6. *N. coriaceum.*
Leaflets 4 to 6, 3 to 5in. long, ovate-lanceolate, entire or sinuate-toothed, narrowed into a distinct petiolule, foveolate in the axils of the primary nerves. Capsules tomentose, deeply divided into 2 to 4 ovoid lobes . . 7. *N. foveolatum.*
Leaflets 2 to 6, 3 to 5in. long, variable both in size and shape, entire or rarely with a few deep serratures; petiolule very short, usually glabrous. Capsule glabrous, with distinct globular lobes 4 or 5 lines diameter. . . . . 8. *N. leiocarpum.*
Leaflets about 10 on the adult trees, about 5in. long, with entire or obscurely toothed margins towards the apex, linear-lanceolate. Capsules glabrous, prominently stipitate, 1in. diameter, 3 or sometimes reduced to 1, globular . . . . 9. *N. Lautererianum.*
Leaflets 4 or sometimes only 2, 2 to 4in. long, entire, petiolulate, oval-oblong. Capsule glabrous, sessile or nearly so, of from 1 to 3 ovoid lobes, indehiscent or splitting longitudinally . . . . 10. *N. divaricatum.*
Leaflets 2 rarely 1, ½ to 1½in. long, ovate or obovate, obtuse, entire, not petiolulate. Capsule glabrous, almost sessile, of 1 or 2 rarely 3 ovoid lobes, splitting like the last . . . . 11. *N. microphyllum.*
Leaflets 2 or sometimes 1 at the end of a short common petiole, 2 or 3in. long, narrowed to a short petiolule, obovate. Capsule renate-obcordate, about ½in. broad, on a short stipes, flattish . . . . 12. *N. distyle.*

Leaflets alternate, 5 to 13, oblong-lanceolate, 5 to 9in. long, puberulent . 13. *N. Callarrie.*

1. **N. semiglaucum** (referring to light colour on under side of leaf), *F. v. M. in Trans. Phil. Soc. Vict.* iii. 25; *Fragm.* iv. 158. Usually a small slender tree, bark light-coloured, smooth. Leaflets 2 to 6, oblong-elliptical, or from almost obovate to nearly lanceolate, obtuse or rarely almost acute or mucronate, 2 to over 4in. long, entire, narrowed into a short petiolule, thin-coriaceous, glabrous and somewhat shining above, more or less glaucous underneath. Panicles either small and axillary or terminal and much-branched, but shorter than the leaves, glabrous or minutely pubescent. Pedicels short. Sepals orbicular, ciliate, otherwise glabrous, the larger inner ones about 1 line diameter. Petals shorter, with two cuneate hairy scales as long as the petals. Stamens exserted. Ovary glabrous, 3-lobed. Capsule 4 to 5 lines diameter, glabrous, very shortly attenuated at the base, with divaricate compressed lobes. Seeds smooth and shining, with a thin arillus.—*Arytera semiglauca*, F. v. M.; *Cupania semiglauca*, F. v. M. Herb., Benth. in. Fl. Austr.

I agree with Baron Mueller in placing this species in *Nephelium.*

Hab.: Many of the south Queensland scrubs; also as far north as Rockingham Bay, *F. v. M.*
Wood of a light-pinkish colour, tough, close-grained, and very prettily marked.—*Bailey's Cat. Ql. Woods No.* 90.

2. **N. connatum** (carpels united), *F. v. M. Herb.; Benth. Fl. Austr.* i. 465. A tree of 20 to 40ft., with a smooth grey bark, the young shoots and inflorescence minutely hoary-tomentose. Leaflets 2 to 6, from obovate to oblong-lanceolate, obtuse 2½ to 4in. long, narrowed at the base, but scarcely petiolulate, quite entire or very obscurely sinuate, thinly coriaceous, glabrous and shining above, somewhat glaucous or minutely tomentose underneath. Flowers small and numerous, in pyramidal panicles rarely exceeding the leaves. Calyx 5-lobed, about 1 line diameter. Petals about ¼ line long, the inner scale as long as the lamina. Filaments short; anthers exserted, oblong, pubescent. Ovary 3-celled; style thickened at the base. Fruit 3-furrowed or 3-lobed, but not deeply so, mucronate,

and not depressed in the centre, somewhat inflated, scarcely coriaceous, hoary, indehiscent or splitting irregularly. Seeds small, shining, black, in a bright red cupular arillus.—*Spanoghea connata*, F. v. M. in Trans. Vict. Inst. iii. 26.

Hab.: Rockingham Bay, Keppel Bay, Brisbane River, Moreton Bay, Port Denison. This is certainly the *Sapindus cinereus*, A. Cunn., referred to by A. Gray, in Bot. Amer. Expl. Exped. i. 258; but the plant from Hunter River more especially described by A. Gray, with coarsely serrate leaves and glabrous bracts, is probably different.—*Benth.*

Wood light-coloured, close in grain, hard and tough.—*Bailey's Cat. Ql. Woods No.* 99.

3. **N. semicinereum** (leaflets grey beneath), *F. v. M. Fragm.* iv. 158. A small glabrous tree with a smooth grey bark; branchlets dark-coloured. Leaves 1, 2, or 3-jugate. Leaflets chartaceous, grey on the under side, shining on the upper side, lanceolate, 4 to 7in. long, 1½ to 3in. broad, lateral parallel nerves erecto-patent, 10 to 18 on each side of the costule, reticulate veins numerous; petiolules rather long. Panicles axillary, 2 or 3in. long, on peduncles of about the same length, with loose spreading branches, bearing small pedicellate flowers near their extremities. Capsules of from 1 to 3 globose lobes, about 3 lines diameter, velvety, bursting irregularly. Seeds black, about half enclosed by a red arillus.

Hab.: Rockingham Bay, *J. Dallachy* (F. v. M., l.c.)

4. **N. subdentatum** (somewhat dentate), *F. v. M.* (as a var. of *N. connatum*), *Benth. Fl. Austr.* i. 465; *Fragm.* ix. 99. A tall shrub or small tree, the young shoots and inflorescence slightly pubescent with minute appressed hairs. Leaflets 2 to 6, ovate or ovate-lanceolate, obtuse or scarcely acute, irregularly sinuate-toothed or rarely almost entire, coriaceous, glabrous on both sides and shining above. Panicles short, little branched. Pedicels short. Calyx truncate or shortly and broadly lobed. Petals none. Filaments very short; anthers oblong, scarcely pubescent. Ovary tomentose, 2 or 3-celled; fruit truncate at the top, slightly hoary with a minute tomentum; capsule lobes usually 2 only, compressed-globular, united to the top, hard and indehiscent.

Hab.: Rockhampton, *P. O'Shanesy* (F. v. M., l.c.); Rockingham Bay, *J. Dallachy* (F. v. M., l.c.) Also Tringilburra Creek, *Bellenden Ker Exped.* 1889.

Wood light-coloured, close-grained, and tough.—*Bailey's Cat. Ql. Woods No.* 99b.

F. v. Mueller thinks that this may be a glabrescent form of *N. connatum*, but there is a considerable difference in general aspect; the calyx is more open and less lobed, I can find no petals, and the fruit is differently shaped.—*Benth.*

5. **N. tomentosum** (tomentose), *Benth. Fl. Austr.* i. 466; *F. v. M. in Trans. Vict. Inst.* ii. 64, *Fragm.* ix. 99. A tree of 20 to 30ft. or more, the young branches and petioles clothed with a soft rust-coloured velvety tomentum. Leaflets 4 to 8, from oval-oblong to oblong-lanceolate, acute or rarely obtuse, 2 to 4in. long, acutely toothed or rarely almost entire, thinly coriaceous, pubescent above or at length glabrous, tomentose-pubescent underneath. Flowers small, crowded, on short slightly-branched tomentose panicles, sometimes reduced to simple racemes. Pedicels very short. Calyx nearly 1 line long, the lobes rather deep and acute. Petals none. Filaments very short; anthers oblong, exserted, glabrous or slightly pubescent. Ovary tomentose, 2 or 3-lobed; style short, with spreading stigmas. Fruit softly tomentose-villous, depressed at the top, of 2 or rarely 3 globular slightly compressed lobes, united to the top, 4 or 5 lines diameter, rather hard, indehiscent. Seeds half immersed in a yellowish arillus.

Hab.: Bremer River, Moreton Bay, *A. Cunningham, W. Hill, F. v. Mueller*; Rockingham Bay, *J. Dallachy*; and Rockhampton, *O'Shanesy* (F. v. M., l.c.)

Wood of a light-yellow colour, close-grained, and hard.—*Bailey's Cat. Ql. Woods No.* 99a.

6. **N. coriaceum** (coriaceous), *Benth. Fl. Austr.* i. 466. Young branches slightly hoary with a very minute tomentum. Leaflets in our specimens always 2, obovate-oblong or elliptical, 2½ to 4in. long, very obtuse, quite entire, coriaceous, glabrous and shining above, pale or glaucous underneath, rounded at the base, on a short petiolule. Flowers not seen. Fruiting panicle branched, shorter than the leaves. Calyx small, with rather acute lobes. Fruit hoary-tomentose, mostly 3-lobed, much depressed in the centre, the lobes nearly globular, coriaceous, indehiscent.

Hab.: Brisbane River, *Fraser*.

7. **N. foveolatum** (foveolate), *F. v. M. Fragm.* ix. 99; *Benth. Fl. Austr.* i. 466. A tree of considerable size, the young branches and inflorescence rusty-tomentose. Leaflets 4 to 6, ovate-lanceolate, or almost ovate, obtuse or acuminate, 3 to 5in. long, entire or sinuate-toothed, narrowed into a distinct petiolule of 1 to 3 lines, thinly coriaceous, glabrous or rarely slightly pubescent underneath, having frequently a cup-shaped cavity in the axils of the primary veins. Panicles in the upper axils broad and many-flowered but shorter than the leaves, the flowers in little clusters or cymes along the principal branches. Calyx tomentose, deeply divided into lanceolate lobes of nearly 1 line, valvate in the bud. Petals minute or rudimentary. Filaments nearly as long as the calyx; anthers oblong, pubescent. Fruit tomentose, deeply divided into 2, 3, or sometimes 4 ovoid lobes, attaining sometimes ½in., opening in 2 thickly coriaceous valves. Seeds completely enveloped in the arillus.—*Arytera foveolata*, F. v. M. in Trans. Vict. Inst. iii. 24.

Hab.: Moreton Bay, *W. Hill, F. v. Mueller;* Isis scrub, *Mrs. Helms*.

8. **N. leiocarpum** (fruit smooth), *Benth. Fl. Austr.* i. 467; *F. v. M. Fragm.* ix. 98. A tall tree, with smooth pale variegated bark; usually glabrous except a very slight pubescence on the young leaves and shoots, and sometimes on the panicles. Leaflets 2 to 6, mostly oblong-elliptical, ovate-lanceolate or lanceolate, acuminate or obtuse, 3 to 4 or even 5in. long, but more variable in size and shape than in most species, entire or rarely with a few deep serratures, narrowed into a very short petiolule, not coriaceous; the rhachis angular. Panicles loose, not much branched, usually glabrous. Calyx about 1 line diameter, with very short broad teeth. Petals broad and short but variable, the scale usually nearly as long as the lamina. Filaments often exceeding the calyx; anthers oblong, glabrous or nearly so. Fruit sessile or nearly so, glabrous, with distinct globular lobes of 4 to 5 lines diameter, coriaceous, indehiscent or opening irregularly in a longitudinal slit, or breaking off transversely. Seeds deeply enclosed in the arillus.—*Spanoghea nephelioides*, F. v. M. in Trans. Vict. Inst. iii. 25.

Hab.: Brisbane River, *F. v. Mueller;* Curtis Island, *Henne* (a var. with smaller, more obtuse and more coriaceous leaflets).

9. **N. Lautererianum** (after Dr. J. Lauterer), *Bail. Bot. Bull.* iv. 8 and xiv. 9. A tall erect tree with umbrageous head; trunk attaining a diameter of 1½ to 2ft., bark smoothish. Leaves alternate, pinnate, glossy, narrow-lanceolate in outline, the young growth more or less viscid and of a bright purplish red; leaflets very irregular, alternate, about 10 on adult trees, linear-lanceolate, about 5in. long, 1in. wide, larger on the young trees; margins entire or obscurely toothed in the upper half; petiolule very short and enlarged where it joins the rhachis; midrib prominent, the lateral veins numerous, almost horizontal from the midrib, with small pits at the axils. Flowers minute in slender, widely branching panicles near the summit of the branchlets, inserted shortly above the axils, or angle formed by the leaf and the branch, panicle branches spike-like, bearing short branchlets which produce at their extremities clusters of from 2 to 4, or at times solitary flowers, which when expanded scarcely exceed 1 line in

breadth. Calyx of 5 ovate lobes. Petals white, larger than the calyx lobes, angular toothed, scales ciliate. Disk lobed. Stamens 8; filaments hairy; anthers glabrous. Capsule glabrous, depressed, prominently stipitate, attaining to a diameter of 1in., of 3 rounded lobes, which are by abortion sometimes reduced to 1 or 2, glabrous inside except for a small dense tuft of white hairs immediately below where the ovules are attached. Seeds compressed, smooth, angular; testa thin, light-brown coloured, 3 or 4 lines broad, entirely concealed in a thick fleshy arillus of an amber colour, very juicy and of sharp acid flavour.

Hab.: Eudlo scrubs, *Field Naturalists*, Nov. 1891 (flowering specimens, *J. H. Simmonds* and *J. F. Bailey*, May 1896).

The nearest Australian ally to the above is probably *N. leiocarpum*.
The sharp, acid, thick arillus closely resembles that in the fruit of *Diploglottis*, and might be used in a similar manner. Doubtless an excellent and palatable jelly might be made from it.
Outer wood grey, inner wood brown, close-grained, hard and heavy, but easily worked; useful for mallets and tool handles.—*Bailey's Cat. Ql. Woods No.* 100.

10. **N. divaricatum** (referring to the few straggling branches of panicles), *Benth. Fl. Austr.* i. 467; *F. v. M. Fragm.* ix. 98. A handsome tree of considerable height, the bark of the trunk smooth, grey or here and there reddish, the young shoots and panicles slightly hoary with a minute tomentum, otherwise glabrous. Leaflets 4 or rarely 2, oval-oblong, elliptical or oblong-lanceolate, obtuse or acuminate, 2 to 3 or rarely 4in. long, entire, narrowed into a petiolule of 2 or 3 lines, thinly coriaceous. Panicles loose, with few divaricate branches, the flower-cymes shortly pedunculate. Calyx very open, about ½ line long, pubescent, divided to the middle into 5 or rarely 4 broad obtuse lobes. Petals small, the inner scale short or in some females nearly as long as the lamina. Filaments short; anthers oblong, pubescent. Ovary tomentose. Fruit glabrous, sessile or nearly so, with 1, 2, or 3 ovoid or nearly globular lobes, nearly ½in. long, indehiscent or splitting longitudinally, more or less villous inside. Seed ovate, 3 to 5 lines long, nearly enveloped in the arillus.—*Arytera divaricata*, F. v. M. in Trans. Vict. Inst. iii. 25.

Hab.: Brisbane River, Moreton Bay, *A. Cunningham. W. Hill, F. v. Mueller*; Pine River, *Fitzalan*; and from thence to Rockingham Bay.

11. **N. microphyllum** (small leaved), *Benth. Fl. Austr.* i. 468 (in part). Glabrous or the young shoots minutely hoary. Leaflets 2 or rarely 1 only, ovate or obovate, obtuse, ½ to 1½in. long, entire, narrowed at the base but not petiolulate, somewhat coriaceous, the primary veins numerous and fine, not distant and raised as in other species. Fruiting panicles short and rather dense. Inflorescence puberulent, almost reduced to racemes, solitary or in pairs, in the axils of the leaves, very slender, 1 to 2in. long, with a few very short branchlets near the base. Pedicels filiform, about 1 line long. Calyx small, 5-lobed, scarcely ¼ line long, hairy. Stamens erect, seldom more than 6, hairy, exceeding 1 line beyond the calyx. Fruit glabrous, almost sessile, with 1, 2, or rarely 3 ovoid lobes, about 5 lines long, splitting irregularly like those of *N. divaricatum*, hirsute inside.

Hab.: Wide Bay, *Bidwill*; Isis scrub, *Mrs. Helms*.

12. **N. distyle** (two-styled), *F. v. M. Fragm.* ix. 99. A tree of considerable size, glabrous except the inflorescence, and sometimes the young shoots. Bark smooth. Leaflets 2, or sometimes reduced to 1 at the end of a short common petiole, from obovate-oblong to elliptical or lanceolate, obtuse or shortly acuminate, 2 to 3in. long, narrowed into a short petiolule, thinly coriaceous, reticulate. Panicles small, pubescent, with minute appressed hairs, the female often reduced to a simple raceme. Calyx small, broad, shortly 5-toothed. Petals minute, orbicular, or sometimes narrow, with a hairy scale at the base. Filaments rather short, particularly in the female flowers. Anthers oval-oblong, obtuse,

x

pubescent. Ovary broadly obcordate, strigose-pubescent. Styles divided to the base, revolute. Capsule renate-obcordate, about ½in. broad, attenuated into a somewhat short stipes, flattish, 2-celled, twice as broad as high by the dimidiate lobes, glabrous outside, lanuginous-tomentose inside; pericarp crustaceous. Seeds ovate, 3 or 4 lines long, the arillus almost enclosing them. Hilum not very large. Cotyledons very unequal, oblique, superposed. Radicle lateral, short. *Ratonia distylis*, F. v. M. in Fl. Austr. i. 462, from further examination removed by Baron Mueller to *Nephelium*.

Hab.: Brisbane River, and Bunya Mountains in the south to Port Denison in the north.

18. **N. callarrie** (aboriginal name of tree at the Upper Barron River), *Bail. Ql. Agri. Journ.* v. A graceful erect tree about 50ft. high, in all parts except the upper side of leaflets thinly covered with a light-coloured pulverulence. Leaves pinnate; leaflets petiolulate, oblong-lanceolate, alternate, 5 to 13, thin, coriaceous, 5 to 9in. long, 1½ to 2in. broad, primary nerves very close, but faint, not more prominent than the close reticulation, which is plain on either side, upper side green, glabrous, under side almost white, margins entire, wavy. Panicles in the upper axils, erect, 6 to 9in. long, with few spreading racemose branches, the flowers small in cymes or clusters of about 3 on short branchlets, almost sessile; calyx silky-hairy, about 2 lines diameter; lobes 5, broad. Petals 5, rudimentary, obtuse, tapering each way from the centre. Stamens 8, slightly exserted; filaments hairy, broadening towards the base, very narrow under the anther; ovary silky, tapering from base to the 2 recurved stigmatic lobes. The position in the genus doubtful until fruit is obtained.

Hab.: Mulgrave River, *Bellenden Ker Expedition*, 1889; Upper Barron River, *J. F. Bailey*, June, 1899.

10. EUPHORIA, Juss.

(From *euphoros*, fertile.)

Flowers regular, polygamous. Sepals 5, distinct, imbricate or valvate in the bud. Petals none or as many as sepals, with or without a scale inside. Disk annular. Stamens 6 to 10, inserted within the disk; filaments short. Ovary 2 or 3-celled, usually lobed, with 1 ovule in each cell; style deeply 2 or 3-lobed, or divided to the base into distinct styles. Fruit deeply 2 or 3-lobed, or reduced to a single carpel, the lobes usually indehiscent, tuberculate. Seeds enclosed in a pulpy arillus; testa coriaceous; cotyledons thick.—Trees, with the young shoots usually pubescent. Leaves pinnate; leaflets, as in *Nephelium*, with the primary pinnate veins raised underneath. Flowers small, in terminal panicles.

The genus extends over tropical Asia, especially the Archipelago, with one Australian endemic species. It is very nearly allied to *Nephelium*, differing chiefly in the distinct sepals (in which respect *N. Beckleri* comes very near to *Euphoria*), and from the Australian *Nephelia* in the tuberculate fruit.—*Benth.*

1. **E. Leichhardtii** (after L. Leichhardt), *Benth. Fl. Austr.* i. 468. Young branches, petioles, and inflorescence rusty-tomentose. Leaflets about 6, from obovate-oblong to ovate-lanceolate, obtuse or acuminate, 2 to 3in. long, entire, rather thin, glabrous or nearly so above, tomentose or pubescent underneath, narrowed into a short petiolule. Panicles terminal, sessile, rather large, the flowers in little dense cymes along its branches. Sepals about 1 line long, tomentose, imbricate. Petals rather shorter, without any scale, but hairy inside, glabrous outside in the typical form. Filaments longer than the calyx; anthers

ovoid. Ovary 3-celled. Style rather thick, with 3 divergent lobes. Young fruit deeply divided into 3 globular lobes, very tomentose and tuberculate, but not seen fully formed.

Hab.: (?) *Leichhardt* (Herb. F. v. M.)

Var. *hebepetala.* Calyx rather smaller. Petals pubescent outside. "Nurrum Nurrum," *Leichhardt* (Herb. F. v. M.)

Baron Mueller places this species with the *Nephetiums.* I think, however, until better known it may as well be left as placed by Mr. Bentham.

11. HETERODENDRON, Desf.

(Various tree, parts.)

Flowers regular, usually hermaphrodite. Calyx broadly cup-shaped, very shortly and irregularly toothed. Petals none. Disk small. Stamens 6 to 15, inserted within or upon the disk; anthers nearly sessile, longer than the calyx. Ovary 2 to 4-lobed, 2 to 4-celled, with 1 ovule in each cell; style short, with an obtuse lobed stigma. Fruit of 1 or 2, rarely 3 or 4 coriaceous or hard lobes, indehiscent. Seed half immersed in an arillus; testa crustaceous; cotyledons thick, flexuose.—Shrubs. Leaves simple, entire or lobed. Flowers small, in short terminal, slightly-branched panicles, often reduced to simple racemes.

The genus is limited to Australia.

Leaves entire, coriaceous, linear, oblong or rarely obovate, usually above
2in. long . . . . 1. *H. oleæfolium.*
Leaves entire, mucronate, toothed or pinnatifid, scarcely coriaceous, rarely
2in. long . . . . 2. *H. diversifolium.*

1. **H. oleæfolium** (Olive-leaved), *Desf. in Mem. Mus. Par.* iv. 8, *t.* 3; *Benth. Fl. Austr.* i. 469. A tall shrub, the young shoots hoary or glaucous with a minute silky pubescence. Leaves linear, lanceolate or narrow-oblong, rarely almost obovate, acute or obtuse, 2 to 4in. long, quite entire, narrowed into a very short petiole, coriaceous and sometimes very rigid. Panicles usually few-flowered and much shorter than the leaves. Calyx broadly cup-shaped, varying from 1½ to nearly 3 lines diameter. Ovary usually 3 or 4-celled, densely tomentose. Fruit of 1, 2, or very rarely 3 or 4 nearly globular lobes, 3 or 4 lines diameter.—DC. Prod. ii. 92; F. v. M. Pl. Vict. i. 90; *Nephelium oleifolium,* F. v. M. Fragm. x. 82.

Hab.: Burdekin River, *F. v. Mueller;* Bowen River and Connor's Creek, *Leichhardt;* common in the interior.

The Queensland specimens have smaller and more glabrous flowers than the more southern ones, with the ovary usually 2-carpellary.—*Benth.*

Wood with the outer yellow, the inner dark-brown, hard and close-grained; suitable for engraving or any purpose to which the box-wood is applied.—*Bailey's Cat. Ql. Woods No.* 100a.

2. **H. diversifolium** (various-leaved), *F. v. M. Fragm.* i. 46; *Benth Fl. Austr.* i. 469. A shrub, the young branches tomentose, pubescent, or perfectly glabrous. Leaves from linear-cuneate to oblong-cuneate or almost obovate, rarely 2in. long and often under 1in., usually mucronate with an almost pungent point, either enitre or with a few sharp teeth or lobes towards the end, or pinnatifid with the triangular pungent lobes rigid and sometimes coriaceous, but less so than *H. oleæfolium.* Flowers few, in short panicles, pubescent or glabrous. Ovary 2-celled. Fruit-lobes very divaricate, ovoid, glabrous or tomentose.—*Nephelium diversifolium,* F. v. M. Fragm. x. 82.

Hab.: Keppel Bay, *R. Brown;* thickets at the foot of the dividing range, *A. Cunningham;* Rockhampton, *Thozet;* Warwick, *Beckler;* Comet River, *Leichhardt.*

There are two forms, one perfectly glabrous, the other with the young shoots and flowers pubescent, the fruit densely pubescent or tomentose.

Wood pinkish except the centre, close in grain, hard and tough; useful for engraving and many other purposes.—*Bailey's Cat. Ql. Woods No.* 101.

12. HARPULLIA, Roxb.
(Name of a species of Chittagong.)

Flowers regular, polygamous. Sepals 4 or 5. Petals as many, without any scale, but sometimes with inflected auricles at the base of the lamina. Disk inconspicuous. Stamens 5 to 8. Ovary 2-celled, with 2 ovules in each cell; style short, or elongated and spirally twisted. Capsule coriaceous, somewhat compressed, with 2 turgid lobes opening loculicidally in 2 valves. Seeds 1 or 2 in each cell, with or without an arillus; cotyledons thick.—Trees. Leaves pinnate; leaflets usually large, the primary veins prominent underneath. Flowers in loose terminal little-branched panicles, sometimes reduced to simple racemes. Capsules usually large, red or orange-coloured.

Besides the Australian species, which are endemic, there are two or three others, natives of tropical Asia and Madagascar.—*Benth.*

Calyx persistent. Petals not auriculate.
  An erect single-stemmed shrub. Leaf rhachis winged. Leaflets entire . . 1. *H. frutescens.*
  A small tree.
    Petiole winged. Leaflets coarsely toothed . . . . 2. *H. alata.*
    Petiole not winged. Leaflets entire. Leaflets coriaceous, very obtuse . . 3. *H. Hillii.*
Calyx deciduous. Petals with inflected auricles . . . . 4. *H. pendula.*

1. **H. frutescens** (shrub-like), *Bail. Rep. Bellenden Ker Exped.*, 1889. A slender usually single-stemmed shrub, so far as observed never attaining more than 5ft., slightly rusty-pubescent. Leaf-rhachis prominently winged to the base, and extending beyond the last pinnæ; pinnæ nearly opposite, sessile, lanceolate, with attenuated points, 4 to 8in. long, 1½ to 2½in. wide in the centre, membranous, of a dark green and always quite entire. Inflorescence a terminal erect panicle with but a few short branches near the base, the female or hermaphrodite flowers near the base, the end or upper ones male, white and fragrant, on short pedicels; sepals imbricate, 4 or 5, linear-obtuse, 3 or 4 lines long, rusty-tomentose. Petals 4, lanceolate, 6 or 7 lines long, recurved. Stamens 7 or 8, nearly as long as the petals, surrounded by a ring of short, erect, obtuse glands. Capsule 2 or 3-celled, forming spreading lobes 1¼in. wide and 1in. deep, the outside clothed with a deep crimson tomentum, 2 seeds in each cell, enveloped in a cup-shaped yellow arillus.

Hab.: Bellenden Ker, at an altitude of over 2000ft., and very frequently met with in the scrubs bordering the rivers on the lower lands of the district, *Bellenden Ker Exped.*, 1889. In Fragm. ix. 89, July, 1875, Baron von Mueller notices a plant, probably this, as occurring at Rockingham Bay, as *H. alata*, F. v. M., having entire leaflets and flowering as a shrub. I have, however, during my recent visit to Bellenden Ker, made a point of hunting for any plant of the species with serrated leaflets, or that attains the size of even a large shrub, and, having failed, think it better to give this northern plant specific rank.

2. **H. alata** (winged), *F. v. M. Fragm.* ii. 108; *Benth. Fl. Austr.* i. 470. A tall tree, the young branches and panicles minutely tomentose, otherwise glabrous. Leaflets usually 6 to 10, oblong elliptical or lanceolate, acutely acuminate and coarsely toothed, almost lobed, 3 to 6in. long, or more in the large leaves of barren shoots, rather rigid, green and much veined on both sides, the common petiole broadly winged. Panicles short, loose. Flowers few, larger than in the other species, on short pedicels. Sepals persistent, about 3 lines long, shortly tomentose. Petals about 4 lines long, oblong-cuneate, narrowed at the base, and not auricled. Stamens 7 or 8, about as long as the sepals in the males, shorter in the females. Capsule 1 to 1½in. broad, coriaceous, nearly glabrous inside. Seeds enveloped in a yellowish arillus.

Hab.: Southern scrubs.

3. **H. Hillii** (after W. Hill), *F. v. M. in Trans. Vict. Inst.* iii. 26, *and Fragm.* ii. 104; *Benth. Fl. Austr.* i 470. A tree of 60 to 80ft., the young branches and inflorescence rusty with a close tomentum, otherwise glabrous. Leaflets usually

5 to 11, broadly-oblong or oval-oblong, very obtuse, 8 to 5in. long, or more in the large leaves of barren shoots, thinly coriaceous, shining, the common petiole not winged. Panicles loose, little branched, shorter than the leaves. Pedicels 2 to 3 lines long. Sepals persistent, broadly-ovate, 2 to 3 lines long. Petals oblong, 3 to 4 lines long, without auricles. Male flowers not seen. Stamens in the females 5 or 6, with very short filaments and acute anthers, probably imperfect. Capsule 1½in. broad, fulvous, thick-chartaceous, shortly stipitate, slightly tomentose outside, the turgid lobes divaricate, hirsute inside. Seeds ovate, 4 to 6 lines long, in the young state showing no arillus, but, according to Beckler, of an orange-yellow when ripe and enclosed in a rich red membrane.—Benth.

Hab.: Moreton to Trinity Bay.
Wood of a close grain and light colour.—*Bailey's Cat. Ql. Woods No.* 103.

4. **H. pendula** (pendulous), *Planch.; F. v. M. in Trans. Vict. Inst.* iii. 26, *Fragm.* ii. 104 ; *Benth. Fl. Austr.* i. 471. Tulipwood. A tall tree, having a sinuous trunk and a pale-coloured bark, glabrous or the young shoots and panicles minutely hoary tomentose. Leaflets 3 to 6, or rarely more, from ovate to elliptical-oblong, obtusely acuminate, 3 to 5in. long, membranous. Panicles loose and slender. Pedicels in flower 3 to 4 lines, in fruit ½ to 1in. long, slender. Sepals deciduous, about 2 lines long. Petals ovate, nearly 3 lines long, with inflected ciliate auricles at the base, representing the inner scales of many other *Sapindaceæ*. Stamens 5 to 7, much longer than the calyx, with slender filaments in the males, small and short in the females. Ovary tomentose, with a long style twisted at the top. Capsule chartaceous, glabrous or slightly pubescent, 1 to 1½in. broad, the lobes inflated. Seeds apparently without any arillus.

Hab.: Moreton Bay, *Fraser. A. Cunningham ;* Wide Bay, *C. Moore ;* Port Denison, *Fitzalan ;* Broadsound, *Thozet ;* and Rockingham Bay.

Wood of a light colour, of some trees showing a more or less quantity of a beautifully figured and coloured dark wood towards the centre. The outer or light wood very tough, easily worked, and might suit for engraving purposes. This outer wood is considered the best we have for lithographers' scrapers.—*Bailey's Cat. Ql. Woods No.* 104.

13. AKANIA, Hook. f.

(Leaves thorny.)

Flowers regular, hermaphrodite (or polygamous?). Calyx campanulate, with 5 short lobes, imbricate in the bud. Petals 5, without any inner scale. Disk adnate to the base of the calyx. Stamens 5 to 10, inserted within the disk. Ovary 3-celled, contracted into a thickish style, with a capitate stigma ; ovules 2 in each cell. Capsule globose-ovate, shortly umbonate at the apex, 8 to 12 lines long, 3-celled, rarely 2-valvate ; valves almost coriaceous, glabrous inside, maturing 1 or 2 ovate-globose a little angular seeds. Testa bony-crustaceous, putaminous. Arillus scarcely any.—Tree. Leaves pinnate. Panicles loose, axillary or terminal.

The genus is limited to a single species, endemic in Australia, allied to *Harpullia*, but very different in the calyx and disk.—*Benth.*

1. **A. Hillii** (after Walter Hill), *Hook. f. in Benth. and Hook. Gen. Pl.* 409 ; *Fragm.* ix. 91 ; *Benth. Fl. Austr.* i. 471. Turnip-wood. An elegant tree of 30 to 40ft., glabrous except the panicle. Leaves often above 2ft. long ; leaflets numerous, lanceolate, acutely acuminate, often above 8in. long, bordered with acute often pungent serratures, rounded at the base and shortly petiolulate, coriaceous, light green, shining above, marked underneath (in the dried state) within each areola of the smaller reticulations with 3 or 4 round ovate or reniform dots. Panicles long, loose, and little branched. Pedicels long and slender. Calyx tomentose, about

2 lines long, the lobes rounded, with thin edges. Petals inserted near the base of the calyx outside the disk ; cuneate-oblong, 4 to 6 lines long. Filaments capillare, 2 lines long. Anthers oblong. Style 8 lines long.—*Cupania lucens*, F. v. M. Fragm, iii. 44.

Hab.: Moreton Bay, *Leichhardt*; Pine River, *W. Hill*; Nerang Creek.
Wood of a light colour, close grained, prettily marked; warps much in drying, particularly if young trees are cut down when full of sap.—*Bailey's Cat. Ql. Woods No.* 106.

14. BLEPHAROCARYA, F. v. M.

(Name referring to the nuts being ciliated.)

Involucre hard-coriaceous, deeply divided into from 20 to 80 partly coherent or connate laciniæ. Flowers sparsely inserted in the base and on the laciniæ of the involucre, sessile. Sepals 4 to 5, lanceolate. Petals of the same number as the sepals or less. Stamens 8, shortly exserted; anthers cordate-ovate, dorsifixed; dehiscing longitudinally. Male flowers : Ovary rudimentary, pilose ; style persistent, setaceous ; stigma simple, capitellate. Female flowers : Ovary 1-celled, 1-ovulate, surrounded by an undulated annular disk. Fruit indehiscent, reniform, strongly compressed, densely ciliate. Pericarp very thin. Seed filling the cavity ; testa membranaceous ; albumen none. Cotyledons straight, reniform, externally slightly convex. Radicle situated at the end of the cotyledon and somewhat shorter, slightly curved and accumbent.—A tree. Leaves abruptly pinnate, pinnæ lanceolate, entire, chartaceous. Panicle with opposite branches. Involucre valvately closed to form a globule like a humectatio. Flowers diœcious, very small. Fruit nestling within the involucre, small and numerous. —F. v. M. in Fragm. xi., and Trim. Journ. Bot. xvii.

Nearly allied to *Dobinea*, an Indian genus.

1. **B. involucrigera** (fruit nestling in a sort of involucre), *F. v. M. Fragm.* xi. 15, 16, 187, *and Bot. Journ.* xvii. 116. "Chargir," Bally Gum, of Martintown Sawmills, *J. F. Bailey*. A tree of moderate size, with a smooth grey back. Monœcious if not diœcious. Branches angular, often somewhat flattened, rusty-tomentose, and speckled with whitish-brown lenticels. Leaves opposite, petioles rusty-pubescent and expanded at the base. Leaflets from 5 to 9 pairs ; the rhachis semi-terete and shortly pubescent, opposite, 3 to 7in. long and 1 to 1½in. broad, more or less gradually contracted into an obtuse point, or obtuse and somewhat emarginate ; petiolules about 8 or 5 lines. Peduncles 2 to 3in. long, bearing erect rather narrow panicles, rather longer than the peduncles, the branches opposite, finely puberulent. The male flowers numerous in small globose heads. Bracts deltoid-cordate 1 to 1½ lines, sepals 4, rhomboid, or lanceolate-ovate, pilose. Petals 4, imbricate in the bud, oval, sessile, membranous, white, 1-nerved. Stamens 8, shortly exserted ; filaments capillare-linear. Anthers pale-yellow, cordate-ovate, scarcely ¼ line long. Pollen granular, smooth. Ovary rudimentary without a style, pilose. Female peduncle and panicles as in the male. Involucre tawny-velvety, with the ultimate branchlets ternate, and when full grown ½in. or less, rather thick. Involucre about 1in., campanulate or hemispherical, composed of broad, linear, gradually attenuated bracts, wholly connate below but not forming a defined receptacle, and united to a various extent above, internally smooth and glabrous, externally sulcate and increased in size by adnate scales. The laciniæ are connate in twos, threes, or fours, but likely before the opening of the flowers the whole are connate, with their united ends twisted together, but during fruiting they are ascending-patent. Flowers rather numerous but scattered within the involucre, in which they occupy the lower part of the laciniæ above the bottom. Sepals ⅔ line long, persistent, externally puberulent, rather imbricate in æstivation. Petals or inner whorl of sepals rather less than the outer, more glabrous and acutely narrowed at the base.

Disk glabrous. Style about ⅔ line long. Stigma rather broader than the style. Ovary compressed, with a silky beard at the apex. Fruit sessile, light-brown, 1½ line high and 3 to 4 lines broad, with a very slight groove at the apex, having a glabrous surface, but ciliated marginally by long, fulvous, silky hairs; testa glabrous, smooth, light-brown; embryo starchy, almost 3 lines broad. Radicle terete, ascending, appressed, scarcely 1 line long.—*F. v. M.*, l.c.

Hab.: Rocky Mountains about the Endeavour River and Coen River; Herberton district, *J. F. Bailey*.

Wood of a light-red colour, close-grained, soft, and easy to work.—*Bailey's Cat. Ql. Woods No.* 106a.

15. DODONÆA, Linn.
(After R. Dodoens.)
(Empleurosma, *Bartl.*)

Flowers polygamous or unisexual, often dioecious. Sepals 5 or sometimes fewer, valvate in the bud. Petals none. Disk small or inconspicuous. Stamens usually 8, sometimes fewer, rarely 10; filaments very short, anthers ovoid or linear-oblong. Ovary 3 or 4, rarely 5 or 6-celled, with 2 ovules in each cell; style short, or in some flowers very long, shortly lobed at the end. Capsule membranous or coriaceous, opening septicidally in as many valves as cells, each valve with a dorsal angle often produced into a vertical wing, and in falling off leaving the dissepiment attached to the persistent axis, or rarely the dissepiment splitting and remaining attached to the valves, thus closing the carpels and leaving only the central filiform axis persistent. Seeds 1 or 2, nearly globular or more frequently compressed, with a thickened funicle, but not arillate; testa crustaceous; embryo spirally curled.—Shrubs, often tall, but scarcely truly arborescent; the young shoots usually viscid, and often the whole plant. Leaves simple or pinnate, with small leaflets, with or without a terminal odd one. Flowers terminal or axillary by the abortion of the flowering branches, solitary, clustered, or in short racemes or panicles.

With the exception of *D. viscosa*, which is widely dispersed over almost all hot countries, and possibly one distinct Sandwich Island species, one from S. Africa, and one or two from Mexico, the *Dodonæas* are all endemic in Australia, and very difficult to distinguish by positive characters. The form of the wings of the capsule, which has been much relied on, is as variable as that of the leaves, and the species, which at first sight appear the most distinct, often pass one into the other by the most insensible gradations. Even the exceptional dehiscence of the capsule, in those species where the dissepiments are carried off with the valves, appears sometimes to be not quite constant, and is at most a purely artificial character separating species in all other respects very closely allied. Several species have in some, occasionally in nearly all the female flowers, a remarkably long style, sometimes ¾ to 1in., whilst other female flowers on the same specimens, or on other specimens of the same species, have no style at all, the stigma or stigmatic surface sessile on the ovary.—*Benth.*

SERIES I. **Cyclopteræ.**—*Leaves entire, toothed, or rarely lobed. Wings of the capsule extending from the base to the style or nearly so, each carpel, including its wing, nearly orbicular or longer than broad.*

Leaves flat, elliptical, oblong-lanceolate or spathulate or, if linear, not filiform, entire or obscurely sinuate, usually above 2in. long, rarely between 1 and 2in.
Young branches very angular. Seed smooth and shining. Leaf-veins indistinct. . . . . 1. *D. triquetra.*
Sepals minute. Anthers linear . . . . 2. *D. lanceolata.*
Sepals 1 to 1½ line long, from half as long to as long as the anthers . 3. *D. petiolaris.*
Young branches terete or slightly angular. Seeds opaque.
Leaves oval-oblong, on a rather long petiole, rounded at the base . .
Leaves narrowed into the petiole, the lateral veins more or less conspicuous.
Leaves elliptical-oblong, lanceolate or spathulate, rarely almost linear-cuneate . . . . 4. *D. viscosa.*
Leaves narrow, linear-cuneate or long and linear . . . . 5. *D. attenuata.*

Leaves flat, more or less cuneate, entire or toothed at the end, rarely
  exceeding 1½in., and usually under 1in.
Much-branched, erect or divaricate shrubs. Terminal flowers clustered
  or shortly racemose.
  Leaves broad-cuneate, rounded or truncate at the end . . . . . . 6. *D. cuneata.*
  Leaves narrow-cuneate, rather acute, acuminate or 3-toothed at the end 7. *D. peduncularis.*
Leaves linear-filiform, heath-like or pine-like, 1 to 3in. long, not crowded 8. *D. filifolia.*
Branches terete or nearly so. Leaves linear or linear-cuneate, obtuse,
  mostly under 1½in. long . . . . 9. *D. lobulata.*

SERIES II. **Platypteræ.**—*Leaves quite entire, flat. Wings of the capsule very divergent or divaricate, not reaching to the style nor to the base, each carpel, including its wing, broader than long, transversely ovate or oblong.*

Leaves linear or lanceolate. Branches very angular. Dissepiments
  persisting on the axis . . . . 10. *D. truncatiales.*
Dissepiments splitting and coming off with the valves.
  Leaves broadly, somewhat bluntly, lanceolate . . . . 11. *D. Hansenii.*
  Leaves narrow-linear . . . . 12. *D. stenophylla.*

SERIES III. **Apteræ.**—*Leaves entire or toothed. Capsule without wings, or the angles slightly and irregularly dilated into very narrow wings.*

Sepals lanceolate. Buds ovoid or globular. Dissepiments persisting on
  the axis of the fruit. Branches scarcely angled. Leaves obovate,
  cuneate, or triangular, glabrous or pubescent. Flowers mostly axillary.
  Sepals narrow, short . . . . 13. *D. triangularis.*

SERIES IV. **Pinnatæ.**—*Leaves all pinnate or very rarely a few simple ones at the base of the branches. Capsule of the Cyclopteræ, except in D. oxyptera and D. inæquifolia, where it approaches that of the Platypteræ, and in D. humilis, where it is apterous.*

Tall shrubs or small trees. Leaflets flat, oblong, lanceolate or obovate, not
  coriaceous. Racemes or panicles terminal, loose.
  Leaflets usually numerous, lanceolate or oblong. Capsule not inflated,
    the wings broad.
    Leaflets ½ to 1in.; rhachis broadly winged. Sepals 1 to 1¼ lines . . 14. *D. megazyga.*
    Leaflets 5 to 27, flat, broad-linear, ¼ to 1in. long; rhachis winged,
      almost ¾ line long . . . . 15. *D. macrozyga.*
  Leaflets few, obovate or oblong. Capsule large and inflated . . . . . 16. *D. physocarpa.*
Much-branched, leafy shrubs. Pedicels solitary or clustered (racemose in
  D. multijuga and *D. pinnata*).
  Leaflets obovate, cuneate or oblong, often toothed, the margins usually
    recurved or revolute. Plant usually pubescent or villous (except *D.
    humilis*).
    Capsule winged, hirsute at least when young.
      Villous. Leaflets 7 to 20 or more; rhachis winged. Sepals
        acuminate. Capsule-wings rounded.
        Pedicels long, clustered . . . . 17. *D. vestita.*
      Pubescent. Leaflets 3 to 7; rhachis angular. Pedicels short.
        Sepals obtuse. Capsule wings acutangular . . . . 18. *D. oxyptera.*
      Plant shortly hairy. Leaves very small, terminal leaflet undivided
        or bifid . . . . 19. *D. Macrossani.*
    Capsule winged, glabrous or very sparingly pubescent. Plant
      pubescent or rarely glabrous. Leaflets usually under 11. Pedicels
      short, clustered . . . . 20. *D. boroniæfolia.*
  Leaflets narrow-linear, convex underneath.
    Capsule-wings rounded; dissepiments splitting and coming off with
      the valves. Leaflets under 15 . . . . 21. *D. adenophora.*
    Differing from the last in elongated pedicels, dissepiments never
      separating from their axes, cells longer and the wing narrower . . 22. *D. tenuifolia.*

In the following 9 species, great as is the diversity in the size of the capsule and the precise shape of the wings, these differences afford no specific characters, and are often very difficult to class as varieties, even when perfectly ripe and well-formed capsules are obtained; and the shape of the wing often alters much during growth, or is apparently affected by the manner in which the capsule has ripened. The very shining seeds distinguish two species, but where they are usually opaque they sometimes are somewhat shining. There remains little but the very uncertain characters derived from foliage to separate all these species, which are yet much too constantly dissimilar to be united into one.—*Benth.*

1. **D. triquetra** (referring to angles of branches), *Andr. Hot. Rep. t.* 280. "Kinginga Kilamul," Moreton Bay, *Watkins.* Erect, usually tall, glabrous, not very viscid, the young branches flattened or angular. Leaves from oval-elliptical to oblong-lanceolate, acuminate, 2 to 3 or rarely 4in. long, the pinnate and reticulate veinlets few and fine, usually scarcely conspicuous. Pedicels slender, in short, oblong, compact panicles or racemes. Sepals minute, rarely ½ line long. Anthers linear, often 1½ line long. Styles, when long, attaining ½in. Capsule of *D. viscosa*, usually middle-sized. Seeds brown, very smooth and shining.—DC. Prod. i. 617; F. v. M. Fragm. i. 75, and Pl. Vict. i. 226.—*D. laurina*, Sieb. in Spreng. Syst. Cur. Post. 152.—*D. longipes*, G. Don, Gen. Syst. i. 674 (from the character given).

Hab.: Brisbane River, Moreton Bay to Rockingham Bay.
The Fiji Island plant referred by A. Gray and Seemann to *D. triquetra* appears to me to be one of the common forms of *D. viscosa.*—*Benth.*
Wood of a light colour, except near the centre, which is dark, close-grained, tough, and nicely marked.—*Bailey's Cat. Ql. Woods* No. 107.

2. **D. lanceolata** (lanceolate), *F. v. M. Fragm.* i. 78; *Benth. Fl. Austr.* i. 475. Very closely allied to *D. triquetra*, with the same angular branches, smooth, almost veinless leaves, slender pedicels, and very shining seeds, and scarcely distinguishable except by the sepals, which are from 1 to 1½ line long. The leaves are perhaps generally rather narrower, and the capsule-wings broader, but neither of these characters can be relied upon.

Hab.: Gladstone, *C. Hedley*; islands of the Gulf of Carpentaria, Northumberland Islands, *R. Brown*; Cape Cleveland, *A. Cunningham*; Sunday Island, *M'Gillivray*; Palm Island, *Henne*; Port Denison, *Fitzalan*.

3. **D. petiolaris** (petiolate), *F. v. M. Fragm.* iii. 13; *Benth. Fl. Austr.* i. 475; *Fragm.* ix. 89. Leaves rounded at the base, here and there truncate, veined as in *D. viscosa*, but sometimes narrowed at the base, on petioles of 2 or 3 lines. Sepals 5, narrow. Stamens 10, 3 times as long as the calyx. Anthers apiculate. The single capsule seen not yet full-grown, but, in that state, does not appear at all different from the larger varieties of *D. viscosa*, of which this plant may probably prove to be a variety.—*Benth.* (in part).

Hab.: Bulloo, *Mrs. Spencer* (F. v. M.)

4. **D. viscosa** (viscid), *Linn.; DC. Prod.* i. 616; *Benth. Fl. Austr.* i. 475. Hop Bush. "Tecan," St. George, *Wedd.* A shrub, sometimes low and stunted, more frequently tall, glabrous, and usually more or less viscid, the young branches frequently compressed or somewhat triangular, but much less so than in *D. triquetra.* Leaves simple, varying from broadly oblong-lanceolate, acute or acuminate, and 3 or 4in. long, to narrow-lanceolate, or oblong-cuneate and very obtuse or almost linear-cuneate, always narrowed into a more or less distinct petiole, entire or obscurely sinuate, or rarely almost 3-toothed at the end, the pinnate veins usually rather numerous and very divergent, sometimes scarcely conspicuous. Panicles or racemes usually short and terminal, or reduced to axillary clusters. Sepals ovate, usually as long as or rather longer than the oblong obtuse anthers. Style rarely lengthened out. Capsule very variable in size, the wings continued from the base to the style, or nearly so, either equally rounded at the top and at the base or more contracted at the base. Seeds rather large, dark-coloured or black, opaque or scarcely shining.—Hook. f. Fl. Tasm. i. 55; F. v. M. Pl. Vict. i. 85.

Hab.: Cumberland Islands, *R. Brown*; Endeavour River, *Banks*; Rodd's Bay and Rockingham Bay, *A. Cunningham*; Cape Upstart and Port Curtis, *M'Gillivray*; Rockhampton, *Thozet*; Moreton Bay, *Fraser, A. Cunningham*, and others; Condamine River, and many other localities.
The form growing in India is said to be used for engraving, turning, tool-handles, and walking-sticks. Has a white sapwood and a hard dark-brown heartwood, which is close-grained. The leaves of a form of this species are said to be used in some parts of India as a febrifuge.
Wood of a brown colour, close-grained, and hard.—*Bailey's Cat. Ql. Woods No.* 108.

The species is abundantly distributed over tropical America, Africa, and Asia, extending to the Pacific Islands, and southward, beyond the tropics, to S. Africa and New Zealand. It includes probably the whole of the extra-Australian described *Dodonæas*, except, perhaps, the *D. eriocarpa* from the Sandwich Islands, *D. Thunbergiana*, Eckl. and Zeyh., from S. Africa, and one or two Mexican ones, which, whether varieties or species, do not occur in Australia. The almost protean forms the species assumes in Australia, even after deducting *D. attenuata*, *D. cuneata*, and *D. megazyga*, which F. v. Mueller unites with it, are very difficult to distribute into definite varieties, although at least the three following are usually considered as species.—*Benth.*

a. vulgaris. Usually tall. Leaves large, obovate-oblong, broadly lanceolate or lanceolate, acuminate or rarely obtuse, the pinnate veins usually numerous and prominent. Capsules large, with rather broad wings, much rounded above and at the base, the terminal sinus (between 2 opposite wings) narrow, each carpel, including its wing, longer than broad.—*D. viscosa*, Linn., and *D. Burmanniana*, D.C.; Griseb. Fl. Brit. W. Ind. 127, with the synonyms adduced; A. Gray, Gen. Ill. t. 182; Wight, Illustr. t. 52.—The most common form in America and tropical Africa, extending in Asia as far north as Scinde and Afghanistan, also in the Pacific Islands; and to this form belong most of the tropical Australian species as well as some from Hastings River, Beckler. Some specimens from Endeavour River, both in the Banksian and in Cunningham's collections, are remarkable for their thick, obscurely veined leaves.—*Benth.*

b. angustifolia. Leaves narrow-lanceolate, mostly long and acutely acuminate, much narrowed at the base, the veins usually conspicuous. Capsules small, with very broad wings, leaving the terminal sinus very open and sometimes narrowed at the base, each carpel, including its wing, orbicular or rather broader than long, although much less so than in the *Platyptera*. - *D. angustifolia*, Swartz; Griseb. Fl. Brit. W. Ind. 128, with the synonyms adduced; Lam. Ill. t. 304, n. 2, and consequently *D. salicifolia*, DC. Prod. i. 617, supposed to be from New Holland; *D. neriifolia*, A. Cunn. in A. Gray, Bot. Am. Expl. Exped. i. 262.—This variety has nearly the same range within the tropics as the large-fruited one, and occasionally is found to pass into it. In Australia it includes many Queensland specimens, and is the common form in N. S. Wales collections. It occurs also in W. Australia, but in Victoria, S. Australia, and Tasmania, as in N. Zealand, it tends rather to pass into the spathulate-leaved form. *D. umbellata* and *D. Kingii*, G. Don, Gen. Syst. i. 674, from the characters given, belong probably to this variety.—*Benth.*

c. spathulata. Usually a more bushy and not so tall a shrub as the preceding varieties, often very viscid. Leaves shorter (although much longer than in *D. cuneata*), obovate-oblong, oblong-cuneate, spathulate, oblanceolate or broadly linear-cuneate, usually obtuse or sometimes truncate, the lateral veins usually conspicuous, but in some thick-leaved specimens scarcely more so than in *D. cuneata*. Capsules very variable, but generally intermediate between those of the var. *vulgaris* and *angustifolia*, but nearer to the former.—*D. spathulata*, Sm. in Rees, Cycl. xii.; DC. Prod. i. 616; *D. conferta*, G. Don, Gen. Syst. i. 674; *viscosa*, var. *asplenifolia*, Hook. f. Fl. Tasm. i. 55.—This is the commonest, perhaps the only form, in Victoria, Tasmania, and S. Australia, and I have seen N. S. Wales specimens from Port Jackson, and northward to New England, Mount Mitchell, and Mount Aiton. It is the prevalent form in New Zealand, and some of the Sandwich Island specimens can be precisely matched with Australian. *D. oblongifolia*, Link, as figured in Bot. Reg. t. 1051, appears to represent rather a short-leaved form of this variety than a long-leaved *D. cuneata*. *D. asplenifolia*, Rudge, in Trans. Linn. Soc. xi. 297, t. 20, DC. Prod. i. 617, judging from N. S. Wales specimens agreeing with the figure, although not authentically named, is an apparently rare form with linear-cuneate, 3-toothed leaves, resembling those of luxuriant drawn-up shoots of *D. cuneata*, but longer.—*Benth.*

5. **D. attenuata** (attenuated), *A. Cunn. in Field, N. S. Wales*, 353; *Benth. Fl. Austr.* i. 477. A viscid shrub, closely resembling the narrowest-leaved forms of *D. viscosa* on the one hand and almost passing into *D. lobulata* on the other. Leaves linear or narrowly linear-cuneate, obtuse, often slightly sinuate-toothed, rather thick and rigid, 1-nerved, the lateral veins inconspicuous, 1½ to 2½in. long in the original form but sometimes longer. Flowers and ovate sepals of *D. viscosa*, in short usually simple racemes. Capsule of *D. viscosa*, usually intermediate between the extremes of the varieties *a* and *b* of that species. Seeds opaque.—Bot. Mag. t. 2860; *D. Preissiana*, Miq. in Pl. Preiss. i. 226; F. v. M. Fragm. i. 72.

Hab.: Brookfield, Stanthorpe, Mitchell's Pinch, Leichhardt district.
Var. *linearis*. Leaves long, narrow-linear, mostly acute, rigid, the margins often recurved. Capsule (only seen in few specimens) rather small, but with the terminal sinus between the wings narrow. Hab.: N. S. Wales border, *J. Ivory; Wilson's River, J. C. Weald*.

6. **D. cuneata** (cuneate), *Rudge, in Trans. Linn. Soc.* xi. 296, t. 19; *Benth. Fl. Austr.* i. 477. A much-branched bushy shrub, glabrous, and usually viscid. Leaves obovate or cuneate, usually ½ to 1in. long and rather broad, rarely narrow-cuneate, attaining 1½in., rounded, truncate, emarginate or 3-toothed at the end.

otherwise entire or rarely obscurely toothed, gradually narrowed into a very short petiole, thin or coriaceous; the lateral veins rarely conspicuous. Racemes short, terminal, scarcely branched, with slender pedicels, or the flowers few in axillary clusters. Sepals ovate-oblong, and capsules of *D. viscosa*, the wings usually not very broad and rather rigid, with the terminal sinus open.—DC. Prod. i. 617.

Hab.: Brisbane and Burnett Rivers, Moreton Bay, and Main Range.

7. **D. peduncularis** (flowers stalked), *Lindl. in Mitch. Trop. Austr.* 361; *Benth. Fl. Austr.* i. 478. A very much-branched glabrous and viscid shrub, closely allied to *D. cuneata*, the smaller branches terete, slender but rigid. Leaves from linear-cuneate to broadly spathulate, either acute or very shortly acuminate or rounded or truncate at the end, and often 3-toothed, ¼ to ½in., or very rarely (when narrow) 1in. long, coriaceous and rigid, 1-nerved, the margins often thickened, the lateral veins inconspicuous. Peduncles sometimes very short. Pedicels rather slender, mostly axillary, solitary or clustered, or in short terminal racemes. Sepals ovate, thicker than in *D. cuneata*. Style 1 to 2 lines long, deeply free. Capsule of *D. viscosa*.—*D. pubescens*, Lindl. in Mitch. Trop. Austr. 342 (the supposed pubescence apparently a mistake).—Benth., but this form is referred to by Baron Mueller.

Hab.: Near Lindley's Range and on the Maranoa, *Mitchell*; Moreton Bay, *S. H. Eaves*; Darling Downs, *Dr. G. Bennett*; Springsure, *Clevelt*; St. George, *J. Wedd*.

8. **D. filifolia** (leaves thread-like), *Hook. in Mitch. Trop. Austr.* 241; *Benth. Fl. Austr.* i. 478. Erect, glabrous, and slightly viscid; branches slender, terete or scarcely angular. Leaves narrow-linear, almost filiform, terete or slightly flattened, often incurved, obtuse or scarcely mucronate, 1 to 3in. long, quite entire. Racemes very few-flowered, the pedicels rather long. Sepals lanceolate, about as long as the anthers. Capsule of *D. viscosa*.—*D. acerosa*, Lindl. in Mitch. Trop. Austr. 273; F. v. M. Fragm. i. 71.

Hab.: Newcastle Ranges, between the Suttor and Burdekin Rivers, *F. v. Mueller*; stony gullies near Mount Mudge, *Mitchell*.

9. **D. lobulata** (leaves shortly lobed), *F. v. M. in Linnæa*, xxv. 372; *Benth. Fl. Austr.* i. 479. Closely allied on the one hand to *D. attenuata* and on the other to *D. ptarmicifolia*, glabrous and viscid, the branchlets scarcely angular. Leaves linear or linear-cuneate, obtuse, mostly 1 to 2in. long, obtusely serrate or pinnatifid with short obtuse callous lobes, coriaceous and rigid, the midrib scarcely conspicuous. Flowers few, in short racemes, the pedicels rather slender. Sepals thin, broadly ovate. Capsule of the smaller forms of *D. viscosa*, the wings not very broad. Seeds smooth and shining.

Hab.: Has been obtained at a few inland western localities.

10. **D. truncatiales** (somewhat truncate), *F. v. M. Fragm.* ii. 148, and *Pl. Vict.* i. 226; *Benth. Fl. Austr.* i. 479. A tall glabrous shrub, scarcely viscid, the younger branches acutely angular. Leaves narrow-lanceolate or linear, rather acute, 2 to 4 or even 5in. long, narrowed into a short petiole, entire or obscurely sinuate-toothed, the lateral veins little conspicuous. Racemes and flowers of *D. viscosa*. Sepals ovate, usually broad and nearly as long as the anthers. Anthers obtuse or very minutely apiculate. Capsule 4 or rarely 3-lobed, flat at the top, the wings oblong, very diverging, not extending to the base of the carpels. Dissepiments remaining attached to the axis as in all the preceding species, or occasionally deciduous, but not splitting as in the two following species. —*D. calycina*, A. Cunn. Herb.; A. Gray, Bot. Amer. Expl. Exped. i. 262.

Hab.: Brisbane River; Condamine, *Rev. B. Scortechini*; Mitchell's Pinches, in the Leichhardt.

11. **D. Hansenii** (after Lars Hansen), *F. v. M. Vict. Nat.*, 1891. A shrub of about 12ft. high, glabrous, hardly viscid. Leaves attaining the length of 2in., slightly shining, on rather conspicuous petioles, chartaceous, broadly and somewhat bluntly lanceolate, but more gradually narrowed into the base than into the apex, subtle-venulated. Flowers unknown. Racemes when fruiting below the leaves corymbose, few-flowered or reduced to 3 or two flowers; pedicels rather long. Sepals early deciduous. Capsule usually 4-celled, its capsular portion hardly as long as broad, its wing-appendages ascendingly divergent, venulous, and from 4 to nearly 6 lines broad, considerably broader than high, rounded-blunt at the upper end, ceasing before the base and before the middle summit of the valves; dissepiments seceding from the axis, closing permanently the carpels. Young seeds longer than broad, almost truncate and also turgescent around the hilum. Ripe seed not obtained.—F. v. M. l.c.

Hab.: Stuart's River, *Stephen Johnson*.

Among the few species with fruit dissepiments seceding from the axis this comes nearest to *D. platyptera*, but the leaves are of larger size, of darker green, of thinner texture, and not of conspicuously glandular punctuation; further, the appendages of the fruit are perceptibly larger, and turn almost diagonally upwards, while those of *D. platyptera* remain nearly at a level with the vertex of the cells. The flowers and mature seeds may also yet show specific differences. From *D. pachyneura* our new plant is also distinguished by much larger leaves, with fainter and more divergent venulation, also by the greater extension of the fruit appendages. The shape of the fruit is much like that of *D. macrozyga* and *D. megazyga*, but its dehiscence, as well as the foliage, are very different. The leaves resemble those of *D. lanceolata* and *D. triquetra.— F. v. M., l.c.*

12. **D. stenophylla** (slender leaves), *F. v. M. Fragm.* i. 72; *Benth. Fl. Austr.* i. 480. Glabrous and viscid. Leaves narrow-linear, rigid, 2 to 6in. long and almost 2 lines broad, the margins usually thickened and entire. Flowers of *D. viscosa*, in short loose racemes or almost cymose panicles. Sepals ovate. Capsule small, the wings broadly oblong or obovate, diverging, not reaching to the style nor to the base of the carpels; dissepiments splitting and falling off with the valves, leaving only the filiform axis persistent. Seeds shining.

Hab.: Broadsound, *R. Brown*; Burdekin River, *F. v. Mueller*; Comet River, *Leichhardt*; and near Moreton Bay.

In flower, this species is scarcely to be distinguished from *D. attenuata*, var. *linearis*; but the fruit is very different.—*Benth*.

13. **D. triangularis** (triangular), *Lindl. in Mitch. Trop. Austr.* 219 (male plant); *Benth. Fl. Austr.* i. 480. An erect shrub of 3 to 4ft., glabrous, pubescent or softly villous. Leaves obovate-cuneate or almost triangular, rounded-truncate or 3-toothed at the end, or very rarely elliptical-oblong, $\frac{1}{2}$ to 1in. or rarely 1$\frac{1}{2}$in. long, coriaceous, 1-nerved, the lateral veins quite inconspicuous. Flowers axillary, solitary or clustered, on short pedicels. Sepals narrow-lanceolate, rather thick. Anthers as in *D. triquetra*, narrow, acuminate, exceeding the calyx. Capsule glabrous or pubescent, 3 or 4-angled, the angles rarely dilated towards the top into very narrow wings; dissepiments remaining attached to the axis, or very rarely deciduous but not splitting.—*D. mollis*, Lindl. in Mitch. Trop. Austr. 212 (with pubescent capsules); *D. trigona*, Lindl. l.c. 286 (with glabrous capsules); *D. Lindleyana*, F. v. M. Pl. Vict. i. 88.

Hab.: Suttor River, *F. v. Mueller*; near Mount Owen, Mount Faraday, and Mantuan Downs, *Mitchell*; near Brisbane and Ironbank forest, *Leichhardt*; and Stanthorpe, *Alex. McPherson*.

14. **D. megazyga** (pairs of leaflets large), *F. v. M. Benth. Fl. Austr.* i. 483; *Fragm.* ix. 86. A tall shrub or tree of 30ft., glabrous and slightly viscid, the young branches acutely angled. Leaves mostly pinnate, the rhachis conspicuously winged; leaflets usually numerous, sometimes above 30, lanceolate, acute, $\frac{1}{2}$ to 1in. long; in some specimens the lower leaves of the branches reduced to very few

leaflets or to a simple linear-lanceolate leaf. Flowers rather large, in short axillary racemes or terminal panicles, the pedicels slender. Sepals ovate. Anthers somewhat broad, the connectivum barbellate, apiculate. Capsules small, with broad obovate or orbicular diverging wings of 3 or 4 lines.

Hab.: Nerang Creek (a tree 20ft. high), *H. Schneider;* near the southern border, on the Condamine River.

15. **D. macrozyga** (pairs of leaflets long), *F. v. Fragm.* iv. 135, ix. 86. A viscid erect shrub, subglabrous, the branchlets compressed-triquetrous. Leaves pinnate with rarely some simple ones; rhachis dilated, with petiole 2 to 3in. long. Leaflets 5 to 27, broad-linear, flat, entire, obtuse, chartaceous, 6 to 12 lines long, ⅔ to 1 line broad, narrowed to both apex and base. Male flowers unknown. Female flowers few, in corymbose racemes, axillary. Pedicels under 1 line long. Bracteoles short. Sepals 4, lanceolate-ovate, almost ¾ line long, deciduous. Capsule acute and very broad, tetrapterous, valves about 2¼ lines high, and the wings almost 4 lines broad.

Hab.: Source of the Cape River, *E. Bowman* (F. v. M., l.c.).

16. **D. physocarpa** (Bladdery appearance of capsule), *F. v. M. Fragm.* i. 74, ix. 86; *Benth. F. Austr.* i. 484. A tall shrub, the flowering-branches short, nearly terete, and as well as the leaves slightly pubescent as in *D. polyzyga,* but much less viscid. Leaves pinnate, the rhachis angular but scarcely dilated; leaflets rarely more than 10 and often only 4 to 6, obovate or oblong, obtuse or mucrohate, mostly 3 to 4 lines long, entire or rarely obscurely 2 or 3-toothed, flat, 1-nerved, sometimes rather thick, but not coriaceous. Racemes terminal, short, loosely few-flowered. Sepals lanceolate, obtuse, nearly 2 lines long. Anthers short, obtuse. Style often elongated. Capsule large, somewhat inflated, often 5 or 6-celled or at times 4, the axis above ½in. long; wings not very broad, rounded above and below, but much injured in our specimens. Seeds shining.

Hab.: Norman River, *T. Gulliver* (F. v. M., l.c.); Gilbert River, *Daintree* (F. v. M., l.c.); and Walsh Range, *R. C. Burton.*

17. **D. vestita** (referring to the clothing of hairs), *Hook. in Mitch. Trop. Austr.* 265; *Benth. Fl. Austr.* i. 484. A much-branched shrub, densely villous, hirsute or pubescent, the hairs sometimes long and almost golden. Leaves pinnate, the rhachis winged; leaflets varying from few broadly obovate-cuneate and 2 or 3 lines long, to above 20, narrow-oblong, and 4 or 5 lines long, entire or rarely 2 or 3-toothed, the margins always much recurved. Pedicels usually in clusters of 3 or 4, about ½in. long. Sepals lanceolate, acute, attaining 3 lines. Anthers 8 to 10, linear, hirsute, spirally twisted as they fade. Capsule when young hirsute with long hairs, the wings broadly orbicular, when far advanced the hairs mostly disappear and the wings are much narrower in proportion to the carpels.—*D. paulliniæfolia,* A. Cunn. Herb.; Steud. Nom. Bot. ed. 2.

Hab.: Belyando River, *Mitchell* (very hirsute specimens, with few, small, broad leaflets, and broadly winged, very hirsute, young fruits); Endeavour River, *Banks, A. Cunningham* (scarcely more than pubescent, with numerous narrow leaflets and narrow-winged, scarcely hirsute, old fruits); Castle Creek and head of Boyd River, *Leichhardt* (leaves and indumentum intermediate, and on one specimen the young fruit, like Mitchell's, on one branch, and an old capsule, like Cunningham's, on another branch).—*Benth.*

18. **D. oxyptera** (wings acute), *F. v. M. Fragm.* i. 74; *Benth. Fl. Austr.* i. 484. A shrub of several feet, the branches virgate, terete, pubescent as well as the leaves and more or less viscid. Leaves pinnate, the rhachis angular but scarcely dilated; leaflets usually 5 to 11, narrow-oblong or oblong-cuneate, obtuse, 2 to 4 lines or rarely ½in. long, the margins recurved. Flowers small,

sessile, or very shortly pedicellate. Sepals broad, acute, about 1 line long. Anthers obtuse, not exceeding the calyx, often hirsute. Capsule small, slightly hairy, the axis 2 to 3 lines long, the wings rigid, divergent, almost triangular and acute.

Hab.: Islands of the Gulf of Carpentaria, *R. Brown.*

Several of R. Brown's specimens have numerous male flowers and fruits on the same individual.—*Benth.*

19. **D. Macrossanii** (after the Hon. J. M. Macrossan), *F. v. M. and Rev. B. Scortechini, Chem. and Drug., Jan.* 7, 1882. Plant clothed with short hairs. Leaves very small, pinnate; leaflets linear or oval-lanceolate, usually 2 pairs, undivided, or some bifid, margins flat; rhachis somewhat dilated. Pedicels of the female flowers solitary, very short. Sepals 3 or 4, ovate-lanceolate, nearly half as long as the very small roundish-triangular or quadrangular fruit. Valves with neither horn-like nor wing-like appendages, seceding from the persistent dissepiments.

Hab.: Near Miles, *Rev. B. Scortechini* (staminate flowers and ripe fruit not obtained).

The only other species which combines pinnate leaves with inappendiculate fruits is *D. humilis,* from which this species differs in its much more conspicuous general hairiness, in the very small not toothed leaflets of only 2 or 1 pairs, in the smaller fruits, the hairs of which are not gland-tipped.—*F. v. M., l.c.*

20. **D. boroniæfolia** (Boronia-leaved), *G. Don, Gen. Syst.* i. 674; *Benth. Fl. Austr.* i. 485. A much-branched shrub, usually pubescent or shortly hirsute, rarely glabrous, often viscid. Leaves pinnate, the rhachis more or less dilated; leaflets 5 to 9 or rarely more, obovate or cuneate-oblong, obtuse or truncate, and usually toothed at the end, 2 to 3 lines long or rarely more, coriaceous, with recurved margins. Pedicels clustered on very short lateral branches, those of the males very short, of the females often 3 to 4 lines long. Sepals ovate-lanceolate, about 1 line long. Anthers almost sessile, short, obtuse, not apiculate. Capsule of *D. viscosa,* glabrous, usually rather small, the wings not very broad, rounded at the top and at the base.—*D. Caleyana,* G. Don, Gen. Syst. i. 674 (from the character given); *D. hirtella,* Miq. in Linnæa, xviii. 94; F. v. M. Pl. Vict. i. 89.

Hab.: On the Maranoa, and Mt. Maria, Warrego.

21. **D. adenophora** (gland-bearing), *Miq. in Linnæa,* xviii. 95; *Benth. Fl. Austr.* i. 486. A rigid shrub, glabrous and usually very viscid, the young branches angular. Leaves pinnate, the rhachis scarcely dilated; leaflets 3 to 9 or rarely 11, linear or slightly cuneate, obtuse and often callous at the tips, 2 to 4 lines long, very rarely slightly toothed at the end, convex or keeled underneath, flat above, rather thick and rigid. Pedicels slender, clustered. Sepals ovate, acute, or very shortly racemose, rather more than 1 line long. Anthers short, very obtuse. Capsule small, the wings rather broad, rounded at the top and at the base; dissepiments splitting and coming off with the valves, leaving only the filiform axis persistent as in *D. platyptera, D. stenophylla,* and *D. bursarifolia.*—*Thouinia (?) adenophora,* Miq. in Pl. Preiss. i. 224; *D. tenuifolia,* Lindl. in Mitch. Trop. Austr. 248 (the Queensland and N. S. Wales specimens).

Hab.: Condamine River, Rosewood, and Killarney.

22. **D. tenuifolia** (slender-leaved), *Lindl. in Mitch. Trop. Austr.* 248; *F. v. M. Fragm.* ix. 85. Plant glabrous, the branchlets a little angular. Leaves paucijugate; leaflets short; broadish-linear, rhachis narrowly winged. Capsule on elongated pedicels, with cells two or three times longer than broad; dissepiments never separating from their axis.—*F. v. M., l.c.*

Hab.: Belyando, Condamine River, Darling Downs.

XL. SAPINDACEÆ.

16. DISTICHOSTEMON, F. v. M.
(Referring to two series of stamens.)

Characters of *Dodonæa* except that the sepals vary from 5 to 8, and the stamens are indefinite, usually above 20, closely packed in 2 or more series.— Pubescent shrub. Leaves simple. Inflorescence more nearly an interrupted spike than in any *Dodonæa*.

The genus is limited to a single species, endemic in Australia, scarcely sufficiently distinct from *Dodonæas*.

1. **D. phyllopterus** (referring to the erect wings of the capsule), *F. v. M. in Hook. Kew Journ.* ix. 306; *Benth. Fl. Austr.* i. 487, *and Fragm.* ix. 89. A tall shrub, softly tomentose-pubescent or villous in all its parts. Leaves very shortly petiolate, oblong or rarely obovate, very obtuse, 1 to 3in. long, soft and velvety on both sides (or the leaves sometimes crenate-dentate and nearly smooth, *F. v. M. Fragm.* ix. 89), the veins prominent underneath. Flowers nearly sessile, in terminal leafless interrupted spikes or racemes of 1 to 3in., rarely branching into oblong panicles. Sepals most frequently 6, but in some specimens almost all 5. Stamens although usually above 20, yet occasionally only 12 to 15, and often above 30; anthers oblong-linear, crowded, with very short filaments as in *Dodonæa*. Styles occasionally elongated as in some *Dodonæas*. Capsule more or less tomentose, obovoid-triquetrous, the angles more or less produced into herbaceous erect wings, usually ovate, very obtuse, and only on the upper outer half of the carpels, but occasionally, especially in the Banksian specimens, not so broad, and continued almost to the base. Seeds very shining, usually 2 in each cell.—*Dodonæa hispidula*, Endl. Atakt. t. 80.

Hab.: Cape River, *Bowman*; Rockingham Bay, *J. Dallachy*; Somerset (where on the sandy wet land there is a short dense growth of it down to the water's edge); Islands of the Gulf of Carpentaria, *R. Brown, Henne*.

Order XLI. ANACARDIACEÆ.

Flowers unisexual, polygamous or hermaphrodite, usually regular. Calyx of 3 to 5 lobes or distinct sepals. Petals 3 to 7, rarely none. Disk usually annular or broad. Stamens of the same number or twice as many as petals, very rarely indefinite, inserted round the disk or rarely upon it; filaments free; anthers versatile. Ovary superior, usually 1-celled, with 1 to 3 styles, or in the *Spondieæ* 2-celled, or very rarely of 2 or more distinct carpels, or in male flowers reduced to 4 or 5 rudimentary style-like carpels. Ovules solitary in the ovary or in each of its cells, pendulous or broadly adnate to the side of the cavity, or suspended from a free funicle erect from the base of the cavity, with a dorsal raphe and inferior micropyle; very rarely in genera not Australian erect, with a ventral raphe and inferior micropyle. Fruit superior or rarely half inferior, free or adnate at the base to the enlarged calyx-tube or disk, 1-celled or (in *Spondieæ*) several-celled, usually drupaceous and indehiscent. Seed erect, horizontal or pendulous; albumen none or very thin. Embryo straight or incurved, cotyledons usually fleshy; radicle short, inferior or more frequently turned upwards or superior.—Trees or shrubs, the bark often exuding a caustic, balsamic or gummy juice. Leaves alternate or very rarely opposite, without real stipules, simple or ternately or pinnately compound, usually without glandular dots. Inflorescence various, usually paniculate, with small flowers. Flesh of the drupes usually oily or full of caustic juice.

The Order is abundantly distributed over the tropical regions of the New and the Old World, more rare in temperate climates. Of the 7 Australian genera, one is common to the New and the Old World, three are Asiatic, and the other three are endemic.

XLI. ANACARDIACEÆ.

Tribe I. **Anacardieæ.**—*Ovary 1-celled, or if 2-celled with one cell early suppressed.*

A. Ovules pendulous from a basal funicle.
 *Sepals and petals not accrescent.
  Calyx 4 to 5-partite. Petals 4 to 6. Stamens 4 to 10. Leaves alternate, usually compound . . . . 1. Rhus.
  Calyx 5-lobed, imbricate. Petals 5, erect, imbricate. Stamens 10. Styles 3, free, spreading. Leaves pinnate. Drupe globose. Seeds ovoid, compressed . . . . 2. Rhodosphæra.
  Calyx 4 to 5-partite. Petals 4 to 5. Stamens 1 to 5. Style filiform. Leaves alternate, simple . . . . 3. Mangifera.
  Calyx 3 to 5-lobed. Petals 3 to 5. Stamens 10. Carpels 5 to 6, one only perfect. Style short. Leaves alternate, simple . . . . . 4. Buchanania.
B. Ovules pendulous from the top of the cell or from the walls of the ovary above the middle.
  Petals imbricate. Stamens 10. Styles 3. Drupes compressed, evittate. Leaves pinnate . . . . 5. Euroschinus.
  Petals imbricate. Stamens 5. Styles 3. Drupe on a much-enlarged peduncle . . . . 6. Semecarpus.

Tribe II. **Spondieæ.**—*Ovary 2 to 5 or more-celled. Ovules pendulous. Leaves pinnate.*
 Polygamous-diœcious. Calyx 5-partite. Petals 5. Stamens 10. Ovary usually 12-celled . . . . 7. Pleiogynium.

1. RHUS, Linn.
(From the red colour of fruit.)

Flowers polygamous. Calyx small, of 4 to 6, usually 5, imbricate sepals. Petals as many as sepals, imbricate in the bud. Disk broad, flat or annular. Stamens as many as petals or rarely 10, inserted round the base of the disk. Ovary 1-celled; styles 3, free or connate, with simple or capitate stigmas; ovule suspended from an erect filiform funicle. Drupe globular or compressed, usually small. Seed inverted or transverse, the radicle turned upwards.—Trees or shrubs. Leaves pinnate, 3-foliolate, or in species not Australian simple. Flowers small, in terminal or axillary panicles.

The species are numerous in the warmer extratropical regions of both the northern and southern hemispheres, especially in S. Africa, more rare within the tropics.—*Benth.*

1. **R. rufa** (red), *Teysman and Binnesidyk in Naturk. Tijd.* xxvii. 52, *F. v. M. in Cens. Austr. Plants.* An evergreen, unarmed and small tree with a milky sap and smooth bark, the branchlets, peduncles and rhachis thinly pubescent. Leaves impari-pinnate, 8 to 9-jugate; leaflets 3 to 5in. long, 1 to 1¼in. broad, flat, chartaceous, ovate or oblong-lanceolate, quite entire, the upper side glabrous, pilose and opaque on the under side. Panicles racemose; flowers small, pentandrous. Sepals imbricate ciliate, ⅓ line diameter. Petals ovate, scarcely 1 line long, white, glabrous outside and inside from middle to base pilose. Stamens 5, the sterile ones very short. Disk crenulate, glabrous. Ovary pilose. Stigmas almost sessile. Drupe small, about 2 or 3 lines diameter, black, glabrous, oblique, rotund, moderately compressed. Pericarp dry, fragile, shining. Putamen fulvo-lividum, very hard. Albumen none. Cotyledons flattish, pale. Radicle curved cylindric.—*R. panaciformis,* F. v. M. Fragm. vii. 22.

Hab.: About Rockingham Bay and Valley of Lagoons, *J. Dallachy* (F. v. M., l.c.)

2. RHODOSPHÆRA, Engl.
(Colour and shape of fruit.)

Flowers polygamo-diœcious. Calyx-segments 5, imbricate. Petals 5, erect, imbricate. Stamens 10 (short in the female flowers?), filaments subulate, anthers obtuse at both ends, of equal length and dehiscing longitudinally. Disk short, cupuliform, slightly 10-crenate. Ovary sessile, subglobose; ovules solitary,

Rhodosphæra rhodanthema, Engl.

shortly suspended from the top of basally fixed, erect funicle. Styles 3, free, patent; stigmas capitate. Drupe globose, epicarp chartaceous, very smooth, mesocarp thick woody, endocarp thin bony, compressed. Seed ovoid, compressed, testa thin membranous. Embryo exalbuminous, cotyledons plain, radicle superior, very short.—A tree of medium size. Leaves impari-pinnate, sub-coriaceous, upper side very minutely puberulent; leaflets shortly petiolulate. Flowers small, red, numerous in dense pyramidal-panicles, axillary or terminal.—DC. Mono. Phane. iv. 324; M. Ad. Englers Monogr. of the Order.

1. **R. rhodanthema** (red-flowered), *Engl. l.c.* A tree of medium size, bark of trunk shedding in rather thick scaly pieces. Leaves pinnate, the common petiole terete. Leaflets usually from 7 to 9, 2 to 3in. or more long, oblong, obtusely-acuminate, usually with tufts of hairs along the midrib on the under side, shortly petiolulate, the pinnate veins prominent underneath; the leaflets of young plants much larger and frequently bluntly lobed. Panicles pyramidal or broadly thyrsoid, dense. Flowers diœcious, red, very shortly pedicellate. Sepals broadly ovate, very obtuse, about 1 line long. Petals ovate, recurved, 1½ line long. Stamens 10. Ovary broad. Styles 3, short, thick, diverging, with capitate stigmas; ovule nearly globular, suspended from an erect funicle. Drupe globular, shining, about ½in. diameter, putamen (endocarp) thick and woolly, striate outside, lined with a separable cartilaginous layer inside. Seeds orbicular, flat, testa membranous.—*Rhus rhodanthema*, F. v. M. in Fl. Austr. i. 189, where Mr. Bentham remarks that the species differs from the greater part of the genus *(Rhus)* in its large red flowers, 10 stamens and larger globular drupes.

Hab.: Creek and river sides in the southern parts of the colony.

I have some specimens belonging to this plant which were collected by A. Cunningham, labelled *Trichilia scabra*, A. Cunn.

Wood very beautiful, and greatly in demand for cabinet-work; the heartwood of a deep glossy brown, soft, and fine-grained.—*Bailey's Cat. Ql. Woods No. 109.*

*8. **MANGIFERA**, Linn.

(Mango-bearing.)

Calyx 4 to 5-partite; segments imbricate, deciduous. Petals 4 to 5, free or adnate to the disk, imbricate; nerves thickened, sometimes ending in excrescences. Stamens 1 to 5, rarely 8, inserted just within the disk or on it, 1 rarely more perfect and much longer than the others, the others with imperfect or smaller anthers, or reduced to teeth. Ovary sessile, 1-celled, oblique. Style lateral. Ovule pendulous, funicle basal, inserted on the side of the cell above the base, rarely horizontal. Drupe large, fleshy; stem compressed, fibrous. Seeds large, compressed; testa papery. Cotyledons plano-convex, often unequal and lobed. —Trees. Leaves alternate, petiolate, quite entire, coriaceous. Flowers small, polygamous in terminal panicles. Pedicels articulate. Bracts deciduous.—Hook. Fl. Brit. Ind. ii. 13.

1. **M. indica** (Indian), *Linn.* Mango. A large tree, glabrous except the panicle. Branches widely spreading. Leaves oblong or linear-oblong, or elliptic or obovate-lanceolate, obtuse, acute or acuminate, 6 to 16in. long, very variable in breadth; crowded at the ends of the branches, often shining; margins often undulate, the young growth usually coloured—a glossy purplish-brown. Petioles 1 to 4in. long, swelled at base. Panicles more or less pubescent. Bracts elliptic, concave. Flowers yellow, odorous, subsessile, rarely pedicellate, male and female in the same panicle. Sepals ovate, oblong, concave. Petals twice as long, ovate,

Y

ridges 3—5, orange. Disk fleshy, 5-lobed. Stamens 1, inserted upon the disk; filaments subulate; anthers purple. Ovary glabrous. Drupe, size, form and colour various.

Hab.: This Indian fruit-tree is met with here, and as a stray from cultivation.—*J. S. Gamble, in Manual of Indian Timbers.*

The wood is used for planking, doors and window-frames, packing cases, boxes, canoes, and boats. The bark gives a gum. and the seeds contain gallic acid. The average weight of a cubic foot of the wood is 41lbs. Wood soft, somewhat spongy; no heart-wood.

The fruit of this tree is often attacked by the blight fungus *Glœosporium Lagenarium,* Pass.

4. BUCHANANIA, Roxb.
(After Dr. Buchanan Hamilton.)

Flowers hermaphrodite. Calyx short, obtusely 3 to 5-toothed. Petals 5, imbricate in the bud. Disk orbicular, crenate. Stamens 10, inserted round the disk. Gynœcium of 5 or 6 distinct carpels, of which one only perfect, the others rudimentary and style-like; style of the perfect one short, with a truncate stigma; ovule suspended from an erect filiform funicle. Drupe small, the putamen crustaceous or bony, 2-valved. Seed with thick cotyledons and a superior radicle.— Trees. Leaves alternate, simple, entire, coriaceous. Flowers small, white, in terminal or axillary panicles.

The genus extends over tropical Asia and the islands of the Pacific, the two Australian species being endemic.

Anthers sagittate.
  Leaves large, lanceolate-oblong, cuneate at the base, the point obtuse or absent, the secondary lateral nerves parallel . . . . 1. *B. mangoides.*
  Leaves smaller, the secondary lateral nerves not parallel. Drupe shortly pilose, apex excentric . . . . 2. *B. Muelleri.*

1. B. mangoides (Mango-like), *F. v. M. Fragm.* vii. 23; *Engler. in DC. Mono. Phane.* iv. 187. A tree about 40ft. in height, the branchlets and peduncles tomentose. Leaves large, attaining 1ft. in length, oblong-lanceolate, narrowed to rather long petiole, thin-coriaceous, bright-green on the upper side, pale beneath; parallel nerves about 16 on each side of midrib, with intermediate ones. Panicles terminal, dense, ferruginous-pilose, 6 or 7in. long. Branches 1 to 1½in. long. Pedicels short. Calyx glabrescent; lobes semiorbicular, ⅔ line long. Petals oblong, 1¼ line long, glabrous. Filaments subulate. Anthers sagittate; lobes oblong-cylindrical, divergent. Styles 5, very short. Ovary strigose-pilose.

Hab.: Family Island, J. Dallachy (F. v. M., l.c)

2. B. Muelleri (after Baron von Mueller), *Engler. in DC. Mono. Phane.* iv. 191. A tree of medium height and widely-spreading head; the young branchlets silky-pilose, at length glabrous. Leaves 4 to 6in. long, and 1½ to 2in. broad, subcoriaceous, ovate-oblong or lanceolate, narrowing from the middle to a petiole of a few lines, swelled at the base, the apex very obtuse or acute or sometimes emarginate. Inflorescence usually below the terminal leaves of the shoots. Peduncles sometimes 2in. long, bearing an elongated rather straggling panicle with few branches, but usually the panicles are almost terminal, shorter, and more dense. Pedicels short. Calyx glabrous; lobes semiorbicular. Petals oblong, 1 or 2 lines long, white. Stamens with subulate filaments about ⅔ line long. Anthers sagittate. Ovary strigose-pilose. Drupe purple, compressed, about 5 lines long and nearly as broad, the apex nearly terminal.

Hab.: Endeavour River, Cape York, and the islands of Torres Straits and Gulf of Carpentaria. Wood of a pinkish colour, close-grained, tough, and easily worked.—*Bailey's Cat. Ql. Woods* No. 109a (by mistake given as *B. mangoides*).

XLI. ANACARDIACEÆ. 323

5. EUROSCHINUS, Hook. f.

(The southern *Schinus*.)

Flowers polygamous, diœcious, or hermaphrodite. Calyx small, 5-lobed, semi-orbicular, imbricate. Petals 5, oblong or oblong-ovate, imbricate in the bud. Disk orbicular, deeply crenate. Stamens 10, inserted round the disk, filaments short. Anthers oblong, dorsifixed. Ovary 1-celled, with 3 thick short styles, or in the males of 3 or 4 linear style-like rudiments; ovule pendulous from the top of the cavity. Drupe small, more or less compressed, the putamen coriaceous. Seeds compressed, with flat cotyledons; the radicle turned upwards.—Tree. Leaves pinnate. Flowers rather small, in terminal or lateral panicles.

The genus is limited to a single species endemic in Australia. It is closely allied to the American genus *Schinus*, but with a rather different habit, a gamosepalous calyx, and the putamen of the fruit does not appear to contain the oily receptacles so conspicuous in that genus.—*Benth.*

1. **E. falcatus** (falcate), *Hook. f. in Benth. and Hook. Gen. Pl.* 422. Maiden's Blush Wood. "Punburra," Moreton Bay, *Watkins*; "Kokare," Barron River, *J. F. Bailey*. A low tree, the branchlets rough with lenticels, glabrous or the young shoots minutely hoary. Petioles 2 or 3in. long, slender. Leaflets 4 to 10, irregular, sometimes opposite, very oblique or falcate, ovate to lanceolate, shortly acuminate, 2 to 4in. long, all but the terminal one very unequal at the base, on petiolules of 1 to 3 lines, penninerved and reticulate, the common petiole terete. Panicles divaricate, many-flowered, not exceeding the leaves. Flowers almost sessile, clustered along the branches, about 1 line long and glabrous. Calyx-lobes obtuse, slightly imbricate. Petals twice as long, oblong, very spreading. Drupes broadly and obliquely ovate, 3 or 4 lines long.

Hab.: Brisbane River; sources of the Burdekin, *F. v. Mueller*; Sunday Island, *M'Gillivray*.

Var. *angustifolius*. Leaves falcate-lanceolate, much acuminate. Flowers rather larger.—Northumberland Islands, *R. Brown*; Rockhampton, *Thozet*. The two forms run the one into the other, and in most of the Queensland coast scrubs.

The wood of each form is very similar—pinkish or white, very soft, light and tough; perhaps might serve for making oars.—*Bailey's Cat. Ql. Woods No.* 111 and 111a.

6. SEMECARPUS, Linn. f.

(From *semeion*, a mark, and *karpos*, fruit: juice of tree used to mark clothes.)

Flowers polygamous or diœcious. Calyx small, 5-lobed. Petals 5, imbricate in the bud. Disk orbicular, slightly lobed or crenate. Stamens 5, inserted round the disk. Ovary 1-celled, with 3 styles, and somewhat club-shaped stigmas; ovule suspended from the top of the cavity. Drupe or nut reniform, seated on the much-enlarged, thick, succulent, fleshy pedicel, cupular, or turbinate base of the calyx; pericarp thick, hard, filled with resinous cells. Seed pendulous, the testa coriaceous, somewhat fleshy inside; embryo thick, with plano-convex cotyledons and a very short superior radicle.—Trees. Leaves alternate. Flowers small, in terminal or lateral panicles.

The genus ranges over tropical Asia, the species most numerous in Ceylon.—*Benth.*

1. **S. australiensis** (Australian), *Engler. DC. Mono. Phane.* iv. 482. Marking-Nut Tree; "Jaln-ba," Annan River, *Roth*. A spreading-headed medium or large tree. Branches thick, with grey bark, the branchlets rough with lenticels. Leaves broadly obovate, very obtuse, 8 to 6in. long, entire or lanceolate and 9in. long, cuneate at the base, on very short petioles, glabrous above, hoary or white underneath but scarcely tomentose, the pinnate veins and reticulate veinlets conspicuous on both sides, and a very prominent nerve-like

margin. Male panicles pyramidal, shorter than or as long as the leaves. Flowers very small, sessile, and clustered. Bracts ¼ to 1 line long. Calyx very short. Petals scarcely 1 line long. Ovary minute and rudimentary or reduced to a tuft of hair. Female flowers sometimes larger than the male calyx-lobes, ½ line long. Petals 1½ line long and almost 1 line broad. Stamens in female flowers shorter than the petals. Drupe oblique-conoid, compressed, 1 to 1¼in. diameter, and about ½in. thick, rugose, supported upon a large, fleshy, succulent pedicel, the nut surrounded inside near the margins by a row of oil-cells.—*S. anacardium*, in Fl. Austr. i. 491.

Hab.: Trinity Bay to Cape York, and the islands of Torres Straits and the Gulf of Carpentaria.

Wood yellow with brown markings, easy to work, strong and tough; might be used in cabinet-work.—*Bailey's Cat. Ql. Woods No.* 110.

7. PLEIOGYNIUM, Engler.

(Alluding to its many female parts. From the specific name given to it by Baron Mueller when placing it in *Spondias*.)

Flowers diœcious or polygamous. Calyx deeply 5-parted, segments ovate, imbricate. Petals 5, imbricate, twice as long as the calyx, obovate, inserted below an annulate crenate disk. Stamens 10, inserted with the petals; filament filiform-subulate. Anthers ovate, versatile, cells longitudinal, their dehiscence introrse. Ovary abortive in the male flowers, in the female flowers of from 5 to 10 or 12 carpels. Ovary depressed, 5 to 10 or 12-celled; ovule 1 in each cell, pendulous; micropyle superior. Styles short, divergent; stigma spathulate, at length patent or reflexed. Drupes slightly depressed, broadly turbinate, with the summit a little elevated and the lower part slightly angular; epicarp fleshy; endocarp thick, woody, shiny on the inside, 5 to 12-celled; cells securiform. Seeds compressed, oblong, and slightly outwardly curved. Embryo with oblong plano-convex cotyledons and short superior radicle.—A tree, the branchlets terete, the foliage clustered at their extremities. Leaves impari-pinnate; petiole and rhachis slightly angular; leaflets membranous, oval, obtuse, often oblique and narrowed at the base; lateral nerves patent. Panicles axillary.—From Ad. Engler, diagnosis in De Candolle's Monogr. Phanera. iv. 255, t. vii., fig. 1, 10.

1. **P. Solandri** (after Dr. D. C. Solander), *Engler, l.c.* Sweet Plum, Burdekin Plum. "Noongi," Port Curtis; "Bungya," Bundaberg, *Keys;* "Rancooran," Rockhampton, *Thozet*. A tree of medium size, frequently unisexual, the branchlets and foliage from velvety pubescent to nearly glabrous. Leaves 3 to 7in. long, on petioles of about 1 or 2in. long; slightly angular as well as the rhachis; leaflets petiolulate, of from 2 to 5 pairs, and a terminal odd one, oval-oblong, 2 to 4in. long, and except the end one, which has a rather long cuneate base, very oblique at the base; the primary veins branched and often glandular, and forming a cavity in the axil at the midrib. Panicles (the male about 5in. long with abortive ovary), slender, narrow and drooping, the flowers with a few short branches, in small clusters along the branches, female usually in spikes not exceeding 2in., with short stamens and sterile anthers. Bracteoles ovate, minute, sessile. Calyx-segments roundish, scarcely exceeding ¼ line long. Petals yellowish-green, about 1 line long, subovate, veined, glabrous above. Filaments linear-subulate, 1 line long. Styles conico-subulate, stellately

Pleiogynium Solandri, Engl.

spreading, recurved, about ½ line long. Fruit turbinate, from 1 to 1½in. or more in diameter, often of a deep purple when ripe, and then with a juicy sarcocarp; putamen hard, rugose, outside, 12-celled, containing 1 seed in each cell.—From Mueller's and Engler's description (in part). *Spondias acida*, Soland; *S. Solandri*, Benth.; *Owenia cerasifera* F. v. M.; *S. pleiogyna*, F. v. M.

Hab.: On the borders of tropical creeks and rivers.

The tree of which the above is a botanic description is that known in Queensland as the Burdekin Plum, or Sweet Plum, and by the Rockhampton natives as "Rancooran," and at Port Curtis as "Noongi." In the 1st vol. of the Flora Australiensis, Mr. Bentham placed it in Meliaceæ as (?) *Owenia cerasifera*, as published previously by Baron v. Mueller in *Hooker's Kew Journal*, with a note that "until the flowers have been seen, this species must remain, in some measure, doubtful," the fruit specimens alone being then known. At page 492 of this same volume, Mr. Bentham describes the same tree as *Spondias Solandri*, changing it from *S. acida*, as named by Solander in the Banksian Herbarium. These specimens were not in fruit, but he tells us that in the Banksian collection was a packet of drupes, named as belonging to this species, and described as such by R. Brown, and from this description there is little doubt but what they were the fruit of the above *Spondias*, and that it and *Owenia cerasifera* are identical. Baron v. Mueller, it would seem, has come to that conclusion, for we find *S. pleiogyna*, under which he described it in his Fragm. iv. and v., now merged into *S. Solandri*, in the last edition of his Census of Australian Plants. It seems to me better that the name given in De Candolle's work, l.c., should be used for our Burdekin Plum. A change of name is always to be avoided where possible, but in the present instance it seems necessary. By an unfortunate oversight C. de Candolle, Monogr. Phaner., i. 596, gives a description from the Flora Austr., i. 386, omitting the note of interrogation, of *Owenia cerasifera*, F. v. M., which name had been changed to *Spondias pleiogyna* twelve or fourteen years before.

Wood hard, dark-brown, with red markings (resembling American Walnut), the grain pretty close; splits straight. An excellent wood for the joiner or cabinetmaker; also for turnery.— *Bailey's Cat. Ql. Woods* No. 112 (under the name of *Spondias pleiogyna*, F. v. M.)

INDEX OF GENERA AND SPECIES.

The synonyms and species incidentally mentioned are printed in italics.

	Page		Page		Page
Abelmoschus		franguloides, A. Gray	271	Billardiera	74
albarubens, F. v. M.	124	zizyphoides, A. Gray	271	angustifolia, DC.	75
divaricatus, Walp.	126	Alyssum	45	canariensis, Wendl.	75
ficulneus, W. & Arn.	124	linifolium, Steph.	45	grandiflora, Putterl.	75
Manihot, Walp.	124	Amoora	283	lutifolia, Putterl.	75
rhodopetalus, F. v. M.	124	nitidula, Benth.	233	Biophytum	
splendens, Walp.	128	Ampelacissus		Apodiscias, Turcz.	180
Abroma	145	acetosa, Planch.	282	Blennodia	46
fastuosa, R. Br.	146	Ancana	22	canescens, R. Br.	46
Abutilon	116	stenopetala, F. v. M.	22	cardaminoides, F. v. M.	47
asiaticum, G. Don	118	Apodytes	248	Cunninghamii, Benth.	48
auritum, G. Don	118	brachystylis, F. v. M.	248	eremigera, Benth.	47
Cunninghamii, Benth.	120	Apophyllum	61	lasiocarpa, F. v. M.	47
Fraseri, Hook.	120	anomalum, F. v. M.	61	nasturtioides, Benth.	47
graveolens, W. & Arn.	119	Arabis		trisecta, Benth.	46
indicum, G. Don	118	gigantea, Hook	44	Blepharocarya	310
leucopetalum, F. v. M.	117	Arenaria		involucrigera, F. v. M.	310
micropetalum, Benth	117	media, Linn.	89	Bocagea	
Mitchelli, Benth.	117	rubra, Linn.	89	pisocarpa	24
muticum, G. Don	119	Argemone	41	Bombax	183
otocarpum, F. v. M.	118	mexicana, Linn.	41	heptaphyllum, Cav.	183
oxycarpum, F. v. M.	119	ochroleuca	41	malabaricum, DC.	133
subviscosum, Benth.	118	Argyrodendron		Boronia	186
tubulosum, Hook	117	trifoliolatum, F. v. M.	140	alulata, Soland.	186
Acronychia	208	Aristotelia	160	anemonifolia, A. Cunn.	188
acidula, F. v. M.	210	australasica, F. v. M.	161	anethifolia, A. Cunn.	188
Baueri, Schott	208	megalosperma, F. v. M	161	arborescens, F. v. M.	186
Cunninghamii, Hook	202	Arytera		artemisiæfolia, F. v. M.	186
Hillii, F. v. M.	208	divaricata, F. v. M.	305	bipinnata, Lindl.	188
imperforata, F. v. M	209	foveolata, F. v. M.	304	Bowmani, F. v. M.	187
lævis, Forst.	209	semiglauca, F. v. M.	302	eriantha, Lindl.	186
laurina, F. v. M.	209	Ascyrum		falcifolia, A. Cunn.	188
melicopoides, F. v. M.	209	humifusum, Labill.	101	floribunda, Sieb.	187
Scortechini, Bail	210	involutum, Labill.	101	granulata, F. v. M.	185
vestita, F. v. M.	210	Asterolasia	195	hirsuta, F. v. M.	184
Adeliopsis	81	Woombye, Bail.	195	hyssopifolia, Sieb.	188
decumbens, Benth.	31	Atalantia	213	lævigata, F. v. M.	188
Adrastæa	17	glauca, Hook. f.	214	lanceolata, F. v. M.	187
salicifolia, DC.	17	recurva, Benth.	214	ledifolia, J. Gay	186
Aglaia	232	Atalaya	299	microphylla, Sieb.	187
elæagnoidea, Benth.	232	hemiglauca, F. v. M.	300	minutiflora, F. v. M.	184
odoratissima, Benth.	232	multiflora, Benth.	300	paleifolia, Endl.	188
Ailanthus	216	variifolia, F. v. M.	300	paradoxa, DC.	187
glandulosa, Desf.	218			parviflora, Sm.	188
imberbiflora, F. v. M.	217	Berchemia		pilonema, Labill.	188
malabarica, DC.	217	ecorollata, F. v. M.	207	pinnata, Sm.	187
punctata, F. v. M.	207	Bergia	99	platyarhachis, F. v. M.	187
Akania	309	ammannioides, Roth.	100	polygalifolia, Sm.	188
Hillii, Hook. f.	309	perennis, F. v. M.	100	tetrathecoides, DC.	188
Alphitonia	270	Berrya	152	Bosistoa	196
excelsa, Reissek	270	Ammonilla, Roxb.	152	sapindiformis, F. v. M.	196

i

INDEX OF GENERA AND SPECIES.

	Page		Page		Page
Brachychiton		tomentosum, *Wight*	104	Cheiranthera	75
acerifolium, F. v. M.	138	*Campylanthera*		cyanea, Brongn.	75
Bidwillii, Hook.	138	ericoides, Lindl.	73	linearis, *A. Cunn.*	75
Delabecchi, F. v. M.	139	Cananga	22	Cissampelos	34
discolor, F. v. M.	138	odorata, *H. f. & T.*	22	pareira, *Linn.*	34
diversifolium, R. Br.	139	Canarium	223	*Cissus*	
luridum, F. v. M.	138	australasicum, *F. v. M.*	224	acetosa, F. v. M.	282
paradoxum, Schott	137	Muelleri, *Bail.*	224	adnata, Roxb.	280
platanoides, R. Br.	138	Cansjera	246	antarctica, Vent.	279
populneum, R. Br.	139	leptostachya, *Benth.*	246	australasica, F. v. M.	283
ramiflorum, R. Br.	137	Capparis	56	carnosa, Lam.	281
Brackenridgea	221	canescens, *Banks*	59	cinerea; Lam.	281
australiana, *F. v. M.*	222	humistrata, *F. v. M.*	61	cordata, Roxb.	280
Brasenia	37	lasiantha, *R. Br.*	57	crenata, Vahl.	281
peltata, *Pursh*	37	lorianthifolia, *Lindl.*	60	geniculata, Blume	281
Brassica	49	lucida, *R. Br.*	60	glandulosa, Poir	289
(Sinapis) nigra, *Boiss.*	50	Mitchelli, *Lindl.*	60	hypoglauca, A. Gray	283
Brathys		nobilis, *F. v. M.*	59	opaca. F. v. M.	283
Billardieri, Spach	101	nummularia, *DC.*	58	Citriobatus	73
Forsteri, Spach	101	ornans, *F. v. M.*	58	lancifolius, *Bail.*	74
humifusa, Spach	101	quiniflora, *DC*	57	multiflorus, *A. Cunn.*	73
Brombya	202	sarmentosa, *A. Cunn.*	58	pauciflorus, *A. Cunn.*	74
platynema, *F. v. M.*	202	Shanesiana, *F. v. M.*	59	Citrus	214
Brotera		Thozetiana, *F. v. M.*	61	australis, *Planch.*	215
Lepricurii, Guill. et Perr.	144	uberiflora, *F. v. M.*	58	australasica, *F. v. M.*	215
ovata, Cav.	144	umbellata, *R. Br.*	57	inodora, *Bail.*	215
Brucea	218	umbonata, *Lindl.*	60	Clausena	213
sumatrana, *Roxb.*	218	Capsella	50	brevistyla, *Oliv.*	213
Buchanania	322	Andræana, *F. v. M.*	50	Clematis	4
mangoides, *F. v. M.*	322	Bursa-pastoris, Mœnch.	50	aristata, *R. Br.*	5
Muelleri, *Engl.*	322	humistrata, *F. v. M.*	51	glycinoides, *DC.*	5
Buettneria		Carapa	237	linearifolia, Steud.	6
dasyphylla, J. Gay	147	moluccensis, *Lam.*	237	microphylla, *DC.*	6
pannosa, DC.	147	Cardamine	44	stenophylla, Fras.	6
Bursaria	72	debilis, Banks	45	stenosepala, DC.	5
diosmoides, Putterl.	73	divaricata, Hook. f.	44	Cleome	54
incana, *Lindl.*	72	eustylis, *F. v. M.*	45	flava, Banks	55
procumbens, Putterl.	73	hirsuta, *Linn.*	44	oxalidea, *F. v. M.*	54
spinosa, *Cav.*	72	parviflora, Linn.	45	pungens, *Willd.*	55
Stuartiana, Klatt.	73	paucijuga, Turcz.	45	tetrandra, *Banks*	55
tenuifolia, *Bail.*	72	stylosa, *DC.*	44	*Clypea*	
Bursera	223	Cardiopteris	250	hernandifolia, W. & Arn.	33
australasica, *Bail.*	223	lobata, *R. Br.*	251	Coatesia	
Busbeckia		moluccana, Blume	251	paniculata, F. v. M.	206
arborea, F. v. M.	59	Cardiospermum	286	*Cocculus*	
corymbiflora, F. v. M.	60	Halicacabum, *Linn.*	286	Moorei, F. v. M.	29
Mitchelli, F. v. M.	60	microcarpum, H. B. & K.	286	Cochlospermum	65
nobilis, *Endl.*	59	Caryospermum	258	Gregorii, *F. v. M.*	66
		arborescens, *F. v. M.*	259	Gillivræi, *Benth.*	65
Cadellia	219	*Catha*		Colletia	
monostylis, *Benth.*	220	Cunninghamii. Hook.	255	Cunninghamii, Fenzl	277
pentastylis, *F. v. M.*	219	*Ceanothus*		pubescens, Brongn.	277
Calandrinia	95	discolor, Vent.	273	Colubrina	269
balonensis, *Lindl.*	96	laniger, Andr.	272	asiatica, *Brongn.*	270
calyptrata, *Hook. f.*	97	Cedrela	237	excelsa, Fenzl	270
pleopetala, *F. v. M.*	97	australis, F. v. M.	238	vitiensis, Seem.	269
pogonophora, *F. v. M.*	98	Toona, *Roxb.*	237	Comesperma	79
polyandra, *Benth.*	97	Celastrus	253	acutifolium, Steetz	82
ptychosperma, *F. v. M.*	98	australis, *Harv. & F.v.M.*	254	coridifolium, A. Cunn.	82
pumila, *F. v. M.*	98	bilocularis, *F. v. M.*	254	defoliatum, *F. v. M.*	82
pusilla, *Lindl.*	97	Cunninghamii, *F. v. M.*	255	ericinum, *DC*	81
spergularina, *F. v. M.*	98	dispermus, *F. v. M.*	254	gracile, Paxt.	80
uniflora, *F. v. M.*	96	montanus, Roxb.	255	latifolium, Steetz	82
volubilis, *Benth.*	97	senegalensis, Lam.	255	lineariæfolium, A. Cunn.	82
Calophyllum	103	Cerastium	86	nudiusculum, *DC.*	82
australianum, *Vesq.*	104	viscosum, Linn.	87	patentifolium, F. v. M.	82
costatum, *Bail.*	104	vulgatum, *Linn.*	87	præselsum, *F. v. M.*	81
inophyllum, *Linn.*	104	*Charissa*, Miq.	250	retusum, *Labill.*	81

INDEX OF GENERA AND SPECIES. iii

	Page
secundum, *Banks*	81
sphærocarpum, *Steetz*	80
sylvestre, *Lindl.*	81
tortuosum, Steetz	80
volubile, *Labill.*	80
Commersonia	147
dasyphylla, Andr.	147
echinata, *Forst.*	148
Fraseri, J. *Gay*	148
Leichhardtii, *Benth.*	148
Cookia	
australis, F. v. M.	207
Corchorus	157
acutangulus, Lam.	158
Cunninghamii, *F. v. M.*	158
hygrophilus, *A. Cunn.*	157
pumilio, *R. Br.*	158
sidoides, *F. v. M.*	158
tomentellus, *F. v. M.*	158
trilocularis, *Linn.*	159
Correa	195
speciosa, *Ait.*	196
Crowea	189
latifolia, Lodd.	189
saligna, *Andr.*	189
scabra, Grah.	190
Cryptandra	274
amara, *Sm.*	276
campanulata, Schlech.	276
capitata, Sieb.	275
ericifolia, *Sm.*	275
largiflora, F. v. M.	276
longistamines, F. v. M.	276
nervata, Reiss.	276
pyramidalis, R. Br.	275
spinescens, Sieb.	275
Cupania	289
anacardioides, *Rich.*	290
anodonta, F. v. M.	290
Bidwilli, *Benth.*	293
Cunninghamii, Hook.	287
curvidentata, *Bail.*	292
diphyllostegia, F. v. M.	287
erythrocarpa, *F. v. M.*	294
flagelliformis, *Bail.*	291
foveolata, *F. v. M.*	292
lucens, F. v. M.	310
Mortoniana, *F. v. M.*	294
nervosa, *F. v. M.*	293
pleurophylla, *F. v. M.*	293
pseudorhus, *A. Rich.*	292
punctulata, *F. v. M.*	297
Robertsonii, *F. v. M.*	291
semiglauca, *F. v. M.*	302
sericolignis, *Bail.*	294
serrata, *F. v. M.*	291
Shirleyana, *Bail.*	290
stipata, F. v. M.	297
tenax. A. Cunn.	299
tomentella. *F. v. M.*	292
Wadsworthii, *F. v. M.*	290
xylocarpa, *A. Cunn.*	293
Cyminosma	
oblongifolium, A. Cunn.	209
Dallachya	268
vitiensis, *F. v. M.*	268

	Page
Delabechea	
rupestris, Lindl.	139
Denhamia	257
heterophylla, F. v. M.	258
obscura, *Meisn.*	257
oleaster, *F. v. M.*	257
pittosporoides, *F. v. M.*	258
xanthosperma, F. v. M.	258
viridissima, *Bail.* and *F. v. M.*	258
Dichoglottis	
australis, Schlecht.	85
tubulosa, Jaub. & Spach	85
Dillenia	
scandens, Willd.	16
speciosa, Bot. Mag.	16
volubilis, Vent.	16
Dimerosporium	
Tetracera, Cke.	10
Diploglottis	286
Cunninghamii, *Hook.. f.*	287
Discaria	276
australis, *Hook.*	277
Distichostemon	319
phyllopterus, *F. v. M.*	319
Dodonæa	311
acerosa, Lindl.	315
adenophora, *Miq.*	318
angustifolia, Sw.	314
asplenifolia, Rudge	314
attenuata, *A. Cunn.*	314
boronizefolia, *G. Don*	318
Burmanniana, DC.	314
calycina, A. Cunn.	315
conferta, G. Don	314
cuneata, *Rudge*	314
filifolia, *Hook.*	315
Hansenii, *F. v. M.*	316
hirtella, Miq.	318
hispidula, Endl.	319
Kingii, G. Don	314
lanceolata, *F. v. M.*	313
laurina, Sieb.	313
Lindleyana, F. v. M.	316
lobulata, *F. v. M.*	315
longipes, G. Don	313
Macrossani, *F. v. M.*	316
megazyga, *F. v. M.*	316
mollis, Lindl.	316
macrozyga, *F. v. M.*	317
neriifolia. A. Cunn.	314
oblangifolia, Link	314
oxyptera, *F. v. M.*	317
paulliniæfolia, A. Cunn.	317
peduncularis, *Lindl.*	315
petiolaris, *F. v. M.*	313
physocarpa, *F. v. M.*	317
Preissiana, *Miq.*	314
pubescens, Lindl.	315
salicifolia, DC.	314
spathulata, Sm.	314
stenophylla, *F. v. M.*	316
tenuifolia, Lindl.	318
triangularis, *Lindl.*	316
trigona, Lindl.	316
triquetra, *Andr.*	313
truncatiales, *F. v. M.*	315
umbellata, G. Don	314

	Page
vestita, *Hook.*	317
viscosa, *Linn.*	313
Drimys	
dipetala, *F. v. M.*	18
Haweana	19
membranea, *F. v. M.*	18
semecarpoides, *F. v. M.*	19
rivularis.	19
Drymaria	89
diandra, *Blume.*	89
Dysoxylon	227
arborescens, *Miq.*	228
Becklerianum, C.DC.	231
cerebriforme, *Bail.*	232
Fraseranum, *Benth.*	230
Klanderi, *F. v. M.*	229
latifolium, *Benth.*	228
Lessertianum, *Benth.*	230
Muelleri, *Benth.*	230
Nernstii, *F. v. M.*	231
oppositifolium, *F. v. M.*	229
Pettigrewianum, *Bail.*	230
rufum, *Benth.*	231
Schiffneri, *F. v. M.*	228
Echinocarpus	
australis, Benth.	159
Elæocarpus	161
arnhemicus, *F. v. M.*	162
Bancroftii, *Bail.* and *F. v. M.*	162
cyaneus, *Ait.*	163
eumundi, *Bail.*	163
foveolatus, *F. v. M.*	164
Grahami, *F. v. M.*	164
grandis, *F. v. M.*	164
Kirtonii, *F. v. M.*	163
obovatus, *G. Don*	163
parviflorus, A. Rich.	163
pauciflorus, Walp.	163
ruminatus, *F. v. M.*	164
reticulatus, Sm.	163
sericopetalus, *F. v. M.*	165
Elæodendron	259
australe, *Vent.*	259
integrifolium, G. Don	259
maculosum, Lindl.	248
melanocarpum, *F. v. M.*	260
Elatine	99
americana, *Arn.*	99
ammanioides, Wight	100
gratioloides, A. Cunn.	99
minima, Fisch. & Mey.	99
perennis, F. v. M.	100
Emmenospermum	271
alphitonioides, *F. v. M.*	271
Cunninghamii, *Benth.*	271
Eriostemon	189
Banksii, *A. Cunn.*	190
Croweí, F. v. M.	189
cuspidatus, A. Cunn.	191
difformis, *A. Cunn.*	191
hispidulus, *Sieb.*	191
intermedius, Hook.	191
lanceolatus, Gærtn.	191
lancifolius, F. v. M.	191
Leichhardtii, F. v. M.	211
lepidatus, Spreng.	193

INDEX OF GENERA AND SPECIES.

	Page		Page		Page
myoporoides, *DC.*	191	Mazlini, *Bail.*	240	latifolia, *F. v. M.*	154
neriifolius, Sieb.	191	maculosa, *F. v. M.*	248	multiflora, *Juss.*	153
paradoxum, Sm.	187	Oxleyana, *F. v. M.*	289	orientalis, *Linn.*	153
parvifolius, *R. Br.*	192	Pimenteliana, *F. v. M.*	241	pleiostigma, *F. v. M.*	154
rhombeus, Lindl.	191	pubescens, *Boil.*	242	polygama, *Roxb.*	154
rotundifolius, A. Cunn.	193	Schottiana, *F. v. M.*	241	prunifolia, A. Gray	153
salicifolius, Sm.	190	Strzeleckiana, *F. v. M.*	248	Richardiana, Hook.	154
salsolifolius, Sm.	194	Frankenia	88	scabrella, *Benth.*	154
sediflorus, *F. v. M.*	193	fruticulosa, DC.	84	sepiaria, Roxb.	153
Smithianus, Hill	191	pauciflora, *DC.*	88	*Guilfoylia*	
squameus, Labil.	194	pulverulenta, var., DC.	88	monostylis, F. v. M.	220
Erodium	178	scabra, Lindl.	88	Gymnosporia	255
cicutarium, *L'Hér.*	178	serpyllifolia, Lindl.	83	montana, *W. & Arn.*	255
cygnorum, Nees	178	Fugosia	181	Gynandropsis	55
Erpetion		australis, *Benth.*	132	Muelleri, *Benth.*	56
hederaceum, G. Don	63	Fumaria		pentaphylla, *DC.*	56
petiolare, G. Don	63	parviflora, *Linn.*	42	Gypsophila	85
reniforme, Sweet	63			tubulosa, *Boiss.*	85
spathulatum, G. Don	63	Galbulimima	19		
Erysimum		baccata, *Bail.*	19	Halfordia	210
blennodioides, F. v. M.	47	Ganophyllum	224	drupifera, *F. v. M.*	210
nasturtium, F. v. M.	47	falcatum, *Blume*	225	scleroxyla, *F. v. M.*	211
Erythroxylon	167	Garcinia, *Linn.*	102	Hannafordia	150
australe, *F. v. M.*	168	Cherryi, *Bail.*	103	Shanesia, *F. v. M.*	151
ellipticum, *R. Br.*	168	Mestoni, *Bail.*	102	Haplostichanthus	24
Eunomia		Warrenii, *F. v. M.*	103	Johnsoni, *F. v. M.*	24
cochlearina, F. v. M.	53	Garuga	222	Harpullia	308
Euonymus	253	floribunda, *Dene.*	222	alata, *F. v. M.*	308
australiana, *F. v. M.*	253	Geijera	205	frutescens, *Bail.*	308
Euphoria	306	Helmsiæ, *Bail.*	206	Hillii, *F. v. M.*	308
Leichhardtii, *Benth.*	306	latifolia, Lindl.	206	pendula, *Planch.*	309
Eupomatia	26	Muelleri, *Benth.*	206	Wadsworthi, F. v. M.	290
Bennettii, *F. v. M.*	26	parviflora, *Lindl.*	206	Harrisonia	221
laurina, *R. Br.*	27	pendula, Lindl.	207	Brownei, *A. Juss.*	221
laurina, Hook.	27	salicifolia, *Schott*	206	*Hartighsea*	
Euroschinus	328	Geococcus	49	acuminata, Miq.	228
falcatus, *Hook. f.*	328	pusillus, *J. Drumm*	49	Fraserana, A. Juss.	230
Evodia	199	Geranium	177	Lessertiana, A. Juss.	230
accedens, *Blume*	201	australe, Nees	177	rufa, A. Rich.	231
acronychoides, F. v. M.	210	carolinianum, Linn.	177	Hearnia	234
alata, *F. v. M.*	200	dissectum, *Linn.*	177	sapindina, *F. v. M.*	234
Bonwickii, *F. v. M.*	200	parviflorum, Willd.	177	Hedraianthera	255
Cunninghamii, F. v. M.	202	philonothum, DC.	177	porphyropetala, *F. v. M.*	256
Elleryana, F. v. M.	201	pilosum, Forst.	177	Helicteres	142
erythrococca, F. v. M.	198	potentilloides, L'Hér.	177	semiglabra, *F. v. M.*	142
haplophylla, F. v. M.	209	Glycosmis	211	spicata, *Colebr.*	142
littoralis, *Endl.*	201	citrifolia, Lindl.	211	*Hemistemma*	
micrococca, F. v. M.	200	crenulata, Turcz.	213	Banksii, R. Br.	12
neurococca, F. v. M.	197	pentaphylla, *Corr*	211	candicans, Hook. f.	12
octandra, F. v. M.	198	*Glæosporium*		Heritiera	
pentacocca, F. v. M.	196	Alphitoniæ, Cke. & Mass.	271	littoralis, *Ait.*	141
vitiflora, *F. v. M.*	201	carpophilum (not phyllum),		Heterodendron	307
xanthoxyloides, *F. v. M.*	200	Mass.	82	diversifolium, *F. v. M.*	307
		Lagenarium, Pass.	322	oleæfolium, *Desf.*	307
Fawcettia	29	Gomphandra	247	Hibbertia	11
tinosporoides, *F. v. M.*	29	australiana, *F. v. M.*	247	acicularis, *F. v. M.*	14
Fitzalania	21	polymorpha, var.	248	Banksii, *Benth.*	12
heteropetala, *F. v. M.*	21	Gossypium	132	Bennettii, *Bail.*	17
Flindersia	238	australe, F. v. M.	132	Billardieri, *F. v. M.*	13
australis, *R. Br.*	239	Sturtii, *F. v. M.*	133	Brownei, *Benth.*	12
australis, F. v. M.	240	Gouania	277	camphorosma, A. Gray	15
Bennettiana, *F. v. M.*	240	australiana, *F. v. M.*	277	candicans, *Benth.*	12
Bourjotiana, *F. v. M.*	241	Hillii, *F. v. M.*	277	canescens, Sieb.	15
Brayleyana, *F. v. M.*	241	*Grævesia*		dentata, *R. Br.*	16
Chatawaiana, *Bail.*	240	cloisocalyx, F. v. M.	122	diffusa, *R. Br.*	16
collina, *Bail.*	242	Grewia	152	fasciculata, *R. Br.*	15
Ifflaiana, *F. v. M.*	240	asiatica, Linn.	154	glaberrima, *F. v. M.*	16
Leichhardtii, *C.DC.*	239	lævigata, Vahl.	153	longifolia, *F. v. M.*	16

INDEX OF GENERA AND SPECIES.

	Page
lepidota, *R. Br.*	14
linearis, *R. Br.*	15
melhanoides, *F. v. M.*	14
Millari, *Bail.*	13
obtusifolia, DC.	15
œnotheroides, *F. v. M.*	17
prostrata, Hook.	15
salicifolia, *F. v. M.*	17
stricta, *R. Br.*	13
synandra, *F. v. M.*	13
velutina, *R. Br.*	14
vestita, *A. Cunn.*	14
virgata, *R. Br.*	15
virgata, Hook.	15
volubilis, *Andr.*	16
Hibiscus	122
Beckleri, *F. v. M.*	127
brachychlænus, *F. v. M.*	125
brachysiphonius, *F. v. M.*	125
Burtonii, *Bail.*	125
divaricatus, *Grah.*	126
diversifolius, *Jacq.*	127
Elsworthii, *F. v. M.*	126
ficulneus, *Linn.*	123
Fitzgeraldi, *F. v. M.*	126
geranioides, *A. Cunn.*	128
grandiflorus, Salisb.	127
heterophyllus, *Vent.*	127
Krichauffianus, *F. v. M.*	129
leptocladus, *Benth.*	128
Lindleyi, Wall.	126
Manihot, *Linn.*	124
Margeriæ, *A. Cunn.*	127
microchlænus, *F. v. M.*	125
Normani, *F. v. M.*	129
Notho-Manihot, *F. v. M.*	124
panduriformis, *Burm.*	129
Patersoni, DC.	131
Patersonius, *Andr.*	131
phyllochlænus, *F. v. M.*	130
Pinonianus, Miq.	126
radiatus, *Cav.*	126
rhodopetalus, *F. v. M.*	124
Richardsoni, Sweet	125
setulosus, *F. v. M.*	128
solanifolius, *F. v. M.*	125
splendens, *Fras.*	127
Sturtii, Hook.	130
tiliaceus, *Linn.*	130
tridactylites, Lindl.	125
trionioides, G. Don	125
trionum, *Linn.*	124
tubulosus, *Cav.*	129
vitifolius, *Linn.*	129
zonatus, *F. v. M.*	128
Hippocratea	260
barbata, *F. v. M.*	260
macrantha, Korth.	260
obtusifolia, *Roxb.*	260
Hugonia	167
Jenkinsii, *F. v. M.*	167
Husemannia	36
protensa, *F. v. M.*	36
Hydropeltis	
purpurea, Mitch.	87
Hymenosporum	71
flavum, *F. v. M.*	71
Hypericum	101
gramineum, *Forst.*	101
involutum, Chois.	101
japonicum, *Thunb.*	101
pedicellare, Endl.	101
pusillum, Chois.	101
Hyptiandra	
Bidwilli, Hook. f.	219
Hypserpa	31
Hypsophila	256
Halleyana, *F. v. M.*	256
oppositifolia, *F. v. M.*	257
Ilex	251
peduncularis, *F. v. M.*	252
Ionidium	63
aurantiacum, *F. v. M.*	64
filiforme, *F. v. M.*	64
linarioides, Presl.	64
suffruticosum, Ging.	63
Vernonii, *F. v. M.*	64
Itea spinosa, Andr.	72
Ixiosporus	
spinescens, F. v. M.	74
Kayea, Wall.	105
Larnachiana, *F. v. M.*	105
racemosa	105
Keraudrenia	149
adenolasia, *F. v. M.*	150
Hillii, *F. v. M.*	150
Hookeri, F. v. M.	150
Hookeriana, *Walp.*	150
integrifolia, Hook.	150
lanceolata, *Benth.*	149
Kleinhovia, *Linn.*	140
hospita, *Linn.*	141
Lagunæa	
Patersonia, Bot. Mag.	131
squamea, Vent.	131
Lagunaria	131
Patersoni, *Ait.*	131
Lasiopetalum	
ledifolium, Vent.	187
Lavatera	107
Behriana, Schlecht.	108
plebia, Sims	108
Lawsonia	
acronychia, Linn. f.	209
Leea	284
Brunoniana, *C. B. Clark*	284
sambucina, *Willd.*	284
staphylea, Roxb.	284
Leichhardtia	32
clamboides, *F. v. M.*	33
Legnephora	29
Moorii, Miers	29
Lepidium	52
hyssopifolium, Desv.	53
monoplocoides, *F. v. M.*	52
papillosum, *F. v. M.*	52
puberulum, Bunge	53
ruderale, *Linn.*	53
strongylophyllum, *F. v. M.*	52
Lepigonum	
anceps, Bartl.	89
brevifolium, Bartl.	89
laxiflorum, Bartl.	89
rubrum, Fries.	89
Leucocarpon	
obscurum, A. Rich	258
Limnacia	30
Selwynia, *F. v. M.*	30
Limonia	
parviflora, Hook.	211
australis, A. Cunn.	218
Linum	166
angustifolium, DC.	166
gallicum, *Linn.*	166
marginale, *A. Cunn.*	166
suædæfolium, Planch.	166
trigynum	167
usitatissimum	166
Lychnis	86
Githago, Linn.	86
Macgregoria	265
racemegera, *F. v. M.*	265
Macintyria	
octundra, F. v. M.	82
Malva	108
Behriana, Schlecht	108
brachystachya, F. v. M.	109
ovata, Cav.	109
parviflora, *Linn.*	109
Preissiana, Miq.	108
rotundifolia, *Linn.*	109
spicata, Linn.	109
sylvestris, *Linn.*	108
tinoriensis, DC.	109
verticillata, *Linn.*	108
tricuspidata, Ait.	110
Malvastrum	109
spicatum, *A. Gray*	109
tricuspidatum, *A. Gray*	109
Mangifera	321
indica, *Linn.*	321
Marianthus	73
lineatus, *F. v. M.*	75
pictus, Lindl.	75
procumbens, *Benth.*	74
Medicosma	202
Cunninghamii, Hook. f.	202
Melhania	143
abyssinica, *A. Rich.*	144
incana, Heyne	143
Leprieurii, Webb.	144
ovata, Boiss.	144
oblongifolia, F. v. M.	144
Melia	226
australasica, A. Juss.	227
composita, *Willd.*	227
Melicope	196
australasica, *F. v. M.*	198
Broadbentiana, *Bail.*	198
choorechillum, *Bail.*	198
erythrococca, *Benth.*	198
Fareana, *F. v. M.*	197
neurococca, *Benth.*	197
pubescens, *Bail.*	199
Melycitus	
oleaster, Lindl.	257
Melochia	144
concatenata, Linn.	145
corchorifolia, *Linn.*	145
nodiflora, Sw.	145

INDEX OF GENERA AND SPECIES.

	Page		Page		Page
pyramidata, *Linn.*	144	aphylla, *R. Br*...	246	glandulosum, *Hook.*	193
supina, *Linn.*	145	retusa, *F. z. M.*	245	Nottii, *F. v. M.*..	193
Melodorum	24	stricta, *R. Br.*	246	retusum, *Hook.*..	194
Leichhardtii, *Benth.*	25	Opilia	246	rotundifolium, *Benth.*	193
Maccreai, *F. v. M.*	25	amentacea, *Roxb.*	247	sediflorum, F. v. M.	193
Uhrii, *F. v. M.*..	25	javanica, Miq. ..	247	squamuligerum, *Hook.*	193
Meniocus		Ornitrophe		Philotheca...	194
australasicus, Turcz.	45	serrata, Roxb. ..	288	australis, *Rudge*	194
linifolius, DC. ..	45	Owenia	235	ciliata, *Hook.* ..	194
serpyllifolius, Desv.	45	acidula, *F. v. M.*	235	calida, *F. v. M.*..	194
Micromelum	211	cepiodora, *F. v. M.*	236	Phlebocalymna ..	249
glabrescens, Benth.	212	cerasifera, F. v. M.	325	lobospora, *F. v. M.*	249
pubescens, *Blume*	212	reticulata, *F. v. M.*	236	Phomà	
Mitrephora, *Blume*	23	venosa, *F. v. M.*	236	diploglottidis, Cke. and	
Froggattii, *F. v. M.*	23	vernicosa, *F. v. M.*	235	Mass...	287
reticulata, H. f. & T.	24	xerocarpa, F. v. M.	236	Phyllosticta	
Modiola, *Mœnch.* ..	120	Oxalis	179	*Evodiæ*, Cke. ..	201
caroliniana, Linn.	120	bipunctata, Grah.	180	neurospilea, Sacc. and	
multifida, *Mœnch.*	120	cognata, Steud...	180	Berk...	279
Murraya ..	212	corniculata, *Linn.*	180	uvariæ, Berk. ..	21
crenulata, *Oliv.*..	213	corymbosa, *DC.*	180	Pigea	
exotica, *Linn.* ..	212	Marteana, Zucc.	180	Banksiana, DC.	64
paniculata, Jack.	212	microphylla, Poir.	180	filiformis, DC. ..	64
Myosurus ..	6	perennans, Haw.	180	Pittosporum ..	68
minimus, *Linn.*	6	Petersii, Klotz. ..	180	acacioides, A. Cunn.	71
aristatus, Gey. ..	6	Preissiana, Steud.	180	angustifolium, Lodd.	71
australis, F. v. M.	6	sessilis, *Hamilt.*	180	ferrugineum, *Ait.*	70
		Oxleya		flavum, Hook. ..	71
Nasturtium	43	xanthoxyla, A. Cunn.	239	fulvum, Rudge ..	70
officinale, *Linn.*	44			hirsutum, Link ..	70
palustre, DC. ..	43	Pachygone..	34	lanceolatum, A. Cunn...	71
semipinnatifidum, Hook.	44	Hullsii, *F. v. M.*	34	ligustrifolium, A. Cunn.	71
terrestre, Br. ..	44	longifolia, *Bail.*	34	linifolium, A. Cunn.	70
Nelumbium	40	pubescens, Benth.	32	longifolium, Putterl.	71
speciosum, *Willd.*	40	Pagetia ...	208	melanospermum, *F. v. M.*	69
Nemedra		medicinalis, F. v. M.	203	nanum, Hook. ..	78
elæagnoidea, A. Juss.	232	monostylis, *Bail.*	203	oleæfolium, A. Cunn. ..	71
Nephelium ..	301	Papaver ...	41	ovatifolium, F. v. M.	70
callarrie, *Bail.* ..	306	gariepinum, DC.	41	phillyræoides, DC.	71
connatum, *F. v. M.*	302	horridum, DC. ..	41	procumbens, Hook.	73
coriaceum, *Benth.*	304	rhœas, *Linn.* ...	41	revolutum, *Ait.*...	69
divaricatum, *F. v. M.*..	305	Paritium		rhomhifolium, *A. Cunn.*	68
distyla, *F. v. M.*	305	tiliaceum, St. Hil.	130	Roëanum, Putterl.	71
diversifolium, F. v. M...	307	Pavonia ..	121	. ruhiginosum, *A. Cunn.*	70
Lautererianum, *Bail.* ..	304	hastata, *Cav.* ..	122	salicinum, Lindl.	71
leiocarpum, *F. v. M.*	304	Pelargonium	179	setigerum, *Bail.*	69
microphyllum, *Benth.*...	305	australe, *Willd.*	179	tinifolium, A. Cunn.	70
oleifolium, F. v. M.	307	crinitum, Nees ..	179	tomentosum, Bonpl.	70
semicinereum, *F. v. M.*	303	Drummondii, Turcz.	179	undulatum, *Vent.*	69
semiglaucum, *F. v. M.*...	302	glomeratum, Jacq.	179	venulosum, *F. v. M.*	70
subdentatum, *F. v. M.*...	303	inodorum, Willd.	179	Wingii, *F. v. M.*	70
tomentosum, *F. v. M.*...	303	littorale, Hueg...	179	Plagianthus	110
Nitraria ..	174	stenanthum, Turcz.	179	glomeratus, Benth.	110
Billiardieri, DC.	174	Pennantia ..	249	microphyllus, *F. v. M.*...	110
Olivieri, Jaub. & Spach.	174	Cunninghamii, *Miers* ..	249	Platynema	
Schoberi, *Linn.*...	174	Pentaceras	207	laurifolium, W. & Arn.	170
Nymphæa ..	38	australis, *Hook. f.*	207	Pleiococca ..	207
Brownii, *Bail.* ..	39	Pericampylus	30	Wilcoxiana, *F. v. M.*	208
cœrulea ..	39	incanus, *Miers* ..	30	Pleogyne ...	35
flava, *Leit.* ..	39	Pestalozzia		australis, *Benth.*	36
gigantea, *Hook.*...	38	Guepini, Desm....	271	Pleiogynium ..	324
lotus, *var.* australis, *Bail.*	88	Phebalium	192	solandri, *Engl.*...	324
minima, Bail. ..	39	aureum, A. Cunn.	193	Pleurandra	
pygmœa, *Ait.* ..	89	Billardieri, *A. Juss.*	194	acicularis, Labill.	14
stellata, *F. v. M.*	89	elæagnifolium, A. Juss.	193	camforosma, Sieb.	13
tetragona, *Georgi*	39	elæagnoides, Sieb.	194	cistiflora, Sieb. ..	13
		elatius, *Benth.* ..	192	cricifolia, DC. ..	13
Olax ..	245	elatum, A. Cunn.	194	ovata, Labill. ..	14

INDEX OF GENERA AND SPECIES.

	Page		Page		Page
riparia, R. Br.	13	*nova-guineusis*	35	Bidwilli, *Benth.*	26
stricta, R. Br.	13	*tumefacta*	35	Brahei, *F. v. M.*	26
Pleuropetalum	250			Salacia	260
Samoense, A. Gr.	250	Ranunculus	6	prinoides, DC.	261
suaveolens, Blume	250	*collinus*, R. Br.	8	Salmalia	
Pœcilodermis		*colonorum*, Endl.	7	*malabarica*, Schott.	133
populnea, Schott.	139	*discolor*, Steud.	7	Salomonia	77
Polanisia	55	*glabrifolius*, Hook.	8	*oblongifolia*, DC.	77
viscosa, DC.	55	*hirtus*, Banks	7	*obovata*, Wight	77
Polyalthia	22	*incisus*, Hook. f.	8	Samadera	218
Armitiana, F. v. M.	23	*inundatus*, R. Br.	8	*Baileyana*, *Oliver*	219
Holtzeana	23	*lappaceus*, *Sm.*	7	Bidwillii, *Oliver*	219
Moonii	24	*leptocaulis*, Hook.	8	Sapindus	300
nitidissima, *Benth.*	23	*parviflorus*, *Linn.*	8	*australis*, *Benth.*	301
Polycarpæa	90	*pilulifer*, Hook.	8	*cinereus*, A. Cunn.	303
breviflora, F. v. M.	92	*plebeius*, R. Br.	7	Saponaria, Linn.	85
Burtoni, *Bail.*	91	*pumilio*, R. Br.	8	*vaccaria*, Linn.	85
corymbosa, *Lam.*	92	*rivularis*, *Banks*	7	Sarcopetalum	32
longiflora, F. v. M.	91	*sessiliflorus*, R. Br.	8	*Harveyanum*, F. v. M.	32
spirostyles, F. v. M.	91	Ratonia	295	Saurauja, *Willd.*	106
synandra, F. v. M.	91	*anodonta*, *Benth.*	296	*Andreana*, *Oliv.*	106
Polycarpon	90	*alphandi*, F. v. M.	288	Schistocarpæa	269
tetraphyllum, *Linn. f.*	90	*Cordieril*, F. v. M.	299	*Johnsoni*, F. v. M.	269
Polygala	77	*Dæmeliana*, F. v. M.	298	Schleichera	
arvensis, *Willd.*	79	*distylis*, F. v. M.	306	*ptychocarpa*, F. v. M.	229
japonica, *Houtt.*	78	*exangulata*, F. v. M.	297	Schmidelia	288
leptalea, DC.	78	*grandissima*, F. v. M.	296	*anodonta*, F. v. M.	296
oligophylla, DC.	78	*lachnocarpa*, F. v. M.	296	*Cobbe*, Linn.	288
persicarisfolia, DC.	78	*Lessertiana*, *Benth.*	298	*pyriformis*, F. v. M.	296
rhinanthoides, *Soland.*	79	*Martyana*, F. v. M.	298	*serrata*, DC.	288
veronicea, F. v. M.	78	*Nugentii*, *Bail.*	297	*timoriensis*, DC.	288
Pomaderris	271	*O'Shanesiana*, F. v. M.	298	Scolopia	66
acuminata, Link	273	*punctulata*, F. v. M.	297	*Brownii*, F. v. M.	66
andromedæfolia, A. Cun.	274	*pyriformis*, *Benth.*	296	Selwynii	30
discolor, DC.	273	*stipitata*, *Benth.*	297	Semecarpus	323
elliptica, *Labill.*	273	*tenax*, *Benth.*	299	*anacardium*, Linn.	324
ferruginea, *Sieb.*	272	Reinwardtia	167	*australiensis*, *Engl.*	323
intermedia, *Sieb.*	273	*trigyna*, *Planch.*	167	Senebiera	51
lanigera, *Sims*	272	Rhamnus	268	*didyma*, *Pers.*	51
malifolia, *Sieb.*	273	*Napeca*, Linn.	268	*integrifolia*, DC.	51
multiflora, Fenzl	273	*Œnoplia*, Linn.	268	*linoides*, DC.	51
obscura, *Sieb.*	272	*vitiensis*, *Benth.*	268	*pinnatifida*, DC.	51
phillyræoides, *Sieb.*	273	*sizyphoides*, Soland	271	Seringia	148
phillyreæfolia, Fenzl	274	Rhodosphæra	320	*corollata*, Steetz	150
prunifolia, A. Cunn.	274	*rhodanthra*, *Engl.*	321	*lanceolata*, Steetz	149
viridirufa, *Sieb.*	273	Rhus	320	*platyphylla*, *J. Gay*	149
Wendlandiana, Don.	273	*rhodanthema*, F. v. M.	321	Sida	110
Popowia		*panaciformis*, F. v. M.	320	*argentea*, *Bail.*	112
australis, *Benth.*	24	*rufa*, Tysm.	320	*asiatica*, Linn.	118
Portenschlagia		Rhytidosporum		*aurita*, Wall.	119
integrifolia, Tratt.	259	*procumbens*, F. v. M.	73	*carpenoides*, DC.	110
australis, Tratt.	259	Riedleia		*compressa*, DC.	115
Portulaca	93	*corchorifolia*, DC.	145	*cordifolia*, *Linn.*	115
Armitii, F. v. M.	95	Rœpera		*corrugata*, *Lindl.*	111
australis, *Endl.*	94	*aurantiaca*, Lindl.	176	*cryphiopetala*, F. v. M.	113
bicolor, F. v. M.	95	*Billardieri*, A. Juss.	176	*filiformis*, A. Cunn.	112
digyna, F. v. M.	94	*fabagifolia*, A. Juss.	176	*Fraseri*, Hook.	120
filifolia, F. v. M.	94	*latifolia*, Hook. f.	175	*graveolens*, Roxb.	119
napiformis, F. v. M.	94	Rœperia		*humillima*, F. v. M.	112
oleracea, Linn.	93	*cleomoides*, F. v. M.	56	*inclusa*, *Benth.*	115
oligosperma, F. v. M.	94	Rulingia	146	*indica*, Linn.	118
Pterospermum	143	*pannosa*, R. Br.	147	*intersians*, F. v. M.	112
acerifolium, *Willd.*	143	*rugosa*, Steetz	147	*intricata*, F. v. M.	112
Pycnarrhena	35	*salvifolia*, *Benth.*	146	*leucopetala*, F. v. M.	117
australiana, F. v. M.	35	Ryssopterys	169	*macropoda*, F. v. M.	118
lucida	35	*timorensis*, *Blume*	169	*micropetala*, R. Br.	118
manillensis	35	Saccopetalum	25	*mutica*, Del.	120
				nematopada, F. v. M.	112

INDEX OF GENERA AND SPECIES.

	Page
oxycarpa, F. v. M.	119
pedunculata, A. Cunn.	112
petrophila, F. v. M.	113
philippica, DC.	115
platycalyx, F. v. M.	115
pleiantha, F. v. M.	114
retusa, Linn.	115
rhombifolia, Linn.	114
rhomboidea, Roxb.	115
Spenceriana, F. v. M.	112
spinosa, Linn.	114
subspicata, F. v. M.	114
trichopoda, F. v. M.	112
tubulosa, A. Cunn.	117
virgata, Hook.	113
Silene	86
anglica, Linn.	86
cerastoides, Linn.	86
gallica, Linn.	86
lusitanica, Linn.	86
quinquevulnera, Linn.	86
Sinapis	
nigra	50
Siphonodon	261
australe, Benth.	261
membranaceum, Bail.	262
pendulum, Bail.	262
Sisymbrium	45
eremigerum, F. v. M.	47
nasturtioides, F. v. M.	47
officinale, Scop.	46
trisectum, F. v. M.	47
Sloanea, Linn.	159
australis, F. v. M.	159
Langii, F. v. M.	160
Macbrydei, F. v. M.	160
Woodsii, F. v. M.	160
Spanoghea	
connata, F. v. M.	303
nephelioides, F. v. M.	304
Spergula	88
arvensis, Linn.	88
Spergularia	89
rubra, Pers.	89
rupestris, Fenzl.	89
Spondias	
acida, Soland	325
Solandri, Benth.	325
pleiogyna, F. v. M.	325
Stackhousia	263
aspericoccca, Schuch.	263
dorypetala, Schuch.	264
elata, F. v. M.	264
Gunniana, Schlecht.	263
Gunnii, Hook. f.	263
intermedia, Bail.	264
linariæfolia, A. Cunn.	263
maculata, Sieb.	263
monogyna, Labill.	263
monogyna, Sieb.	264
Muelleri, Schuch.	263
muricata, Lindl.	264
nuda, Lindl.	264
obtusa, Lindl.	263
spathulata, Sieb.	263
viminea, Sm.	264
Stellaria	87
angustifolia, Hook.	88

	Page
flaccida, Hook.	88
glauca, With.	87
media, Linn.	88
media, var., Hook. f.	88
pungens, Brongn.	87
squarrosa, Hook.	87
Stenanthemum	274
Scortechini, F. v. M.	274
Stenopetalum	18
lineare, R. Br.	49
nutans, F. v. M.	49
velutinum, F. v. M.	48
Stephania	33
aculeata, Bail.	33
hernandiæfolia, Walp.	33
lætificata, Miers	33
Sterculia	134
acerifolia, A. Cunn.	138
Bidwilli, Hook.	137
caudata, Hew.	139
discolor, F. v. M.	138
diversifolia, G. Don	139
Garrawayæ, Bail.	136
laurifolia, F. v. M.	136
lurida, F. v. M.	138
quadrifida, R. Br.	136
ramiflora, Benth.	137
rupestris, Benth.	129
Trichosiphon, Benth.	138
vitifolia, Bail.	137
Sturtia	
gossypioides, R. Br.	133
Suriana	220
maritima, Linn.	220
Synoum	233
glandulosum, A. Juss.	234
Muelleri, C. DC.	234
Talinum	95
patens, Willd.	95
polyandrum, Hook.	97
Tarrietia	140
argyrodendron, Benth.	140
actinophylla, Bail.	141
Tasmannia	
dipetala, R. Br.	18
insipida, R. Br.	18
monticola, A. Rich.	18
Tetracera	9
Cowleyana, Bail.	9
Dæmeliana, F. v. M.	10
Nordtiana, F. v. M.	9
Wuthiana	10
Tetrapasma	
juncea, G. Don	277
Tetratheca	76
thymifolia, Sm.	76
Thespesia	132
populnea, Corr.	132
Thlaspi	53
cochlearinum, F. v. M.	53
Thouinia	
adenophora, Miq.	318
hemiglauca, F. v. M.	300
varifolia, F. v. M.	300
Thylacium lucidum, DC.	60
Tinospora	28
smilacina, Benth.	28

	Page
Tribulopis	
angustifolia, R. Br.	173
pentandra, R. Br.	173
Solandri, R. Br.	173
Tribulus	171
acanthococcus, F. v. M.	172
angustifolius, Benth.	173
cistoides, Linn.	172
hystrix, R. Br.	172
lanuginosus, Linn.	172
leptophyllus, Bail.	173
minutus, Leichh.	173
occidentalis, R. Br.	172
pentandrus, Benth.	172
Solandri, F. v. M.	173
terrestris, Linn.	171
Trichilia	
arborescens, Spreng.	228
glandulosa, Sm.	234
Trichosiphon	
australe, Schott.	188
Triphasia	
glauca, Lindl.	214
Tripterococcus	
spathulatus, F. v. M.	268
Tristellateia	170
australasica, A. Rich.	170
Triumfetta	155
nigricans, Bail.	156
pilosa, F. v. M.	156
procumbens, Forst	155
rhomboidea, Jacq.	156
Winneckeana, F. v. M.	156
Tristichocalyx	31
diffusus, Miers	32
pubescens, F. v. M.	31
Turræa	226
Billardieri, A. Juss.	226
concinna, Benn.	226
pubescens, Hellen	226
Unona	
fulgens, Lahill.	23
Leichhardtii, F. v. M.	25
nitidissima, Dun.	23
nitens, F. v. M.	23
Urena	121
Armitiana, F. v. M.	121
lobata, Linn.	121
sinuata, Linn.	121
Uromyces	
diploglottidis, Oke. and Mass.	287
Uvaria	20
Goezeana, F. v. M.	21
heteropetala, F. v. M.	22
membranacea, Benth.	21
odorata, Linn.	22
Ventilago	266
ecorollata, F. v. M.	267
viminalis, Hook.	267
Villaresia	250
Moorei, F. v. M.	250
Samoensis, A. Gr.	250
Smythii, F. v. M.	250
suaveolens, Benth.	250
Viola	62

INDEX OF GENERA AND SPECIES. ix

	Page		Page		Page
betonicæfolia, *Sm.*	62	Waltheria	145	granulata, *C. Moore*	184
hederacea, *Labill.*	63	americana, *Linn.*	145	*hirsuta,* DC. ..	184
longiscapa, DC.	62	*indica,* Linn. ..	145	*lævigata, Sm.* ..	183
phyteumæfolia, DC.	62	Wormia	10	*lanceolata,* R. Br.	184
Sieberiana, Spreng.	63	*alata, R. Br.* ..	10	*obcordata, A. Cunn.*	184
Vitis	278			*pauciflora,* Sm...	184
acetosa, F. v. M.	282	Xanthophyllum ..	82	*pilosa, Rudge* ..	183
acris, F. v. M. ..	281	Macintyrii, *F. v. M.*	82	*revoluta,* A. Cunn.	183
adnata, Wall. ..	280	Ximenia	244	*Smithii, Andr...*	184
antarctica, Benth.	279	americana, *Linn.*	245	Zizyphus	267
brachypoda, F. v. M.	280	*elliptica,* Forst...	245	*celtidifolia,* DC.	268
cardiophylla, F. v. M.	280	*exarmata, F. v. M.*	245	*jujuba, Lam.* ..	268
carnosa, W. & Arn.	281	*laurina,* Delile...	245	*œnoplia, Mill.* ..	268
clematidea, F. v. M.	282	Xylocarpus		*pomaderroides,* Fenzl	270
cordata, Wall. ..	279	Granatum, Koen	237	*rufula,* Miq. ..	268
Gardineri, Bail.	283	Xylosma	66	Zygophyllum ..	174
hypoglauca, F. v. M.	282	*orbiculatum,* Forst.	67	*apiculatum, F. v. M.*	175
japonica, Willd.	281	*ovatum, Benth...*	67	*aurantiacum, F. v. M.*	176
nitens, F. v. M.	280			*australasicum,* Miq.	174
oblonga, Benth...	279	Zanthoxylum ..	204	*Billardieri,* DC.	176
opaca, F. v. M...	283	*brachyacanthum, F. v. M.*	204	*crenatum, F. v. M.*	175
pedata, Wall. ..	281	*parviflorum, Benth.*	205	*fruticulosum,* DC.	176
penninervis, F. v. M.	283	*torvum, F. v. M.*	205	*glaucescens, F. v. M.*	175
psoralifolia, F. v. M.	281	*veneficum, Bail.*	204	*glaucum, F. v. M.*	175
saponaria, Seem.	280	Zieria	182	*iodocarpum, F. v. M.*	175
sterculifolia, F. v. M.	283	*aspalathoides, A. Cunn.*	183	*prismatothecum, F. v. M.*	175
trifolia, Linn. ..	281	*cytisoides, Sm...*	184	*terminale,* Turcz.	175

VERNACULAR NAMES.

	Page		Page		Page
Aguridal ..	262	Bumble	60	Crow's Apple	236
Alexandrian Laurel	104	Bungya	324	Crow's Ash	239
Ane	270	Bunji-bunji ..	241		
An-gi-ur	137	Burdekin Plum ..	324	Dentie	139
Aquaie	40	Buttercup	7	Dilly Boolen ..	285
Arnurna	38			Dooja	215
Arrago	240			Doomba-tree ..	104
Arsenic Plant ..	17	Cairns Hickory ..	240	Dunanya	270
		Callarrie	306		
Bally Gum of Martintown	310	Caloon	164	Ebonyheart ..	162
Baragara	270	Caltrops, not Catrops	171	Eeger	60
Batham	127	Cannon-ball Tree ..	237	Elemi-tree (Queensland)	224
Beech (Red) ..	240	Cardwell Maple ..	240	Emu Apple.. ..	235
Binkey	139	Carrot-wood ..	223		
Bladder Ketmia ...	124	Catchfly	86	Figo del Inferno ..	41
Blue Waterlily ..	38	Cat's-head	172	Flame Tree ..	138
Bogum bogum ..	240	Cattle-bush	300		
Boiong	140	Cedar (red) ..	237	Ghittoe	211
Bombyx cynthia Tree	218	Cedar (white) ..	227	Gotoobah	245
Boodyarra	234	Chambin	297	Gruee	235
Boogoobi	201	Chargir	310		
Boolboora	237	Convavola	136	Hickory (Cairns) ..	240
Boorbal	300	Cooreenyan ..	128	Hop Bush	212
Boorbalmin ..	300	Coorunyan ..	123		
Bottle Tree (broad leaf)	138	Corn Cockle ..	86	Indian Tulip Tree ..	132
" (narrow leaf)	139	Cow-herb	85	Ironbark (scrub, Cairns)	230

VERNACULAR NAMES.

Name	Page	Name	Page	Name	Page
Irrpo	38	Momin	237	Silk-wood	294
Ithnee	126	Mondo	60	Spider-flower	55
Ivory Wood	261, 262	Mondolen	59	Spotted Tree	243
		Moorgun	164	Stave Wood	140
Jaln-ba	323	Moorjung	292	Sundri	141
Jimmie-Jimmie	233	Morna	136	Sweet Plum	324
Jujube	268	Mouse-ear Chickweed	87		
		Mouse-tail	6	Urgullathy	38
Kalan	139	Murgurpul	237		
Kam-doo-thal	60	Murtilam	268	Wanga	237
Kame	224	Mustard	50	Watercress	44
Kaooroo	38	Mycoolon	88	Watershield	87
Karoom	154			White Beech (of Bunya Mountain)	163
Kary	70	Nankeen Dye-wood	206		
Kedgy Kedgy	280	Narroo	240	White Cedar	227
Keena	104	Ngar-Golly	126	White Wood	300
Kel-lan	139	Noongi	324	Wilga	206
Ketey	138			Woolah	163
Kilamul	313	Orange (Native)	215	Woota	237
Kinginga	313	Orange Thorn	78	Wyjelah	57
Kokare	323	Ouraie	154		
Kooline	154			Yako-kalo	38
Kooraloo	290	Pam-mo	154	Yarra	127
Koral-ba	136	Peirir	140	Yellow Plum	245
Kou-nung	154	Pigweed	93	Yellow-wood	289
Kuman	136	Pink Waterlily	40	Yellow Wood-sorrel	180
Kumquat (Queensland)	214	Pomegranate (Native)	60		
Kurrajong	139	Poon Spar Tree	104		
		Poppy	41	Tacamahac	104
Leopard-tree	243	Poppy (prickly)	41	Talwalpin	130
Lime (N.Q.)	215	Punburra	323	Tchaln-ji	224
Lime (Finger)	215			Tecan	313
Logan Apple	210	Quandong	164	Teeweeree	9
Longullah	58			Thal-ango-thera	198
		Rancooran	324	Thandorah	267
Maiden's Blush Wood	323	Rarum	59	Thindah	38
Mango	321	Red Ash	270	Thoogeer	60
Maraguigi	291	Red Beech (Johnstone Riv.)	140	Thoolambool	38
Maple (Cardwell)	240	Red Beech (Herberton)	240	Thoongoon	38
Marking-nut Tree	323	Red Cedar	237	Thukouro	98
Mee-a-mee	270			Thulla-kurbin	57
Merangara	25	Sacred Lotus	40	Towra	196
Mijah	58	Satin-wood	294	Tree of Heaven	218
Mille	38	Satin-wood (Cairns)	230	Tuckeroo	290
Mock Orange	69	Scortechini's Crab	210	Tulip-wood Tree	309
Moi-u	38	Silk-cotton Tree	133	Turnip-wood	309

www.ingramcontent.com/pod-product-compliance
Lightning Source LLC
Chambersburg PA
CBHW051248300426
44114CB00011B/943